physica status solidi (b)

basic research

Volume 133 Number 1 January 1986

AKADEMIE-VERLAG BERLIN

VCH PUBLISHERS INC. · DEERFIELD BEACH/FLORIDA

ISSN 0370-1972 phys. stat. sol. (b), Berlin 133 (1986) 1, 1—432, K1—K76, A1—A

International Classification System for Physics*)

60. Condensed matter: structure, mechanical and thermal properties
(see also 68.20. Solid surface structure, 71. Electron states)
   - 61. Structure of liquids and solids; crystallography
   (see also 68.20. Solid surface structure, 71. Electron states)
   - 62. Mechanical and acoustical properties of condensed matter
   (see also 61.70. Defects in crystals, 68.30. Surfaces and interfaces)
   - 63. Lattice dynamics and crystal statistics
   (see also 65. Thermal properties, 66.70. Thermal conduction, 68.30. Dynamics of surface and interface vibrations, 78.30. Infrared and Raman spectra)
   - 64. Equations of state, phase equilibria, and phase transitions
   - 65. Thermal properties of condensed matter
   (see also 63. Lattice dynamics; for thermodynamic properties of quantum fluids, see 67.40; for thermal properties of solid helium, see 67.80)
   - 66. Transport properties of condensed matter (nonelectronic)
   - 67. Quantum fluids and solids; liquid and solid helium
   - 68. Surfaces and interfaces; thin films and whiskers
   (for impact phenomena, see 79; for crystal growth, see 61.50)

70. Condensed matter: electronic structure; electrical, magnetic, and optical properties
   - 71. Electronic states
   (see also 63. Lattice dynamics, 73. Electronic structure and electrical properties of surfaces, interfaces, and thin films)
   - 72. Electronic transport in condensed matter
   (for surfaces, interfaces, and thin films, see 73)
   - 73. Electronic and electrical properties of surfaces, interfaces, and thin films
   - 74. Superconductivity
   - 75. Magnetic properties and materials
   - 76. Magnetic resonances and relaxation in condensed matter; Mössbauer effect
   - 77. Dielectric properties and materials
   (for conductivity phenomena, see 72.20 and 72.80)
   - 78. Optical properties and condensed-matter spectroscopy and other interactions of matter with particles and radiation
   (for phonon spectra, see 63)
   - 79. Electron and ion emission by liquids and solids; impact phenomena

*) Excerpt; reproduced with permission of International Council for Scientific and Technical Information (ICSTI).

(The Substance Classification is given on cover three)

Author Index

physica status solidi (b)

basic research

Volume 133 · Number 1 · Pages 1 to 432, K1 to K76, and A1 to A8

January 1, 1986

AKADEMIE-VERLAG · BERLIN

and

VCH PUBLISHERS INC.
DEERFIELD BEACH/FLORIDA

physica status solidi (b), ISSN 0370-1972,

is published monthly by Akademie-Verlag Berlin, GDR. Annual subscription price in USA and Canada US $ 695.00 including air speed delivery.

Second class postage is paid at Jamaica, NY 11431.

USA Postmasters: Send address changes to physica status solidi (b), VCH Publishers, Inc., 303 NW 12th Avenue, Deerfield Beach, FL 33442 – 1705.

Air freight and mailing in the USA by Publications Expediting Inc., 200 Meacham Avenue, Elmont, NY 11003,

Enquiries, subscriptions, and orders for single copies from customers in the USA and CANADA should be directed to:

VCH Publishers, Inc., 303 NW 12th Avenue, Deerfield Beach, FL 33442 – 1705, USA;

from all other countries to:

Akademie-Verlag Berlin, Leipziger Str. 3 – 4, DDR-1086 Berlin.

Editorial Note:

"physica status solidi (b)" undertakes that an original paper *accepted for publication* before the 8th of any month will be published within 50 days of this date unless the author requests a postponement. In special cases there may be some delay between *receipt* and *acceptance* of a paper due to the review and, if necessary, revision of the paper.

Schriftleiter und verantwortlich für den Inhalt:
Professor Dr. Dr. h. c. P. Görlich, DDR-1086 Berlin, Leipziger Str. 3—4, Postfach 1233 bzw. DDR-6900 Jena, Schillbachstr. 24.

Verlag:
Akademie-Verlag Berlin, DDR-1086 Berlin, Leipziger Straße 3—4; Fernruf: 2236221 und 2236229; Telex-Nr.: 114420; Bank: Staatsbank der DDR, Berlin, Kto.-Nr.: 6836-26-20712.

Chefredakteur:
Dr. H.-J. Hänsch.

Redaktionskollegium:
Prof. Dr. E. Gutsche, Dr. H.-J. Hänsch, Dr. H. Lange, Dr. S. Oberländer.

Anschrift der Redaktion:
DDR-1086 Berlin, Leipziger Str. 3—4, Postfach 1233; Fernruf: 2236279.

Veröffentlicht unter der Lizenznummer 1310 des Presseamtes beim Vorsitzenden des Ministerrates der Deutschen Demokratischen Republik.

Gesamtherstellung:
VEB Druckerei „Thomas Müntzer", DDR-5820 Bad Langensalza.

Erscheinungsweise:
Die Zeitschrift „physica status solidi (b)" erscheint jeweils am 1. eines jeden Monats. Jährlich erscheinen 6 Bände zu je 2 Heften.

Bestellnummer dieses Bandes: 1068/133.

Contents

Defects, atomistic aspects

Magnetism

Extended electronic states and transitions

Localized electronic states and transitions

Electric transport

Erratum

Pre-Printed Titles

Systematic List

Subject classification:	Corresponding papers begin on the following pages (pages given in italics refer to the principle subject classification):

Contents of Volume 133 Continued on Page 435

International Classification System for Physics *)

60. Condensed matter: structure, mechanical and thermal properties

61. Structure of liquids and solids; crystallography
    (see also 68.20. Solid surface structures, 71. Electron states)
    - 61.10. X-ray determination of structures
      (for specific determinations, see 61.55 to 61.80)
    - 61.12. Neutron determination of structures
      (for specific determinations, see 61.55 to 61.80)
    - 61.14. Electron determination of structures
      (for specific determinations, see 61.55 to 61.80)
    - 61.16. Other determination of structures
      (for specific determinations, see 61.55 to 61.80)
    - 61.20. Classical, semiclassical, and quantum theories of liquid structure
      (for electronic states, see 71; for liquid helium, see 67)
    - 61.25. Studies of specific liquid structures
    - 61.30. Liquid crystals
    - 61.40. Amorphous and polymeric materials
    - 61.50. Crystalline state (including molecular motions in solids)
      (for magnetic structure and spin systems, see 75.25)
    - 61.55. Specific structures of elements and alloys
    - 61.60. Specific structures of inorganic compounds
    - 61.65. Specific structures of organic compounds
    - 61.70. Defects in crystals
      (see also 61.80. Radiation damage, 62. Mechanical and acoustical properties, 71.55. Impurities and defect levels, 76.30. EPR of color centers and other defects, 78.50. Impurity and defect absorption in solids)
    - 61.80. Radiation damage and other irradiation effects
      (for techniques of structure determination, see 61.10 to 61.16; for electron and ion impact phenomena, see 79.20)
    - 61.90. Other topics in structure of liquids and solids

62. Mechanical and acoustical properties of condensed matter
    (see also 61.70. Defects in crystals, 68.30. Surfaces and interfaces)
    - 62.10. Mechanical properties of liquids
      (for viscosity of liquids, see also 66.20)
    - 62.20. Mechanical properties of solids (related to microscopic structure)
    - 62.30. Mechanical and elastic waves
    - 62.40. Anelasticity, internal friction, and mechanical resonances
      For thermomechanical effects, see 65.70
      For magnetomechanical effects, see 75.80
      For piezoelectric effects, see 77.60
      For elasto-optical effects, see 78.20
    - 62.50. High-pressure and shock-wave effects in solids
    - 62.60. Acoustical properties of liquids
      For lattice dynamics, phonons, see 63
      For second sound in quantum fluids, see 67.40
    - 62.65. Acoustical properties of solids
      For magnetoacoustic effects in metals, see 75.80
      For acoustoelectric effects, see 72.50
      For acousto-optical effects, see 78.20
    - 62.80. Ultrasonic relaxation
      (see also 74.30. Ultrasonic attenuation in superconductors)
    - 62.90. Other topics in mechanical and acoustical properties of condensed matter

---

*) Excerpt, reproduced with permission of International Council for Scientific and Technical Information (ICSTI).

63. Lattice dynamics and crystal statistics
(see also 65. Thermal properties, 66.70. Thermal conduction, 68.30. Dynamics of surface and interface vibrations, 78.30. Infrared and Raman spectra)
63.10. General theory
63.20. Phonons and vibrations in crystal lattices
63.50. Vibrational states in disordered systems
63.70. Statistical mechanics of lattice vibrations
(see also 65. Thermal properties of condensed matter, and 66.70. Thermal conduction)
63.75. Statistical mechanics of displacive phase-transitions
For order–disorder and statistical mechanics of model systems, see 64.60
For crystallographic aspects of polymorphic and order–disorder transformations, see 61.50
63.90. Other topics in lattice dynamics and crystal statistics

64. Equations of state, phase equilibria, and phase transitions
64.10. General theory of equations of state and phase equilibria
64.30. Equations of state of specific substances
(see also 65.70. Thermal expansion)
64.60. General studies of phase transitions
(for magnetic, superconducting, and quantum fluid critical phenomena, see 75, 74, and 67, respectively)
64.70. Phase equilibria, phase transitions, and critical points of specific substances
64.75. Solubility, segregation, and mixing
64.80. Other phase properties of systems
64.90. Other topics in equations of state, phase equilibria, and phase transitions

65. Thermal properties of condensed matter
(see also 63. Lattice dynamics; for thermodynamic properties of quantum fluids, see 67.40; for thermal properties of solid helium, see 67.80)
65.20. Heat capacities of liquids
65.40. Heat capacities of solids
(for specific heat of superconductors, see 74.30; for specific heat of magnetic systems, see 75.40)
65.50. Thermodynamic properties and entropy
65.70. Thermal expansion and thermomechanical effects
(see also 64.30. Equations of state)
For thermal conduction in nonmetallic liquids, see 66.60; for nonmetallic solids, see 66.70
For electronic thermal conduction, see 72.10 and 72.20
For thermal conductivity of superconductors, see 74.30
For pyroelectric and electrocaloric effects, see 77.70
65.90. Other topics in thermal properties of condensed matter

66. Transport properties of condensed matter (nonelectronic)
66.10. Diffusion and ionic conduction in liquids
66.20. Diffusive momentum transport
(for viscosity of liquids, see also 62.10)
66.30. Diffusion in solids
66.60. Thermal conduction in nonmetallic liquids
(for thermal conduction in liquid metals, see 72.15)
66.70. Nonelectronic thermal conduction and heat-pulse propagation in nonmetallic solids
(for thermal conduction in solid metals, see 72.15; for statistical mechanics of lattice vibrations, see 63.70)
66.90. Other topics in nonelectronic transport properties

67. Quantum fluids and solids; liquid and solid helium
    67.20. Quantum effects on the structure and dynamics of nondegenerate fluids
    67.40. Boson degeneracy and superfluidity of helium-4
    67.50. Fermi fluids; liquid helium-3
    67.60. Mixed systems; liquid helium-3, -4 mixtures
    67.70. Films (including physical adsorption)
    67.80. Solid helium and related quantum crystals
    67.90. Other topics in quantum fluids
    (e.g., neutron-star matter)

68. Surfaces and interfaces; thin films and whiskers
    (for impact phenomena, see 79; for crystal growth, see 61.50)
    68.10. Fluid surfaces and fluid–fluid interfaces
    68.15. Liquid thin films
    68.20. Solid surface structures
    68.25. Mechanical and acoustical properties of solid surfaces and interfaces
    68.30. Dynamics of solid surfaces and interface vibrations
    68.40. Surface energy of solids; thermodynamic properties
    68.45. Solid–fluid interface processes
    68.48. Solid–solid interfaces (including bicrystals)
    (for grain boundaries, see 61.70)
    68.55. Thin film growth, structure, and epitaxy
    68.60. Physical properties of thin films: nonelectronic
    68.70. Whiskers and dendrites: growth, structure, and nonelectronic properties
    68.90. Other topics in the structure and nonelectronic properties of surfaces and thin films

70. Condensed matter: electronic structure; electrical, magnetic, and optical properties

71. Electronic states
    (see also 63. Lattice dynamics, 73. Electronic structure and electrical properties of surfaces, interfaces, and thin films)
    71.10. General theories and computational techniques
    71.20. Electronic density of states determinations
    (including energy states of liquid semiconductors)
    (see also 65.40. Electronic heat capacity)
    71.25. Nonlocalized single-particle electronic states
    71.30. Metal–insulator transitions
    71.35. Excitons and related phenomena (including electron–hole drops)
    71.36. Polaritons
    (including photon–phonon and photon–magnon interactions)
    71.38. Polarons and electron–phonon interactions
    (see also 63.20. phonon–electron interactions in lattices)
    71.45. Collective effects
    71.50. Localized single-particle electronic states (excluding impurities)
    71.55. Impurity and defect levels
    71.60. Positron states
    (see also 78.70. Positron annihilation)
    71.70. Level splitting and interactions
    (see also 75.10. — in magnetic phenomena, 75.30. Exchange and superexchange interactions, 73.20. Electronic surface states)
    71.90. Other topics in electron states

72. Electronic transport in condensed matter
(for surfaces, interfaces, and thin films, see 73)
72.10. Theory of electronic transport; scattering mechanism
72.15. Electronic conduction in metals and alloys
72.20. Conductivity phenomena in semiconductors and insulators
(for nonelectronic conduction, see 66.70)
72.30. High-frequency effects; plasma effects
72.40. Photoconduction and photovoltaic effects; photodielectric effects
72.50. Acoustoelectric effects
72.55. Magnetoacoustic effects
72.60. Mixed conductivity and conductivity transitions
72.70. Noise processes and phenomena
72.80. Conductivity of specific semiconductors and insulators
72.90. Other topics in electronic transport in condensed matter

73. Electronic structure and electrical properties of surfaces, interfaces, and thin films
73.20. Electronic surface states
(for emission and impact phenomena, see 79)
73.25. Surface conductivity
73.30. Surface double layers, Schottky barriers, and work functions
73.40. Interfaces
73.60. Electronic properties of thin films
73.90. Other topics in electrical properties of surfaces, interfaces, and thin films

74. Superconductivity
74.10. Occurrence, critical temperature
74.20. Theory
74.30. General properties
74.40. Fluctuations and critical effects
74.50. Tunneling phenomena, Josephson effect, and proximity effects
74.55. Type-I superconductivity
74.60. Type-II superconductivity
74.70. Superconducting materials
74.90. Other topics in superconductivity

75. Magnetic properties and materials
75.10. General theory and models of magnetic ordering
(see also 71.25. Nonlocalized single-particle electronic states, 71.70. Level splitting and interactions)
75.20. Diamagnetism and paramagnetism
75.25. Spin arrangements in magnetically ordered materials
(neutron studies, etc.)
75.30. Magnetically ordered materials, other intrinsic properties
(for critical point effects, see 75.40)
75.40. Critical-point effects, specific heats, short-range order
(see also 65.40. Heat capacities)
75.50. Studies of specific magnetic materials
75.60. Domain effects, magnetization curves, and hysteresis
75.70. Magnetic films and plates
75.80. Magnetomechanical and magnetoelectric effects, magnetostriction
For galvanomagnetic effects, see 72.15 and 72.20
For magneto-optical effects, see 78.20
75.90. Other topics in magnetic properties and materials

76. Magnetic resonances and relaxation in condensed matter; Mössbauer effect
  - 76.20. General theory of resonances and relaxation
  - 76.30. Electron paramagnetic resonance and relaxation
  - 76.40. Diamagnetic and cyclotron resonances
  - 76.50. Ferromagnetic, antiferromagnetic, and ferrimagnetic resonances; spin-wave resonance (see also 75.30. Spin waves)
  - 76.60. Nuclear magnetic resonance and relaxation
  - 76.70. Magnetic double resonances and cross effects
  - 76.80. Mössbauer effect; other γ-ray spectroscopy
  - 76.90. Other topics in magnetic resonances and relaxation

77. Dielectric properties and materials
(for conductivity phenomena, see 72.20 and 72.80)
  - 77.20. Permittivity
  - 77.30. Polarization and depolarization effects
  - 77.40. Dielectric loss and relaxation
  - 77.50. Dielectric breakdown and space-charge effects
  - 77.55. Dielectric thin films
  - 77.60. Piezoelectricity and electrostriction (for piezo-optical effects, see 78.20)
  - 77.70. Pyroelectric and electrocaloric effects
  - 77.80. Ferroelectricity and antiferroelectricity
  - 77.85. Electrical resonances
  - 77.90. Other topics in dielectric properties and materials

78. Optical properties and condensed-matter spectroscopy and other interactions of matter with particles and radiation
(for phonon spectra, see 63)
  - 78.20. Optical properties and materials
  - 78.30. Infrared and Raman spectra and scattering
  - 78.35. Brillouin and Rayleigh scattering
  - 78.40. Visible and ultraviolet spectra
  - 78.45. Stimulated emission
  - 78.50. Impurity and defect absorption in solids
  - 78.55. Photoluminescence
  - 78.60. Other luminescence and radiative recombination
  For photoconduction and photovoltaic effects, see 72.40
  - 78.65. Optical properties of thin films
  - 78.70. Other interactions of matter with particles and radiation
  - 78.90. Other topics in optical properties of condensed matter and other interactions of matter with particles and radiation

79. Electron and ion emission by liquids and solids; impact phenomena
  - 79.20. Impact phenomena, including electron spectra and sputtering
  - 79.40. Thermionic emission
  - 79.60. Photoemission and photoelectron spectra
  - 79.70. Field emission and field ionization
  - 79.75. Exoelectron emission
  - 79.80. Resonance tunneling
  - 79.90. Other topics in emission and impact phenomena in condensed matter

Substance Classification (according to physica status solidi)

S1. Transition metals and their alloys and compounds
(for oxides, see S.10)
- S1.1. Fe
- S1.2. Other 3d metals
- S1.3. Noble metals
- S1.4. Lanthanides
- S1.5. Actinides
- S1.61. Carbides, silicides, germanides
- S1.62. Nitrides, phosphides, arsenides
- S1.63. Sulfides, selenides, tellurides

S2. Alkali metals and their alloys and compounds
(for alkali halides, see S9.11)

S3. Alkaline earth metals and their alloys and compounds
(for alkaline earth halides, see S9.12)

S4. Other metals and their alloys

S5. Nonmetallic elements
- S5.11. Silicon
- S5.12. Germanium
- S5.2. Solidified Gases

S6. Group IV compounds
(for transition metal compounds, see S1.61)

S7. Group V compounds
(for transition metal compounds, see S1.62)
- S7.1. III–V compounds
- S7.11. Phosphides
- S7.12. Arsenides
- S7.13. Antimonides
- S7.15. Mixed crystals
- S7.16. Higher compounds

S8. Chalcogenides
(for transition metal compounds, see S1.63)
- S8.1. II–VI compounds
- S8.11. Sulfides
- S8.12. Selenides
- S8.13. Tellurides
- S8.15. Mixed crystals
- S8.16. Higher compounds

S9. Halides
- S9.11. Alkali halides
- S9.12. Alkaline earth halides
- S9.15. Mixed crystals
- S9.16. Higher compounds

S10. Oxides
- S10.1. Simple oxides
- S10.15. Mixed crystals

S11. Other higher compounds
- S11.1. Salts of oxy-acids
- S11.2. Spinels, garnets, ferrites

S12. Organic compounds
- S12.1. Quasi-one-dimensional systems

Review Article

phys. stat. sol. (b) **133**, 17 (1986)

Subject classification: 71.55; 72.20; 72.40; S8; S8.15

Institut für Experimentalphysik der Universität Linz[1])

Deep Level Defects in Narrow Gap Semiconductors

By

K. Lischka

Contents

1. Introduction

The investigation of *deep* level defects in *narrow* gap semiconductors became a vital field in semiconductor physics. It has been realized that these defects have an important influence on the performance of devices made from narrow gap semiconductors and that some basic physical properties of deep defects can be studied very effectively in connection with narrow gap semiconductors.

Traditionally defects in semiconductors were classified according to their binding energy rather than to their physical properties. Deep defect levels were considered to be at least 100 meV separated from the next band edge. Binding energies of this magnitude are comparable with the band gap energy of narrow gap semiconductors and therefore deep level defects were considered to be of minor importance in narrow gap semiconductors.

However, a recent theoretical model [1] showed clearly that the localization of defect states in semiconductors is not correlated with their binding energy. Adopting this concept, deep defect states are bound by the central cell potential of the defect (and therefore cannot be described by the hydrogen effective mass theory); deep

[1]) A-4040 Linz-Auhof, Austria.

levels do not follow the next band edge when the host band structure is varied either by alloy mixing, the application of hydrostatic pressure or a high magnetic field, or simply by the variation of the lattice temperature.

It is the purpose of this article to review experimental and theoretical studies of defect states in narrow gap semiconductors which are deep due to the above definition, irrespective of their binding energy. We will not only include defects which have bound states in the forbidden gap but also those defects which form quasilocalized states resonant with a band continuum. The latter are believed to play an important role for the doping of narrow gap semiconductors. However, this paper is not intended to cover all the numerous investigations of deep level defects in narrow gap semiconductors. We will emphasize some selected defects in IV–VI and II–VI narrow gap compounds which were found to have either interesting physical properties or are important for the technological application of these semiconductors.

In Chapter 2 we start with a description of experimental methods for deep level investigations which were applied to narrow gap semiconductors. Chapter 3 deals with deep level theories and their application to defects in narrow gap semiconductors. Experimental results concerning defects in IV–VI compounds are summarized in Chapter 4. Intrinsic defects and impurity states in these semiconductors were found to have quite unusual properties. Chapter 5 is concerned with defects in the $Hg_{1-x}\cdot Cd_xTe$ system. Defect states in zero gap $Hg_{1-x}Cd_xTe$ have been studied quite extensively and revealed the first examples of resonant defect levels in semiconductors. Their properties were described successfully by approaches which assumed a localized defect potential. Deep defect states in semiconducting $Hg_{1-x}Cd_xTe$, an important material for the production of infrared detectors, have been characterized by deep level methods. Their influence on device performance has been confirmed. In Chapter 6 we give a conclusive summary and outlook.

2. Experimental Methods

Experimental methods for the investigation of deep level defects in semiconductors were designed to measure the binding energy of defect states, the thermal and optical cross-sections for carrier capture and emission, as well as the density of deep defects. Beside the determination of physical parameters a fundamental problem of defect investigations is the identification of the chemical nature of the defect under consideration. Deep level methods have been described in a number of excellent reviews and books [2 to 7]. We will describe only those experimental methods which have been successfully applied to narrow gap semiconductors and discuss briefly the reasons why some important methods have not been used with narrow gap semiconductors so far.

The determination of the free carrier concentration and mobility by Hall effect and conductivity measurements as a function of sample temperature is still one of the simplest and most effective methods for the determination of the concentration and the binding energy of defects in semiconductors. It has been argumented that this method is more suitable for shallow than for deep defect levels since for the latter one has to go to rather high temperatures in order to ionize the defect levels. This argument, definitely right for standard wide gap semiconductors does not necessarily hold for narrow gap semiconductors since binding energies of deep states in these compounds are comparable with that of shallow states in wide gap semiconductors. Actually a number of deep trap states and also quasilocalized resonant levels in narrow gap semiconductors were observed for the first time by Hall measurements.

The sensitivity of this method is limited, however, by the concentration of shallow donors or acceptors. A current approach to eliminate the undesired free carriers is the investigation of trap states in the depletion region of a reverse biased p–n junction or a Schottky barrier. A number of junction techniques were developed which are capable to reveal a set of defect parameters [4 to 6]. Their applicability to narrow gap semiconductors is constrained to those materials which allow the production of junctions with good reverse characteristics. A frequently quoted figure of merit for diodes is the dynamic resistance–area product (R_0A) at zero bias voltage. Typical values of R_0A for $Hg_{1-x}Cd_xTe$ diodes exceed that of diodes from IV–VI materials with comparable gap energy by a factor of 10^3 to 10^4. For this reason, junction techniques have been successfully used with II–VI compounds, but no investigations with IV–VI semiconductors are reported. Deep levels in $Hg_{1-x}Cd_xTe$ were observed by deep level transient spectroscopy (DLTS) [8], admittance spectroscopy [9], and by the thermally stimulated current (TSC) technique [10].

The DLTS, developed by Lang [4], is probably the most popular capacitance method for the study of deep level defects. Both minority and majority carrier traps can be investigated by this method. Basically a DLTS apparatus consists of a fast capacitance bridge which allows to monitor the capacitance of the sample which may be a diode, a Schottky barrier, or any structure in which a depletion region can be set up. The device is kept at reverse bias producing a depletion region with empty traps. Short voltage pulses are used to modulate the depletion region and bring to the junction region free electrons and holes which can be captured by empty traps. After the end of the forward voltage pulse the trapped carriers are emitted thermally giving rise to a transient in the junction capacitance which is monitored by the bridge. From the transients at various sample temperatures the thermal emission rate e, the defect binding energy E_D, and the concentration N_D can be evaluated. The capture cross-section σ is found by varying the length of the fill pulse. Concentration profiles $N_D(z)$ are measured by varying the reverse bias voltage or the voltage of the filling pulse. Both minority and majority carrier traps can be investigated by this method.

In admittance spectroscopy the conductance and capacitance of a junction is measured applying a small ac voltage to the sample and detecting the signal in phase and the signal 90° out of phase with a double lock-in amplifier. Free carriers are swept into and out of the depletion region. At high temperatures and low frequencies defect levels at the edge of the junction region can fill and empty at the applied frequency. This produces a capacitance signal proportional to the charge of trapped carriers. At low temperatures the capture and emission processes cannot follow the applied field and the capacitance signal goes to zero. From the step in the capacitance signal N_D can be obtained. The emission rate e can be measured from the conductance signal which has a maximum at approximately that temperature where $2\pi f = e$, with f being the frequency of the ac voltage.

The thermally stimulated capacitance (TSCAP) and thermally stimulated current (TSC) methods are single-shot deep level techniques. A reverse biased junction is cooled to low temperatures where the thermal emission of trapped carriers becomes negligible. Then the bias is reduced to a low reverse level for a short time and traps in the depletion region fill rapidly. Following the fill pulse the reverse bias is re-established and the sample temperature is increased with a constant heating rate. The emission of trapped carriers is observed then by monitoring the junction capacitance and current allowing the trap energy and concentration to be evaluated.

Narrow gap semiconductors find their main application in the field of infrared opto-electronic devices. Consequently numerous measurements of the excess carrier lifetime were performed. Beside important information on intrinsic band–band recombination, these experiments revealed some data on deep defect levels. Carrier recombination was studied by transient photoconductivity (TPC) and the photoelectromagnetic effect (PEM) in bulk samples.

In transient photoconductivity experiments an excess carrier concentration is generated by short light pulses. The transient of the conductivity after the end of the light pulse is a direct measure of the excess carrier lifetime. The temperature variation of the lifetime reveals the position of the defect levels in the gap. When the density of the defects is known capture cross-sections can be also obtained. The same information can be obtained from steady state photoconductivity in combination with the PEM effect yielding the minority carrier lifetime in the sample. The PEM ʌoltage drop across a sample results from the diffusion of photogenerated carriers in a magnetic field perpendicular to the direction of the light propagation.

All experimental methods described up to now are able to give thermal binding energies and thermal cross-sections of the defect levels. However, for a thorough characterization of a deep defect both thermal and optical energies and cross-sections arc important. According to the well-known Frank-Condon principle, the coupling between localized defect states and the lattice vibrations shows up in a difference between thermal and optical energies. The optical parameters are usually measured in absorption, spectral impurity photoconductivity, and photoluminescence experiments.

Standard absorption experiments in the sub-band-gap (impurity absorption) region suffer from the fact that the optical cross-sections of localized defect states are small giving a small signal-to-noise ratio. The sensitivity is improved significantly by the population modulation spectroscopy [11]. In this experiment the transmitted intensity of a tunable dc probe beam with a photon energy $\hbar\omega \leqq E_g$ is modulated by a fixed wavelength chopped pump beam with a photon energy $\hbar\omega_p \geqq E_g$.

In photoluminescence experiments the radiative recombination of excess electron–hole pairs generated by the intense light of a laser beam ($\hbar\omega \geqq E_g$) is observed. It is the advantage of this method that transitions involving defect states show up in relatively sharp structures and it has been used widely for the investigation of wide gap semiconductors. However, the spectral range of photoluminescence from narrow gap semiconductors is the middle and far infrared, where the sensitivity of photon petectors is small compared to the visible and near infrared region, and due to the small binding energy of the defects competing nonradiative processes are effective. Photoluminescence involving gap levels in II–VI and IV–VI narrow gap compounds was reported recently [12, 13].

In impurity photoconductivity experiments, the photoconductivity response of a sample for light in the sub-band-gap region is measured. For transitions from deep states to band states the observed structures in the spectral response are relatively broad and featureless. This limits the accuracy of the impurity photoconductivity method for the determination of the photoionization energy of deep traps. Serious advantage results from the application of high magnetic fields. Resonant transitions between gap states and Landau bands give sharp structures in the magneto-photoconductivity response. Recently resonant structures in the magneto-photovoltaic response of a Schottky barrier on PbTe were observed [14] which were due to transitions between a gap state and Landau levels of both conduction and valence bands, giving not only the binding energy of thc defect state with high accuracy but also valuable data on the influence of a high magnetic field on localized defect states.

All experimental methods described so far are able to provide data of the defect under investigation but do not lend themselves to investigate the chemical nature of the defect, allowing to identify certain defect levels with specific impurity atoms or native defects. Magnetic resonance methods, e.g. electron paramagnetic resonance

(EPR) and optically detected magnetic resonance (ODMR), proved to be useful in completing our understanding of the chemical nature of defects in semiconductors. However, to the author's knowledge, no published result of defect identifications using these techniques exists for narrow gap semiconductors. The reasons for that are most likely the difficulties with the preparation of bulk samples with sufficiently low carrier concentration.

The lack of ESR data is a major limitation of deep level investigation in narrow gap semiconductors, since strictly speaking, the chemical nature of any defect centre in these materials has been identified. However, there were some conclusive suggestions drawn from the experimental data which shed some light on the identification of defects.

3. Theoretical Models for Deep Level Defects in Narrow Gap Semiconductors

The recent developments in the theoretical understanding of deep level defects in semiconductors were summarized in a number of reviews [15 to 17].

Two complementary concepts were used to handle the deep defect problem: a) the first-principle theories [18 to 21] and b) semiempirical models [1, 22, 23]. The first principle type of models employ self-consistent pseudopotential techniques to calculate the wave function and binding energy of defect states with a high degree of accuracy. Disappointingly, the calculations involved are rather laborious. Consequently, the application of first-principle methods has been limited to some selected defect–host systems only; narrow gap semiconductors were not treated.

Semiempirical models employ a modern version of the tight binding theory augmented by the Koster-Slater Green's function technique. They allow to investigate qualitatively the properties of a great number of different defects in various semiconductors. It was possible to show by relative simple calculations how the defect level energy of a certain impurity will change as the host changes or to calculate the ordering of the energy levels of different impurities in a single host.

It was a main advantage of the tight binding model that a number of approximate rules were set up allowing a new definition of deep defect states. According to the tight binding picture a defect level is deep regardless of its binding energy, if it does not follow a nearby band edge when the host band gap is varied by the application of pressure or due to alloy mixing. The wave function of deep levels is a host-like state, localized at the dangling bonds of the atoms surrounding the defect. As a consequence, different impurities at the same site of a semiconductor will have similar properties and chemical trends.

Tight binding calculations of deep level impurities in narrow gap semiconductors were published for the $Hg_{1-x}Cd_xTe$ system [24]. Results are shown in Fig. 1. The trends of defect levels from different impurity atoms occupying the same position in the lattice are similar. A number of impurities which have deep defect levels in the forbidden zone of wide gap CdTe form resonant states in the narrow gap compound, demonstrating the necessity to include resonant levels in a review on deep level defects in narrow gap semiconductors.

Defects in IV–VI compounds have not been subject to tight binding calculations. However, an early attempt to understand the lack of thermal freeze out of free carriers at low temperatures involved the calculation of the electronic energy levels associated with vacancies in PbTe [25, 26] by the Green's function method of Koster and Slater [27]. The effect of the vacancies was treated as a time independent localized potential. The perturbed wave functions were expanded in terms of free carrier wave functions of the unperturbed lattice. One important result of this calculation was that both Pb and Te vacancies do not produce bound states in the gap. The localized defect

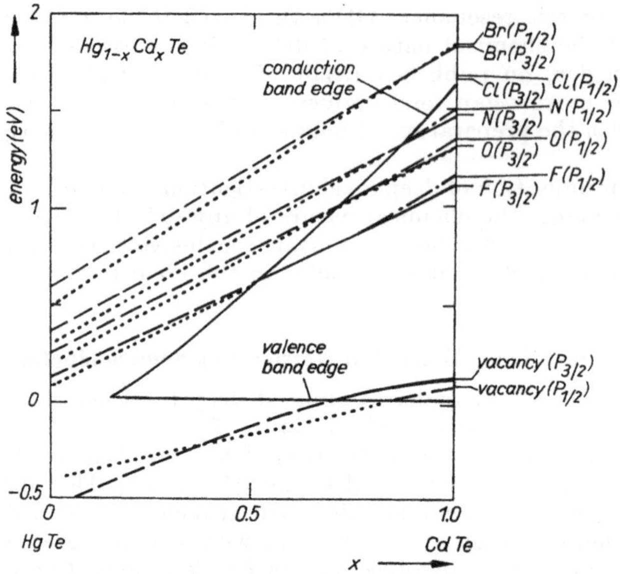

Fig. 1. Substitutional cation site impurity levels in $Hg_{1-x}Cd_xTe$ as a function of x. The level energies were calculated by the tight binding method (after [24])

potential of a vacancy splits off electronic levels from the valence band to the conduction band. As shown in Fig. 2 a Te vacancy promotes four spin degenerate electron states to the conduction band. The neutral Te atom leaving the crystal takes six valence states to the conduction band. The neutral Te atom leaving the crystal takes six valence electrons with it. Consequently two electrons in the crystal cannot find a free state in the valence band. They occupy the lowest conduction band states and contribute to the electrical conductivity. A Pb vacancy was found to contribute two free holes. The results of [25, 26] were corroborated by cluster calculations for intrinsic defects in the Pb salts [28].

Recently the donor or acceptor behaviour of group III elements substituting Pb in PbTe has been calculated [29]. It was assumed that the balance between a) the energy needed to promote the s-level electrons of the impurity element above the top of the valence band of the host crystal and b) the lattice energy gained as a result of the increased charge and smaller radius of the impurity are responsible for its doping properties. The lattice distortion energy and the polarization were calculated assuming an ionic lattice model. The calculations revealed an acceptor action of Tl and Ga and a donor action of In and Al.

The influence of lattice distortion has also been incorporated in a recent model of

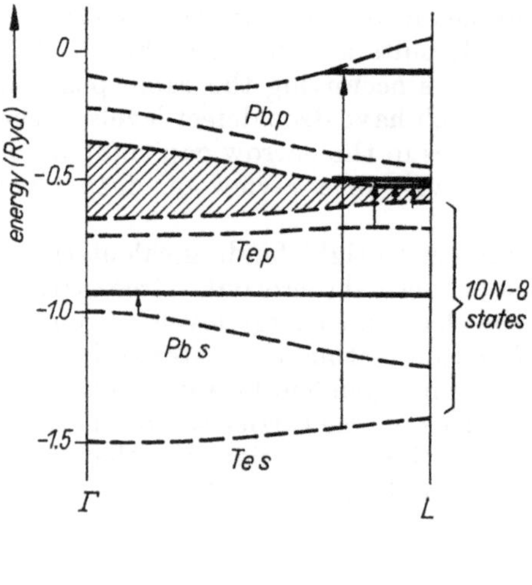

Fig. 2. Sketch of the PbTe band structure as perturbed by a short-range Te vacancy potential. Localized states are split from the valence bands leaving $10N - 8$ states for $10N - 6$ electrons forcing two electrons into the conduction band (after [25])

the Te vacancy in PbTe [30]. The defect states were calculated assuming that the interatomic bonds in IV–VI semiconductors are constructed primarily of atomic p-orbitals. According to [30] a Te vacancy forms a singly (Γ_{6-}) and a doubly (Γ_{8-}) degenerate state within the conduction band. A change of the charge state of the defect centre due to electron capture will lift the degeneracy of the Γ_{8-} level by Jahn-Teller distortion, yielding a gap level and a level in the conduction band. The Jahn-Teller model has been used for the interpretation of experimental data from In doped $Pb_{1-x}Sn_xTe$ and will further be discussed in connection with these data in Chapter 5.

4. Deep Level Defects in IV–VI Compounds

Among the semiconductors of technological relevance, the IV–VI compounds are the only system with ten valence electrons per unit cell. They crystallize in the rocksalt structure with an octahedral coordination of the next neighbour atoms [31]. The band structure has a direct gap at the L point of the Brillouin zone. Some ternary alloys exhibit a zero gap transition which is due to an accidental degeneracy of bands and not symmetry induced as in II–VI compounds. An unusual property of the lead chalcogenides is the high electrical conductivity of undoped material. The type and density of free carriers present depends on the deviation from stoichiometry of the compound.

The main application of the lead chalcogenides has been shifted from the production of infrared sensitive photovoltaic detectors [32, 33] to tunable infrared emitting laser diodes [34].

In the following sections we will summarize experimental investigations of defects in nominally undoped and in doped lead chalcogenides.

4.1 Intrinsic defect levels in IV–VI compounds

Crystalline lead chalcogenides in thermal equilibrium at elevated temperatures contain significant concentrations of lattice point defects like vacancies and interstitials which produce deviations from the stoichiometric composition. It has been realized quite early [35] that these defects, which can be quenched by rapid cooling, are electrically active and contribute free carriers. Crystals with excess metal usually show n-type conductivity, crystals with excess of the non-metal element are p-type. The coexistence region of various lead chalcogenides was obtained from anneal experiments under metal and non-metal overpressure. A typical diagram is shown in Fig. 3. The carrier concentration obtained from Hall measurement at 77 K for PbTe and $Pb_{1-x}Sn_xTe$ is plotted versus the reciprocal isothermal annealing temperature. Intrinsic defects were associated with donors and acceptors. The measured carrier concentration was assumed to be equal to the deviation from stoichiometry of the compound. The bound states of intrinsic defects were assumed to be hydrogen-like states with a relative small ionization energy due to the extremely large static dielectric constant of the lead chalcogenides. However, magnetic freeze-out experiments did not reveal any evidence for the existence of hydrogen-like intrinsic defect states [36 to 38].

In the model of Parada and Pratt [25, 26], which was described in Chapter 3, the properties of vacancies and interstitials in the lead chalcogenides were discussed in a localized defect potential approach. A Pb vacancy was found to be the source of holes and a chalcogenide vacancy a source of two electrons. These defects have no trap levels in the gap, but resonant states deep in the valence or conduction band and freeze out of free carriers cannot occur according to this model.

The resonant defect states have been the subject of experimental investigations carried out with samples containing a high density of intrinsic point defects. The defects were generated by bombarding the crystals with high energy ions [39 to 41].

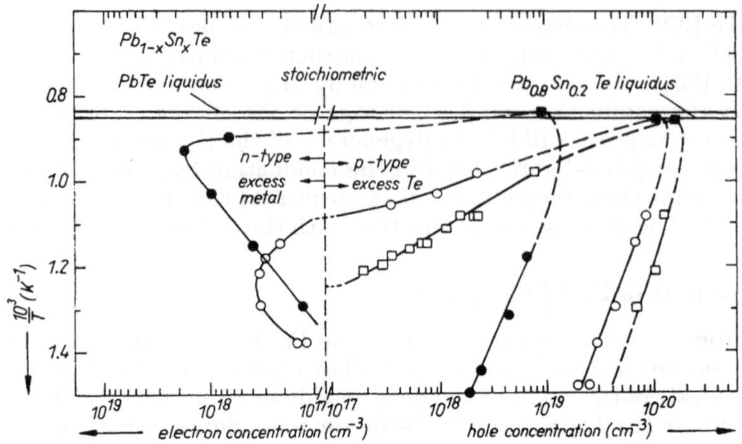

Fig. 3. Carrier concentration obtained by Hall measurements at 77 K as a function of the isothermal annaeling temperature for several $Pb_{1-x}Sn_xTe$ alloy compositions (after [35]). ● $x = 0$, ○ 0.13, □ 0.20

Ion implantation in IV–VI compounds was recently reviewed [42]. We will summarize here only the most important results. Ion implantation of the host lattice elements (e.g. Te and Pb in PbTe) and subsequent annealing of the lattice damage revealed a single donor action of implanted Pb atoms [43]. Implanted Te atoms were found to be electrically inactive. It was assumed that most of the implanted atoms sit at interstitial sites after the anneal process at moderate (≈ 300 °C) temperatures. The resulting donor action of Pb interstitials is in agreement with calculations of interstitial defect levels which showed that Pb interstitials are donors and Te interstitials are either neutral or weak acceptors in PbTe [44].

It is well established experimentally [45 to 47] that the ion implantation damage in PbTe reveals n-type conduction, irrespective of the ion species used for implantation. The free electron density saturates with increasing damage concentration (which was obtained in Rutherford backscattering measurements [39]). In Fig. 4 the electron concentration in the damage region of PbTe samples irradiated with 300 keV Pb^+ ions is plotted as a function of the implantation dose. Independent of the type and the density of free carriers in the samples prior to the ion bombardment, the carrier concentration in the damage region was equal for all samples when the dose of implanted ions exceeded about 10^{13} ions/cm². It is important to note that the saturation of the

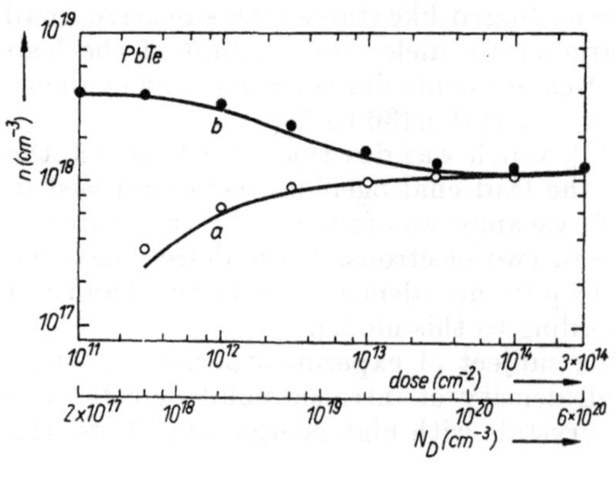

Fig. 4. The room temperature electron concentration, n, of ion implantation damage in two PbTe samples. Different doses of 300 keV Pb^+ ions were implanted. Points are experimental values, the curves are calculated with the intrinsic defect model described in the text. N_D is the assumed density of Frenkel defects (after [41])

electron concentration occurred well below those fluences (10^{14} ions cm^{-2}) where saturation of the lattice damage was observed in Rutherford backscattering experiments.

The curves in Fig. 4 were calculated using a model [41] which was based on the following assumptions: a) The region damaged by ion implantation is crystalline at least for a dose below 10^{14} ions cm^{-2}, containing a large number of point defects. The point defects govern the electrical properties of the damage region. b) The densities of anion and cation vacancy–interstitial pairs (Frenkel defects) are equal. This is not a stringent point. Slightly different densities of anion and cation Frenkel defects do not change the basic features of the model. c) According to [25, 26, 43] Te vacancies are assumed to contribute two electrons, Pb vacancies two holes, and Pb interstitials are single donors. The net effect of one anion–cation Frenkel pair is to create one free electron giving n-type conduction in the damage region. d) Tellurium vacancies create two defect levels close to the conduction band edge. For simplicity the first and second ionization energies of Te vacancies were assumed to be equal.

According to this model each pair of Frenkel defects contributes one free electron but two (resonant) trap levels, resulting in a pinning of the Fermi level when the concentration of defects is increased (curve a in Fig. 4). In case the samples have an electron concentration prior to implantation, which exceeds the saturation carrier density (this was realized by impurity doping), the concentration of free electrons decreases with increasing density of Frenkel defects, as shown by curve b in Fig. 4.

The above model has been successfully applied for the interpretation of the dose and temperature dependence of the free carrier concentration in ion bombarded PbTe [40], $Pb_{1-x}Sn_xTe$ [41, 48], and $PbTe_{1-x}Se_x$ [49]. The only adjustable parameter was the energetic position of the anion vacancy levels. The position of the levels at 0 K was obtained from the saturated carrier concentration at low temperature. Fig. 5 shows the defect level position at 0 K for the $PbTe_{1-x}Se_x$ and $Pb_{1-x}Sn_xTe$ systems.

According to [25, 26] the defect states in the intrinsic defect model were assumed to be due to anion vacancies. However, no direct experimental identification of the defects was possible. It has been noted [49] that an anion vacancy in $PbTe_{1-x}Se_x$ has the same neighbour atoms for all values of x, in contrast to $Pb_{1-x}Sn_xTe$, where the cation element is varying with composition. Since vacancy states can be regarded as built up mainly by the dangling bonds of the next neighbours [1] it can be understood qualitatively why no substantial shift of the intrinsic (vacancy) level is observed in $PbTe_{1-x}Se_x$, while in $Pb_{1-x}Sn_xTe$ a pronounced shift from the conduction band down to the valence band of SnTe is observed.

It has been pointed out in [48] that the intrinsic defect model has important consequences for the interpretation of phase diagrams particulary of the $Pb_{1-x}Sn_xTe$ system.

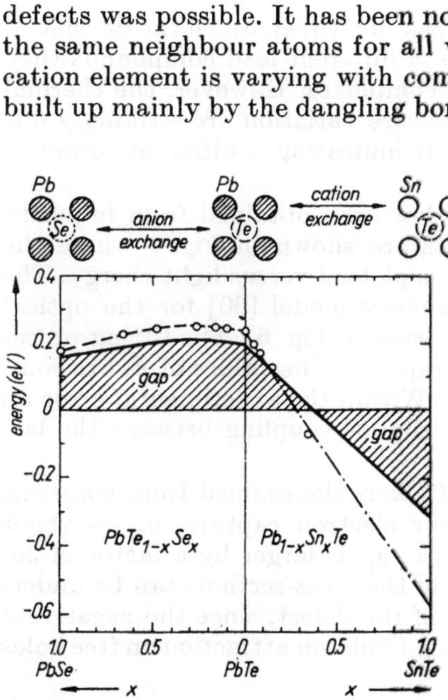

Fig. 5. The position of ion implantation induced defect levels in $PbTe_{1-x}Se_x$ and $Pb_{1-x}Sn_xTe$ (after [49]) at $T = 0$ K

The phase diagrams were obtained by annealing $Pb_{1-x}Sn_xTe$ crystals in metal or Te-rich atmosphere. The resulting carrier concentration was determined by Hall measurements. The concentration of free carriers was correlated with the density of vacancies assuming that they form donors or acceptors with zero ionization energy. This concept does not hold, however, if one accepts that the anion vacancy level moves down to the valence band with increasing x in $Pb_{1-x}Sn_xTe$. Anion vacancies in $Pb_{1-x}Sn_xTe$ have to be considered either neutral or doubly ionized depending on the position of the Fermi level. Consequently, the left branch of the phase diagram in Fig. 3 for $x \geqq 0$ shows the compensation of anion and metal vacancies and does not reveal the deviation from stoichiometry.

Conclusively, the intrinsic defect model is able to explain the electric properties of ion implantation damage in the lead chalcogenides. It has been successfully applied to explain the asymmetry of the phase diagram of $Pb_{1-x}Sn_xTe$ in a quite natural way. The assumptions which were made are in good agreement with theoretical models of intrinsic point defects in the lead chalcogenides, for instance the donor or acceptor character of vacancies. However, the chemical nature of the defects which bring about the pinning of the Fermi level has not been identified although some evidence exists that the assumption of anion vacancies is favourable. The main question that arises in connection with the intrinsic defect level model is how a resonant state can bind free carriers. It seems that lattice relaxation around the defect has to be considered as important.

Lead chalcogenides which were not doped by adding intentionally impurities during or after the crystal growth process have found considerable technological applications. Experiments which revealed the existence of deep trap levels in the gap of undoped lead compounds have been reviewed recently [50]. We will, therefore, only shortly summarize the properties of these traps and add new data which have not been considered.

The binding energy of the traps E_T, and its temperature variation, dE_T/dT, was obtained by different experiments. The values of E_T and dE_T/dT from traps in different lead compounds are listed in Table 1.

The chemical nature of these traps has not yet been identified. Since the density of the traps was within the same order of magnitude in different lead compounds they may be associated with intrinsic defects or defect complexes. However, the thermal activation energy of the trap levels and its temperature variation are strikingly different from that of damage induced resonant levels indicating a different origin of trap and resonant states.

The optical ionization energy of traps in PbTe has been obtained from impurity photoconductivity measurements [56]. The results are shown in Fig. 6 where the photoresponse of PbTe in the sub-band-gap region is plotted versus light energy. The experimental data points were fitted by the Lucovsky model [60] for the optical ionization of localized defects in semiconductors (curves in Fig. 6). The fitting procedure revealed the optical ionization energy of the traps, E_i. The values of E_i are compared with the thermal activation energy in Fig. 7. Within the experimental error no significant difference was observed indicating a negligible coupling between the lattice and the trap states.

The capture cross-sections of trap levels in PbTe were determined from transient photoconductivity data [56]. The cross-section for electron capture, σ_n, is about 10^{-20} cm^2, the capture cross-section for hole capture, σ_p, is larger by a factor of 30. It was pointed out in [56] that the relative values of the cross-sections can be understood by the assumption of an acceptor character of the defect, since the negatively charged trap centre occupied by an electron exerts a Coulomb attraction on free holes giving $\sigma_p \geqq \sigma_n$.

Table 1

The binding energy E_T of trap levels in lead chalcogenides (the energy scale zero is the conduction band edge)

compound	x	E_T at 0 K (meV)	dE_T/dT (meV/K)	experimental method*)	ref.
PbS	—	−210	—	H, TPC	[51]
PbSe	—	− 20	−0.20	H, SPR	[52, 53]
	—	−130	−0.32	H	[52, 53]
$PbTe_{1-x}Se_x$	0.26	− 16	0.05	H	[49]
	0.48	− 12	0.03	H	[49]
	0.66	− 4	0.12	H	[49]
PbTe	—	− 75	0.38	H, TPC, SPR	[54 to 56]
	—	−160	0.91	H	[57]
$Pb_{1-x}Sn_xTe$	0.02	− 60 ± 10	—	TPC	[56]
	0.03	− 45 ± 10	—	TPC	[56]
	0.05	− 30 ± 10	—	TPC	[56]
	0.08	− 40 ± 10	—	TPC	[56]
	0.11	− 25 ± 10	—	TPC	[56]
	0.18	− 20	—	H	[58]
	0.17 to 0.22	− 30 ... −50	—	PC, PEM	[59]

*) Experimental methods: Hall effect (H), photoconductivity (PC), transient photoconductivity (TPC), photoelectromagnetic effect (PEM), spectral photoresponse (SPR).

The application of absorption methods for the determination of the ionization energy E_i of deep traps is limited due to the relatively broad structures which are observed. This is a consequence of the spread of the trap state wave functions in k-space, having

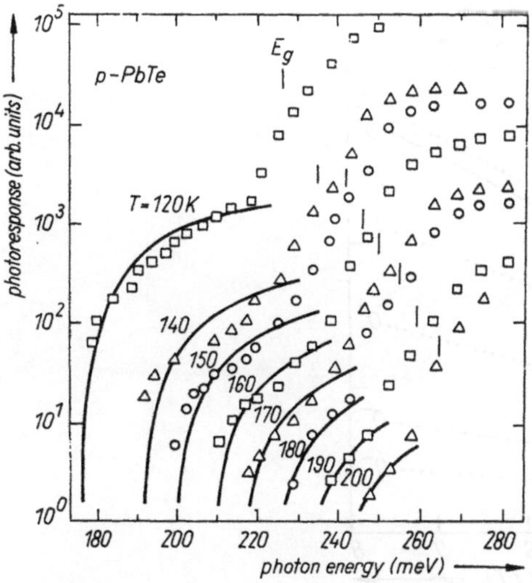

Fig. 6

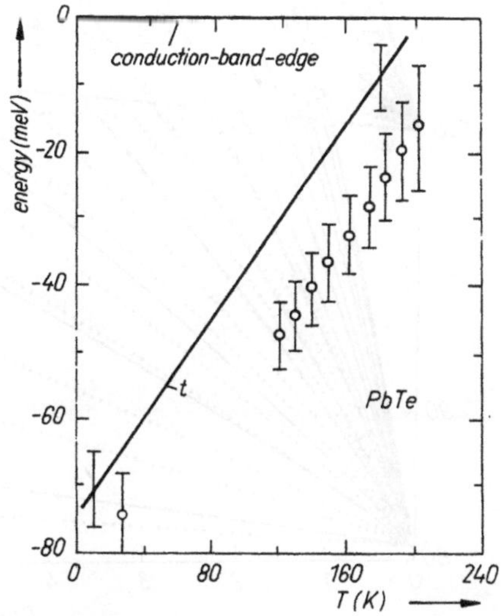

Fig. 7

Fig. 6. Spectral photoresponse of p-PbTe at various temperatures. The data in the sub-band-gap region are fitted by Lucovsky's model of deep level photoionization (full curves) (after [56])

Fig. 7. Optical (o) and thermal (t) activation energies of a deep trap level in PbTe vs. lattice temperature (after [56])

a maximum density at those values of $\boldsymbol{k}$ where the density of band states has a minimum. It can qualitatively be understood by the density-of-state picture that transitions between a trap level and a band occur in a wide range of energies with a maximum transition rate at about $2E_i$ [2]. The application of a high magnetic field B modifies this picture considerably. The band states are split into Landau sub-bands which have a pronounced maximum of the density of states close to the band edge. Resonant optical transitions between trap states and the extrema of the Landau bands will occur. Narrow gap semiconductors provide suitable conditions for the observation of these resonant transitions. The effective mass of free carriers is small yielding a relatively large splitting of Landau levels. The mobility of free carriers is high giving full quantization even at low magnetic fields. Resonant structures in the photoconductivity of InSb due to transitions between a mid-gap level and the conduction band Landau levels were observed in [61]. Illumination of a Schottky barrier on p-PbTe with sub-band-gap light gave resonant structures in the magneto-photovoltaic effect [14].

Experimental results for PbTe are shown in Fig. 8. The magnetic field positions of the maxima of the photovoltaic response are indicated by dots. The curves are the

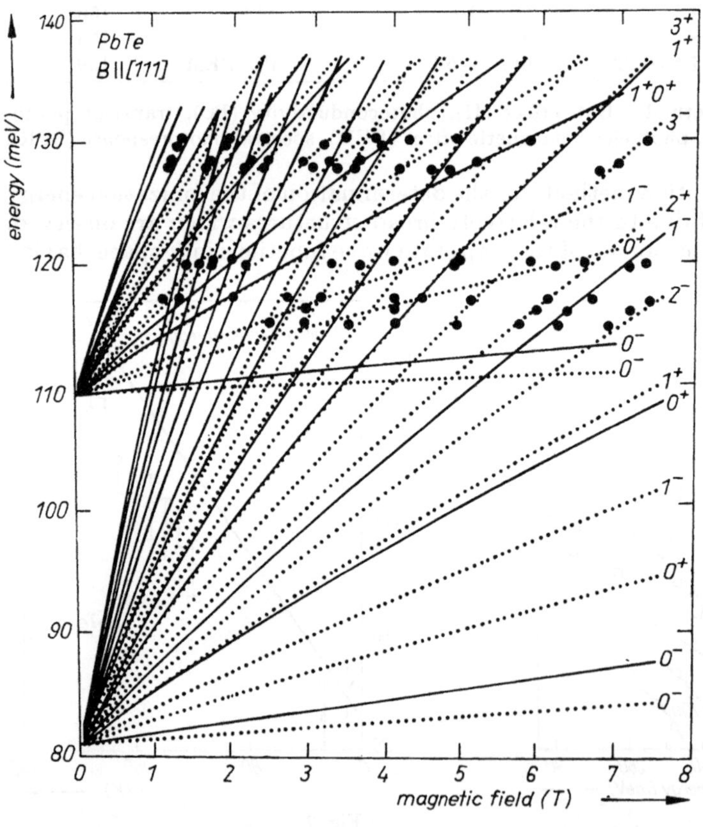

Fig. 8. Resonant structures of the magneto-photovoltaic effect of a Schottky barrier on p-PbTe obtained with different CO_2 laser emission lines (dots). The curves are calculated conduction band Landau levels (lower set) and valence band Landau levels (upper set) of PbTe. Full lines stand for ellipsoids with the long axis along [111]. Dotted lines for the remaining three equivalent ellipsoids with the long axis along ⟨111⟩ directions (after [14])

Fan-chart of the valence band (upper part) and the conduction band (lower part) in PbTe. Assuming that the resonant structures originate from transitions between a deep trap level in the space charge region of the Schottky barrier and the Landau levels of conduction and valence band, respectively, the energy of intersection at $B = 0$ gives the ionization energy of the traps. The trap level is situated 81 meV below the conduction band edge in agreement with impurity photoconductivity data [56]. The good fit of the experimental data by the Fan chart approach indicates that the deep trap level is not effected by the magnetic field. This conclusion is in agreement with a theoretical model for the influence of an external magnetic field B on the binding energy of deep level defects [62]. It was an essential conclusion in [62] that the ground state of a defect which is bound by a short-range potential is actually not affected by the magnetic field and the ground level will not follow the lowest Landau band with increasing B. In contrast, shallow effective-mass-like bound states will closely follow the lowest Landau band edge. Therefore, high magnetic fields may be used to discriminate localized defect states from delocalized, effective-mass-like impurity states, irrespective of their binding energy. The above-mentioned experimental investigations provided first data in support of the theoretical model, showing that the investigation of deep level defects in narrow gap semiconductors is not only important for the technological application of these semiconductors but can add significantly to our understanding of basic deep level physics.

4.2 Deep level impurities in IV–VI compounds

Doping of lead chalcogenides by impurity elements has been studied in a number of experiments. A comprehensive summary of the doping properties of various elements has been given in [63]. It has been suggested that most of the impurity atoms form levels degenerated with a band continuum. Evidence for hydrogen-like impurity states has been found only in a few cases [64, 65]. In this article we will concentrate on group III impurities, particularly In and Ga in PbTe and $Pb_{1-x}Sn_xTe$, since these impurities were found to have a number of interesting and unusual physical properties.

More than a decade ago it was realized [66] that the Hall coefficient of indium doped PbTe crystals with an indium concentration between 0.1 and 6 at% exhibits an unusual temperature dependence as shown in Fig. 9. The Hall data and the dependence of the thermoelectric power on an external magnetic field were explained by the assumption of a quasilocalized twofold degenerated indium defect level above the conduction band edge which pins the Fermi level when the In concentration is high. The pressure coefficient of the indium level, or more exactly, of the pinned Fermi level, was determined to be $dE_{In}/dP = 7 \times 10^{-6}$ eV/bar. Alloying of PbTe with SnTe resulted in a decrease of the pinned Fermi energy [67]. Both, hydrostatic pressure

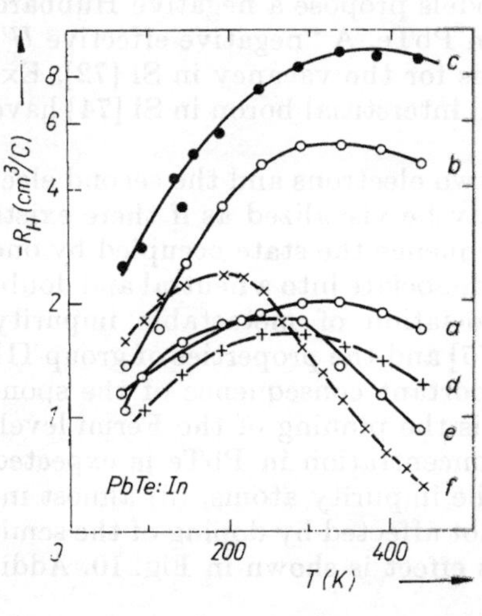

Fig. 9. Temperature dependence of the Hall coefficient, R_H, of PbTe doped with indium: (a) 0.1, (b) 0.5, (c) 0.6, (d) 1, (e) 3, and (f) 6 at% (after [66])

experiments and experiments with ternary alloys indicated a strong localization of the bound state wave function since the associated levels did not show any correlation with a nearby band edge.

It was pointed out by Andreev et al. [68] that a simple defect model for the indium impurity cannot explain the apparent lack of any EPR signal from electrons with unpaired spins in half filled In levels and the absence of any effect of Coulombic repulsion between electrons at the impurity sites. A "variable valence model" (VVM) was proposed for the indium impurity. In this model the bonds in PbTe were regarded as mainly ionic. Indium atoms at Pb sites were considered to be in the In^{2+} state giving two electrons to the conduction band. The essential feature of the VVM was that the In^{2+} states are unstable and decay into In^{+} and In^{3+} by the following process:

$$2\, In^{2+} \rightarrow In^{+} + In^{3+} . \tag{1}$$

The driving force for this reaction is the energy ΔE which is gained by the polarization of the lattice around the impurity ions. In ionic crystals ΔE is given by the expression

$$\Delta E = eV_3 - eV_2 - \frac{e^2}{2R_+}\left(1 - \frac{1}{\varepsilon_0}\right) - \frac{e^2}{2R_{3+}}\left(1 - \frac{1}{\varepsilon_0}\right). \tag{2}$$

V_3 and V_2 are the third and second ionization potentials of In, R_{3+} and R_+ are the radii of In^{3+} and In^{+}, respectively. Using reasonable values for these parameters a negative ΔE was calculated in [68]. Although the degree of the ionicity of the lead salts is still controversial their complex band structure is hardly compatible with a pure ionic model. This idea led Weiser et al. [69] to propose an "autocompensation" model (ACM) for the indium impurity. The ACM, conceptually similar to the VVM differed from the latter by the fact that the relaxation of the lattice around an indium impurity was considered to be important and leads to the dissociation of two neutral indium atoms into a positive and a negative charged state. In a subsequent paper by Weiser [29] two effects, the electronic relaxation and the relaxation of the lattice around group III impurities in PbTe, were considered on the basis of an ionic bond model. It was concluded that lattice polarization and lattice distortion are important in order to explain the energy of group III impurity levels in PbTe. It is difficult to decide, which of the proposed models is to be preferred since the nature of the bonds in IV–VI compounds is still under discussion. However, the common feature of both, the VVM and the ACM, merits discussion. Both models propose a negative Hubbard correlation energy U [70] for the indium impurity in PbTe. A "negative effective U" system [71] was first proposed on a theoretical basis for the vacancy in Si [72]. Experimentally the isolated vacancy in silicon [73] and interstitial boron in Si [74] have been identified as negative U systems so far.

A defect has negative U properties if it can bind two electrons and the second electron is bound more strongly than the first. This may be visualized as if there exists an attraction between the two electrons. As a consequence the state occupied by one electron will be thermodynamically metastable and dissociate into a neutral and doubly charged state. Recently the spontaneous dissociation of metastable impurity states has been reviewed by Drabkin and Moishes [75] and the properties of group III elements in PbTe were discussed in detail. An important consequence of the spontaneous dissociation of metastable impurity states is the pinning of the Fermi level. For large indium concentrations the free carrier concentration in PbTe is expected to be (i) considerably smaller than the density of the impurity atoms, (ii) almost independent of the density of In impurities, and (iii) not affected by doping of the semiconductor by other donor or acceptor centres. This effect is shown in Fig. 10. Addi-

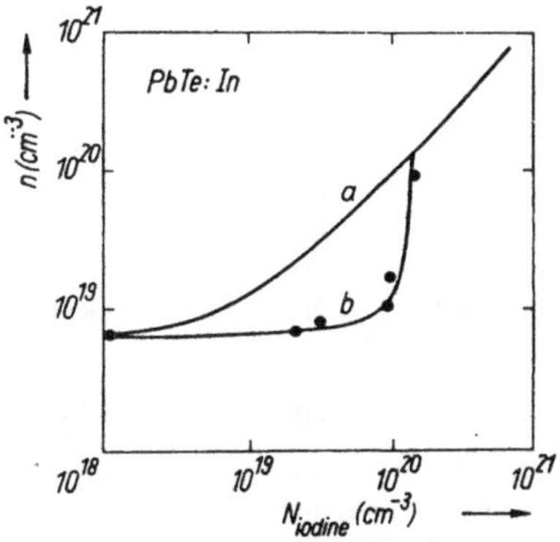

Fig. 10. The electron concentration of In doped PbTe additionally doped by single iodine donors. Dots are experimental data. The curves are calculated assuming (a) one impurity state per indium atom, (b) two impurity states per indium atom (after [66])

tional doping of indium doped PbTe by iodine donors [76] did not affect the free carrier concentration in a wide range of the iodine donor concentration, N_{iodine}.

The curves in Fig. 10 were calculated assuming that each indium impurity can bind a) one and b) two electrons. Noteworthy an effect of the first ionization stage of the In impurities was not observed. These results are corroborated by measurements of the magnetic susceptibility of PbTe doped with 10^{20} cm^{-3} In [77] and indium doped $Pb_{1-x}Sn_xTe$ [78]. Both investigations showed that the density of states with unpaired spins is some orders of magnitude less than the density of the In impurities. Measurements of the low temperature specific heat (1 to 4.2 K) of heavily doped PbTe:In in a high magnetic field [79] did not show the Schottky anomaly [80] as would be expected if impurity atoms with uncompensated spins are present in the samples. All these experimental data are in support of a negative U model for the indium impurity in PbTe and $Pb_{1-x}Sn_xTe$. However, the value of the negative U energy was not measured so far.

The observation of impurity band conduction [81] which was interpreted to be due to an overlap of the wave functions of indium impurity states in heavily doped $Pb_{1-x}Sn_xTe$ ($N_{\text{In}} \geqq 0.2$ at%) does not necessarily contradict the negative U model since (i) the splitting of the impurity levels was not determined and may be small compared to kT at elevated temperatures and (ii) all experimental data in support of a negative U impurity were obtained with samples containing a much smaller indium concentration than those samples showing impurity band conduction.

The interest in PbTe and $Pb_{1-x}Sn_xTe$ doped with indium or gallium was further enhanced by the observation of various anomalous effects which are summarized in Table 2. It is the common feature of these anomalies that they occur only at temperatures below a "threshold" temperature T_0 which is included in Table 2.

About ten years ago a sharp peak of the mobility at about 20 K was observed in transport investigations [82] of $Pb_{0.8}Sn_{0.2}Te$ for the first time. As shown in Fig. 11, this effect was only observed in In doped samples. The data were interpreted as possibly being due to a low temperature phase transition which is favoured by the incorporation of In atoms into the PbSnTe lattice.

The resistivity of $Pb_{0.78}Sn_{0.22}Te$ samples doped with more than 0.4 at% In is plotted in Fig. 12 [83]. The full lines are experimental results obtained with samples facing background radiation from a radiation shield at 4.2 K. Illumination of the samples decreased the low temperature resistivity by some orders of magnitude. The samples remained in the high conductivity state after the light was switched off (persistent photoconductivity). In order to return to the high resistivity state the samples had to be heated above the threshold temperature T_0 of about 20 K. The application of a high electric field resulted in a quench of the high conductance state. The mobility

Table 2

Low temperature anomalies in indium doped $Pb_{1-x}Sn_xTe$

parameter, showing ananomalous behaviour	x	indium concentration (at%)	"threshold" temperature T_0 (K)	ref.
Hall mobility	0.2	0.05 to 0.1	≈20	[82]
Hall mobility, resistivity, persistent photoconductivity	0.22	0.7 to 2.7	≈20	[83]
magnetic susceptibility	0.2 0.25	0.2 to 0.5 0.2 to 0.5	≈30 ≈20	[78]
spontaneous voltage	0.22	0.6	15 ± 1	[86
carrier concentration, Hall mobility spontaneous voltage	0.22 to 0.26	exceeding 0.5	≈15	[87]
persistent photoconductivity	0	≈0.01	≈20	[85]
negative photoconductivity	0.23	0.50 to 1	20 to 40	[89]

of free carriers showed a pronounced maximum at T_0. It was concluded in [83] that these experimental results indicate a phase transition which occurs in $Pb_{1-x}Sn_xTe$ samples doped with indium. Pronounced steps were found in the relaxation of the persistent photoconductivity under the influence of a high electric field [84]. These steps were attributed to the successive relaxation of ferroelectric domains in the samples.

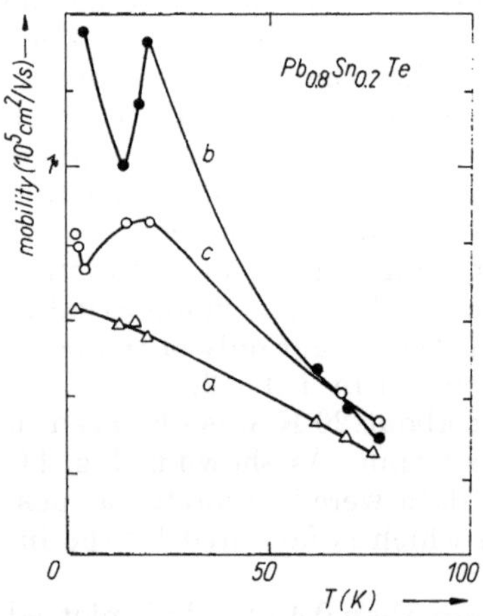

Fig. 11

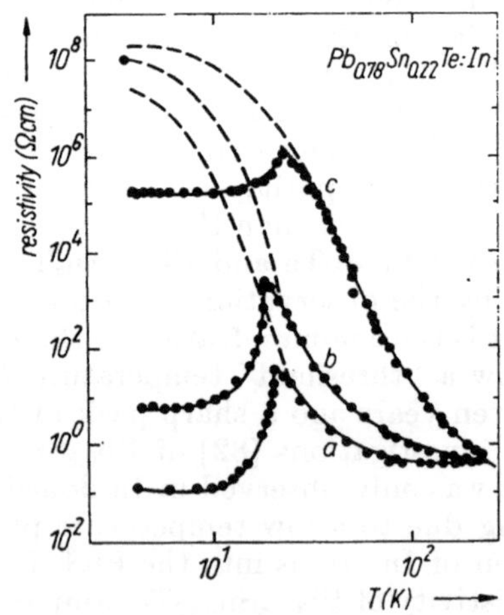

Fig. 12

Fig. 11. Temperature dependence of the mobility of $Pb_{0.8}Sn_{0.2}Te$. (a) Not doped with indium, (b) doped with 0.05 at% indium, (c) doped with 0.1 at% indium (after [82])

Fig. 12. Temperature dependence of the resistivity of In doped $Pb_{0.78}Sn_{0.22}Te$ illuminated with light (full curves), without illumination (dashed curves). The indium concentration is (a) 0.74, (b) 0.97, (c) 2.7 at% (after [83])

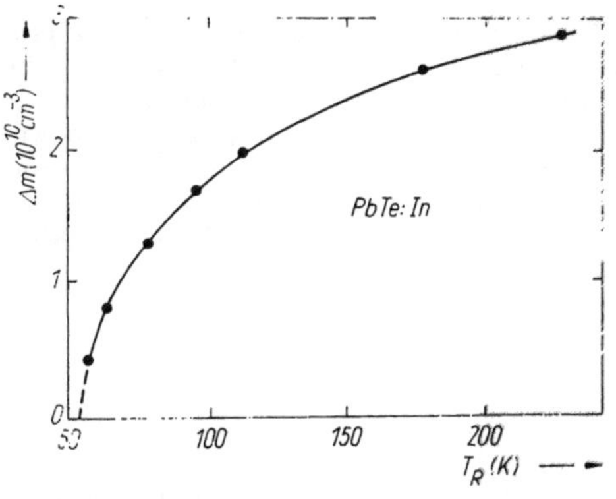

Fig. 13

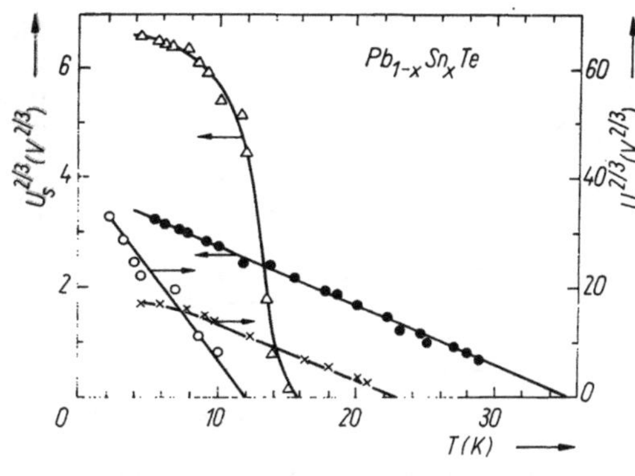

Fig. 14

Fig. 13. The density of excess photocarriers Δm generated by illumination of an In doped PbTe sample with the infrared emission from a carbon resistor heated to a temperature T_R. The sample temperature is 4.2 K (after [85])

Fig. 14. Temperature dependence of the spontaneous voltage U_s in PbTe: Ga (●), $Pb_{0.78}Sn_{0.22}Te$: In (△), $Pb_{0.78}Sn_{0.22}Te$: In (×), $Pb_{0.74}Sn_{0.26}Te$: In (○). The Ga and In concentration is about 0.5 at% (after [87])

Persistent photoconductivity was also observed in PbTe doped with 0.01 at% indium at temperatures below 20 K [85]. When the samples were exposed to a thermal radiation source with a temperature T_R exceeding 50 K, the free electron density was increased. This is shown in Fig. 13 where the density of photocarriers Δm is plotted as a function of T_R.

Herrmann and co-workers [86, 87] observed a spontaneous voltage U_s in In doped $Pb_{0.78}Sn_{0.22}Te$ and also in PbTe doped with gallium. As shown in Fig. 14, for most of the investigated samples U_s was proportional to $(T_0 - T)^{3/2}$ as it would be expected for a displacive phase transition.

An increase of the resistivity of p-$Pb_{0.75}Sn_{0.25}Te$ doped with 5×10^{18} cm^{-3} indium was observed [88] when the samples were irradiated by CO_2 laser light. The negative photoconductivity (NPC) signal appeared at temperatures below 21 K and its rise and fall times were in the order of milliseconds. It was assumed in [81] that the NPC and other low temperature transport anomalies are due to the negative U properties of the indium defect centres rather than to a phase transition. The configuration coordinate diagram of an indium impurity which was presented in [88] is shown in Fig. 15. It was assumed that the following processes will lead to NPC:

$$2p + In^- \xrightarrow{\text{optical absorption}} 2p + e + In^0 , \qquad (3)$$

$$2p + e + In^0 \xrightarrow{\text{electron-hole recombination}} p + In^0 , \qquad (4)$$

$$p + In^0 \xrightarrow{\text{hole capture}} In^+ . \qquad (5)$$

Process (4) is rapid in $Pb_{1-x}Sn_xTe$, process (5) is assumed to be rapid too since the sample is degenerate p-type below 21 K. The thermal emission of holes from In^+ to return the system to equilibrium in the dark is assumed to be slow because of the thermal activation energy required for emission of the first hole. No parameters of the negative U impurity were reported in [88].

The kinetics of NPC in $Pb_{1-x}Sn_xTe$ were studied in detail by Zasavitskii et al. [89]. They used n- and p-type $Pb_{0.77}Sn_{0.23}Te$ samples with an indium concentration between

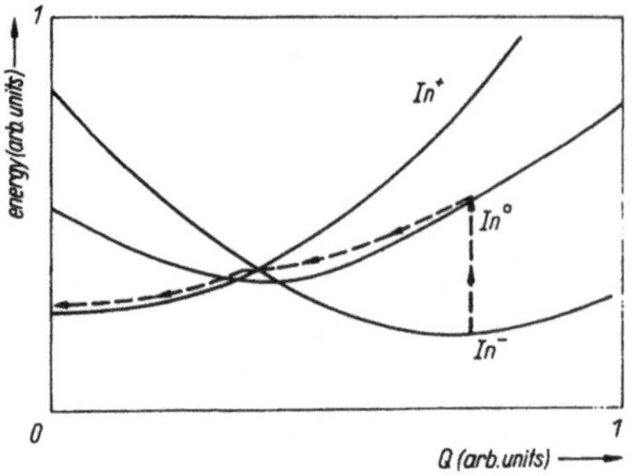

Fig. 15. A configuration coordinate model proposed for indium impurities in $Pb_{1-x}Sn_xTe$. The total energy of the indium impurity in three different charge states (In^+, In^0, In^-) is shown. Dashed line indicates a process which was assumed to explain the negative photoconductivity in $Pb_{0.75}Sn_{0.25}Te$ doped with 5×10^{18} cm^{-3} In (after [88])

1×10^{18} and 10^{20} cm^{-3}. They found that a) a minimal photon energy $h\nu$ of the exciting laser radiation of about 0.34 eV and b) the presence of 300 K background radiation is required in order to get NPC. Only n-type samples showed NPC at temperatures below a threshold temperature T_0 which depended on the indium concentration and was in the range between 20 and 40 K. At temperatures above T_0 only positive photoconductivity was observed. The relaxation times τ_- and τ_+ for negative and positive photoconductivity, respectively, are plotted versus inverse temperature in Fig. 16 showing an activated temperature dependence. The activation energy E_{act} of τ_+ and the threshold temperature T_0 were both decreasing with increasing indium concentration N_{In} of the samples. This is shown in Fig. 17.

The interpretation of the experimental data presented in [89] is based on the Jahn-Teller defect model by Volkov and Pankratov [30]. According to [30] a Te vacancy will form a singly (Γ_{6-}) and a doubly (Γ_{8-}) degenerate state within the conduction band, as shown in the left part of Fig. 18. Electron capture lifts the degeneracy of the Γ_{8-} level by Jahn-Teller distortion (dashed line in Fig. 18). Including a relaxation of the lattice around the defect the total energy of an occupied vacancy is given by curve E_1 in Fig. 18. The unoccupied state of the defect, E_0, can be recovered only by

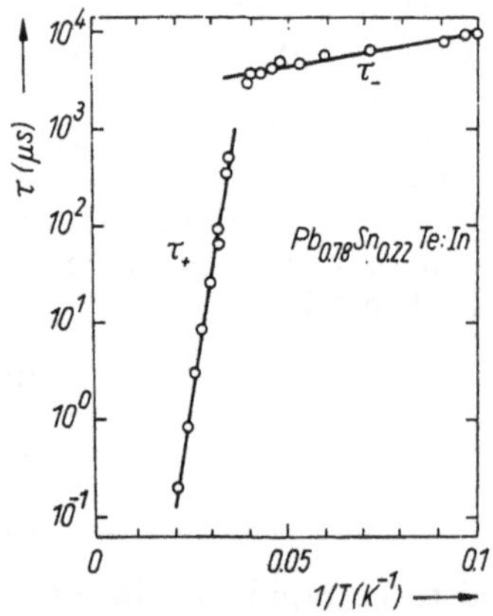

Fig. 16

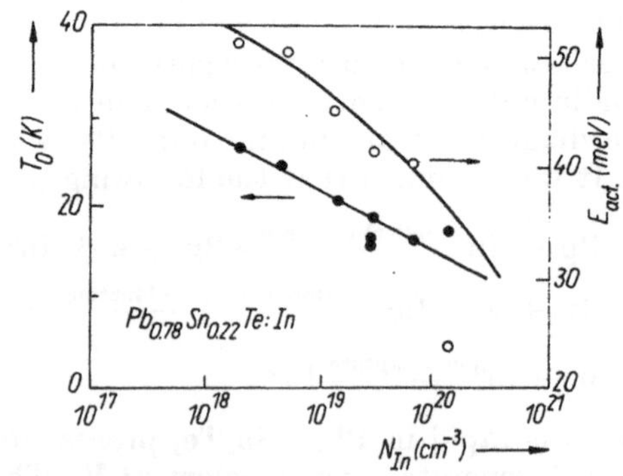

Fig. 17

Fig. 16. Temperature dependence of the relaxation times of positive (τ_+) and negative (τ_-) photoconductivity of In doped $Pb_{0.78}Sn_{0.22}Te$. The In concentration is 5×10^{18} cm^{-3} (after [89])

Fig. 17. The effect of In doping on the threshold temperature T_0 for negative photoconductivity, and the activation energy, E_{act}, of the positive photoconductivity in $Pb_{0.78}Sn_{0.22}Te$ (after [89])

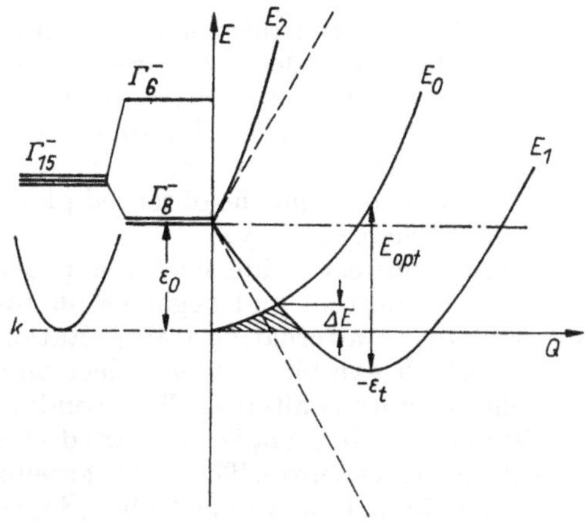

Fig. 18. Configuration coordinate model proposed for a Te vacancy in PbTe. The dashed lines indicate the splitting of the doublet Γ_8-state as a result of a Jahn-Teller instability. The curves E_0, E_1, and E_2 correspond to the total energy of the centre for different charge states (after [30])

thermal activation of the trapped electron across the barrier ΔE, by tunneling through the potential barrier or by photoexcitation indicated by the arrow E_{opt}. It was pointed out in [89] that the main effect of indium doping is to fill the metal vacancies, which act as acceptors in $Pb_{1-x}Sn_xTe$ and in this way, to change the conductivity of the samples from p- to n-type [90]. For n-type samples containing a reasonable concentration of Jahn-Teller centres the mechanism which leads to NPC has been described in [89] as follows. At low sample temperature the thermal activation of carriers from the Jahn-Teller centre to the conduction band is blocked by the barrier ΔE. Background radiation can optically excite trapped electrons by supplying an optical energy E_{opt}. This process leads to the observed increase of the sample conductivity due to background radiation at low temperatures. Negative PC is due to the quenching of background-radiation-excited PC by light with a minimum $h\nu$ of $\varepsilon_0 + E_g$. This energy is necessary to transfer electrons from the valence band to the Jahn-Teller centre. The photogenerated excess holes can recombine relatively fast ($\approx 10^{-8}$ s) with free electrons by band–band recombination processes and the conductivity drops below the dark value. Two parameters of the defect centre were obtained. From the minimum $h\nu$ for NPC excitation the value of ε_0 was determined to be about 0.24 eV. The barrier energy ΔE was obtained from the temperature dependent relaxation of the positive photoconductivity to be about 30 to 40 meV.

No direct experimental evidence exists for the nature of the defect centre. It may be an In impurity or a native point defect, possibly a Te vacancy. Recent experimental investigations of undoped $Pb_{1-x}Sn_xTe$ showed a clear evidence for NPC [91]. This observation corroborates the assumption that the Jahn-Teller centre is an intrinsic point defect, possibly a Te vacancy.

Concluding this chapter we shall point out that the experiments with In doped PbTe and $Pb_{1-x}Sn_xTe$ did not, so far, result in *one* consistent model for that impurity. However, some interesting properties emerged.

1. It seems agreeable that doping of $Pb_{1-x}Sn_xTe$ with In results in the appearance of localized defect levels. These defects exhibit the distinguished features of a negative U system, in which the Jahn-Teller relaxation overcomes the Coulomb energy costs when a second electron is added to an occupied defect. Although no unequivocal experimental evidence exists, this has been inferred from the observation of the pinning of the Fermi level as well as from transport and magnetic susceptibility data. Disappointingly, the value of U has not been determined experimentally.

2. Direct evidence (e.g. the appearance of a spontaneous voltage across the sample surfaces) for a ferroelectric phase transition at low temperatures was obtained for

$Pb_{1-x}Sn_xTe$:In and PbTe:Ga [92] containing rather high concentrations of impurities (more than 0.5 at%). Undoped $Pb_{1-x}Sn_xTe$ ($x \leqq 0.25$) shows a tendency towards a phase transition with a negative Curie temperature [93]. The phase transition in doped $Pb_{1-x}Sn_xTe$ was associated with an increase of the Curie temperature due to a lattice softening by impurity atoms [92]. However, for impurity concentrations exceeding 0.5 at%, the formation of a mixed crystal cannot be ruled out and the observed phase transition may be due to a lattice instability of the PbSnInTe alloy.

Samples which were lightly doped revealed only indirect evidence for a phase transition. Several anomalous transport effects (e.g. persistent and negative photoconductivity) were observed at low temperatures. In an alternative interpretation these effects were associated with defect centres which exhibit a strong electronic (Jahn-Teller) and lattice relaxation when their charge state is altered. Photoconductivity experiments with lightly doped ($N_{In} \leqq 10^{19}$ cm^{-3}) $Pb_{1-x}Sn_xTe$ supported this model and revealed some important parameters of the defect centre. Therefore it seems more favourable to assume that the transport anomalies in lightly doped $Pb_{1-x}Sn_xTe$ are due to the capture and release of carriers in Jahn-Teller defect centres rather than to a phase transition.

5. Deep Level Defects in II–VI Compounds

The $Hg_{1-x}Cd_xTe$ compounds are the best studied narrow gap alloys of group II and group VI elements. $Hg_{1-x}Cd_xTe$ ($x \approx 0.2$) found an important application in infrared-detector technology, and the outstanding physical properties of the alloy system, e.g. a symmetry induced semiconductor–semimetal (SC–SM) transition, have been investigated thoroughly. The physics and applications of the $Hg_{1-x}Cd_xTe$ system were summarized in a number of review papers and books [94 to 98]. Since investigations of defects were extensively discussed in these reviews, we shall be concerned mainly with the question if the defects have deep level properties according to the definition of a deep level given in the Introduction.

CdTe is a direct wide gap semiconductor ($E_g = 1.6$ eV). The gap of $Hg_{1-x}Cd_xTe$ decreases continuously with an increasing admixture of HgTe until it becomes zero for $x = 0.14$. In the semimetallic region ($x \leqq 0.14$) the band structure of the alloy is inverted with the Γ_6 level below the Γ_8 level. In contrast to the zero gap IV–VI compounds, where the degeneracy of conduction and valence bands is accidential, the zero gap transition in $Hg_{1-x}Cd_xTe$ is induced by the symmetry of the crystal structure. The interesting effects which are due to the symmetry induced gapless state have been reviewed in [99]. Particular attention has been paid to defect states in symmetry induced zero gap semiconductors. Due to the absence of a real gap between the conduction and valence band defect states must be degenerate with band states. Obviously, defect states will be resonant rather than localized bound states.

In the first part of this chapter we shall discuss the "deep" level properties of resonant acceptor states in SM $Hg_{1-x}Cd_xTe$, viewing the zero gap state as the ultimate limit of a narrow gap semiconductor. Recent investigations of deep level defects in narrow gap $Hg_{1-x}Cd_xTe$ ($x \approx 0.2$ to 0.3) will be summarized in the second part of this chapter.

5.1 *Defects in zero gap $Hg_{1-x}Cd_xTe$*

Experimental evidence for resonant defect levels in the conduction band of HgTe and zero gap $Hg_{1-x}Cd_xTe$ was gathered from a number of transport [100 to 104] and magneto-optic [105, 106] investigations.

The Hall coefficient and the mobility of free carriers in HgTe exhibit a nonmonotonic temperature variation [100]. Dips of the conductivity were observed at 1 K

(A_0), 9 K (A_1), and 32 K (A_2). These transport anomalies have been interpreted as being due the enhanced scattering of conduction electrons by acceptor resonant states when the Fermi energy coincides with such a level. From the temperature variation of the Fermi energy in HgTe the positions of the acceptor levels was determined: $E_{A_0} = 0.7$ meV, $E_{A_1} = 2.25$ meV, and $E_{A_2} = 9.5$ meV. Optical transitions involving two of these levels, A_0 and A_1, were observed in magneto-optic [105, 106] investigations. The fact that A_2 was not observed in optical experiments led to an alternative interpretation of the A_2 anomaly. It was proposed [107] that the transport effects are due to a resonant scattering by optical phonons when the Fermi energy is equal to the optical phonon energy. It has been pointed out [108] that the shift of the A_2 energy, which is observed when the alloy composition is altered, is hardly compatible with the optical phonon model, since the energy of optical phonons in HgTe and CdTe differs only by about 10%. However, a quantitative interpretation of transport data from $Hg_{1-x}Cd_xTe$ ($x = 0.095$ and 0.12) showing the A_2 anomaly was given using the optical phonon model [109]. A further argument in support of the resonant level model is the strong influence of annealing conditions on the amplitude of the A_1 and A_2 structures [104]. Experimental data are shown in Fig. 19. Two HgTe samples were annealed under different conditions and showed pronounced different A_2 anomalies, although the resistivity of the samples was almost identical in the relevant temperature range.

The position of the A_0 and A_1 level has been measured in $Hg_{1-x}Cd_xTe$ alloys with different composition [110] and as a function of hydrostatic pressure [102]. In Fig. 20 the energy of A_1 is plotted as a function of the gap energy. A variation of the gap by a change of the alloy composition or the application of hydrostatic pressure yielded a pronounced shift of the A_1 level. For a small positive gap the A_1 level remains in the conduction band, its energy with respect to the conduction band edge decreased with $dE_{A_1}/dE_g \approx 0.5$. The A_0 level shows a small shift and is essentially attached to the valence band.

No direct identification of the defect to which the resonant acceptor levels belong to has been reported. The results of rather indirect attempts to find out the nature of these defects by annealing and doping experiments are summarized in Table 3. Tentatively the A_1 level has been assigned to a Hg vacancy.

The theoretical models of defect states in zero gap semiconductors have been reviewed in [99, 111]. We shall briefly outline the basic features of these models. Two groups of approaches can be distinguished: a) model calculations assuming a Coulomb impurity potential yielding "shallow" effective-mass-like resonant states and b) calculations of the resonant level energy and width assuming a localized short-range defect potential. The latter models bear the distinct features of "deep" level theories since the coupling of different bands by the defect potential is essential.

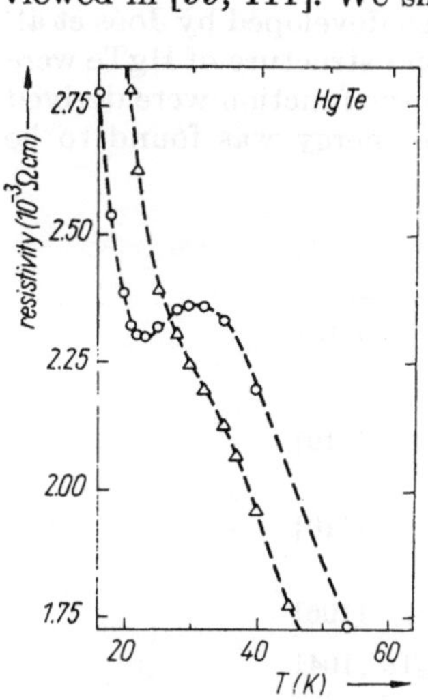

Fig. 19. Temperature dependence of the resistivity of HgTe samples annealed at $T = 225$ (○) and 300 °C (△) under saturated vapour pressure of Hg (after [110])

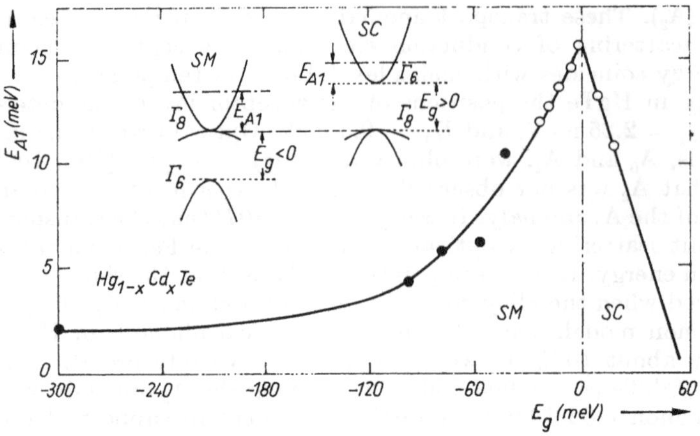

Fig. 20. Energy difference between the acceptor level A_1 and the bottom of the conduction band as a function of the gap energy E_g of $Hg_{1-x}Cd_xTe$. Dots are data from [110]. Circles are data from hydrostatic pressure experiments with $Hg_{0.86}Cd_{0.14}Te$ (after [116])

An early attempt to understand defect states in zero gap semiconductors was that by Liu and Brust [112]. The defect potential was assumed to be a statically screened Coulomb potential plus a central-cell potential localized at the site of the impurity ion. Expressions for the position and the width of resonant impurity states were derived. Since the detailed band structure of the zero gap semiconductor was not considered in [112] the results may be regarded as qualitative.

A more quantitative analysis of acceptor levels due to a Coulomb impurity potential has been performed by Gelmont and Dyakonov [113]. They found that in case the ratio of electron and hole mass, $m_e/m_h = \gamma$, becomes zero, the acceptor level lies $\frac{9}{4}$ hole Rydberg above the valence band edge. The width of the acceptor resonance, Γ, is related to the resonance energy, E_R, by

$$\frac{\Gamma}{E_R} \sim \gamma^{3/2} . \tag{6}$$

A refined model of effective mass impurities in HgTe has been developed by Joós et al. [114]. Screening of the Coulomb potential and a realistic band structure of HgTe were incorporated. The radial components of the acceptor state wave function were derived in a variational treatment of the problem. The resonance energy was found to be

Table 3

Resonant defect levels in $Hg_{1-x}Cd_xTe$

acceptor	assigned defect	experiment	ref.
A_0	Cu	analysis of ingot	[110]
A_1, A_2	Hg_V	annealing under Hg vapour pressure	[110]
A_0, A_1	Hg_V	annealing under reduced Hg vapour pressure	[106]
A_2	Hg_V	annealing of n-type $Hg_{1-x}Cd_xTe$	[104]

sensitive to screening while the relative width, Γ/E_R, is only weakly influenced. Within the range of band parameters reported for HgTe, either the 0.7 meV (A_0) or the 2.2 meV (A_1) resonance was reproduced, corresponding to hole masses of $0.35m_0$ or $0.58m_0$, respectively. An estimate of E_R at the SM–SC transition gave A_0 of about 6.5 meV and A_1 of 11 meV. Experimental evidence for a resonant level at 5.4 meV in $Hg_{0.845}Cd_{0.155}Te$ ($E_g = -7.1$ meV) was reported [115], favouring the assumption that the A_0 level is associated with a charged impurity.

It was suggested that the A_1 and A_2 resonances are associated with intrinsic defects, probably Hg vacancies [104, 110]. Models for vacancy levels in zero gap $Hg_{1-x}Cd_xTe$ were developed, assuming a short-range defect potential which couples states from different bands [108, 116]. In a generalization of the one-band, one-site Green's function approach of Koster and Slater [27] two bands, the conduction and the valence band, were included in [116]. Three bands, the Te p-band (Γ_8 valence band), the Hg s-band (Γ_6 valence band in the SM configuration), and the conduction band were taken into account in [108]. Wannier functions for these bands were used as a basis to set up a matrix representation of the one-particle Green's functions G and the vacancy potential V. Assuming that Wannier functions located at different lattice sites do not overlap, G and V were reduced to 2×2 or 3×3 matrices, respectively. The position of the vacancy resonant level was obtained from

$$\det [I - GV] = 0 \tag{7}$$

with I being the unit matrix. In order to simplify the calculation a model density of band states, $N(E)$, which was suggested by Hubbard [117], has been used:

$$N(E) = \frac{2}{\pi W^2} (-2EW - E^2)^{1/2}, \qquad -2W \leqq E \leqq 0 . \tag{8}$$

$N(E)$ depends only on the halfwidth W of the symmetric band. Near the band edges $N(E)$ is appropriate to reproduce a parabolic band. All parameters beside the gap energy E_g were assumed to be constant when the alloy composition is varied or hydrostatic pressure is applied.

The vacancy "deep level" models [108, 116] were used to fit experimental data of the resonance energy E_{A_1}. The full line in Fig. 20 was calculated using the two-band vacancy model [116]. Both the increase of E_{A_1} in the the SC region and the decrease of E_{A_1} in the SM region have been well reproduced by the model. A linear extrapolation of the data in the SC region gives a real bound state for the A_1 resonance for compounds with $x \geqq 0.2$.

5.2 Deep level defects in narrow gap $Hg_{1-x}Cd_xTe$

Semiconducting $Hg_{1-x}Cd_xTe$ ($x = 0.2$ to 0.3) has become the most important infrared detector material for applications in the 3 to 5 and 8 to 14 μm spectral region. The presence of deep defect levels in the gap can have a stringent influence on the device quality. If the lifetime of excess carriers is governed by trap levels which present efficient pathways for nonradiative recombination, the performance of detectors can be improved significantly by decreasing the density of traps. In a series of investigations, Polla and co-workers [8 to 12, 118 to 121] have applied a set of deep defect characterization techniques to $Hg_{1-x}Cd_xTe$. Deep level transient spectroscopy (DLTS) which was applied for the first time to a narrow gap semiconductor [8, 10], admittance spectroscopy (AS) [9], thermally stimulated current (TSC) [10], population modulation spectroscopy (PMS) [11], and infrared photoluminescence [12] were used to measure the deep trap parameters in $Hg_{1-x}Cd_xTe$. A summary of the data is given in Table 4. All samples were undoped p-type $Hg_{1-x}Cd_xTe$. The carrier concentration has been

Table 4
Data of deep defects in $Hg_{1-x}Cd_xTe$ (after [10])

sample No.	x-value	trap energy DLTS $E_T - E_V$ (eV)	trap energy AS $E_T - E_V$ (eV)	trap density DLTS (cm^{-3})	trap density AS (cm^{-3})	electron capture cross-section, σ_e (c,m²)	hole capture cross-section, σ_p (cm^2)
1	0.207	0.037	0.044	3.4×10^{15}	5.3×10^{15}	1.0×10^{-15}	1.2×10^{-17}
2	0.215	0.035	0.046	9.1×10^{14}	2.3×10^{15}	2.1×10^{-15}	3.1×10^{-17}
3	0.219	0.052		1.4×10^{15}		1.2×10^{-15}	1.1×10^{-18}
4	0.271	0.081	0.070	4.4×10^{14}	8.0×10^{14}	8.0×10^{-16}	1.0×10^{-19}
5	0.285	0.092	0.081	9.9×10^{14}	1.2×10^{15}	3.8×10^{-17}	9.4×10^{-18}
6	0.321	0.134		1.1×10^{15}		5.6×10^{-17}	
7	0.305	0.208	0.161	8.7×10^{13}	4.4×10^{14}	1.5×10^{-16}	8.5×10^{-13}
8	0.320	0.218	0.206	1.1×10^{16}	2.2×10^{16}	1.4×10^{-16}	1.5×10^{-16}

varied by deviation from stoichiometry and the net acceptor concentration is believed to be due to cation (Hg, Cd) vacancies [122]. The measured deep level energies versus alloy composition of $Hg_{1-x}Cd_xTe$ are plotted in Fig. 21. Two deep levels were observed, one slightly below midgap (D_1) and one at approximately $\frac{3}{4} E_g$ (D_2). Since the density of D_1 was direct proportional to the free hole concentration it was tentatively associated with a cation vacancy related defect. This assumption is corroborated by recent tight binding calculations [24]. Results were shown in Fig. 1. The dotted line in Fig. 21 shows the shift of the cation vacancy $P_{3/2}$ level as calculated in [24]. The tight binding results reproduce the trend of the experimental data well. The absolute position of the vacancy level is about 0.3 eV below the experimentally observed D_1 level.

It has been observed [118, 119] that deep traps in undoped p-type $Hg_{1-x}Cd_xTe$ have a profound influence on the lifetime of excess carriers and, consequently, on the

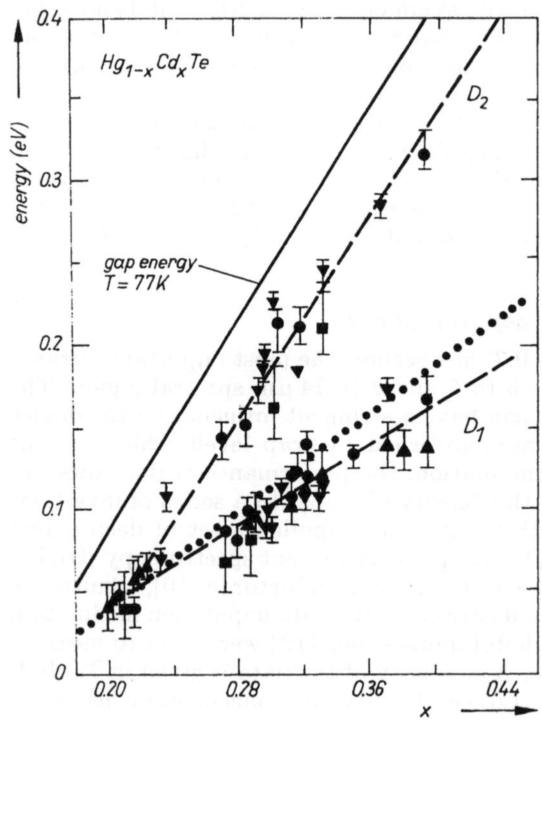

Fig. 21. Energy of deep levels D_1 and D_2 in $Hg_{1-x}Cd_xTe$ versus alloy composition x. Data were obtained by various techniques: DLTS (●), diode step recovery (▲), admittance spectroscopy (■), photoluminescence (●), population modulation spectroscopy (▼) (after [121]). The dotted line shows the shift of the cation vacancy $P_{3/2}$ level calculated by the tight binding method [24] but shifted by about 0.3 eV in energy to fit experimental data for D_1

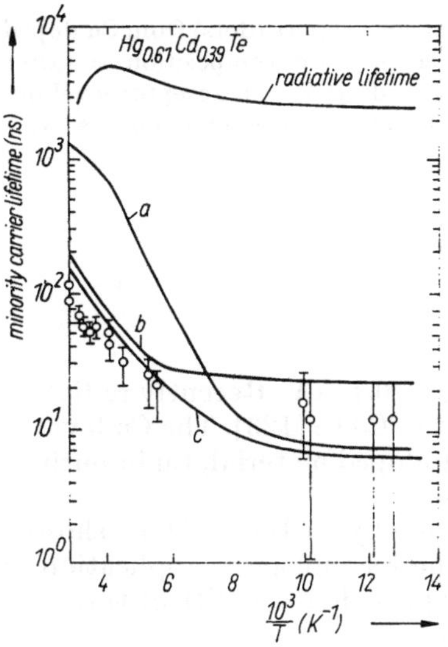

Fig. 22

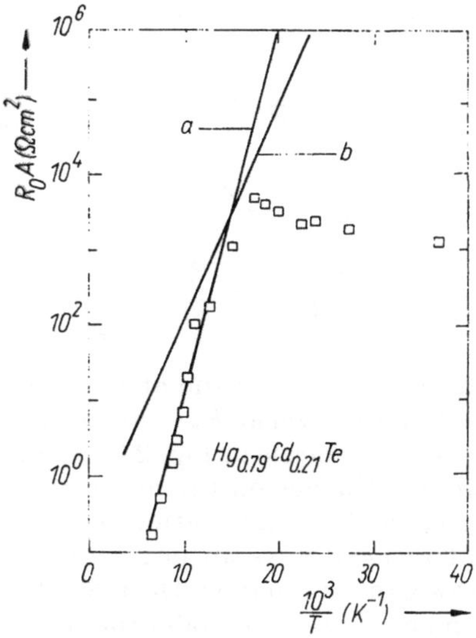

Fig. 23

Fig. 22. Temperature dependence of the minority carrier lifetime in $Hg_{0.61}Cd_{0.39}Te$. Circles are experimental data, curves are calculated using the DLTS data of (a) trap D_2, (b) trap D_1, and (c) a combination of both (after [118])

Fig. 23. Temperature dependence of the zero bias resistance-area product, R_0A, of a n^+ on p-$Hg_{0.79}Cd_{0.21}Te$ photodiode. Curves are calculated using DLTS trap level data; (a) diffusion limited R_0A, (b) generation recombination limited R_0A (after [119])

response of n^+–p photodiodes. The deep traps provide an effective recombination channel for electrons and holes in the p-type region, thus decreasing the number of minority carriers reaching the junction. This has been inferred from the correlation of experimental data of the minority carrier lifetime with DLTS data in terms of a Shockley-Read model [118]. An example is shown in Fig. 22. The data points are the minority carrier lifetime obtained by the pulse recovery method. The curves were calculated assuming a Shockley-Read process involving deep level D_2 (curve a), deep level D_1 (curve b), and the effect of both trap levels (curve c). The trap energy E_T, the trap density, and the capture cross-sections σ_e and σ_p, which were measured by DLTS, have been used in the calculation.

Deep traps can provide centres for the generation of excess carriers, thus having a remarkable effect on the noise currents of n^+–p photodiodes in $Hg_{1-x}Cd_xTe$ and the related figure of merit, the zero bias resistance-area product (R_0A) [118, 119]. Fig. 23 compares experimental data with theoretical values of R_0A. The curves were calculated using the DLTS parameters of deep traps assuming a) R_0A to be limited by the diffusion of carriers produced by generation centres on the p-side of the diode and b) R_0A limited by generation–recombination centres in the depletion region of the diode. The influence of the traps on the diode performance is evident from the good fit of calculated and experimental data in a wide range of R_0A. The saturation of the experimental data at low temperatures is due to a third noise current mechanism which was not considered in [118, 119].

Copper is a common impurity in tellurium and is sometimes used as contact material for p-$Hg_{1-x}Cd_xTe$. DLTS of copper doped $Hg_{1-x}Cd_xTe$ revealed two hole traps close to the valence band edge [122]. The copper levels are indicated in Fig. 24. The levels

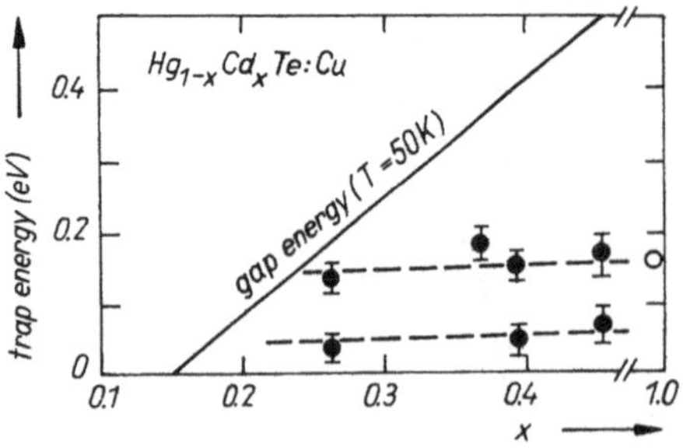

Fig. 24. DLTS trap energies (dots) from Cu doped $Hg_{1-x}Cd_xTe$ versus alloy composition x (after [122]). The circle indicates a Cu acceptor level obtained by photoluminescence experiments with CdTe:Cu (after [123])

are pinned to the valence band at $E_v + 0.06$ eV and $E_v + 0.15$ eV. Recently radiative transitions to a Cu level at $E_v + 0.15$ eV were observed in CdTe [123]. The Cu level in CdTe is included as circle in Fig. 24. Like D_1 and D_2 in undoped material, Cu impurities have a strong influence on the minority carrier lifetime.

Evidently, the first application of deep level spectroscopy to $Hg_{1-x}Cd_xTe$ showed that deep traps can have an important influence on the technological application of these narrow gap compounds. Deep level investigations provided a wealth of trap parameters which allowed to model the device performance.

6. Conclusions

We have summarized experimental investigations of defect states in narrow gap semiconductors. The physical properties of these defects allowed to classify them as deep level defects. However, their binding energy is small, generally less than 100 meV. It has variously been demonstrated that these deep level defects have a profound influence on the minority carrier lifetime and can strongly effect the efficiency of infrared devices made with narrow gap II–VI and IV–VI compounds. Beside the apparent technological importance, the investigation of deep defect levels in narrow gap semiconductors can significantly add to our understanding of basic deep level physics, e.g. the influence of a strong external magnetic field on the binding energy of a deep level.

A number of defects in narrow gap semiconductors have resonant levels. The properties of these states attracted much experimental and theoretical interest. Resonant levels in II–VI and IV–VI compounds have quite different properties. Resonant states in narrow and zero gap $Hg_{1-x}Cd_xTe$ act as scattering centres, affecting the mobility of carriers when the Fermi level coincides with the resonant level. Resonant levels in IV–VI compounds act as bound states. It seems that lattice relaxation in terms of a configuration coordinate model has to be considered in order to understand their properties.

Several theoretical models for defects in narrow gap semiconductors have been proposed. Early considerations were mainly concerned with intrinsic defects, e.g. vacancies, assuming an extremely localized defect potential. Remarkably, these models were among the first theoretical attempts in semiconductor physics dealing with a localized defect potential. A recently developed tight binding model for deep defects in II–VI compounds provided a good fit of the chemical trend of deep trap energies in narrow gap $Hg_{1-x}Cd_xTe$.

The identification of defects in narrow gap semiconductors is still meagre. Some levels were tentatively ascribed to doping elements which were incorporated during the growth process. On the other hand, many experiments revealing deep level data have been carried out with undoped material, and it was concluded that at least some of the observed levels are due to intrinsic point defects. However, these results must be considered as qualitative leaving a large task for future investigations with narrow gap semiconductors.

Acknowledgements

The author is indebted to a number of scientists for valuable discussions, comments, and the stimulation of this work: H. Heinrich, W. Jantsch, L. Palmetshofer, A. Lopez-Otero, H. Sitter, P. Vogl, R. Enderlein, K. H. Herrmann, R. S. Allgaier, J. H. Langer, and I. I. Zasavitskii. The expert help of C. Leitner and U. Hannesschläger with the preparation of the manuscript is acknowledged.

References

[1] H. P. Hjalmarson, P. Vogl, D. J. Wolford, and J. D. Dow, Phys. Rev. Letters 44, 810 (1980).
[2] H. J. Queisser, Festkörperprobleme 11, 45 (1971).
[3] A. G. Milnes, Deep Impurities in Semiconductors, Wiley, New York/London 1973.
[4] D. V. Lang, J. appl. Phys. 45, 3023 (1974).
[5] C. T. Sah, Solid State Electronics 19, 975 (1976).
[6] H. G. Grimmeiss, Annu. Rev. Mater. Sci. 7, 341 (1977).
[7] J. Bourgoin and M. Lannoo, Point Defects in Semiconductors II, Springer Series in Solid-State Sciences, Vol. 35, Ed. M. Cardona, P. Fulde, and H. J. Queisser, Springer-Verlag, Berlin 1983.
[8] D. L. Polla and C. E. Jones, Solid State Commun. 36, 809 (1980).
[9] D. L. Polla and C. E. Jones, J. appl. Phys. 51, 6233 (1980).
[10] D. L. Polla and C. E. Jones, J. appl. Phys. 52, 5118 (1981).
[11] D. L. Polla, R. L. Aggarwal, J. A. Mroczkowski, J. F. Shanley, and M. B. Reine, Appl. Phys. Letters 40, 338 (1982).
[12] D. J. Polla and R. L. Aggarwal, Appl. Phys. Letters 44, 775 (1984).
[13] B. Kaufmann, phys. stat. sol. (a) 81, K31 (1984).
[14] J. Niewodniczanska-Zawadzka and K. Lischka, Solid State Commun. 50, 721 (1984).
[15] S. T. Pantelides, Rev. mod. Phys. 50, 797 (1978).
[16] P. Vogl, Festkörperprobleme 21, 191 (1981).
[17] P. Vogl, Adv. Electronics and Electron Phys. 62, 101 (1983).
[18] G. A. Baraff and M. Schlüter, Phys. Rev. B 19, 4965 (1979).
[19] M. Jaros, C. O. Rodriguez, and S. Brand, Phys. Rev. B 19, 3137 (1979).
[20] J. Bernhole, N. O. Lipari, and S. T. Pantelides, Phys. Rev. B 21, 3545 (1980).
[21] U. Lindefelt and A. Zunger, Phys. Rev. B 24, 5913 (1981).
[22] O. F. Sankey, H. P. Hjalmarson, J. D. Dow, D. J. Wolford, and B. G. Streetman, Phys. Rev. Letters 45, 1656 (1980).
[23] T. L. Reinecke and B. W. Henris, Proc. XVI. Internat. Conf. Phys. Semicond., Montpellier 1982; Physica 117/118B, 194 (1983).
[24] Akiko Kobayashi, O. F. Sankey, and J. D. Dow, Phys. Rev. B 25, 6367 (1982).
[25] N. J. Parada and G. W. Pratt, Phys. Rev. Letters 22, 180 (1969).
[26] N. J. Parada, Phys. Rev. B 3, 2042 (1971).
[27] G. F. Koster and J. C. Slater, Phys. Rev. 95, 1167 (1954).
[28] L. A. Hemstreet, Phys. Rev. B 11, 2260 (1975).
L. A. Hemstreet, Phys. Rev. B 12, 1212 (1975).
[29] K. Weiser, Phys. Rev. B 23, 2741 (1981).
[30] B. A. Volkov and O. A. Pankratov, Dokl. Akad. Nauk SSSR 255, 93 (1980); Soviet Phys.—Doklady 25, 922 (1981).
[31] Yu. I. Ravich, B. A. Efimova, and I. A. Smirnov, Semiconducting Lead Chalcogenides, Ed. L. S. Stilbans, Plenum Press, New York/London 1970; an updated, translated version of the original Russian edition, Izd. Nauka, Moscow 1968.
[32] I. Melnailis and T. C. Harman, in: Semiconductors and Semimetals, Vol. 5, Ed. R. K. Willardson and A. C. Beer, Academic Press, New York/London 1970 (p. 111).
[33] H. Holloway, in: Physics of Thin Films, Vol. 11, Academic Press, New York 1980 (p. 105).
[34] H. Preier, Appl. Phys. 20, 189 (1979).
[35] T. C. Harman, in: Physics of IV–VI Compounds, Ed. S. Rabii, Gordon & Breach, New York 1974 (p. 141).

[36] P. J. STILES, E. BURSTEIN, and D. N. LANGENBERG, J. appl. Phys., Suppl. 32, 2174 (1961).
[37] D. M. FINLAYSON and I. A. JOHNSON, phys. stat. sol. (b) 71, 395 (1975).
[38] G. BAUER, M. BURKHARD, H. HEINRICH, and A. LOPEZ-OTERO, J. appl. Phys. 47, 1721 (1976).
[39] L. PALMETSHOFER, H. HEINRICH, O. BENKA, and W. RESCHENEDER, Appl. Phys. Letters 30, 557 (1977).
[40] K. H. GRESSLEHNER, W. JANTSCH, K. LISCHKA, L. PALMETSHOFER, and H. HEINRICH, Proc. III. Internat. Conf. Phys. Narrow-Gap Semicond., Warszawa 1977, Ed. I. RAULUSZKIEWICZ, M. GORSKA, and E. KACZMAREK, Polish Sci. Publ., Warsazawa 1978 (p. 205).
[41] K. H. GRESSLEHNER and L. PALMETSHOFER, J. appl. Phys. 51, 4735 (1980).
[42] L. PALMETSHOFER, Appl. Phys. A34, 139 (1984).
[43] H. HEINRICH, A. LOPEZ-OTERO, L. PALMETSHOFER, and L. D. HAAS, in: Lattice Defects in Semiconductors 1974, Conf. Ser. No. 23, The Institute of Physics, London 1975 (p. 264).
[44] G. W. PRATT, J. Nonmetals 1, 103 (1973).
[45] J. P. DONNELLY, T. C. HARMAN, and A. G. FOYT, Appl. Phys. Letters 18, 259 (1971).
[46] D. M. STAUDTE and F. J. BRYANT, Solid State Electronics 26, 83 (1983).
[47] J. P. DONNELLY, A. R. CALAWA, T. C. HARMAN, A. G. FOYT, and W. T. LINDLEY, Solid State Electronics 15, 403 (1972).
[48] K. H. GRESSLEHNER, L. PALMETSHOFER, H. HEINRICH, and N. C. SHARMA, in: Physics of Narrow-Gap Semiconductors, Lecture Notes Phys., Vol. 152, Springer-Verlag, Berlin/Heidelberg/New York 1983 (p. 444).
[49] G. BRUNTHALER, K. LISCHKA, and L. PALMETSHOFER, Proc. IV. Internat. Conf. Phys. Narrow-Gap Semicond., Linz 1981, Ed. E. GORNIK, H. HEINRICH, and L. PALMETSHOFER, Springer-Verlag, Berlin/Heidelberg/New York 1982 (p. 458).
[50] K. LISCHKA, Appl. Phys. A 29, 177 (1982).
[51] K. Y. DUH and J. N. ZEMEL, Thin Solid Films 26, 165 (1975).
[52] K. LISCHKA and H. PREIER, phys. stat. sol. (b) 101, K129 (1980).
[53] K. LISCHKA, L. PALMETSHOFER, K. H. GRESSLEHNER, and R. GRISAR, Proc. XV. Internat. Conf. Phys. Semicond., Kyoto 1980, Ed. SHAJI TANAKA and YUTAKA TOYOZAWA, J. Phys. Soc. Japan 49, Suppl. A, 791 (1980).
[54] K. LISCHKA and W. HUBER, J. appl. Phys. 48, 2632 (1977).
[55] K. LISCHKA and W. HUBER, Solid State Electronics 21, 1509 (1978).
[56] K. LISCHKA, R. DURSTBERGER, G. LINDEMANN, and H. STAUDINGER, phys. stat. sol. (b) 123, 319 (1984).
[57] B. SCHLICHT, R. DORNHAUS, G. NIMTZ, L. D. HAAS, and T. JAKOBUS, Solid State Electronics 21, 1509 (1978).
[58] F. F. SIZOV, V. B. ORLETSKII, and M. V. RADCHENKO, Fiz. Tekh. Poluprov. 14, 2117 (1980); Soviet Phys. — Semicond. 14, 1258 (1980).
[59] P. BERNDT, D. GENZOW, and K. H. HERRMANN, phys. stat. sol. (a) 38, 497 (1976).
[60] G. LUCOVSKY, Solid State Commun. 3, 299 (1965).
[61] D. G. SEILER, M. W. GOODWIN, and K. H. LITTLER, Proc. XVI. Internat. Conf. Phys. Semicond., Montpellier 1982; Physica 117/118B, 167 (1983).
[62] W. PÖTZ and P. VOGL, Solid State Commun. 48, 249 (1983).
[63] G. NIMTZ and B. SCHLICHT, Springer Tracts mod. Phys. 98, 1 (1983).
[64] F. KUCHAR, J. C. RAMAGE, R. A. STRADLING, and A. LOPEZ-OTERO, J. Phys. C 10, 5101 (1977).
[65] A. V. LEWIS, F. KUCHAR, R. J. NICHOLAS, J. C. RAMAGE, and L. PALMETSHOFER, Phys. Rev. B 25 (1982).
[66] A. A. AVERKIN, V. I. KAIDANOV, and R. B. MELNIK, Fiz. Tekh. Poluprov. 5, 91 (1971); Soviet Phys. — Semicond. 5, 75 (1971).
[67] B. A. AKIMOV, L. I. RYABOVA, O. B. YATSENKO, and S. M. CHUDINOV, Fiz. Tekh. Poluprov. 13, 752 (1979); Soviet Phys. — Semicond. 13, 441 (1979).
[68] YU. V. ANDREEV, K. I. GEIMANN, I. A. DRABKIN, A. V. MATVEENKO, E. A. MOZHAEV, and B. YA. MOIZHES, Fiz. Tekh. Poluprov. 9, 1873 (1975); Soviet Phys. — Semicond. 9, 1235 (1976).
[69] K. WEISER, A. KLEIN, and M. AINHORN, Appl. Phys. Letters 34, 607 (1979).

[70] J. HUBBARD, Proc. Roy. Soc. A276, 238 (1963).
[71] P. W. ANDERSON, Phys. Rev. Letters 34, 953 (1975).
[72] G. A. BARAFF, E. O. KANE, and M. SCHLÜTER, Phys. Rev. B 21, 5662 (1980).
[73] G. D. WATKINS and J. R. TROXELL, Phys. Rev. Letters 44, 593 (1980).
[74] R. D. HARRIS, J. L. NEWTON, and G. D. WATKINS, Phys. Rev. Letters 48, 1271 (1982).
[75] I. A. DRABKIN and B. YA. MOIZHES, Fiz. Tekh. Poluprov. 15, 625 (1981); Soviet Phys. – Semicond. 15, 357 (1981).
[76] V. I. KAIDANOV, R. B. MELNIK, and I. A. CHERNIK, Fiz. Tekh. Poluprov. 7, 759 (1973); Soviet Phys. – Semicond. 7, 522 (1973).
[77] I. A. DRABKIN, M. A. KVANTOV, and V. V. KOMPANIETS, Fiz. Tekh. Poluprov. 13, 2064 (1979); Soviet Phys. – Semicond. 13, 1206 (1979).
[78] K. I. GEIMAN, I. A. DRABKIN, M. A. KVANTOV, and A. V. MATVEENKO, Fiz. Tekh. Poluprov. 14, 814 (1980); Soviet Phys. – Semicond. 14, 478 (1980).
[79] S. N. LYKOV and I. A. CHERNIK, Fiz. Tekh. Poluprov. 14, 1861 (1980); Soviet Phys. – Semicond. 14, 1112 (1980).
[80] C. KITTEL, Introduction to Solid State Physics, 5th edition, Wiley & Sons, New York 1976 (p. 454).
[81] S. N. LYKOV, YU. I. RAVICH, and I. A. CHERNIK, Fiz. Tekh. Poluprov. 11, 1731 (1977); Soviet Phys. – Semicond. 11, 1016 (1977).
[82] YU. V. ANDREEV, K. I. GEIMAN, I. A. DARBKIN, A. V. MATVEENKO, E. A. MOZHAEV, and B. YA. MOIZHES, Fiz. Tekh. Poluprov. 9, 1873 (1975); Soviet Phys. – Semicond. 9, 1235 (1976).
[83] B. M. VUL, I. D. VORONOVA, G. A. KALYUZHNAYA, T. S. MAMEDOV, and T. SH. RAGIMOVA, Zh. eksper. teor. Fiz., Pisma 29, 21 (1979); Soviet Phys. – J. exper. theor. Phys., Letters 29, 18 (1979).
[84] K. H. HERRMANN, and K.-P. MÖLLMANN, phys. stat. sol. (b) 76, K67 (1983).
[85] B. A. AKIMOV, N. B. BRANDT, K. R. KURBANOV, L. I. RYABOVA, A. T. KHASANOV, and D. R. KHOKHLOV, Fiz. Tekh. Poluprov. 17, 1604 (1983); Soviet Phys. – Semicond. 17, 1021 (1983).
[86] K. H. HERRMANN, G. A. KALYUZHNAYA, K.-P. MÖLLMANN, and M. WENDT, phys. stat. sol. (a) 71, K21 (1982).
[87] K. H. HERRMANN and K.-P. MÖLLMANN, phys. stat. sol. (a) 80, K101 (1983).
[88] C. M. PENCHINA, A. KLEIN, and K. WEISER, Proc. XV. Internat. Conf. Phys. Semicond., Kyoto 1980, Ed. SHAJI TANAKA and YUTAKA TOYOZAWA, J. Phys. Soc. Japan 49, Suppl. A, 783 (1980).
[89] I. I. ZASAVITSKII, A. V. MATVEENKO, B. N. MATSONASHVILI, and V. T. TROFIMOV, Fiz. Tekh. Poluprov. 17, 2184 (1983); Soviet. Phys. – Semicond. 17, 1396 (1984).
[90] I. I. ZASAVITSKII, Fiz. Tehk. Poluprov. 15, 2275 (1981); Soviet Phys. – Semicond. 15, (1981).
[91] I. I. ZASAVITSKII, K. LISCHKA, and H. HEINRICH, to be published.
[92] K. P. MÖLLMANN, K. H. HERRMANN, and R. ENDERLEIN, Proc. XVI. Internat. Conf. Phys. Semicond., Montpellier 1982; Physica 117/118B, 582 (1983).
[93] W. JANTSCH, Springer Tracts mod. Phys. 99, 1 (1983).
[94] T. C. HARMAN, in: Physics and Chemistry of II–VI Compounds, Ed. M. AVEN and J. S. PRENER, North-Holland Publ. Co., Amsterdam 1967 (p. 767).
[95] D. LONG and J. L. SCHMIT, see [32] (p. 175).
[96] T. C. HARMAN and I. MELNGAILIS, in: Applied Solid State Science, Vol. 4, Ed. R. WOLFE, Academic Press, New York 1974 (p. 1).
[97] R. K. WILLARDSON and A. C. BEER (Ed.), Mercury Cadmium Telluuride, Semiconductors and Semimetals, Vol. 18, Academic Press, New York/London 1981.
[98] R. DORNHAUS and G. NIMTZ, Springer Tracts mod. Phys. 98, 119 (1983).
[99] M. AVEROUS, phys. stat. sol. (b) 95, 9 (1979).
[100] S. OTEMEZGUINE, F. RAYMOND, G. WEILL, and C. VÉRIÉ, Proc. X. Internat. Conf. Phys. Semicond., Cambridge 1970, Ed. S. P. KELLER, J. C. HENSEL, and F. STERN, USA EC, Conf. 700801, 1970, (p. 536).
[101] C. T. ELLIOTT, J. MELNGAILIS, T. C. HARMAN, J. A. KAFALAS, and W. C. KERNAN, Phys. Rev. B 5, 2985 (1972).

[102] A. MAUGER, S. OTMEZGUINE, C. VÉRIÉ, and J. FRIEDEL, Proc. XII. Internat. Conf. Phys. Semicond., Stuttgart 1974, Ed. M. PILKUHN, Teubner, Stuttgart 1974 (p. 1166).
[103] C. FAU, J. CALAS, M. AVEROUS, and B. LOMBOS, see [40] (p. 183).
[104] C. VÉRIÉ, F. RAYMOND, F. BAILLY, I. VACQUIÉ, G. WEILL, A. KAZACKI, and J. RIOUX, Proc. Internat. Conf. Phys. Semicond., Edinburgh 1978, Inst. Phys. Conf. Ser. No. 43, 1978 (p. 241).
[105] J. TUCHENDLER, M. GRYNBERG, Y. COUDER, H. THOME, and R. LE TOULLEC, Phys. Rev. B 8, 3884 (1973).
[106] G. BASTARD, Y. GULDNER, C. RIGAUX, NGUYEN HY HAU, J. P. VIEREN, M. MENANT, and A. MYCIELSKI, Phys. Letters A 46, 99 (1973).
[107] W. WALUKIEWICZ, Phys. Rev. Letters 33, 650 (1974).
[108] L. LIU and C. VÉRIÉ, Phys. Rev. Letters 37, 453 (1976).
[109] T. DIETL, J. J. DUBOWSKI, and W. SZYMANSKA, see [104] (p. 245).
[110] C. FINCK, S. OTMEZGUINE, G. WEILL, and C. VÉRIÉ, Proc. XI. Internat. Conf. Phys. Semicond., Warszawa 1972, Ed. Polish Academy of Sciences, PWN-Pol. Scientific Publ., Warszawa 1972 (p. 944).
[111] L. LIU, see [40] (p. 151).
[112] L. LIU and D. BRUST, Phys. Rev. 157, 627 (1967).
[113] B. L. GELMONT and M. I. DYAKONOV, Zh. eksper. teor. Fiz. 62, 713 (1972); Soviet Phys. — J. exper. theor. Phys. 35, 377 (1972).
[114] B. JOOS, A. K. DAS, and P. R. WALLACE, Phys. Rev. B 18, 5693 (1978).
[115] A. KOZACKI, S. OTMEZGUINE, G. WEILL, and C. VÉRIÉ, Proc. XIII. Internat. Conf. Phys. Semicond,. Rome 1976, Ed. F. G. FUMI, North-Holland Publ. Co., Amsterdam 1976 (p. 467).
[116] A. MAUGER and J. FRIEDEL, Phys. Rev. B 12, 2412 (1975).
[117] J. HUBBARD, Proc. Roy. Soc. A281, 401 (1964).
[118] C. E. JONES, JIJAYAN NAIR, and D. L. POLLA, Appl. Phys. Letters 39, 248 (1981).
[119] D. L. POLLA, M. B. REINE, and C. E. JONES, J. appl. Phys. 52, 5132 (1981).
[120] D. L. POLLA, S. P. TOBIN, M. B. REINE, and A. K. SOOD, J. appl. Phys. 52, 5182 (1981).
[121] D. L. POLLA, W. SCOTT, and C. E. JONES, see [49] (p. 434).
[122] C. E. JONES, VIJAYAN NAIR, and J. LINDQUIST, and D. L. POLLA, J. Vacuum Sci. Technol. 21, 187 (1982).
[123] J. P. CHAMONAL, E. MOLVA, and J. L. PAUTRAT, Solid State Commun. 43, 801 (1982).

(Received April 1, 1985)

Original Papers

phys. stat. sol. (b) 133, 47 (1986)

Subject classification: 1.1 and 17; 22.1.1; 22.1.2

All-Union Research Institute of Scientific Instrument Engineering, Leningrad[1])

Determination of the Photoemission Generation Depth with Use of Experiments on the Dynamic Scattering of X-Rays

By

M. V. Kruglov, I. K. Solomin, and A. V. Lunev

The angular dependences of photoemission are measured in Bragg diffraction of X-rays on germanium and silicon crystals covered with amorphous layers of various thickness. The function $K(z)$ — the integral of electron escape probability versus generation depth — is obtained from the experimental data. The escape depth is found to be substantially lower than the estimates often used whereas $K(z)$ is satisfactorily described by Liljequist's theory. In the photoeffect technique, the depth of the explored layer depends on the primary-to-secondary electron recording efficiency ratio. The capabilities of two methods of deep scanning — by changing the initial energy of electrons by a variation of the radiation wavelength and by means of the escape energy selection of electrons — are compared.

Измерены угловые зависимости фотоэмиссии при брэгговской дифракции рентгеновских лучей на кристаллах германия и кремния, покрытых аморфными слоями различной толщины. Из экспериментальных данных восстановлена функция $K(z)$ — первообразная вероятности выхода электронов в зависимости от глубины зарождения. Глубина выхода значительно меньше часто используемых оценок, в то же время, $K(z)$ удовлетворительно описывается теорией Лильеквиста. Толщина зондируемого слоя в методе фотоэффекта зависит от соотношения эффективностей регистрации первичных и вторичных электронов. Сравниваются возможности двух методов глубинного сканирования: изменением начальной энергии электронов за счет варьирования длины волны излучения и с помощью селекции электронов по энергиям вылета.

1. Introduction

X-ray-generated photoelectrons and Auger electrons scattering elastically and nonelastically in solids suffer a gradual energy loss and change the direction of their motion. Part of the electrons escapes from the sample. The function $P(z)$ describing the distribution of the electrons versus depth of their generation is a characteristic of the electron yield. The analytic calculation of this function is complicated for an initial energy of electrons of several keV because the scattering results from several different mechanisms of interaction. That is why the calculation is usually performed semi-empirically, with rather rough approximations [1, 2]. Another way is to use the numerical simulation of the process by the Monte-Carlo technique. For the energy range under consideration, substantial progress has been recently achieved in this direction in connection with the development of the Mössbauer spectroscopy technique for conversion electrons. [3]. However, the results of these calculations strongly differ from the older estimates used often in works on the X-ray-generated photoelectric effect.

---

[1]) Stakhanovtsev 1, 195112 Leningrad, USSR.

The data of $P(z)$ acquire particular importance in connection with the ever increasing use of the technique of studying lattice disturbances in surface layers of crystals from angular dependences of photoemission $\varkappa(\theta)$ under the conditions of X-ray diffraction [4, 5]. In this technique, $P(z)$ determines the contribution of layers arranged at various depth to the general signal.

It should be emphasized that the direct information about $P(z)$ can be obtained from the dependences $\varkappa(\theta)$ measured for model samples — single crystals with surface amorphous film of known thickness [6]. The utilization of the results obtained in such a manner is more preferable than the use of calculated data since the real $P(z)$ depends on the detection parameters: the angular and energy interval, the recording efficiency versus energy and escape angle, etc., which are not sufficiently exactly known or are not monitored.

Howerer, results of determination of the X-ray-generated photoemission electron escape depth with the use of this technique are not numerous and are contradictory. Other experimental data are insufficiently informative (this subject is discussed in detail below). These are apparently the reasons why various authors employ insufficiently substantiated estimates. Thus, most works on development and utilization of the technique of the photoelectric effect in the dynamic diffraction of X-rays employ the function $P(z)$ first proposed in [7] and derived on the basis of rough approximations in [1]:

$$P(z) = (1/2)\,(1 - z/L)\,, \tag{1}$$

where L is identified with the so-called practical path of electrons L_i. At the same time, experimental values of the escape depth that are several times less than L_i were obtained in our preliminary investigation [8] and in [9] (under the conditions of recording electrons escaping directly into a gas-filled chamber of a proportional counter). The absence of generally accepted, trustworthy data on the depth of X-ray photoemission generation and the importance of this subject, especially in studying lattice disturbances, stimulated the present experimental investigation of the function $P(z)$ (more exactly, its integral — see below). Two more subjects that have not been investigated earlier are considered: the influence of the recording conditions on measurement results and the comparison of the deep scanning efficiency by two methods — by changing the energy of radiation quanta and by the energy separation of electrons.

2. Investigation Technique

The method of the photoelectric effect in dynamic diffraction of X-rays allows for the direct experimental study of the function $P(z)$, as first shown in [6]. This possibility is based on a sharp difference between the photoemission angular dependences $\varkappa(\theta)$ for the crystal and the disordered layer.

Under the diffraction conditions, $\varkappa(\theta)$ for a certain electron shell is given by (in relative units)

$$\varkappa(\theta) = (1/B) \int_0^\infty \mathrm{d}z\, P(z)\, \overline{|E(\theta, z)|^2}\,, \tag{2}$$

where z is the coordinate read along the internal normal to the surface; $\overline{|E(\theta, z)|^2}$ is the intensity of the diffraction wave field, averaged over the distribution of the electron density in the elementary cell, as normalized to the square of the incident wave amplitude;

$$B = \int_0^\infty \mathrm{d}z\, P(z)\, \mathrm{e}^{-\mu z}$$

is the normalization factor.

Let the radiation diffract on a thick crystal base covered with an amorphous film having thickness t. Assume that the radiation is practically not absorbed by the film, i.e. $t \ll 1/\mu$, where μ is the normal absorption factor of the radiation per unit length along z outside the diffraction region. Assume also that $P(z)$ is similar for the crystal and the amorphous layer, i.e. neglect the orientation effects of electron passage.

We represent $\varkappa(\theta)$ as two components: emission of electrons generated in the film and generated in the base:

$$\varkappa(\theta) = \varkappa_{\mathrm{am}}(\theta) + \varkappa_{\mathrm{cr}}(\theta) = (1/B)\,[\int_0^t \mathrm{d}z\, P(z)\, \overline{|E_{\mathrm{am}}(\theta)|^2} + \int_0^\infty \mathrm{d}z\, P(z)\, \overline{|E_{\mathrm{cr}}(\theta, z)|^2}]\,. \tag{3}$$

It is obvious that $\overline{|E_{\mathrm{am}}|^2}$ is equal to the sum of squares of component amplitudes because of the random arrangement of atoms in the film:

$$\overline{|E_{\mathrm{am}}(\theta)|^2} = 1 + \overline{|E_{\mathrm{h}}(\theta)|^2}\,. \tag{4}$$

In the crystal base, the field for each state of polarization is defined by the known formula for the perfect crystal:

$$\overline{|E_{\mathrm{cr}}(\theta, z')|^2} = |E_{\mathrm{cr}}(\theta)|^2_{z'=0}\, \mathrm{e}^{-\sigma(\theta) z}$$

$$|E_{\mathrm{cr}}(\theta)|^2_{z'=0} = 1 + |E_{\mathrm{h}}(\theta)|^2 + 2C\, \frac{\chi_{ih}}{\chi_{i0}}\, |E_{\mathrm{h}}(|\theta)|^2 \cos \zeta_{\mathrm{h}}\,. \tag{5}$$

Here $z' = z - t$ is the coordinate read from the crystal–amorphous film interface while the other symbols correspond to [10].

Passing, for the sake of simplicity of presentation, to the case of symmetrical diffraction where $|E_{\mathrm{h}}(\theta)|^2 = R(\theta)$ is the reflection factor measured experimentally, which is the most widespread in experiments, we obtain from (3), (4), and (5)

$$\varkappa_{\mathrm{am}}(\theta) = [1 + R(\theta)\,(1/B) \int_0^t \mathrm{d}z\, P(z)\,; \tag{6}$$

$$\varkappa_{\mathrm{cr}}(\theta) = |E_{\mathrm{cr}}(\theta)|^2_{z'=0}\,(1/B) \int_0^\infty \mathrm{d}z'\, P(z' + t)\, \mathrm{e}^{-\sigma(\theta) z'}\,. \tag{7}$$

As can be seen from (7), $\varkappa_{\mathrm{cr}}(\theta)$ is proportional to the curve for the crystal without the film $\varkappa_{\mathrm{perf.}}(\theta)$ if $P(z + t)$ is substituted by $P(z)$. Thus, $\varkappa(\theta)$ consists of two different functions: $1 + R(\theta)$ that can be interpreted as the emission induced by the passing and reflected rays without their interaction and a feebly distorted function $\varkappa_{\mathrm{perf.}}(\theta)$ for the crystal without film. The relative contribution of these two functions depends on both the film thicknesses and $P(z)$.

Assume further that

$$\mathrm{e}^{-\sigma(\theta) z} = 1 \quad \text{at all } z \text{ where } P(z) \neq 0\,. \tag{8}$$

Denote $(1/B) \int_0^t \mathrm{d}z\, P(z) \equiv K(t)$ ($B = \int_0^\infty \mathrm{d}z\, P(z)$ since $\mathrm{e}^{-\mu z} = 1$ in (2) if (8) is true); then it is evident from (7) that $\varkappa_{\mathrm{cr}}(\theta) = \varkappa_{\mathrm{perf.}}(\theta)\,[1 - K(t)]$, and we obtain finally

$$\varkappa_t(\theta) = [1 + R(\theta)]\, K(t) + [1 - K(t)]\, \varkappa_{\mathrm{perf.}}(\theta)\,. \tag{9}$$

$K(t)$ has the meaning of the ratio of the number of electrons escaping from the surface layer having the thickness t to the total number of escaping electrons. The function $K(t)$ can be reconstructed by measuring $\varkappa_t(\theta)$ for the crystal with amorphous film of various thicknesses.

The curves $\varkappa_{perf.}(\theta)$ and $R(\theta)$ can be calculated [4, 11]. However, as shown in [9], it is preferable to measure $R(\theta)$ simultaneously with $\varkappa_t(\theta)$ and $\varkappa_{perf.}(\theta)$ before the film is applied.

The error of $K(t)$ determination sharply increases at angles close to the precise Bragg angle [9]; when $K(t)$ is calculated using (9). This neighbourhood should be excluded from the consideration. Note also that as the numerical estimations carried out and the computer experiment (calculation of $P(z)$ using (1) with $L = 200$ nm) showed, (9) based on the approximation of (8) could be used with an accuracy up to $|\Delta K| < 0.02$ at all angles, except $-0.6 \leqq y \leqq 0.4$ (y is the standard angular deviation [10]). These estimations also showed that the procedure of fitting the calculated curves with varied $P(z)$ to experiment [11] is hardly reasonable since the direct reconstruction of $P(z)$ is possible from experimental data.

3. Experiment and Data Processing

The functions $\varkappa(\theta)$ and $R(\theta)$ were measured on a double-crystal apparatus with a sample vacuum chamber. Strongly asymmetrical monochromators were utilized. This allowed the apparatus distortions to be practically avoided [12]. The experiment was carried out at two values of the X-ray quantum energy: 8.05 and 4.95 keV (CuK_α and VK_α lines). The samples were [110] germanium and [111] silicon slices. The measurement was performed in the symmetrical diffraction geometry. The $\varkappa_t(\theta)$ and $R(\theta)$ curves were measured simultaneously while the $\varkappa_{perf.}(\theta)$ curve was measured for each slice before the film was applied.

Amorphous layers were applied by cathode sputtering. Film thicknesses $t \geqq 220$ nm were measured by means of an optical microinterferometer while thicknesses $t < 220$ nm were determined from the time of deposition and from the specially measured deposition rate. The error of thickness determination was assessed from the straggling of several rate measurements and was 10 and 15% for silicon and germanium, respectively.

Electrons were recorded by a VEU-6 multiplier in the pulse count mode. The function $\varkappa(\theta)$ was measured at three different voltages U_c at the flat grid between the sample and the multiplier: (i) an accelerating potential $U_c = -250$ V in accordance with the common measurement procedure, (ii) a retarding potential $U_c = +50$ V providing the cut-off of true secondary electrons [13], (iii) $U_c = +1.3$ kV for germanium and $U_c = +1.9$ kV for silicon to cut off Auger electrons and, thus, to separate electrons having predominantly similar initial energy E_0: L photoelectrons of germanium ($E_0 = 6.7$ keV) or K photoelectrons of silicon ($E_0 = 6.2$ keV). This point should be dwelt upon in more detail.

Emitted electrons are known to consist of fast electrons — photoelectrons and Auger electrons — and slow truly-secondary electrons, the number of secondary electrons being of the same order as or more than the number of fast electrons [14]. In the event of the accelerating potential $U_c = -250$ V the best conditions are achieved for recording of secondary electrons: their almost complete collection over angles and the optimum energy (≈ 300 eV) from the viewpoint of counting by the multiplier are achieved. However, such a weak field does practically not influence the efficiency of recording of primary electrons since only those electrons are recorded whose direction of escape allows them to get into the multiplier window. Furthermore, the faster electrons reaching the multiplier are less likely counted. As a result, over 80% of the electrons recorded with our procedure are secondaries.

At the same time, the $P(z)$ calculated for fast electrons are used, as a rule, in works on the photoelectric effect, the results being usually represented as integral of

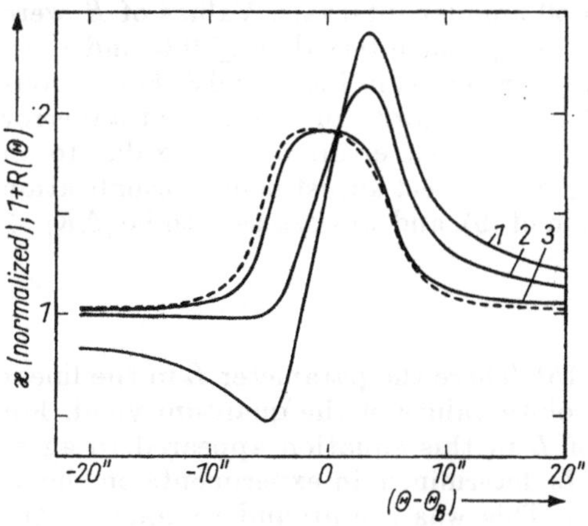

Fig. 1. Experimental angular functions of photoemission $\varkappa(\theta)$ for germanium crystals with amorphous germanium film of various thickness ((1) $t = 0$, (2) 18, (3) 108 nm) and function $1 + R(\theta)$ (dashed line). (220) reflection, CuK$_\alpha$ radiation

$P(z)$ over final energies and escape angles. Such an assumption would be valid in the event where each escaping slow electron were accompanied by a primary one by virtue of which the recording of a secondary electron were equivalent to the recording of a primary one. In order to test this assumption, $K(z)$'s were determined at $U_c = -250$ V and $U_c = +50$ V for comparison in this work. Finally, the results at $U_c = +1.3$ (1.9) kV could be compared with the calculated ones for electrons of one and the same initial energy.

Fig. 1 shows several experimental dependences $\varkappa_t(\theta)$ for germanium as an example. It is well seen how the shape of $\varkappa_t(\theta)$ changes from the typical curve with a maximum and a minimum to a curve resembling the function $1 + R(\theta)$ as the film thickness t is increased in accordance with (9).

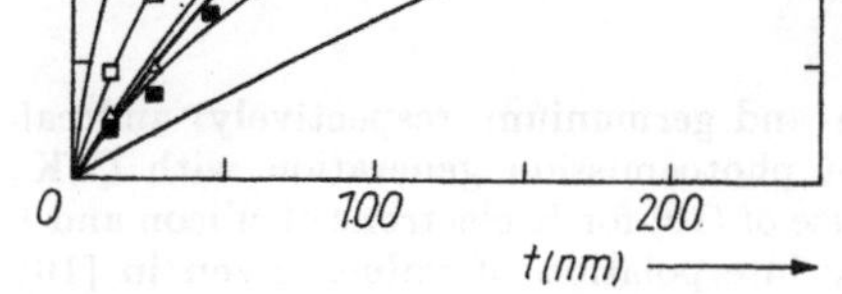

Fig. 2

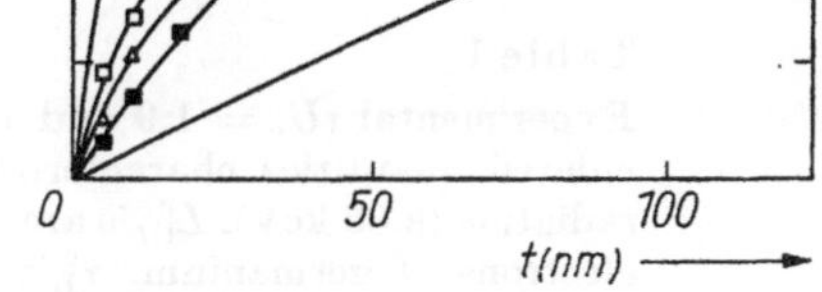

Fig. 3

Fig. 2. Integral of electron escape probability versus depth of electron generation for silicon. CuK$_\alpha$ radiation: □ $U_c = -250$ V, △ +50 V, ■ +1.9 kV; VK$_\alpha$ radiation: ○ $U_c = -250$ V. Calculation (solid lines): CuK$_\alpha$ radiation: (1) with use of (11), $r_B = 715$ nm; (2) with use of (1), $L_i = 660$ nm

Fig. 3. The same as in Fig. 2, but for germanium. ■ $U_c = 1.3$ kV, $r_B = 364$ nm (CuK$_\alpha$), $L_i = 374$ nm (CuK$_\alpha$)

Each experimental curve was divided into 50 angular intervals. Values of K were calculated with the use of (9) for each interval, except the interval $|y| \leq 0.6$, and averaged. The results obtained in such a manner are represented in Fig. 2 and 3. For various intervals, the standard deviation was < 0.005 in the majority of cases. The total error depended, in addition to statistical errors, on systematic errors occurring due to an inaccurate matching of curves as to the angles, an incorrect adjustment, normalization errors, the approximation used in the derivation of (9), and was assessed to be $|\Delta K| < < 0.03$.

4. Discussion

Let us compare the results obtained with [1, 15]. There the parameter L in the linear dependence (1) was determined from the absolute values of the quantum yield. For various substances and energies the value of L in this equation appeared to agree rather well with the so-called practical path L_i determined in experiments on shooting-through of thin films with electrons [13]. This was the ground to consider the approximation (1) to be valid and to identify the parameter L having the meaning of the mean distance from the point of generation to the point of stopping with L_i. Further, such a viewpoint was adopted in most works on the technique of the photoelectric effect in diffraction of X-rays.

The empirical formula

$$L_i = \frac{6 \times 10^{-6} A}{\varrho Z} E_0^{1.3 \text{ to } 1.5} \text{ [cm]}, \tag{10}$$

where Z is the atomic number, A is the atomic weight, E_0 is the initial energy of the electron, ϱ is the substance density, is known to be suitable for calculation of L_i [13].

Values of L_i calculated with the use of (10) for the most numerous group of electrons — K electrons of silicon and L electrons of germanium — are represented in Table 1. The results of calculation of L on the basis of the experimental $K(t)$'s (Fig. 2 and 3) are given in the same table. As can be seen, the values obtained in the experiment are several times less than the respective values of L_i.

As we think, the cause of the difference is that the physical model leading to the approximation (1) describes the electron transport process with an extreme simplification, not taking into consideration the statistical nature of the phenomenon. This model introduces the quantity L defined as the mean distance of electrons from the point of excitation to the point of stopping and depending only on the initial energy E_0. It is assumed that the straggling of the distances is substantially smaller than L. This assumption cannot be considered to be correct since it is known [13] that the straggling

Table 1

Experimental ($U_c = 1.9$ and 1.3 kV for silicon and germanium, respectively) and calculated quantities characterizing the depth of photoemission generation with CuK_α radiation (8.05 keV). L^c_{lin}'s are calculated with use of (10) for K electrons of silicon and L electrons of germanium. r^c_B's are obtained by interpolation of values given in [16]. L^e_{lin}'s and r^e_B's are mean results for various thicknesses of amorphous film as calculated from experimental $K(t)$'s (Fig. 2 and 3): L^e_{lin}'s with use of (1), r^e_B's with use of (11). The mean deviation for various thicknesses is given

material	L^e_{lin} (nm)	L^c_{lin} (nm)	r^e_B (nm)	r^c_B (nm)
Si	368 ± 10%	666	959 ± 2%	715
Ge	136 ± 3%	374	353 ± 5%	364

is comparable with or even larger than the mean distance itself because of the diffusive nature of motion on a considerable portion of the path. Furthermore, the fact of equal values of L and L_i contradicts the physical meaning of these quantities — as it follows from the procedure of their determination, L_i should be close not to the mean, but to the maximum distance of electrons. The equal values are apparently accidental and result from a model inaccuracy and possible experimental errors that might stem from an inaccurate determination of absolute values of quantum and electron flows in measurements of the quantum yield. These experimental data should be probably considered to be insufficiently accurate as well as the numerical results of [6]. In the last of the above mentioned papers, the large errors might be caused by apparatus distortions, absence of mathematical processing, and insufficient set of the thicknesses employed.

Let us compare the results obtained with the data [16]. This paper presents a calculation of the distribution function of Mössbauer conversion electrons over the depth of their generation, the energies, and the escape directions. The calculation was performed by a computer simulation of the random process of interaction of electrons with substance particles with due allowance made for the fairly strict notions of the interaction mechanisms. For the entire set of escaping electrons, the calculation is very well approximated by the formula:

$$P(z) = \begin{cases} 0.74 - 2.7z/r_B + 2.5(z/r_B)^2 & \text{at} \quad z \leqq 0.55r_B \\ 0 & \text{at} \quad z > 0.55r_B \,. \end{cases} \tag{11}$$

Here r_B is the so-called Bethe radius (the full path of electrons in the approximation of continuous slowing-down, which depends only on E_0 and the density).

As can be seen from Fig. 2 and 3, the experimental $K(t)$'s obtained with the cut-off of secondary and Auger electrons well agree with the dependences calculated with the use (11). The values of r_B obtained by interpolating the data of [16] were utilized in the calculation.

The values of r_B^e calculated using (11) and the experimental $K(t)$'s are given in Table 1. The close values of r_B^e and r_B^c is an evidence that (11) satisfactorily describes the function $P(z)$. Unfortunately, the accuracy of the approximation (11) cannot be strictly quantitatively assessed due to errors in the determination of $K(t)$ caused chiefly by the inaccurate measurement of amorphous film thicknesses, due to possible errors of interpolation in the determination of r_B^c, as well as due to the fact that the detection method used does not provide complete collection of electrons (electrons are recorded in a limited solid angle with a lower energy limiting). However, it may be ascertained that the utilization of this approximation to describe $P(z)$ is much more preferable than the approximation (1).

In order to assess the influence of the detection conditions on the depth from which the information is received in the technique of the photoelectric effect, let us compare the experimental $K^{U_c}(t)$'s obtained at various potentials between the sample and the control grid (see Fig. 2 and 3). As can be seen, for all thicknesses the values of K^{+50V} are slightly higher than those of $K^{+1.3kV}$ (Ge) or $K^{+1.9kV}$ (Si). This is undoubtedly connected with the appearance of the low-energy groups — Auger and fluorescent electrons — in the general flow as the retarding potential is decreased. As can be also seen from Fig. 2 and 3, the values of K^{-250V} obtained under the conditions of recording of both the primary and secondary electrons are higher than those of K^{+50V} obtained in the recording of only primary electrons. In the region of small thicknesses t the compared K's differ by a factor of 2 to 3. Thus, in the event of recording of secondary electrons the information is fed from substantially lower depths than in event of recording of primary ones. From this it follows that in the technique of the photoelec-

Table 2

Experimental K^{U_c} obtained in this work for CuK_α and VK_α lines at $U_c = -250$ V and those obtained in [9] with CuK_α radiation for various energies of detected electrons: F fast, S slow, material: silicon

t (nm)	K^{-250V}		K [9]	
	CuK_α	VK_α	S	F
29	0.310	0.58	0.15	0.45
46	0.519	0.85	0.35	0.68

tric effect the depth of the explored layer depends on a number of conditions influencing the primary-to-secondary recorded electron number ratio. In most works devoted to the development or utilization of the technique, this circumstance was not taken into account.

Let us dwell upon the possible causes of the difference of the compared K's. One of the explanations can be as follows. If the secondary electron is excited by a primary one knocked-out from an atom very close to the surface in the zone of escape of secondary electrons (several nm's [13]), part of such secondary electrons can go out into vacuum, not accompanied by the primary ones that initiated them, but went into the sample. Formally, this is described by an increase in the probability of escape of the electron formed near the surface when secondary electrons are predominantly recorded ($U_c = -250$ V) as compared to the case of recording of only primary ones ($U_c = +50$ V). For the explanation of the difference observed it is necessary of course to assume a noticeable contribution of such a process. The other possible cause is connected with an increase in the secondary emission coefficient when the energy of the electrons exciting the secondary ones decreases [13]. Because of this fact, low-energy Auger electrons excite secondary ones with a higher probability as compared to photoelectrons, and their contribution to the general signal at $U_c = -250$ V becomes larger than the true ratio of photoelectrons to Auger electrons.

Further, let us assess the capabilities of two different methods of deep scanning in the technique of the photoelectric effect: by changing the initial energy of electrons (first method) and by means of selection of the final energy of electrons at a constant initial one (second method). To do this, we compare the values of K obtained for silicon at various radiation wavelengths — CuK_α and VK_α — with those obtained for similar samples in [9] for various portions of the energy spectrum of detected electrons. As can be seen from Table 2, in both cases comparable variations of K are observed. This fact and the circumstance that in the first case there is the principle possibility to change the radiation wavelength and the electron escape depth together with it within limits exceeding substantially those under consideration, whereas the capabilities of escape depth variation by the second method are very likely practically exhausted give an evidence of the higher efficiency of the first method of deep scanning. However, in the cases where there is no necessity to vary the escape depth within wide limits for solution of a particular problem, but its small variations (by a factor of 2 to 3) are sufficient the use of the second method providing the simultaneous information from various depths and, therefore, being prompter is more preferable. Furthermore, the second method is more preferable from the viewpoint of a simpler analysis of experimental results since the parameters of the internal wave field slightly change with variation of the radiation wavelength.

References

[1] W. T. Ogier and D. V. Ellis, J. appl. Phys. **36**, 3788 (1965).
[2] O. Lopez, J. appl. Phys. **46**, 2504 (1975).
[3] D. Liljequist, Electron Penetration in Solids and Its Application to Mössbauer Spectroscopy, USIP-Reports, Stockholm 1979.
[4] V. N. Shchemelev and M. V. Kruglov, Fiz. tverd. Tela **14**, 3556 (1972).
[5] M. V. Kruglov, V. N. Shchemelev, and G. G. Kareva, phys. stat. sol. (a) **46**, 343 (1978).
[6] M. V. Kruglov, E. A. Sozontov, V. N. Shchemelev, and B. G. Zakharov, Kristallografiya **22**, 693 (1977).
[7] H. Sauer, Helv. phys. Acta **23**, 381 (1965).
[8] I. K. Solomin and M. V. Kruglov, Heads of Reports XIIIth All-Union Conf. Application of X-Rays to Material Investigation, Chernogolovka 1982.
[9] A. I. Tsumakov, G. V. Smirnov, M. V. Kruglov, and I. K. Solomin, Fiz. tverd. Tela **26**, 746 (1984).
[10] Z. G. Pinsker, Dynamical Scattering of X-Rays in Perfect Crystals, Springer-Verlag 1978 and Izd. Nauka, Moscow 1974 (pp. 20, 27, 77, 185) (in Russian).
[11] E. Kh. Mukhamedzhanov, V. A. Maslov, A. I. Tsuzo, and R. M. Imamov, Poverkhnost **3**, 54 (1984).
[12] M. V. Kruglov, E. A. Sozontov, and I. K. Solomin, Fiz. tverd. Tela **25**, 653 (1983).
[13] I. M. Bronstein and B. S. Fraiman, Secondary Electron Emission, Izd. Nauka, Moscow 1969 (pp. 88, 94, 170) (in Russian).
[14] V. N. Shchemelev and M. A. Rumsh, Fiz. tverd. Tela **4**, 2795 (1962).
[15] L. G. Eliseenko, V. N. Shchemelev, and M. A. Rumsh, Fiz. tverd. Tela **8**, 3649 (1966).
[16] D. L. Liljequist, T. Ekdahl, and U. Bäverstam, Nuclear Instrum. and Methods **155**, 529 (1978).

(Received June 3, 1985)

phys. stat. sol. (b) 133, 57 (1986)

Subject classification: 78.30; S1.62

Department of Solid State Physics, Polish Academy of Sciences, Zabrze[1] *(a) and Laboratoire de Physique des Solides, Université Paul Sabatier, Toulouse*[2] *(b)*

Some Aspects of Raman Scattering in Cd_3As_2 Single Crystals

By

J. WESZKA (a), M. RENUCCI (b), and A. ZWICK (b)

Raman scattering measurements in the tetragonal Cd_3As_2 single crystals are carried out at temperatures of 300 and 77 K. The spectra obtained are interpreted in terms of phonon–electron and electron–electron interactions. The features in the low-frequency interval are due to contributions from phonon and single particle scattering. The electronic RS involving intra- and interband excitations is dominant at frequencies higher than 80 cm^{-1}. The resonant enhancement of the band at 300 cm^{-1} arises from the incident light frequency being in the vicinity of $\Sigma_1 - \Sigma_4$.

La diffusion Raman dans les mono-cristaux de Cd_3As_2 quadratique a été étudiée à 300 et 77 K. Les spectres Raman obtenus sont interprétés en termes des interactions un phonon–électron et électron–électron. Des spectres en région de basses fréquences sont attribués à la diffusion à un phonon et à une particule. Cependant la diffusion Raman faisant intervenir des excitations électroniques intra- et interbandes est dominant aux fréquences au-dessus de 80 cm^{-1}. Une croissance résonante d'une bande à 300 cm^{-1} resulte du voisinage des énergies de raie incidente et un seuil d'excitation $\Sigma_1 - \Sigma_4$.

1. Introduction

The Raman scattering (RS) technique has recently been widely utilized for studying a quite large variety of electron–electron and phonon–electron interactions in solids [1]. These effects are of a particular interest in narrow-gap semiconductors, where intra- and interband as well as phonon excitation energies can overlap. Since the tetragonal α-Cd_3As_2 is a nearly zero-gap semiconductor, permanently degenerated within a quite large temperature interval with very high electron densities [2, 3], the afore-said phenomena are expected to play a rather important role in RS in these crystals. For these reasons, the authors of this paper aimed at giving insight into these effects for the lattice dynamics and other properties of the α-Cd_3As_2 crystals to be better understood. Subsequent sections contain a short information on physical properties and structure of the crystal under consideration (Sections 2 and 3). They are followed by the factor group analysis (Section 4) and the presentation of the experimental arrangement (Section 5). The obtained results and their discussion are presented in Sections 6 and 7, respectively.

2. Some Physical Properties of α-Cd_3As_2

Single crystals of the tetragonal modification of Cd_3As_2 have been grown from the vapour phase in evacuated quartz-sealed ampoules placed in a two-zone resistivity furnace by the method described elsewhere [3]. The inverted forbidden gap of this

---

[1]) Ul. Kawalca 3, 41-800 Zabrze, Poland.

[2]) 118 Route de Narbonne, 31062 Toulouse Cédex, France.

crystal was found to be of 0.09 eV at 77 K [3]. The electron density in as-grown n-type crystals is as high as 2×10^{24} m^{-2} [4], though Lovett [2] and Blom and Gelten [5] reported on samples with electron density as low as 2.8×10^{23} m^{-3} obtained after a thermal treatment. Though their densities being such high, the electron mobilities also attain rather high values ranging from 0.3 to 2.0 m^2/Vs at 300 K, while approaching 28 m^2/Vs at 4.2 K [2, 3]. This material is permanently degenerate over a large temperature range. From the optical measurements 0.14 eV band gap was reported [4]. Some authors completed the reflectivity studies covering the photon energy interval from 0 up to 30 eV [6]. There have been observed maxima in the reflectivity spectrum situated at 1.8 to 2.2, 2.8, and 4.0 eV and so on [7, 8]. The higher frequency maxima coincide with those observed in the thermoreflectivity spectra [8] and their localization appears to be in rather good agreement with the energy values arising from Lin Chung's band model [9].

3. Crystal Structure

The crystalline structure of the tetragonal α-Cd_3As_2 has the C_{4v}^{12} space group symmetry and its body-centred unit cell ($a = 1.267$ nm, $c = 2.548$ nm, and $c/a \approx 2.01$) contains as much as 160 atoms [10]. According to Steigmann and Goodyear [10] this structure can be seen as closely packed f.c.c. arrangement of arsenic atoms with six of eight tetrahedral sites occupied by cadmium atoms. These non-occupied sites are arranged along the diagonal of such an arsenic cube. The arrangement of these non-occupied sites in the vicinal arsenic f.c.c. "unit cells" determines the crystal structure of the II_3–V_2 compounds [11]. The α-Cd_3As_2 crystal structure being appreciably complex, would be considerably simplified when treated as a distorted fluorite-type one with two tetrahedral vacancies [9]. In the real α-Cd_3As_2 structure Cd atoms are displaced from their ideal positions towards the non-occupied sites. These displacements being rather large for Cd atoms (0.025 nm) are small enough for As atoms (0.005 nm). As a consequence the atoms are tetrahedrally coordinated but the tetrahedra are slightly distorted.

4. Factor Group Analysis

According to the group theory the primitive unit cell must be used as the basis of a factor group analysis [12]. If the Bravais unit cell in the reciprocal space is not primitive, its volume has to be divided by the factor equal to how many times it is larger than the primitive unit cell for a given symmetry class. Thus, the volume of such a primitive unit cell for the α-Cd_3As_2 crystal contains only 80 atoms of the constituent elements. The factor group analysis for this crystal has been carried out in accordance with the correlation method of Fateley and McDevitt [12], in which site symmetries of the constituent elements are taken as a starting point. The distribution of the lattice modes among the irreducible representations of the crystal factor group are found through the correlations among the irreducible representations of the crystal factor group and those of the site symmetry groups of the constituent elements. According to Steigmann and Goodyear [10], the Cd atoms occupy six sets of equivalent positions of C_1 symmetry (each of multiplicity 8), while the As atoms form six sets of C_2 symmetry positions (multiplicity 4) and one set of C_1 symmetry. Then the correlations among the irreducible representations of both the site groups and the crystal factor group must be found. For the Cd atoms the correlation between

C_1 and C_{4v} groups is as below

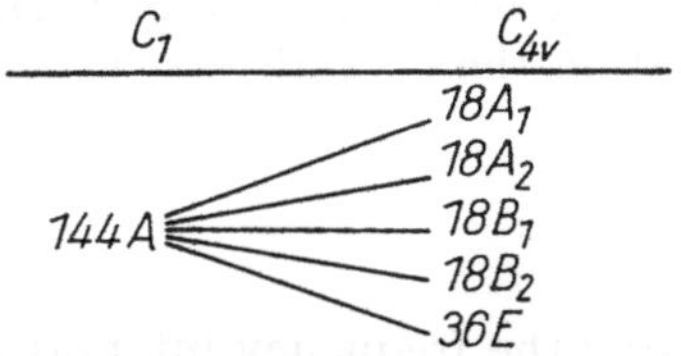

The 144 degrees of vibrational freedom of Cd atoms are represented by

$$\Gamma^{Cd} = 18\,A_1 + 18\,A_2 + 18\,B_1 + 18\,B_2 + 36\,E\,.$$

For As atoms at C_1 sites the same correlation results in the following representation:

$$\Gamma_1^{As} = 3\,A_1 + 3\,A_2 + 3\,B_1 + 3\,B_2 + 6\,E\,.$$

The correlation between C_2 and C_{4v} groups for As atoms at the sites of C_2 symmetry

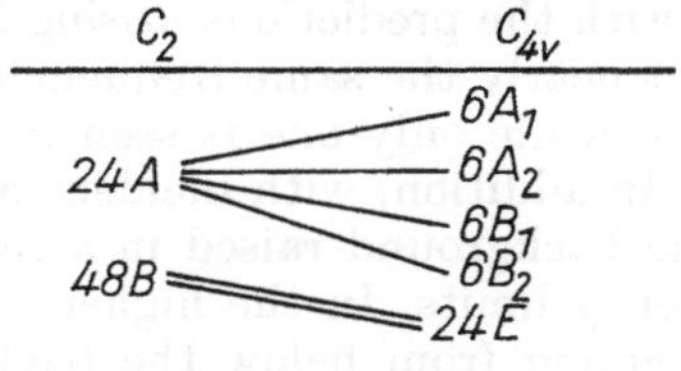

results in the representation

$$\Gamma_2^{As} = 6\,A_1 + 6\,A_2 + 6\,B_1 + 6\,B_2 + 24\,E\,.$$

Thus, the total distribution of the α-Cd_3As_2 lattice modes among the irreducible representations of the factor group is as follows:

$$\Gamma^{tot} = 27\,A_1 + 27\,A_2 + 27\,B_1 + 27\,B_2 + 66\,E\,.$$

Then, the representation of Raman active modes is found to be

$$\Gamma^{R} = 26\,A_1 + 27\,B_1 + 27\,B_2 + 65\,E\,,$$

while that of IR-active modes

$$\Gamma^{IR} = 26\,A_1 + 65\,E\,.$$

The components of the Raman tensor for this crystallographic class are

$$\underset{A_1(z)}{\begin{pmatrix} a & \cdot & \cdot \\ \cdot & a & \cdot \\ \cdot & \cdot & b \end{pmatrix}} \quad \underset{B_1}{\begin{pmatrix} c & \cdot & \cdot \\ \cdot & -c & \cdot \\ \cdot & \cdot & \cdot \end{pmatrix}} \quad \underset{B_2}{\begin{pmatrix} \cdot & d & \cdot \\ d & \cdot & \cdot \\ \cdot & \cdot & \cdot \end{pmatrix}} \quad \underbrace{\underset{(x)}{\begin{pmatrix} \cdot & \cdot & e \\ \cdot & \cdot & \cdot \\ e & \cdot & \cdot \end{pmatrix}} \quad \underset{(y)}{\begin{pmatrix} \cdot & \cdot & \cdot \\ \cdot & \cdot & e \\ \cdot & e & \cdot \end{pmatrix}}}_{E}\,.$$

5. Experimental

The RS measurements in the tetragonal α-Cd_3As_2 single crystals were carried out in the backscattering configuration. The radiation of the Ar^+ ion laser was used as a source of light. To avoid the crystal deterioration the power of the incident radiation was kept at the level of 300 mW, though there were runs it was raised up to 400 mW. All the Raman spectra had been recorded with the Coderg triple-grating Raman spectrometer. The spectral resolution was changed from 1 to 3 cm^{-1}. The crystals examined were mounted onto the cold finger in the sample chamber of the cryostat. While measuring this chamber was filled in with helium. The RS spectra for various symmetry species were recorded at (100), (001), (112) crystal faces at 77 and 300 K. The polarization of the scattered light was always kept horizontally. Thus, when changing the scattered light polarization the crystal was turned around a given axis.

6. Results

The RS spectra from the (100) crystal face covering the frequency interval from the Raylaigh line up to 400 cm^{-1} are shown in Fig. 1. They were recorded at room temperature with the use of the $\lambda_0 = 476.5$ nm line of 400 mW power. The spectral resolution was as large as 3 cm^{-1}. The spectra of $A_1 + B_1$, A_1, and E species were being recorded in (yy), (zz), and (yz) polarization geometries, respectively. The spectra $z(xx)\,\bar{z}$, $z(yy)\,\bar{z}$, and $z(x+y, x+y)\,\bar{z}$ recorded at 77 K are given in Fig. 2. To excite them radiation of $\lambda_0 = 488$ nm wavelength and 300 mW power was used. In this case the spectral resolution has been equal to 2 cm^{-1}. There are a few features well pronounced being hardly seen in the obtained spectra, in contrast with the predictions arising from the factor group analysis. These features appeared at nearly the same frequencies in the spectra of (xx), (yy), (zz) polarization geometries, while only one is seen in the (xy) and (xz) spectra, independently of temperature. In addition, with incident and scattered light polarization parallel one observes the background raised in a rather appreciable fashion within the 0 to 80 cm^{-1} frequency limits. In the higher frequency area rather weak and broad bands slightly emerging from below the background, with an exception of the band at about 300 cm^{-1}, are seen in these spectra. For perpendicular polarization geometries the intensities of the RS spectra are much smaller than for the parallel ones and relatively flat and smooth background is observed. The main features observed in the spectra $z(xx)\,\bar{z}$, $z(yy)\,\bar{z}$, and $x(zz)\,\bar{x}$ appeared at 27 and 46 cm^{-1}, while in (xy) and (xz) spectra the only feature has been observed at 27 and 46 cm^{-1}, respectively (Fig. 1). When the temperature is decreased the band at 27 cm^{-1} undergoes a slight shift downward by 2 cm^{-1}, the others being unshifted. The low temperature spectra were found to be strongly dependent on the frequency of the incident radiation. However, this dependence has appeared to be quite different for the features within the frequency range up to 80 cm^{-1} (Fig. 3) and for the band

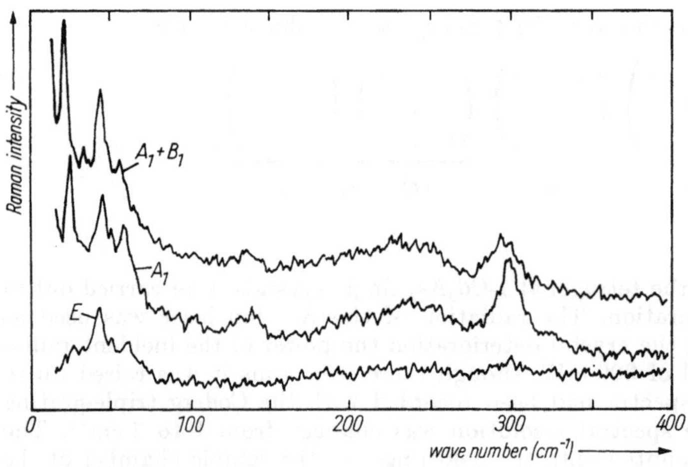

Fig. 1. Raman spectra recorded on (100) face at 300 K

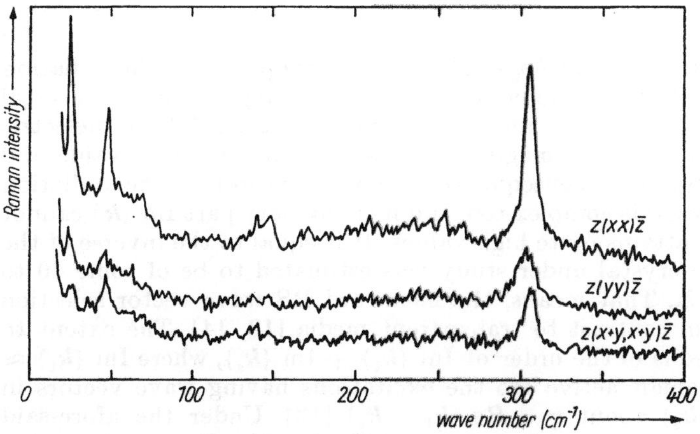

Fig. 2. Raman spectra recorded on (001) face at 77 K

at about 300 cm^{-1} (Fig. 4). The $z(xx)\,\bar{z}$ spectra, shown in Fig. 3 and 4, in which the lattice modes of both A_1 and B_1 species are allowed, were recorded with 2 cm^{-1} spectral resolution. The power of the incident radiation was kept equal to 200 mW.

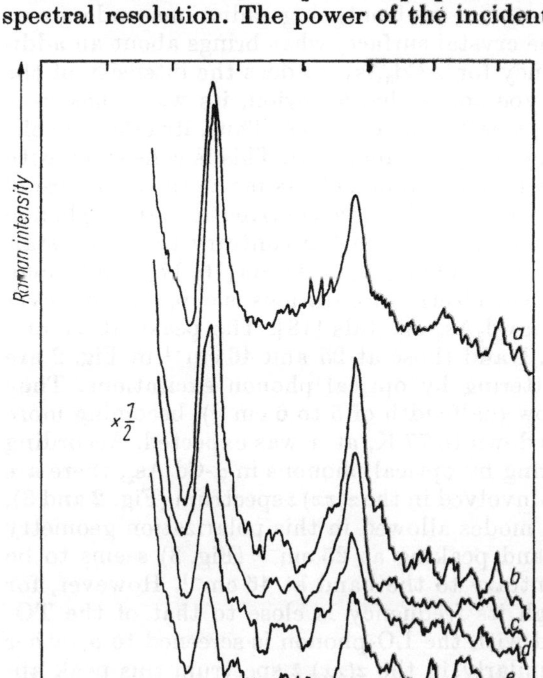

Fig. 3

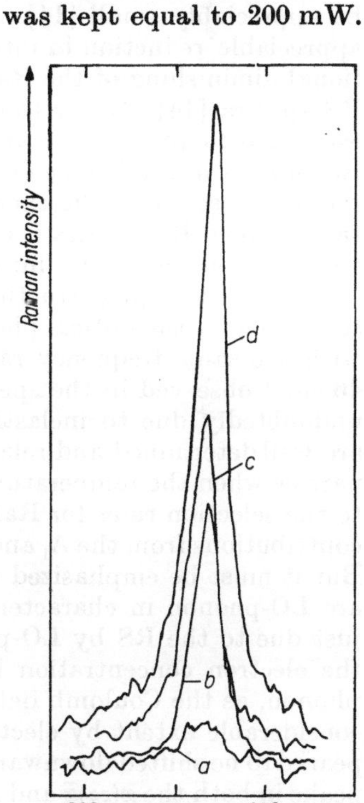

Fig. 4

Fig. 3. Raman spectra, $z(xx)\,\bar{z}$ polarization at 77 K. Incident light wavelengths: (a) 514.5, (b) 496.5, (c) 488, (d) 476.5, (e) 457.9 nm

Fig. 4. Resonance enhancement of the band at 300 cm^{-1}, $z(xx)\,\bar{z}$, 77 K. Incident light wavelengths: (a) 501.7, (b) 488, (c) 476.5, (d) 457.9 nm. The power of the incident radiation was equal to 200 mW

7. Discussion

Being narrow-gap in character the α-Cd_3As_2 single crystals are opaque for the radiation of the Ar^+ ion laser which is strongly attenuated while penetrating through the crystal surface into its interior [14]. At the frequencies in question the α-Cd_3As_2 dielectric function is complex and its real and imaginary parts are comparable in value and much higher than unity [6]. As a consequence, the wave vector of the radiation propagating through the crystal is complex too, and its imaginary part Im $\{\boldsymbol{k}\}$ cannot be neglected any more as it attains quite high values. It is equal to the inverse of the skin-depth δ, which for the crystal under study was estimated to be of some 40 to 50 nm, small compared to λ_0. That means, the backward RS wave vector selection rules fail to be conserved in contrast to transparent media [13, 14]. The extent to which they are not conserved is of the order of $\text{Im}\{\boldsymbol{k}_i\} + \text{Im}\{\boldsymbol{k}_s\}$, where $\text{Im}\{\boldsymbol{k}_i\} \approx \approx \text{Im}\{\boldsymbol{k}_s\} \approx \delta^{-1}$. Thus, Raman active are the excitations having wave vectors in the range $\text{Im}\{\boldsymbol{k}_i\} + \text{Im}\{\boldsymbol{k}_s\}$ about $\boldsymbol{q} = \text{Re}\{\boldsymbol{k}_i - \boldsymbol{k}_s\}$ [13]. Under the afore-said conditions the range of the momentum transfer contributing the Raman scattering is spread over the wave-vector range from 0 up to 10^6 cm^{-1}, while the real part of the momentum transfer being of the same order of the magnitude. Once the wave vector range is so large the efficiency of inelastic light scattering in α-Cd_3As_2 should be appreciably small [14]. Moreover, both in- and outgoing radiations undergo an appreciable reduction in intensity at the crystal surface, what brings about an additional diminishing of the Raman efficiency for α-Cd_3As_2, so does the intensity of the RS spectra [14]. As for the Schottky-type space charge region, its width has been estimated to amount to some 4 to 5 nm, small compared to δ. Thus, its effect on the RS process in α-Cd_3As_2 should be merely of significance [15]. This is consistent with the nearly semimetallic character of the crystal examined. As far as the structure of the obtained RS spectra is concerned, they would involve contributions from phonon and electron intra- and interband excitations. But, on the contrary to the crystals having much simpler structure and larger forbidden gap, i.e. GaAs [16, 17], the Raman scattering by the optical phonons as well as electron excitations is supposed to cover with the same frequency range in the α-Cd_3As_2 crystals [18]. The peaks at 27 and 46 cm^{-1} observed in the spectra in Fig. 1 and those at 25 and 46 cm^{-1} in Fig. 2 are undoubtedly due to inelastic light scattering by optical phonon excitations. They are well determined and relatively narrow (halfwidth of 5 to 6 cm^{-1}), becoming more narrow when the temperature decreases down to 77 K, at it was expected. According to the selection rules for Raman scattering by optical phonons in α-Cd_3As_2, there are contributions from the A_1 and B_1 species involved in the $z(xx)\bar{z}$ spectrum (Fig. 2 and 3). But it must be emphasized that the A_1 modes allowed in this polarization geometry are LO-phonon in character and the band peaking at 25 cm^{-1} (Fig. 3) seems to be just due to the RS by LO-phonons, contrary to the band at 46 cm^{-1}. However, for the electron concentration being so high its frequency is close to that of the TO-phonon, as the Coulomb field associated with the LO-phonon is screened to a rather considerable extent by electrons [1]. Similarly, in the $z(xx)\bar{z}$ spectrum this peak appeared to be shifted downward by 2 cm^{-1} with respect to the frequencies of the relevant peaks in both the $x(zz)\bar{x}$ and $x(yy)\bar{x}$ spectra. But this shift can be understood as in both these spectra the A_1 modes are TO-phonon ones. On the other hand, since the α-Cd_3As_2 crystal is nearly semimetallic with very high density of conduction electrons, the LO-phonon–plasmon interactions should be noticeable in the $z(xx)\bar{z}$ spectrum. Furthermore, as the plasma frequency is equal to about 330 cm^{-1}, the LO-phonon–plasmon coupling should bring about two such coupled modes to occur, the low frequency LO-phonon one and the high frequency one of plasmon character. Then, for the above-mentioned reasons the band at 25 cm^{-1} in the $z(xx)\bar{z}$ spectrum is thought to be due to just the LO-phonon low frequency coupled mode, while that observed at about 27 cm^{-1} in the $x(zz)\bar{x}$ and $x(yy)\bar{x}$ spectra is the TO-phonon mode. However, for the momentum transfer to be quite sizeable the plasmon mode as well as the high frequency coupled plasmon mode cease to exist as being totally screened by single free electron excitations [15]. On the contrary, the LO-phonon low frequency coupled mode is only partially damped. Then, while changing the temperature the electron density can undergo rather slight variations which consequently make it shift slightly towards lower frequencies. Though these changes are thought to be negligible in the temperature regime used [18], nevertheless they may appear to be meaningful for the LO-phonon–plasmon interactions. The small value of this shift seems not to be in contradiction with that. The peaks in question are seen to be grown on the raised background. This effect could be attributed to single free electron scattering [19]. As the momentum transfer $\boldsymbol{q} \approx \boldsymbol{q}_F$, the electrons involved can be treated as a free electron gas. Under the conditions of the experiment this effect should be too weak to be observed [16]. However, the non-parabolicity of the conduction band [5] would make it sufficiently strong to be observed due to energy fluctuations, which are not screened as they are relatively little affected by Coulomb interactions as compared with the charge density fluctuations [19]. Then, the spectrum within the low frequency limit (up to about 80 cm^{-1}) would be due to a simple summation of the non-interacting contributions, i.e. phonon scattering and single free electron scattering. The cut-off frequency $\omega \approx qv_F$ has been estimated to be about 75 cm^{-1} in good agreement with the experimental data.

Additionally to the inelastic light scattering mechanisms discussed above, some contribution coming from the excitations involving intra- and interband electron transitions should be taken into account for the narrow-gap character of α-Cd_3As_2. Moreover, the intermediate states are now real states of the crystal. Then, the RS spectrum should follow the joint density of electron–hole pair states [15]. For incident light being far from the maxima in the joint density of e–h pair states, rather broad structureless bands are expected to appear in the RS spectrum [15]. Such features are seen in the present spectra at frequencies higher than 80 cm^{-1}. As has been mentioned above, the most pronounced features in the presented spectra were found to be dependent on the incident light wavelength at 77 K [20]. However, this effect seems to take place in different ways for the features within the low frequency range and for the band at 300 cm^{-1} (Fig. 3 and 4, respectively). That points to the rather different origins of these resonant effects. As is clearly seen in Fig. 3, the low frequency peaks at 25 and 46 cm^{-1} behave in some opposite fashion when the frequency of the incident light increases. For $\lambda_0 = 457.9$ nm the relative intensities of both peaks are nearly the same. The resonant-like behaviour of the bands at 25 and 46 cm^{-1} would be related to the maxima in the jointdensity of e–h pair states observed at 2.1 and 2.8 eV [7, 8] and predicted by the simplified Lin Chung model [9]. For the energy of the incident radiation between of them, some cancellation of the resonant terms in the expression for the RS cross section can result in the fluctuations of their intensities in a way as observed in the spectra in Fig. 3. On the other hand, the resonant enhancement of the band at about 300 cm^{-1}, when the incident light frequency increased, would arise from the interband electronic RS process in which the frequency of the incident light is in the vicinity of the maximum in the joint density of e–h pair states at 2.8 eV [7, 8, 15]. This maximum is attributed to the Σ_1–Σ_4 transition. To give an account for the 300 cm^{-1} band behaviour the following three-band model is proposed. The incident radiation excites the electron from the valence band into the higher conduction band and then, the electron undergoes a transition to the lower conduction band. Thus, the frequency at which this band is observed in the RS spectrum corresponds with the gap between the valence band and the lower conduction band. The first of these transitions is the Σ_1–Σ_4 one, which explains the increase in intensity of the band at 300 cm^{-1} when the incident light photon energy approaches 2.8 eV.

8. Concluding Remarks

The obtained RS spectra of the α-Cd_3As_2 crystals have been interpreted in terms of phonon– and electron–electron interactions. The low frequency area of the spectrum is due to summation of the phonon and single free particle scattering contributions. At higher frequencies the intra- and interband electron excitations appear to be dominant. The resonance enhancement of the band at 300 cm^{-1} is interpreted in terms of the three-band electron RS with the Σ_1–Σ_4 transition involved.

Acknowledgements

One of the authors, J. W., is indebted to Prof. J. B. Renucci for his support and valuable discussions, and to Prof. W. Żdanowicz for his support and aid. We wish to thank to Dr. K. Kloc for supplying us with the crystals.

References

[1] P. M. PLATZMAN and P. A. WOLF, Waves and Interactions in Solid State Plasmas, Academic Press, Inc., 1973.
[2] D. LOVETT, Phys. Letters A **30**, 90 (1969).
[3] W. ŻDANOWICZ, Acta phys. Polon. **25**, 663 (1967).
[4] L. M. ROGERS, R. M. JENKINS, and A. J. CROCKER, J. Phys. D **4**, 793 (1971).
[5] F. A. P. BLOM and M. J. GELTEN, Phys. Rev. B **8**, 2411 (1979).
[6] M. ZIVITZ and J. R. STEVENSON, Phys. Rev. B **10**, 3457 (1974).
[7] V. V. SOBOLEV, N. N. SYRBU, T. A. ZYUBINA, and J. A. UGAI, Fiz. Tekh. Poluprov. **5**, 327 (1971).
[8] V. P. BHOLA, J. Phys. Chem. Solids **38**, 1237 (1977).
[9] P. J. LIN CHUNG, Phys. Rev. **188**, 1271 (1969).
[10] G. A. STEIGMANN and J. GOODYEAR, Acta cryst. **B24**, 1062 (1968).
[11] A. D. IZOTOV, V. P. SANYGIN, and V. F. PONOMAREV, Kristallographiya **23**, 764 (1978).
[12] G. A. FATELEY and L. MCDEVITT, J. appl. Optics **25**, 155 (1971).
[13] E. BURSTEIN and A. PINCZUK, in: Light Scattering in Solids, Ed. M. CARDONA, Springer-Verlag, 1975 (p. 22).
[14] D. L. MILLS, A. A. MARADUDIN, and E. BURSTEIN, Ann. Phys. (N.Y.) **56**, 504 (1970).
[15] E. BURSTEIN, A. PINCZUK, and R. F. WALLIS, in: Phys. Semimet. and Narrow-Gap Semicond., Proc. Conf., Dallas (Texas) 1970, Ed. D. L. CARTER and R. T. BATE, Pergamon Press, London/Oxford 1971 (p. 251).
[16] A. MOORADIAN, in: Light Scattering in Solids, Ed. G. B. WRIGHT, Springer-Verlag, 1978 (p. 285).
[17] A. MOORADIAN and G. B. WRIGHT, Phys. Rev. Letters **16**, 999 (1966).
[18] P. L. RADOLF and S. G. BISHOP, Phys. Rev. B **5**, 442 (1972).
[19] P. A. WOLF, in: Light Scattering Spectra of Solids, Ed. G. B. WRIGHT, Springer-Verlag, New York/Heidelberg/Berlin 1968 (p. 273).
[20] J. WESZKA, M. RENUCCI, and J. B. RENUCCI, Acta phys. Polon. **A67**, 227 (1985).

(Received April 30, 1985)

phys. stat. sol. (b) **133**, 65 (1986)

Subject classification: 71.45 and 77.20; S9.16

Sektion Physik der Humboldt-Universität zu Berlin[1]*), Bereich 04*

Dielectric Tails in K_2ZnCl_4: No Evidence for an Intrinsic Chaotic State

By

T. NATTERMANN

It is shown, that the increasing dielectric tail found recently for decreasing impurity concentration N_i in the incommensurate system K_2ZnCl_4 *does not* indicate the existence of an intrinsic chaotic state. Using a simple model for the pinning of solitons by impurities (but more realistic than those used previously in this context) it is shown, that the dielectric tails grow with growing soliton density n_s but with dropping N_i. The precise dependence on N_i depends on the pinning mechanism (i.e. weak or strong pinning).

Es wird gezeigt, daß der anwachsende dielektrische Ausläufer, der kürzlich für abnehmende Störstellenkonzentration N_i in dem inkommensurablen System K_2ZnCl_4 gefunden wurde, nicht die Existenz eines intrinsischen chaotischen Zustands anzeigt. Mit einem einfachen Modell für die Verankerung von Solitonen durch Störstellen (jedoch realistischer als die, die früher in diesem Zusammenhang benutzt wurden) wird gezeigt, daß die dielektrischen Ausläufer mit wachsender Solitonendichte n_s, jedoch mit fallendem N_i wachsen. Die genaue Abhängigkeit von N_i hängt vom Verankerungsmechanismus (d. h. schwache oder starke Verankerung) ab.

1. Introduction

In a recent paper, Zhang et al. [1] considered the excess dielectric constant $\Delta\chi = \chi - \chi_0$ or dielectric tail below the transition temperature T_c to the commensurate (C) phase of K_2ZnCl_4. With decreasing impurity concentration N_i the tail became longer, and its relaxation at constant temperature became slower. In order to explain their data, Zhang et al. [1] assume a linear dependence of $\Delta\chi$ on the discommensuration (soliton) density n_s, as has been proposed in [2]. Although n_s should be zero below T_c, pinning by a) impurities or b) the (intrinsic) Peierls lattice pinning potential [3] will keep the system in a metastable chaotic state with $n_s > 0$. Since $\Delta\chi$ exhibited an increase with decreasing impurity concentration Zhang et al. concluded that the chaotic state is intrinsic rather than impurity induced. In the present communication we argue that this is not the case.

The paper is organized as follows. In Section 2 we summarize the main results of previous investigations on the influence of impurities on the soliton lattice of the incommensurate phase. They are the framework for our further discussion.

In Section 3 the dielectric susceptibility is related to the behaviour of the soliton lattice. A previous treatment of pinning effects in the susceptibility is critically illustrated.

Section 4 contains our new treatment of those effects both for the case of weak and strong pinning as well as some concluding remarks.

2. Soliton Phenomenology and Impurity Pinning

We consider first the influence of impurities on the soliton lattice of the incommensurate (I) phase. In the thermodynamic equilibrium, the free-energy density $\mathcal{F}(l)$ of the soli-

[1]) Invalidenstr. 42, DDR-1040 Berlin, GDR.

ton lattice of spacing l is

$$\mathcal{F}(l) = \frac{1}{l}\,[-\sigma\tau + U(l)]. \tag{1}$$

Here $\tau \approx \tau_1(T - T_c)/(T_I - T_c)$ is a reduced temperature, T_I is the transition temperature between the disordered (D) and the I-phase, and $\sigma > 0$. $-\sigma\tau$ has the meaning of the free energy per unit area of the soliton, $U(l)$ of a repulsive interaction between the solitons. In equilibrium $\partial\mathcal{F}/\partial l|_{l=l_s} = 0$, which yields the soliton density $n_s = \xi/l_s$. ξ denotes the width of the solitons. Neglecting fluctuations $U(l) = U_0(l) \approx 4\sigma\, e^{-l/\xi}$ [3], which yields $n_s \sim \ln^{-1}(T_s - T)$ for $T \to {}_+T_c$ and $n_s = 0$ for $T < T_c$.

Discussing the influence of impurities, one has to distinguish several cases. For brevity we will restrict ourselves to three-dimensional systems from the very beginning.

It is convenient to introduce the mean impurity distance D ($N_i = D^{-3}$) and the quantity $w_s = (\vartheta/g)^{1/2}$. g denotes the interface stiffness. w_s has the meaning of the maximal distortion of the soliton which is still compensated by the energy gain $-\vartheta$ from the interaction with a *single* impurity. In three dimensions, w_s is independent of the linear extension of the soliton. Depending on the relative size of D, ξ, l_s, and w_s, various pinning regions may be introduced. For simplicity, we distinguish here only the case of *weak pinning* where $D \ll \xi$ or $w_s \ll \xi$ if $D > \xi$ from that of *strong pinning* where $w_s > \xi$ if $D > \xi$.

a) In the weak-pinning case, soliton distortions arise due to the spatial fluctuations in the impurity density. In this case one has to distinguish further between impurities, breaking or preserving the local order parameter symmetry, respectively [4 to 6]. A phenomenological summary is given in [7].

The main results for the three-dimensional weak-pinning case are:

(i) Impurities change the interaction law $U(l)$ between solitons, which obtains a power-law contribution $U_\lambda(l) \approx Bg(\lambda/l)^{1/\bar\beta}$. The length scale λ measures the local fuzziness of the soliton due to the disorder. In particular, $\lambda = \vartheta^2 N_i/g$, $\bar\beta = 1$ and $\lambda = (\vartheta^2 N_i \xi/g^2)^{1/2}$, $\bar\beta = \frac{1}{2}$ for symmetry breaking and symmetry preserving impurities, respectively. B is a numerical factor. It follows that in approaching T_c for $T_c < T < T_\lambda$, $U_\lambda(l)$ dominates the soliton interaction resulting in $n_s \approx (\xi/\lambda)\,(\sigma\tau/g)^{\bar\beta}$. T_λ is determined implicitly from the equality of energies $U_\lambda(l_s) \approx U_0(l_s)$. The transition to the C-phase remains of second order.

(ii) The impurities destroy the long-range order of the I-phase, i.e. the whole I-phase is "chaotic" and there is no longer a sharp transition at T_I. Practically, the correlation length L_c is very large, if the impurity concentration is low. Note, however, that in the region $T_c < T < T_\lambda$, L_c is of the order of the intersoliton distance l_s [7].

More important, once prepared in a non-equilibrium state, impurities may keep the system there for a long time. These pinning effects occur already at temperatures $T_\lambda^* > T_\lambda$. T_λ^* is determined now from the equality of the *driving force* $-\partial U/\partial l|_{l=l_s}$ and the *pinning force* E_c. We have estimated the pinning force E_c by $E_c \approx U_\lambda(\xi)/\xi$ [7, 8]. Thus, the soliton lattice starts to be frozen in a non-equilibrium state if the soliton density approaches the value $n_s^* = n_s(T_\lambda^*)$,

$$n_s^* \approx \ln\left[\frac{Bg}{4\sigma}\left(\frac{\lambda}{\xi}\right)^{1/\bar\beta}\right] \sim \ln N_i\,. \tag{2a}$$

From a practical point of view one has to consider T_λ^* as the onset temperature of the chaotic phase, which has an initial soliton density n_s^*.

Thermal hopping will lead to a temporal decrease of the pinning force [8]. The mechanism for the approach of n_s to equilibrium is not fully clear, but the formation and

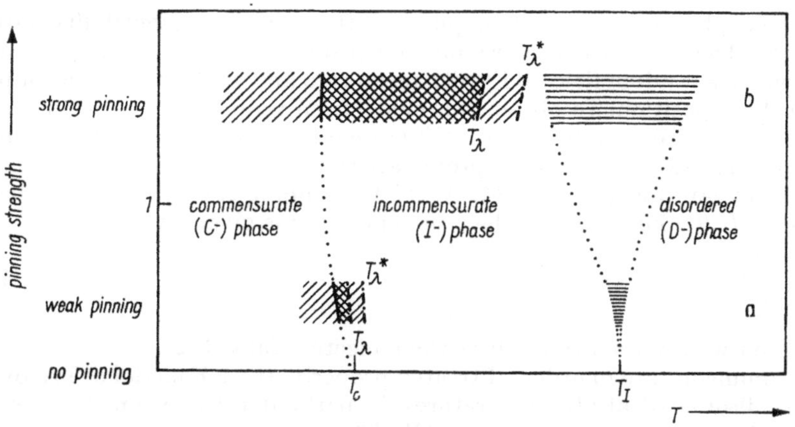

Fig. 1. Schematic phase diagram for weak and strong pinning. Even weak pinning destroys the sharp transition between the D- and the I-phase at T_{I}. Principally, the system is disordered for $T > T_c$. Practically, the transition region (horizontally hatched) can be very tight and the correlation length L_c can be large. In equilibrium, the transition to the C-phase at T_c remains sharp. For $T_c < T < T_\lambda$ (cross-hatched region) the influence of the pinning is remarkable, leading to a "chaotic" state with L_c of the order of the intersoliton distance l. However, already for $T < T_\lambda^*$ metastable soliton configurations occur, which reach deeply into the C-phase (diagonally hatched region)

growth of nuclei or anti-nuclei, formed by p solitons that terminate in a deperiodization loop ($p = 6$ for K_2ZnCl_4), seems to be the decisive mechanism [7, 9].

b) In the strong-pinning case, the soliton tries to overlap with a maximal number of impurities, each lowers its energy by $-\vartheta$. The qualitative behaviour is expected to be the same as in the weak-pinning region. However, the correlation length L_c is probably of the order of the impurity distance D, for $T < T_\lambda^*$ with T_λ, T_λ^* of the order of T_{I}. It is reasonable to consider

$$n_s^* \approx \frac{\xi}{D} \sim N_i^{1/3} \tag{2b}$$

as a lower bound for the initial non-equilibrium soliton density.

The different temperature regions are schematically scetched in Fig. 1.

3. Dielectric Susceptibility

In incommensurate ferroelectrics like K_2ZnCl_4, the polarization P is a second-order parameter which couples essentially to the phase φ of the first-order parameter such that in equilibrium solitons separate domains of alternating up and down polarization. Using the exponential wall interaction $U_0(l)$ one finds a Curie-Weiss law for χ if one approaches T_c from the I-phase [10 to 12]. An obvious interpretation of this result has been given in [13]: an applied electric field E increases the total electric moment mainly by shifting domain walls such that domains with $\boldsymbol{P}$ parallel (antiparallel) to $\boldsymbol{E}$ become larger (smaller). These shifts are limited by the repulsive interaction $U(l)$. In this way we got for arbitrary $U(l)$

$$\chi - \chi_0 = \frac{P_c^2}{l_s U''(l_s)} \sim \frac{\partial n_s^{-1}}{\partial T}, \tag{3}$$

where P_c is the spontaneous polarization just below T_c.

A first attempt to explain non-equilibrium pinning effects in the susceptibility data has been made in [2]. The authors consider a one-dimensional model where impurities are assumed to produce a random field (RF) with mean zero and width σ_E, which couples linearly to the soliton position. To obtain this model from a similar three-dimensional one, one has to assume an infinite soliton stiffness g or a homogeneous RF perpendicular to the modulation axis. Using an approximate treatment, Prelovsek and Blinc [2] find a Curie-Weiss law for χ if $T > T_c^*$. At T_c^* the pinning energy of the RF becomes comparable to $U_0(l)$. For $T < T_c^*$ their susceptibility is given by

$$\chi - \chi_0 = \frac{P_c}{\pi \sigma_E} n_s \,. \tag{4}$$

Equation (4) has been widely used to analyse the dielectric data [1, 2].

However, some comments are in order. If treated properly, the RF model of Prelovsek and Blinc [2] is disordered at all temperatures. In particular, there is no C-phase. This is due to entropic contributions from the RFs. The same statement applies even to a more realistic three-dimensional model, where the RF is inhomogeneous and the soliton stiffness is finite [14]. In reality, the influence of disorder is more complicated and weaker such that there is a stable C-phase in three dimensions [4 to 7].

Considering metastable states, as necessary for the explanation of the dielectric tails, one is forced to perform approximations. In the following we consider a simple model to explain the dependence of $\Delta\chi$ on the soliton density n_s and defect concentration N_i. Contrary to [2], the equilibrium properties of our model permit a stable C-phase.

4. Dielectric Tails and Impurities

In the strong-pinning case, to begin with, the solitons behave as elastic membranes with fixed distortions at the impurity positions. Switching on a small external electric field E the soliton will be distorted by δl between the impurities, in order to increase the total polarization (see Fig. 2). The corresponding increase $\delta\sigma$ of the energy of a singled-out soliton, which is subjected to the interaction with neighbouring solitons in a distance l_s is given by

$$\delta\sigma = \frac{1}{2} U(l_s + \delta l) + \frac{1}{2} U(l_s - \delta l) + \frac{1}{2} g \frac{\delta l^2}{D^2} - 2EP_c \delta l \,.$$

Minimizing $\delta\sigma$ with respect to δl and using $\delta P = 2P_c \delta l/l_s$ for the change of the polariza-

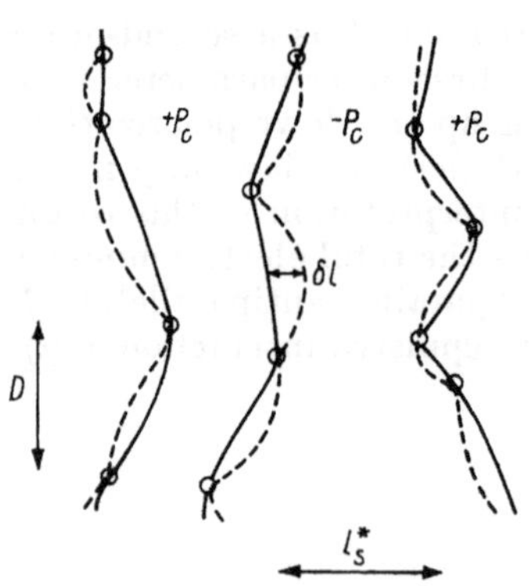

Fig. 2. Schematic representation of a strongly pinned soliton configuration. Under the influence of an external field E parallel to the $+P_c$ directions solitons become distorted by δl (marked with dashed lines)

tion, we get for the excess susceptibility

$$\chi - \chi_0 = \frac{\partial\, \delta P}{\partial E} = \frac{P_c^2}{U''(l_s) + g/(4D^2)}\, \frac{1}{l_s} \sim \frac{P_c^2}{g\xi}\, \frac{n_s}{N_i^{2/3}}. \tag{5}$$

As in (3) $\Delta\chi$ is proportional to the non-equilibrium soliton density $n_s = \xi/l_s$. However, for sufficiently large l such that $U''(l) \ll g/D^2$, $\Delta\chi$ is proportional to $D^2 = N_i^{-2/3}$. Thus, decreasing the impurity concentration N_i, $\Delta\chi$ increases, as observed by Zhang et al. [1]. This conclusion remains valid even if we admit that the non-equilibrium soliton density $n_s \approx n_s^*$, entering (5) will decrease as $1/D \sim N_i^{1/3}$ for increasing impurity distance (cf. (2b)).

A similar result is obtained for the weak-pinning case. Here the largest pinning forces result from a length scale $L_{\min} = \xi^{3/2} g/\vartheta N_i^{1/2}$ [6, 15]. Since on scales $L \ll L_{\min}$ the domain wall is essentially unpinned, we assume that (5) remains valid in the weak-pinning case provided D is replaced by $L_{\min}$.

In the region where we can neglect the soliton interaction we get from (5) for the weak-pinning case

$$\Delta\chi \approx \frac{P_s^2 \xi^2 g}{\vartheta^2}\, \frac{n_s}{N_i}. \tag{6}$$

The initial value of n_s is now given approximately in (2a). Again, $\Delta\chi$ increases for $N_i \to 0$.

Thus, both for strong and weak pinning we find increasing dielectric tails for decreasing impurity concentrations, in agreement with the findings in [1]. Although additional pinning due to the underlying lattice may increase n_s, it will also render more difficult the wall motion and hence decrease the response on an external field. In this respect the considerations made in [1] seem to be partly inconsistent. Indeed, a lower N_i will also lead to a lower metastable soliton density n_s, even if there is an additional pinning due to the lattice. Using then (4), this would mean a decrease of χ! On the contrary, according to our above calculation, the decrease of n_s is overcompensated by the N_i dependence of the denominator of (5), (6).

For n_s low enough Zhang et al. [1] found empirically

$$\frac{\partial n_s}{\partial t} = -C n_s^2 \tag{7}$$

for the time behaviour of the soliton density, suggesting that solitons decay by pair annihilation. Since the initial value n_s^* of n_s decreases with N_i ((2a), (2b)) according to (7) the approach to equilibrium at a fixed temperature should be slower for lower impurity concentration, in agreement with [1]. The strong increase of C (cf. (7)) for decreasing T ($< T_c$) is consistent with the fact that the (impurity induced) chaotic state is highly metastable and that the approach to equilibrium becomes faster for increasing supercooling [7, 9].

We add a further remark on (7). In terms of walls separating domains of alternating magnetization (7) describes the rate of disappearance of such domains. In terms of the original phase walls, where φ changes by $2\pi/p$ ($p = 6$), the collision of two walls means the creation of a $4\pi/p$ wall, which has in general a higher energy than that of the two separate $2\pi/p$ walls. Possibly, such a state is a saddle-point which the system has to pass to reach the equilibrium.

References

[1] G. Zhang, S. L. Qiu, M. Dutta, and H. Z. Cummins, Solid State. Commun. (1985), in the press.

[2] P. Prelovsek and R. Blinc, J. Phys. C 17, 577 (1984).
R. Blinc, P. Prelovsek, A. Levstik, and C. Filipic, Phys. Rev. B 29, 1508 (1984).

[3] P. BAK, Rep. Progr. Phys. **45**, 187 (1982).
[4] J. VILLAIN, J. Physique Lettres **43**, L551 (1982).
J. VILLAIN, B. SEMERIA, F. LANCON, and L. BILLARD, J. Phys. C **16**, 6153 (1983).
[5] T. NATTERMANN, J. Phys. C **16**, 4125 (1983).
[6] T. NATTERMANN, phys. stat. sol. (b) **119**, 209 (1983); J. Phys. C **18** (1985), in the press.
[7] T. NATTERMANN, J. Phys. C**18** (1985), in the press.
[8] T. NATTERMANN, phys. stat. sol. (b) **129**, 153 (1985).
[9] V. JANOVEC, G. GODEFROY, and L. R. GODEFROY, Ferroelectrics **53**, 333 (1984).
K. KAWASAKI, J. Phys. C **16**, 6911 (1983).
P. PRELOVSEK, Ferroelectrics **54**, 29 (1984).
[10] D. G. SANNIKOV, J. Phys. Soc. Japan, Suppl. B **49**, 75 (1980).
[11] T. NATTERMANN and S. TRIMPER, J. Phys. C **14**, 1603 (1981).
[12] A. LEVSTIK, P. PRELOVSEK, C. FILIPIC, and B. ZEHS, Phys. Rev. B **25**, 3416 (1982).
[13] T. NATTERMANN, Solid State Commun. **44**, 869 (1982).
[14] T. NATTERMANN, J. Phys. C **16**, 4113 (1983).
[15] T. M. RICE, S. WHITEHOUSE, and P. LITTLEWOOD, Phys. Rev. B **24**, 2751 (1981).

(Received July 15, 1985)

phys. stat. sol. (b) 133, 71 (1986)

Subject classification: 64.70; 65.50; S5.11; S5.12

Department of Physics, Himachal Pradesh University, Simla[1])

The Structural Disorder Model: Pressure Induced Transitions in the Semiconductors Si and Ge

By

P. K. DIXIT, B. A. VAID, and K. C. SHARMA

The structural disorder model, recently proposed by the authors to explain the thermodynamic properties near the transitions of first order, is generalized to include the pressure induced transition in the semiconductors Ge and Si. The calculated values show at least a in order of magnitude agreement with the experimental data.

Das kürzlich von den Autoren vorgeschlagene Fehlordnungsmodell zur Erklärung der thermodynamischen Eigenschaften in der Nähe eines Übergangs erster Ordnung wird verallgemeinert und schließt den druckinduzierten Übergang in den Halbleitern Ge und Si ein. Die berechneten Werte befinden sich in größenordnungsmäßiger Übereinstimmung mit den experimentellen Daten.

1. Introduction

In recent years, there has been a significant interest in structural phase transformations in simple solids particularly in semiconductors in the form of semiconducting to metallic transitions under pressure, both experimentally [1 to 5] and theoretically, and subsequently in the study of equilibrium and transport properties in different phases. Nevertheless, the study of thermodynamic properties appears to have been neglected, inspite of the fact that it plays an important role in the calculation of electronic and optical properties, in phases which are obtained in pressure induced transitions.

Recently in tetrahedral semiconductors under high pressures, various structural sequences in Si [6, 7] and in Ge [7] have been reported. In Si a systematic increase in coordination number from 4 (Si-I) → 6 (Si-II) → 8 (Si-V) → 12 (Si-VII) was found at high pressure. The first transformation, i.e. I → II, was observed [7] between the range 8.8 to 13.0 GPa. The structural behaviour of Ge under pressure was found [7] to be much simpler as there was only one transition from Ge-I (tetrahedral, α-Sn) to Ge-II (tetragonal, β-Sn) between the range 10.3 to 11.4 GPa. So far as this range of pressures for structural transformation is concerned, there has been a slight variation in recent literature, too. Associated with these transitions, a variation in reduced volume was also observed in a systematic manner. Similar pressure induced transitions have also been reported for other elemental and alloy semiconducting systems in the literature.

At the theoretical level, Van Vechten [8] calculated the phase diagrams with pressure for Si, Ge, and III–V compounds using simplified models for semiconducting and metallic phases and predicted for the first time the semiconductor to metal

[1]) Summer Hill, Simla 171005, India.

transition in some cases. This work encouraged many subsequent studies. Papers [9, 10] investigated the problem for Sn under pressure, using the first principle pseudopotential approach and reproduced the structural sequence α-Sn $\rightarrow$ β-Sn $\rightarrow$ b.c.t. and predicted a b.c.t. $\rightarrow$ h.c.p. transition at pressures above the known experimental range. Similar calculations for Ge [11], Si [12], and C [13] yielded the structure dependent energy $U_S = U_E + U_{BS}$ as a function of reduced volume where U_E is the Ewald energy and U_{BS} is the band structure energy. They reproduced the well known α–β transition in Ge and Si and predicted for C a transition from the α-Sn structure to a simple cubic structure around 2300.0 GPa. In Si, additional phase transitions at higher pressures were also predicted which were later supported by experiments. However, only little work has so far been done on the studies of thermodynamic properties explicitly in different phases particularly near the phase transitions, although much work on electron transport as a function of pressure in semiconducting systems exists since the paper of Minomura and Drickamer [14].

Recently, the authors have presented a model [15] (referred to as I) known as the structural disorder model (SDM), for calculating thermodynamic properties near the transition. This model was applied successfully to explain the thermodynamic properties of liquid metals [15], amorphous semiconductors [16], and liquid semiconductors [17] near the transition. Essential features of SDM are: (a) the transition has to be of first order exhibiting a quasi-discontinuous behaviour in volume which is an extensive quantity. (b) The transition temperature T_c has been assumed to be a point where spontaneous production of "defects" takes place, which is an activated process. (c) Finally the thermodynamic theory of defects becomes applicable and various properties are calculated in "additive approximation". (d) The "defect" is to be characterised by its activation energy and activation free volume rather than by its well defined geometrical arrangement. (e) The basic input is the change in packing fraction (PF) or volume fraction of defects f associated with the transition.

In this paper we shall generalize the above concepts of SDM to the situation where pressure is the main variable and the transition from one phase to another takes place at some critical value of the pressure p_t. Such a situation of pressure-induced transitions is exhibited in an interesting manner while studying the structural [6, 7] or the electron transport [18] behaviour in semiconducting alloy systems. In these transitions we observe that the transformation from one phase to another is almost a discontinuous one at some p_t, which can be considered analogous to that observed in melting of solids.

2. Basic Expressions

From the thermodynamic theory of point defects in thermal equilibrium, we recall

$$x = n_{\text{def}}/N = [1 + e^{E/kT}]^{-1}, \tag{1}$$

where n_{def} is the number of defects, N is the total number of molecules on the lattice sites. E is the activation energy needed to create a defect, given by

$$E = U_{\text{def}} + pV_{\text{def}} - TS_{\text{def}}, \tag{2}$$

in general, which has the meaning of Gibb's function. Assuming $TS_{\text{def}} \ll U_{\text{def}} \ll pV_{\text{def}}$ at higher pressures, for simplification, we have, $E \approx pV_{\text{def}}$, where $V_{\text{def}} = fV_b$, with V_b as the building unit of phase I. E is to be calculated from the change in volume during transition. The total volume of defects is a certain fraction "f" of V^{I} corresponding to phase I,

$$\Delta V/V^{\text{I}} = fx.$$

The activation energy E at the transition pressure p_t may be expressed as

$$E = p_t V_{\text{def}} = kT \ln\left(\frac{1}{x} - 1\right), \tag{3}$$

where T corresponds to room temperature. In order to have expressions for the change in heat capacity ΔC_p and thermal expansion $\Delta\alpha$, we have to know dx/dT in which E is assumed to be temperature independent and get, in additive approximation

$$\Delta C_p = \frac{R(E/kT)^2 \exp(E/kT)}{[1 + \exp(E/kT)]^2}, \tag{4}$$

$$\Delta\alpha = \frac{f(E/kT)\exp(E/kT)}{T[1 + \exp(E/kT)]^2}, \tag{5}$$

where R is the gas constant. To obtain an expression for the change in compressibility $\Delta\chi$, we calculated dx/dp and get

$$\Delta\chi = \frac{f(E/kT)\exp(E/kT)}{[1 + \exp(E/kT)]^2}\frac{1}{p_t} + \chi^{\mathrm{I}} f x . \tag{6}$$

3. Calculation and Results

For the calculation of thermodynamic properties from the above expressions, the volume change at p_t may be taken from the data given by Olijnyk et al. [7] and the volume fraction of defects f may be taken from the table given in paper I, depending upon the structure/coordination number of the new phase after transition. With this, one can calculate the activation energy, $\Delta\alpha$, ΔC_p, and $\Delta\chi$.

It is easy to consider the pressure induced structural transition in Ge from phase I → II at p_t = 11.4 GPa at which the transition to phase II (β-Sn) becomes complete. In case of Si the situation is not so easy. Due to lack of data of various thermodynamic quantities near the transition, in different phases, it is not simple to justify the validity of calculations of this paper. However, from the data of V/V_0 versus p [7], it is possible to extract the values of compressibility near the transition I → V. This requirement guides us to choose in Si the transition I → V having coordination numbers 4 and 8, respectively. Moreover, the phase structure of Si–V has the lowest energy [7] in the observed range of interest. Thus mainly the energy considerations, so far as the stability of a particular phase at a particular pressure is concerned, have been the guiding factor in choosing the transformation Si I → Si V in this paper. $\Delta V/V_0$ can be directly taken from the data for Ge and Si. Depending on the coordination number of the new phase, we can, in some approximation at least, have an idea of the volume fraction of defects f from paper I. The relevant values of $\Delta V/V_0$, p_t, f, and χ^{I} and other calculated quantities of E/kT and of thermodynamic properties are presented in Table 1.

4. Discussion

We observe from Table 1 that the calculated values of $\Delta\chi$ at transition under pressure are at least in order-of-magnitude agreement with the data. It is possible to calculate the change in entropy, once the change in compressibility and the $(dp/dT)^{-1}$ data are available. However, in absence of detailed data we are unable to compare the calculations except that of compressibility with those extracted from the V/V_0 versus p curves. It can be further observed that $\Delta\alpha$ and ΔC_p are of the same order as those

Table 1

Thermodynamic properties of Si and Ge near the pressure-induced transition at pressure p_t and at room temperature T

element	transition	$\Delta V/V_0$	f	E/kT	exp (E/kT)	p_t (GPa)	χ^I (10^{-10} m²/N)	$\Delta\alpha$ (10^{-5} K⁻¹)	ΔC_p (J/kg)	$\Delta\chi$ (10^{-10} m²/N)
Si	I → V	−0.22	−0.50	0.240	1.27	16.0	0.062	−9.86	117.2	−0.032
	(data for comparison)									−0.021
Ge	I → II	−0.16	−0.35	0.172	1.187	11.4	0.072	−4.97	355.9	−0.024
	(data for comparison)							—	—	−0.009

$\Delta V/V_0$ is the change in volume at the transition, f is the volume fraction of defects. E is the activation energy. χ^I and $\Delta\chi$ represent the compressibility of phase I and change in compressibility during transition. $\Delta\alpha$ and ΔC_p are the change in thermal expansion coefficient and in heat capacity, respectively. Approximate experimental values of $\Delta\chi$ are given for Si and Ge.

in crystalline state which, for comparison, should be available near the transition pressure p_t. It suggests detailed experiments in this direction on various semiconducting systems.

Furthermore, some of the concepts developed in this paper may also be extended to understand cohesive energies and the transport properties in semiconductors near the transition. In spite of the success of SDM to interpret various thermodynamic properties in liquid metals, amorphous semiconductors, pressure induced transitions, there is a need of relating SDM with the microscopic first principle approaches like the pseudopotential technique.

References

[1] J. F. Cannon, J. phys. Chem. Ref. Data **3**, 797 (1974).
[2] M. C. Gupta and A. L. Ruoff, J. appl. Phys. **51**, 1072 (1980).
[3] A. Werner, J. A. Sanjurjo, and M. Cardona, Solid State Commun. **44**, 155 (1982).
[4] M. Baublitz, Jr. and A. L. Ruoff, J. appl. Phys. **53**, 5669 (1982).
[5] S. B. Quadri, E. F. Skelton, and A. W. Webb, J. appl. Phys. **53**, 3609 (1982).
[6] J. Z. Hu and I. L. Spain, Solid State Commun. **51**, 263 (1984).
[7] H. Olijnyk, S. K. Sikka, and W. B. Holzapfel, Phys. Letters **103**, 137 (1984).
[8] J. A. Van Vechten, Phys. Rev. B **7**, 1479 (1973).
[9] J. Hafner, Phys. Rev. B **10**, 4151 (1974).
[10] J. Ihm and M. L. Cohen, Phys. Rev. B **23**, 1576 (1981).
[11] M. T. Yin and M. L. Cohen, Solid State Commun. **38**, 625 (1981).
[12] M. T. Yin and M. L. Cohen, Phys. Rev. Letters **45**, 1004 (1980).
[13] M. T. Yin and M. L. Cohen, Phys. Rev. Letters **50**, 2006 (1983).
[14] S. Minomura and H. G. Drickamer, J. Phys. Chem. Solids **23**, 451 (1962).
[15] B. A. Vaid, K. C. Sharma, and V. K. Syal, phys. stat. sol. (b) **119**, 703 (1983).
[16] B. A. Vaid and K. C. Sharma, phys. stat. sol. (b) **122**, 423 (1984).
[17] P. K. Dixit, B. A. Vaid, and K. C. Sharma, unpublished.
[18] G. Parthasarathy, K. J. Rao, and E. S. R. Gopal, Solid State Commun. **52**, 867 (1984).

(Received July 11, 1985)

phys. stat. sol. (b) 133, 75 (1986)

Subject classification: 62.80 and 76.80

Department of Physics, Belorussian State University, Minsk[1])

Mössbauer Radiation Dynamical Diffraction in Crystals Being Subjected to the Action of External Variable Fields

By

V. G. Baryshevskii and V. V. Skadorov

A dynamical theory is developed of the Mössbauer radiation diffraction by crystals being subjected to an variable external field action. Equations describing the dynamical diffraction by nonstationary crystals are obtained. It is shown that the resonant interaction between Mössbauer radiation and shift field induced in the crystal by a variable external field giving rise to an effective conversion of the incident wave into a wave with changed frequency.

Развита теория динамической дифракции мессбауэровского излучения на кристаллах, подверженных действию внешних переменных полей. Получены уравнения описывающие динамическую дифракцию γ-квантов на нестационарных кристаллах. Показано, что резонансное взаимодействие между мессбауэровским излучением и полем смещений, индуцируемым внешним переменным полем, приводит к эффективному преобразованию падающей волны в волну с измененной частотой.

1. Introduction

The dynamical theory of the Mössbauer radiation diffraction by deformed crystals was developed in [1], where the external source resulting in a shift field was assumed to be time-independent. Recently, the problem of γ-quanta diffraction by crystals in variable external fields has attracted attention of both theoreticians and experimentalists [2 to 7].

A similar problem for X-ray radiation when the crystal was vibrated by an ultrasonic wave was discussed in [8, 9]. In these papers it was assumed that the acoustic shift field generated in instant superlattice with a period equal to the ultrasonic wavelength and the quasi-static approximation was used. Such an approximation supposes that the ultrasonic lattice oscillation can be neglected during the time of γ-quantum transmission through the crystal. The diffraction by a static shift wave is treated in terms of the multiple wave dynamical diffraction equations. Although for X-ray radiation such an approach is appropriate, for the Mössbauer radiation having an extremely small Γ-width of Mössbauer transitions in nuclei the interaction time of γ-quanta with a crystal is of the order of or longer than an ultrasonic vibration period. As a result the quasi-static approximation is not applicable [4]. The Mössbauer radiation dynamical diffraction should be considered as diffraction by a moving lattice which is produced by the shift field induced by an external variable field (the scattering of photons by a moving ultrasonic lattice is known in optics as forced Mandelstam-Brillouin scattering [10, 11]).

---

[1]) Minsk 220080, USSR.

2. Calculation of the Current Induced in Crystals by γ-Quanta

The γ-quantum field inside the crystal without external currents and charges obeys the equation

$$\mathrm{rot}\,\mathrm{rot}\,\boldsymbol{E}(\boldsymbol{r};t) + \frac{1}{c^2}\frac{\partial^2}{\partial t^2}\boldsymbol{E}(\boldsymbol{r};t) + \frac{4\pi}{c^2}\frac{\partial}{\partial t}\boldsymbol{I}(\boldsymbol{r};t) = 0\,. \tag{1}$$

The current density $\boldsymbol{I}(\boldsymbol{r};t)$, induced by the incident γ-quanta in the crystal whose nuclei are vibrated by the external variable field, is defined as

$$\boldsymbol{I}(\boldsymbol{r};t) = \sum_m \langle\psi_e(t)|\,\hat{\boldsymbol{j}}_e(\boldsymbol{r}-\boldsymbol{r}_m-\boldsymbol{u}(\boldsymbol{r}_m;t))\,|\psi_e(t)\rangle + \\ + \sum_m \langle\psi_N(t)|\,\hat{\boldsymbol{j}}_N((\boldsymbol{r}-\boldsymbol{r}_m-\boldsymbol{u}(\boldsymbol{r}_m;t))\,|\psi_N(t)\rangle\,, \tag{2}$$

where $|\psi_e(t)\rangle$ and $|\psi_N(t)\rangle$ are the wave functions of the electron shell and the nucleus of the m-th atom in the crystal at the instant t; $\boldsymbol{r}_m = \boldsymbol{R}_e + \boldsymbol{r}_j + \boldsymbol{u}_j$ is the radius vector of the m-th atom in the absence of external fields, $\boldsymbol{R}_e$ is the radius vector of the elementary cell containing the atom, $\boldsymbol{r}_j$ is the radius vector of the m-th atom inside the elementary cell, $\boldsymbol{u}_j$ is the displacement of the atom from its equilibrium value due to the thermal motion; $\boldsymbol{u}(\boldsymbol{r}_m;t)$ is the displacement of the m-th atom from its equilibrium position caused by the external variable field; $\boldsymbol{j}_e$ and $\boldsymbol{j}_N$ are the operators of the current densities induced in the electron shell and the nucleus of the atom by γ-quanta. Let the shift field produced in the crystal by an external variable field have the form

$$\boldsymbol{u}(\boldsymbol{r},t) = \boldsymbol{a}\sin(\boldsymbol{\varkappa}\boldsymbol{r} - \Omega t + \varphi)\,, \tag{3}$$

where $\boldsymbol{a}$ is the amplitude of the forced nuclear vibration in the crystal under the crystal excitation, $\boldsymbol{\varkappa}$ is the wave vector of the sound (light) wave, $\Omega = \Omega(\boldsymbol{\varkappa})$ is the nuclear vibration frequency which is a rather complex function of $\boldsymbol{\varkappa}$ (e.g. in the case of ultrasonic pumping $\Omega = \varkappa v$, where v is the sound velocity), φ is the phase of the forced vibration of the nucleus. Making use of perturbation theory from relations (2), (3) one obtains [5] that the current density induced in the crystal by the γ-quanta in the presence of an external variable field is defined by the expression

$$\boldsymbol{I}(\boldsymbol{r};t) = \int \mathrm{d}^3r'\,\mathrm{d}t'\Big(\sum_{\boldsymbol{\tau}}\sum_{p=-\infty}^{\infty}\hat{\sigma}_{\boldsymbol{\tau}+p\boldsymbol{\varkappa}}(\boldsymbol{r}-\boldsymbol{r}';t-t')\times \\ \times\exp\{-i(\boldsymbol{\tau}+p\boldsymbol{\varkappa})\,\boldsymbol{r}'\}\exp\{ip\Omega t'\}\Big)\,\boldsymbol{E}(\boldsymbol{r};t')\,, \tag{4}$$

where

$$\hat{\sigma}_{\boldsymbol{\tau}+p\boldsymbol{\varkappa}}(\boldsymbol{r};t) = \int\frac{\mathrm{d}^3k}{(2\pi)^3}\frac{\mathrm{d}\omega}{(2\pi)}\,\hat{\sigma}(\boldsymbol{k},\boldsymbol{k}+\boldsymbol{\tau}+p\boldsymbol{\varkappa};\omega)\exp\{i(\boldsymbol{k}\boldsymbol{r}-\omega t)\}$$

is the $(\boldsymbol{\tau}+p\boldsymbol{\varkappa})$ component of the crystal conductivity tensor and

$$\hat{\sigma}(\boldsymbol{k},\boldsymbol{k}+\boldsymbol{\tau}+p\boldsymbol{\varkappa},\omega) = \\ = (-1)^p\,\mathrm{e}^{ip\varphi}\Big[\mathrm{J}_p((\boldsymbol{\tau}+p\boldsymbol{\varkappa})\,\boldsymbol{a})\,\sigma_{\boldsymbol{\tau}}^{e}(\omega) + \frac{i\eta}{h\omega V(2Z_0+1)}\times \\ \times\sum_j\exp\Big\{-\frac{1}{2}Z_j(\boldsymbol{k}) - \frac{1}{2}Z_j(\boldsymbol{k}+\boldsymbol{\tau})\Big\}\exp\{i\boldsymbol{\tau}\boldsymbol{r}\}\times \\ \times\sum_{M_0,M}\langle M_0|\,\hat{\boldsymbol{j}}_N^j(\boldsymbol{k})\,|M\rangle\otimes\langle M|\,\hat{\boldsymbol{j}}_N^j(\boldsymbol{k}+\boldsymbol{\tau}+p\boldsymbol{\varkappa})\,|M_0\rangle\times \\ \times\sum_{n=-\infty}^{\infty}\frac{\mathrm{J}_n(\boldsymbol{k}\boldsymbol{a})\,\mathrm{J}_{n+p}((\boldsymbol{k}+\boldsymbol{\tau}+p\boldsymbol{\varkappa})\,\boldsymbol{a})}{\omega-[\omega(M,M_0)+n\Omega]+i\Gamma/2}\Big] \tag{5}$$

is the crystal conductivity tensor in the $(\boldsymbol{k}; \omega)$ representation, where $\sigma_{\boldsymbol{\tau}}^{e}$ is the cyrstal conductivity coefficient ot the electronic subsystem in the absence of external perturbation, $J_p(\ldots)$ is the Bessel function of the first kind; M_0, M, $|M_0\rangle$, $|M\rangle$ are the magnetic quantum numbers and spin wave functions of the nucleus in the ground and in the excited states; $\otimes$ means the tensor product; J_0 is the nuclear spin in the ground state; η is the relative concentration of the Mössbauer nuclei in the crystal.

3. Dielectric Tensor of the Crystal in the Presence of a Variable External Field

Equation (1) describing the γ-quantum propagation through the crystal takes the following form in the $(\boldsymbol{k}; \omega)$ representation:

$$\left(k^2 - \frac{\omega^2}{c^2} - \boldsymbol{k} \otimes \boldsymbol{k}\right) \boldsymbol{E}(\boldsymbol{k}, \omega) = \frac{4\pi i\omega}{c^2} \boldsymbol{I}(\boldsymbol{k}, \omega) .$$

Making used (4) and (5) we can obtain for the current density induced in the crystal by the γ-quanta

$$\left(k^2 - \frac{\omega^2}{c^2} [1 + \hat{\chi}(\boldsymbol{k}, \boldsymbol{k}; \omega)] - \boldsymbol{k} \otimes \boldsymbol{k}\right) \boldsymbol{E}(\boldsymbol{k}; \omega) =$$

$$= \frac{\omega^2}{c^2} \sum_{\boldsymbol{\tau}} \sum_{p=-\infty}^{\infty} \hat{\chi}(\boldsymbol{k}, \boldsymbol{k} + \boldsymbol{\tau} + p\boldsymbol{\varkappa}; \omega) \boldsymbol{E}(\boldsymbol{k} + \boldsymbol{\tau} + p\boldsymbol{\varkappa}, \omega + p\Omega) , \tag{6}$$

where

$$\hat{\chi}(\boldsymbol{k}, \boldsymbol{k} + \boldsymbol{\tau} + p\boldsymbol{\varkappa}; \omega) = \frac{4\pi i}{\omega} \hat{\sigma}(\boldsymbol{k}, \boldsymbol{k} + \boldsymbol{\tau} + p\boldsymbol{\varkappa}; \omega) \tag{7}$$

is the $(\boldsymbol{\tau} + p\boldsymbol{\varkappa})$ component of the crystal polarizability, which is connected with the conductivity tensor by the relation

$$\hat{\chi}(\boldsymbol{r}, \boldsymbol{r}'; \omega, t') = \frac{4\pi i}{\omega} \hat{\sigma}(\boldsymbol{r}, \boldsymbol{r}'; \omega, t') =$$

$$= \frac{4\pi i}{\omega} \sum_{\boldsymbol{\tau}} \sum_{p=-\infty}^{\infty} \hat{\sigma}_{\boldsymbol{\tau}+p\boldsymbol{\varkappa}}(\boldsymbol{r} - \boldsymbol{r}'; \omega) \exp\{-i(\boldsymbol{\tau} + p\boldsymbol{\varkappa}) \boldsymbol{r}'\} \exp\{i(\omega + p\Omega) t'\} . \tag{8}$$

Since the elements of the tensor $\hat{\chi}(\boldsymbol{k}, \boldsymbol{k} + \boldsymbol{\tau} + p\boldsymbol{\varkappa}; \omega)$ can be interpreted as the scattering amplitudes of the γ-quanta by the unit cell, then, from (6) it follows that the electromagnetic wave describing the γ-quantum in the crystal in the presence of the variable external field contains not only waves with wave vectors $\boldsymbol{k}$, $(\boldsymbol{k} + \boldsymbol{\tau})$ and frequency ω but also waves with wave vectors $(\boldsymbol{k} + p\boldsymbol{\varkappa})$, $(\boldsymbol{k} + \boldsymbol{\tau} + p\boldsymbol{\varkappa})$ and frequencies $(\omega + p\Omega)$, where $p = \pm 1, \pm 2, \ldots$ The appearance of the γ-quanta with such wave vectors and frequencies under the action of a variable external field can be understood if the shift wave (3) produced by the variable external field in the crystal is considered as moving grating with the period $\Lambda = 2\pi/\varkappa$. As a result of diffraction by this moving grating the γ-quanta with wave vectors $(\boldsymbol{k} + p\boldsymbol{\varkappa})$, $(\boldsymbol{k} + \boldsymbol{\tau} + p\boldsymbol{\varkappa})$ and frequencies $(\omega + p\Omega)$ arise. The effect for the ultrasonic excitation of the crystal lattice is well known in the optical wave range as forced Mandelstam-Brillouin scattering phenomenon [10, 11]. In the case of vibration by the action of light wave we have the phenomenon of the parametric conversion of γ-quanta [4, 5]. Thus, under the action of a variable external field on the Mössbauer crystal it is necessary to take into account both the γ-quantum diffraction by the crystal lattice and the diffraction by a moving grating produced by the variable field. The contributions of the γ-quanta diffracted both by the crystal lattice and by the moving grating are represented by the terms on the right side of (6). The dielectric tensor of the crystal is defined by the relation

$$\hat{\varepsilon}(\boldsymbol{r}, \boldsymbol{r}'; t, t') = 1 + \hat{\chi}(\boldsymbol{r}, \boldsymbol{r}'; t, t') =$$
$$= \sum_{\boldsymbol{\tau}} \sum_{p=-\infty}^{\infty} \hat{\chi}_{\boldsymbol{\tau}+p\boldsymbol{\varkappa}}(\boldsymbol{r} - \boldsymbol{r}'; t - t') \exp\{-i(\boldsymbol{\tau} + p\boldsymbol{\varkappa})\,\boldsymbol{r}'\} \exp\{-ip\Omega t'\}, \quad (9)$$

where

$$\hat{\chi}_{\boldsymbol{\tau}+p\boldsymbol{\varkappa}}(\boldsymbol{r}; t) = \int \frac{\mathrm{d}^3k}{(2\pi)^3} \frac{\mathrm{d}\omega}{2\pi} \hat{\chi}(\boldsymbol{k}, \boldsymbol{k} + \boldsymbol{\tau} + p\boldsymbol{\varkappa}; \omega) \exp\{i(\boldsymbol{kr} - \omega t)\}$$

and $\hat{\chi}(\boldsymbol{k}, \boldsymbol{k} + \boldsymbol{\tau} + p\boldsymbol{\varkappa}; \omega)$ is defined by (7). As a result, the relation between the induction $\boldsymbol{D}(\boldsymbol{r}; t)$ and the electric strength $\boldsymbol{E}(\boldsymbol{r}; t)$ of the γ-quantum in the $(\boldsymbol{k}, \omega)$ representation

$$\boldsymbol{D}(\boldsymbol{k}; \omega) = \sum_{\boldsymbol{\tau}} \sum_{p=-\infty}^{\infty} \hat{\varepsilon}(\boldsymbol{k}, \boldsymbol{k} + \boldsymbol{\tau} + p\boldsymbol{\varkappa}; \omega)\, \boldsymbol{E}(\boldsymbol{k} + \boldsymbol{\tau} + p\boldsymbol{\varkappa}; \omega + p\Omega) \quad (10)$$

is nonlocal not only in the wave vector space but also in the frequency space because the crystal becomes a nonstationary medium in this case.

The characteristic resonance width of the Mössbauer nuclei is of the order of 10^6 to $10^7\ \mathrm{s}^{-1}$ and we can neglect the forced vibration frequency Ω in the resonance factor of (10) only when the frequency Ω is much smaller than Γ, i.e. when Ω is smaller than $10^5\ \mathrm{s}^{-1}$. The value of the wave vectors of the forced nuclear vibrations in the crystal for this frequency range is smaller than $1\ \mathrm{cm}^{-1}$, therefore, the wave vector can be neglected in (5). Using the addition theorem of the Bessel functions

$$\sum_{n=-\infty}^{\infty} \mathrm{J}_n(\boldsymbol{ka})\, \mathrm{J}_{n+p}((\boldsymbol{k} + \boldsymbol{\tau})\,\boldsymbol{a}) = (-1)^p\, \mathrm{J}_p(\boldsymbol{\tau a})$$

one can obtain the $(\boldsymbol{\tau}, p)$ component of the crystal polarizability tensor under the action of the variable external field whose frequency is much smaller than the width of the Mössbauer resonance transition from (5) and (8)

$$\hat{\chi}_p(\boldsymbol{k}, \boldsymbol{k} + \boldsymbol{\tau}; \omega) = (-1)^p\, \mathrm{e}^{ip\varphi}\, \mathrm{J}_p(\boldsymbol{\tau a})\, \hat{\chi}(\boldsymbol{k}, \boldsymbol{k} + \boldsymbol{\tau}; \omega), \quad (11)$$

where $\hat{\chi}(\boldsymbol{k}, \boldsymbol{k} + \boldsymbol{\tau}; \omega)$ is the $\boldsymbol{\tau}$ component of the Mössbauer crystal polarizability tensor without external perturbation. Thus, for a Mössbauer crystal the quasi-static approximation is applicable only in the sonic frequency range of the forced vibration. Neglecting the Mössbauer crystal anisotropy the Mössbauer radiation optics in this frequency range is similar to the X-ray optics and, in the absence of hyperfine splitting, it completely coincides with the X-ray optics. Therefore, we suppose below that the variable external field frequency is greater than the resonance transition width of Mössbauer nuclei.

4. The Equations of γ-Quantum Dynamical Diffraction

The relation (4) obtained above for the current density induced by γ-quanta allows us to write the equation for the γ quantum field in the crystal where the shift wave caused by the external perturbation is propagating

$$\operatorname{rot}\operatorname{rot} \boldsymbol{E}(\boldsymbol{r}; t) +$$
$$+ \frac{1}{c^2} \frac{\partial^2}{\partial t^2} \boldsymbol{E}(\boldsymbol{r}; t) + \frac{4\pi}{c^2} \frac{\partial}{\partial t} \int \mathrm{d}^3r'\, \mathrm{d}t' \Big(\sum_{\boldsymbol{\tau}} \sum_{p=-\infty}^{\infty} \hat{\sigma}_{\boldsymbol{\tau}+p\boldsymbol{\varkappa}}(\boldsymbol{r} - \boldsymbol{r}'; t - t') \times$$
$$\times \exp\{-i(\boldsymbol{\tau} + p\boldsymbol{\varkappa})\,\boldsymbol{r}'\} \exp\{ip\Omega t'\}\Big)\, \boldsymbol{E}(\boldsymbol{r}'; t') = 0\,. \quad (12)$$

As it has been noted above, the external perturbation produces a moving grating with the period $\Lambda = 2\pi/\varkappa$ in the crystal. Consequently, in the crystal a certain new lattice appears which is a superposition of the ordinary crystal lattice and the moving grating generated by the external variable field. The basic vectors of this new lattice are functions of the crystal lattice base vectors and $\boldsymbol{\Lambda} = \Lambda\boldsymbol{\varkappa}/\varkappa$, i.e. $\boldsymbol{a}_i = \boldsymbol{f}_i(\boldsymbol{a}_1, \boldsymbol{a}_2, \boldsymbol{a}_3, \boldsymbol{\Lambda})$. The set of vectors $(\boldsymbol{\tau} + p\boldsymbol{\varkappa})$ defines the reciprocal lattice of this new lattice and the condition $(\boldsymbol{\tau} + p\boldsymbol{\varkappa})\,(l_1\boldsymbol{a}_1' + l_2\boldsymbol{a}_2' + l_3\boldsymbol{a}_3') = (\boldsymbol{\tau} + p\boldsymbol{\varkappa})\,\boldsymbol{C}_l = 2\pi N$ is satisfied (l_1, l_2, l_3, and N are integers). It is easily verified that (12) being not translation invariant with respect to the ordinary crystal lattice and to the moving grating separately, has translation invariance with respect to the new complex lattice. Besides, (12) is also invariant under $t \to t + T$ transformation, where $T = 2\pi/\Omega$ is the period of the forced nucleus vibrations. Thus, (12) is invariant under the transformation $\boldsymbol{r} \to \boldsymbol{r} + \boldsymbol{C}_l$ and $t \to t + T$. Therefore, according to the Bloch theorem for the γ-quantum incident wave packet $\boldsymbol{E}_0(\boldsymbol{r}; \omega) \exp\{i(\boldsymbol{kr} - \omega t)\}$ with the wave vector $\boldsymbol{k}$ and the frequency ω the solution of (12) should be found in the form of an expansion in Bloch waves,

$$\boldsymbol{E}(\boldsymbol{r}; t) = \exp\{i(\boldsymbol{kr} - \omega t)\} \sum_{\boldsymbol{\tau}} \sum_{p=-\infty}^{\infty} \boldsymbol{E}_{\boldsymbol{\tau}+p\boldsymbol{\varkappa}}(\boldsymbol{r}; \omega + p\Omega) \times \\ \times \exp\{i(\boldsymbol{\tau} + p\boldsymbol{\varkappa})\boldsymbol{r}\} \exp\{-ip\Omega t\}\,. \tag{13}$$

Substituting (13) into (12) and expanding the Bloch wave amplitudes in the integrand of (12) in a Taylor series, one obtains [1] the following set of equations for the dynamical diffraction of γ-quanta by a crystal, in the presence of a shift wave (3) caused by the variable external field

$$\left(\frac{(\boldsymbol{k} + \boldsymbol{\tau} + p\boldsymbol{\varkappa})^2}{k^2} - 1\right) \boldsymbol{E}_{\boldsymbol{\tau}+p\boldsymbol{\varkappa}}(\boldsymbol{r}; \omega + p\Omega) - \frac{2i}{k^2}\left((\boldsymbol{k} + \boldsymbol{\tau} + p\boldsymbol{\varkappa})\Delta\right) \boldsymbol{E}_{\boldsymbol{\tau}+p\boldsymbol{\varkappa}}(\boldsymbol{r}; \omega + p\Omega) = \\ = \sum_{\boldsymbol{\tau}'} \sum_{p'=-\infty}^{\infty} \hat{\chi}(\boldsymbol{k} + \boldsymbol{\tau} + p\boldsymbol{\varkappa}, \boldsymbol{k} + \boldsymbol{\tau} + \boldsymbol{\tau}' \mid p'\boldsymbol{\varkappa}; \omega)\, \boldsymbol{E}_{\boldsymbol{\tau}+\boldsymbol{\tau}'+p'\boldsymbol{\varkappa}}(\boldsymbol{r}; \omega + p'\Omega)\,. \tag{14}$$

The tensor $\hat{\chi}(\boldsymbol{k} + \boldsymbol{\tau} + p\boldsymbol{\varkappa}; \boldsymbol{k} + \boldsymbol{\tau}' + p'\boldsymbol{\varkappa}; \omega)$ may be interpreted as the tensor amplitude of the scattering of an incident γ-quantum with wave vector $(\boldsymbol{k} + \boldsymbol{\tau}' + p'\boldsymbol{\varkappa})$ and frequency $(\omega + p'\Omega)$ into a γ-quantum with wave vector $(\boldsymbol{k} + \boldsymbol{\tau} + p\boldsymbol{\varkappa})$ and frequency $(\omega + p\Omega)$ by a crystal element and is determined by

$$\hat{\chi}(\boldsymbol{k} + \boldsymbol{\tau} + p\boldsymbol{k}, \boldsymbol{k} + \boldsymbol{\tau} + \boldsymbol{\tau}' + p'\boldsymbol{\varkappa}; \omega) = \\ = (-1)^{(p-p')}\, \mathrm{e}^{-i(p-p')\varphi}\, \mathrm{J}_{(p'-p)}(\boldsymbol{\tau}'\boldsymbol{a})\, \hat{\chi}^{\mathrm{e}}_{\boldsymbol{\tau}}(\omega) + \\ + \sum_{j} \sum_{M_0, M} \hat{\mu}^j(\boldsymbol{k} + \boldsymbol{\tau} + p\boldsymbol{\varkappa}, \boldsymbol{k} + \boldsymbol{\tau} + \boldsymbol{\tau}' + p'\boldsymbol{\varkappa}; M_0 M) \times \\ \times \sum_{n=-\infty}^{\infty} \frac{(i\Gamma/2)\,(-1)^{(p-p')}\, \mathrm{e}^{-i(p-p')\varphi}\, \mathrm{J}_{p+n}((\boldsymbol{k} + \boldsymbol{\tau})\,\boldsymbol{a})\, \mathrm{J}_{n+p'}((\boldsymbol{k} + \boldsymbol{\tau} + \boldsymbol{\tau}')\,\boldsymbol{a})}{\omega - [\omega^j(M, M_0) + n\Omega] + i\Gamma/2}, \tag{15}$$

where the tensor $\hat{\mu}^j$ is determined by

$$\hat{\mu}^j(\boldsymbol{k} + \boldsymbol{\tau} + p\boldsymbol{\varkappa}, \boldsymbol{k} + \boldsymbol{\tau} + \boldsymbol{\tau}' + p'\boldsymbol{\varkappa}; M_0, M) = \\ = \frac{4\pi i\eta}{\hbar\omega^2 V(2J_0 + 1)} \exp\left\{-\frac{1}{2} Z_j(\boldsymbol{k} + \boldsymbol{\tau}) - \frac{1}{2} Z_j(\boldsymbol{k} + \boldsymbol{\tau} + \boldsymbol{\tau}')\right\} \times \\ \times \exp\{i\boldsymbol{\tau}'\boldsymbol{r}_j\} \frac{\langle M_0|\, \hat{\boldsymbol{j}}_{\mathrm{N}}^{i}(\boldsymbol{k} + \boldsymbol{\tau} + p\boldsymbol{\varkappa})\,|M\rangle \otimes \langle M|\, \hat{\boldsymbol{j}}_{\mathrm{N}}^{i}(\boldsymbol{k} + \boldsymbol{\tau} + \boldsymbol{\tau}' + p'\boldsymbol{\varkappa})\,|M_0\rangle}{\Gamma/2}. \tag{16}$$

5. The Modulation of the Mössbauer Radiation by a Variable External Field

In this section we will discuss the diffraction by a moving grating induced by the variable external field in a crystal. Assuming that the crystal is far from the exact Bragg condition the diffraction by the crystal lattice may be neglected. It follows from (14) that the diffraction by the moving grating obeys the set of equations

$$\alpha_p \boldsymbol{E}_{p\varkappa}(\boldsymbol{r}; \omega + p\Omega) - \frac{2i}{k} (\boldsymbol{n}_p \nabla) \boldsymbol{E}_{p\varkappa}(\boldsymbol{r}; \omega + p\Omega) =$$

$$= \sum_{p'=-\infty}^{\infty} \hat{\chi}(k\boldsymbol{n}_p, k\boldsymbol{n}_{p'}; \omega) \boldsymbol{E}_{p'\varkappa}(\boldsymbol{r}; \omega + p'\Omega), \tag{17}$$

where $\boldsymbol{n}_p = (\boldsymbol{k} + p\boldsymbol{\varkappa})/k$ is the normal to the phase surface of the Bloch wave $\boldsymbol{E}_{p\varkappa}(\boldsymbol{r}; \omega + p\Omega) \exp \{i[(\boldsymbol{k} + p\boldsymbol{\varkappa}) \boldsymbol{r} - (\omega + p\Omega) t]\}$ in a crystal. The quantities $\alpha_p = 2p (\varkappa/k) \cos \psi + p^2(\varkappa/k)^2$, where ψ is the angle between the vectors $\boldsymbol{\varkappa}$ and $\boldsymbol{k}$, determine the number of the Bloch waves which are excited in the crystal in the presence of a variable external field. Since $|\mathrm{J}_p(\boldsymbol{ka})| \leqq 1$, then $|\chi_{\mathrm{N}}^{\beta\gamma}(k\boldsymbol{n}_p, k\boldsymbol{n}_{p'}; \omega)| \leqq |\mu_{\beta\gamma}|$ where $\chi_{\mathrm{N}}^{\beta\gamma}$ are the matrix elements of the polarizability tensor of the crystal's nuclear subsystem and $\mu_{\beta\gamma} = \sum_{j, M_0, M} \hat{\mu}_{\beta\gamma}^j(\boldsymbol{k}; M_0, M)$ are the matrix elements of the tensor $\hat{\mu} = \sum_j \sum_{M_0, M} \hat{\mu}^j(\boldsymbol{k}; M_0, M)$ that is defined by (15) for $\boldsymbol{\tau} = \boldsymbol{\tau}' = 0$ and $p = p' = 0$. Consequently, the γ-quanta with wave vectors $(\boldsymbol{k} + p\boldsymbol{\varkappa})$ and frequencies $(\omega + p\Omega)$ interact with the moving grating only when the condition $|\alpha_p| \leqq |\mu_{\beta\gamma}|$ is satisfied. For $|\alpha_p| > |\mu_{\beta\gamma}|$ the interaction between the moving grating and the γ-quanta with wave vectors $(\boldsymbol{k} + p\boldsymbol{\varkappa})$ and frequencies $(\omega + p\Omega)$ may be neglected. As a result, the γ-quantum field in the crystal consists of the $(2m + 1)$ Bloch waves that interact with the moving grating, where the integer m may be defined by the condition

$$m = |\mu_{\beta\gamma}| \Big/ 2 \left(\frac{\varkappa}{k}\right) |\cos \psi| , \tag{18}$$

and an infinite set of the Bloch waves that do not interact with the nuclear subsystem of the crystal. Since $(\boldsymbol{n}_p\boldsymbol{n}) = 1 + 2p(\varkappa/k) \cos \psi$, where $\boldsymbol{n} = \boldsymbol{k}/k$ is the normal to the phase surface of the Bloch wave $\boldsymbol{E}(\boldsymbol{r}; \omega) \exp \{i(\boldsymbol{kr} - \omega t)\}$, then $\boldsymbol{n}_p$ may be considered to be equal to $\boldsymbol{n}$ for $|p| \leqq m$ and the $(2m + 1)$ Bloch waves interacting with the moving grating are orthonormal to the phase normal $\boldsymbol{n} = \boldsymbol{k}/k$ up to $|\mu_{\beta n}|^2$. The Bloch wave amplitudes are described by the equations

$$\alpha_p \boldsymbol{E}_{p\varkappa}(\boldsymbol{r}; \omega + p\Omega) - \frac{2i}{k} \frac{\mathrm{d}}{\mathrm{d}\xi} \boldsymbol{E}_{p\varkappa}(\boldsymbol{r}; \omega + p\Omega) =$$

$$= \sum_{p'=-m}^{m} \hat{\chi}_{p,p'}(\boldsymbol{k}; \omega) \boldsymbol{E}_{p'\varkappa}(\boldsymbol{r}; \omega + p'\Omega), \tag{19}$$

where $\xi = (\boldsymbol{nr})$ is the coordinate along the normal $\boldsymbol{n}$. The amplitude $\hat{\chi}_{p,p'}(\boldsymbol{k}; \omega)$ for the scattering of a γ-quantum with frequency $(\omega + p'\Omega)$ and wave vector $(\boldsymbol{k} + p'\boldsymbol{\varkappa})$ into a γ-quantum with wave vector $(\boldsymbol{k} + p\boldsymbol{\varkappa})$ and frequency $(\omega + p\Omega)$ by a unit cell for $|p|, |p'| \leqq m$ becomes

$$\hat{\chi}_{p,p'}(\boldsymbol{k}; \omega) = \delta_{p,p'} \chi_0^{\mathrm{e}}(\omega) + \sum_j \sum_{M_0, M} \hat{\mu}^j(\boldsymbol{k}; M_0, M) \times$$

$$\times \sum_{n=-\infty}^{\infty} \frac{(i\Gamma/2) (-1)^{(p-p')} \mathrm{e}^{-i(p-p')\varphi} \mathrm{J}_{p+n}(\boldsymbol{ka}) \mathrm{J}_{n+p'}(\boldsymbol{ka})}{\omega - [\omega^j(M, M_0) + n\Omega] + i\Gamma/2}, \tag{20}$$

where the tensor $\hat{\mu}^j(\boldsymbol{k}; M_0, M)$ is defined by

$$\hat{\mu}^j(\boldsymbol{k}; M_0, M) = \frac{4\pi i\eta}{\hbar\omega^2 V(2J_0 + 1)} \times$$
$$\times \exp\{-Z_j(\boldsymbol{k})\} \frac{\langle M_0| \boldsymbol{j}_{\mathrm{N}}^j(\boldsymbol{k}) |M\rangle \otimes \langle M| \hat{\boldsymbol{j}}_{\mathrm{N}}^{j*}(\boldsymbol{k}) |M_0\rangle}{\Gamma/2} . \quad (21)$$

For Bloch waves which do not interact with the nuclear subsystem of the crystal the equations are given by

$$-\frac{2i}{k} (\boldsymbol{n}_p\nabla) \boldsymbol{E}_{p\varkappa}(\boldsymbol{r}; \omega + p\Omega) = \chi_0^{\mathrm{e}}(\omega) \boldsymbol{E}_{p\varkappa}(\boldsymbol{r}; \omega + p\Omega) . \quad (22)$$

The quantity $(\varkappa/k)$ varies from 10^{-7} to 10^{-4} under ultrasonic pumping. As follows from (18), the number of interacting Bloch waves varies as a function of the γ-quantum wavelength and the concentration of Mössbauer isotopes from a few tens to three. The number of interacting Bloch waves increases sharply at the angles ψ for which $|(\pi/2) - \psi| \leqq 10^{-2}$. Thus, under ultrasonic pumping the γ-quantum diffraction by the moving grating is a multiwave one. A similar result can be obtained for radio-frequency and ultrahigh frequency fields and light waves.

To write the solution of the set (19) it is convenient to introduce a $2(2m + 1) \times \times 2(2m + 1)$ matrix $\hat{\chi}_{\mathrm{N}}$ with the matrix elements

$$(\hat{\chi}_{\mathrm{N}})_{p,p'}^{\beta,\gamma} = \sum_j \sum_{M_0,M} \hat{\mu}_{\beta,\gamma}^j \sum_{n=-\infty}^{\infty} \frac{(i\Gamma/2)(-1)^{(p-p')} \mathrm{e}^{-i(p-p')\varphi} \mathrm{J}_{p+n}(\boldsymbol{ka}) \mathrm{J}_{n+p'}(\boldsymbol{ka})}{\omega - [\omega^j(M, M_0) + n\Omega] + i\Gamma/2} -$$
$$- \delta_{\beta,\gamma}\delta_{p,p'}\alpha_p =$$
$$- \sum_j \sum_{M_0,M} \hat{\mu}_{\beta,\gamma}^j(\boldsymbol{k}; M_0, M) \left(\hat{g}^j(\boldsymbol{k}; M_0, M)\right)_{p,p'} - \delta_{\beta\gamma}(\hat{\alpha})_{p,p'} , \quad (23)$$

where $\beta, \gamma = 1, 2$; $p = m, \ldots, -m$; $\mu_{\beta,\gamma}^j = (\boldsymbol{e}_\beta^* \hat{\mu}^j \boldsymbol{e}_\gamma)$ are matrix elements of the tensor $\hat{\mu}^j$; $\boldsymbol{e}_\beta$ are orthonormal vectors on the plane which is orthogonal to the phase normal $\boldsymbol{n} = \boldsymbol{k}/k$. The solution of the equation system (19) can be written in the following form:

$$E_{p\varkappa}^\beta(\boldsymbol{r}; \omega + p\Omega) = \exp\{i\tfrac{1}{2}\chi_0^{\mathrm{e}}(\omega) k\boldsymbol{n}(\boldsymbol{r} - \boldsymbol{r}_0)\} \times$$
$$\times \sum_{\gamma=1,2} \sum_{p'=-m}^{m} \left(\exp\{i\tfrac{1}{2}\hat{\chi}_{\mathrm{N}} k\boldsymbol{n}(\boldsymbol{r} - \boldsymbol{r}_0)\}\right)_{p,p'}^{\beta,\gamma} \times$$
$$\times E_{p'\varkappa}^\gamma(\boldsymbol{r}, \omega + p'\Omega) , \quad (24)$$

where $E_{p\varkappa}^\beta(\boldsymbol{r}; \omega + p\Omega) = \left(\boldsymbol{e}_\beta^* \boldsymbol{E}_{p\varkappa}(\boldsymbol{r}; \omega + p\Omega)\right)$ is the projection of $\boldsymbol{E}_{p\varkappa}$ on the vector $\boldsymbol{e}_\beta$. The solution of (22) is given by

$$\boldsymbol{E}_{p\varkappa}(\boldsymbol{r}; \omega + p\Omega) = \exp\{i\tfrac{1}{2}\chi_0^{\mathrm{e}}(\omega) (\boldsymbol{k} + p\boldsymbol{\varkappa}) (\boldsymbol{r} - \boldsymbol{r}_0)\} \boldsymbol{E}_{p\varkappa}(\boldsymbol{r}_0; \omega + p\Omega) \quad (25)$$

for $|p| > m$. The expressions (24) and (25) connect the state of the γ-quantum electromagnetic field at the crystal point $\boldsymbol{r}_0$ with that at the point $\boldsymbol{r}$. In order to obtain the γ-quantum field in the crystal in the case when the γ-quantum wave packet $\boldsymbol{E}_0(\boldsymbol{r}; \omega) \exp\{i(\boldsymbol{kr} - \omega t)\}$ is incident on the incoming crystal surface it is necessary to supplement (19) and (22) by boundary conditions. In the present paper we suppose that the angle between the wave vector $\boldsymbol{k}$ and the normal $\boldsymbol{N}$ to the incoming surface of a crystal is small. Since the elements of the matrix $\hat{\chi}_{\mathrm{N}}$ have values of the order of

10^{-4} in this case, the boundary condition is given by

$$\boldsymbol{E}^{\mathrm{I}} = \boldsymbol{E}^{\mathrm{II}}\,, \tag{26}$$

where $\boldsymbol{E}^{\mathrm{I}}$ is the incident γ-quantum field on the incoming surface of the crystal; $\boldsymbol{E}^{\mathrm{II}}$ is the γ-quantum field inside the crystal on its incoming surface. Since the shift wave (3) caused by the variable external field is propagating into the crystal, then the crystal incoming surface at t is defined by the radius vector

$$\boldsymbol{r}(t) = \boldsymbol{r}_0 + \boldsymbol{a}\sin(\boldsymbol{\varkappa r}_0 - \Omega t + \varphi)\,, \tag{27}$$

where $\boldsymbol{r}_0$ is the radius vector of an arbitrary point on the incoming surface in the absence of the variable external field. The field $\boldsymbol{E}^{\mathrm{I}}$ is given by

$$\boldsymbol{E}^{\mathrm{I}} = \boldsymbol{E}^0(\boldsymbol{r}(t);\omega)\exp\{i(\boldsymbol{kr}(t) - \omega t)\}\,,$$

where $\boldsymbol{r}(t)$ is defined by (27). As it has been shown in Section 4, the interaction between the γ-quanta and the nuclear subsystem of the crystal is essential only when the forced vibration amplitude has a magnitude of the order of the γ-quantum wavelength. Consequently, we can write $\boldsymbol{E}^0(\boldsymbol{r}_0 + \boldsymbol{a}\sin(\boldsymbol{\varkappa r}_0 - \Omega t + \varphi);\omega) = \boldsymbol{E}^0(\boldsymbol{r}_0;\omega)$, since within distances of the order of the γ-quantum wavelength the wave packet amplitude $\boldsymbol{E}^0(\boldsymbol{r};\omega)$ changes slowly. Expanding $\exp\{i(\boldsymbol{ka})\sin(\boldsymbol{\varkappa r}_0 - \Omega t + \varphi)\}$ into a series of Bessel functions and substituting the field $\boldsymbol{E}^{\mathrm{II}} = \exp\{i(\boldsymbol{kr}_0 - \omega t)\} \times \times \sum_{p=-\infty}^{\infty} \boldsymbol{E}_{p\varkappa}(\boldsymbol{r}_0;\omega + p\Omega)\exp\{ip(\boldsymbol{\varkappa r}_0 - \Omega t)\}$ into the boundary condition (26) one obtains the Bloch wave amplitudes in a crystal on the incoming surface:

$$\boldsymbol{E}_{p\varkappa}(\boldsymbol{r}_0;\omega + p\Omega) = \mathrm{e}^{ip\varphi}\,\mathrm{J}_p(\boldsymbol{ka})\,\boldsymbol{E}^0(\boldsymbol{r}_0;\omega)\,. \tag{28}$$

Using (13), (24), (25), and (28) one obtains the field of the γ-quanta, which were emitted by a source at time t and were detected by a detector behind the absorber at time $t > t_0$,

$$\begin{aligned}
\boldsymbol{E}(\boldsymbol{r};t,t_0) = {} & \sum_{p,p'=-m}^{m} \exp\{ip(\boldsymbol{\varkappa r} - \Omega t)\rangle \int_{-\infty}^{\infty} \exp\{ikn_{\mathrm{e}}(\omega)\,\boldsymbol{n}(\boldsymbol{r} - \boldsymbol{r}_0)\} \times \\
& \times \left(\hat{S}(\boldsymbol{r} - \boldsymbol{r}_0;\omega)\right)_{p,p'} \mathrm{e}^{ip'\varphi}\,\mathrm{J}_{p'}^{\mathrm{T}}(\boldsymbol{ka})\,\boldsymbol{E}_0(\boldsymbol{r}_0,\omega) \times \\
& \times \exp\{i[\boldsymbol{kr}_0 - \omega(t - t_0)]\}\,\frac{\mathrm{d}\omega}{2\pi} + \\
& + \int_{-\infty}^{\infty} \exp\{ikn_{\mathrm{e}}(\omega)\,\boldsymbol{n}(\boldsymbol{r} - \boldsymbol{r}_0)\}\exp\{ik\varphi(\boldsymbol{r},\boldsymbol{r}_0;t)\} \times \\
& \times \boldsymbol{E}^0(\boldsymbol{r}_0;\omega)\exp\{i[\boldsymbol{kr}_0 - \omega(t - t_0)]\}\,\frac{\mathrm{d}\omega}{2\pi}\,,
\end{aligned} \tag{29}$$

where $\hat{S}(\boldsymbol{r} - \boldsymbol{r}_0;\omega) = \left(\exp\{i\tfrac{1}{2}k\hat{\chi}_{\mathrm{N}}\boldsymbol{n}(\boldsymbol{r} - \boldsymbol{r}_0)\} - 1\right)$ is the evolution operator of the γ-quantum electromagnetic field in a crystal; $n_{\mathrm{e}}(\omega) = \left(1 - \tfrac{1}{2}\chi_0^{\mathrm{e}}(\omega)\right)$ is the refraction index of the crystal's electronic subsystem; $\varphi(\boldsymbol{r},\boldsymbol{r}_0;t) = (\boldsymbol{na})\sin[\boldsymbol{\varkappa r} + \tfrac{1}{2}\chi_0^{\mathrm{e}}(\omega) \times \times \boldsymbol{\varkappa}(\boldsymbol{r} - \boldsymbol{r}_0) - \Omega t + \varphi]$ is the phase. Then the intensity of the γ-quanta transmitted through the absorber is given by

$$\begin{aligned}
I(\boldsymbol{r}) = \int_{-\infty}^{\infty} & \exp\{-\mu^{\mathrm{e}}(\omega)\,\boldsymbol{n}(\boldsymbol{r} - \boldsymbol{r}_0)\}\,\{\mathrm{J}_0[(\boldsymbol{ka})\,\mathrm{sh}\left(\mathrm{Im}\,\chi_0^{\mathrm{e}}(\omega)\,\boldsymbol{\varkappa}(\boldsymbol{r} - \boldsymbol{r}_0)\right)] \times \\
& \times I_0(\boldsymbol{r}_0;\omega) + \mathrm{Sp}\left(\hat{S}(\boldsymbol{r} - \boldsymbol{r}_0;\omega)\,[\hat{g}_0\hat{\varrho}_0(\boldsymbol{r}_0;\omega)]\,\hat{S}^+(\boldsymbol{r} - \boldsymbol{r}_0;\omega)\right) + \\
& + \mathrm{Im}\,\mathrm{Sp}\left(\hat{C}(\boldsymbol{r} - \boldsymbol{r}_0)\,\hat{S}(\boldsymbol{r} - \boldsymbol{r}_0;\omega)\,[\hat{g}_0\hat{\varrho}_0(\boldsymbol{r}_0;\omega)]\right)\}\,\frac{\mathrm{d}\omega}{2\pi}\,,
\end{aligned} \tag{30}$$

where $\hat{C}(\boldsymbol{r}-\boldsymbol{r}_0)$ is the diagonal $(2m+1)\times(2m+1)$ matrix with the elements $(\hat{C})_{p,p'}=\delta_{p,p'}\exp\{-i\frac{1}{2}p\chi_0^{e}\boldsymbol{\varkappa}(\boldsymbol{r}-\boldsymbol{r}_0)\}$, $\hat{\varrho}_0(\boldsymbol{r}_0;\omega)$ is the density matrix of an incident γ-quantum with the frequency ω; $\hat{g}_0$ is the $(2m+1)\times(2m+1)$ matrix with the elements $(\hat{g}_0)_{p,p'}=e^{i(p-p')\varphi}\,J_p(\boldsymbol{ka})\,J_{p'}(\boldsymbol{ka})$; $\mu^{e}(\omega)$ is the absorption coefficient of the electron subsystem of the crystal. It should be emphasized that the intensity is space modulated with the period $4\pi\cos\theta/\mathrm{Re}\,\chi_0^{e}\cos\vartheta$, where θ is the angle between $\boldsymbol{k}$ and the incoming surface normal $\boldsymbol{N}$, ϑ is the angle between $\boldsymbol{\varkappa}$ and $\boldsymbol{N}$.

Suppose that the crystal is a plane-parallel plate with thickness L, the concentration of the Mössbauer isotope is small and the condition $k\,|\mu_{\beta\gamma}|\,L/\cos\theta\ll 1$ is satisfied (thin resonance absorber). Then the evolution operator $\hat{S}(\boldsymbol{r}-\boldsymbol{r}_0;\omega)$ can be written in the following form:

$$\hat{S}(\boldsymbol{r}-\boldsymbol{r}_0;\omega)=\tfrac{1}{2}\,ik\boldsymbol{n}(\boldsymbol{r}-\boldsymbol{r}_0)\,\hat{\chi}_{\mathrm{N}}\,, \tag{31}$$

where the matrix $\hat{\chi}_{\mathrm{N}}$ has been defined above. The intensity of the γ-quanta passed through the thin absorber is given by

$$\begin{aligned}I(\boldsymbol{r})=\int_{-\infty}^{\infty}\exp\left\{-\frac{\mu^{e}(\omega)\,L}{\cos\theta}\right\}\Bigg\{&J_0\left[(\boldsymbol{ka})\,\mathrm{sh}\left(\mathrm{Im}\,\chi_0^{e}(\omega)\,\frac{\varkappa L\cos\vartheta}{\cos\theta}\right)\right]I_0(\boldsymbol{r}_0;\omega)-\\ &-\frac{kL}{\cos\theta}\,\mathrm{Im}\,[\sum_{j}\sum_{M_0,M}\mathrm{Sp}\,(\hat{\mu}^{j}(\boldsymbol{k};M_0,M)\,\hat{\varrho}_0(\boldsymbol{r}_0;\omega))\times\\ &\times\mathrm{Sp}\,(\hat{C}(L)\,\hat{g}^{j}(\boldsymbol{k};M_0,M)\,\hat{g}_0)-\mathrm{Sp}\,(\hat{C}(L)\,\hat{\alpha}\hat{g}_0)\,I_0(\boldsymbol{r}_0;\omega)]\Bigg\}\frac{d\omega}{2\pi},\end{aligned} \tag{32}$$

where the matrices $C(\hat{L})$, $\hat{g}_0$, $\hat{\alpha}$, and $\hat{g}^{j}(\boldsymbol{k};M_0,M)$ have been obtained above. As follows from (32), in such ordinary experiments the intensity of the γ-quanta is proportional to the product of four Bessel functions. It should be emphasized once more, that the intensity and the polarization vector oscillate as a function of the crystal thickness with the period $4\pi\cos\theta/(\mathrm{Re}\,\chi_0^{e}(\omega))\,\varkappa\cos\vartheta$.

6. The Resonant Interaction between γ-Quanta and the Shift Field Induced by the Variable External Field

Consider now the case when the forced vibration amplitude is small compared to the γ-quantum wavelength, i.e. $|\boldsymbol{ka}|\ll 1$. In this case $J_p(\boldsymbol{ka})\approx(J/p!)\,(\boldsymbol{ka}/2)^{p}$. Since $(\hat{\chi}_{\mathrm{N}})_{p,p'}\sim(1/|p-p'|\,!)\,(\boldsymbol{ka}/2)^{(p-p')}$, only terms linear in $(\boldsymbol{ka})$ are kept in the matrix $\hat{\chi}_{\mathrm{N}}$. Consequently, as a first approximation only the matrix elements

$$\begin{aligned}(\hat{\chi}_{\mathrm{N}})_{p,p}&=\hat{\chi}_{\mathrm{N}}(\omega+p\Omega)-\alpha_p\,,\\ (\hat{\chi}_{\mathrm{N}})_{p,p\pm1}&=\mp e^{\pm i\varphi}\left(\frac{\boldsymbol{ka}}{2}\right)(\hat{\chi}_{\mathrm{N}}(\omega+p\Omega)-\hat{\chi}_{\mathrm{N}}(\omega+(p\pm1)\,\Omega))\end{aligned} \tag{33}$$

are nonzero, where $\hat{\chi}_N(\omega+p\Omega)$ are the tensor amplitudes of the forward scattering of the γ-quanta with frequencies $(\omega+p\Omega)$ by a unit cell in the absence of a variable external field. It is easy to show that the Bloch wave amplitudes $\boldsymbol{E}_{p\varkappa}(\boldsymbol{r};\omega+p\Omega)$ are proportional to $(\boldsymbol{ka})^{p}\,\boldsymbol{E}_0(\boldsymbol{r}_0;\omega)$, where $\boldsymbol{E}^{0}(\boldsymbol{r}_0;\omega)$ is the wave packet amplitude of an incident γ-quantum at the point $\boldsymbol{r}_0$ on the incoming surface of a crystal. Hence only three interacting Bloch waves should be taken into account: the basic wave with amplitude $\boldsymbol{E}(\boldsymbol{r};\omega)$ and the two first satellites with the amplitudes $\boldsymbol{E}_{+\varkappa}(\boldsymbol{r};\omega+\Omega)$ and

$E_{-\varkappa}(\boldsymbol{r}; \omega - \Omega)$. For simplicity the hyperfine splitting in the crystal is assumed to be absent. As a result, the linear operator $\hat{\chi}_N$ takes the following form;

$$\hat{\chi}_N = \begin{vmatrix} (\chi_{+1} - \alpha_{+1}); & -\left(\frac{\boldsymbol{ka}}{2}\right) e^{-i\varphi} (\chi_0 - \chi_{+1}); & 0 \\ -\left(\frac{\boldsymbol{ka}}{2}\right) e^{i\varphi} (\chi_0 - \chi_{+1}); & \chi_0; & \left(\frac{\boldsymbol{ka}}{2}\right) e^{-i\varphi} (\chi_0 - \chi_{-1}) \\ 0; & \left(\frac{\boldsymbol{ka}}{2}\right) e^{i\varphi} (\chi_0 - \chi_{-1}); & (\chi_{-1} - \alpha_{-1}) \end{vmatrix} . \quad (34)$$

Here the following notations are used: χ_p are the forward scattering amplitudes of the γ-quanta with frequencies $(\omega + p\Omega)$ by the nuclear subsystem of the crystal unit cell in the absence of the external field,

$$\chi_p = \frac{\mu(\omega)\,(i\Gamma/2)}{\omega + p\Omega - \omega_0 + i\Gamma/2},$$

where the quantity $\mu(\omega)$ takes the following form for the M1-transition [12]:

$$\mu(\omega) = \frac{6\pi\eta i}{\hbar V k^3} \frac{2J + 1}{2J_0 + 1} \frac{1}{1 + \alpha} \sum_j \exp\{-Z_j(\boldsymbol{k})\};$$

$\alpha_p = 2p(\varkappa/k) \cos\psi$; $p = 0, \pm 1$. For the resonant absorber being a plane-parallel plate with thickness L the Bloch wave amplitudes on the outgoing surface of the crystal are represented through the wave packet amplitude of the incident γ-quanta by the expression

$$\begin{pmatrix} \boldsymbol{E}_{+\varkappa}(\boldsymbol{r}; \omega + \Omega) \\ \boldsymbol{E}(\boldsymbol{r}; \omega) \\ \boldsymbol{E}_{-\varkappa}(\boldsymbol{r}; \omega - \Omega) \end{pmatrix} = $$

$$= \exp\left\{i\chi_0^{e}(\omega) \frac{kL}{2\cos\theta}\right\} \exp\left\{ik\hat{\chi}_N \frac{L}{2\cos\theta}\right\} \begin{vmatrix} e^{i\varphi}\left(\frac{\boldsymbol{ka}}{2}\right) \\ 1 \\ -e^{-i\varphi}\left(\frac{\boldsymbol{ka}}{2}\right) \end{vmatrix} \boldsymbol{E}^0(\boldsymbol{r}_0; \omega)\,, \quad (35)$$

where θ is the angle between the vectors $\boldsymbol{k}$ and $\boldsymbol{N}$; $\boldsymbol{r} = \boldsymbol{r}_0 + \boldsymbol{n}(L/\cos\theta)$. The eigenvalues and the projection operators of matrix $\hat{\chi}_N$ can be given explicitly because the characteristic equation for the matrix $\hat{\chi}_N$ is a cubic one which can be solved in radials. However, the expression is too complicate and can be simplified using the fact that non-diagonal matrix elements of χ_N are proportional to $\boldsymbol{ka}$. Thus, making use of perturbation theory [13] one obtains in the first order

$$\boldsymbol{E}_{+\varkappa}(\boldsymbol{r}; \omega + \Omega) =$$

$$= \exp\left\{i\chi_0^{e}(\omega) \frac{kL}{2\cos\theta}\right\} e^{i\varphi} \left(\frac{\boldsymbol{ka}}{2}\right) \left[\exp\left\{i\lambda_1^0 \frac{kL}{2\cos\theta}\right\} + \right.$$

$$\left. + \frac{\chi_0 - \chi_{+1}}{\lambda_2^0 - \lambda_1^0} \left(e^{i\lambda_1^0 \frac{kL}{2\cos\theta}} - e^{i\lambda_2^0 \frac{kL}{2\cos\theta}}\right)\right] \boldsymbol{E}^0(\boldsymbol{r}_0; \omega)\,,$$

$$E(r;\omega)=\exp\left\{i\chi_0^e(\omega)\frac{kL}{2\cos\theta}\right\}\exp\left\{i\lambda_2^0\frac{kL}{2\cos\theta}\right\}E^0(r_0;\omega),$$

$$E_{-\varkappa}(r;\omega-\Omega)=$$

$$=-\exp\left\{i\chi_0^e(\omega)\frac{kL}{2\cos\theta}\right\}e^{-i\varphi}\left(\frac{ka}{2}\right)\left[\exp\left\{i\lambda_2^0\frac{kL}{2\cos\theta}\right\}+\right.$$

$$\left.+\frac{\chi_0+\chi_{+1}}{\lambda_2^0-\lambda_3^0}\left(e^{i\lambda_3^0\frac{kL}{2\cos\theta}}-e^{i\lambda_2^0\frac{kL}{2\cos\theta}}\right)\right]E^0(r_0;\omega), \tag{36}$$

where

$$\lambda_1^0=(\chi_{+1}-\alpha_{+1});\qquad \lambda_2^0=\chi_0;\qquad \lambda_3^0=(\chi_{-1}-\alpha_{-1})$$

are the eigenvalues of the matrix $\hat{\chi}_N$ in the first order of perturbation theory; they are equal to the diagonal elements of this matrix. The corrections to the eigenvalues of the matrix $\hat{\chi}_N$ arise in the second order of perturbation theory; they are proportional to $(ka)^2$. The Bloch satellite amplitudes in the crystal are proportional to (ka) and, in general, they are small compared to the basic Bloch amplitude. If, however, the conditions $\mathrm{Re}\,(\lambda_2^0-\lambda_1^0)=0$ and $(ka/2)\,|\chi_0-\chi_{+1}|/\mathrm{Im}\,(\lambda_2^0-\lambda_1^0)\approx 1$ or $\mathrm{Re}\,(\lambda_2^0-\lambda_3^0)=0$ and $(ka/2)\,|\chi_0-\chi_{-1}|/\mathrm{Im}\,(\lambda_2^0-\lambda_3^0)\approx 1$ are satisfied then one of the Bloch wave amplitudes $E_{+\varkappa}(r;\omega+\Omega)$ or $E_{-\varkappa}(r;\omega-\Omega)$ becomes comparable with the basic wave amplitude. For example, for ultrasonic pumping the condition $\mathrm{Re}\,(\lambda_2^0-\lambda_3^0)=0$ is equivalent to

$$\varkappa=\frac{k\,\mathrm{Re}\,\chi_0}{2\cos\psi}+\frac{\Delta\omega}{v} \tag{37}$$

and the condition $\mathrm{Re}\,(\lambda_2^0-\lambda_1^0)=0$ is equivalent to

$$\varkappa=-\frac{k\,\mathrm{Re}\,\chi_0}{2\cos\psi}-\frac{\Delta\omega}{v}. \tag{37 a}$$

The conditions $(ka/2)\,|\chi_0-\chi_{\pm1}|/\mathrm{Im}\,(\lambda_2^0-\lambda_{1,3}^0)\approx 1$ are satisfied for

$$(ka)\Delta\omega/\Gamma\approx 1. \tag{37 b}$$

Here $\Delta\omega=\omega-\omega_0$, v is the sound velocity in the crystal. Hence under the conditions (37) and (37b) or (37a) and (37b) the system of two waves with amplitudes $E(r;\omega)$ and $E_{-\varkappa}(r;\omega-\Omega)$ or $E(r;\omega)$ and $E_{+\varkappa}(r;\omega+\Omega)$ satisfies the parametric resonance conditions in the crystal, i.e. the conditions of the resonance interaction with the crystal shift field.

It is obvious that, in case the resonance conditions (37) to (37b) are satisfied, perturbation theory cannot be applied. In order to describe the γ-quantum diffraction by the moving grating under these conditions we should take into account the fact that only two of these three waves interact resonantly. Therefore, the interaction between these two waves should be described exactly and the third wave can be accounted for by perturbation theory. As a result, we have

$$E_{+\varkappa}(r;\omega+\Omega)=$$

$$=\exp\left\{i\chi_0^e(\omega)\frac{kL}{2\cos\theta}\right\}e^{i\varphi}\left(\frac{ka}{2}\right)\left[\exp\left\{i\lambda_1\frac{kL}{2\cos\theta}\right\}+\right.$$

$$+\frac{\chi_0-\chi_{+1}}{\lambda_3-\lambda_2}\left(\frac{\chi_{-1}-\alpha_{-1}-\lambda_2}{\lambda_2-\lambda_1^0}\exp\left\{i\lambda_2\frac{kL}{2\cos\theta}\right\}+\frac{\chi_{-1}-\alpha_{-1}-\lambda_3}{\lambda_3-\lambda_1^0}\exp\left\{i\lambda_3\frac{kL}{2\cos\theta}\right\}-\right.$$

$$\left.\left.-\frac{(\lambda_3-\lambda_2)(\chi_{-1}-\alpha_{-1}-\lambda_1^0)}{(\lambda_3-\lambda_1^0)(\lambda_2-\lambda_1^0)}\exp\left\{i\lambda_1\frac{kL}{2\cos\theta}\right\}\right)\right]E^0(r_0;\omega),$$

$$\boldsymbol{E}(\boldsymbol{r};\omega) = \exp\left\{i\chi_0^e(\omega)\,\frac{kL}{2\cos\theta}\right\}\left[\frac{\lambda_3-\chi_{-1}+\alpha_{-1}}{\lambda_3-\lambda_2}\exp\left\{i\lambda_3\,\frac{kL}{2\cos\theta}\right\} - \right.$$
$$-\frac{\lambda_2-\chi_{-1}+\alpha_{-1}}{\lambda_3-\lambda_2}\exp\left\{i\lambda_2\,\frac{kL}{2\cos\theta}\right\} - \left(\frac{\boldsymbol{ka}}{2}\right)\frac{\chi_0-\chi_{-1}}{\lambda_3-\lambda_2}\times$$
$$\left.\times\left(\exp\left\{i\lambda_3\,\frac{kL}{2\cos\theta}\right\} - \exp\left\{i\lambda_2\,\frac{kL}{2\cos\theta}\right\}\right)\right]\boldsymbol{E}^0(\boldsymbol{r}_0;\omega)\,,$$

$$\boldsymbol{E}_{-\varkappa}(\boldsymbol{r};\omega-\Omega) = \exp\left\{i\chi_0^e(\omega)\,\frac{kL}{2\cos\theta}\right\}\mathrm{e}^{-i\varphi}\left(\frac{\boldsymbol{ka}}{2}\right)\times$$
$$\times\left[\frac{\chi_0-\chi_{-1}}{\lambda_3-\lambda_2}\left(\exp\left\{i\lambda_3\,\frac{kL}{2\cos\theta}\right\} - \exp\left\{i\lambda_2\,\frac{kL}{2\cos\theta}\right\}\right) + \right.$$
$$+\frac{\chi_0-\lambda_2}{\lambda_3-\lambda_2}\exp\left\{i\lambda_2\,\frac{kL}{2\cos\theta}\right\} -$$
$$\left. -\frac{\chi_0-\lambda_3}{\lambda_3-\lambda_2}\exp\left\{i\lambda_3\,\frac{kL}{2\cos\theta}\right\}\right]\boldsymbol{E}^0(\boldsymbol{r}_0;\omega)\,, \tag{38}$$

where

$$\lambda_{3,2} = \frac{\chi_0+\chi_{-1}-\alpha_{-1}}{2} \mp \sqrt{\left(\frac{\chi_0-\chi_{-1}+\alpha_{-1}}{2}\right)^2 + \left(\frac{\boldsymbol{ka}}{2}\right)^2(\chi_0-\chi_{-1})^2}\,,$$
$$\lambda_1 = \lambda_1^0 - \left(\frac{\boldsymbol{ka}}{2}\right)^2\frac{(\chi_0-\chi_{+1})\,(\chi_{-1}-\alpha_{-1}-\lambda_1^0)}{(\lambda_1^0-\lambda_2)\,(\lambda_1^0-\lambda_3)}\,, \tag{39}$$
$$\lambda_1^0 = \chi_{+1}-\alpha_{+1}$$

are the eigenvalues of the matrix χ_N in the given approximation. Since the interaction between the basic wave with the amplitude $\boldsymbol{E}(\boldsymbol{r};\omega)$ and the satellite with the amplitude $\boldsymbol{E}_{-\varkappa}(\boldsymbol{r};\omega-\Omega)$ is accounted for exactly (38) and (39) describe the γ-quantum field in the crystal also under parametric resonance. As follows from (38) and (39), the resonant scattering of the γ-quantum with frequency into the γ-quantum with frequency $(\omega-\Omega)$ occurs over the length $L_{\text{res}} = 1/k\,\mathrm{Re}\,\Delta$, where

$$\Delta = \sqrt{\left(\frac{\chi_0-\chi_{-1}-\alpha_{-1}}{2}\right)^2 + \left(\frac{\boldsymbol{ka}}{2}\right)^2(\chi_0-\chi_{-1})^2}\,.$$

Hence the yield of the γ-quanta with frequency $(\omega-\Omega)$ out of the crystal will be considerable only if this length will be smaller than the γ-quantum absorption depth $L_{\text{abs}} = 1/k\,\mathrm{Im}\,\lambda_{3,2}$. One can see that for crystals with $|\mu(\omega)| \approx 10^{-5}$; $(\boldsymbol{ka}/2) \approx 3 \times 10^{-1}$ when the conditions (37) and (37b) are satisfied (condition (37b) means that $\Delta\omega \approx 2\Gamma$), $\mathrm{Im}\,\lambda_{2,3} \approx \mathrm{Im}\,(\chi_0/2)$ and $\mathrm{Re}\,\Delta \approx (2/\sqrt{3})\,\mathrm{Im}\,\chi_0$, i.e. $L_{\text{res}}/L_{\text{abs}} \approx 1/\sqrt{3} \approx 0.6$. Since the γ-quantum absorption depth is greater than L_{res}, then the intensity of the γ-quanta with frequency $(\omega-\Omega)$ passing through the crystal and the intensity of the γ-quanta with frequency ω will be comparable for a crystal thickness $L = L_{\text{res}}$. In particular, for a crystal 90% enriched by ^{57}Fe $L_{\text{res}} = 0.2 \times 10^{-4}$ cm, whereas the absorption depth equals 1.53×10^{-4} cm for $\Delta\omega = 2\Gamma$.

If the conditions (37a) and (37b) are fulfilled the interaction between the basic wave with the amplitude $\boldsymbol{E}(\boldsymbol{r};\omega)$ and the satellite with the amplitude $\boldsymbol{E}_{+\varkappa}(\boldsymbol{r};\omega+\Omega)$ should be described exactly, whereas for interaction with the satellite with amplitude $\boldsymbol{E}_{-\varkappa}(\boldsymbol{r}; \omega - \Omega)$ can be accounted within perturbation theory. The expressions for the Bloch wave amplitudes, in this case, are analogous to the expressions (38) and (39), but χ_{-1} is replaced by χ_{+1}, α_{+1} by α_{-1}, $\boldsymbol{E}_{-\varkappa}$ by $\boldsymbol{E}_{+\varkappa}$ and $e^{i\varphi}$ by $-e^{i\varphi}$.

Thus, in the simplest case when diffraction by the crystal lattice is absent, the interaction between the Mössbauer γ-quanta and the shift field (3) induced by the variable external field in the crystal differ radically from the interaction between X-ray and ultrasonic wave. Besides, it should be noted that because the parametric resonance width is defined by the ratio $k \operatorname{Im} \chi_0/2 \, |\cos \psi|$, i.e. of the order of 10^3 cm^{-1}, the resonant interaction between the γ-quanta and the forced nuclear vibrations can be easily observed experimentally.

7. Conclusion

The dynamical theory of Mössbauer radiation diffraction by crystals in variable external fields has been developed. The scattering radiation intensity in thin and thick crystals has been obtained. It is shown that the resonant interaction between the Mössbauer radiation and the ultrasound wave gives rise to an effective conversion of the incident wave into a wave with changed frequency.

References

[1] V. G. Baryshevskii and V. V. Skadorov, phys. stat. sol. (b) **108**, 631 (1981).
[2] D. I. Bolef and S. I. Ruby, Phys. Rev. Letters **5**, 5 (1960).
[3] L. Pheifer, N. Heiman, and J. Walker, Phys. Rev. **136**, 74 (1972).
[4] V. G. Baryshevskii, Dokl. Akad. Nauk BSSR **26**, 329 (1982).
[5] V. G. Baryshevskii, Phys. Letters A **96**, 439 (1983).
[6] A. R. Mkrtchyan, G. A. Arutyunyan, A. R. Arakelyan, and R. G. Gabrielyan, phys. stat. sol. (b) **92**, 23 (1979).
[7] I. A. Dubovcev, P. S. Zearumov, and N. P. Philippova, Zh. eksper. teor. Fiz. **65**, 1058 (1979).
[8] I. R. Entin, Zh. eksper. teor. Fiz. **47**, 214 (1979).
[9] K. G. Assur and I. R. Entin, Fiz. tverd. Tela **24**, 2122 (1982).
[10] M. Born and E. Wolf, Principles of Optics, Pergamon Press, London 1968.
[11] S. A. Athmanov and N. I. Koroteev, Methods of Nonlinear Optics in Scattering Light Spectroscopy, Izd. Nauka, Moscow 1981.
[12] A. M. Afanasev and Yu. Kagan, Zh. eksper. teor. Fiz. **48**, 327 (1965).
[13] T. Kato, Perturbation Theory for Linear Operators, Springer-Verlag, Berlin 1966.

(Received November 15, 1984)

phys. stat. sol. (b) 133, 89 (1986)

Subject classification: 63.20 and 65.70; 62.20; S 12

Department of Physics, Panjab University, Chandigarh[1]) *(a)*
and Physikalisches Institut der Universität Bayreuth[2]) *(b)*

Calculation of Thermal Expansion and Phonon Frequency Shift in Deuterated Naphthalene

By

V. K. JINDAL (a) and J. KALUS (b)

A calculation of the implicit shift to the phonon energies for phonons in deuterated naphthalene at $q = 0$ is presented, the "explicit shift" having been already described in an earlier paper. With this the calculation of the shift in the phonon energies due to temperature becomes complete and consistent, ready for direct comparison with the measured results of the shift obtained by neutron scattering experiments. The results for the total calculated shift in the phonon energies, arising from both, the explicit anharmonic terms and the implicit (i.e. the volume dependent) shift are presented and compared with the experimental neutron scattering results. The calculations of thermal expansion, compressibility, and the Grüneisen constants which naturally enter the calculation of the implicit shift, are also described and a comparison with the experimental neutron scattering results of these also is made. All the results are very satisfactory, strengthening the faith in the assumption for the model of the molecular crystals.

Eine Berechnung der impliziten Verschiebung zu den Phononenenergien in deuteriertem Naphtalen bei $q = 0$ wird durchgeführt; die „explizite" Verschiebung wurde bereits in einer früheren Arbeit beschrieben. Damit wird die Berechnung der Verschiebung in den Phononenenergien infolge der Temperatur vollständig und konsistent und geeignet für einen direkten Vergleich mit den gemessenen Ergebnissen der Verschiebung, die mit Neutronenstreuung erhalten werden. Die Ergebnisse für die berechnete Gesamtverschiebung der Phononenenergien, die sowohl von den expliziten anharmonischen Termen als auch von der impliziten (d. h. der volumenabhängigen) Verschiebung herrühren, werden angegeben und mit experimentellen Neutronenstreuergebnissen verglichen. Die Berechnungen der thermischen Ausdehnung, der Kompressibilität und der Grüneisenkonstanten, die in die Berechnung der impliziten Verschiebung eingehen, werden ebenfalls beschrieben und ein Vergleich mit experimentellen Neutronenstreuergebnissen durchgeführt. Alle Ergebnisse sind sehr befriedigend und bestätigen die Richtigkeit der Annahmen für das Modell der Molekülkristalle.

1. Introduction

The calculation of anharmonic phonon frequencies and bandwidths of molecular crystals is quite involved, mainly due to the complexities of the theoretical and computational procedures. Due to this reason, a calculation of this type has so far been reported only for limited cases of molecular crystals. The anharmonic phonon frequencies differ from the harmonic phonon frequencies by a shift from their harmonic, low-temperature frequencies. This shift, normally described as the phonon frequency shift depends upon temperature and consists of two parts from calculational point of view. The part arising due to the explicit anharmonic terms in the crystal potential energy, has been termed as explicit shift. The other part arises due to the implicit

[1]) Chandigarh 160014, India.
[2]) Postfach 3008, D-8580 Bayreuth, FRG.

temperature dependence of the phonon frequencies and is obtained from the volume dependence of the phonon frequencies. The latter contribution has been termed as the implicit shift.

In an earlier paper of Jindal and Kalus 1983 [1], hereinafter referred to as I, a calculation of the explicit shift was presented for deuterated naphthalene for which experimental results from neutron scattering were available [2 to 4]. In [1], the explicit shift was calculated by retaining cubic and quartic anharmonic terms in the crystal potential. For comparison with experimental results, the calculated explicit shifts were compared with the "measured explicit shift". The "measured explicit shift" was extracted from the measured shift using the measured results of Grüneisen constants and the thermal expansion. In fact, about the same time, we noticed quite a similar type of calculation having been attempted by Della Valle et al. [5] for protonated naphthalene. The results of [1] and the work by Della Valle et al. [5] were similar qualitatively. The quantitative differences have been attributed to slightly different calculational details and the different systems taken.

In the present work, we calculate the implicit contribution to the shift in the phonon frequencies of all the phonons at the Γ point for deuterated naphthalene and present the combined results of the explicit and implicit shifts. These are then compared directly with the experimental results of phonon frequency shifts obtained by neutron scattering experiments. The calculation of the implicit shift requires a calculation of thermal expansion, the compressibility, and the Grüneisen constants. These are therefore also calculated and compared with the results from neutron scattering experiments [2, 6]. The procedure and assumptions regarding the model of our molecular crystal are similar to those of [1].

The paper is organized as follows: The theoretical procedure for the calculation of implicit phonon frequency shift is given in Section 2. Expression for thermal expansion for its subsequent use is also obtained in this section. In Section 3, the numerical calculations for compressibility, thermal expansion, implicit shift, etc. are described and results presented. Comparison of the calculated and experimental results is made and discussed in Section 4. Concluding remarks are presented in Section 5. Finally, the expression for thermal expansion in higher order is given in the Appendix.

2. Theoretical Procedure

The implicit shift in the phonon frequencies arises due to the thermal expansion and is obtained from the volume dependence of the phonon frequencies. The implicit temperature dependence of the phonon frequencies is therefore written as

$$\left(\frac{\partial \omega_{\boldsymbol{q}j}}{\partial T}\right)_{\text{im}} = \frac{\partial \omega_{\boldsymbol{q}j}}{\partial V}\,\frac{\partial V}{\partial T} \tag{1}$$

which when expressed in terms of the Grüneisen constants $\gamma_{\boldsymbol{q}j}$ becomes

$$\left(\frac{\partial \omega_{\boldsymbol{q}j}}{\partial T}\right)_{\text{im}} = -\gamma_{\boldsymbol{q}j}\omega_{\boldsymbol{q}j}\beta\,, \tag{2}$$

where β is the volume thermal expansion coefficient and $\gamma_{\boldsymbol{q}j}$ are defined as

$$\beta = \frac{1}{V}\,\frac{\partial V}{\partial T} \tag{3}$$

and

$$\gamma_{\boldsymbol{q}j} = -\,\frac{\partial \ln \omega_{\boldsymbol{q}j}}{\partial \ln V}\,. \tag{4}$$

Rewriting (2) and integrating from $T = 0$ to T, we obtain the expression for the fractional implicit phonon frequency shift as

$$\left(\frac{\Delta\omega_{qj}}{\omega_{qj}}\right)_{\text{im}} = \exp\left[-\gamma_{qj}\varepsilon(T)\right] - 1\,, \tag{5}$$

where $\varepsilon(T) = \int\limits_0^T \beta \,\mathrm{d}T$ is the volume thermal expansion. In obtaining equation (5), we assume that the Grüneisen constants γ_{qj} do not depend on temperature.

As is evident (equation (5)) a calculation of the implicit shift to the phonon frequencies requires a calculation of (i) the Grüneisen constants and (ii) the volume thermal expansion. The calculation of these will be discussed in the following sections.

Expressions for the calculation of thermal expansion, especially for cubic crystals exist already in literature [7, 8]. We shall obtain approximate expressions valid for our molecular crystal by minimizing the free energy of the crystal.

To obtain the lowest-order contribution to the thermal expansion, we start with the expression for quasiharmonic free energy [7],

$$F_{\text{qh}} = \Phi(V) + kT \sum_{qj} \ln\left(2 \sinh \frac{\hbar\omega_{qj}}{2kT}\right), \tag{6}$$

where $\Phi(V)$ is the static lattice contribution and ω_{qj} are the harmonic phonon frequencies.

Expressing the strained lattice volume $V = V(T)$ in terms of unstrained volume $V_0 = V(0)$ and the volume strain parameter ε, i.e.

$$V = V_0(1 + \varepsilon)\,, \tag{7}$$

we write the static lattice contribution to the free energy as

$$\Phi(V) \approx \Phi(V_0) + \frac{1}{2}\,\varepsilon^2 V_0^2 \left(\frac{\partial^2\Phi}{\partial V^2}\right)_{V_0}, \tag{8}$$

where we assume $(\partial\Phi/\partial V)_{V_0} = 0$, which holds if we start with a structure that gives the minimum of potential energy. Furthermore we assume here that the temperature dependence of ε is "regular". ε depends on the lattice parameters a, b, c and the angle β between the a–b and b–c planes. (The angles α and γ are $\pi/2$ and do not change with temperature). Regular means, that $\mathrm{d}a/\mathrm{d}T =$ const. $\mathrm{d}b/\mathrm{d}T$, and so on, where the constant is temperature independent. Equation (8) can be rewritten as

$$\Phi(V) = \Phi(V_0) + \tfrac{1}{2}\,\varepsilon^2 V_0 B\,, \tag{9}$$

where we use the definition of bulk modulus B,

$$B = V_0 \left(\frac{\partial^2\Phi}{\partial V^2}\right)_{V_0}. \tag{10}$$

Therefore, the condition for the minimum of free energy,

$$\frac{\partial F_{\text{qh}}}{\partial\varepsilon} = \varepsilon V_0 B - \frac{1}{2} \sum_{qj} \gamma_{qj}\hbar\omega_{qj} \coth \frac{\hbar\omega_{qj}}{2kT} = 0\,,$$

gives

$$\varepsilon = \frac{1}{2V_0B} \sum_{qj} \gamma_{qj}\hbar\omega_{qj} \coth \frac{\hbar\omega_{qj}}{2kT} \tag{11}$$

which is the lowest-order volume thermal expansion. The volume expansion coefficient β is obtained as

$$\beta = \frac{k}{V_0 B} \sum_{qj} \gamma_{qj} \left(\frac{\hbar \omega_{qj}}{kT}\right)^2 \frac{e^{\hbar \omega_{qj}/kT}}{(e^{\hbar \omega_{qj}/kT} - 1)^2}. \tag{12}$$

The expression for thermal expansion in the next higher order can be approximately evaluated by replacing ω_{qj} in (6) by the renormalized phonon frequency $\tilde{\omega}_{qj}$ where $\tilde{\omega}_{qj} = \omega_{qj} + \Delta\omega_{qj}$; $\Delta\omega_{qj}$ being the explicit anharmonic shift. An expression for the thermal expansion thus obtained is described in the Appendix.

3. Numerical Calculations and Methods

For the calculation of thermal expansion and implicit phonon frequency shift (11), (5), and (A12), we need to calculate (i) the phonon frequencies, (ii) the Grüneisen constants, and (iii) the bulk modulus. We now discuss a numerical calculation of these for deuterated naphthalene. We assume as in I, the molecules to behave like rigid units and obtain molecule–molecule interactions from atom–atom interaction summations. The atom–atom interactions are assumed to be of the 6-exp type,

$$V(r) = -\frac{A}{r^6} + B \exp(-\alpha r), \tag{13}$$

where A, B, and α are constants depending on the nature of atom pairs involved. We again use for this calculation the potential parameters A, B, and α provided by Kitaigorodski.

3.1 Phonon frequencies and Grüneisen constants

The harmonic phonon frequencies ω_{qj} for values of $\boldsymbol{q}$ in the Brillouin zone were obtained by the procedure described in I.

For a calculation of the Grüneisen constants (4), we need the volume dependence of the phonon frequencies. This is obtained by computing the "new volume" of the crystal under hydrostatic pressure p. For this, one minimizes the potential energy function

$$\Phi_T = \Phi(V) + p\,\Delta V \tag{14}$$

by varying the crystal structure parameters [2]. This new structure gives the new volume and the dynamical matrix is again diagonalized corresponding to this structure. From the changes in the phonon frequency (say at a pressure of 0 and 2×10^8 Pa), the Grüneisen parameters are easily calculated. We present the results for the Grüneisen parameters for various phonon branches at Γ point in Table 1. It may

Table 1
Calculated and measured Grüneisen constants

phonon branch	γ_{qj} (calculated)	γ_{qj} (measured)
4 $A_u\Gamma_2$	4.22	3.7 ± 0.7
5 $B_u\Gamma_4$	4.15	—
6 $B_g\Gamma_3$	4.05	4.0 ± 0.9
7 $A_g\Gamma_1$	4.21	3.6 ± 0.6
8 $B_g\Gamma_3$	4.84	3.4 ± 0.8
9 $A_g\Gamma_1$	4.27	3.5 ± 0.6
10 $A_u\Gamma_2$	4.37	2.9 ± 0.5
11 $A_g\Gamma_1$	3.96	3.3 ± 0.4
12 $B_g\Gamma_3$	4.08	2.6 ± 0.5

be mentioned here that it is by no means evident that the pattern of the strains related with pressure is identical to the pattern of strains created by the thermal expansion. Therefore, for the same relative volume changes (remembering that the lattice constants may be different for the pressure or temperature effect) the Grüneisen parameters γ_{qj} may be different. The assumption is that because of the summation over all phonons in (12) this difference is washed out to some extent.

3.2 Bulk modulus

For the calculation of the bulk modulus (10), the potential energy of the crystal as a function of volume is required. The total potential energy of the crystal of N molecules is given by

$$\Phi = \tfrac{1}{2} N \sum_{l'k'}{}' \varphi(lk, l'k') . \tag{15}$$

Expressing the intermolecular potential energy $\varphi(lk, l'k')$ in terms of the atom–atom potentials, (15) becomes

$$\Phi = \tfrac{1}{2} N \sum_{l'k'}{}' \sum_{ij} V(r_{ij}^{lk,l'k'}) , \tag{16}$$

where i and j run through the number of atoms in a molecule and $r_{ij}^{lk,l'k'}$ represents the atom–atom distance between "i" atom of lk molecule and "j" atom of $l'k'$ molecule. Using (13), a part of the summation in (16) denoted by Φ_{near} has been carried out exactly for atom contacts within a radius of $R = 0.55$ nm. The remaining "distant potential" part, Φ_{dist} can be approximated by neglecting the short-range repulsive part and assuming the molecules to be uniformly distributed outside a sphere of radius R. In this way, one obtains

$$\Phi_{\text{dist}} \approx -\frac{1}{2} N(\sum_{ij} A_{ij})\, n \int_R^\infty \frac{4\pi r^2 \, \mathrm{d}r}{r^6} , \tag{17}$$

where n is the number of molecules per unit volume. For our case $n = 2/V_c$. V_c being the volume of a unit cell. Therefore,

$$\Phi_{\text{dist}} \approx -\frac{4\pi}{3} N \frac{A_T}{V_c R^3} , \tag{18}$$

$$A_T = \sum_{ij} A_{ij} = N_C^2 A_{CC} + 2N_C N_D A_{CD} + N_D^2 A_{DD} , \tag{19}$$

N_C and N_D being the number of "C" and "D" atoms per molecule. From these two contributions, the total potential energy of the crystal was obtained.

To calculate the potential energy at various volumes, the same procedure as described in Section 3.1 was used. The new volume corresponding to an applied pressure p was obtained by minimizing $\Phi + p\,\Delta V$ with respect to the crystal structure parameters. From this data Φ at various volumes was obtained. This is tabulated in Table 2. A plot of Φ for various cell volumes is given in Fig. 1. We also fitted this

Table 2
The total potential energy at various unit cell volumes and pressures

p (10^8 Pa)	−2	0	2	4
V_0 $(0.1\text{ nm})^3$	328.1903	322.9244	318.4898	314.5457
Φ (10^{-4} J/mol)	−7.93498	−7.95118	−7.93825	−7.90391

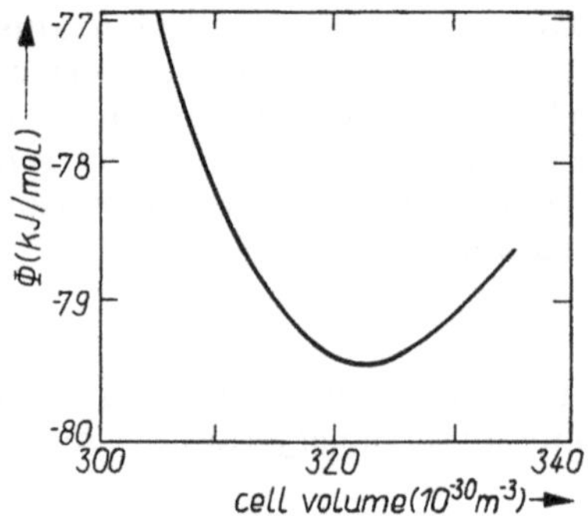

Fig. 1. The total crystal potential energy plotted as function of cell volume

curve to the equation

$$\Phi = a + bV_0 + cV_0^2 + dV_0^3\,, \tag{20}$$

where Φ is in J/mol and V_0 is the volume of a unit cell in $(0.1\ \text{nm})^3$.

This equation facilitates the evaluation of the bulk modulus B, or the compressibility $K = 1/B$. The coefficients a, b, c, and d are presented in Table 3. The calculated results for K are also presented in this table.

Table 3

Coefficients of the potential energy Φ and volume relationship and results of calculated and measured compressibility of deuterated naphthalene. See text for the units of a, b, c

a	b	c	d	K (Pa^{-1})	
				calculated	exp.
3.2848×10^6	-2.8045×10^5	84.300	-8.0572×10^{-2}	7.585×10^{-11}	9.991×10^{-11} (100 K) 7.2×10^{-11} (extrapolated to 0 K)

3.3 Thermal expansion

The calculation of thermal expansion can now be easily carried out (11), once the bulk modulus, the Grüneisen constants, and the phonon frequencies are computed as described earlier. The Brillouin-zone summation was carried out, again as in I, using a mesh size of 4 for the $\boldsymbol{q}$-vectors. The calculated results for lowest order ε and β are given in Fig. 2 and 3, respectively.

The calculation of the higher-order thermal expansion and its coefficient is also important and has been carried out using (A12) of the Appendix. Here one requires (i) $(\partial^3\Phi/\partial V^3)_{V_0}$, which can be immediately obtained from equation (20) and (ii) $\tilde{\gamma}_{\boldsymbol{q}j}$, the "renormalized" Grüneisen constants. A calculation of the latter is quite cumbersome and time consuming as for a calculation of these ((A8) and (A9)), $\gamma_{\boldsymbol{q}j}^{\text{an}}$ has to be evaluated and this involves the computation of the total explicit shift at two volumes of the crystal. We attempted this for phonons corresponding to branches 4 and 7 at Γ point. These are tabulated in Table 4 at various temperatures along with the quantity $[(\Delta\omega/\omega)\,\gamma_{\boldsymbol{q}j}^{\text{an}}]$ where $\Delta\omega/\omega$ is the relative explicit shift. Fortunately these contributions are much too small in comparison with "unrenormalized" $\gamma_{\boldsymbol{q}j}$. Therefore, we just replaced $\tilde{\gamma}_{\boldsymbol{q}j}$ by $\gamma_{\boldsymbol{q}j}$ in our calculation of the higher-order thermal expansion. In

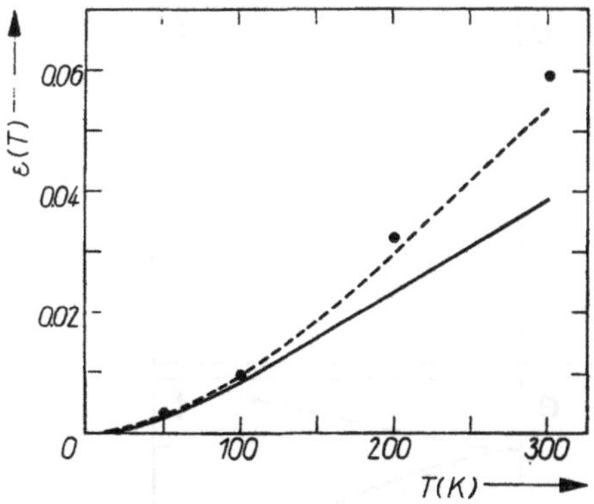

Fig. 2

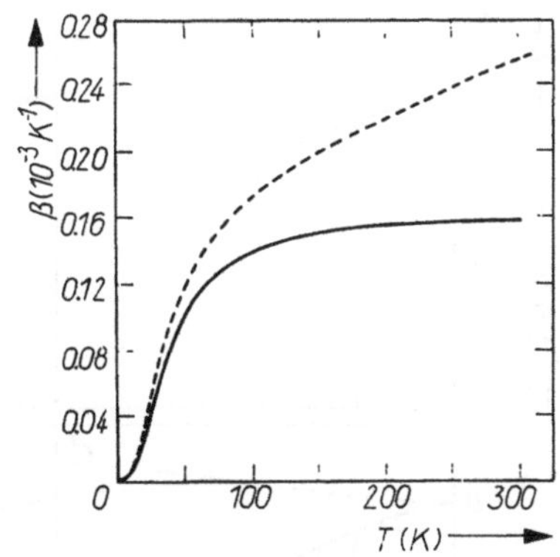

Fig. 3

Fig. 2. Volume thermal expansion $\varepsilon(T)$ as a function of temperature. The dashed curve represents the calculation from the higher-order expression for thermal expansion. The experimental points are obtained from [6]. The 50 K measurement is from [9]

Fig. 3. Volume thermal expansion coefficient β as a function of temperature. The dashed curve represents the higher-order results

fact it turns out that the second term in (A12) is the dominant contribution. The third term involving $\tilde{\gamma}_{qj}$ is negligibly small. The results thus calculated for ε and β are also presented in Fig. 2 and 3.

Table 4

γ_{qj}^{an} at various temperatures for two phonons

T (K)	for branch 4		for branch 7	
	$-\gamma^{an}$	$-\frac{\Delta\omega}{\omega}\gamma^{an}$	$-\gamma^{an}$	$-\frac{\Delta\omega}{\omega}\gamma^{an}$
0	0.5	0.009	0.4	0.003
10	0.58	0.01	0.4	0.003
100	2.87	0.08	3.53	0.06
200	3.64	0.17	4.16	0.13
300	3.83	0.27	4.29	0.21

3.4 Implicit shift

It is now a simple matter to calculate the implicit shift (5), once the essential ingredients have been calculated as described. We have calculated the implicit shift using higher-order thermal expansion results. As the explicit shift had already been calculated in I, we present here the combined results for the total relative shift in Fig. 4.

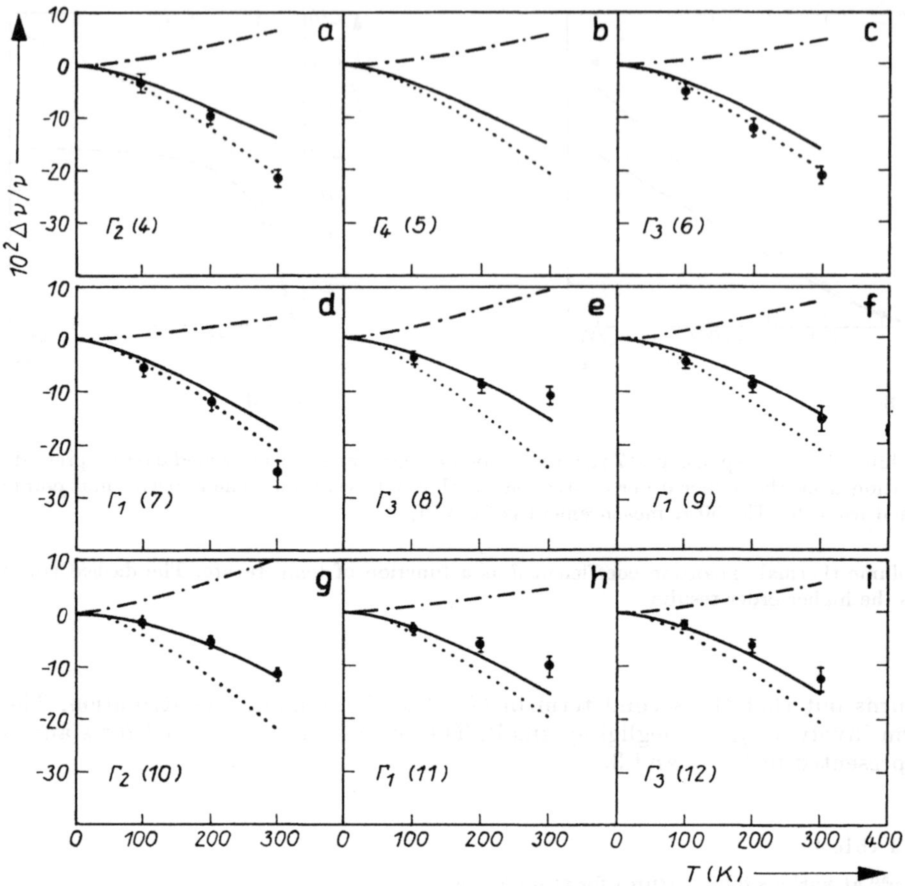

Fig. 4. Calculated and experimental results of the relative shift in phonon frequencies of all the nine optic translational and librational modes at the Γ point as a function of temperature. In parts a) to i), the branch numbers are given in parentheses. The dash-dotted curves represent the calculated explicit shift, whereas the dotted curves give the calculated implicit shift. The full curves represent the total calculated relative shifts

4. Comparison with Experimental Results and Discussion

The results obtained here for the Grüneisen constants, compressibility, thermal expansion, and phonon frequency shift have been compared with the experimental results obtained from neutron scattering experiments. The experimental results of Grüneisen constants from [2] and [6] are given in Table 1, for the compressibility from [6], in Table 3. From various measurements of [6], we have extrapolated the compressibility at 0 K for proper comparison with our calculated results. This extrapolation was done by fitting a curve for the three measurements. The curve was almost linear. Experimental results for thermal expansion [6, 9] are presented in Fig. 2 along with the calculated results. The experimental phonon shifts presented in Fig. 4 are from [4].

A quick comparison of the calculated and experimental results of all these quantities reveals very satisfactory agreement except for the Grüneisen constants, where the calculated results are systematically higher for all phonons. The measured Grüneisen constants reported here are at 100 K. In fact if one takes into account the temperature dependence of the Grüneisen constants one finds from recent measurements [10] that it is by no means small for all the modes. According to these results, γ_{qj} are likely to increase generally with temperature by even 20% in the temperature range between 0 and 300 K. Thus the calculated γ_{qj} tend to improve the agreement with measured γ_{qj} in the high-temperature region. Indeed if the same set of calculated γ_{qj} is used to obtain the implicitly measured shift in I, the results of this paper and that of I tend to coincide, which is not unexpected as the calculated and measured thermal expansion results already agree. The good agreement of the calculated quantities here with measured values supports the simple ideas of the used models for the molecular potential.

Similarly the results for the relative phonon frequency shifts are in quite good agreement with the observed shifts, as presented in Fig. 4.

5. Concluding Remarks

In this work, we have calculated the implicit shift to the phonon frequencies for deuterated naphthalene for all the phonons at Γ point. As the explicit shift had already been calculated for these phonons, this work therefore completes the calculation undertaken. The results for the total calculated shift presented here are in fairly close agreement with experimental results. Similarly, results of compressibility and thermal expansion of deuterated naphthalene which had to be calculated for obtaining the implicit shifts, when compared with the experimental results, are found to be in good agreement.

The calculation of the physical quantities for deuterated naphthalene presented here, has been made using the atom–atom interactions valid for all C–C, D–D, and C–D type interactions, where C and D stand for carbon and deuterium atom, respectively. Results obtained support very favourably this idea of the use of atom–atom potentials to form molecule–molecule potentials, It can be hoped that such potential would work for other molecular crystals, too.

Acknowledgements

A major part of this work was done while VKJ was also at Physikalisches Institut, Universität Bayreuth, FRG, on a position offered by Bundesministerium für Forschung und Technologie (BMFT). The authors gratefully acknowledge the support of BMFT for this work.

Appendix

Thermal expansion in higher order

For considering the higher-order contribution, we modify the expression for free energy (6) by replacing the harmonic phonon frequency ω_{qj} with the renormalized phonon frequency, $\tilde{\omega}_{qj} = \omega_{qj} + \Delta\omega_{qj}$; $\Delta\omega_{qj}$ being the explicit anharmonic shift. As a result we write for the free energy

$$F = \Phi(V) + \sum_{qj} F_{qj}\,, \tag{A1}$$

where

$$F_{qj} = kT \ln\,(2 \sinh \tilde{k}_{qj}) \tag{A2}$$

and

$$\tilde{k}_{qj} = \frac{\hbar \tilde{\omega}_{qj}}{2kT}. \tag{A3}$$

Hereinafter, we omit the subscript in $\tilde{k}_{qj}$. In order to obtain ε we have to minimize the free energy. For that we first obtain $\partial F_{qj}/\partial \varepsilon$, which is written as

$$\frac{\partial F_{qj}}{\partial \varepsilon} = \frac{\partial F_{qj}}{\partial \tilde{k}} \frac{\partial \tilde{k}}{\partial \varepsilon}. \tag{A4}$$

Expanding $\partial F_{qj}/\partial \tilde{k}$ and $\partial \tilde{k}/\partial \varepsilon$ around the unstrained values $\tilde{k}_0 = \tilde{k}_0(V_0)$, in powers of ε, we rewrite (A4) as

$$\frac{\partial F_{qj}}{\partial \varepsilon} \approx \left[\left(\frac{\partial F_{qj}}{\partial \tilde{k}}\right)_0 + \varepsilon \left(\frac{\partial \tilde{k}}{\partial \varepsilon}\right)_0 \left(\frac{\partial^2 F_{qj}}{\partial \tilde{k}^2}\right)_0\right] \left[\left(\frac{\partial \tilde{k}}{\partial \varepsilon}\right)_0 + \varepsilon \left(\frac{\partial^2 \tilde{k}}{\partial \varepsilon^2}\right)_0\right] \tag{A5}$$

or

$$\frac{\partial F_{qj}}{\partial \varepsilon} \approx \left(\frac{\partial \tilde{k}}{\partial \varepsilon}\right)_0 \left(\frac{\partial F_{qj}}{\partial \tilde{k}}\right)_0 + \varepsilon \left[\left(\frac{\partial \tilde{k}}{\partial \varepsilon}\right)_0^2 \left(\frac{\partial^2 F_{qj}}{\partial \tilde{k}^2}\right)_0 + \left(\frac{\partial^2 \tilde{k}}{\partial \varepsilon^2}\right)_0 \left(\frac{\partial F_{qj}}{\partial \tilde{k}}\right)_0\right], \tag{A6}$$

where terms up to order ε are retained. Various derivations of $\tilde{k}$ and F_{qj} in (A6) are easily obtained from (A2) and (A3) as

$$\left(\frac{\partial F_{qj}}{\partial \tilde{k}}\right)_0 = kT \coth \tilde{k}_0 , \tag{A7a}$$

$$\left(\frac{\partial^2 F_{qj}}{\partial \tilde{k}^2}\right)_0 = -kT \operatorname{cosech}^2 \tilde{k}_0 , \tag{A7b}$$

$$\left(\frac{\partial \tilde{k}}{\partial \varepsilon}\right)_0 = \tilde{\gamma} -_{qj} \tilde{k}_0; \qquad \tilde{\gamma}_{qj} = -\frac{\partial \ln \tilde{\omega}_{qj}}{\partial \ln V}, \tag{A7c}$$

and

$$\left(\frac{\partial^2 \tilde{k}}{\partial \varepsilon^2}\right)_0 = [\tilde{\gamma}_{qj}^2 + \tilde{\gamma}_{qj} - \tilde{\gamma}'_{qj} V_0]\, \tilde{k}_0 , \tag{A7d}$$

where the "renormalised Grüneisen constant" $\tilde{\gamma}_{qj}$ is written as

$$\tilde{\gamma}_{qj} = \gamma_{qj} + \frac{\Delta\omega}{\omega} \gamma_{qj}^{\mathrm{an}} , \tag{A8}$$

$$\gamma_{qj}^{\mathrm{an}} = -\frac{\partial \ln \Delta\omega_{qj}}{\partial \ln V}, \tag{A9}$$

and

$$\tilde{\gamma}'_{qj} = \frac{\partial \tilde{\gamma}_{qj}}{\partial V}. \tag{A10}$$

Similarly, expanding static lattice contribution $\Phi(V)$ as Taylor series expansion in ε up to how ε^3 gives

$$\frac{\partial \Phi}{\partial \varepsilon} \approx \varepsilon V_0 B + \frac{\varepsilon^2}{2} V_0^3 \left(\frac{\partial^3 \Phi}{\partial V^3}\right)_0 . \tag{A11}$$

Therefore, from the minimum of free energy (A1), using equations (A6) to (A8), (A10) and (A11) gives

$$\varepsilon = \varepsilon_0 \left[1 - \varepsilon_0 \frac{1}{2} \frac{V_0^3(\partial^3\Phi/\partial V^3)_0}{V_0 B} + \frac{kT}{V_0 B} \sum_{qj} \tilde{\gamma}_{qj}^2 [k_0^2 \operatorname{cosech}^2 k_0 - k_0 \coth k_0]\right], \tag{A12}$$

where $k_0 = \hbar\omega_{qj}/(2kT)$, ω_{qj} being the harmonic phonon frequency at $V = V_0$. In obtaining (A12), we have neglected $\tilde{\gamma}'_{qj}$ as the volume dependence of Grüneisen constants is very weak. Also ε_0 is the lowest-order thermal expansion.

References

[1] V. K. Jindal and J. Kalus, J. Phys. C 16, 3061 (1983).
[2] U. Schmelzer, E. L. Bokhenkov, B. Dorner, J. Kalus, G. A. Mackenzie, I. Natkaniec, G. S. Pawley, and E. F. Sheka, J. Phys. C 14, 1025 (1981).
[3] J. Kalus, B. Dorner, V. K. Jindal, N. Karl, I. Natkaniec, G. S. Pawley, W. Press, and E. F. Sheka, J. Phys. C 15, 6533 (1982).
[4] E. F. Sheka, E. L. Bokhenkov, B. Dorner, J. Kalus, G. A. Mackenzie, I. Natkaniec, G. S. Pawley, and U. Schmelzer, J. Phys. C 17, 5893 (1984).
[5] R. G. Della Valle, P. F. Fracassi, R. Righini, and S. Califano, Chem. Phys. 74, 179 (1983).
[6] J. F. J. Jordan, A. Axmann, H. Egger, and J. Kalus, phys. stat. sol. (a) 71, 457 (1982).
[7] A. A. Maradudin, P. A. Flinn, and R. A. Coldwell Horsfall, Ann. Phys. (U.S.A.) 15, 6533 (1961).
[8] R. A. Cowley, Rep. Progr. Phys. 31, 123 (1968).
[9] G. S. Pawley and E. Baharie, in: Ann. Rep., Annexe, ILL, Grenoble 1978 (p. 210).
[10] W. Häfner and W. Kiefer, private communication (1985).

(Received May 29, 1985)

phys. stat. sol. (b) **133**, 101 (1986)

Subject classification: 63.90; 65.50; S2; S9.11; S9.12; S10.1

Physics Department, University of Roorkee[1])

On the Pressure Dependence of Grüneisen Parameter in Solids

By

M. Kumari and N. Dass

A simple theory is developed to study the pressure dependence of Grüneisen parameter, γ. The final results show that γ varies linearly with the pressure, that is, $\gamma(p, T) = \gamma(0, T) - \eta p$ where η is found to be a pressure-independent but temperature-dependent parameter. Moreover, it is shown that η is related to some other thermodynamic parameters and hence can have positive or negative magnitude which in turn will decide whether $\gamma(p, T)$ decreases or increases with pressure. The calculated values of $\gamma(p, T)$ in the whole range of pressure and temperature are found in very good agreement with the experimental results in case of all the solids studied like NaCl, CaF_2, MgO, quartz, Na, K, and Li. Further, it is shown that the second Grüneisen parameter, q, and the Debye temperature, θ, can also be calculated as a function of pressure with the help of the present theory.

Eine einfache Theorie zur Untersuchung der Druckabhängigkeit des Grüneisenparameters γ wird entwickelt. Das Endresultat zeigt, daß γ linear mit dem Druck variiert, d. h. $\gamma(p, T) = \gamma(0, T) - \eta p$, wobei η ein druckunabhängiger jedoch temperaturabhängiger Parameter ist. Darüberhinaus wird gezeigt, daß η mit einigen anderen thermodynamischen Parametern verknüpft ist und positive oder negative Größe haben kann, was umgekehrt entscheidet, ob $\gamma(p, T)$ abnimmt oder zunimmt mit dem Druck. Die berechneten Werte von $\gamma(p, T)$ im gesamten Druck- und Temperaturbereich sind in guter Übereinstimmung mit den experimentellen Ergebnissen im Falle aller untersuchten Festkörper: NaCl, CaF_2, MgO, Quarz, Na, K und Li. Darüberhinaus wird gezeigt, daß der zweite Grüneisenparameter q und die Debye-Temperatur θ auch als Funktion des Druckes mit Hilfe der vorgestellten Theorie berechnet werden können.

1. Introduction

On surveying the literature, it is found out that the Grüneisen parameter γ decreases in case of solids [1, 2] and increases in case of liquids [3] with the increase in pressure. Therefore, at present, there is no theory which can tell why $(\partial\gamma/\partial T)_p$ should be negative in solids and positive in case of liquids. Furthermore, the problem becomes somewhat complicated when it is found experimentally that alkali metal liquids like Li, Na, and K behave like that of a solid [4] as far as the behaviour of γ with pressure is concerned. Moreover, accurate values of γ as a function of pressure are required in obtaining (i) isothermal data from shock wave experiments, (ii) the second Grüneisen parameter, q, and (iii) the pressure dependence of the Debye temperature θ which is needed for studying the recoilless fraction of γ-ray emission or absorption in Mössbauer effect, particularly to account for the effect of an impurity in the host materials [25, 26]. Thus, it is clear that the study of the pressure dependence of γ is an interesting problem from the theoretical as well as from the practical point of view, particularly, due to the lack of a proper theory in this regard.

---

[1]) Roorkee 247667, India.

Now, the pressure dependence of γ can be obtained by various methods. The simplest method, to get the pressure dependence of γ, may be with the help of thermodynamic formulation like $\gamma = \alpha V \beta_T / C_v$ where accurate values of the thermal expansion α, heat capacity C_v and isothermal bulk modulus B_T, are required as functions of pressure. But this method is quite unable to give the physical reasoning for the decrease or the increase of γ with pressure.

The second method to find γ versus pressure can be with the help of the lattice theory based on different models [5, 6]. But, the most unfortunate situation is that the results obtained from these methods may be in sharp disagreement with each other. For example, Ho and Ruoff [5] have found an increase in γ with pressure whereas Boehler et al. [2] have found the reverse trend in case of CaF_2. Hence, a theory for studying the pressure dependence of γ becomes more important under these conditions, however simple the theory may be.

Therefore, the purpose of the present paper is to develop a simple theory, based on thermodynamic principles, to study $\gamma(p, T)$ as a function of pressure and that at different temperatures, too. The final result obtained indicates that $\gamma(p, T)$ varies linearly with pressure, at least to a good approximation, that is,

$$\gamma(p, T) = \gamma(0, T) - \eta p \,, \tag{1}$$

where $\gamma(0, T)$ is the value of the Grüneisen parameter at some reference pressure, taken as zero here, and temperature T. η is a pressure-independent but temperature-dependent parameter and is shown to be related with some other thermodynamic parameters.

2. Theory

2.1 Pressure dependence of Grüneisen parameter

The theory which we are going to develop in this paper is based on two basic assumptions. These are

(i) the product of the volume coefficient of thermal expansion α and the isothermal bulk modulus B_T is a pressure-independent quantity for a given substance, that is

$$\alpha(p, T)\, B_T(p, T) = \alpha(0, T)\, B_T(0, T) = \xi; \tag{2}$$

(ii) the isothermal as well as the adiabatic bulk modulus varies linearly with pressure, that is

$$B_T(p, T) = B_T(0, T) + \beta_T p \tag{3}$$

and

$$B_S(p, T) = B_S(0, T) + \beta_S p \,. \tag{4}$$

The first assumption represented by (2) has been shown to be valid by Dass and Kumari [9, 10], Boehler and Kennedy [8], and Yagi [7]. Further, it may be mentioned that Dass and Kasir [11] and Dass and Mustafa [12] have used this assumption successfully in representing the temperature dependence of isothermal compressibilities K_T in terms of the pressure dependence of K_T in alkali halides. Furthermore, Dass and Kumari [9] have found that this assumption gives rise to Murnaghan's equation of state which was originally obtained on the basis of finite strain theory and is extensively used in the literature for studying the compression of solids and liquids.

As far as the validity of the second assumption in the form of (3) is concerned, Anderson [13] has applied (3) successfully to estimate the compression at high pressures from the ultrasonic measurements done under moderate pressures. The most spectacular success of (3) was in case of MgO as shown by Anderson [14]. Thus, in conclusion, it may be said that (3) holds throughout the pressures found in the earth's deep interiors as argued by Ramsay [15]. After establishing the validity of (3), we are justified to extend this idea in the case of adiabatic bulk modulus as given in (4).

Now, to establish that $\gamma(p, T)$ varies linearly with pressure as given in (1), we subtract (3) from (4). The result is

$$B_S(p, T) - B_T(p, T) = [B_S(0, T) - B_T(0, T)] - (\beta_T - \beta_S)\, p\,. \tag{5}$$

However, the thermodynamic relation between $B_S(p, T)$ and $B_T(p, T)$ can be written with the help of (2) as

$$\begin{aligned} B_S(p, T) - B_T(p, T) &= T\gamma(p, T)\,\alpha(p, T)\, B_T(p, T) \\ &= T\xi\gamma(p, T)\,. \end{aligned} \tag{6}$$

Combination of (5) and (6) gives (1) with

$$\eta = (\beta_T - \beta_S)/(T\xi)\,. \tag{7}$$

Thus, it is clear from (7) that η is a pressure-independent but temperature-dependent parameter because β_T, β_S, and ξ are pressure-independent but temperature-dependent quantities. Furthermore, β_T, β_S, and ξ are positive quantities. Hence, η will depend eventually on the condition (a) $\beta_T > \beta_S$ or (b) $\beta_T < \beta_S$.

In case $\beta_T > \beta_S$, η will have a positive magnitude and hence $\gamma(p, T)$ will decrease with pressure. This result is in accordance with the experimental results in case of solids [1, 2]. Moreover, the condition that $\beta_T > \beta_S$ is also in accordance with the calculations of Overtone [16] in solids. Thus, it is clear that the decrease of $\gamma(p, T)$ with pressure is connected with the condition that $\beta_T > \beta_S$.

In case $\beta_T < \beta_S$, η will be negative as other quantities involved are positive and hence $\gamma(p, T)$ will increase with pressure. This result is in accordance with the experimental results of H_2O and Hg [3]. The results of Davis and Gordon [27] on the compression of Hg show that $\beta_T < \beta_S$ and hence $\gamma(p, T)$ must increase with pressure and in reality an increase has been found experimentally by Boehler [3], but the two results are in better agreement at low pressures than at high pressures. This is the reason that the results for Hg are not given here. Moreover, such a condition is also found to exist in case of CaF_2 which will be discussed later on. Therefore, for the first time, this simple theory is capable of predicting the increase of $\gamma(p, T)$ with pressure which was not available so far and the condition under which it can happen is $\beta_T < \beta_S$.

2.2 Second Grüneisen parameter

Equation (1) is capable of giving the value of the second Grüneisen parameter q, as a function of pressure. To obtain q, differentiation of (1) with respect to pressure gives

$$\left(\frac{\partial\gamma}{\partial p}\right)_T = -\eta\,. \tag{8}$$

Equation (8) gives easily the desired relation for q as

$$q = \left[\frac{\partial \ln \gamma}{\partial \ln V}\right]_T = \frac{B_T(p, T)}{\gamma(p, T)}\,\eta\,. \tag{9}$$

Thus, it is clear from (9) that q will increase with pressure, as $B_T(p, T)$ increases but $\gamma(p, T)$ decreases with pressure, for those substances for which η has a positive magnitude, i.e. $\beta_T > \beta_S$. The reverse trend will follow for those substances for which η is negative, i.e. $\beta_T < \beta_S$.

2.3 Pressure dependence of Debye temperature

It will be interesting to obtain a relation for the Debye temperature as a function of pressure with the help of the present theory. This is outlined below:

The Grüneisen parameter can also be defined in terms of Debye temperature as

$$\gamma = -\frac{\partial \ln \theta}{\partial \ln V}. \tag{10}$$

Equation (10) can be written alternatively as

$$\gamma(p, T) = \frac{B_T(p, T)}{\theta(p, T)} \left[\frac{\partial \theta(p, T)}{\partial p}\right]_T. \tag{11}$$

Equation (11) can be integrated with the help of (1) and (3) and the result so obtained is

$$\ln\left[\frac{\theta(p, T)}{\theta(0, T)}\right] = -\frac{\eta p}{\beta_T} + \delta \ln\left[1 + \frac{\beta_T p}{B_T(0, T)}\right], \tag{12}$$

where

$$\delta = \frac{1}{\beta_T}\left[\gamma(0, T) + \frac{B_T(0, T)}{\beta_T}\right] \tag{13}$$

is a constant for a given substance.

However, (12) can be simplified on the assumption that the pressure is low enough and hence the term $\eta p/\beta_T$ and $\eta B_T(0, T)/\beta_T^2$ can be neglected in comparison with other terms. Thus, the simplified result is

$$\theta(p, T) = \left[1 + \frac{\gamma(0, T)}{B_T(0, T)} p\right] \theta(0, T). \tag{14}$$

Equation (14) is generally used whenever there is a need to obtain the values of θ as a function of pressure, for example, in calculating the recoilless fraction of γ-ray emission or absorption in Mössbauer effect [26].

It will be shown later that (12) and (14) give the same results up to a pressure of 10 kbar. However, within an error of 5%, (14) can be used instead of (12) up to 25 kbar. But beyond this pressure, the error involved is quite high and hence no physical justification can be given for the use of (14) instead of (12).

3. Calculations and Discussion

Calculations for getting $\gamma(p, T)$ as a function of pressure have been done in many solids to see the validity of (1) combined with (7). The calculations for q and $\theta(p, T)/\theta(0, T)$ are also done wherever possible.

3.1 Sodium chloride

The values of β_T and β_S at 300, 550, and 800 K in the pressure range 0 to 10 kbar are obtained by the least-square fit from (3) and (4), respectively, using the experimental data for $B_T(p, T)$ and $B_S(p, T)$ from Spetzler et al. [17]. The values of β_T and β_S so obtained are reported in Table 1 along with the values of $\alpha(0, T)$ and $B_T(0, T)$. For all the data being available, the values of η are calculated and are reported in Table 1.

With the help of η values, the values of $\gamma(p, T)$ at different temperatures are calculated as function of pressure with the help of (1). The calculated values of $\gamma(p, T)$

Table 1

Input data used in calculations

substance	T (K)	$\alpha(0, T)$ (10^{-5} K^{-1})	$B_T(0, T)$ (kbar)	β_T	β_S	$\gamma(0, T)$	η (10^{-3}/kbar)
MgO	298	3.15 [13]	1622.0 [13]	4.54 [13]	4.50 [13]	1.387 [1]	2.627
Li	298	15.8 [4]	119.8 [20]	3.55 [20]	3.51 [20]	0.98 [4]	7.336
Na	298	20.8 [4]	64.0 [22]	3.62 [21]	3.60 [13]	1.07	5.041
K	298	24.6 [4]	33.8 [13]	3.98 [13]	3.946 [13]	1.08	13.693
quartz	298	3.56 [13]	374.2 [13]	6.34 [13]	6.33 [13]	0.675 [2]	2.565
NaCl [17]	300	11.752	238.41	4.999	4.896	1.6275	12.25
	550	14.252	197.86	5.362	5.186	1.6851	11.35
	800	17.190	157.39	5.728	5.509	1.7591	10.12
CaF_2	194.15	4.584 [23]	851.8 [5]	4.693 [5]	4.563 [5]	1.736	−17.12
	273.15	5.490 [23]	830.9 [5]	5.832 [5]	5.570 [5]	1.723	−21.03
	295.5	5.625 [23]	824.8 [5]	6.172 [5]	5.861 [5]	1.679	−22.68

are compared with the experimental results in Table 2. The overall agreement is very good in the whole range of temperature and pressure as the maximum error is not more than 2% in any case.

Table 2

Calculated values of γ and q as functions of pressure and temperature in sodium chloride

p (kbar)	300 K			550 K			800 K		
	γ [17] (exp.)	γ (calc.)	q	γ [17] (exp.)	γ (calc.)	q	γ [17] (exp.)	γ (calc.)	q
0	1.6275	1.6275	1.794	1.6851	1.6851	1.332	1.7591	1.7591	0.905
2	1.6095	1.6030	1.898	1.6661	1.6624	1.424	1.7363	1.7389	0.983
4	1.5915	1.5785	2.005	1.6470	1.6397	1.518	1.7143	1.7186	1.062
6	1.5732	1.5540	2.116	1.6277	1.6170	1.615	1.6923	1.6984	1.143
8	1.5545	1.5295	2.230	1.6077	1.5943	1.714	1.6698	1.6781	1.225
10	1.5347	1.5050	2.347	1.5867	1.5716	1.816	1.6464	1.6579	1.310

Further, the calculated values of q are also reported in Table 2. Unfortunately, no experimental data of q are available as a function of pressure and temperature for comparison in this case. However, $q = 1.44$ has been reported by Roberts and Ruppin [6] at 298 K and atmospheric pressure, which is much less than the value obtained in the present calculation at 300 K and zero pressure.

The calculations for $\theta(p, T)/\theta(0, T)$ are also done in case of sodium chloride in the pressure range 0 to 100 kbar and in the temperature range 300 to 800 K as all the parameters needed for computation are available from Table 1. The results so obtained are reported in Table 3. The ratio $\theta(p, T)/\theta(0, T)$ is found to increase with pressure as well as with temperature. As no experimental data are available for comparison, therefore, the calculations are done using (12) and (14). The results of (12) and (14) are the same up to 10 kbar, within 5% error up to 25 kbar, and beyond this pressure there is no physical justification to use (14).

Table 3

$\theta(p, T)/\theta(0, T)$ as a function of pressure and temperature in NaCl

p (kbar)	300 K		550 K		800 K	
	from (12)	from (14)	from (12)	from (14)	from (12)	from (14)
2	1.013	1.013	1.017	1.017	1.022	1.022
4	1.026	1.027	1.032	1.034	1.042	1.045
6	1.038	1.041	1.048	1.051	1.061	1.067
8	1.050	1.054	1.062	1.068	1.074	1.089
10	1.061	1.068	1.076	1.085	1.097	1.112
15	1.088	1.102	1.107	1.128	1.137	1.167
20	1.112	1.136	1.136	1.170	1.172	1.223
25	1.133	1.171	1.162	1.213	1.204	1.279
30	1.153	1.204	1.185	1.255	1.233	1.335
50	1.215	1.341	1.259	1.426	1.323	1.559
75	1.264	1.512	1.318	1.639	1.399	1.838
100	1.290	1.683	1.353	1.852	1.446	2.118

Further, to test (4) still to high pressure, the calculations of $B_S(p, T)$ are done in the pressure range 2 to 32 kbar using $B_S(0, T) = 252.09$ kbar at 300 K [17]. The present calculated values of $B_S(p, T)$ are compared in Table 4 with the values of Hardy and Karo [18] as reported by Boehler et al. [19]. The agreement is excellent. However, it may be mentioned here that Hardy and Karo obtained the values of $B_S(p, T)$ as a function of pressure on the basis of a lattice dynamic model.

Thus, it is clear that (1) is quite successful in representing the pressure dependence of $\gamma(p, T)$ and that at different temperatures, too. Moreover, it is also clear that (4) can be used for high pressures like (3).

3.2 Fluorite

Fluorite appears to be a controversial case in the sense that Ho and Ruoff [5] had found that $\gamma(p, T)$ increases with pressure whereas Boehler et al. [2] found the reverse trend. Therefore, the real type of behaviour of $\gamma(p, T)$ with pressure is not known and hence there is a scope for experimental work in this case. However, the theory developed here favours the results of Ho and Ruoff.

Table 4

Values of B_S (kbar) as functions of pressure in sodium chloride

p (kbar)	B_S [19]	B_S (calc.)	p (kbar)	B_S [19]	B_S (calc.)	p (kbar)	B_S [19]	B_S (calc.)
2.05	260.00	262.12	9.78	299.5	299.97	22.06	361.5	360.10
2.80	264.00	265.00	11.09	306.5	306.39	23.46	368.5	366.95
3.46	267.0	269.03	12.49	313.5	313.24	24.90	375.5	374.00
4.19	270.5	272.60	13.90	323.0	320.14	26.37	383.0	381.02
4.92	274.5	276.18	15.35	328.0	327.24	27.76	390.0	388.00
5.63	278.0	279.65	16.77	335.0	334.20	29.18	397.5	394.95
6.29	282.0	282.88	18.00	341.5	340.22	30.45	402.5	401.17
7.01	285.5	286.41	19.33	348.0	346.73	31.89	409.5	408.22
8.35	292.0	292.97	20.76	355.0	353.73			

The values of β_T and β_S at 194.15, 273.15, and 295.5 K are obtained by the least square fit using the experimental values of $B_T(p, T)$ and $B_S(p, T)$ from Ho and Ruoff [5]. The values of β_T and β_S so obtained are reported in Table 1 along with the values of $\alpha(0, T)$, $B_T(0, T)$, and η. It is clear from Table 1 that $\beta_T < \beta_S$ in the whole temperature range and hence η will have a negative magnitude indicating that $\gamma(p, T)$ will increase with pressure as found by Ho and Ruoff. Therefore, the calculated values of $\gamma(p, T)$ from (1) are compared in Table 5 with the values of Ho and Ruoff. The overall agreement is very good in the whole range of pressure and temperature although our results are slightly higher, but the error is not more than 2.5% in any case.

Table 5

Pressure and temperature dependence of γ in CaF_2

T = 194.15 K			273.15 K			295.5 K		
p (kbar)	γ [5]	γ (calc.)	p (kbar)	γ [5]	γ (calc.)	p (kbar)	γ [5]	γ (calc.)
0.001	1.736	1.736	0.001	1.723	1.723	0.001	1.679	1.679
0.241		1.740	0.265		1.728	0.264		1.685
0.574		1.746	0.578		1.735	0.582		1.692
0.924	1.745	1.752	0.911	1.733	1.742	0.950	1.689	1.700
1.273		1.758	1.281		1.750	1.269		1.708
1.615		1.764	1.741		1.760	1.669	1.697	1.717
2.129	1.757	1.772	2.026	1.745	1.766	2.126		1.727
2.677		1.782	2.520		1.776	2.629	1.708	1.739
3.189	1.768	1.790	3.030	1.756	1.787	3.139	1.713	1.750
3.732		1.800	3.505		1.797			
4.164	1.778	1.807	4.028	1.768	1.808			

3.3 Magnesium oxide and quartz

All the data needed for the calculations of $\gamma(p, T)$ are available as reported in Table 1. Thus, taking the value of η from Table 1, the calculations of $\gamma(p, T)$ are done in both cases and the results so obtained are compared with the experimental data in Table 6. The agreement found is very good in both solids in the pressure range 0 to 30 kbar.

Table 6

Pressure dependence of Grüneisen parameter in alkali metals, MgO, and quartz

p (kbar)	Li		Na		K		MgO		quartz	
	γ [4] (exp.)	γ (calc.)	γ [4] (exp.)	γ (calc.)	γ [4] (exp.)	γ (calc.)	γ [1] (exp.)	γ (calc.)	γ [2] (exp.)	γ (calc.)
0	0.98	0.98	1.21	1.07	1.27	1.08	1.387	1.387	0.675	0.675
5	0.90	0.94	1.05	1.05	1.02	1.02	—	1.374	0.662	0.662
10	0.85	0.90	1.00	1.02	0.92	0.95	1.364	1.361	0.648	0.649
15	0.84	0.86	0.98	0.99	0.86	0.88	—	1.347	0.632	0.636
20	0.82	0.83	0.96	0.97	0.81	0.81	1.328	1.334	0.617	0.624
25	0.80	0.79	0.94	0.94	0.76	0.74	—	1.321	0.603	0.610
30	0.76	0.75	0.94	0.92	0.73	0.67	1.307	1.308	0.594	0.598
40							1.273	1.282	—	0.572
50							1.235	1.256	—	0.547

Table 7

Pressure and temperature dependence of Grüneisen parameter γ in $Mg_2Si_2O_4$, SiO_2, quartzite, and MgO

p (kbar)	$Mg_2Si_2O_4$						SiO_2		MgO				p (kbar)	quartzite	
	298 K		673 K		973 K		298 K		673 K		973 K			298 K	
	γ [1]	γ (calc.)	γ [1]	γ (calc.)	γ [1]	γ (calc.)	γ [1]	γ (calc.)	γ [1]	γ (calc.)	γ [1]	γ (calc.)		γ [2]	γ (calc.)
0	1.154	1.154	1.187	1.187	1.206	1.206	0.567	0.567	1.386	1.386	1.384	1.384	0	0.651	0.651
10	1.140	1.131	1.163	1.153	1.197	1.187	0.568	0.560	1.374	1.367	1.374	1.367	5	0.645	0.642
20	1.120	1.107	1.132	1.119	1.180	1.168	0.566	0.554	1.356	1.348	1.360	1.352	10	0.638	0.632
30	1.091	1.084	1.095	1.085	1.160	1.150	0.552	0.547	1.336	1.329	1.344	1.336	15	0.627	0.623
40	1.061	1.061	1.051	1.051	1.132	1.131	0.541	0.541	1.310	1.310	1.320	1.320	20	0.615	0.614
50	1.027	1.038	1.003	1.017	1.099	1.112	0.526	0.534	1.283	1.291	1.294	1.304	25	0.603	0.604
													30	0.592	0.595
													35	0.585	0.586
η (10^{-3}/kbar)	2.324		3.407		2.691		6.454		1.893		1.607			1.863	

3.4 Alkali metals

The alkali metals to be considered here are Li, Na, and K. In these solids, too, all the data required for getting the value of η are available as reported in Table 1. The calculated values of $\gamma(p, T)$ are compared with the experimental results of Boehler [4] in Table 6. The agreement is very good in the whole pressure range except the value of $\gamma(0, T)$.

3.5 Other solids

Once it is established that $\gamma(p, T)$ varies linearly with pressure, then (1) can be used in those cases in which the values of $B_T(p, T)$, and $B_S(p, T)$ as functions of pressure are not available, but on the other hand, the experimental values of $\gamma(p, T)$ as functions of pressure are available. This is what exactly has been done in cases like SiO_2 (298 K), Mg_2SiO_4 (298, 873, and 973 K) quartzite (298 K), and MgO (673 and 973 K). The results so obtained are compared with the experimental results in Table 7 along with the values of η. The agreement is excellent in all cases in the whole range of pressure and temperature.

4. Conclusions

From Tables 2 to 7, the following conclusions are obvious:

(i) The present simple theory clearly exhibits that the Grüneisen parameter varies linearly with pressure obeying (1). Hence, the calculations done for $\gamma(p, T)$ as a function of pressure in all the cases are in very good agreement with the available experimental results without any exception.

(ii) For the first time, the present theory explains the reason why $\gamma(p, T)$ should decrease or increase with pressure in a given system. If $\beta_T > \beta_S$, $\gamma(p, T)$ will decrease with pressure, whereas it will increase if $\beta_T < \beta_S$. Thus, the accurate values of β_T and β_S will decide about the nature of variation of $\gamma(p, T)$ with pressure.

(iii) To avoid the controversy in case of CaF_2, more experimental work is needed to be done to arrive at one definite conclusion.

(iv) q increases with pressure and decreases with temperature as shown in case of NaCl.

(v) The ratio of $\theta(p, T)/\theta(0, T)$ increases with pressure as well as with temperature as shown in case of NaCl. Further, use of (14) beyond 25 kbar pressure cannot be justified on physical grounds.

(vi) From Table 4, it is clear that (4) can also be used to very high pressures like (3) for studying the pressure dependence of $B_S(p, T)$.

(vii) The present theory exhibits that once the accurate values of β_T, β_S, $\alpha(0, T)$, and $B_T(0, T)$ become available one can easily compute the pressure dependence of $B_T(p, T)$, $B_S(p, T)$, $\gamma(p, T)$, q, and $\theta(p, T)$.

Acknowledgements

The authors are thankful to the Dean of Research and Industrial Liasion, University of Roorkee, for providing financial support for this work. One of us (MK) is also thankful to the University Grants Commission, New Delhi, India, and Physics Department, University of Roorkee, for the award of a research fellowship.

References

[1] R. BOEHLER, J. geophys. Res. 87, 5501 (1982).
[2] R. BOEHLER, A. SKOPROPONOV, D. O. MORE, and G. C. KENNEDY, J. geophys. Res. 84, 3527 (1979).
[3] R. BOEHLER and G. C. KENNEDY, J. appl. Phys. 48, 4183 (1977).
[4] R. BOEHLER, Phys. Rev. B 27, 6754 (1983).
[5] P. S. HO and A. L. RUOFF, Phys. Rev. 161, 864 (1967).
[6] R. W. ROBERTS and R. RUPPIN, Phys. Rev. B 4, 204 (1971).
[7] T. YAGI, J. Phys. Chem. Solids 39, 563 (1978).
[8] R. BOEHLER and G. C. KENNEDY, J. Phys. Chem. Solids 41, 1019 (1980).
[9] N. DASS and M. KUMARI, phys. stat. sol. (b) 127, 103 (1985).
[10] N. DASS and M. KUMARI, phys. stat. sol. (b) 124, 531 (1984).
[11] N. DASS and F. A. KASIR, phys. stat. sol. (b) 106, K99 (1981).
[12] N. DASS and B. M. MUSTAFA, phys. stat. sol. (b) 119, K27 (1983).
[13] O. L. ANDERSON, J. Phys. Chem. Solids 27, 547 (1966).
[14] O. L. ANDERSON, Bull. Seismol. Soc. Amer. 56(3), 725 (1966).
[15] W. H. RAMSAY, Monthly Notices Roy. Astron. Soc., Geophys. Suppl. 6, 42 (1950).
[16] W. C. OVERTONE, J. chem. Phys. 37, 116 (1962).
[17] H. SPETZLER, C. G. SAMMIS, and R. J. O. CONNELL, J. Phys. Chem. Solids 33, 1727 (1973).
[18] R. J. HARDY and A. M. KARO, J. Phys. Chem. Solids 38, 335 (1977).
[19] R. BOEHLER, I. C. GETTING, and G. C. KENNEDY, J. Phys. Chem. Solids 38, 233 (1977).
[20] R. A. EELICE and J. TRIVISONNE, Phys. Rev. B 16, 5773 (1977).
[21] KEH-JIM DUNN, Phys. Rev. B 12, 3497 (1975).
[22] S. N. VAIDYA, I. C. GETTING, and G. C. KENNEDY, J. Phys. Chem. Solids 32, 2545 (1971).
[23] D. N. BATCHELDER and R. O. SIMMONS, J. chem. Phys. 41, 2324 (1968).
[24] R. BOEHLER, J. geophys. Res. 86, 7159 (1981).
[25] W. M. VISSCHER, Phys. Rev. 129, 28 (1963).
[26] S. S. NANDWANI and S. P. PURI, phys. stat. sol. 41, 199 (1970).
[27] L. A. DAVIS and R. B. GORDON, J. chem. Phys. 46, 2650 (1967).

(Received June 3, 1985)

phys. stat. sol. (b) **133**, 111 (1986)

Subject classification: 61.80; S1.3

M. I. Kalinin Leningrad Polytechnical Institute[1])

Defect Clusters and Subcascades

II. The Splitting Threshold and Subcascade Dimensions[2])

By

A. I. MELKER, S. N. ROMANOV, and N. L. TARASENKO

Energy thresholds of splitting displacement cascades into subcascades are calculated for some materials on the basis of the theory developed earlier. The group representation of splitting is constructed and the density of subcascade size distribution is obtained. A simple method for analytical estimation of temperature influence on splitting is suggested.

Energieschwellen der Aufspaltung von Verlagerungskaskaden in Subkaskaden werden für einige Materialien auf der Grundlage der früher entwickelten Theorie berechnet. Die Gruppendarstellung der Aufspaltung wird konstruiert und die Dichte der Subkaskaden-Größenverteilung wird erhalten. Es wird eine einfache Methode für eine analytische Berechnung des Temperatureinflusses auf die Aufspaltung vorgeschlagen.

1. Introduction

In the previous paper [1] the theory of displacement cascade splitting was developed and the formulas for calculating the mean number of subcascades produced by PKA were obtained. However, to compare the theory with experiment it is necessary to know the threshold of cascade splitting and some other parameters. In this paper, methods of calculating these quantities are considered and a comparison with experiment is given.

2. Splitting Threshold

From mechanics it is known that the distance between colliding particles at the moment of their nearest approach corresponding to a head-on collision $r = b$ can be obtained from the condition, which expresses the equality of the repulsion potential energy and the kinetic energy of relative motion,

$$\Phi(r)|_{r=b} = \frac{m_2 E}{m_1 + m_2} = \frac{E}{2}\bigg|_{m_1=m_2} . \tag{1}$$

Here m_1 and m_2 are the mass of the moving and resting particle, respectively, E the kinetic energy of the moving particle, and b the collision diameter.

Consider a collision of an arriving atom A, which moves along a close-packed direction, with atoms of a solid [2]. Put the origin of the coordinate system at the arriving atom. Then

$$\sum_{i=1}^{N} \Phi(r_{Ai}) = \Phi(b) + \Phi(r_0 - b) + \sum_{i=3}^{N} \Phi(r_i > r_0 - b) , \tag{2}$$

[1]) Ul. Polytekhnicheskaya 29, 195251 Leningrad, USSR.

[2]) Part I see phys. stat. sol. (b) **132**, 425 (1985).

where N is the number of atoms interacting with atom A, r_0 the interatomic distance of the crystal lattice. For the LNS (Lindhard-Nielsen-Scharff) potential [3]

$$\Phi(r) = \frac{Z_1 Z_2 e^2}{r} k_s s \left(\frac{a}{r}\right)^{s-1}, \tag{3}$$

where Z_1, Z_2 are the charge numbers, e the electron charge, a the screening constant, k_s and s are constants, we have in the case of a single constituent $Z_1 = Z_2 = Z$, $m_1 = m_2 = m$, and the following relation between E and b:

$$E = 4k_s s Z^2 E_R \frac{a_0}{a} \left(\frac{a}{b}\right)^s, \tag{4}$$

where E_R is the Rydberg energy, a_0 the radius of the first electron orbit in a hydrogen atom.

If $b \ll r_0$, the terms with the argument $r_i > r_0 - b$ almost do not influence the collision result, because the corresponding forces are near to zero. Denote

$$\frac{\Phi(r_0)}{\Phi(b)} = \delta \tag{5}$$

and take for the criterion of cascade splitting the following condition: if δ becomes less than a small given quantity, there occurs a transition to collisions with large-angle scattering and the cascade splits. It follows from (3) and (5) that

$$\delta = \left(\frac{b}{r_0}\right)^s. \tag{6}$$

Table 1

Threshold energy of splitting of cascades into subcascades

element	Z	r (nm)	E (keV)	
			$\delta = 0.1$	$\delta = 0.05$
Al	13	0.286	0.9	1.8
Si	14	0.235	2.3	4.7
Ti	22	0.289	3.3	6.6
V	23	0.262	4.3	8.6
Cr	24	0.250	5.1	10.1
Fe	26	0.248	5.9	11.8
Ni	28	0.249	6.6	13.2
Cu	29	0.256	6.6	13.2
Ga	31	0.244	8.1	16.3
Ge	32	0.245	8.5	17.0
As	33	0.316	5.4	10.8
Zr	40	0.317	7.4	14.8
Nb	41	0.286	9.4	18.9
Mo	42	0.272	10.9	21.8
Ag	47	0.289	11.6	23.2
Au	79	0.288	27.8	55.6
Pb	82	0.350	20.0	40.0
U	92	0.275	39.3	78.6

In the PKA energy interval considered in [1] $s = 2$, therefore using (4) and (6), one can write the kinetic energy of the arriving atom, above which the cascade splits, as

$$E_s = \frac{6.65}{\delta} Z^2 E_R \frac{a_0}{a} \left(\frac{a}{r_0}\right)^2 . \tag{7}$$

In Table 1 the splitting energy for some elements calculated at $\delta = 5$ and 10% is given. Except for a few metals almost nothing is known about the experimental value of this energy. For gold the splitting energy is equal to ≈ 50 keV [4]. This value is in good agreement with the theoretical one calculated at $\delta = 5\%$. This tolerance also permits to obtain the coincidence of the theoretical splitting energy for iron with that deduced from computer experiments [5], the latter being equal approximately to 10 keV. Probably, the 5% tolerance is optimal for experimental and theoretical values of the splitting energy to agree.

3. Subcascade Dimensions

In general, the subcascade size depends on the energy of the arriving atom, its free path during collisions with large-angle scattering, and the number of subcascades in a cascade. Strictly speaking the subcascade size distribution must be obtained as a folding of the distributions of these quantities. However, such an approach leads to big mathematical difficulties, so we consider a simpler way to estimate the size distribution.

Let us denote by β the number of high-energy branches in a cascade. Analogous to [6] the energy of the arriving atom A will be measured in β-units. If $\beta = 1$, the arriving atom produces one cascade, but if $\beta = 3$, the created cascade splits into three subcascades. Fig. 1 shows the graphs whose arcs are the high-energy branches of the cascade generated by arriving atom A with an energy β. These graphs define all the possible ways of splitting the cascade into a given number of subcascades, the graphs being a particular case of the diagrams suggested to describe the formation of athermal vacancy clusters [6]. (In terms of [6], these diagrams, except the first two, correspond to the subgroup of branchings, L).

Let S_k denote the set of all the processes leading to splitting of a cascade into k subcascades. Using the diagrams shown in Fig. 1 we can construct analogous to [6] the group representation of these processes. In the compact system of denotations introduced in [6] the group representations have the form

$$\begin{aligned} S_3 &= S_1 \cdot S_2 = L_3 , \\ S_5 &= S_3 \cdot S_2 = L_3 \cdot S_2 = 2L_5 , \\ S_7 &= S_5 \cdot S_2 = 2L_5 \cdot S_2 = 5L_7 , \\ S_9 &= S_7 \cdot S_2 = 5L_7 \cdot S_2 = 14L_9 , \\ S_{11} &= S_9 \cdot S_2 = 14L_9 \cdot S_2 = 45L_{11} , \end{aligned} \tag{8}$$

where

$$\begin{aligned} L_5(A1, A2) &= L_3(A1)\, S_2(A4, 25) , \\ L_5(A1, 13) &= L_3(A1)\, S_2(14, 35) , \end{aligned} \tag{9}$$

$$\begin{aligned} L_7(A1, A2, A4) &= L_5(A1, A2)\, S_2(A6, 47) , \\ L_7(A1, A2, 25) &= L_5(A1, A2)\, S_2(26, 57) , \\ L_7(A1, A2, 13) &= L_5(A1, A2)\, S_2(16, 37) , \\ L_7(A1, 13, 35) &= L_5(A1, 13)\, S_2(36, 57) , \\ L_7(A1, 13, 14) &= L_5(A1, 13)\, S_2(16, 47) , \end{aligned} \tag{10}$$

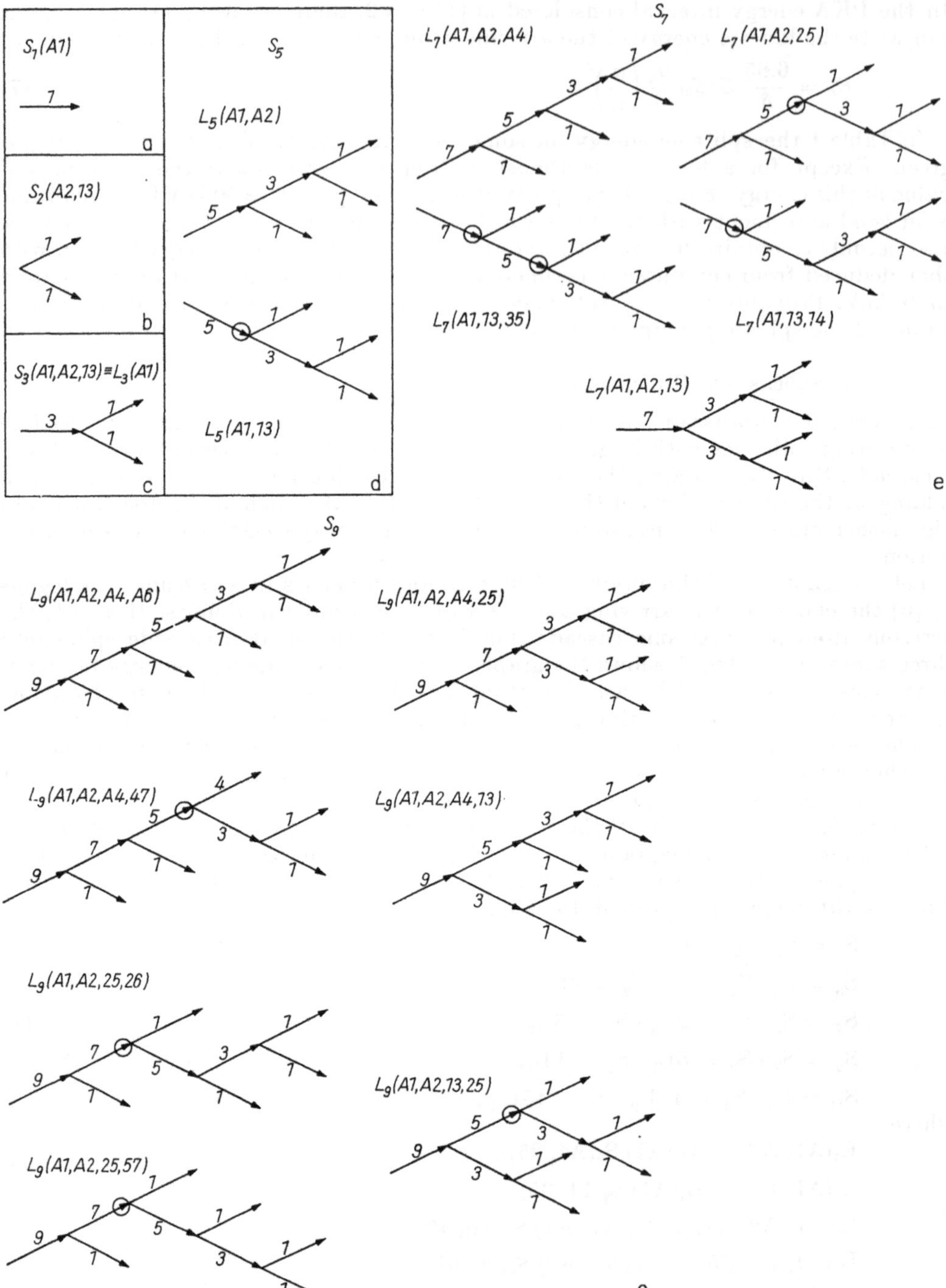

Fig. 1. Formation of high-energy branches during splitting of a cascade; a) S_1, b) S_2, c) S_3, d) S_5, e) S_7, f) S_9; for S_9 only half the graphs is given, the other half is symmetrical with respect to the horizontal axis; the numbers denote the energies of arriving atoms in β-units

$$
\begin{aligned}
L_9(A1, A2, A4, A6) &= L_7(A1, A2, A4)\, S_2(A8, 69)\,, \\
L_9(A1, A2, A4, 47) &= L_7(A1, A2, A4)\, S_2(48, 79)\,, \\
L_9(A1, A2, A4, 25) &= L_7(A1, A2, A4)\, S_2(28, 59)\,, \\
L_9(A1, A2, A4, 13) &= L_7(A1, A2, A4)\, S_2(18, 39)\,, \\
L_9(A1, A2, 25, 26) &= L_7(A1, A2, 25)\, S_2(28, 69)\,, \\
L_9(A1, A2, 25, 57) &= L_7(A1, A2, 25)\, S_2(58, 79)\,, \\
L_9(A1, A2, 13, 25) &= L_7(A1, A2, 13)\, S_2(28, 59)\,,
\end{aligned} \tag{11}
$$

and so on. Here

$$
\begin{aligned}
L_3(A1) &\equiv L_3(A1, A2, 13)\,, \\
L_5(A1, A2) &\equiv L_5(A1, A2, 13, A4, 25)\,, \\
L_7(A1, A2, A4) &\equiv L_7(A1, A2, 13, A4, 25, A6, 47)\,, \\
L_9(A1, A2, A4, A6) &\equiv L_9(A1, A2, 13, A4, 25, A6, 47, A8, 69)\,,
\end{aligned} \tag{12}
$$

and so forth, double symbols A1, A2, 13, A4, 25, ... correspond to the processes of producing the high-energy branches, the first symbol denoting the arriving atom and the second one the branch number.

Formulae (8) do not contain cascades with even k numbers. Generally speaking there is no chance to split a cascade into an even number of subscascades, because it means that the length of the first arc or one of the terminals must be equal to zero; however, such defect clusters are observed in experiments. The reason consists in the following. Probably, in this case one high-energy branch has a length which is less than the device resolution. Therefore, an "experimental" set S_{2k} is incorporated into a "theoretical" set S_{2k+1}.

Let $q(\beta)$ be the mean free path distribution for an arriving atom with an energy β (for details see Part 1). By counting on the diagrams of Fig. 1 the general arc number and the number of arcs with a given β, it is not difficult to obtain the contribution of the $q(\beta)$ distribution to the general one. Consequently, the density of subcascade size distribution $f(D)$ can be written as follows:

$$
\begin{aligned}
&\beta = 1\,, && q(1)\,, \\
&\beta = 3\,, && \tfrac{1}{3}\,[2q(1) + q(3)]\,, \\
&\beta = 5\,, && \tfrac{1}{5}\,[3q(1) + q(3) + q(5)]\,, \\
&\beta = 7\,, && \tfrac{1}{35}\,[20q(1) + 6q(3) + 4q(5) + 5q(7)]\,, \\
&\beta = 9\,, && \tfrac{1}{63}\,[35q(1) + 10q(3) + 6q(5) + 5q(7) + 7q(9)]\,, \\
&\beta = 11\,, && \tfrac{1}{495}\,[270q(1) + 76q(3) + 46q(5) + 30q(7) + \\
& && \quad + 28q(9) + 45q(11)]\,.
\end{aligned} \tag{13}
$$

Here, we have limited ourselves to cascades containing eleven subcascades whose appearance was observed in silver irradiated by 14 MeV neutrons [7]. For silver $\beta = 11$ corresponds to $E = 253$ keV, which is nearly half the maximum energy transferred to a silver PKA by a 14 MeV neutron. It can be easily seen that the subcascade size

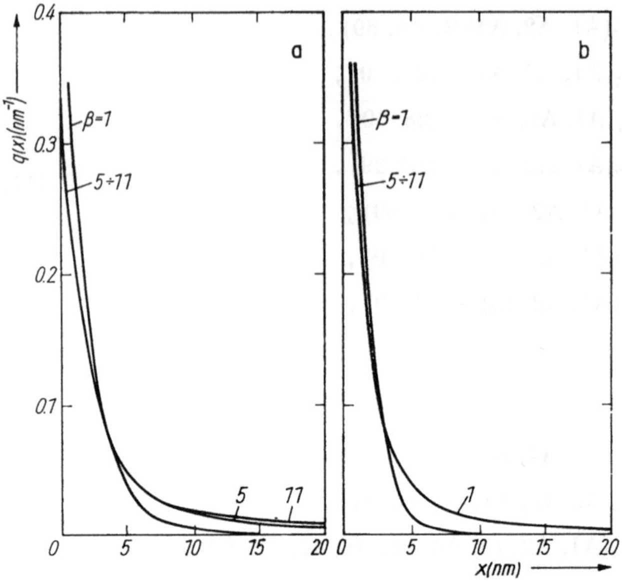

Fig. 2. Displacement subcascade size distribution for a) gold and b) silver for $\beta = 1, 5, 11$; $\theta_b = 0.03$

distribution is more than half determined by the mean free path distribution at the minimum energy set equal to splitting energy (for $\beta = 1, 3, 5, 7, 9, 11$ the $q(1)$ contribution is 1, 2/3, 3/5, 4/7, 5/9, and 6/11, respectively). This result explains the following regularity found in [4], the average size of subcascades in gold is almost constant over a wide range of incident ion energies.

Fig. 2a and b show the subcascade size distributions calculated for gold and silver. The existing experimental data for gold [4] are in satisfactory agreement with the theoretical distribution corrected for the resolution of transmission electron microscopy (TEM) which is equal to 2.5 nm (Fig. 3). The discrepancy at large sizes of subcascades can be caused by taking no account of the energy losses during elastic collisions with small-angle scattering as well as of the influence of temperature (see Section 4). We suppose that the method considered will describe precisely the experimental data for other materials. As the charge numbers of silver and refractories proposed as the first wall materials for thermonuclear reactors are close, the data of Fig. 2b can be used to estimate the radiation damage structure in these materials.

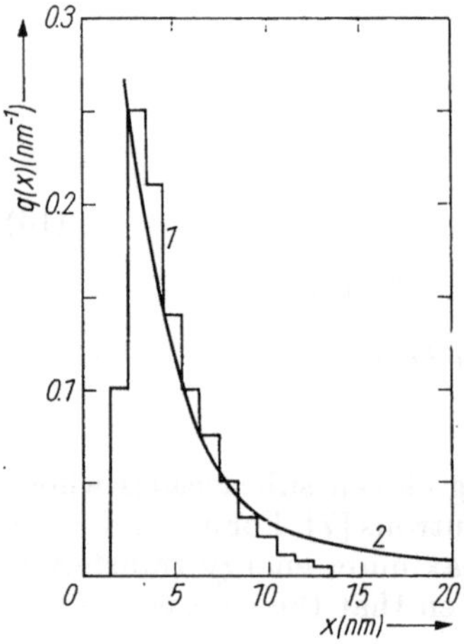

Fig. 3. Comparison of the experimental and corrected theoretical distribution; (1) TEM experiment for 250 keV Au^+ irradiation of gold [4]; (2) $\theta_b = 0.03$, $\beta = 5$ (initial energy 278 keV)

4. Influence of Temperature

According to [8, 9] the temperature of a material does not practically influence the high-energy branches of a cascade, at the same time increasing the temperature decreases the dimensions of the low-energy branches initiated by the secondary knock-out atoms with momenta normal to the PKA trajectory [1]. The influence of temperature is investigated mainly for focused collisions [10] which take place if the energy of an arriving atom does not exceed ≈ 100 eV. The formulae permitting to estimate the influence of thermal vibrations on binary collisions at higher energies were obtained in [2].

Let us consider analogous to [2] how the temperature influences the localization of the cascade low-energy branches. It is known [11] that the mean square root displacement of an atom from a crystal lattice site which corresponds to zero vibrations is equal to

$$\langle u_0^2 \rangle \approx \hbar^2/(mk\theta_D) , \tag{14}$$

where $\hbar = h/(2\pi)$, h is the Planck constant, m the atom mass, k the Boltzmann constant, and θ_D the Debye temperature. If the temperature increases, the displacements also grow and at $T > \theta_D$ they are [12]

$$\langle u_D^2 \rangle \approx \frac{9\hbar^2 T}{mk\theta_D^2} . \tag{15}$$

It follows from (14), (15) that one can realize the displacement $\langle u_T^2 \rangle$ even at zero temperature, if the atom mass has decreased.

Suppose that increasing the temperature of a solid is equivalent to decreasing the atom mass of this solid, and replace the real atom mass by an effective one which depends on the temperature as

$$m(T) = \frac{m\theta_D}{9T} . \tag{16}$$

This means that now one must use the effective collision diameter, because (1) takes the form

$$\Phi(r)|_{r=b(T)} = \frac{m_2(T)}{m_1 + m_2(T)} E . \tag{17}$$

Hence, for the LNS potential at $m_1 = m_2 = m$

$$b(T) = b \left[\frac{m + m(T)}{2m(T)} \right]^{1/s} \tag{18}$$

and, because $\lambda = 1/(n_0 \pi b^2)$, at $s = 2$

$$\lambda(T) \approx \frac{2\lambda}{1 + 9T/\theta_D} . \tag{19}$$

For iron [13] $\theta_D = 432$ K, therefore $\lambda(T)|_{T=242\,K} \approx \lambda/3$. Computer experiments [9] have shown that not taking into consideration the thermal vibrations, a $\langle 111 \rangle$ crowdion generated by a PKA with energy 50 eV in α-iron has a range of $22r_0$, but at $T = 242$ K its value decreases to $8r_0$. This is in good agreement with the theoretical estimate made above.

It is known [14] that a monocascade averaged statistically is an rotation ellipsoid whose large axis is directed along a PKA trajectory. For that reason if the temperature

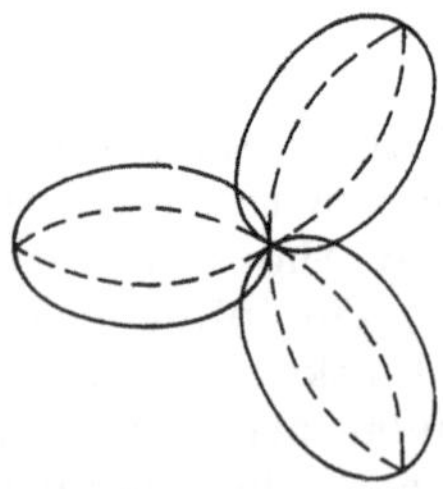

Fig. 4. Subcascade localization under the influence of temperature

increases, the large axis does not change, at the same time the small axes is drastically contracted, which leads to the localization of subcascades (Fig. 4). As a result, the subcascade structure of radiation damage is visualized with better resolution [4, 8, 13].

5. Summary and Conclusion

In the present paper the theory developed earlier for the description of splitting displacement cascades in a solid has been applied to the calculation of the energy threshold of splitting in some materials. The calculated energy thresholds are in good agreement with experiments. A diagram technique permitting to construct the group representation of splitting is discussed. On this basis the density of subcascade size distribution is obtained, the results being also in agreement with experiment. A simple method for an analytical estimation of the influence of temperature on splitting is suggested. It is shown that temperature growth decreases the cross section of subcascades.

References

[1] A. I. Melker, S. N. Romanov, and N. L. Tarasenko, phys. stat. sol. (b) **132**, 425 (1985).
[2] A. I. Melker, in: Radiatsionnye defekty v metallakh, Ed. A. T. Lukyanov, Izd. Nauka, Alma-Ata 1981 (p. 15).
[3] J. Lindhard, V. Nielsen, and M. Scharff, Kong. Danske Vid. Selsk., mat.-fys. Medd. **36**, No. 1 (1968).
[4] P. P. Pronko and K. L. Merkle, in: Applications of Ion Beams to Metals, Ed. S. T. Picraux, E. P. EerNisse, and F. L. Vook, Plenum Press, New York 1974 (p. 481).
[5] J. R. Beeler, Jr., J. nuclear Mater. (Amsterdam) **53**, 207 (1974).
[6] A. I. Melker and S. N. Romanov, phys. stat. sol. (b) **126**, 133 (1984).
[7] R. L. Lyles and K. L. Merkle, in: Radiat. Eff. and Tritium Technol. Fusion React., Proc. Internat. Conf., Gatlinburg (Tenn.) 1975, Vol. 1, Oak Ridge 1976 (p. 191).
[8] V. V. Kirsanov, A. L. Suvorov, and Yu. V. Trushin, Protsesy radiatsionnogo defektoobrazovaniya v metallakh, Energoatomizdat, Moskva 1985.
[9] V. V. Kirsanov, in: Defekty v kristallakh i ikh modelirovanie na EVM, Ed. Yu. A. Osipyan, Izd. Nauka, Leningrad 1980 (p. 134).
[10] M. W. Thompson, Defects and Radiation Damage in Metals, Cambridge University Press, 1969.
[11] R. Guyer, Solid State Phys. **23**, 413 (1969).
[12] J. M. Ziman, Principles of the Theory of Solids, Cambridge University Press, 1972.
[13] L. Kaufman, E. V. Clougherty, and R. J. Weiss, Acta metall. **11**, 323 (1963).
[14] A. I. Melker and S. N. Romanov, Fiz. i Khim. Obrabotki Mater. **6**, 12 (1979).

(Received August 15, 1985)

phys. stat. sol. (b) 133, 119 (1986)

Subject classification: 66.30; S1

Natuurkundig Laboratorium, Vrije Universiteit, Amsterdam[1])

Electromigration in Niobium

A Finite-Cluster-Model Study

By

M. G. E. BRAND and A. LODDER

The windforce acting upon a host atom and the impurities Ta, Cr, and Ni in niobium is calculated in a finite cluster model. In this model all strong and multiple scattering effects up to $l_{max} = 2$ are preserved. Special attention is given to the influence of the vacancy, the cluster size, and of charge transfer. A positive windforce is found, in accordance with experiment, so that this appears to be not directly related to a hole current contribution, contrary to what is believed commonly.

La force due au vent électronique agissant sur un atome du réseau ou sur une impureté substitutionelle (Ta, Cr, Ni) dans le niobium est calculée à l'aide d'un modèle de cluster de dimension finie. Tous les effects de diffusion forte et multiple ($l_{max} = 2$) sont incorporés dans ce modèle. L'influence de la lacune, de la taille du cluster ainsi que du transfert de charge est considérée en détail. La force de vent électronique est parallèle au champs électrique en accord avec les résultats expérimentaux. Contrairement aux interprétations habituelles cette force de vent positive n'est pas directement reliée à une conduction de trous dans le niobium.

1. Introduction

When describing electromigration one generally distinguishes two contributions to the driving force, a direct force and a wind force,

$$\boldsymbol{F} = Ze\boldsymbol{E} + \boldsymbol{F}_{\text{wind}} \equiv (Z + Z_{\text{wind}})\, e\boldsymbol{E} \equiv Z^* e\boldsymbol{E}\,. \tag{1}$$

The direct electrostatic force has been the subject of many discussions: in early ballistic theories [1] Z is the valence of the impurity, while according to Bosvieux and Friedel [2] there is no direct force contribution because the impurity is completely screened. Recent investigations [3] starting from linear response theory have led to the conclusion that the presence of $Ze\boldsymbol{E}$ in (1) is correct.

Only recently some theories have been proposed for the wind force in the strongly scattering noble and transition metals [4 to 6]. A complete alloy calculation being still quite involved although feasible [6], we made a finite-cluster-model (FCM) study, like in [5], of interstitial migration in the transition-metal hydrides V(H), Nb(H), Ta(H), Cu(H), and Pd(H) [7]. The results obtained stimulated us to study substitutional Nb alloys treated earlier by Gupta et al. [8]. In their treatment, although electronic band effects were taken into account, backscattering effects being important in electromigration were neglected and the formalism used did not allow for a vacancy at the end of the migration path. A FCM treatment does not incorporate band effects, but all (multiple) scattering effects including backscattering are accounted for exactly and the influences of the presence of a vacancy, of charge transfer, and lattice distortion can be studied explicitly.

[1]) Postbus 7161, 1007 MC Amsterdam, The Netherlands.

In Section 2 the model is presented, Section 3 gives some computational remarks, and in Section 4 the results are presented with a discussion.

2. The Model

In the FCM approach, the alloy to be studied is modelled by a finite cluster of muffin-tin host potentials around the migrating atom. The remote host environment is replaced by the muffin-tin zero of the host potential. Since the impurity and, in the case of substitutional migration, the vacancy will always influence the host potentials, at least in the first surrounding shell, one would like to choose the cluster as large as possible.

For studying the substitutional Nb alloys we have used a b.c.c. configuration shown in Fig. 1. The jump path is along the line from the starting position 1 to the vacancy position 16.

As starting point for a FCM expression of the wind force we have taken a complete alloy formulation [6]. The FCM formula for the wind force so obtained has the following form [7]:

$$\begin{aligned} \boldsymbol{F}^{\mathrm{FCM}}_{\mathrm{wind}} = &- \frac{eE\tau}{m} \cdot \frac{16\varepsilon_{\mathrm{F}}}{3} \Big[\sum_{\substack{Lm_1 \\ (l<l_{\max})}} \sum_{\substack{p'L' \\ p''L''}} (\cot \delta^1_{l+1} - \cot \delta^1_l)\, \boldsymbol{C}_{L,l+1,m_1} \times \\ &\times \operatorname{Im} \{ T^{1p'*}_{l+1,m_1,L'} T^{1p''}_{LL''} \boldsymbol{U}^{p'p''}_{L'L''} \cdot \hat{\boldsymbol{E}} \} + \sum_{mm_1} \sum_{p'L'} \boldsymbol{C}_{l_{\max} m, l_{\max}+1, m_1} \times \\ &\times \operatorname{Im} \{ T^{1p'*}_{l_{\max} m, L'} (\boldsymbol{U}^{1p'}_{l_{\max}+1, m_1, L'} + \\ &+ \sum_{\substack{p''L'' \\ p'''L'''}} B^{1p''}_{l_{\max}+1, m_1, L''} T^{p''p'''}_{L''L'''} \boldsymbol{U}^{p'''p'}_{L'''L'}) \cdot \hat{\boldsymbol{E}} \} \Big] , \end{aligned} \tag{2}$$

where ε_{F} is the Fermi energy. The vectorial matrix $\boldsymbol{C}$ is a Gaunt coefficient,

$$\boldsymbol{C}_{LL'} = \sqrt{\frac{3}{4\pi}} \int \mathrm{d}\hat{\boldsymbol{r}}\; Y_L(\hat{\boldsymbol{r}})\, \hat{\boldsymbol{r}}\, Y_{L'}(\hat{\boldsymbol{r}}) . \tag{3}$$

The vectorial matrix $\boldsymbol{U}$ defined by

$$\boldsymbol{U}^{pp'}_{LL'} = i^{l-l'-1} \sqrt{\frac{3}{4\pi}} \int \mathrm{d}\hat{\boldsymbol{k}}\; Y_L(\hat{\boldsymbol{k}})\, \hat{\boldsymbol{k}} \exp{(i\boldsymbol{k} \cdot \boldsymbol{R}_{pp'})}\; Y_{L'}(\hat{\boldsymbol{k}}) \tag{4}$$

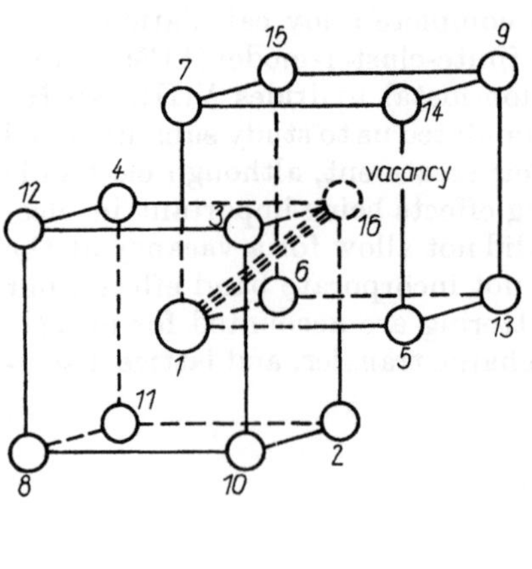

Fig. 1. The cluster corresponding to a complete first host shell around the migration path, according to the b.c.c. structure of the Nb alloys. A cluster of N atoms ($N < 15$) incorporated the numbered atoms up to N and the vacancy was renumbered to $N + 1$

is real and antisymmetric for exchange of (pL) and $(p'L')$. The scattering effects are fully present in the cluster t-matrix,

$$T = t(1 - Bt)^{-1}\,, \tag{5}$$

where the matrix t of the individual scatterers is diagonal,

$$t_l^j = -\sin \delta_l^j \exp (i\delta_l^j)\,. \tag{6}$$

δ_l^j is the phase shift of the j-th scatterer for angular momentum l, and B is a commonly used [5] geometrical matrix,

$$B_{LL'}^{pp'} = 4\pi i^{l-l'-1} \sum_{L''} i^{l''} C_{LL'L''} h_{L''}^{+}(\boldsymbol{R}_{pp'}) \tag{7}$$

containing Gaunt coefficients C. For more detailed definitions we refer to [5].

3. Computational Remarks

We point out that the windforce (2) is proportional to the electron drift velocity $-eE\tau/m$. In view of (1) Z_{wind} now can be calculated if the transport relaxation time τ is known. Like Gupta [4, 8] we will use for it the measured specific resistivity ϱ which is related to τ as

$$\varrho = m/(ne^2\tau)\,. \tag{8}$$

Z_{wind} is calculated from the tensor force expression (2) by equating it to the ratio $\boldsymbol{F}_{\text{wind}}^{\text{FCM}} \cdot \hat{\boldsymbol{E}}/eE$ when $\hat{\boldsymbol{E}}$ is taken to be along the migration path.

Muffin-tin potentials were used and constructed in a standard way [7]. The vacancy potential was obtained by adding the tails of the surrounding potentials. The angular momentum cut-off value was $l_{\max} = 2$, which does justice to the strong d-scattering characteristic of transition metals. In Table 1 the phase shifts for the different scatterers are given for the studied alloy clusters Nb(Nb), Nb(Ta), Nb(Cr), Nb(Ni).

Mainly calculations are done for the clusters consisting of nine and fifteen atoms with a vacancy. These clusters, being symmetric with respect to the migration path, allowed to restrict the average of the windforce to the first half of the jump path.

Table 1
Phase shifts of the different scatterers in the alloys investigated. The second host shell phase shifts are not given but they were still different from the unperturbed host ones

alloy	scatterer	δ_0	δ_1	δ_2
Nb(Nb)	vacancy	−0.9505	−0.3006	−0.0527
	first host shell	−0.9166	−0.3496	1.0122
	unperturbed host	−0.8937	−0.3399	1.1406
Nb(Ta)	Ta	−0.9722	−0.4048	0.6377
	vacancy	−0.9527	−0.3013	−0.0529
	first host shell	−0.9042	−0.3444	1.0760
Nb(Cr)	Cr	−0.6547	−0.1527	0.6632
	vacancy	−0.9629	−0.3056	−0.0539
	first host shell	−0.9094	−0.3466	1.0520
Nb(Ni)	Ni	−0.4159	−0.0677	−0.3666
	vacancy	−0.9732	−0.3096	−0.0548
	first host shell	−0.9157	−0.3493	1.0142

The calculations, particularly those for the larger clusters, were performed on a CDC 205 vector processor. In our case the accuracy was limited by the inversion routine necessary to obtain T in (5). For $N > 20$ the accuracy was lost and the results became unreliable.

4. Results and Discussion

We first present results which reveal the different aspects and possibilities of the model, like the influence of the cluster size, the vacancy and charge transfer. We calculated the properties of Nb(Nb) for that purpose. Other alloys show similar features.

The detailed cluster-size dependence of Z_{wind} is given in Fig. 2, using the phase shifts for Nb of Gupta et al. [8]. Since these authors give only host phase shifts all scatterers in the clusters are equal and the vacancy is represented by just omitting a scatterer at that position. In addition to the full potential results, i.e. $l_{\max} = 2$, connected by a solid line, we performed calculations for the same clusters using only the s phase shift (dotted line) and the s and p phase shifts (dashed line). Note the difference in scale, for the s and sp lines given along the right vertical axis.

The strong d-scatterer character of the transition metal atom is clear from the large negative Z_{wind} values for clusters consisting of few scatterers. It is interesting to note that for cluster sizes of 7 to 13 scatterers the d contribution more than cancels the sp contribution, leading to positive Z_{wind} values. This is surprising because a positive wind force commonly is attributed to holes, being related to a reversal of the sign of the effective mass and therefore to the band structure. In the model, however, band structure is not accounted for. We must conclude that a positive Z_{wind} is due to the large (multiple) scattering effects in transition metals which manifest themselves already locally, in a relatively small cluster. This conclusion is supported by earlier findings in interstitial alloys of transition metals, with hydrogen as the migrating atom [7]. A different kind of support is given by Verbruggen [9] who concludes from a very recent tight-binding model analysis of the expression of the windforce, that its sign is not merely a band effect but follows directly from the value of the matrix element of the potential of the migrating atom.

It is seen from Fig. 2 that the clusters accounting only for first surrounding shell atoms are too small yet to achieve convergence. However, while all first shell atoms are equivalent and the atoms are strong scatterers, remark that the cluster-size

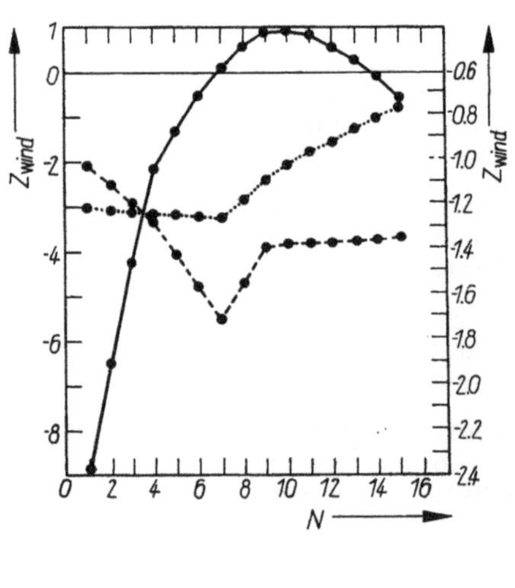

Fig. 2. The values of Z_{wind} for clusters of 1 to 15 atoms, connected by lines as a guide for the eye. The solid line connects values corresponding to the full potential. The right vertical axis holds for the values obtained by using only the s phase shift (dotted line) and the s and p phase shifts (dashed line)

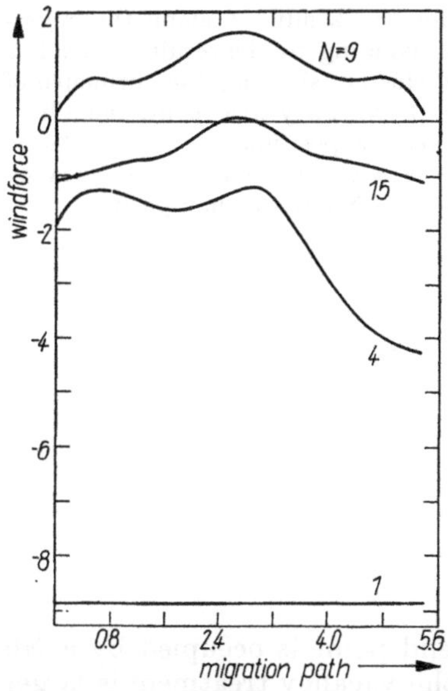

Fig. 3. The windforce as a function of the position along the migration path for clusters of 1, 4, 9, and 15 atoms

dependence for $N < 5$ is clearly larger than for the larger clusters. This trend comes out even clearer if atoms in the second shell are added. Some test calculations for $N > 15$ did show that the values no longer vary much while they are again positive (≈ 0.15). Of course, a complete saturation cannot be expected.

Fig. 3 shows in more detail the cluster size dependence. For $N = 1$ and 4, and the symmetric configurations $N = 9$ and 15 the windforce is plotted as a function of the position of the moving atom along the migration path. Only for symmetric clusters the function is symmetric. For $N = 1$ the force is of course position independent. The $N = 4$ curve shows nicely the local influence of Nb atoms around the migrating atom (see Fig. 1).

The effect of charge transfer and the vacancy (as a void or as a scatterer) is illustrated by Table 2. To that end the constructed potentials and the corresponding phase shifts of Table 1 are used. In order to avoid the strong effects which occur when two scatterers approach each other the windforce was averaged over the first half of the migration path also in these cases. Between parentheses values are given obtained by the common practice [8] of averaging starting point and midway wind force values. Even in the symmetric cases (first two lines) this practice is seen to be not always reliable. The values in this and the following table are corrected for the value of the wind force at the origin of a cluster with a host atom at the vacancy position, which should be zero.

Comparison of the first and second lines shows the influence of charge transfer while at the vacancy position there is no scatterer. In the third line the vacancy is

Table 2

The effect of charge transfer and vacancy treatment on Z_{wind} for clusters of 9 and 15 atoms. The values between parentheses are obtained according to the saddle point method

	$N = 9$	$N = 15$
host scatterers, vacancy as void	1.66(1.66)	−0.35(−0.25)
charge transfer, vacancy as void	1.44(1.73)	−0.23(−0.29)
charge transfer, vacancy as scatterer	2.06(1.75)	−0.05(−0.50)
host scatterers, Nb at vacancy position	0.32(−0.63)	−0.19(−0.87)

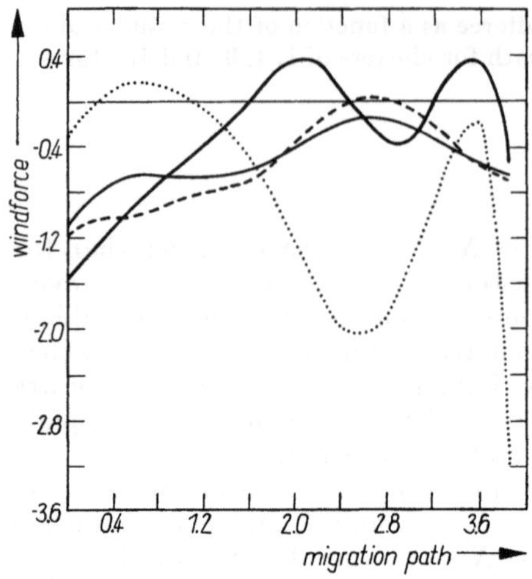

Fig. 4. Curves for the windforce along the migration path corresponding to the results of Table 2 for the largest cluster, showing the influence of charge transfer, the vacancy treatment, and a host atom at the vacancy position. ---- no charge transfer, ——— charge transfer, ——— vacancy as scatterer, ······ Nb at vacancy position

accounted for as a scatterer. In the fourth line the end point is occupied by a Nb atom. The influence of charge transfer as well as of the vacancy treatment is larger for the smaller cluster than for $N = 15$. In the calculation of Gupta et al. the vacancy treatment is unclear, because the expression used to obtain the midway force value is based on an argument and not on a derivation. Although for the wave function a host atom occupies the vacancy position (our fourth line), for the scattering the vacancy is treated as a void (first line). Also the saddle point method was used to obtain Z_{wind}. One can conclude from Table 2 that results are very sensitive to the approach used. Obviously the vacancy is crucial in substitutional migration. The way in which we account for it as a scatterer may be questionable, because it is not a static scatterer. Its scattering properties change during the process while a second vacancy develops at the starting position. To simulate this we have made some calculations with the migrating atom halfway, surrounded by two smaller vacancies. The results did not change significantly so the method used seems to be justified.

In Fig. 4 the wind force is given as a function of the position of the migrating atom, for $N = 15$, corresponding to the four cases of Table 2. Uncorrected windforce values are used: it is seen that for the cluster with 16 atoms the windforce value at the origin is negative (-0.33) and so not exactly zero, which is to be found for the unperturbed host. For the two cases with a scatterer at the end point it is not possible to show the curves for positions much farther from the origin. Heavy oscillations start to occur due to overlap of scatterers.

A beginning of it is already seen for the case of Nb at the vacancy position. Only the curves for a vacancy as a void are symmetric with respect to the midway position 2.701.

Now we turn to the different Nb alloys studied by Gupta et al. [8]. Results are presented in Table 3, together with the results by Gupta et al. and some experimental values. The values for $N = 9$ are surprisingly good. The values for $N = 15$ are slightly larger than those by Gupta et al. for Nb (Ta) and Nb(Ni), and for Nb(Nb) and Nb(Cr) they are slightly worse. For the four cluster sizes larger than $N = 15$ accessible by our computer program the values depend on the positions of the additional atoms, but they become significantly positive and do not vary strongly. Typical values are those for a $N = 18$ cluster shown in the Table. In any effective valence calculation, in a final comparison with experiment, one should consider a possible

Table 3

Different calculated values for Z_{wind} and values derived from the measured effective valences Z^* [8], for the alloys indicated

alloy	present theory			previous theory [8]	experiment [8]
	$N = 9$	$N = 15$	$N = 18$		
Nb(Nb)	2.06	−0.065	0.71	0.096	1.5
Nb(Ta)	1.70	0.12	0.68	0.094	2.0 to 16.0
Nb(Cr)	1.32	−0.11	0.66	0.043	1.0
Nb(Ni)	1.04	0.25	0.67	0.083	—

contribution of a phonon-wind term, to be added on the right-hand side of (1) according to recent work by Bohn and Simon [10] and which may be considerable for the class of fast diffusers [11]. In the present case of slowly diffusing Nb alloys, however, this term, for the time being, is assumed to be negligibly small.

In conclusion, the present study confirms a finding of an earlier study on interstitial alloys [7] that a positive sign of the wind force in a transition metal is already found in a finite-cluster model and therefore it does not require long-range effects leading to a description in terms of holes. Further, the results are sensitive to the treatment of the vacancy. On the other hand, effects which may originate from the real Fermi surface of niobium, being rather different from a free electron sphere, require a complete alloy calculation [6]. Comparison of such a calculation with experiment would be very interesting, because it could be conclusive in the fundamental question whether a muffin-tin approach is sufficient in electromigration, being a complicated dynamical process.

References

[1] V. B. Fiks, Soviet Phys. — Solid State **1**, 14 (1959); H. B. Huntington and A. R. Grone, J. Phys. Chem. Solids **20**, 76 (1961).
[2] C. Bosvieux and J. Friedel, J. Phys. Chem. Solids **23**, 123 (1962).
[3] R. S. Sorbello, Phys. Rev. B **31**, 798 (1985) and references therein.
[4] R. P. Gupta, Phys. Rev. B **25**, 5188 (1982).
[5] R. S. Sorbello, A. Lodder, and S. J. Hoving, Phys. Rev. B **25**, 6178 (1982).
[6] A. Lodder, J. Phys. F **14**, 2943 (1984).
[7] A. Lodder and M. G. E. Brand, J. Phys. F **14**, 2955 (1984).
[8] R. P. Gupta, Y. Serruys, G. Brebec, and Y. Adda, Phys. Rev. B **27**, 672 (1983).
[9] A. H. Verbruggen, Ph. D. Thesis, Vrije Universiteit, Amsterdam 1985.
[10] H.-J. Bohn and G. Simon, Z. Naturf. **37a**, 1327 (1982).
[11] H.-J. Bohn and G. Simon, Z. Naturf. **36a**, 10 (1981).

(Received August 21, 1985)

Note added in proof

We are grateful that Dr. Bohn and Prof. Simon have been willing to estimate the phonon wind valency for the ions Ta, Cr and Ni. Following their theory they find 0.91, 0.47 and 0.25 resp., with an uncertainty of a factor of 2. Apparently the measured difference between Ta and Cr is reproduced better including this contribution.

phys. stat. sol. (b) 133, 127 (1986)

Subject classification: 76.20

Institute of Physics, Academy of Sciences of the Georgian SSR, Tbilissi (a) and Faculty of Physics, Tbilissi State University[1]) *(b)*

Cross-Relaxation of Nuclear Spins at Low Temperatures

By

L. L. BUISHVILI (a), N. P. FOKINA (b), and L. T. CHELIDZE (b)

A system of equations is derived for the evolution under action of cross-relaxation (CR) of the dipole–dipole reservoir (DDR) and Zeeman subsystems, redefined taking into account the fact, that the part of secular dipole–dipole interaction H_d', which causes homogeneous displacement of spin resonant frequencies at low temperatures (LT) must be subtracted from H_d'. The case is considered, when the influence of the change of LT displacement due to CR on the energy, absorbed by DDR, is stronger, than that of the usual high-temperature term. It is shown, that in this situation both, DDR and spins under CR acquire a negative temperature in spite of the fact, that the frequency of saturated spins is smaller than the frequency of spins under CR. This behaviour differs from the high-temperature result.

Выведена система уравнений для эволюции под действием кросс-релаксации (КР) диполь–дипольного резервуара (ДДР) и зеемановских подсистем, переопределенных с учетом того, что часть секулярного d–d взаимодействия H_d' вызывающая при низких температурах (НТ) однородный сдвиг резонансных частот спинов, должна быть вычтена из H_d'. Рассмотрен случай, когда влияние изменения НТ сдвига в процессе КР на величину энергии, поглощаемой ДДР, сильнее, чем влияние обычного высокотемпературного члена. Показано, что в этой ситуации ДДР и спины, нагреваемые посредством КР от насыщенных спинов, приобретают отрицательную температуру несмотря на то, что частота насыщенных спинов меньше частоты нагреваемых. Такое поведение отличается от высокотемпературного результата.

1. Introduction

Recently low-temperature (LT) situations have gained special attention in connection with researches made in nuclear dynamic polarization and nuclear relaxation [1]. Quite often samples in which it is necessary to obtain high polarization of nuclei of one kind, also contain nuclei of another kind and therefore their cross-relaxation (CR) becomes possible. Consideration of CR at low temperatures is of certain interest.

Using the method developed in [2, 3] we choose statistically independent subsystems [4] at LT.

Demagnetizing fields produced by the spin orientation cause a homogeneous displacement of resonant frequencies and change the Zeeman energy of spins. The corresponding term (the energy of spin waves with zero wave vector) is to be subtracted from the secular part of the dipole–dipole (d–d) interaction. The remainder represents the fluctuative term that causes quasi-equilibrium in all subsystems.

Note that the CR problem at low temperatures is considered in [5] without taking into account homogeneous displacement at resonances and redefinition of the subsystems.

---

[1]) Chavchavadze pr. 3, 380028 Tbilissi, USSR.

2. Basic Equations

The Hamiltonian of the system, consisting of two kinds of spins with d–d interaction located in the constant magnetic field $\boldsymbol{H}_0 \parallel z$, is

$$\mathcal{H} = \omega_I I^z + \omega_S S^z + \mathcal{H}'_{\rm d} + \mathcal{H}_{\rm CR}\,, \qquad (1)$$

where $\mathcal{H}'_{\rm d}$ is the part of d–d interaction that is secular with respect to Zeeman energies $\omega_I I^z$ and $\omega_S S^z$; ω_I and ω_S are Larmor frequencies of I and S spins;

$$I^z = \sum_{\alpha=1}^{N_I} I_\alpha^z \quad \text{and} \quad S^z = \sum_{i=1}^{N_S} S_i^z$$

are projections of operators of the full spin on the z-axis, N_I and N_S are the number of I and S spins, respectively. $\mathcal{H}'_{\rm d}$ is equal to

$$\mathcal{H}'_{\rm d} = \frac{1}{2} \sum_{\substack{i,j\\(i\neq j)}}^{N_S N_S} [U_{ij}^{zz} S_i^z S_j^z + \tfrac{1}{2}\, U_{ij}^{+-} (S_i^+ S_j^- + S_i^- S_j^+)] +$$
$$+ \frac{1}{2} \sum_{\substack{\alpha,\beta\\(\alpha\neq\beta)}}^{N_I N_I} [U_{\alpha\beta}^{zz} I_\alpha^z I_\beta^z + \tfrac{1}{2}\, U_{\alpha\beta}^{+-} (I_\alpha^+ I_\beta^- + I_\alpha^- I_\beta^+)] + \sum_{i\alpha}^{N_S N_I} U_{i\alpha}^{zz} S_i^z I_\alpha^z\,, \qquad (2)$$

where

$$S_i^\pm = S_i^x \pm i S_i^y\,; \qquad I_\alpha^\pm = I_\alpha^x \pm i I_\alpha^y\,,$$

$$U_{ij}^{zz} = -2U_{ij}^{+-} = \frac{\gamma_S^2 \hbar (1 - 3\cos^2\theta_{ij})}{r_{ij}^3}\,,$$

$$U_{\alpha\beta}^{zz} = -2U_{\alpha\beta}^{+-} = \frac{\gamma_I^2 \hbar (1 - 3\cos^2\theta_{\alpha\beta})}{r_{\alpha\beta}^3}\,,$$

$$U_{i\alpha}^{zz} = -2U_{i\alpha}^{+-} = \frac{\gamma_I \gamma_S \hbar (1 - 3\cos^2\theta_{id})}{r_{i\alpha}^3}\,.$$

γ_I and γ_S are the gyromagnetic ratios of I and S spins, lattice positions of S spins are denoted by Latin indexes, the ones of I spins are denoted by Greek indexes, distances between lattice positions are denoted by letter r with corresponding indexes, θ is the angle between the corresponding vector $\boldsymbol{r}$ and the z-axis.

$\mathcal{H}_{\rm CR}$ is responsible for the CR and is regarded as a perturbation,

$$\mathcal{H}_{\rm CR} = \frac{1}{4} \sum_{i\alpha}^{N_S N_I} U_{i\alpha}^{+-} (S_i^+ I_\alpha^- + S_i^- I_\alpha^+)\,. \qquad (3)$$

Zeeman Hamiltonians taking into account demagnetizing fields [2] are equal to

$$\mathcal{H}_I^{Z^*} = (\omega_I + M_{1I})\, I^z \equiv \omega_I^* I^z\,,$$
$$\mathcal{H}_S^{Z^*} = (\omega_S + M_{1S})\, S^z \equiv \omega_S^* S^z\,,$$

where $M_{1,I,S}$ are the first moments of the lineshapes of I and S spins, respectively. Like in [2, 3], it can be shown that the fluctuative part of $\mathcal{H}'_{\rm d}$ is equal to

$$\mathcal{H}_{\rm d}^* = \frac{1}{2} \sum_{\substack{i,j\\(i\neq j)}}^{N_S N_S} [(U_{ij}^{zz} - U_{0SS}^{zz})\, S_i^z S_j^z + \tfrac{1}{2}\,(U_{ij}^{+-} - U_{0SS}^{+-})\,(S_i^+ S_j^- + S_i^- S_j^+)] +$$
$$+ \frac{1}{2} \sum_{\substack{\alpha,\beta\\(\alpha\neq\beta)}}^{N_I N_I} [(U_{\alpha\beta}^{zz} - U_{II}^{zz})\, I_\alpha^z I_\beta^z + \tfrac{1}{2}\,(U_{\alpha\beta}^{+-} - U_{0II}^{+-})\,(I_\alpha^+ I_\beta^- + I_\alpha^- I_\beta^+)] +$$
$$+ \sum_{i\alpha}^{N_S N_I} (U_{i\alpha}^{zz} - U_{0SI}^{zz})\, S_i^z I_\alpha^z\,, \qquad (4)$$

where

$$U_{0SS}^{zz,+-} = \frac{1}{N_S^2} \sum_{ij} U_{ij}^{zz,+-}, \qquad U_{0II}^{zz,+-} = \frac{1}{N_I^2} \sum_{\alpha\beta} U_{\alpha\beta}^{zz,+-},$$

$$U_{0SI}^{zz} = \frac{1}{N_S N_I} \sum_{i\alpha} U_{i\alpha}^{zz}.$$

$\mathcal{H}_d^*$ will play the role of a LT dipole–dipole reservoir (DDR). Subsystems with Hamiltonians $\mathcal{H}_I^{z*}$, $\mathcal{H}_S^{z*}$, $\mathcal{H}_d^*$ achieve a state of internal equilibrium at the time T_2 due to the $\mathcal{H}_d^*$ interaction. After that temperatures β_I^{-1}, β_S^{-1}, and β_d^{-1} can be ascribed to these subsystems, respectively.

Zubarev's nonequilibrium statistical operator [6] enables us to obtain the following set of equations for the evolution of inverse temperatures β_S, β_I, and β_d under the action of CR:

$$\frac{dp_S}{dt} = -\frac{1-p_S p_I}{2N_S} \tanh\left(\frac{\omega_S^*\beta_S - \omega_I^*\beta_I - \Delta^*\beta_d}{2}\right) W^{CR}(\Delta^*),$$

$$\frac{dp_I}{dt} = \frac{1-p_S p_I}{2N_I} \tanh\left(\frac{\omega_S^*\beta_S - \omega_I^*\beta_I - \Delta^*\beta_d}{2}\right) W^{CR}(\Delta^*), \tag{5}$$

$$\frac{d}{dt}(\omega_d^{*2}\beta_d) = \frac{\Delta^*(1-p_S p_I)}{N_I + N_S} \tanh\left(\frac{\omega_S^*\beta_S - \omega_I^*\beta_I - \Delta^*\beta_d}{2}\right) W^{CR}(\Delta^*),$$

where

$$p_m = \tanh\left(\frac{\omega_m^*\beta_m}{2}\right); \qquad m = I, S$$

and

$$W^{CR}(\Delta^*) = \pi \sum_{i\alpha} (U_{i\alpha}^{+-})^2 f_{i\alpha}^{CR}(\Delta^*)$$

is the probability of the transition induced by CR;

$$f_{i\alpha}^{CR}(\Delta^*) = \frac{1}{2\pi} \int_{-\infty}^{\infty} \frac{\langle \{J_{i\alpha}(t)\, J_{i\alpha}^+\}\rangle_Z}{\langle \{J_{i\alpha} J_{i\alpha}^+\}\rangle_Z}\, e^{i\Delta^* t}\, dt;$$

$$J_{i\alpha} = S_i^+ I_\alpha^-, \qquad J_{i\alpha}(t) = e^{i\mathcal{H}_d^* t} J_{i\alpha} e^{-i\mathcal{H}_d^* t},$$

$$\{AB\} = \tfrac{1}{2}(AB + BA).$$

$\langle\ldots\rangle_Z$ denotes the averaging by the Zeeman statistical operator

$$\varrho_Z = (\mathrm{tr}\, e^{-C})^{-1} e^{-C}; \qquad C = \sum_m \beta_m \mathcal{H}_m^{Z*}.$$

Since our analysis is fulfilled for the disordered state of the spin system, we assume $\omega_d^*\beta_d \ll 1$, where ω_d^* is the mean quantum of the low-temperature DDR, and for calculating the specific heat of DDR we use the high-temperature approximation with respect to the DDR temperature,

$$\langle(\mathcal{H}_d^*)^2\rangle_Z \approx \frac{N_I + N_S}{4}\, \omega_d^{*2}.$$

ω_d^* is connected with ω_{dS} and ω_{dI} (the corresponding values calculated for $(\mathcal{H}'_d)^{SS}$ and $(\mathcal{H}'_d)^{II}$ at infinite Zeeman temperatures) in the following way:

$$\omega_d^{*2} = \frac{N_S(1-p_S^2)}{N_I+N_S}\left[\left(1-\frac{2}{3}\,p_S^2\right)\omega_{dS}^2 + p_S^2\overline{\Delta_{SS}^2} + p_I^2\overline{\Delta_{IS}^2} + 2p_Sp_I\overline{\Delta_{SS}\Delta_{IS}}\right] +$$

$$+ \frac{N_I(1-p_I^2)}{N_I+N_S}\left[\left(1-\frac{2}{3}\,p_I^2\right)\omega_{dI}^2 + p_I^2\overline{\Delta_{II}^2} + p_S^2\overline{\Delta_{SI}^2} + 2p_Sp_I\overline{\Delta_{II}\Delta_{SI}}\right] +$$

$$+ \frac{(1-p_S^2)\,(1-p_I^2)}{4(N_I+N_S)}\sum_{i\alpha}(U_{i\alpha}^{zz})^2\,,$$

where

$$\Delta_{SS} = a_{Sj} - \overline{a_{SS}}\,,\qquad \Delta_{SI} = \Delta_{S\alpha} - \overline{a_{SI}}\,,$$

$$\Delta_{II} = a_{I\alpha} - \overline{a_{II}}\,,\qquad \Delta_{IS} = \Delta_{Ij} - \overline{a_{IS}}\,,$$

$$\overline{a_{mn}} = \frac{N_m}{2}\,U_{0mn}^{zz}\,;\qquad a_{Sj} = \frac{1}{2}\sum_j U_{ij}^{zz}\,;\qquad a_{Ij} = \frac{1}{2}\sum_\alpha U_{j\alpha}^{zz}\,,\ \dots\,.$$

Resonant frequency displacements with respect to the Larmor values are equal to the first moments of the functions of the lineshapes of spins S and I, taken with $\mathcal{H}'_d$ interaction,

$$\begin{aligned} M_{1S} &= -\tfrac{3}{2}\,p_S\overline{a_{SS}} - p_I\overline{a_{IS}}\,,\\ M_{1I} &= -\tfrac{3}{2}\,p_I\overline{a_{II}} - p_S\overline{a_{SI}}\,. \end{aligned} \tag{6}$$

Equations (6) are identical with those for the first moments given in [1]. Using (6), the difference of resonant frequencies for two kinds of spins, Δ^*, can be written in the form

$$\Delta^* = \omega_S^* - \omega_I^* = -\Delta - p_S(\tfrac{3}{2}\,\overline{a_{SS}} - \tfrac{1}{2}\,\overline{a_{SI}}) + p_I(\tfrac{3}{2}\,\overline{a_{II}} - \tfrac{1}{2}\,\overline{a_{IS}})\,;$$

$$\Delta = \omega_I - \omega_S\,.$$

Calculating the first moment of $f_{i\alpha}^{CR}(\Delta^*)$ with the $\mathcal{H}_d^*$ interaction, we get nonzero $M_{1i\alpha}^*$ in spite of leaving in $\mathcal{H}_d^*$ only the fluctuative part of the secular d–d interaction,

$$M_{1i\alpha}^* = -p_S[(a_{Si} - \overline{a_{SS}}) - (a_{S\alpha} - \overline{a_{SI}})] + p_I[(a_{I\alpha} - \overline{a_{II}}) - (a_{Ii} - \overline{a_{IS}})]\,.$$

The cause of this is that mean local fields are subtracted from $\mathcal{H}'_d$,while individual spins subjected to local fields different from the mean ones participate in the elementary CR act.

The nonzero value of the first moment of the cross-relaxation function leads to the fact that the maximum of $f_{i\alpha}^{CR}$ is located at $\Delta^* = M_{1i\alpha}$. This fact will cause a complicated behaviour of CR: if at $M_{1i\alpha} = 0$ DDR compensates the difference of resonant frequencies of flip-flopping spins, then at $M_{1i\alpha} \neq 0$ DDR also compensates the difference of local fields at every spin from the mean local field. The flip of each individual spin cannot occur without involving DDR, that leads to "the mixing-up" of evolutions of subsystems $\mathcal{H}_S^{Z*}$, $\mathcal{H}_I^{Z*}$, and $\mathcal{H}_d^*$ during CR. This phenomenon is similar to "the mixing-up" of the Zeeman subsystem and DDR during the spin–lattice relaxation at LT pointed out in [7]. Further on the case will be considered when the local field coincides with the mean one, i.e. $M_{1i\alpha} = 0$. Such a situation is realized if spins are crystallographically equivalent and the sample has ellipsoidal form [1]. Then the maximum of the $f_{i\alpha}^{CR}$-function is located at $\Delta^* = 0$ unlike the results of [5], where the cross-relaxation function is maximum at $\Delta = 0$. The equations of [5] coincide with (5), if LT displacements are introduced in the former equations and it is assumed that $\langle \mathcal{H}_k \mathcal{H}_{k'} \rangle - \langle \mathcal{H}_k \rangle \langle \mathcal{H}_{k'} \rangle = 0$ (where $k, k' = I, S, \mathrm{d}$), which is fulfilled automatically for the redefined subsystems if, as is accepted, $\omega_\mathrm{d}^* \beta_\mathrm{d} \ll 1$.

3. Solutions of (5) in Particular Cases and Discussion

The set of equations (5) has two first integrals

$$N_S p_S + N_I p_I = \text{const}\,, \tag{7}$$

$$-2N_S(\Delta - a_I p_I)\, p_S - N_S a_S p_S^2 + \tfrac{1}{2}[-2N_I(\Delta + a_S p_S)\, p_I + N_I a_I p_I^2 + (N_I + N_S)\, \omega_\mathrm{d}^{*2} \beta_\mathrm{d}] = \text{const}\,, \tag{8}$$

where

$$a_I = \tfrac{3}{2}\overline{a_{II}} - \tfrac{1}{2}\overline{a_{IS}}\,; \qquad a_S = \tfrac{3}{2}\overline{a_{SS}} - \tfrac{1}{2}\overline{a_{SI}}\,.$$

The quasi-stationary solution of (5) under arbitrary initial conditions is obtained by the joint solution of (7), (8), and the equation

$$\omega_S^* \beta_S - \omega_I^* \beta_I - \Delta^* \beta_\mathrm{d} = 0\,. \tag{9}$$

Further on the frequency shift will be accounted for only in Δ^*. The obtained set of equations is transcendental and in the general case its solution seems impossible; that is why the following particular cases will be considered:

(i) The case of insignificant deviations from the initial state. It is assumed that at the initial moment of time spins S are in the saturated state ($\beta_S = 0$), while spins I and DDR are in equilibrium with the lattice,

$$\beta_I(0) = \beta_\mathrm{d}(0) = \beta_\mathrm{L}\,.$$

We consider the first state of evolution of the spin system under the action of CR, when deviations from the initial values are insignificant. In that case, after a certain interval of time τ, the values of inverse temperatures can be presented in the following manner:

$$\beta_S \approx \frac{p_{S\mathrm{L}}}{N_S \omega_S} W^\mathrm{CR}(-\Delta + a_I p_{I\mathrm{L}})\,\tau\,,$$

$$\beta_I \approx \beta_\mathrm{L} - \frac{p_{S\mathrm{L}}}{N_I \omega_I (1 - p_{I\mathrm{L}}^2)} W^\mathrm{CR}(-\Delta + a_I p_{I\mathrm{L}})\,\tau\,,$$

$$\beta_\mathrm{d} \approx \beta_\mathrm{L} + \frac{(\Delta - a_I p_{I\mathrm{L}})\, p_{S\mathrm{L}}}{(N_I + N_S)\, \omega_\mathrm{d}^{*2}\Big|_{\substack{p_S = 0 \\ p_I = p_{I\mathrm{L}}}}} W^\mathrm{CR}(-\Delta + a_I p_{I\mathrm{L}})\,\tau\,,$$

where

$$p_{m\mathrm{L}} = \tanh\left(\frac{\omega_m \beta_\mathrm{L}}{2}\right).$$

Let us compare deviations of β_d from β_L in the case of low and high temperatures,

$$\frac{|\beta_\mathrm{d} - \beta_\mathrm{L}|^\mathrm{LT}}{|\beta_\mathrm{d} - \beta_\mathrm{L}|^\mathrm{HT}} \sim \frac{\tanh\,(\omega_S \beta_\mathrm{L}/2)^\mathrm{LT}}{(\omega_S \beta_\mathrm{L}/2)^\mathrm{HT}}\,. \tag{10}$$

As known [8, 9], the deviation of DDR from equilibrium occurs as a result of the action of CR. It is evident from (10) that at low temperatures the deviation of DDR is greater than at high temperatures.

(ii) Qualitative remarks about the quasi-stationary solution of the set of equations (5) and its numerical solution. It is assumed that at the initial moment of time spins I

and DDR are in equilibrium with the lattice, spins S are saturated and subsequently are sustained in this state.

The evolution of β_I and β_d occurs due to CR. In that case (8) takes the form

$$\frac{N_I}{a_I}(\Delta - a_I p_I)^2 - (N_I + N_S)\,\omega_d^{*2}|_{p_S=0}\,\beta_d =$$

$$= \frac{N_I}{a_I}(\Delta - a_I p_{IL})^2 - (N_I + N_S)\,\omega_d^{*2}\Big|_{\substack{p_S=0\\ p_I=p_{IL}}}\beta_L\,. \tag{11}$$

To simplify this expression we assume that

$$\frac{N_I}{N_S} \ll 1\,; \qquad \sum_{i\alpha} (U_{i\alpha}^{zz})^2 \ll \omega_{dS}^2$$

(i.e. $\gamma_I \ll \gamma_S$) and denote

$$Z = \frac{\Delta}{a_I} - p_I\,.$$

Then (11) can be approximately written as

$$-Z^3 + Z\left(\frac{\Delta}{a_I} - p_{IL}\right)^2 = 2\,\frac{N_S}{N_I}\,\frac{\omega_{dS}^2}{a_I^2}\,\operatorname{arctanh}\left(Z - \frac{\Delta}{a_I}\right). \tag{12}$$

Let us consider (12) in the region of parameter values, where it has an unique real solution. Analyzing the curves of the left- and right-hand sides of (12), it can be concluded that at $\Delta < 0$ and $\Delta > a_I p_{IL}$ the equation is satisfied by the values of p_I in the interval $0 < p_I < p_{IL}$. If $a_I p_{IL}/2 < \Delta < a_I p_{IL}$, then $0 < p_I < (2\Delta/a_I - p_{IL})$, and β_d is in the interval

$$0 < \beta_d < \frac{N_I}{N_S}\,\frac{a_I p_{IL}}{\omega_{dS}^2}\left(2\,\frac{\Delta}{a_I} - p_{IL}\right).$$

However, if $0 < \Delta < a_I p_{IL}/2$, (12) is satisfied by the values of p_I in the interval $(2\Delta/a_I - p_{IL}) < p_I < 0$. The corresponding values of β_d are also negative and lie in the interval

$$\frac{N_I}{N_S}\,\frac{a_I p_{IL}}{\omega_{dS}^2}\left(2\,\frac{\Delta}{a_I} - p_{IL}\right) < \beta_d < 0$$

in spite of the fact that the source of heating of I spins (the saturated line of S spins) has a smaller frequency than the resonant frequency of I spins established as a result of the action of CR $\big((\omega_I^* - \omega_S) > 0\big)$ (Fig. 1). When the LT displacement of I spins is absent, then $\beta_d > 0$ at $\Delta > 0$. Such a behaviour can be explained by the fact that in the LT case the mean energy absorbed by DDR during the CR takes the form

$$\frac{N_S}{4}\,\omega_{dS}^2(\beta_d - \beta_L) = \frac{N_I \Delta}{2}\,(p_{IL} - p_I) - \frac{N_I a_I}{4}\,(p_{IL}^2 - p_I^2) \tag{13}$$

while only a first term is present in high-temperature approximation.

If $\Delta > a_I p_{IL}/2$, the first term of (13) plays the main role and the qualitative results of high-temperature theory remain valid. At $0 < \Delta < a_I p_{IL}/2$ the mean energy absorbed by DDR is defined mostly by the change of the LT displacement induced by heating of I spins (the second term of (13) having the sign opposite to the sign of the

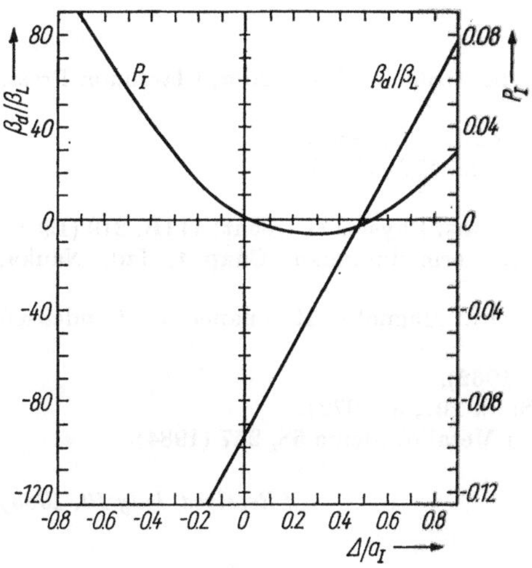

Fig. 1. Dependences $p_I(\Delta/a_I)$ and $\beta_d/\beta_L\,(\Delta/a_I)$ at the following values of parameters:

$$\omega_{dS}^2/a_I^2 = 10^{-1}\,, \quad N_I/N_S = 10^{-2}\,,$$
$$p_{IL} = 1\,, \quad a_I\beta_L = 10^{-3}$$

first term). As a result, β_d becomes negative and so does p_I when equilibrium is established in the coordinate system that rotates with the frequency ω_S round the z-axis, giving

$$\beta_d = \frac{2\,\text{arctanh}\,p_I}{\Delta - a_I p_I}\,. \tag{14}$$

At $\Delta = a_I p_{IL}/2$ the mean energy received by DDR is equal to zero, and it remains in equilibrium with the lattice. Because of that DDR does not suppress CR and saturation is transferred to I spins ($p_I = 0$). At $\Delta = 0$, p_I is also zero, but according to (13) there is a DDR shift from equilibrium. Thus, when the LT displacement is taken into account, this leads to the change of CR dynamics even in the case of a unique solution of (12).

The problem under consideration also has a bistability region (a region of parameter values, where there are two stable solutions of (12) for one value of the parameter) if the condition of strong LT displacement is fulfilled,

$$2\,\frac{N_S}{N_I}\,\frac{\omega_{dS}^2}{a_I^2} \ll 1\,.$$

The behaviour of the solution of (12) in the bistability region is analogous to the behaviour of the quasi-stationary polarization at NMR saturation of one kind of nuclei considered in [10]. The role of the control parameter in the problem of CR can be played by the value of Δ, if the latter can be changed by means of turning the sample.

The dependence of p_I on the difference of ω_S and the equilibrium frequency of I spins, $\omega_{IL}^* = \omega_I - a_I p_{IL}$, will be the same as the dependence of p_q on the value $\Omega - \omega_{IL}^*$ in the case of quasi-stationary saturation of I spins by a varying magnetic field with frequency Ω (Fig. 2 in [10]), i.e. a hysteresis with one point of frustration will take place.

If (12) allows an approximation arctanh $p_I \approx p_I$, then all the formulas from [10] are valid for the CR problem with the replacement

$$\frac{2\omega_d^2}{U^2} \to 2\,\frac{N_S}{N_I}\,\frac{\omega_{dS}^2}{a_I^2}\,.$$

References

[1] A. Abragam and M. Goldman, Nuclear Magnetism: Order and Disorder, Clarendon Press, Oxford 1982.
[2] L. L. Buishvili and N. P. Fokina, Fiz. tverd. Tela **25**, 381 (1983).
[3] L. L. Buishvili and N. P. Fokina, Fiz. tverd. Tela **25**, 1761 (1983).
[4] J. Philippot, Phys. Rev. **133**, A471 (1964).
[5] L. J. de Haas, W. Th. Wenckebach, and N. J. Poulis, Physica (Utrecht) **111B**, 219 (1981).
[6] D. N. Zubarev, Neravnovesnaya statisticheskaya termodinamika, Chap. 4, Izd. Nauka, Moscow 1971.
[7] B. I. Kochelaev, Transactions of All-Union Conf. Magnetic Resonance in Condensed Matter, Chap. 3, Kazan 1984 (p. 40).
[8] B. N. Provotorov, Zh. eksper. teor. Fiz. **42**, 882 (1962).
[9] V. A. Atsarkin and M. I. Rodak, Uspekhi fiz. Nauk **107**, 3 (1972).
[10] L. L. Buishvili and N. P. Fokina, Fiz. Metallov i Metallovedenie 58, 257 (1984).

(Received July 10, 1985)

phys. stat. sol. (b) 133, 135 (1986)

Subject classification: 73.40 and 75.10

Institute of Physics, Academy of Sciences of the Ukrainian SSR, Kiev[1])

Electron Potential Energy near the Ferromagnetic Metal–Vacuum Interface

By

A. M. GABOVICH and A. I. VOITENKO

The charge electrostatic energy $W(z)$ near the surface of a metal with an arbitrary spontaneous magnetization $0 \leqq \zeta \leqq 1$ is calculated using the random-phase approximation for the metal dielectric function. It is shown that the variation of the surface value $W(0)$ under the ferromagnetic transition amounts up to 12% of its original value and depends on free-electron concentration. Such a variation can be observed in LEED experiments.

Вычислена электростатическая энергия $W(z)$ заряда вблизи поверхности металла с произвольной спонтанной намагниченностью $0 \leqq \zeta \leqq 1$, причем диэлектрическая проницаемость металла учитывалась в рамках приближения хаотических фаз. Показано, что изменение поверхностного значения $W(0)$ при ферромагнитном переходе составляет до 12% первоначальной величины и зависит от концентрации свободных электронов. Такое изменение может быть наблюдаемо в экспериментах по дифракции медленных электронов.

1. Introduction

On the basis of the one-electron solid state theory [1] and taking into account many-particle correlations within the framework of the density-functional formalism [2 to 7] it is possible to construct a reasonable electron potential energy $W(z)$ [7 to 10] near the metal–vacuum interface ($z = 0$). In the bulk of metal ($z > 0$) $W(z)$ is periodic in z, and in vacuum ($z < 0$) far from the inhomogeneous surface layer [4, 5, 8 to 11] the $W(z)$ asymptotics, when $z \to -\infty$, coincides with a classical polarization energy of an electron $W_0(z) = -e^2/4\,|z|$ [12] (image force energy). If one is not interested in the anisotropy of surface electron properties [8 to 11], such a behaviour of $W(z)$ can be qualitatively deduced [4 to 6] from the much simpler jellium model [1, 13] where discrete ions are replaced by a homogeneous compensating background. So, when $z \to \infty$, instead of the periodic potential energy we have in the jellium model an averaged internal energy $W(\infty)$ [4].

Even in this case the self-consistent calculations of $W(z)$ are very laborious and complicated. The essence of further simplifications consists of abandoning the self-consistency requirement and making use of the approximation of a sharp interface with infinite potential barrier for metal electrons [14 to 24]. Then $W(z)$ turns out [21, 22] to be a functional of the bulk dielectric function $\varepsilon(\boldsymbol{k})$ of a metal [13] taking into account its spatial dispersion.

The Green's function formulation [21 to 23] allowed the dependence of $W(z)$ on the Fermi surface shape and on the electron–electron interaction in metals to be determined. Calculations for semiconductor surfaces were carried out as well [22, 25, 26].

---

[1]) Prospekt Nauki 144, 252650 Kiev 28, USSR.

Until recently there were no experiments to distinguish between real $W(z)$ and classical $W_0(z)$. (The suggestion [27] that the saturation of $W(z)$ at the interface leads to a violation of the Fowler-Nordheim law for the field emission [28] is not justified [29]). In a series of papers [30 to 32] the potential barrier at the interface was investigated by low-energy electron diffraction (LEED) method. The real part of $W(z)$ was approximated as follows:

$$\mathrm{Re}\, W(z) = \begin{cases} -\dfrac{e^2}{4\,|z - z_0|} + \dfrac{\eta e^2}{4(z - z_0)^2}\,; & z < z_s \\ -\,U_0; & z \geqq z_s\,. \end{cases} \tag{1}$$

Here $z_s > z_0 > 0$, η, and $U_0 = -W(\infty)$ are constants determined from experimental data. Thus, Jennings et al. [30 to 32] stated the LEED data for Cu(001) and W(001) to make possible an estimate of the internal electron energy for these metals. In the subsequent high-precision LEED experiments for Cu(001) surface [33] $W(z)$ was shown (see also calculations [34]) to reach at the interface a finite value $W(0) \neq W(\infty)$, in agreement with the theory [17, 18, 21, 22]. Nevertheless, the trend of $W(z)$ into the metal cannot be determined from these LEED experiments, but the quantity $W(\infty)$ itself can be found e.g. making use of the interference electron spectroscopy [35]. It is of interest that the value $W(0) = -7$ eV found for Cu in [33] amounts to 58% of the internal electron energy for copper $W(\infty) = -12$ eV and is rather close to the universal value $\delta = W(0)/W(\infty) = 2/3$ for the simple Thomas-Fermi model [18, 21].

Thus, an experimental opportunity to determine the surface energy $W(0)$ became available so that any considerable change of the electron gas screening properties may be observed. In contrast to semiconductors, the current carrier concentration in metals cannot be altered by doping or temperature variation. However, as was suggested for the first time in [36] there is another possibility to affect $\varepsilon(\boldsymbol{k})$ (and hence $W(0)$). It consists in investigating metals with a low-temperature transition to the state of a strong or weak ferromagnet. This problem is considered here in detail.

2. Formulation and Results of Calculations

Until now there is no full microscopic theory of ferromagnetic electron gas screening properties which takes into account the spin-dependent many-body interactions (see [37 to 40]). Therefore, we confine ourselves to the random-phase approximation (RPA) [13], where $\varepsilon(k)$ has the form

$$\varepsilon(k) = 1 + \frac{4\pi e^2}{k^2}\,\{\Pi_+(k) + \Pi_-(k)\}\,, \tag{2}$$

where e is the electron charge, $\Pi_\pm(k)$ are the static polarization operators for the appropriate subbands of the spin-polarized electron gas [41]. The quantities $\Pi_\pm$ are defined by ($\hbar = 1$)

$$\Pi_\pm(k) = \frac{m k_{\mathrm{F}\pm}}{2\pi^2}\, f_\mathrm{L}(k, k_{\mathrm{F}\pm})\,; \qquad k_{\mathrm{F}\pm}^3 = 3\pi^2 n\,(1 \pm \zeta)\,, \tag{3}$$

where $f_\mathrm{L}(k, k_\mathrm{F})$ is the Lindhard function

$$f_\mathrm{L}(k, k_\mathrm{F}) = \frac{1}{2} + \frac{k_\mathrm{F}}{2k}\left(1 - \frac{k^2}{4k_\mathrm{F}^2}\right) \ln \left|\frac{k + 2k_\mathrm{F}}{k - 2k_\mathrm{F}}\right|, \tag{4}$$

n is the electron density in the paramagnetic state, the relative magnetization ζ is determined by the relationship

$$\zeta = \frac{|n_+ - n_-|}{n}\,, \tag{5}$$

and $n_\pm$ are electron densities in each of the spin-split subbands $(n = n_+ + n_-)$. So, ferromagnet screening properties are determined by two dimensionless parameters ζ and $\alpha = me^2/\pi k_F$. The values of the electron density parameter α range from 0.2 to 0.9 for various metals [1, 13], whereas ζ, by definition, may rise from 0 for paramagnets to 1 for strong ferromagnets.

According to [21, 22] the charge Coulomb energy near the metal–vacuum interface has the form

$$W(z) = -e^2 \int_0^\infty q\,dq\,D(q, z)\,; \tag{6}$$

$$D(q, z) = \begin{cases} \dfrac{1}{2q}\,e^{2qz}\,\dfrac{1 - qa(q, z)}{1 + qa(q, z)}\,; & z < 0 \\[2ex] \dfrac{qa^2(q, z)}{1 + qa(q, 0)} - \dfrac{1}{2}\,[a(q, 0) + a(q, 2z)] + \dfrac{1}{2q}\,; & z > 0\,, \end{cases} \tag{7}$$

where

$$a(q, z) = \frac{1}{\pi} \int_{-\infty}^{\infty} \frac{\cos k_\perp z\,dk_\perp}{(k_\perp^2 + q^2)\,\varepsilon(\sqrt{k_\perp^2 + q^2})}\,. \tag{8}$$

In the quasiclassical Thomas-Fermi approximation (TFA) [1, 13] the dielectric function $\varepsilon(k)$ is determined by

$$\varepsilon(k) = 1 + \frac{\tilde{\varkappa}^2}{k^2}\,; \tag{9}$$

$$\tilde{\varkappa}^2 = \frac{\varkappa_{TF}^2}{2}\,[(1 + \zeta)^{1/3} + (1 - \zeta)^{1/3}]\,, \tag{10}$$

where $\varkappa_{TF} = (6\pi ne^2/E_F)^{1/2}$ is the inverse Thomas-Fermi length, E_F the electron Fermi energy in the paramagnetic state. In the TFA case one can use Sidyakin's results [18] (see also [19 to 22, 29]) accurate to the substitution of $\varkappa_{TF}$ by $\tilde{\varkappa}$,

$$a(q, z) = \frac{1}{(q^2 + \tilde{\varkappa}^2)^{1/2}} \exp\{-|z|\,(q^2 + \tilde{\varkappa}^2)^{1/2}\}\,, \tag{11}$$

$$W(z) = \begin{cases} \dfrac{e^2}{4z}\left\{1 + \dfrac{\pi}{(\tilde{\varkappa}z)^2}\,[\mathrm{H}_0(2\tilde{\varkappa}z) - \mathrm{N}_0(2\tilde{\varkappa}z)] + \dfrac{\pi}{\tilde{\varkappa}z}\,[\mathrm{H}_1(2\tilde{\varkappa}z) - \mathrm{N}_1(2\tilde{\varkappa}z)]\right\}; & z < 0\,, \quad (12) \\[2ex] -\dfrac{e^2\tilde{\varkappa}}{2}\left\{1 - \dfrac{1}{2\tilde{\varkappa}z}\left[\left(1 + \dfrac{1}{\tilde{\varkappa}z}\right)^2 \exp\{-2\tilde{\varkappa}z\} - 2\mathrm{K}_2(2\tilde{\varkappa}z)\right]\right\}; & z > 0\,, \quad (13) \end{cases}$$

where H_i, N_i, and K_i are the Struve, Neumann, and McDonald functions of the i-th order, respectively. We are particularly interested in the $W(z)$ function values when $z = 0$ and $z \to \infty$, i.e. in surface and internal charge potential energies. From (12) and (13) we have

$$W(0) = -\frac{e^2\tilde{\varkappa}}{3}, \qquad W(\infty) = -\frac{e^2\tilde{\varkappa}}{2}. \tag{14}$$

Hence it follows that in the TFA the ratio δ of the quantities concerned does not depend on magnetization and electron density, being the universal constant: $\delta = W(0)/W(\infty) = 2/3$. Expanding (14) in ζ for a paramagnetic electron gas ($\zeta \to 0$) and in $(1 - \zeta)$ for a strong ferromagnet ($\zeta \to 1$), we obtain

$$W_{\text{para}}(0) = \frac{2}{3} W_{\text{para}}(\infty) = -\frac{e^2\varkappa_{\text{TF}}}{3}\left(1 - \frac{\zeta^2}{18}\right); \qquad \zeta \to 0\,, \tag{15}$$

$$W_{\text{ferro}}(0) = \frac{2}{3} W_{\text{ferro}}(\infty) = -\frac{e^2\varkappa_{\text{TF}}}{3 \times 2^{1/3}}\left[1 + \frac{1}{2^{4/3}}(1 - \zeta)^{1/3}\right]; \quad \zeta \to 1\,. \tag{16}$$

Thus in the TFA the absolute value of metal surface energy reduces approximately by 20%, independent of electron concentration, when a phase transition to the strong ferromagnetic state occurs.

Note that although the TFA for $\varepsilon(\boldsymbol{k})$ is qualitatively correct in considering surface electron properties, they can be substantially affected by the quantum character of electron gas screening. Indeed, first the Lindhard expression for $\varepsilon(k)$ (see (2) to (4) when $\zeta = 0$) has a large-k symptotics different from (9),

$$\varepsilon_{\text{L}}(k) = 1 + \frac{4\varkappa_{\text{TF}}^2 k_{\text{F}}^2}{3k^4}, \tag{17}$$

where k_{F} is the Fermi number. This fact is of importance in the integral (6) where the large-q contribution is very significant for small z, i.e. near the interface. Second the logarithmic singularity of $\mathrm{d}f_{\text{L}}(k, k_{\text{F}})/\mathrm{d}k$ when $k = 2k_{\text{F}}$ and $\zeta = 0$ leads [21, 22] to electron energy Friedel oscillations damping in the metal depth. These oscillations would survive also for a ferromagnet. The weak-ferromagnetism case ($0 < \zeta < 1$) is of special interest, when a nontrivial interference of both spin subband contributions should be observed. In this paper we carried out numerical calculations of $W(z)$ according to (6) to (8) taking into account (2) to (4).

In Fig. 1 the dependences of the surface energy $W(0)$ on magnetization ζ for various $\alpha = 0.1, 0.5$, and 0.9 are shown by dashed (TFA) and solid (RPA) curves. For all α $|W(0)|$ in the RPA is less than that in the TFA, and the dependences of $W(0)$ on ζ become smoother.

In Fig. 2 the ratio $\delta = W(0)/W(\infty)$ calculated in the RPA as a function of ζ is shown for $\alpha = 0.1, 0.5$, and 0.9 (curves 1 to 3). For comparison the universal quasi-classical ratio $\delta = 2/3$ is marked by a dashed line. In the RPA the variation of δ with ζ exists but it disappears provided the electron gas density increases, i.e. for small α (see curve 1). The qualitative similarity of electron gas properties in the TFA and RPA in this limit takes place due to the primary role of the small momentum contribution to the energy [13], when the functions $\varepsilon(k)$ in both approximations coincide. Moreover, many-body correlations are insignificant when $\alpha \to 0$ [13]. Therefore, even for a more realistic electron liquid model one should expect the constancy of δ for high electron densities.

The dependences of $W(0)$ on α in the paramagnetic ($\zeta = 0$, curves 1) and in the ferromagnetic ($\zeta = 1$, curves 2) cases are shown in Fig. 3. Dashed curves correspond to the TFA and solid curves to the RPA. The behaviour of $W(0)$ with variation of α does not change when the ferromagnetic transition occurs. The dependence of $W(0)$ on α provided $\zeta = 0$ and taking into account the many-body interactions in the Taylor

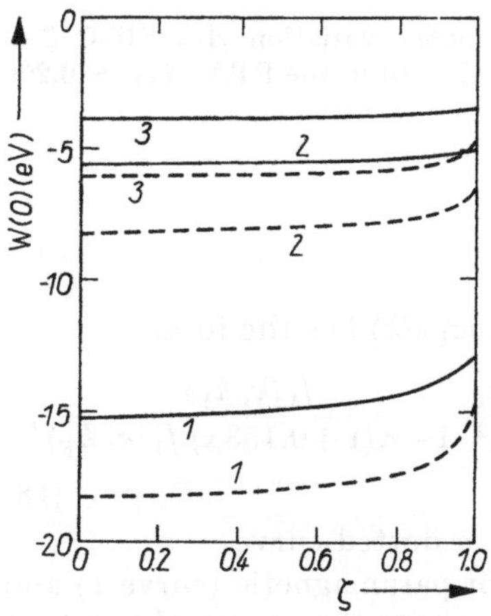

Fig. 1

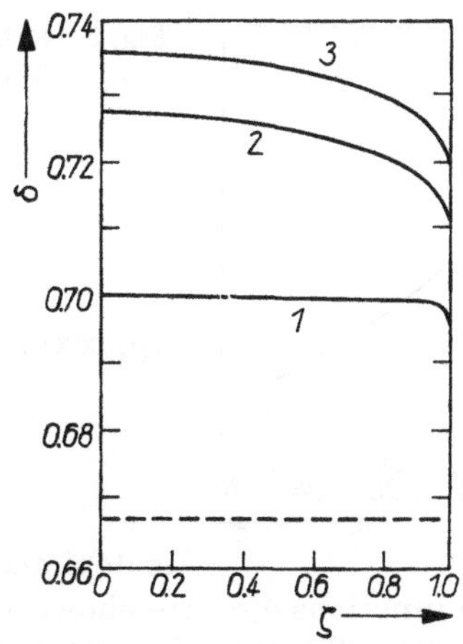

Fig. 2

Fig. 1. The dependences of the surface potential energy $W(0)$ on relative magnetization ζ in the TFA (dashed curves) and the RPA (solid curves). Curves 1, 2, 3 correspond to the values of the electron density parameter $\alpha = me^2/\pi k_F = 0.1, 0.5$, and 0.9, respectively

Fig. 2. The dependences of the ratio $\delta = W(0)/W(\infty)$ on ζ for the RPA and (1) $\alpha = 0.1$, (2) 0.5, and (3) 0.9. Dashed line corresponds to the TFA

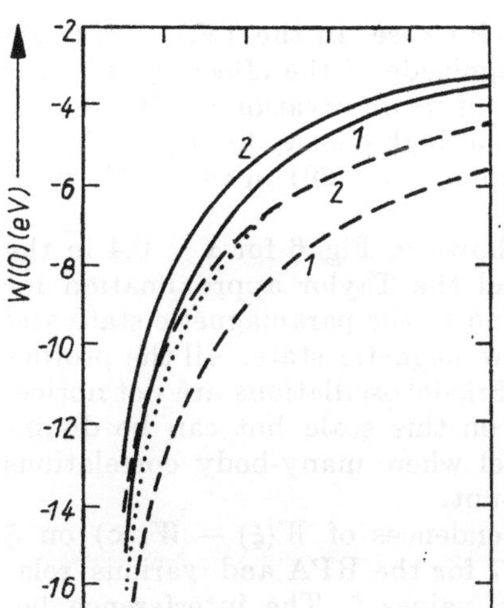

Fig. 3

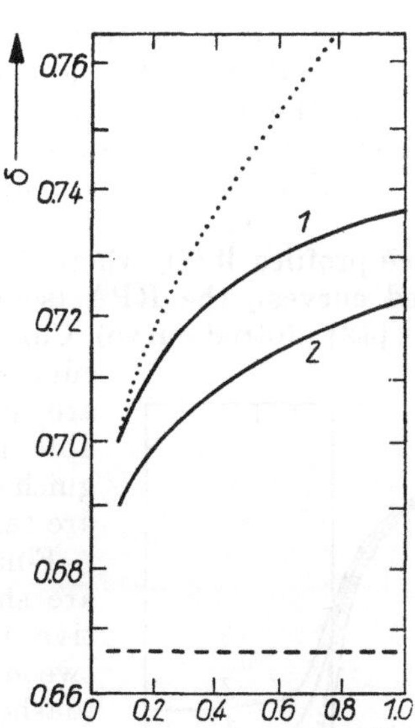

Fig. 4

Fig. 3. The dependences of $W(0)$ on α in the TFA (dashed curves), the RPA (solid curves), and the Taylor model [42] (dotted curve). Curves 1 correspond to the paramagnetic state ($\zeta = 0$) and curve 2 to the strong ferromagnet ($\zeta = 1$). Dotted curve corresponds to $\zeta = 0$

Fig. 4. The curves $\delta(\alpha)$ for the TFA (dashed line, $\zeta = 0$ and $\zeta = 1$), the Taylor model (dotted curve, $\zeta = 0$), the RPA (solid curves, (1) $\zeta = 0$ and (2) $\zeta = 1$)

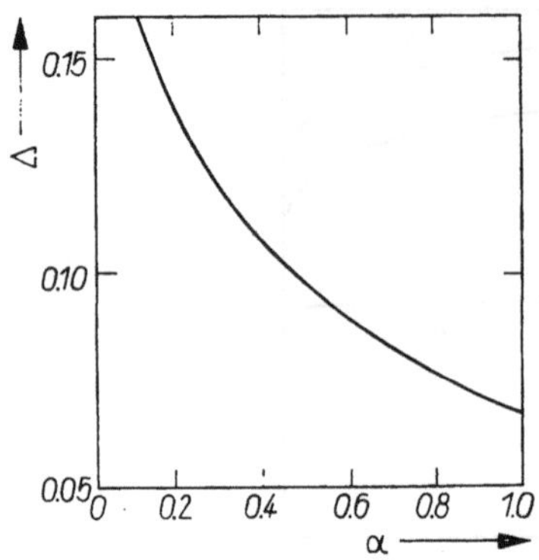

Fig. 5. Relative surface energy variation $\Delta = (W(0, \zeta = 0) - W(0, \zeta = 1))/W(0, \zeta = 0)$ in the RPA. $\Delta_{TF} \approx 0.206$

approximation [42], when $\varepsilon(k)$ has the form

$$\varepsilon(k) = 1 + \frac{\varkappa_{TF}^2}{k^2} \frac{f_L(k, k_F)}{1 - \alpha(1 + 0.153\alpha) f_L(k, k_F)}, \tag{18}$$

is depicted in Fig. 3 by a dotted curve.

In Fig. 4 the functions $\delta(\alpha)$ are shown in the RPA for paramagnetic (curve 1) and ferromagnetic (curve 2) states of a metal. In the TFA we have the universal constant value (dashed line). The dotted curve corresponds to the Taylor model of the paramagnetic metal. One can see that the interelectron correlations lead to a steeper dependence $\delta(\alpha)$.

The most important quantity determined in this paper is the relative variation Δ of the metal surface energy below the ferromagnetic transition,

$$\Delta = \frac{W(0, \zeta = 0) - W(0, \zeta = 1)}{W(0, \zeta = 0)}. \tag{19}$$

Its dependence on α is depicted in Fig. 5 for the RPA case. In the TFA $\Delta \approx 0.206$ and does not depend on α. As one can see, the magnitude of the effect ($\lessapprox 12\%$) is not as large as in the quasiclassical approximation but its observation is still possible by the method of Dietz et al. [33]. Note that in the high-density limit ($\alpha \to 0$) the RPA and TFA values of Δ become close. For example, $\Delta \approx 0.201$ in the RPA when $\alpha = 10^{-3}$.

The surface profiles $W(\xi)$, where $\xi = 2k_Fz$, are shown in Fig. 6 for $\alpha = 0.4$ in the TFA (dashed curves), the RPA (solid curves), and the Taylor approximation for paramagnets [42] (dotted curve). Curves 1 correspond to the paramagnetic state and curves 2 to the ferromagnetic state. All the profiles are quite similar. Friedel oscillations are not noticeable in the RPA on this scale but can be distinguished in the model where many-body correlations are taken into account.

Finally, the dependences of $W(\xi) - W(\infty)$ on ξ are shown in Fig. 7 for the RPA and various relative magnetization values ζ. The interference between spin-split subbands can be seen for $\zeta = 0.5$ (dashed curve) and $\zeta = 0.75$ (dash-dotted curve).

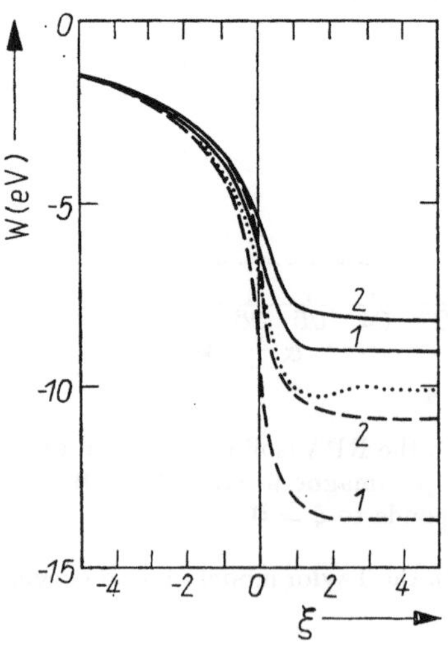

Fig. 6. The profiles $W(\xi)$ near the interface ($\xi = 2k_Fz$) for $\alpha = 0.4$. Dashed curves TFA, solid curves RPA, dotted curve model [42]. Curves 1 correspond to $\zeta = 0$, curves 2 to $\zeta = 1$

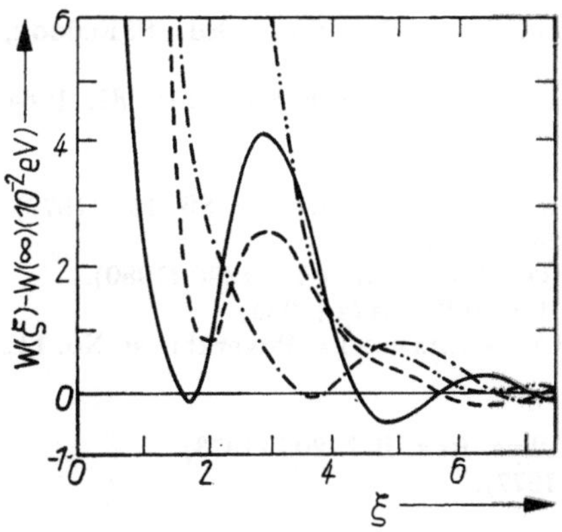

Fig. 7. Friedel oscillations of the difference $W(\xi) - W(\infty)$ in the RPA for $\alpha = 0.4$. $\zeta = 0$ (———), 0.5 (– – –), 0.75 (·—·—·), 1 (—··—··)

Now we are not able to suggest an experimental method to observe such delicate peculiarities of the energy (≈ 0.01 eV) against the large values of $W(z) \approx -10$ eV.

To summarize, our calculations prove the possibility of the ferromagnetic transition impact on the surface electron properties of metals. It can be observed in LEED experiments. For their unambiguous interpretation the surface magnetic structure must coincide with the bulk one.

References

[1] A. O. E. Animalu, Intermediate Quantum Theory of Crystalline Solids, Prentice-Hall, Englewood Cliffs (N.J.) 1977.
[2] P. Hohenberg and W. Kohn, Phys. Rev. **136**, B864 (1964).
[3] W. Kohn and L. J. Sham, Phys. Rev. **140**, A1133 (1965).
[4] N. D. Lang, Solid State Phys. **28**, 225 (1973).
[5] M. B. Partenskii, Uspekhi fiz. Nauk **128**, 69 (1979); **129**, 590 (1979).
[6] M. B. Partenskii, Poverkhnost No. 10, 15 (1982).
[7] E. Zaremba, Lecture Notes Phys. **187**, 167 (1983).
[8] J. A. Appelbaum, Electronic Structure of Solid Surfaces, in: Surface Physics of Materials, Vol. 1, Ed. J. M. Blakely, Academic Press, Inc., 1975 (p. 79).
[9] J. A. Appelbaum and D. R. Hamann, Rev. mod. Phys. **48**, 479 (1976).
[10] J. E. Inglesfield, Rep. Progr. Phys. **45**, 223 (1982).
[11] V. F. Ukhov, R. M. Kobeleva, G. V. Dedkov, and A. I. Temrokov, Elektronnostatisticheskaya teoriya metallov i ionnykh kristallov, Izd. Nauka, Moskva 1982.
[12] J. D. Jackson, Classical Electrodynamics, John Wiley & Sons, New York 1962.
[13] D. Pines and P. Nozieres, The Theory of Quantum Liquids, W. A. Benjamin, Inc., New York 1966.
[14] H. J. Juretschke, Phys. Rev. **92**, 1140 (1953).
[15] T. L. Loucks and P. H. Cutler, J. Phys. Chem. Solids **25**, 105 (1964).
[16] R. W. Davies, Surface Sci. **11**, 419 (1968).
[17] D. M. Newns, J. chem. Phys. **50**, 4572 (1969); Phys. Rev. B **1**, 3304 (1970).
[18] A. V. Sidyakin, Zh. eksper. teor. Fiz. **58**, 573 (1970).
[19] J. Heinrichs, Phys. Rev. B **8**, 1346 (1973).
[20] A. A. Kornyshev, A. I. Rubinshtein, and M. A. Vorotyntsev, phys. stat. sol. (b) **84**, 125 (1977); J. Phys. C **11**, 3307 (1978).
[21] A. M. Gabovich, L. G. Ilchenko, E. A. Pashitskii, and Yu. A. Romanov, Zh. eksper. teor. Fiz. **75**, 249 (1978).
[22] A. M. Gabovich, L. G. Ilchenko, and E. A. Pashitskii, Fiz. tverd. Tela **21**, 1683 (1979).
[23] L. G. Ilchenko, E. A. Pashitskii, and Yu. A. Romanov, Fiz. tverd. Tela **22**, 2700 (1980).
[24] F. Bechstedt, R. Enderlein, and D. Reichardt, phys. stat. sol. (b) **117**, 261 (1983).
[25] A. I. Voitenko, A. M. Gabovich, and L. G. Ilchenko, Fiz. tverd. Tela **23**, 1531 (1981).
[26] A. M. Gabovich and A. I. Voitenko, phys. stat. sol. (b) **110**, 407 (1982).
[27] T. J. Lewis, Phys. Rev. **101**, 1694 (1956).

[28] R. H. GOOD, JR. and E. A. MÜLLER, Field Emission, in: Hdb. Phys., Ed. S. FLÜGGE, Vol. 21, Springer-Verlag, Berlin 1956 (p. 176).
[29] A. I. VOITENKO, A. M. GABOVICH, and V. M. ROZENBAUM, Zh. eksper. teor. Fiz. 87, 1064 (1984).
[30] M. N. READ and P. J. JENNINGS, Surface Sci. 74, 54 (1978).
[31] G. L. PRICE, P. J. JENNINGS, P. E. BEST, and J. C. L. CORNISH, Surface Sci. 89, 151 (1979).
[32] P. J. JENNINGS and G. L. PRICE, Surface Sci. 93, L124 (1980).
[33] R. E. DIETZ, E. G. MCRAE, and R. L. CAMPBELL, Phys. Rev. Letters 45, 1280 (1980).
[34] R. O. JONES, P. J. JENNINGS, and O. JEPSEN, Phys. Rev. B 29, 6474 (1984).
[35] V. N. BYKOV, YU. A. KULYUPIN, S. A. NEPIIKO, and V. G. SHAMONYA, Poverkhnost No. 11, 78 (1984).
[36] A. M. GABOVICH, Fiz. tverd. Tela 25, 3179 (1983).
[37] D. J. KIM, B. B. SCHWARTZ, and H. C. PRADDAUDE, Phys. Rev. B 7, 205 (1973).
[38] J. C. STODDART and P. HANKS, J. Phys. C 10, 3167 (1977).
[39] A. K. RAJAGOPAL, Phys. Rev. B 17, 2980 (1978).
[40] A. H. MACDONALD and R. TAYLOR, Solid State Commun. 38, 995 (1981).
[41] A. K. RAJAGOPAL, H. BROOKS, and N. R. RANGANATHAN, Nuovo Cimento, Suppl. 5, 807 (1967).
[42] R. TAYLOR, J. Phys. F 8, 1699 (1978).

(Received June 13, 1985)

phys. stat. sol. (b) 133, 143 (1986)
Subject classification: 75.10

Department of Physics, Northeast University of Technology, Shenyang (a), Department of Physics and Measurement Technology, University of Linköping[1]) *(b), and Institute of Physics, A. Mickiewicz University, Poznań (c)*

CDW–SDW Transition in the Extended Hubbard Model

By
JIAN ZHONG (a), KAIYI ZHANG (a), K. A. CHAO (b), R. MICNAS (c), and S. ROBASZKIEWICZ (c)

The ground state phase diagram of the extended Hubbard model for arbitrary electron density is investigated with the broken-symmetry Hartree approximation. It is found that for two-sublattice orderings, only the charge-order and the antiferromagnetism exist without the mixed phase.

Das Phasendiagramm des Grundzustands des ausgedehnten Hubbardmodells für beliebige Elektronendichte wird mit der Hartree-Näherung gebrochener Symmetrie untersucht. Es wird gefunden, daß für Zwei-Untergitter-Ordnungszustände nur die Ladungs-Ordnung und der Antiferromagnetismus ohne Mischphase existieren.

1. Introduction

The problem of electron orderings in narrow band systems have attracted much attention in general. One of the Hamiltonians which have been studied extensively in particular is the extended Hubbard model (EHM) Hamiltonian

$$H = \sum_{ij\sigma} t_{ij} c^{\dagger}_{i\sigma} c_{j\sigma} + U \sum_{i} n_{i\uparrow} n_{i\downarrow} + \frac{W}{2} \sum_{ij}{}' \sum_{\sigma\sigma'} n_{i\sigma} n_{j\sigma'} - \mu \sum_{i\sigma} n_{i\sigma}\,, \tag{1}$$

where t_{ij} is the hopping integral, U the intraatomic Coulomb interaction, W the nearest neighbor Coulomb interaction, $n_{i\sigma} = c^{\dagger}_{i\sigma} c_{i\sigma}$, and μ the chemical potential. The primed sum runs only over the nearest neighbor sites.

The thermodynamic properties of (1) depend on the electron density n and the lattice dimensionality d. For $d = 1$, although reliable results have been derived with the weak-coupling renormalization group [1] and with the Monte-Carlo simulation [2], exact solutions were obtained only in the limit $U \to \infty$ [3 to 8]. For the general case of arbitrary dimensionality, the phase diagram of (1) in the limit of zero bandwidth ($t_{ij} = 0$) exhibits the high charge-ordered and the low charge-ordered states, as well as rich critical behaviors [9]. The effect of electron hopping on the CO stability has been examined with the Hartree approximation [10] and the high temperature expansion method [11]. For the special case $n = 1$, the EHM was also studied with the local approach on the electron correlation [12].

Earlier works [13, 14] on the EHM have reached the conclusion that the nearest neighbor Coulomb interaction W does not enhance the weak tendency toward ferromagnetism in the pure Hubbard model. Therefore, one interesting problem is the competition between the charge-order (CO) and the antiferromagnetism (AF) or its modifications [15 to 19]. So far the AF–CO transition in the EHM has been studied mainly for $n = 1$ [12]. In this paper we will investigate the AF–CO transition for

[1]) 58183 Linköping, Sweden.

arbitrary electron density within the framework of the broken-symmetry Hartree approximation. In Section 2 we shall formulate the theory. The analytical results are given in Section 3. In Section 4 we check the stabilities of various phases to determine the ground state properties. A short discussion follows in Section 5.

2. Broken-Symmetry Hartree Theory

Let us rewrite the Hamiltonian (1) in the Bloch representation

$$H = \sum_{k\sigma} (\varepsilon_k - \mu)\, c_{k\sigma}^{\dagger} c_{k\sigma} + \frac{1}{2N} \sum_{kk'q\sigma\sigma'} \Big[\Big(\sum_{\xi} e^{i\boldsymbol{q}\cdot\boldsymbol{\xi}}\Big) W + U\delta_{\sigma',-\sigma}\Big]\, c_{k+q,\sigma}^{\dagger} c_{k\sigma} c_{k'-q,\sigma'}^{\dagger} c_{k'\sigma'}, \tag{2}$$

where ε_k and $c_{k\sigma}$ are the Fourier transforms of t_{ij} and $c_{i\sigma}$ respectively, N is the number of sites, and ξ the nearest neighbor vector. From now on we will restrict ourselves to systems which can be divided into two interpenetrating sublattices A and B. That is, there exists a vector $\boldsymbol{Q}$ such that $\exp(i\boldsymbol{Q}\cdot\boldsymbol{R}_j) = 1$ if $\boldsymbol{R}_j$ belongs to A, and $\exp(i\boldsymbol{Q}\cdot\boldsymbol{R}_j) = -1$ if $\boldsymbol{R}_j$ belongs to B.

If we choose the x-axis as the spin quantization axis, then we can introduce the order parameters

$$M_0 = \frac{2}{N} \sum_j \langle s_j^x \rangle = \frac{2}{N} \sum_j \langle c_{j\sigma}^{\dagger} c_{j,-\sigma} \rangle = \frac{2}{N} \sum_k \langle c_{k\sigma}^{\dagger} c_{k,-\sigma} \rangle \tag{3a}$$

for the ferromagnetic (F) ordering,

$$M_Q = \frac{2}{N} \sum_j \langle s_j^x \rangle \exp(i\boldsymbol{Q}\cdot\boldsymbol{R}_j) = \frac{2}{N} \sum_j \langle c_{j\sigma}^{\dagger} c_{j,-\sigma} \rangle \exp(i\boldsymbol{Q}\cdot\boldsymbol{R}_j) = \frac{2}{N} \sum_k \langle c_{k+Q,\sigma}^{\dagger} c_{k,-\sigma} \rangle \tag{3b}$$

for the antiferromagnetic (AF) ordering, and

$$n_Q = \frac{1}{N} \sum_{j\sigma} \langle n_{j\sigma} \rangle \exp(i\boldsymbol{Q}\cdot\boldsymbol{R}_j) = \frac{1}{N} \sum_{k\sigma} \langle c_{k+Q,\sigma}^{\dagger} c_{k\sigma} \rangle \tag{3c}$$

for the charge-order (CO). Since there is no magnetic ordering along the y-axis and the z-axis, we have the conditions

$$\sum_j \langle s_j^y \rangle = \sum_j \langle s_j^z \rangle = \sum_j \langle s_j^y \rangle \exp(i\boldsymbol{Q}\cdot\boldsymbol{R}_j) = \sum_j \langle s_j^z \rangle \exp(i\boldsymbol{Q}\cdot\boldsymbol{R}_j) = 0\,. \tag{4}$$

Finally, the electron density n is given by

$$n = \frac{1}{N} \sum_{j\sigma} \langle c_{j\sigma}^{\dagger} c_{j\sigma} \rangle = \frac{1}{N} \sum_{k\sigma} \langle c_{k\sigma}^{\dagger} c_{k\sigma} \rangle\,. \tag{5}$$

The broken-symmetry Hartree approximation is introduced via the Bogolyubov variational principle for the free energy F [20]

$$F \leqq F_0 = -\frac{1}{\beta} \ln \operatorname{Tr} \{\exp(-\beta H_0)\} + \langle H - H_0 \rangle_0 + \mu N_e\,, \tag{6}$$

where N_e is the number of electrons, H_0 the trial Hamiltonian, $1/\beta = k_B T$, and $\langle \ldots \rangle_0$ the average with the density matrix $\varrho_0 = \exp(-\beta H_0)/\operatorname{Tr}\{\exp(-\beta H_0)\}$. The broken-symmetry Hartree Hamiltonian H_0 is chosen as

$$H_0 = \sum_k{}' (c^\dagger_{k\uparrow}, c^\dagger_{k\downarrow}, c^\dagger_{k+Q,\uparrow}, c^\dagger_{k+Q,\downarrow}) \times$$

$$\times \begin{pmatrix} \varepsilon_k - \bar\mu & \frac{1}{2} m_0 & -\Delta & \frac{1}{2} m_Q \\ \frac{1}{2} m_0 & \varepsilon_k - \bar\mu & \frac{1}{2} m_Q & -\Delta \\ -\Delta & \frac{1}{2} m_Q & -\varepsilon_k - \bar\mu & \frac{1}{2} m_0 \\ \frac{1}{2} m_Q & -\Delta & \frac{1}{2} m_0 & -\varepsilon_k - \bar\mu \end{pmatrix} \begin{pmatrix} c_{k\uparrow} \\ c_{k\downarrow} \\ c_{k+Q,\uparrow} \\ c_{k+Q,\downarrow} \end{pmatrix}, \tag{7}$$

where $\bar\mu = \mu + A_0$, and A_0, m_0, m_Q and Δ are variational parameters to minimize the trial free energy F_0. The primed summation over k is restricted to the inner half of the Brillouin zone.

The eigensolutions of H_0 can be obtained easily. The energy spectrum consists of four branches

$$E(k, p) = \pm \left[\{\varepsilon_k^2 + (\Delta \pm \tfrac{1}{2} m_Q)^2\}^{1/2} \pm \tfrac{1}{2} m_0 \right] - \bar\mu\,, \tag{8}$$

where $p = 1$ corresponds to the combination of signs $(+, -, +)$, $p = 2$ to $(-, -, -)$, $p = 3$ to $(+, +, -)$, and $p = 4$ to $(-, +, +)$. The trial free energy per site is readily derived as

$$\frac{F_0}{N} = (A_0 + \mu)\, n - \frac{1}{N\beta} \ln \operatorname{Tr} \{\exp(-\beta H_0)\} + \Delta n_Q - \frac{1}{2} m_Q M_Q -$$

$$- \frac{1}{2} m_0 M_0 + \frac{1}{2} Z W (n^2 - n_Q^2) + \frac{1}{4} U (n^2 + n_Q^2 - M - M_Q^2)\,, \tag{9}$$

where Z is the coordination number. If we define $X(kp) = [1 + \exp\{\beta E(k, p)\}]^{-1}$, then the order parameters and the electron density are computed as

$$n = \frac{1}{2N} \sum_{kp} X(kp)\,, \tag{10a}$$

$$n_Q = \frac{1}{4N} \sum_k \left[\frac{\{X(k2) - X(k1)\}(2\Delta - m_Q)}{B_k^-} + \frac{\{X(k4) - X(k3)\}(2\Delta + m_Q)}{B_k^+} \right], \tag{10b}$$

$$M_Q = \frac{1}{4N} \sum_k \left[\frac{\{X(k2) - X(k1)\}(2\Delta - m_Q)}{B_k^-} + \frac{\{X(k3) - X(k4)\}(2\Delta + m_Q)}{B_k^+} \right], \tag{10c}$$

and

$$M_0 = \frac{1}{2N} \sum_k \{X(k1) + X(k2) - X(k3) - X(k4)\}\,, \tag{10d}$$

where $B_k^\pm = \{\varepsilon_k^2 + (\Delta \pm \frac{1}{2} m_Q)^2\}^{1/2}$.

A minimization of the trial free energy F_0 yields the optimum values

$$A_0 = \frac{-n(2ZW + U)}{2}, \tag{11a}$$

$$\Delta = \frac{n_Q(2ZW - U)}{2}, \tag{11b}$$

$$m_Q = -U M_Q\,, \tag{11c}$$

$$m_0 = -U M_0\,, \tag{11d}$$

and

$$\frac{F_0}{N} = (A_0 + \mu)\, n + \frac{1}{4}\,(2ZW + U)\, n^2 + \frac{\Delta^2}{(2ZW - U)} + \\ + \frac{1}{4}\,\frac{(m_0^2 + m_Q^2)}{U} + \frac{1}{2N\beta} \sum_{kp} \ln\,\{1 - X(kp)\}\,. \tag{12}$$

If we substitute (11 a to d) into (10 a to d), we obtain four coupled equations for the chemical potential and the order parameters. For given values of U, W, n and given band structure ε_k, these equations will be solved self-consistently. The ground state properties are derived as the limiting case $\beta \to \infty$.

3. Analytical Solutions

Using a square density of states

$$\varrho(E) = \begin{cases} \dfrac{1}{2D}; & |E| < D \\ 0; & |E| > D\,, \end{cases} \tag{13}$$

the analytical solutions can be obtained for four pure phases, three bi-mixed phases, and one tri-mixed phase. In the following we will list the results.

(i) Pure non-ordered (NO) phase with $\Delta = m_0 = m_Q = 0$.
We have the chemical potential

$$\bar{\mu} = (n - 1)\, D \tag{14a}$$

and the ground state energy per site

$$\frac{E_0(\mathrm{NO})}{N} = \left(\frac{n}{2} - 1\right) nD + \frac{1}{4}\,(2ZW + U)\, n^2\,. \tag{14b}$$

(ii) Pure CO phase with $m_0 = m_Q = 0$ but $\Delta \neq 0$.
We have the chemical potential

$$\bar{\mu} = D\left\{1 - |n - 1| \cosh \frac{2D}{2ZW - U}\right\} \Big/ \sinh \frac{2D}{2ZW - U}\,, \tag{15a}$$

the variational parameter

$$\Delta^2 = \bar{\mu}^2 - (n - 1)^2\, D^2\,, \tag{15b}$$

and the ground state energy per site

$$\frac{E_0(\mathrm{CO})}{N} = D\left\{|n - 1| \Big/ \sinh \frac{2D}{2ZW - U} - \frac{1}{2}\,(n^2 - 2n + 2) \coth \frac{2D}{2ZW - U}\right\} + \\ + \frac{1}{4}\,(2ZW + U)\, n^2\,. \tag{15c}$$

(iii) Pure AF phase with $m_0 = \Delta = 0$ but $m_Q \neq 0$.
We have the chemical potential

$$\bar{\mu} = D\left\{1 - |n - 1| \cosh \frac{2D}{U}\right\} \Big/ \sinh \frac{2D}{U}\,, \tag{16a}$$

the variational parameter

$$\tfrac{1}{4}\, m_Q^2 = \bar{\mu}^2 - (n - 1)^2\, D^2\,, \tag{16b}$$

and the ground state energy per site

$$\frac{E_0(\mathrm{AF})}{N} = D\left\{|n-1|\Big/\sinh\frac{2D}{U} - \frac{1}{2}(n^2-2n+2)\coth\frac{2D}{U}\right\} + \frac{1}{4}(2ZW+U)\,n^2. \tag{16c}$$

(iv) Pure F phase with $m_Q = \Delta = 0$ but $m_0 \neq 0$.
In the region $U < 2D$ where the Hartree theory is reasonable, we found no ferromagnetic solution.

(v) Bi-mixed CO–AF phase with $m_0 = 0$ but $\Delta \neq 0$ and $m_Q \neq 0$.
For $n \neq 1$ this phase does not exist. For $n = 1$ the chemical potential is pinned at $\bar{\mu} = 0$, and the variational parameters satisfy

$$\frac{2\Delta}{2ZW-U} - \frac{m_Q}{U} = \frac{1}{2D}(2\Delta - m_Q)\ln\left|\frac{D + \{D^2 + (\Delta - m_Q/2)^2\}^{1/2}}{\Delta - m_Q/2}\right|, \tag{17a}$$

$$\frac{2\Delta}{2ZW-U} + \frac{m_Q}{U} = \frac{1}{2D}(2\Delta + m_Q)\ln\left|\frac{D + \{D^2 + (\Delta + m_Q/2)^2\}^{1/2}}{\Delta + m_Q/2}\right|. \tag{17b}$$

The ground state energy per site is calculated as

$$\frac{E_0(\mathrm{CO\text{–}AF})}{N} = \frac{1}{4}(2ZW+U) - \frac{1}{4}\left\{D^2 + \left(\frac{\Delta - m_Q}{2}\right)^2\right\}^{1/2} - \frac{1}{4}\left\{D^2 + \left(\frac{\Delta + m_Q}{2}\right)^2\right\}^{1/2}. \tag{17c}$$

(vi) Bi-mixed F–AF phase with $\Delta = 0$ but $m_0 \neq 0$ and $m_Q \neq 0$.
If $n \neq 1$ there is no solution for this phase at any value of $U/2D$. On the other hand, if $n = 1$ there is also no solution in the region $U < 2D$.

(vii) Bi-mixed F–CO phase with $m_Q = 0$ but $m_0 \neq 0$ and $\Delta \neq 0$.
Again we found no such phase if $n \neq 1$, as well as in the region $U < 2D$ if $n = 1$.

(viii) Tri-mixed CO–AF–F phase with $\Delta \neq 0$, $m_0 \neq 0$ and $m_Q \neq 0$.
This phase exists only for two cases. If $m_0/2 - \bar{\mu} + |\Delta - m_Q/2| \geqq 0$, $|\Delta - m_Q/2| \geqq m_0/2 - \bar{\mu}$ and $\Delta + m_Q/2 < m_0/2 + \bar{\mu}$, then the chemical potential and the parameters satisfy

$$m_0 = (n-1)\,U\,, \tag{18a}$$

$$2D(n-1) = [\{\tfrac{1}{2}(n-1)\,U + \bar{\mu}\}^2 - (\Delta + \tfrac{1}{2}m_Q)^2]^{1/2}\,, \tag{18b}$$

$$\frac{2\Delta}{2ZW-U} - \frac{m_Q}{U} = \frac{1}{2D}(2\Delta - m_Q)\ln\left|\frac{D + \{D^2 + (\Delta - \frac{1}{2}m_Q)^2\}^{1/2}}{\Delta - \frac{1}{2}m_Q}\right|, \tag{18c}$$

$$\frac{2\Delta}{2ZW-U} + \frac{m_Q}{U} = \frac{1}{2D}(2\Delta + m_Q)\ln\left|\frac{D + \{D^2 + (\Delta + \frac{1}{2}m_Q)^2\}^{1/2}}{2D(n-1) + \{4D^2(n-1) + (\Delta + \frac{1}{2}m_Q)^2\}^{1/2}}\right|. \tag{18d}$$

The ground state energy per site is then

$$\frac{E_0(\mathrm{CO\text{–}AF\text{–}F})}{N} = \frac{1}{2}(n-1)\,\bar{\mu} + \frac{1}{4}(2ZW+U)\,n^2 - \frac{1}{4}\left\{D^2 + \left(\Delta - \frac{1}{2}m_Q\right)^2\right\}^{1/2} - \frac{1}{4}\left\{D^2 + \left(\Delta + \frac{1}{2}m_Q\right)^2\right\}^{1/2}. \tag{18e}$$

On the other hand, if $m_0/2 - \bar{\mu} + |\Delta - m_Q/2| < 0$ and $\Delta + m_Q/2 \leqq m_0/2 + \bar{\mu}$, then we have to solve the following coupled equations:

$$n - 1 \pm \frac{m_0}{U} = \frac{1}{D} \left\{ \left(\frac{1}{2} m_0 \pm \bar{\mu} \right)^2 - \left(\Delta \pm \frac{1}{2} m_Q \right)^2 \right\}^{1/2}, \tag{19a}$$

$$\frac{2\Delta}{2ZW - U} \pm \frac{m_Q}{U} = \frac{1}{2D} (2\Delta \pm m_Q) \ln \left| \frac{D + \{D^2 + (\Delta \pm \frac{1}{2} m_Q)^2\}^{1/2}}{\bar{\mu} \pm \frac{1}{2} m_0 + \{(\frac{1}{2} m_0 \pm \bar{\mu})^2 - (\Delta \pm \frac{1}{2} m_Q)\}^{1/2}} \right|. \tag{19b}$$

In this case, the ground state energy per site is also given by (18e).

These analytical solutions clearly indicate that the CO and the AF phases exhibit the mirror-image symmetry with respect to $ZW = U$. This symmetry is characteristic of the Hartree approximation.

4. Phase Stability and Ground State Properties

For bi-mixed phases only the CO–AF solution exists under the special condition $n = 1$. When $n = 1$ the pure CO phase and the pure AF phase have very simple solutions. We can then prove without difficulty that the CO–AF phase is unstable against the pure CO phase if $ZW \geqq U$, and is unstable against the pure AF phase if $ZW \leqq U$.

Next, we consider the tri-mixed phase. Under the first set of conditions $m_0/2 - \bar{\mu} + |\Delta - m_Q/2| \geqq 0$, $|\Delta - m_Q/2| \geqq m_0/2 - \bar{\mu}$, and $\Delta + m_Q/2 < m_0/2 + \bar{\mu}$, using (18 a to d) one derives the inequality

$$\frac{E_0(\text{CO–AF–F})}{N} > \frac{1}{2} (n - 1) \bar{\mu} + \frac{1}{4} (2ZW + U) n^2 - \frac{D}{2} \left\{ \cosh \frac{2D}{U} - 2(n - 1) \right\} \Big/ \sinh \frac{2D}{U} \tag{20a}$$

if $ZW < U$, and

$$\frac{E_0(\text{CO–AF–F})}{N} > \frac{1}{2} (n - 1) \bar{\mu} + \frac{n - 1}{\sinh \frac{2D}{2ZW - U}} - \frac{D}{2} \coth \frac{2D}{2ZW - U} + \frac{1}{4} (2ZW + U) n^2 \tag{20b}$$

if $ZW > U$. If $ZW = U$, the tri-mixed phase does not exist. By comparing (20 a to b) to pure phase solutions, we can prove that

$$\begin{aligned} E_0(\text{CO–AF–F}) &> E_0(\text{AF}) \quad \text{if} \quad ZW < U\,, \\ &> E_0(\text{CO}) \quad \text{if} \quad ZW > U\,. \end{aligned} \tag{20c}$$

On the other hand, under the second set of conditions $m_0/2 - \bar{\mu} + |\Delta - m_Q/2| < 0$ and $\Delta + m_Q/2 \leqq m_0/2 + \bar{\mu}$, we can also use (19 a to b) to derive the inequality

$$E_0(\text{CO–AF–F}) > \frac{1}{U} D^2(n - 1)^2 + \frac{1}{4} (2ZW + U) n^2 + \frac{D}{4} (n - 1) \left\{ \frac{1}{\sinh \frac{2D}{U}} + \frac{1}{\sinh \frac{2D}{2ZW - U}} \right\} - \frac{D}{4} \left\{ \coth \frac{2D}{U} + \coth \frac{2D}{2ZW - U} \right\} \tag{21a}$$

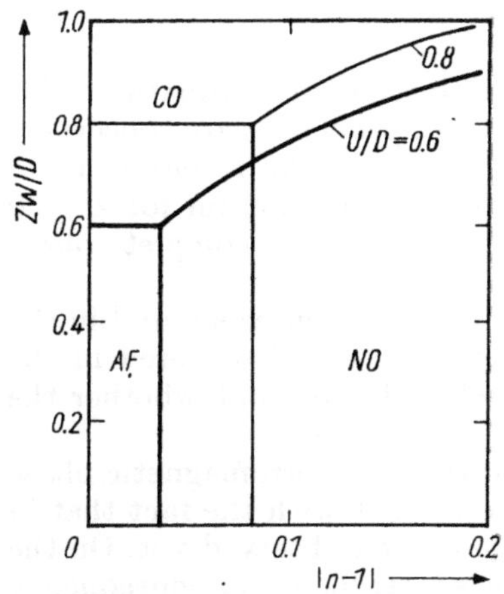

Fig. 1

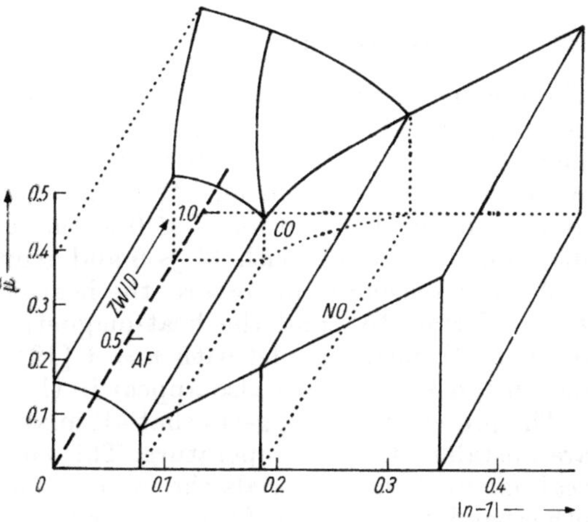

Fig. 2

Fig. 1. Ground state phase diagram for two values of $U/D = 0.6$ (heavy curves) and 0.8 (thin curves)

Fig. 2. Behavior of the chemical potential $\bar{\mu} = \mu - n(2ZW + U)/2$ for $U/D = 0.8$

if $ZW < U$, and

$$E_0(\text{CO–AF–F}) > \frac{1}{U} D^2(n-1)^2 + \frac{D}{4}(n-1)\left\{\frac{1}{\sinh\frac{2D}{U}} + \frac{1}{\sinh\frac{2D}{2ZW-U}}\right\} - \frac{D}{4}\left\{\coth\frac{2D}{U} + \coth\frac{2D}{2ZW-U}\right\} + \frac{1}{4}(2ZW+U)\,n^2 \tag{21b}$$

if $ZW > U$. To proceed further, we have performed a numerical calculation to reach the same conclusion (20c), as well as the non-existence of the tri-mixed phase if $ZW = U$.

Now we need only to consider the pure phases to determine the ground state phase diagram. From (14a to 16c) we found the phase boundaries between the CO and the AF phases, the AF and the NO phases, and the CO and NO phases are given by $ZW = U$, $|n-1| = \exp(-2D/U)$, and $|n-1| = \exp(-2D/(2ZW-U))$, respectively. If $ZW > U$ and $|n-1| < \exp(-2D/(2ZW-U))$, the ground state is charge-ordered. If $ZW < U$ and $|n-1| < \exp(-2D/U)$, the ground state is antiferromagnetic. Finally, if $ZW > U$ and $|n-1| > \exp(-2D/(2ZW-U))$, or $ZW < U$ and $|n-1| > \exp(-2D/U)$, then the ground state exhibits no ordering at all.

Fig. 1 gives the ground state phase diagram in the $ZW/D - |n-1|$ plane for two values of $U/D = 0.6$ and 0.8. For each stable phase we can use the corresponding equations to compute the chemical potential as shown in Fig. 2 for $U/D = 0.8$, as well as the order parameters as shown in Fig. 3 for $U/D = 0.8$. The transitions from ordered phases to non-ordered phase are clearly continuous.

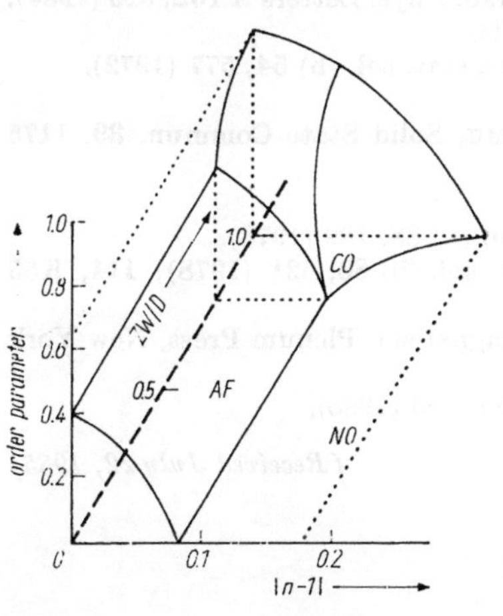

Fig. 3. Behavior of the AF and the CO order parameters for $U/D = 0.8$

5. Discussion

The present analysis is based on the broken-symmetry Hartree approximation. This approximation is quite reasonable as far as the ground state properties are concerned, judging from the comparison with exact results [10] and from the recent work of Nozieres and Schmitt-Rink [21]. Combining the present phase diagram for $U > 0$ and the phase diagram for $U < 0$ given in [10], we have provided a complete phase diagram of the EHM with $W > 0$ and arbitrary electron density.

With increasing temperature, the heat-charge-order phenomenon was found in the $U < 0$ EHM [10], and the heat-magnetization phenomenon was discovered in the ordinary Hubbard model with $n \neq 1$ [19]. It is interesting to find out whether the heat-magnetization can also appear in the EHM.

The present work suggests that all mixed phases and the pure ferromagnetic phase are unstable at zero temperature. This conclusion is consistent with the fact that in real narrow-band materials the mixed orderings have not been observed yet. On the other hand, the absence of the pure ferromagnetic phase is most likely the consequency of using a simple square density of states (13). In our openion, the most serious assumption of this work is the exclusion of all possible incommensurate magnetic and charge orderings. In view of the existing works on the Hubbard model [15, 17, 19], incommensurate orderings are likely to appear in the extended Hubbard model.

References

[1] V. J. Emery, Highly Conducting One-dimensional Solids, Ed. J. Devreese, R. Evrard, and V. van Doven, Plenum Press, New York 1979 (p. 247).
[2] J. E. Hirsch and D. J. Scalapino, Phys. Rev. B **27**, 7169 (1983); **29**, 5554 (1984).
[3] U. Brandt, Z. Phys. B **22**, 397 (1975).
[4] A. Ovchinnikov, Zh. eksper. teor. Fiz. **64**, 342 (1973).
[5] J. F. Kwak and G. Beni, Phys. Rev. B **13**, 652 (1976).
[6] J. Hubbard, Phys. Rev. B **17**, 494 (1978).
[7] J. Bernasconi, M. J. Rice, W. R. Schneider, and S. Strässler, Phys. Rev. B **12**, 1090 (1975).
[8] S. Robaszkiewicz and T. Kostyrko, Physica (Utrecht) B **121**, 134 (1983).
[9] R. Micnas, S. Robaszkiewicz, and K. A. Chao, Phys. Rev. B **29**, 2784 (1984).
[10] S. Robaszkiewicz, R. Micnas, and K. A. Chao, Phys. Rev. B **24**, 4018 (1981); **26**, 3915 (1982).
[11] R. Micnas, S. Robaszkiewicz, and K. A. Chao, J. Magn. Magn. Mat. **31/34**, 365 (1983).
[12] A. M. Oleś, R. Micnas, S. Robaszkiewicz, and K. A. Chao, Phys. Letters A **102**, 323 (1984).
[13] R. Kishore and S. K. Joshi, Phys. Rev. B **3**, 3901 (1971).
[14] J. Schneider, E. Heiner, and W. Haubenreisser, phys. stat. sol. (b) **54**, 577 (1972).
[15] D. R. Penn, Phys. Rev. **142**, 350 (1966).
[16] A. N. Andriotis, P. N. Poulopoulos, and E. N. Economu, Solid State Commun. **39**, 1175 (1981).
[17] A. M. Oleś, J. Phys. C **15**, 1065 (1983).
[18] A. Richter, G. Röpke, and F. Goedsche, phys. stat. sol. (b) **88**, 189 (1978).
[19] W. Schumacher, U. Lindner, and R. Jauch, phys. stat. sol. (b) **86**, 621 (1978); **114**, K55 (1982).
[20] S. V. Tyablikov, Methods in the Quantum Theory of Magnetism, Plenum Press, New York 1967.
[21] P. Nozieres and S. Schmitt-Rink, J. low-Temp. Phys. **59**, 195 (1985).

(Received July 22, 1985)

phys. stat. sol. (b) 133, 151 (1986)

Subject classification: 75.10

Department of Physics, University of Leeds[1])

On SCCF Theory of Magnetic Properties of 1-D Binary Alloys (Magnon Spectrum, Magnetization)[2])

By

J. Šlechta

A detailed SCCF calculation of the magnon densities of states of 1-D binary alloys, both with statistical short range order and purely random ones, is presented and the results are compared with the electronic densities of states of those alloys. Also the demonstration of the higher spatial correlations (HSC) in the shapes of the magnon densities of states is discussed. It is shown that for some constitutions of 1-D binary alloys the long range magnetic order is likely to exist and results for the temperature dependence of the magnetization of those binary alloys is also given.

Отдано детальное исчисление плотностей состояний магнонов в 1D двухкомпонентных сплавов, как чисто случайных так с кореляциями по коротких растояниях, по ССНД методу. Зависимость формы этих плотностей от высших пространственных кореляций тоже объявлена. Показано что для специальных конституций этих сплавов можно ожидать существование магнитного порядка по далеких растояниях и для этих случаев отдана температурная зависимость намагниченности данных сплавов.

1. Introduction

The SCCF method has been shown to be the best method of calculation of the density of states of disordered materials in existence [1]. It has been applied successfully to the transport properties of those materials [2, 3]. The core of that method is same for all the types of elementary excitations of the solid in question [4] and that provides a unique common ground for a comparison between their properties.

In this work the method is applied to investigation of the magnetic properties of 1-D binary alloys studied recently also by different authors [5]. At first there are presented the shapes of the densities of states of magnons.

2. Theory

The equation of motion of the magnon component of the elementary excitations of magnetic materials can be written in the form [6, 7]

$$2\pi \sum_j \{[\omega/\sigma - \tfrac{1}{2} \sum_{k\,\text{near. neigh.}} J_{ik}]\,\delta_{ij} + \tfrac{1}{2} J_{ij}\}\, G_{jl} = \delta_{il}\,. \tag{1}$$

Here ω is the magnon frequency, σ is the magnetization per a spin, J_{ij} is the magnetic exchange interaction, G_{ij} is the Green function of the problem, i, j are the indeces of the locations of those two spins in focus. The equation of motion (1) can be easily

---

[1]) 18, Lidgett Hill, Leeds 8, LS 8 1PE, Great Britain.

[2]) Preliminary presented at the 21th Annual Solid State Physics conference of the Institute of Physics held in Southampton in December 1984.

modified into the form suitable for the application of the SCCF method [4, 8]. Namely it can be written into the form

$$\sum_j K_{ij} G_{jl}(E) = \delta_{il} , \tag{2}$$

where

$$K_{ii} = E_i = E - \pi \sum_{k \atop \text{near. neigh.}} J_{ik} \tag{3}$$

and

$$K_{ij} = \pi J_{ij} \quad \text{for} \quad i \neq j \tag{4}$$

with

$$E = \frac{2\pi\omega}{\sigma} \tag{5}$$

and

$$\tilde{G}_{ij}(E) = G_{ij}\left(\frac{\sigma E}{2\pi}\right). \tag{6}$$

The Green function $\tilde{G}_{ij}(E)$ was calculated by the SCCF method [4].

3. Numerical Results and Discussion

For the sake of continuity of this work with the previous works on application of the SCCF method to 1-D binary alloys [2, 4, 8 to 10] the results presented here were obtained for a similar model of 1-D binary alloys and values of the parameters involved as in the preceding works mentioned. Namely: $\alpha = 1$ — the strength of the A–A magnetic exchange interaction, $\beta = 5$ — the strength of the A–B magnetic interaction and $\gamma = 2$ — the strength of the B–B magnetic interaction. The only larger difference between the magnetic and electronic cases is in the form of the diagonal term K_{ii} in (2) given by (3) which plays here the role of the "atomic" energies in the electronic case [4]. That difference proves responsible for the special features of the magnon densities of states in relation to the electronic ones.

The densities of states of magnons $\tilde{\varrho}(E)$ acquired by SCCF method for selected values of the parameters of relative concentration x and the sticking parameter ε [2 to 4] are given in Fig. 1 a to c. At the first glance they are in various aspects noticeably different to the electronic densities of states of similar binary alloys [4, 9, 10]. Namely:

(i) they show less clearly defined separate bands of atoms A and B, relative to the electronic ones; due to the term (3) in (2). Because of that term we have here three diagonal "pseudoatom" energies, $E_1 = \alpha/2 = 0.5$, $E_2 = \beta/2 = 2.5$, and $E_3 = \gamma/2 = 1$. These formally play the role of the centres of three "bands". However none of those "pseudoatomic" energies plays its role in any of all the representative M-testing clusters [3, 4], consistently. Nor there can be ascribed to it clear values of the off-diagonal elements which define the width of the relevant bands. That is one of the largest differences between the magnon and electronic cases.

(ii) They show only one dominant peak within the interval (0.1), in this case. It actually consists of two close peaks; at $E = 0.1$ and $E = 0.3$. Off those two peaks the one at $E = 0.3$ is higher for $x = 0.1$; the peak at $E = 0.1$ is higher for $x = 0.9$. This existence of the main peak of the density of states of magnons at $E \approx 0$ agrees with the findings of different authors [7, 11]. A second side peaks develops only in the case of the purely random alloys; at $E = 2.3$. It is most intensive for $x = 0.1$.

(iii) They are much more slowly convergent towards the crystalline densities of states of magnons that is for $x \to 0$, or $x \to 1$ (100% of atoms A). In the electronic case,

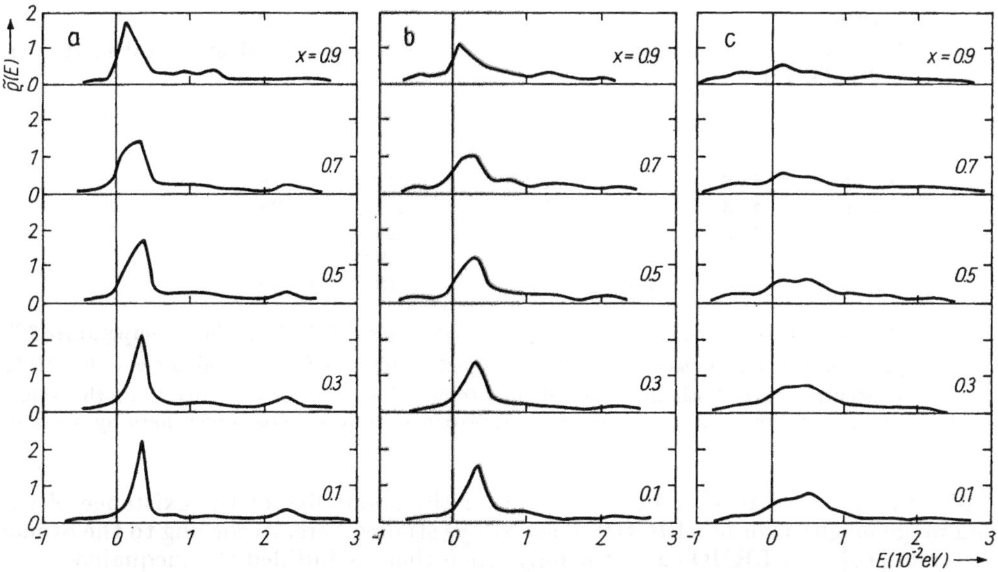

Fig. 1. The concentration dependence of the shape of the magnon densities of states for a) $\varepsilon = 0.1$, separative case; b) $\varepsilon = 0.5$, the separative and alternative cases are equal here; and c) $\varepsilon = 0.1$, alternative case. x is the relative concentration parameter $0 \leqq x \leqq 1$

unlike for magnons, strong van Hove singularities [12] showed themselves already for $x = 0.1$ and $x = 0.9$ [4, 9]. Here they are nonexistent, yet.

(iv) They change less profoundly with the changes of the values of the parameters x and ε which control the higher spatial correlations (HSC) [3].

Because of the fast convergence of SCCF [4, 9, 10] the results for the shapes of the magnon densities of states of the 1-D binary alloys presented above are close to the true shapes of those densities in real (pseudo) 1-D binary alloys and serve for a set of reference densities of states of magnons for that type of alloys. By comparison the experimentally acquired magnon densities of states with a spanning set of computed ones, for a set of parameters x and ε, one can determine those parameters in the given material and thus HSC, especially when simultaneously supplemented by different studies of such a kind [2, 3, 13 to 15].

Table 1

The value of the basic characteristics of the 1-D purely random binary alloys [14, 17]

x	J_{av}	T_{av}^{M}	T_{av}^{M}/J_{av}	f_{T_M}	a
0.1	1.57	3.14	2	0.12	0.038
0.3	2.45	4.89	1.97	0.15	0.031
0.5	2.91	5.81	1.99	0.13	0.025.
0.7	3.08	6.14	1.99	0.12	0.02
0.9	2.4	4.7	1.95	0.08	0.017

J_{av} average exchange interaction; T_{av} average values of the local Curie temperatures T_i^M f_{T_M} relative fluctuations of T_i^M; a magnetic enhance factor.

Table 2

The values of the basic characteristics of 1-D binary alloys with statistical short range order [14, 17]

ε	J_{av}	T_{av}^{M}	T_{av}^{M}/J_{av}	f_{T_M}	a
0.1 alt.	4.02	9.15	2.27	0.17	0.018
0.3 alt.	4.03	7.97	1.97	0.21	0.028
0.5	2.23	4.42	1.98	0.28	0.063
0.3 sep.	2.23	4.42	1.98	0.19	0.042
0.1 sep.	1.84	3.65	1.98	0.18	0.049

J_{av} average exchange interaction; T_{av} average values of the local Curie temperature T_i^M; f_{T_M} relative fluctuations of T_i^M; a magnetic enhance factor; and for $x = 0.5$ (50% concentration of both the types of the atoms). For different concentrations the values of those parameters can be obtained approximately from those given here by a simple scaling.

We shall progress further by evaluation of the possibility of the existence of the long range magnetic order (LRMO) in the alloys studied here. According to the Stoner criterium [16] such LRMO can exist only when there is fulfilled the inequality

$$V > 1/\tilde{\varrho}(E_F)\,, \tag{7}$$

where V is the value of the (average) exchange interaction, and E_F is the Fermi level. Let us give here, first, in Table 1 and 2, the values of the basic characteristics of magnetic alloys [14, 17]. By utilizing the results for the values of E_F given in [2] and the values of $\tilde{\varrho}(E_F)$ which can be extracted from [4], there are given, in Table 3, the values of $1/\tilde{\varrho}(E_F)$ for the three types of compositions of the binary alloys considered in [2], namely: (i) M–M — both atoms A and B are metallic, (ii) M–I — atoms A are metallic while atoms B are those of an insulator, (iii) I–M when atoms A are of an insulator and atoms B are metallic. By the combination of the results in Tables 1, 2 and 3 there can be obtained the information about the cases when there is possible LRMO in 1-D binary alloys. These are marked by "Y", in Table 3. The opposite cases are marked by "N" there. From that there can be seen that the proofs of lack of LRMO in 1-D magnetic systems given in [18, 19] are not strictly conclusive for the disordered materials for within them there had been used: a) the concept of k-space not valid for disordered matter ([4], appendix 2); b) Born-Karman conditions, also not very suitable, although less principly, for the non-crystalline state.

Because of the possibility of the existence of LRMO in 1-D binary alloys there is meaningfull to investigate also the temperature dependence of the magnetization per a spin $\sigma(T)$ on the bases of the modified form of the formula in [6, 7]

$$\frac{1}{2}(1-\sigma) = \frac{2\pi}{\sigma}\int_{-\infty}^{\infty}\tilde{\varrho}\left(\frac{2\pi\omega}{\sigma}\right)\left(\exp\left(\frac{2\pi\omega\beta}{\sigma}\right)-1\right)^{-1}\mathrm{d}\omega\,. \tag{8}$$

Here $\beta = 1/k_0T$, where k_0 is the Boltzmann constant $\tilde{\varrho}(E) \approx \operatorname{Im}\tilde{G}(E)$. From Table 1 and 2 there can be seen that the values of f_{T_M} differ little for all the types of the alloys considered and so the shapes of $\sigma(T)$ can be expected to be nearly the same for all of them, as a consequence of [17]. That really was confirmed by computation. In Fig. 2 there is given a sample $\sigma(T)$ for those alloys studied here. There, similarily to the transport properties of 1-D binary alloys in [2, 3], holds the scaling properties of the temperature and energy scale. Namely when the energy unit in Fig. 1 a to c is

Table 3
The concentration dependence of the values of $1/\tilde{\varrho}(E_F)$ for the three types of the composition of the 1-D binary alloys (see the text) for various values of ε

x	M–M	M–I	I–M
a) $\varepsilon = 0.1$, separative case			
0.1	6.25 N	100 N	12.50 N
0.3	2.7 N	50 N	7.14 N
0.5	3.03 N	14.3 N	12.5 N
0.7	3.03 Y	14.3 N	33.33 N
0.9	1.67 Y	1.9 N	50 N
b) $\varepsilon = 0.3$ separative case			
0.1	3.03 N	4.55 N	3.23 N
0.3	6.25 N	33.33 N	33.33 N
0.5	12.50 N	12.5 N	14.29 N
0.7	3.85 Y	5.26 N	12.5 N
0.9	11.11 N	2.86 Y	16.67 N
c) $\varepsilon = 0.5$			
0.1	5.56 N	16.67 N	12.5 N
0.3	3.23 N	33.33 N	10 N
0.5	7.69 N	12.5 N	11.11 N
0.7	12.5 N	16.67 N	33.33 N
0.9	6.25 N	11.11 N	50 N
d) $\varepsilon = 0.3$ alternative case			
0.1	8.33 N	5.26 N	11.11 N
0.3	3.85 N	25 N	12.5 N
0.5	25 N	14.29 N	14.29 N
0.7	5 N	25 N	33.33 N
0.9	2.78 Y	4 Y	11.11 N
e) $\varepsilon = 0.1$, alternative case			
0.1	9.09 N	20 N	9.09 N
0.3	5.26 N	25 N	10.53 N
0.5	11.76 N	11.76 N	11.11 N
0.7	12.5 N	11.76 N	16.67 N
0.9	4.35 N	8.33 N	50 N
f) purely random case			
0.1	5.55 N	100 N	10 N
0.3	5.55 N	16.6 N	11.11 N
0.5	6.67 N	20 N	33.3 N
0.7	10 N	3.7 N	33.3 N
0.9	11.11 N	6.25 N	9.09 N

In the table there is denoted the case when the long range magnetic order is possible, see (7), by the mark "Y" there, the opposite case by the mark "N".

10^{-2} eV the temperature unit in Fig. 2 is 100 K. Due to that the temperature range in Fig. 2 can be adjusted to any given value of the average interaction J_{av}. We add still that the results for $\sigma(T)$ presented above can be interpreted as to be constructed within the basis of the vectors γ_i [17] the directions of which are determined to satisfy the balance of the local interactions of every spin, for the absolute temperature $T = 0$, for the given disorder. Within such an interpretation $\sigma(T)$ presented here is valid for any

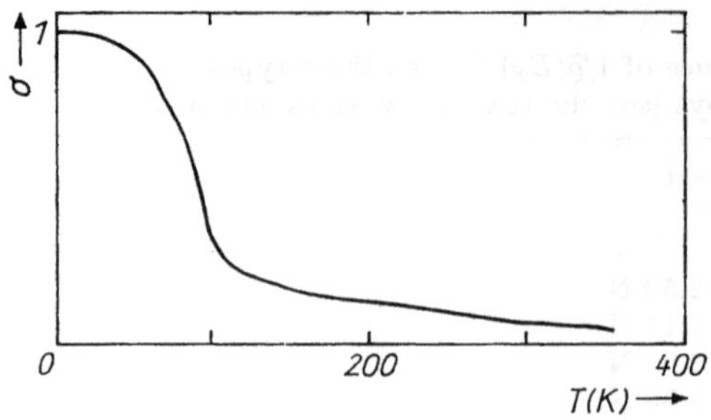

Fig. 2. The representative example of the temperature dependence of the magnetization of a spin of the alloys studied in this work

kind of LRMO of 1-D binary alloys, like ferromagnetic, or antiferromagnetic, or a more general one [17], with the same value of f_{T_M} as in the cases above.

Acknowledgements

The results were computed on the author's Micro Sinclair ZX Spectrum with 48K RAM. He would like to thank the Benevolent Fund of the Institute of Physics, and Dr. L. Cohen, the executive secretary of the Institute of Physics, for their financial and moral support of himself; and the Brotherton Library, especially Mr. D. Cox, its librarian, for their hospitality.

References

[1] J. Šlechta, phys. stat. sol. (b) **120**, 329 (1983).
[2] J. Šlechta, J. Phys. C **12**, 1819 (1979).
[3] J. Šlechta, phys. stat. sol. (b) **127**, 403 (1985).
[4] J. Šlechta, J. Phys. C **10**, 2047 (1977).
[5] D. Lj. Mirjanić, Z. Bundalo, B. S. Tošić, and J. P. Šetrajčić, phys. stat. sol. (b) **128**, 151 (1985).
[6] S. V. Tyablikov, Methods in Quantum Theory of Magnetism, Plenum Press, New York 1967.
[7] C. C. Montgomery, J. I. Krugels, and R. M. Stubbs, Phys. Rev. Letters **25**, 669 (1970).
[8] J. Šlechta, J. Phys. F **4**, 1148 (1974).
[9] J. Šlechta, phys. stat. sol. (b) **71**, K165 (1975).
[10] J. Šlechta, phys. stat. sol. (b) **104**, K143 (1981).
[11] F. Goedsche, phys. stat. sol. (b) **44**, 191 (1971).
[12] L. van Hove, Phys. Rev. **89**, 1189 (1953).
[13] J. Šlechta, J. non-crystall. Solids **22**, 345 (1976).
[14] J. Šlechta, J. Magnetism magnetic Mater. **22**, 113 (1980).
[15] J. Šlechta, J. Physique **42**, C4–229 (1981).
[16] R. Brout, Phase Transitions, W. A. Benjamin, Inc., New York 1965.
[17] J. Šlechta, phys. stat. sol. (b) **67**, 595 (1975); **70**, 531 (1975); **123**, K133 (1984).
[18] N. D. Mermin and H. Wagner, Phys. Rev. Letters **17**, 1133 (1966).
[19] S. Robaszkiewicz and R. Micnar, phys. stat. sol. (b) **73**, K35 (1976).

(Received July 5, 1985)

phys. stat. sol. (b) **133**, 157 (1986)
Subject classification: 75.10; S1.4

L.M.S.E.S., Institut de Physique, Strasbourg[1])

Detailed Low-Field Magnetic Properties of Rare-Earth Ion States in Cubic Crystal Fields

By

G. FISCHER and A. HERR

The effect of a cubic crystal field on rare-earth ions is calculated and presented in a graphical form representing the individual magnetic moment and Van Vleck susceptibility of the lowest-lying sub-state, for $J = 3.5; 4; 4.5; 6; 7.5; 8$. This is a practical guide for analyzing the moderate-field part of the magnetization. The results are given for the three principal directions. In order to analyze susceptibility measurements, the calculated Curie constant (or the Van Vleck susceptibility) of the fundamental multiplet is also represented.

L'effet d'un champ cristallin cubique sur les ions des terres rares est calculé, et les résultats sont donnés sous forme graphique représentant le moment magnétique et la susceptibilité de Van Vleck individuels du sous-état le plus bas, pour les valeurs de J correspondantes: 3,5; 4; 4,5; 6; 7,5; 8. Ces courbes ont une utilité pratique pour l'analyse des courbes d'aimantation dans des champs modérés. Les résultats sont donnés pour les trois directions principales. Dans le but d'exploiter les mesures de susceptibilité, la constante de Curie (ou la susceptibilité de Van Vleck) du multiplet fondamental est aussi représentée.

1. Introduction

After the pioneer work of Stevens [1] who developed the operator equivalent technique in crystal field problems, Lea, Leask, and Wolf [2] (LLW) gave the Hamiltonian a more practical form in the case of cubic symmetry, with the well-known parameters W and x. The first is a simple scale factor and the second reflects the ratio of the fourth and the sixth degree crystal fields. In their fundamental work, the energy of each level and the corresponding state(s) were calculated for all J-values, and their dependence on x was given. The eigenvectors were written in the $|J, J_z\rangle$ basis, the z-axis being one of the fourfold axes. The usefulness of these results when a weak magnetic field is present is straightforward principally for the determination of the magnetic moment(s) of the levels *other than* Γ_8, for they do not depend on the direction of the field and can thus be deduced from the case of a field applied in [001] direction.

Ebina and Tsuya [3, 4] treated the same problem but in presence of finite magnetic fields or exchange fields in the [001] and [111] directions. In addition, the mean value of the magnetic moment in each state was calculated. However, the crystal field parameters they used were different from the W, x of LLW and also not so convenient.

In the same period, Schumacher und Hollingsworth [5] considered the perturbation effect of a weak magnetic field H on each level, and developed the energy up to the H^2-term. They thus obtained the individual moment (from the term in H) and the individual Van Vleck susceptibility (from the term in $H^2/2$) of each state. Only the case when $x = 1$ was considered, with the field parallel to the [001] and [111] directions, and for all J-values relevant to rare-earth ions.

---

[1]) 3, rue de l'Université, 67084 Strasbourg, France.

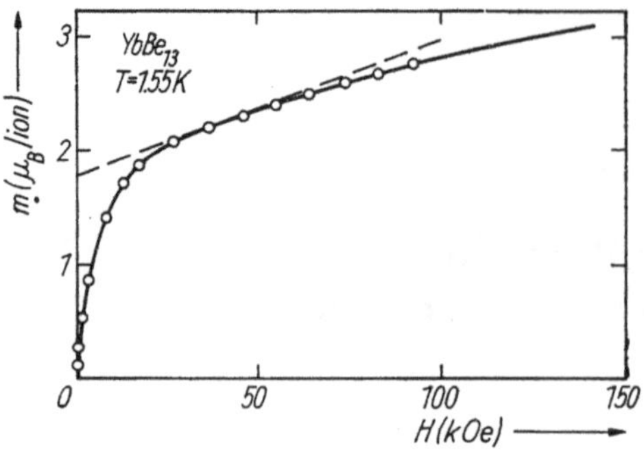

Fig. 1. Magnetization of $YbBe_{13}$, as published by Heinrich et al. [7]. The dashed line represents $m = m_i + \chi_i^{VV} H_{int}$ of the lowest sub-level, for $W = 1.90$ K and $x = 0.45$; the use of the internal field H_{int} is due to the presence of a small exchange field, with a molecular field constant $\lambda = 2.2\ kOe/\mu_B$. This line corresponds to the moderate-field part of the magnetization curve, when the field is just high enough to privilege that sub-level. The measured sample is polycrystalline, but this has little effect since m_i and χ_i^{VV} are isotropic here (see Fig. 2a)

We present here results of new calculations of the effect of a cubic crystal field on rare-earth f-electron states with the widely used LLW formalism, though an improved one has recently been proposed [6]. The domain of the previously quoted works is now extended into an unified area: the variation with x of the eigenstates, the individual magnetic moment, and Van Vleck susceptibility of each state, for the three principal crystallographic directions; in addition, the Curie constant $C_F(x)$ and the Van Vleck susceptibility $\chi_F^{VV}(x)$ of the fundamental multiplet are also reported. These data are useful for the following reasons:

a) When a magnetic field H is present, at low temperature, the splitting of the sub-levels of the fundamental multiplet (if magnetic) makes the lowest one the most important. When H is not too high, i.e. when the behaviour of the sub-levels is not very different from that predicted by perturbation theory, the zero-field individual moment and Van Vleck susceptibility of that lowest sub-level are the essential quantities to be known. Fig. 1 shows the example of the magnetization curve of paramagnetic $YbBe_{13}$, which has been approximately fitted by Heinrich et al. [7] with $W = 1.9$ K and $x = 0.45$. These values yield $m_i = 1.71\mu_B$ for the individual zero-field moment of the lowest sub-level, and $\chi_i^{VV} = 1.14 \times 10^{-5}\ \mu_B/Oe$ (i.e. 0.064 cm^3/mol) for the corresponding individual Van Vleck susceptibility; the perturbation theory gives then a field-dependent moment $m = m_i + \chi_i^{VV} H$ at low temperature. This behaviour is reflected in the quasi-linear part of the experimental magnetization curve (see Fig. 1).

Inversely, from the quasi-linear part observed in many cases, and mainly at low temperature, one can immediately deduce an estimate of m_i and χ_i^{VV}. We report in the following figures the calculated variation of $m_i(x)$ and $\chi_i^{VV}(x)$ for the lowest sub-level, and this can be a very simple guiding tool in searching for W and x.

Such figures have also been given for the individual magnetic moments in the case of hexagonal symmetry by Segal and Wallace [8].

b) In low-temperature susceptibility measurements, one is concerned with the whole fundamental multiplet and the experimental values of C_F (or χ_F^{VV}), when compared to the calculated $C_F(x)$ $\left(\text{or } \chi_F^{VV}(x)\right)$, can easily add information about W and x.

We then restrict the description of these quantities to the case of the lowest sub-levels only.

2. Calculations

The crystal field Hamiltonian has different expressions depending on the direction of the z quantization axis [9].

In the [001] quaternary direction,

$$\mathcal{H}_{\text{CF}}^{(4)} = W\left[\frac{x}{F_4}(O_4^0 + 5O_4^4) + \frac{(1-|x|)}{F_6}(O_6^0 - 21O_6^4)\right],$$

in the [110] binary direction,

$$\mathcal{H}_{\text{CF}}^{(2)} = W\left[\frac{x}{4F_4}(-O_4^0 + 20O_4^2 + 15O_4^4) + \right.$$
$$\left. + \frac{(1-|x|)}{8F_6}\left(-13O_6^0 - \frac{105}{2}O_6^2 + 105O_6^4 - \frac{231}{2}O_6^6\right)\right],$$

in the [111] ternary direction,

$$\mathcal{H}_{\text{CF}}^{(3)} = W\left[\frac{x}{3F_4}(-2O_4^0 + 40\sqrt{2}\,O_4^3) + \frac{(1-|x|)}{9F_6}(16O_6^0 + 140\sqrt{2}\,O_6^3 + 150O_6^6)\right].$$

The diagonalization was performed using a standard program. In order to determine the proper eigenvectors in case of degeneracy, a simple method was the systematic use of a Zeeman term $-g_J\mu_B J_z H_z$, with a low field value (1 Oe) compared to the crystal field (the calculations were performed with $W = 10$ K). This automatically enables us to get the right eigenvectors to be used for perturbation treatment. The error due to that Zeeman term is negligible.

The calculation of the individual magnetic moment m_i of each state and of its Van Vleck susceptibility χ_i^{VV} has been performed using the Hamiltonian $\mathcal{H}_{\text{CF}}^{(4)} - g_J\mu_B H \boldsymbol{J}\cdot\boldsymbol{u}$ ($\boldsymbol{u}$ being the unitary vector in the direction of the field, taken successively in the [001], [110], and [111] directions) with

$$m_i = g_J\mu_B\langle i|\,\boldsymbol{J}\cdot\boldsymbol{u}\,|i\rangle$$

and

$$\chi_i^{\text{VV}} = -2g_J^2\mu_B^2 \sum_{j\neq i} \frac{|\langle i|\,\boldsymbol{J}\cdot\boldsymbol{u}\,|j\rangle|^2}{E_i - E_j}.$$

Care has been taken in the summation to reject the states which, though different from $|i\rangle$ due to the effect of the 1 Oe field, are issued from the same original Zeeman multiplet.

In some cases the m_i were also calculated with the quantization axis along $\boldsymbol{u}$ (with $m_i = g_J\mu_B\langle i^u|\,J_z\,|i^u\rangle$, i^u being the eigenvector of the state i), and χ_i^{VV}, too; the results turned to be the same as above.

The use of $\mathcal{H}_{\text{CF}}^{(4)}$, $\mathcal{H}_{\text{CF}}^{(2)}$, and $\mathcal{H}_{\text{CF}}^{(3)}$ is of practical importance essentially for the representation of the eigenvectors, which in these cases are easy to handle since they are written in a $|J, J_z\rangle$ basis.[2])

The Curie constant C_F of the fundamental multiplet is simply deduced from the mean of the squares of the corresponding individual moments, and its Van Vleck susceptibility χ_F^{VV} (when this multiplet has no moment) is the mean $\bar{\chi}^{\text{VV}}$ of the individual χ_i^{VV} of the states involved in it.

In the following (Fig. 2 to 7) graphical representation of the individual magnetic moments of the lowest sub-level, and of its individual Van Vleck susceptibility is given, with x varying from -1 to 1. The variation of the Curie constant of the fundamental multiplet and — when this multiplet is non-magnetic — of its mean Van Vleck susceptibility is also drawn in the figures.

---

[2]) These eigenvectors are not tabulated here but are available from the authors on request.

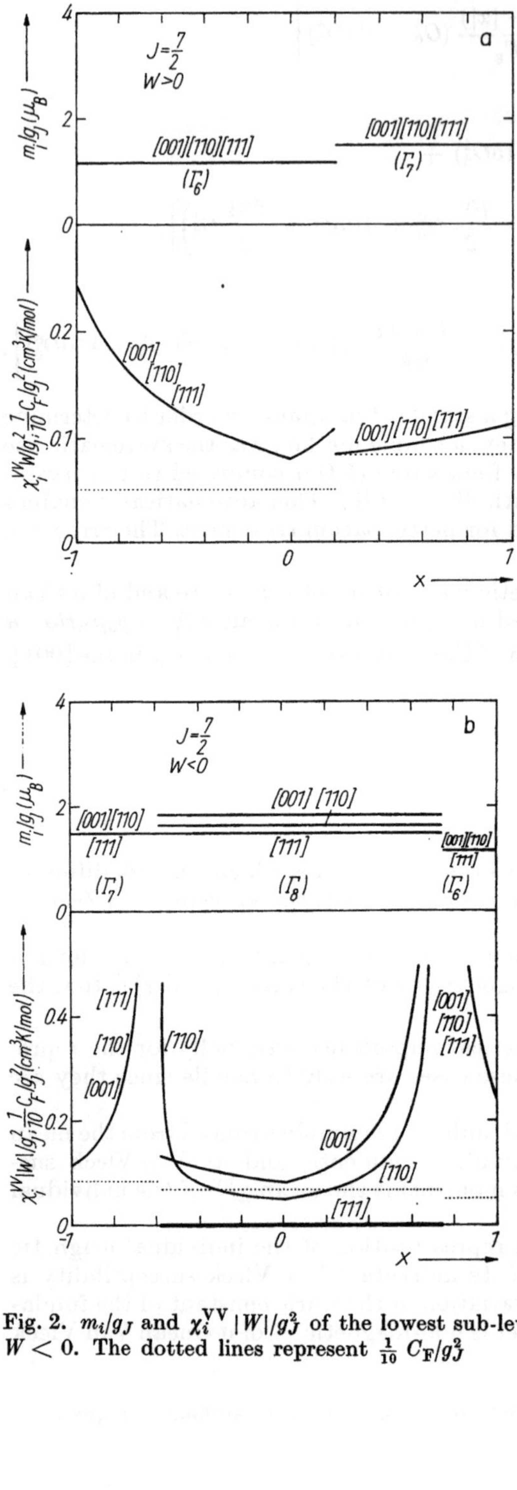

Fig. 2. m_i/g_J and $\chi_i^{VV}\,|W|/g_J^2$ of the lowest sub-level for $J = \frac{7}{2}$, a) when $W > 0$ and b) when $W < 0$. The dotted lines represent $\frac{1}{10}\,C_F/g_J^2$

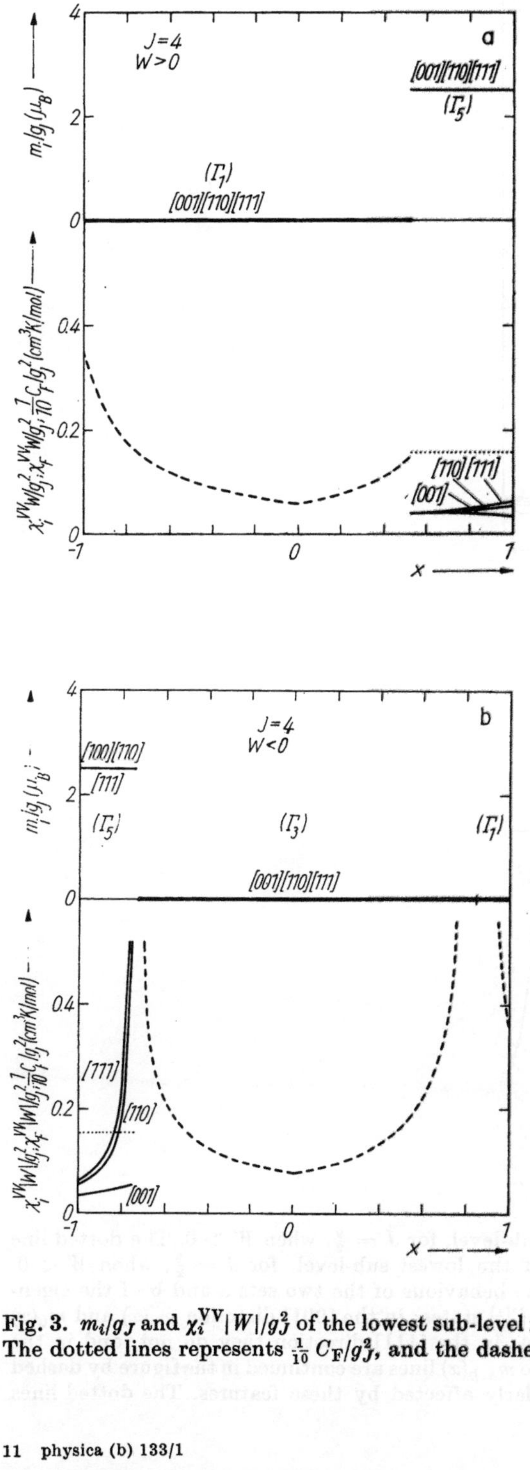

Fig. 3. m_i/g_J **and** $\chi_i^{VV}|W|/g_J^2$ **of the lowest sub-level for** $J = 4$**, a) when** $W > 0$ **and b) when** $W < 0$**. The dotted lines represents** $\frac{1}{10}\,C_F/g_J^2$**, and the dashed line** $\chi_F^{VV}|W|/g_J^2$

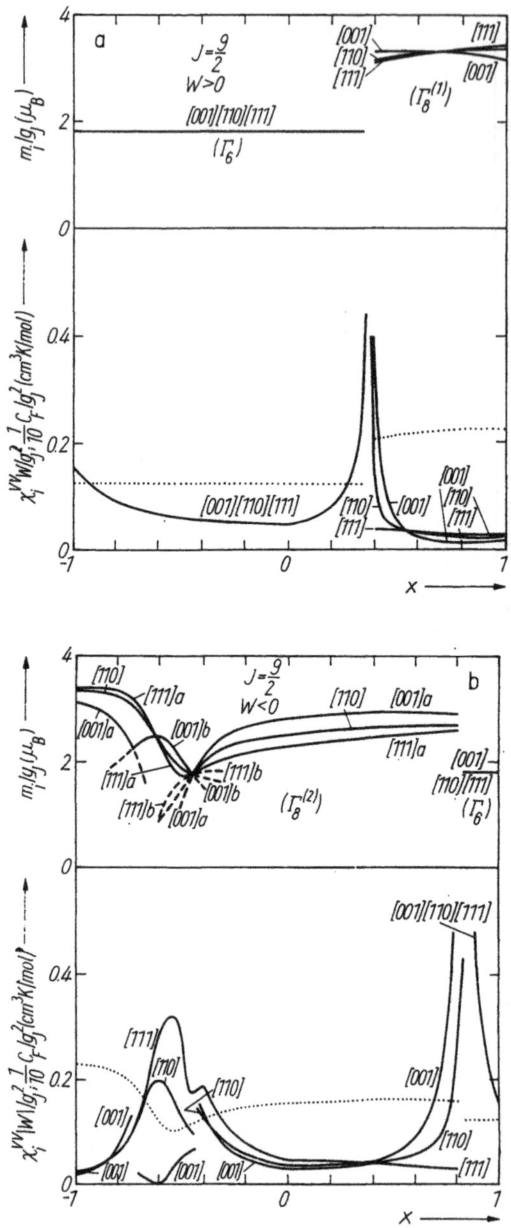

Fig. 4. a) m_i/g_J and $\chi_i^{VV} W/g_J^2$ of the lowest sub-level, for $J = \frac{9}{2}$, when $W > 0$. The dotted line represents $\frac{1}{10} C_F g_J^2$. b) m_i/g_J and $\chi_i^{VV} |W|/g_J^2$ of the lowest sub-level, for $J = \frac{9}{2}$, when $W < 0$. The peaks of $m_i(x)$ are effects of changes in the behaviour of the two sets a and b of the eigenvectors of $\Gamma_8^{(2)}$ caused by the proximity of the $\Gamma_8^{(1)}$ states: in the [001] direction $m_a(x)$ and $m_b(x)$ intersect twice in the range $-0.8 \leq x \leq -0.4$; in the [111] direction they do not, and in the [110] direction a and b are indistinguishable. The $m_{a,b}(x)$ lines are continued in the figure by dashed lines to show their own trend. $\chi_i^{VV}(x)$ is similarly affected by these features. The dotted lines represent $\frac{1}{10} C_F/g_J^2$

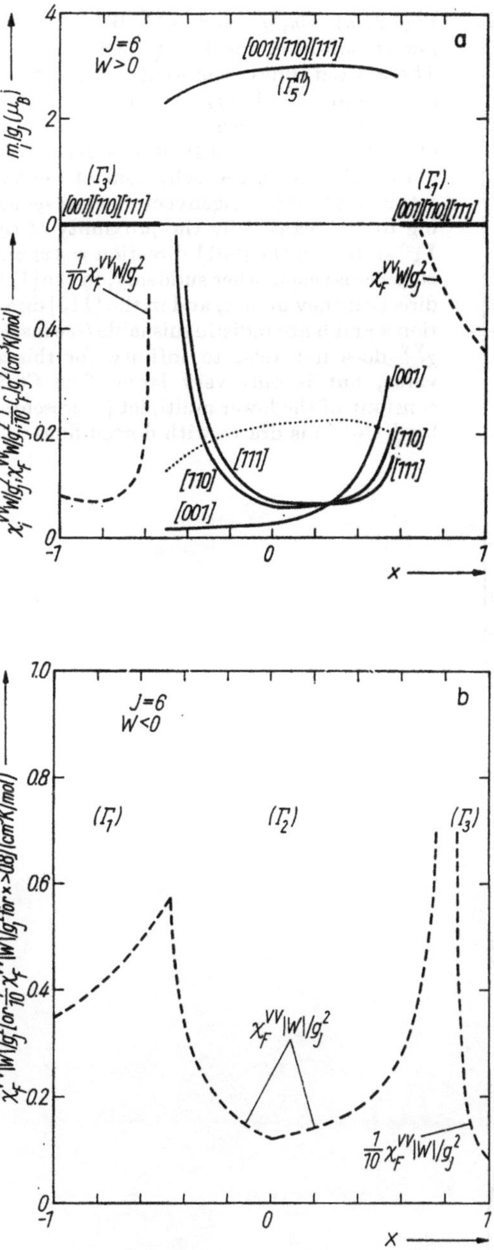

Fig. 5. a) m_i/g_J and $\chi_i^{VV} W/g_J^2$ of the lowest sub-level for $J = 6$, when $W > 0$. For $x \approx 0.6$, the high values of χ_i^{VV} are not due to the proximity $(\Gamma_1\text{–}\Gamma_5^{(1)})$ because there are no mixing effects between these two states, but to Γ_4, and thus χ_i^{VV} does not go to infinity. The dashed lines represent $\chi_F^{VV} W/g_J^2$ (with a reducing factor $\frac{1}{10}$ in the range $-1 \leqq x \lessapprox -0.5$) which also remains finite for $x \approx 0.6$ for the above-mentioned reasons. The dotted line represents $\frac{1}{10} C_F g_J^2$. b) $\chi_F^{VV} \, |W|/g_J^2$ of the lowest level for $J = 6$, when $W < 0$ (with a reducing factor $\frac{1}{10}$ in the range $0.8 \lessapprox x \leqq 1$). The m_i are here zero over the whole range

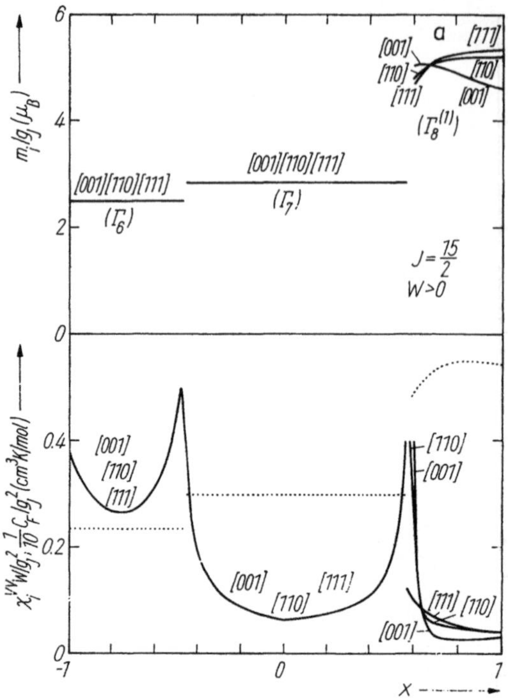

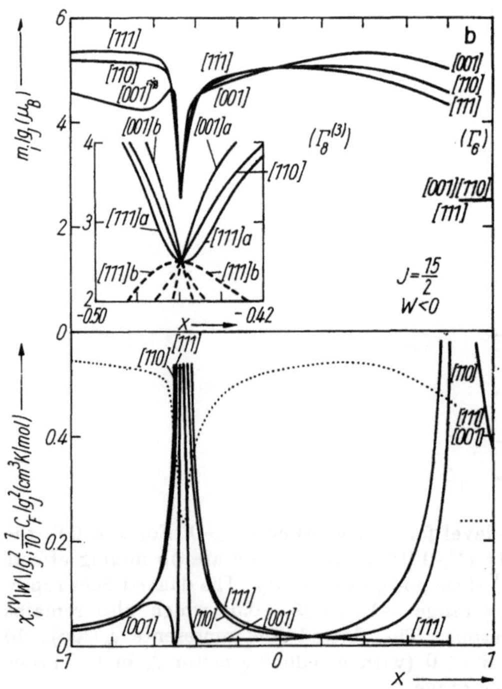

Fig. 6. a) m_i/g_J and $\chi_i^{VV}\, W/g_J^2$ of the lowest sub-level for $J = \frac{15}{2}$, when $W > 0$. The dotted lines represent $\frac{1}{10}\, C_F/g_J^2$. b) m_i/g_J and $\chi_i^{VV}\, |W|/g_J^2$ of the lowest sub-level for $J = \frac{15}{2}$, when $W < 0$. The peak of $m_i(x)$ near $x = -0.46$ is an effect of the abrupt change in the behaviour of the two sets a and b of the eigenvectors corresponding to $\Gamma_8^{(3)}$, caused by the proximity of the $\Gamma_8^{(2)}$ states. In the [001] direction $m_a(x)$ and $m_b(x)$ cross each other suddenly, in the [111] direction they do not, and in the [110] direction a and b are indistinguishable (see inset). χ_i^{VV} does not tend to infinity for this x-value, but is only very large. The Curie constant of the lower multiplet (represented by $\frac{1}{10}\, C_F/g_J^2$) is drawn with dotted lines

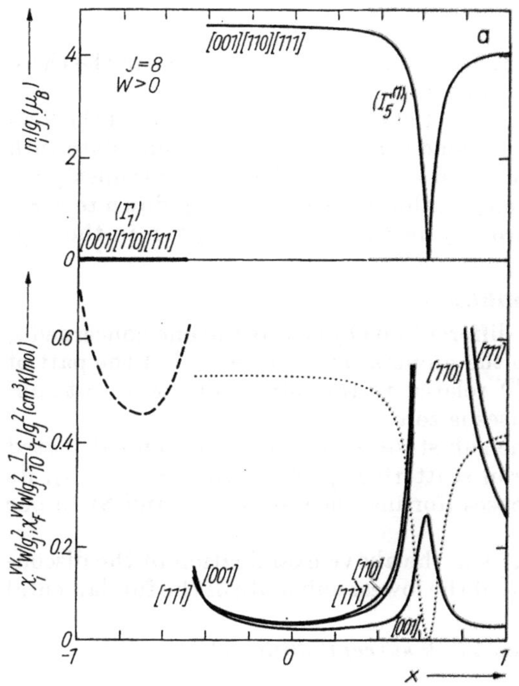

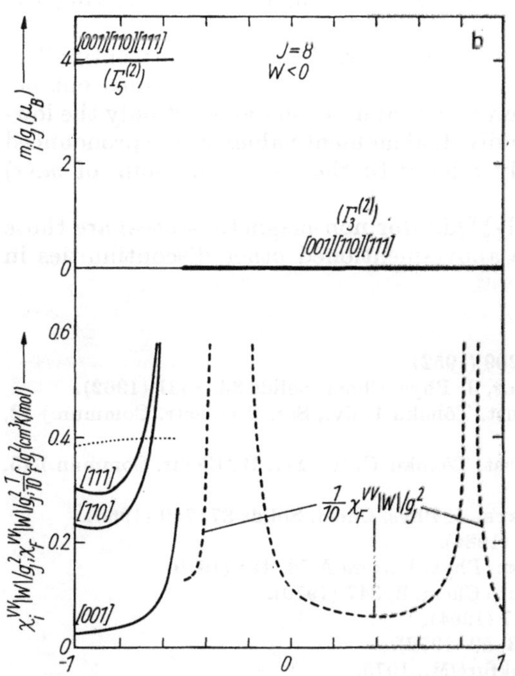

Fig. 7. a) m_i/g_J and $\chi_i^{VV}W/g_J^2$ of the lowest sub-level for $J = 8$, when $W > 0$. The dotted line represents $\frac{1}{10}\,C_F/g_J^2$ and the dashed line $\chi_F^{VV}W/g_J^2$. For $x \approx 0.63$, χ_i^{VV} does not become infinite in the [001] direction because then the magnetic states of $\Gamma_5^{(1)}$ do not mix with the $\Gamma_3^{(1)}$ states. For analogous reasons χ_i^{VV} and χ_F^{VV} do not tend to infinity for $x \approx -0.47$. b) m_i/g_J and $\chi_i^{VV}|W|/g_J^2$ of the lowest sub-level for $J=8$, when $W < 0$. In the [001] direction χ_i^{VV} remains finite for $x \approx -0.55$ because the magnetic states of $\Gamma_5^{(2)}$ do not mix with the $\Gamma_3^{(2)}$ states. The dotted line represents $\frac{1}{10}\,C_F/g_J^2$ and the dashed line $\frac{1}{10}\,\chi_F^{VV}|W|/g_J^2$

3. Comments on the Results

3.1 Moments

Except for the Γ_8 states the individual magnetic moments m_i are the same in the three directions, due to their isotropy, as is well known.

Apparent discontinuities in $\mathrm{d}m_i/\mathrm{d}x$ are the effect of changes in the nature of the lowest sub-state of the fundamental multiplet: beyond certain x-values the sub-state with the highest moment is different from that retained before. For doublets or triplets, due to the opposite values for the extremal moments, this gives peaks going down to zero; for quadruplets, the peaks do not go to zero in general. Details are given in the caption of each figure.

3.2 Individual Van Vleck susceptibilities

They become important in general when a different level j is near the one concerned i, but there are exceptions. Those are due to the numerator $|\langle i|\,\boldsymbol{J}\cdot\boldsymbol{u}\,|j\rangle|^2$ of the part of the individual Van Vleck susceptibility χ_i^{VV} related to the mixing of the lowest substate i with each j-state, which in some cases is zero.

The sum of such numerators over all the sub-states of i and j has been calculated by Birgeneau [10] for the analysis of neutron scattering spectra; when $\boldsymbol{J}\cdot\boldsymbol{u} = J_z$, its behaviour can be partly related in some cases (for instance for $J = \frac{15}{2}$ and 8) to our χ_i^{VV}'s.

Other discontinuities in the χ_i^{VV} are due, as in the above examination of the discontinuities in $\mathrm{d}m_i/\mathrm{d}x$, to changes in the nature of the lowest sub-state of the fundamental multiplet.

3.3 Curie constant C_F and mean Van Vleck susceptibilities χ_F^{VV} of the fundamental multiplet

The $C_\mathrm{F}(x)$ curves agree well with those representing $|\langle i|\,J_z\,|i\rangle|^2$ (which is a sum over all the sub-states of i) as given by Happel [11] and concerning neutron scattering, except for small differences for $J = 8$ and $\frac{9}{2}$.

Except for the cases when the levels cross each other, there are no discontinuities in $\mathrm{d}C_\mathrm{F}/\mathrm{d}x$, in spite of possible discontinuities in $\mathrm{d}m_i/\mathrm{d}x$ which are here ruled out because C_F involves all the sub-states of the fundamental multiplet and not only the lowest one and because of the squaring of the individual moment values. Some pronounced minima nevertheless occur which are only related to the local behaviour of $m_i(x)$ (for $J = \frac{9}{2}$, $\frac{15}{2}$, and 8).

The only discontinuities in $\chi_\mathrm{F}^{\mathrm{VV}}$ and in $\mathrm{d}\chi_\mathrm{F}^{\mathrm{VV}}/\mathrm{d}x$ (for non-magnetic states) are those related to the crossing of the levels (the above-mentioned other discontinuities in χ_i^{VV} only occur for magnetic multiplets).

References

[1] K. W. H. STEVENS, Proc. Phys. Soc. **A65**, 209 (1952).
[2] K. R. LEA, M. J. M. LEASK, and W. P. WOLF, J. Phys. Chem. Solids **23**, 1381 (1962).
[3] Y. EBINA and N. TSUYA, Sci. Rep. Res. Inst. Tôhoku Univ., Ser. B (Electr. Commun.) **12**, 1, 167, 183 (1960); **13**, 1, 25, 43 (1961).
[4] Y. EBINA and N. TSUYA, Sci. Rep. Res. Inst. Tôhoku Univ., Ser. B (Electr. Commun.) **15**, (Suppl.) 1, 47 (1963).
[5] D. P. SCHUMACHER and C. A. HOLLINGSWORTH, J. Phys. Chem. Solids **27**, 749 (1966).
[6] U. WALTER, J. Phys. Chem. Solids **45**, 401 (1984).
[7] G. HEINRICH, J. P. KAPPLER, and A. MEYER, Phys. Letters A **74**, 121 (1979).
[8] E. SEGAL and W. E. WALLACE, J. Solid State Chem. **2**, 347 (1970).
[9] M. T. HUTCHINGS, Solid State Phys. **16**, 227 (1964).
[10] R. J. BIRGENEAU, J. Phys. Chem. Solids **33**, 59 (1972).
[11] H. HAPPEL, Dissertation, Universität Frankfurt/M., 1975.

(Received October 1, 1985)

phys. stat. sol. (b) 133, 167 (1986)

Subject classification: 75.80; S1.1

Laboratorio de Magnetismo, Facultad C. Fisicas, Universitad Complutense, Madrid (a) and Max-Planck-Institut für Metallforschung, Institut für Physik, Stuttgart[1]) *(b)*

Critical Exponents of the Magnetostriction in Amorphous Alloys

By

M. VÁZQUEZ (a), A. HERNANDO (a), and H. KRONMÜLLER (b)

The magnetostriction of some amorphous alloys is investigated in the temperature range of the magnetic phase transition. Measurements are made with the help of a method described previously, and which is based on the observation of the macroscopical torsional strain under a longitudinal electrical current. The magnetostriction near the ferromagnetic phase transition is found to obey a $(T_c - T)^\varkappa$-law with critical exponents depending on the type of alloy and values varying in the range of $\varkappa = 1.1$ to 1.4.

Die Magnetostriktion amorpher Legierungen wird im Temperaturbereich der magnetischen Phasenumwandlung untersucht. Die Messungen werden mit Hilfe einer früher beschriebenen Methode durchgeführt, die auf der Beobachtung makroskopischer Torsionsverzerrungen bei Anregen eines longitudinalen elektrischen Stromes beruht. In der Nähe der magnetischen Phasenumwandlung gehorcht die Magnetostriktion einem $(T_c - T)^\varkappa$-Gesetz, wobei die kritischen Exponenten materialabhängig sind und Werte von $\varkappa = 1{,}1$ bis 1,4 annehmen.

1. Introduction

The dependence of the magnetostriction on temperature has been reported for different amorphous alloys [1 to 4]. In a wide temperature range, the magnetostriction λ increases with the spontaneous magnetization, $M_s(T)$ and decreases with increasing temperature T following a power law as

$$\lambda(T) \sim M_s^n(T)\,, \tag{1}$$

where the reported exponents n [5, 6] depend on the considered alloy and in the range of temperatures considered vary usually between 2 and 3. Nevertheless, the experimental results do not account for the range of temperature near the Curie temperature because of the difficulty of its measurement.

The purpose of the present work has been the determination of λ near the Curie point where the ferromagnetic phase transition takes place and the determination of the critical exponent describing the temperature dependence of λ near T_c.

2. Experimental Procedure and Results

Measurements of λ have been carried out following the method described in detail elsewhere [7, 8]. This method is based on the fact that when a longitudinal and a circular magnetic field are simultaneously applied to the amorphous ribbon, the sample undergoes a torsional deflection. The longitudinal magnetic field was obtained by means of a Helmholtz system and the circular one by a dc current I along the

---

[1]) Heisenbergstr. 1, D-7000 Stuttgart, FRG.

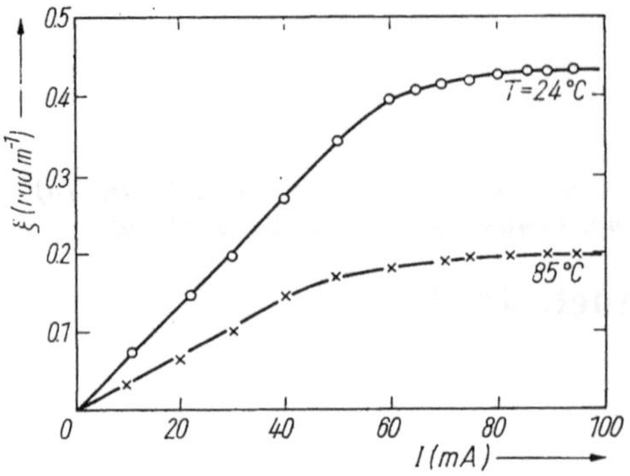

Fig. 1. Dependence of the torsional strain ξ on the dc current flowing along the ribbon ($Fe_{20}Ni_{60} \cdot P_4B_{16}$). The value of the constant longitudinal applied field was 8 Oe

sample. The ribbon was vertically placed and fixed at the top. The lower end of the ribbon was clamped to a light aluminium tube (2g in weight) which was introduced into an electrical bath thus allowing the flow of electrical currents while the sample was free to undergo torsional strain. Moreover, the sample was placed within a furnace so that the torsional deflection could be measured also as a function of temperature. The deflection was measured with the help of a small mirror cemented onto the aluminium tube by focusing a laser beam and measuring the deviations of the reflected beam on a scale disposed far away.

The influence of both applied magnetic fields has been discussed recently [7, 8] and the magnetostriction is roughly given by

$$\lambda = \xi_{\max} a/2 \,, \tag{2}$$

where $a/2$ is the half-thickness of the ribbon and $\xi_{\max}$ the saturation value of the torsional deflection. $\xi_{\max}$ is obtained from the experimental $\xi \sim I$ curve when a fixed longitudinal magnetic field is applied (high enough to produce almost saturation of the magnetization along its direction and being usually of the order of magnitude of a few Oe (see Fig. 1).

Two different samples obtained by quenching from the melt were used for measurements. Their nominal compositions are $Fe_{20}Ni_{58}Si_8B_{14}$ and $Fe_{20}N_{60}P_4B_{16}$. The variation of λ with temperature of both samples is shown in Fig. 2.

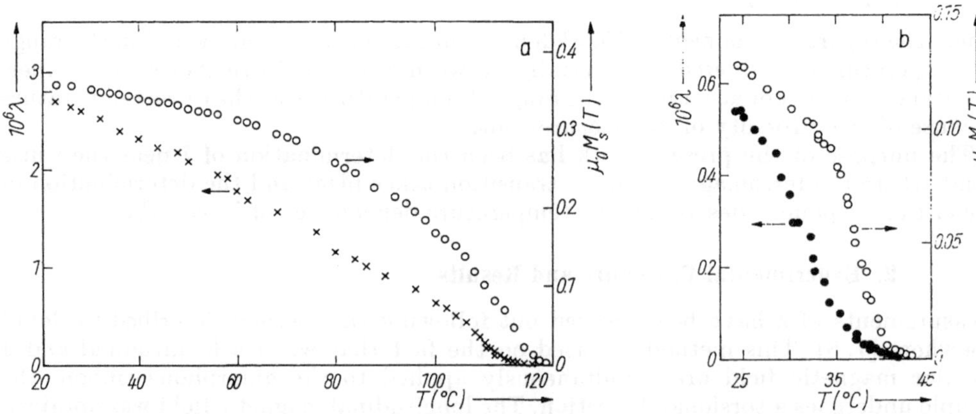

Fig. 2. Temperature dependence of the magnetostriction λ and of the magnetic polarization $\mu_0 M_s$ (8 Oe applied field) for a) $Fe_{20}Ni_{60}P_4B_{16}$ and b) $Fe_{20}Ni_{58} \cdot Si_8B_{14}$ alloys

With the help of an additional furnace, the dependence of the longitudinal component of the magnetization on temperature was determined for an applied longitudinal magnetic field of 8 Oe. These results are represented in Fig. 2.

3. Discussion

In the region close to the ferromagnetic phase transition the dependence of the magnetization $M(H, T)$ is described by the critical exponents α, β, and γ which are defined as

$$M(0, T) \sim (T_c - T)^\beta ; \qquad T < T_c , \tag{3}$$

$$M(H, T) \sim H^{1/\delta} ; \qquad T = T_c , \tag{4}$$

$$\chi_0(0, T) \sim (T - T_c)^{-\gamma} ; \qquad T > T_c , \tag{5}$$

where χ_0 corresponds to the paramagnetic susceptibility for zero field. These coefficients have been intensively studied from both experimental and theoretical viewpoints in amorphous alloys [9, 10].

The material laws (3) to (5) have been found to be valid for amorphous alloys within a narrow temperature range of $|T - T_c|/T_c < 0.04$, and at larger deviations from T_c the exponents β, γ become temperature dependent [9, 10, 11]. Furthermore in a series of experiments it has been shown [9, 12] that the critical exponents of amorphous alloys near T_c correspond to those of the three-dimensional Heisenberg model with $\gamma = 1.39$, $\beta = 0.36$, and $\delta = 4.8$. It is self-suggesting that also the temperature dependence of $\lambda(T)$ near T_c according to

$$\lambda(T) \sim (T_c - T)^\varkappa \tag{6}$$

may be represented by a critical exponent $\varkappa$. The asymptotic critical exponent $\varkappa_c$ now is derived from experimental data by plotting $\ln \lambda(T)$ versus $\ln (T_c - T)$. Similarly the critical exponent β may be derived from $\ln M(T)$ versus $\ln (T_c - T)$ plots.

Fig. 3 shows the results of such plots for $\lambda(T)$ of the amorphous alloys $Fe_{20}Ni_{60}P_4B_{16}$ and $Fe_{20}Ni_{60}Si_8B_{14}$. From the stright lines found for these alloys up to temperatures of 10 K above T_c we derive critical exponents

$$\varkappa_c = 1.12 \quad \text{for} \quad Fe_{20}Ni_{58}Si_8B_{14}$$

and

$$\varkappa_c = 1.40 \quad \text{for} \quad Fe_{20}Ni_{60}P_4B_{16} .$$

These results are summarized in Table 1 together with previous results of these alloys for T_c and β [12]. The critical exponents, $\varkappa_c$, should be considered as approximate values with error limits $\Delta\varkappa = \pm 0.05$. Efrrors may also arise from the longitudinal

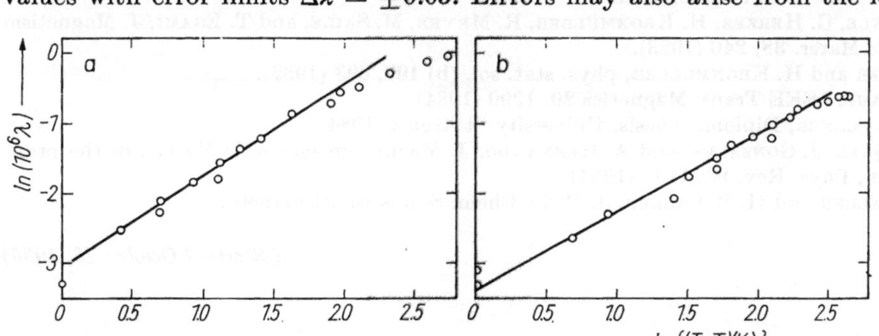

Fig. 3. ln ($10^6\lambda$) vs. ln ($T_c - T$) curve near tho Curie temperature for a) $Fe_{20}Ni_{60}P_4B_{16}$ and b) $Fe_{20}Ni_{58}Si_8B_{14}$ alloys

Table 1
Curie temperature and the critical exponents β and $\varkappa$ of two amorphous alloys

	T_c (K)		β_c		$\varkappa_c$	3β	
	present paper	[12]	present paper	[12]		present paper	[12]
$Fe_{20}Ni_{58}Si_8B_{14}$	311	312.6	0.36	0.38	1.12	1.08	1.14
$Fe_{20}Ni_{60}P_4B_{16}$	389	390.6	0.55	0.45	1.40	1.65	1.35

electrical current which may raise the sample temperature as it has been studied previously [13]. In the present investigation dc currents up to 70 mA have been used. For such low values of the dc current increases of temperature can be disregarded. This is also supported by the very similar values obtained for T_c in the present paper and in [12]. It is of interest to note that within the error limits the empirical relation

$$\varkappa_c \approx 3\beta_c \tag{7}$$

holds, indicating the validity of the well-known relation $\lambda_s \sim M_s^3$ which is expected to apply for magnetic properties of second order in the direction cosines of the spontaneous magnetization [14, 15].

References

[1] R. C. O'HANDLEY, Phys. Rev. B 18, 930 (1978).
[2] R. C. O'HANDLEY and M. O. SULLIVAN, J. appl. Phys. 52, 1841 (1981).
[3] A. HERNANDO, V. MADURGA, C. NUÑEZ DE VILLAVICENCIO, and M. VÁZQUEZ, Appl. Phys. Letters 45, 802 (1984).
[4] J. M. BARANDIARAN, A. HERNANDO, and O. V. NIELSEN, J. Magnetism magnetic Mater. 46, 317 (1985).
[5] R. C. O'HANDLEY, in: Amorphous Metallic Alloys, Ed. F. E. LUBORSKY, Butterworths, London 1983 (p. 280).
[6] K. HANDRICH and S. KOBE, Amorphe Ferro- und Ferrimagnetika, Physik Verlag, Weinheim 1980 (p. 92).
[7] M. LINIERS, V. MADURGA, M. VÁZQUEZ, and A. HERNANDO, Phys. Rev. B 31, 4425 (1985).
[8] C. NUÑEZ DE VILLAVICENCIO, M. VÁZQUEZ, V. MADURGA, and A. HERNANDO, J. Magnetism magnetic Mater., to be published.
[9] M. FÄHNLE, G. HERZER, H. KRONMÜLLER, R. MEYER, M. SAILE, and T. EGAMI, J. Magnetism magnetic Mater. 38, 240 (1983).
[10] R. MEYER and H. KRONMÜLLER, phys. stat. sol. (b) 109, 693 (1982).
[11] S. N. KAUL, IEEE Trans. Magnetics 20, 1290 (1984).
[12] W.-U. KELLNER, Diploma Thesis, University Stuttgart, 1984.
[13] M. VÁZQUEZ, J. GONZALES, and A. HERNANDO, J. Magnetism magnetic Mater., in the press.
[14] C. ZENER, Phys. Rev. 96, 1335 (1954).
[15] E. R. CALLEN and H. B. CALLEN, J. Phys. Chem. Solids 16, 310 (1960).

(Received October 16, 1985)

phys. stat. sol. (b) 133, 171 (1986)

Subject classification: 71.25; S8.16.

Institute of Physics, Academy of Sciences of the Azerbaidzhan SSR, Baku[1])

Band Structure of $TlGaSe_2$ Ternary Layered Crystals

By

S. G. ABDULLAEVA and N. T. MAMEDOV

The band structure in the high-symmetry points of the $TlGaSe_2$ Brillouin zone is calculated by the pseudopotential method. The top of the valence band is found to be located at the Γ point. The bottom of the conduction band is located at the line Γ–Y and the additional extremum of this band — at the point Γ of the Brillouin zone of $TlGaSe_2$. The data obtained are compared with the results of previous band calculations and optical absorption data of $TlGaSe_2$.

Методом псевдопотенциала расчитана зонная структура в высокосимметричных точках зоны Бриллюэна слоистых кристаллов $TlGaSe_2$. Найдено, что потолок валентной зоны расположен в точке Г. Дно зоны проводимости расположено на линии Г–Y, а дополнительный экстремум этой зоны в точке Г. Проведено сравнение полученных данных с результатами предыдущих расчетов зонной структуры к данными по оптическому поглощению.

1. Introduction

Energy structure and exciton states of binary layer semiconductors, in particular, of $A^{III}B^{VI}$ crystals, have been intensively studied in recent years [1, 2]. This is mainly associated with the strong anisotropy of the electron states, which may be expected in crystals with layered structures. However [3, 4], the existence of strongly anisotropic electron states in the near-absorption-edge region of GaSe-type crystals is questionable since the electron densities of the atoms localized at the boundaries of the layers and forming the bands near the absorption edge are sufficiently overlapping. Rather than by this overlapping is caused by the presence of a directed covalent component (along with the van-der-Waals type) in the interlayer bond, experimentally discovered in the gallium selenide analog — GaS [5]. Therefore, the layered crystals with isotropic interlayer bond, in particular, the $A^{III}B^{III}C_2^{VI}$ ternary compounds, are of considerable interest.

An individual layer of the mostly studied ternary compound $TlGaSe_2$, having a thickness (≈ 0.7 nm [6]) smaller than that of GaSe, contains seven atomic sheets which are occupied by the atoms in the Se–Tl–Ga–Se–Ga–Tl–Se sequence. Therefore, the bonding forces within a layer of $TlGaSe_2$ at least are of the same order as those in GaSe. The last-mentioned statement is also true for the interlayer forces because the frequencies of the modes corresponding to interlayer vibrations and registered in the Raman spectra of $TlGaSe_2$ [7] and GaSe [8] crystals are essentially the same. Thus, one can expect that in $A^{III}B^{III}C_2^{VI}$ crystals the interaction between the atoms in different layers are very weak compared with those between the atoms in the same layer as well as in $A^{III}B^{VI}$ layered compounds.

In [9] the band structure of monoclinic $TlGaSe_2$ was obtained on the basis of the consideration of the monoclinic modification of $TlGaSe_2$ as a deformed tetragonal one. The present paper contains the results of a pseudopotential calculation of monoclinic $TlGaSe_2$ crystals.

[1]) Prospekt Narimanova 33, 370143 Baku, USSR.

2. Symmetry Analysis

Calculation of the electronic band structure of monoclinic $TlGaSe_2$ with C_s^4 space group symmetry presents a problem due to the low symmetry (C_s^4) and the great number of atoms within the unit cell (32) [6]. Therefore, we used the coordinates of the atoms in the $TlGaSe_2$ unit cell of space group C_{2h}^6, which are given in [10].

The primitive translational vectors of the $TlGaSe_2$ lattice were chosen as

$$\boldsymbol{t}_1 = \begin{pmatrix} \frac{a}{2} \\ \frac{b}{2} \\ 0 \end{pmatrix}, \quad \boldsymbol{t}_2 = \begin{pmatrix} -\frac{a}{2} \\ \frac{b}{2} \\ 0 \end{pmatrix}, \quad \boldsymbol{t}_3 = \begin{pmatrix} -C \sin\left(-\frac{\beta\pi}{2}\right) \\ 0 \\ C \cos\left(-\frac{\beta\pi}{2}\right) \end{pmatrix}, \tag{1}$$

where $\boldsymbol{a}$, Z, $\boldsymbol{c}$ are the lattice parameters and β is the angle between the a- and c-axes given in [6].

The basis vectors of the reciprocal lattice are, respectively, equal to

$$\boldsymbol{g}_1 = 2\pi \begin{pmatrix} \frac{1}{a} \\ \frac{1}{b} \\ \frac{1}{a} \operatorname{tg}\left(-\frac{\beta\pi}{2}\right) \end{pmatrix}; \quad \boldsymbol{g}_2 = 2\pi \begin{pmatrix} -\frac{1}{a} \\ \frac{1}{b} \\ -\frac{1}{a} \operatorname{tg}\left(-\frac{\beta\pi}{2}\right) \end{pmatrix}; \quad \boldsymbol{g}_3 = 2\pi \begin{pmatrix} 0 \\ 0 \\ \frac{1}{C \cos\left(-\frac{\beta\pi}{2}\right)} \end{pmatrix}. \tag{2}$$

Fig. 1 shows the Brillouin zone of the $TlGaSe_2$ base-centred monoclinic lattice. Positions of high-symmetry k-points and k-lines along with the relationships between our notations and those given in [11] are summarized in Table 1.

The following notations of the symmetry elements are used; E identity; C_2 rotation through $\pi/2$ about y-axis; J inversion, and σ reflection perpendicular to the y-axis. As $C_2 \boldsymbol{g}_2 = \boldsymbol{g}_1$, Table 1 presents only A and C symmetry points because B and D enter the stars of the wave vectors A and C, respectively.

The irreducible representation tables for the Γ, Z, and A points, the Λ-line, and the F-symmetry plane are shown in Tables 2 to 6, respectively.

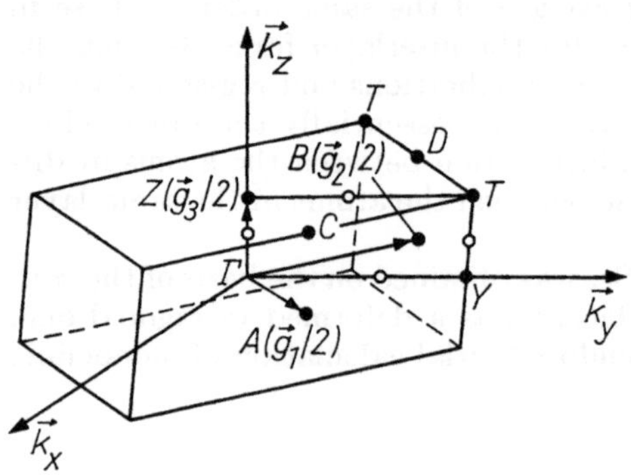

Fig. 1. Brillouin zone of $TlGaSe_2$

Table 1

k-points of high symmetry	notations in accordance with [11]	corresponding point groups	elements of point groups
$\Gamma(0, 0, 0)$	Γ	C_{2h}	E, J, σ, C_2
$Z\left(\frac{g_3}{2}\right)$	Z	C_{2h}	E, J, σ, C_2
$Y\left(\frac{g_1 + g_2}{2}\right)$	V	C_{2h}	E, J, σ, C_2
$T\left(\frac{g_2 + g_3 - g_1}{2}\right)$	—	C_{2h}	E, J, σ, C_2
$A\left(\frac{g_1}{2}\right)$	L	C_i	E, J
$C\left(\frac{g_1 + g_3}{2}\right)$	M	C_i	E, J
$\Lambda(\mu(g_1 + g_2) + \frac{1}{2} g_3)^{*})$	—	C_2	E, J
$\Delta\mu(g_1 + g_2)^{*})$	—	C_2	E, C_2
$F(\mu_1(g_1 - g_2) + \mu_3 g_3)^{*})$	—	C_s	E, σ

*) $\mu, \mu_1, \mu_3 \leqq \frac{1}{2}$

Table 2

All the symmetry elements form the wave vector group of the Γ point (centre of the Brillouin zone). The wave vector group for the Y point is the same as that for the Γ point

Γ	E	$\sigma/\tau^{*})$	J	$C_2/\tau^{*})$
Γ_1	1	1	1	1
Γ_2	1	−1	1	−1
Γ_3	1	−1	−1	1
Γ_4	1	1	−1	−1

*) τ is half of the primitive translational vector along the z-axis ($t_{3/2}$).

Table 3

The wave vector groups for the Z and T points are identical. The electronic states in the Z point are classified only by one (two-dimensional) representation

Z	E	σ/τ	J	C_2/τ
Z_1	$\begin{pmatrix}1 & 0\\ 0 & 1\end{pmatrix}$	$\begin{pmatrix}0 & -1\\ 1 & 0\end{pmatrix}$	$\begin{pmatrix}1 & 0\\ 0 & -1\end{pmatrix}$	$\begin{pmatrix}0 & 1\\ 1 & 0\end{pmatrix}$

Table 4

The groups of the wave vectors for the A and C points contain only two symmetry elements. Tables of irreducible representations for the A and C points correlate

A	E	J
A_1	1	1
A_2	1	−1

Table 5

The Λ-line joints the points Z and T. The Δ-line joints the points Γ and Y. The groups of wave vectors of Λ and Δ are the same. Along the whole Λ-line the electronic states are twofold degenerate because of the time-reversal symmetry

Λ	E	C_2/τ
Λ_1	1	1
Λ_2	1	−1

Table 6

The lines joining the Γ and Z points as well as the Y and T points lie in the F-symmetry plane

F	E	σ/τ
F_1	1	$\exp(-i\mu_3\pi)$
F_2	1	$-\exp(-i\mu_3\pi)$

3. Band Structure and Discussion

The band structure of the monoclinic $TlGaSe_2$ crystals has been calculated by the pseudopotential method [12]. The calculation technique, the values of the atomic potentials, and the form factors do not differ from those used in [9]. The energy spectrum in the symmetry points Γ and Y was calculated with the use of 720 plane waves in the expansion of the electron wave function. The convergence of the levels located near the forbidden gap amounted to 0.2 eV.

The energy spectrum at the symmetry lines and in the other symmetrical points was determined by the *kp*-method [13]. The solutions for the Γ and Y points were used as the zero approximations. The order of the secular determinant of the *kp*-method was equal to 130. The accuracy of calculations done by the mentioned method was checked by comparing the results for the intermediate points on the Γ–Y line obtained under different zero approximations and by checking double degeneration in the Z point and along the whole Λ-line.

The calculated electron spectrum is shown in Fig. 2. The values of the electron energy (E) as a function of wave vector ($\boldsymbol{k}$) are displayed by the points. Such a manner of representing the results of band structure calculation has been previously used by the authors of [14]. The points are not connected due to the complicated band structure. In Fig. 2 one can see that the forbidden gap for the direct transitions is realized in the Γ-point and corresponds to the $\Gamma_2 \leftrightarrow \Gamma_3$ transitions allowed in the $\boldsymbol{e} \parallel \sigma$ orientation ($\boldsymbol{e}$ is the electrical vector of the incident light). The direct energy gap is about 2.1 eV. An additional minimum of the conduction band (Γ_1), with the energy position approximately the same as that of Γ_3, is localized near the absorption edge. Generally speaking, the transitions of $\Gamma_2 \leftrightarrow \Gamma_1$ are forbidden in dipole approximation. These transitions can be shown to become allowed in quadrupole approximation in both $\boldsymbol{e} \parallel \sigma$ and $\boldsymbol{e} \perp \sigma$ light polarizations. Taking into account the double groups, it is easy to show that due to a mixing of states transitions are allowed in all light polarizations, as well. Thus, it is impossible to explain the selection rules for direct optical transitions obtained in [15] neither in quadrupole approximation nor with account of spin–orbit interaction.

Recently, a ferroelectric phase transition was discovered in $TlGaSe_2$ at $T = 107$ K [16]. Since this transition occurs with reducing of symmetry it is reasonable to assume that the high-temperature phase has the C_{2h}^2 space group symmetry and the low-tem-

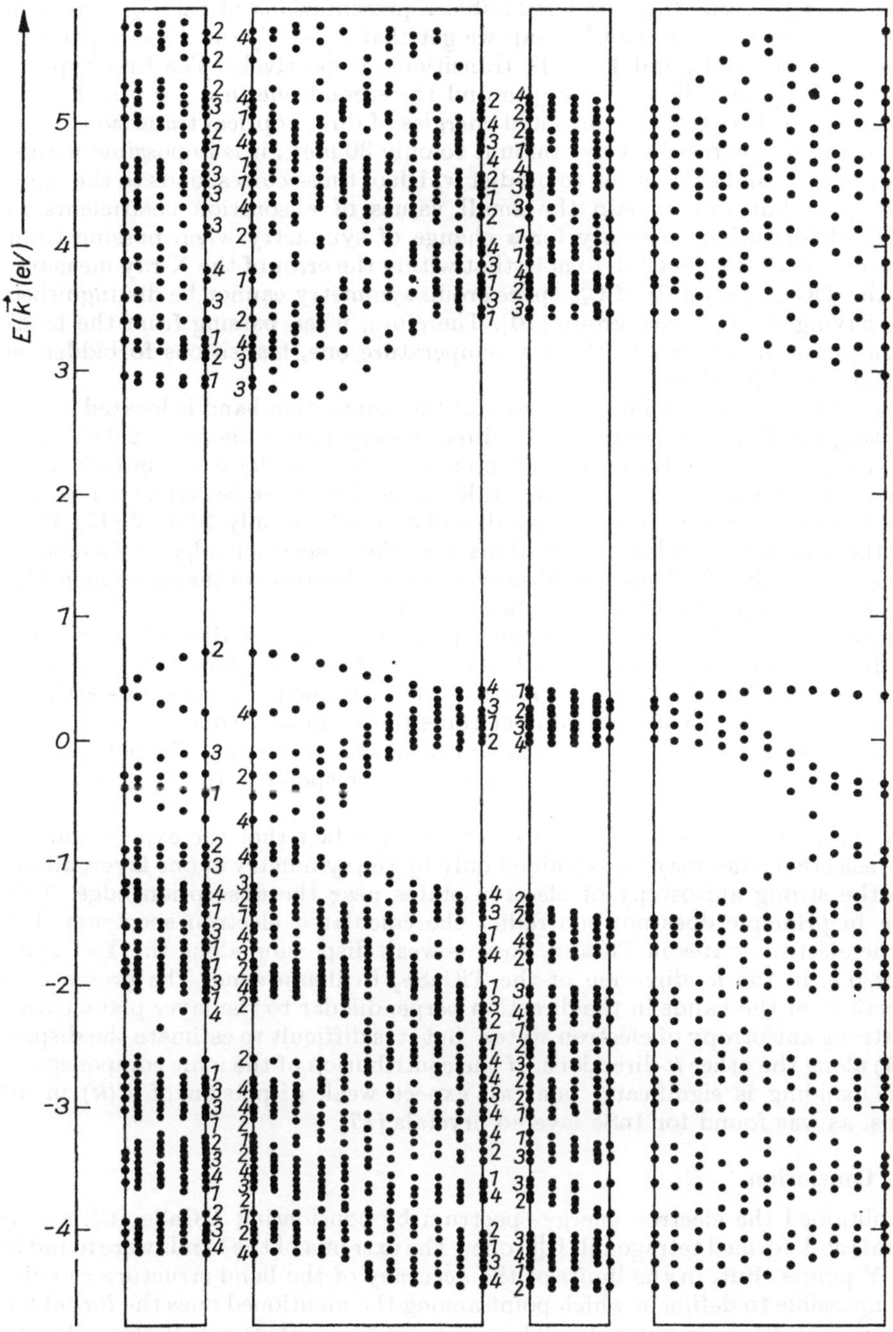

Fig. 2. Band structure of $TlGaSe_2$

perature one — C_s^4. Expanding the irreducible representations of the C_{2h}^6 group in irreducible representations of the C_s^4 group we get that $\Gamma_2 \leftrightarrow \Gamma_3$ and $\Gamma_2 \leftrightarrow \Gamma_1$ transitions transform to $\Gamma_2 \leftrightarrow \Gamma_2$ and $\Gamma_2 \leftrightarrow \Gamma_1$ transitions, respectively. The first type of transitions ($\Gamma_2 \leftrightarrow \Gamma_2$) are allowed in $\boldsymbol{e} \perp \sigma$ and the second one in $\boldsymbol{e} \parallel \sigma$ orientation. Since the difference between the threshold energies of direct optical transitions polarized in $\boldsymbol{e} \parallel \sigma$ and $\boldsymbol{e} \perp \sigma$, respectively, amount to only 30 meV, it is impossible within the accuracy of our band calculation to predict which of those corresponds to the minimum direct gap, but can explain the small values of absorption coefficients in $TlGaSe_2$. The deformation necessary for a change of symmetry when passing from C_{2h}^6 to C_s^4 is very small. It is enough to note that within the error of the X-ray measurements [6], the $TlGeSe_2$ crystals of C_{2h}^6 space group symmetry cannot be distinguished from those having the C_s^4 space group [10]. Therefore, when passing from the high-temperature phase of $TlGeSe_2$ to the low-temperature one, transitions forbidden in nature, become weakly allowed.

As seen from Fig. 2 the absolute minimum of the conduction band is located at the line connecting the Γ and Y points. The indirect energy gap is about ≈ 2.0 eV. The extremum at the Γ point and the bottom of conduction band are located approximately at equal energies (see Fig. 2). The experimentally found difference between the energies of direct and indirect transitions is also small and amounts to only 70 meV [15]. Consequently, the presence of indirect transitions near the absorption edge of $TlGaSe_2$ is confirmed by the results of calculation. However, the explanation of the experimentally obtained polarization peculiarities presents a difficulty.

In compliance with [15] indirect transitions in $TlGaSe_2$ are allowed in $\boldsymbol{e} \perp \boldsymbol{C}^*$ ($\boldsymbol{C}^*$ is the direction perpendicular to the layer plane). It can be shown that the symmetry of one layer of $TlGaSe_2$ is S_4^2. The fourth-order-inversion axis coincides with the $\boldsymbol{C}^*$ direction and selection rules for indirect transitions can be written as: $\boldsymbol{e} \perp S_4$ — the transitions are allowed and $\boldsymbol{e} \parallel S_4$ — the transitions are forbidden. The polarization peculiarities of indirect transitions obtained in [15] correspond to the S_4^2 space group symmetry.

The unit cell of $TlGaSe_2$ contains two layers and the fact that the experimentally established selection rules may be explained only by the symmetry of one layer, allows to assume the strong anisotropy of electron states near the absorption edge. This assumption in principle does not contradict the calculated electron spectrum. For example, the electron states in $TlGaSe_2$ possess weak dispersion along the Y–T symmetry lines, i.e. in the $\boldsymbol{k}_z$ direction of the $TlGaSe_2$ Brillouin zone. The presence of "plane" sections of the bands in the direction perpendicular to the layer plane corresponds to strong anisotropy of electron states. But it is difficult to estimate the dispersion of $E(\boldsymbol{k})$ along the other $\boldsymbol{k}$-directions. If the contribution of the ionic component to interatomic bonding is significant, one can expect weak dispersion of $E(\boldsymbol{k})$ in all $\boldsymbol{k}$-directions, as was found for InSe layered crystals [17].

4. Conclusion

In [9] we obtained the electron energy spectrum by considering $TlGaSe_2$ (C_s^4) monoclinic crystals as deformed tetragonal (D_{4h}^{18}) ones. The extrema of the bands were found in the Γ and Y points. But, in the limits of the accuracy of the band structure calculation, it is impossible to define in which point among the mentioned ones the forbidden gap for direct transitions is minimum. The results of the present calculation allow to consider that the energy gap in the centre of the $TlGaSe_2$ Brillouin zone is minimum.

The comparison of the results of the calculation by the method of pseudopotentials with those given in [9] revealed that the number of bands of a certain symmetry, the total width of the valence bands, and the qualitative picture of the band sequence are in good agreement. But, as the present calculations were made without additional approximations, we believe that they give a better illustration of the band structure of $TlGaSe_2$.

Acknowledgements

The authors express their thanks to G. B. Abdullaev for the support of this work and to F. M. Gashimzade for helpful discussion of the results.

References

[1] E. Mooser (Ed.), Physic and Chemistry of Materials with Layered Structures, D. Reidel Publ. Co., Dordrecht/Boston 1979.
[2] L. N. Bylaevskii, Uspekhi fiz. Nauk **116**, 449 (1975).
[3] M. Shluter, Nuovo Cimento **13B**, 313 (1973).
[4] G. L. Belenkii, M. O. Godzhaev, N. T. Mamedov, E. Yu. Salaev. and R. A. Suleimanov, phys. stat. sol. (a) **53**, 137 (1979).
[5] A. Kuhn, A. Bourdon, J. Rigoult, and A. Rimsky, Phys. Rev. B **25**, 4081 (1982).
[6] V. D. Muller and H. Hahn, Z. anorg. allg. Chem. **438**, 258 (1978).
[7] B. I. Mavrin, Kh. E. Sterin, N. M. Gasanly, Z. D. Khalafov, E. Yu. Salaev, K. R. Allakhverdiev, and R. M. Sardarly, Fiz. tverd. Tela **19**, 2960 (1977).
[8] J. C. Irvin, R. M. Hoff, B. P. Clayman, and R. A. Bromley, Solid State Commun. **13**, 1531 (1973).
[9] S. G. Abdullaeva, N. T. Mamedov, and G. S. Orudzhev, phys. stat. sol. (b) **119**, 41 (1983).
[10] W. Henkel, H. D. Hochhermer, C. Carlone, A. Werner, S. Ves, and H. G. Schnering, Phys. Rev. B. **26**, 3211 (1982).
[11] C. J. Bradley and A. P. Cracknell, The Mathematical Theory of Symmetry in Solids, Clarendon Press, Oxford 1972.
[12] V. Heine, M. Cohen, and D. Weaire, The Pseudopotential Theory, Izd. Mir, Moskva 1970 (in Russian).
[13] A. A. Ovchinnikova (Ed.), The Methods of Calculation in Theory of Solids, Izd. Mir, Moskva 1975 (in Russian).
[14] J. Abbati, L. Braicovich, and G. Ciucci, Nuovo Cimento **39B**, 727 (1977).
[15] S. G. Adbullaeva, G. L. Belenkii, and N. T. Mamedov, Fiz. Tekh. Poluprov. **15**, 949 (1981).
[16] A. A. Volkov, Yu. G. Goncharov, G. V. Kozlov, S. P. Lebedev, A. M. Prokhorov, R. A. Aliev, and K. R. Allakhverdiev, Zh. eksper. teor. Fiz., Pisma **37**, 517 (1983).
[17] Y. Depersing, E. Doni, R. Girlanda, A. Baldereschi, and K. Maschke, Solid State Commun. **27**, 1449 (1978).

(Received March 13, 1985)

phys. stat. sol. (b) 133, 179 (1986)

Subject classification: 71.38; S8

Institute of Physics, Academy of Sciences of the Azerbaidzhan SSR, Baku[1])

Selection Rules for Matrix Elements of Electron–Phonon Scattering Processes in Crystals with TlSe Structure

By

F. M. GASHIMZADE and E. V. TAGIROV

Selection rules for intervalley transitions in TlSe structural crystals are obtained. The results are presented both, without and with regard to the electron spin.

Получены правила отбора для междолинных переходов в кристаллах со структурой TlSe. Приведены результаты как без учёта, так и с учётом спина электрона.

1. Introduction

Intervalley scattering plays a great role in various physical processes. The scattering between equivalent minima explains effects concerned with overpopulation of the valleys, e.g., in high electric field [1, 2] or in acousto-electric effects [2]. The scattering between nonequivalent minima contributes to indirect optical transitions and to the Gunn effect [1, 2].

The first step in considering all these effects is to determine the type of phonons which take part in the process. One can obtain useful information using group theoretical selection rules, as was done for Ge and Si [3].

The present paper is concerned with obtaining the selection rules for matrix elements of the intervalley transitions, with and without taking into account the electron spin, for TlSe-type crystals. Tl Se belongs to the chain structural semiconductors A^3B^6. There are binary compounds TlS and InSe as well as triple semiconductors $TlInSe_2$, $TlInTe_2$, $TlGaTe_2$ with common formula $A^3B^3C_2^6$ [4], which belong to this group. The crystal lattice for all these semiconductors is volume-centred tetragonal with ratio of axes $c/a < 1$, where c is the tetragonal axis. The symmetry space group is D_{4h}^{18}, for which the tables of irreducible representation characters of wave vector groups are given in [5]. The band structure of TlSe and analogous materials is calculated in [6, 7]. The influence of spin–orbit interaction has been studied by Gashimzade and Orudzhev [8].

2. Scattering between Equivalent Valleys

To analyse the phonon-assisted intervalley transitions we have to calculate the sum [3]

$$\sum_{a \in \bar{G}} \chi_k^*(a_0^{-1} a a_0)\, \chi_w(a_0'^{-1} a a_0')\, \chi_k(a)\,, \tag{1}$$

where a_0, a_0' are determined from the quasi-impulse conservation law,

$$-a_0 k + a_0' w + k = K \tag{2}$$

[1]) Prospekt Narimanova 33, 370143 Baku, USSR.

and

$$\bar{G} = G_{a_0 k} \cap G_{a_0' w} \cap G_k \,.$$

Here χ_k is an irreducible representation character of the group of the electron wave vector for the corresponding valley; χ_w is the same for a phonon taking part in the scattering; $\boldsymbol{K}$ is a reciprocal lattice vector. For forbidden processes the sum (1) is equal to zero. However, (1) takes no account of the time reversal symmetry. When the wave function of the electron Ψ and $\hat{K}\Psi$ are linearly dependent, instead of (1) we have to consider the expression [3]

$$\sum_{a \in \bar{G}} \chi_w^*(a_0'^{-1} a a_0') \, \chi_k(a_0^{-1} a a_0) \, \chi_k(a) \pm \sum_{a \in \bar{\bar{G}}} \chi_w^*(a_0'^{-1} a a_0') \, \chi_k(a^2) \,, \tag{3}$$

where

$$\begin{aligned} -a_0' \boldsymbol{w} + a_0 \boldsymbol{k} + \boldsymbol{k} &= \boldsymbol{K} \,, \\ \bar{G} &= G_{a_0 k} \cap G_{a_0' w} \cap G_k \,, \\ \bar{\bar{G}} &= G_{a_0' w} \cap G_k a_0^{-1} \,. \end{aligned} \tag{4}$$

In (3) the upper sign refers to the spinless case and the lower one is used if taking into account the spin. But if Ψ and $\hat{K}\Psi$ are linearly independent, we have to calculate the following sum in addition to (1):

$$\sum_{a \in \bar{G}} \chi_k'^*(a_0^{-1} a a_0) \, \chi_w(a_0'^{-1} a a_0') \, \chi_k(a) \,. \tag{5}$$

Here, the prime marks the conjugate representation degenerate with the considered one because of time reversal symmetry.

There are two essentially different scattering processes between equivalent valleys, first, transitions to the opposite minimum (g-scattering), and the second, transitions to lateral ones (f-scattering). The selection rules with indication of the scattering type are given in Tables 1 to 3. The calculations were carried out by (3) for all points except the representations P_1, P_2, P_3, P_4 and P_6, P_7 (double-group representations) of the point P. For the point P the time reversal symmetry leads, firstly, to a linear dependence of Ψ and $\hat{K}\Psi$ only for the representation P_5, so that for P_1, P_2, P_3, P_4, and P_6, P_7 the selection rules are determined by (1). Secondly, it results in merging of the bands (P_1, P_4), (P_2P_3), and (P_6P_7), so we also take into account the transitions $P_1 \to P_4$, $P_2 \to P_3$, $P_6 \to P_7$ by (5).

We must note that taking into account the electron spin relaxes the selection rules since the Hamiltonian symmetry decreases because of the spin–orbit interaction. Nevertheless, for point A, as is obvious from comparison of Tables 1 and 2, considering the electron spin results in an additional prohibition of the optical phonon A_3 and the acoustic phonon A_4 to interact with electrons. The matter is that the spin–orbit interaction splits the double-degenerate level A_5 at point A. As a result, the selection rules change fundamentally in point A, at least because of decreasing number of states between which scattering is possible.

The absolute minimum of the conduction band in TlSe is placed at point D and belongs to the irreducible representation D_1 [6]. As is obvious from Table 1 three types of intervalley scattering are possible for this point. The longitudinal acoustic phonon A_4 and the optical phonon A_1 take part in the g-process. They are the lowest-frequency phonons according to the phonon spectra calculated for TlSe in [9]. Only the acoustic phonon A_4 takes part in the f-scattering for the transition $(\pi/2a, \pi/2a, \pi/4c) \to$ $\to (\pi/2a, -\pi/2a, -\pi/4c)$ and the high-frequency optical phonons T_1 and T_6 in the f′-process for the transition $(\pi/2a, \pi/2a, \pi/4c) \to (-\pi/2a, \pi/2a, \pi/4c)$. Thus, the g- and

Table 1

Selection rules for a scattering between equivalent valleys for points N, P, A, Δ, D, Σ, G neglecting spin

group of wave vector	N		P		A		Δ		D					Σ		G	
type of scat.	f		g			g	g	f	g	f		f′		g	f	g	f
irr. rep.	N_1 N_3 N_5 N_7	N_2 N_4 N_6 N_8	P_5	P_1 P_2 P_3 P_4	A_1 A_2 A_3 A_4	A_5	Δ_1 Δ_2 Δ_3 Δ_4	Δ_1 Δ_2 Δ_3 Δ_4	D_1 D_2 D_3 D_4	D_1 D_3	D_2 D_4	D_1 D_3	D_2 D_4	Σ_1 Σ_2 Σ_3 Σ_4	Σ_1 Σ_2 Σ_3 Σ_4	G_1 G_2 G_3 G_4	G_1 G_2 G_3 G_4
phonon	T_1	T_3	T_1T_3 T_4T_6 T_8T_9	T_1T_4 T_6T_9	A_1	A_1 A_3 A_4	Δ_1	Σ_1	A_1 A_4	A_4	A_2	T_1 T_6	T_3 T_8	Σ_1	Δ_1	Σ_1	Δ_1

f-processes would be more probable for the intervalley scattering without taking into account the electron spin in n-TlSe. The spin adds (see Table 2) to the f′-transition only the high-frequency phonons T_3, T_5, T_{10} (the phonon T_8 is absent [9]). Hence the g- and f-processes would remain dominant as before.

3. Scattering between Nonequivalent Valleys

In this case we should use the expression [3]

$$\sum_{a \in \bar{G}} \chi_{k'}^{*}(a_0^{-1} a a_0)\, \chi_{w}(a_0'^{-1} a a_0')\, \chi_{k}(a)\,, \tag{6}$$

where

$$-a_0 \boldsymbol{k}' + a_0' \boldsymbol{w} + \boldsymbol{k} = \boldsymbol{K}\,,$$

$$\bar{G} = G_{a_0 \boldsymbol{k}'} \cap G_{a_0' \boldsymbol{w}} \cap G_{\boldsymbol{k}}\,.$$

Table 2

Selection rules for a scattering between equivalent valleys for points N, P, A, Δ, D, Σ, G taking into account spin

group of wave vector	N	P	A	Δ		D			Σ		G	
type of scat.	f	g	g	g	f	g	f	f′	g	f	g	f
irr. rep.	N_9 N_{10}	P_6 P_7	A_6 A_7	Δ_5	Δ_5	D_5	D_5	D_5	Σ_5	Σ_5	G_5	G_5
phonon	T_1 T_3 T_{10}	$T_1T_2T_3T_4$ $T_5T_6T_7T_8$ T_9T_{10}	A_1	Δ_1	Σ_1 Σ_3 Σ_4	A_1 A_4	A_2 A_4 A_5	T_1T_3 T_5T_6 T_8T_{10}	Σ_1	Δ_1 Δ_3 Δ_4	Σ_1	Δ_1 Δ_3 Δ_4

Here $\boldsymbol{k}'$ and $\boldsymbol{k}$ are the wave vectors of the electron in the valleys between which the transition is considered. In TlSe and analogous compounds the minima are arranged at the point T and on the lines A, D, Σ, Δ and belong to the irreducible representations T_4, A_4, D_1, Σ_1, Δ_1, respectively. The selection rules for the transitions between these minima are given in Table 3. One can see from the table that the intervalley scattering in TlSe takes place assisted by the low-frequency acoustic phonon N_5, the optical phonon N_1, and the phonon D_1.

Table 3

Selection rules for a scattering between nonequivalent valleys for points T_4, A_4, Σ_1, D_1, Δ_1 neglecting spin

irr. rep.	T_4	A_4	Σ_1	D_1	Δ_1
T_4		A_4	G_1	D_1	Δ_2
A_4	A_4		Λ_1	$N_1N_5D_1$	L_1
Σ_1	G_1	Λ_1		$F_1\Lambda_1$	$\Delta_1\Delta_2H_1$
D_1	D_1	$N_1N_5D_1$	$F_1\Lambda_1$		—
Δ_1	Δ_2	L_1	$\Delta_1\Delta_2H_1$	—	

The phonons taking part in the scattering are given on the cross of the corresponding columns and lines. The symmetry of the system gives no additional information for the transition $D_1 \rightarrow \Delta_1$, since the wave vector of the transferred phonon is in the common direction.

In Table 3 we do not take into account the electron spin. If the spin-orbit interaction plays an essential role, then the symmetry of the system prohibits only the phonon A_1 among all A-phonons for the transition $T_4 \rightarrow A_4$. And there are no selection rules for the other intervalley transitions.

References

[1] K. SEEGER, Semiconductor Physics, Springer-Verlag, Wien/New York 1973.

[2] M. CONWELL, High-Field Transport in Semiconductors, Academic Press, New York/London 1967.

[3] H. STREITWOLF, Gruppentheorie in der Festkörperphysik, Akademische Verlagsgesellschaft Geest & Portig, Leipzig 1967.

[4] G. D. GUSEINOV, G. B. ABDULLAEV, S. M. BIDZINOVA, F. M. SEIDOV, M. Z. ISMAILOV, and A. M. PASHAEV, Phys. Letters A 33, 421 (1970).

[5] F. M. GASHIMZADE, Fiz. tverd. Tela 2, 3040 (1960).

[6] F. M. GASHIMZADE and G. S. ORUDZHEV, Fiz. Tekh. Poluprov. 15, 1311 (1981).

[7] F. M. GASHIMZADE, G. S. ORUDZHEV, and M. A. NIZAMETDINOVA, Fizicheskie svoistva slozhnykh poluprovodnikov, Elm, Baku 1982 (p. 80).

[8] F. M. GASHIMZADE and G. S. ORUDZHEV, Dokl. Akad. Nauk Azerbaidshan SSR, 27, No. 12 (1981).

[9] M. G. SHAKHTAKHTINSKII, D. A. GUSEINOV, M. A. NIZAMETDINOVA, and V. J. SHTEINSHRAIBER, Fiz. tverd. Tela 23, 1522 (1981).

(Received May 30, 1985)

phys. stat. sol. (b) 133, 183 (1986)

Subject classification: 79.70

Institute of Experimental Physics, University of Wroclaw[1])

The Second-Order Correction to Field-Electron-Emission Formulae for an Extended Temperature Range

By

C. J. WORKOWSKI

Corrected formulae for the total energy distribution, the emission current density, and the inversion temperature are derived under the condition that the field-electron emission takes place from a free-electron metal. The corrections relate to the series truncation approximating the WKB phase integral inherent in the analytic approximation of the electron tunneling probability. The assumed approximation for the phase integral takes into account not only the zero and first order terms, present in the existing theory, but also the quadratic term. The expression for the total energy distribution, obtained on this basis, is in typical circumstances valid for field emission in the electron energy range of $|\varepsilon| \leqq 1.5$ eV with respect to the Fermi energy. The accuracy of the approximations used is also discussed.

Unter der Voraussetzung, daß die Feldemission des Metalls durch das Freie-Elektronen-Modell beschrieben wird, werden korrigierte Formeln für die totale Energieverteilung, die Emissionsstromdichte und die Inversionstemperatur abgeleitet. Die Korrektion erfolgt durch entsprechendes Abbrechen einer Reihe, die das WKB-Phasenintegral approximiert, was immer mit einer analytischen Näherung der Tunnelwahrscheinlichkeit des Elektrons verbunden ist. Die Approximation des Phasenintegrals berücksichtigt nicht nur die Glieder nullter und erster Ordnung, wie in der existierenden Theorie, sondern auch das Glied zweiter Ordnung. Die so erzielte Formel für die gesamte Energieverteilung gilt unter typischen Feldemissionsverhältnissen im Bereich der relativ zur Fermi-Energie gemessenen Elektronenenergie $|\varepsilon| \leqq 1{,}5$ eV. Die Genauigkeit der angewendeten Approximation wird eingehend diskutiert.

1. Introduction

Higher-order corrections to field-electron-emission current formulae have been calculated in an analytic form by Cutler and Good [1]. Nevertheless, inspection of their arguments reveals that these authors have obtained emission current equations starting from the expression for the distribution of the normal component of the electron energy. But it is now well known that the measuring instruments, such as energy analysers, measure the total electron energy distribution: this follows from the properties of the radially symmetric electric field around the field emitter tip [2]. On the other hand, the normal-energy distribution occurs during tunneling of electrons through the insulator in thin film sandwich structures (see e.g. [3] and [4]).

There exist, however, more reasons which justify an elaboration of advanced models for field emission based on a free electron concept. For one reason, some features of a free electron model for field emission seem to be experimentally realizable. It is a well known experimental fact that emitter surface modification by adsorption removes, in some cases, band-structure effects in the energy distribution. Existing experimental data are too few to make generalizations, but a somewhat more extended discussion of this problem has been given by Workowski [5].

[1]) Cybulskiego 36, 50-205 Wrocław, Poland.

The second reason follows from the presence of some effects in field emission, such as: temperature dependence of the work-function, polarization of emitter surface atoms in the external electric field; also the interaction between the emitted electron and a metal surface at short distance leads to exchange and correlation effects, etc.

All the above-mentioned effects are small, but experimentally accessible inasmuch as the sensitivity and accuracy of measurements are sufficient. Analysis of experimental data, of course, requires a knowledge of exact electron-emission models, including the free-electron-like model.

The aim of the present paper is to propose improved formulae for the total energy distribution of field-emitted electrons, for the emission current density, and for the inversion temperature. Calculations are carried out by the use of asymptotic methods.

2. Correction Terms in Tunneling Equations

A convenient starting point for discussion is the following one-dimensional tunneling equation:

$$j'_{\mathrm{TF}}(\varepsilon) = \left(\frac{4\pi m e}{h^3}\right) f(\varepsilon) \int\limits_0^{E_\mathrm{F}+\varepsilon} D(W)\,\mathrm{d}W \tag{1}$$

that describes, in terms of the current density per unit energy range, the electron-energy distribution with respect to the total electron energy ε, i.e. the total energy distribution (TED) of field-emitted electrons from a metal in a free-electron approximation. The variable W represents the energy corresponding to the z component of the electron momentum. Note that the factor multiplying the integral in (1) is usually called the "supply function" and $f(\varepsilon) = (1 + \mathrm{e}^{\varepsilon/kT})^{-1}$ is the Fermi function (k is the Boltzmann constant).

Consider the electric field F applied perpendicularly to the metal surface in the vacuum and let $\varphi + E_\mathrm{F}$ be the unperturbed surface barrier. φ denotes the work function of the metal and E_F the Fermi energy; here all the energies are taken relative to the bottom of the conduction band apart from ε which is measured with reference to E_F. Therefore, the electron tunneling probability or tunneling function $D(W)$ given, in the framework of the first-order WKB approximation, as

$$D(W) = [1 + \mathrm{e}^{A(W)}]^{-1} = \mathrm{e}^{-A(W)}\,[1 + \sum_{n=1}^{\infty} (-1)^n\, \mathrm{e}^{-nA(W)}] \tag{2}$$

introduces into (1) the surface barrier in the following way:

$$A(W) = 2 \int\limits_{z_1}^{z_2} |k(z)|\,\mathrm{d}z\,, \qquad k(z) = \left\{\left(\frac{2m}{\hbar^2}\right) [V(z) - W]\right\}^{1/2}, \tag{3}$$

where the potential energy

$$V(z) = \varphi + E_\mathrm{F} - \left(\frac{e^2}{16\pi\,\varepsilon_0 z}\right) - eFz \tag{4}$$

represents a one-electron static surface barrier modified by an external electric field F. Introducing $V(z)$ into expression (3) and performing integration along the normal z to the metal surface between two turning points z_1 and z_2, we obtain the relation for a phase integral $A(W)$ in the form [2]

$$A(W) = \left(\frac{4}{3eF}\right) \left(\frac{2m}{\hbar^2}\right)^{1/2} [\varphi - (W - E_\mathrm{F})]^{3/2}\, v \left\{\left(\frac{e^3 F}{4\pi\varepsilon_0}\right)^{1/2} \Big/ [\varphi - (W - E_\mathrm{F})]\right\}. \tag{5}$$

The function $v(y')$ of the argument

$$y' = \frac{(e^3F/4\pi\varepsilon_0)^{1/2}}{\varphi - (W - E_f)} \tag{6}$$

arises from the image-force rounding of the barrier appropriate to a metal. The argument y' expresses the relative depression of the potential barrier height due to the action of the image force on the electron with a given energy W. $v(y')$ is an elliptic function well known in the theory of field emission [6].

The image-force potential and thus the barrier potential can be modified for a variety of reasons. Such attempts and their application to analysis of the surface phenomena have been made by many authors [7 to 13]. However, the problems related to the barrier correction are beyond the scope of present paper.

To carry out further analytical calculations the right-hand side of expression (5) has to be expanded into a Taylor series with respect to the small quantity $|W - E_F|/\varphi$. After this operation we can write the series

$$A(W) = c - \left[\frac{W - E_F}{\varphi d_0}\right] + \left[\frac{W - E_F}{\varphi f_0}\right]^2 - \left[\frac{W - E_F}{\varphi g_0}\right]^3 + \dots \tag{7}$$

with the coefficients

$$c = Cv(y)\,, \tag{8}$$

$$\frac{1}{d_0} = \frac{3}{2}\,Ct(y)\,, \tag{9}$$

$$\frac{1}{f_0^2} = \frac{3}{8}\,Cv(y)\,(1 - y^2)^{-1}\,, \tag{10}$$

$$\frac{1}{g_0^3} = \frac{1}{16}\,C[4v(y) - 3t(y)\,(1 - y^2)]\,(1 - y^2)^{-1}\,, \tag{11}$$

where

$$C = \frac{4}{3}\left(\frac{2m}{e^2\hbar^2}\right)^{1/2}\left(\frac{\varphi^{3/2}}{F}\right) = 6.8309\left(\frac{\varphi^{3/2}}{F}\right). \tag{12}$$

The work function φ and the applied electric field F are expressed now in units of eV and V nm^{-1}, respectively; the constants e and m represent the elementary charge and electron mass. The variable y depends on the above quantities

$$y = [y']_{W=E_F} = \frac{(e^3F/4\pi\varepsilon_0)^{1/2}}{\varphi} = \frac{1.2000F^{1/2}}{\varphi}$$

and the function $t(y)$ is connected with $v(y)$ by the simple relation

$$t(y) = v(y) - \frac{2}{3}\,y\,\frac{dv(y)}{dy}.$$

Note that the quadratic and the cubic expansion terms in the series (7) have also been calculated by Murphy and Good [6] and by Stratton [14].

Prior to making the next step in our considerations we have to comment on the series in (7). In field-emission experiments the work function of the emitting surface varies between 2 and 6 eV approximately. Also, the condition of thermodynamic equilibrium for the electron gas in an emitting metal is fulfilled when the extracted emission current density is extremely low. This implies, of course, a relatively low electric field strength near the emitting surface. In consequence, one can accept the following values for the work function and the applied field:

$$\varphi = 4\ \mathrm{eV}\,, \qquad F = 1.6\ \mathrm{V\,nm^{-1}} \tag{13}$$

as the typical field-emission conditions; the latter value corresponds to the emission current density of 0.1 A cm^{-2} in a crude estimation. In these circumstances one can estimate, comparing (5) and (7), the contribution of higher order terms in the series (7). If we take into account the energy range of 1.5 eV below and above the Fermi energy for W, then a maximum error of 1% in the tunneling probability occurs on the upper border of this energy range, and is caused by disregarding all expansion terms higher than the quadratic one. Omission of the summation term in (2) produces an error considerably less than that stated above. Therefore, we retain the first three terms in the series of (7) for further consideration. As will be discussed in Section 5, this approximation seems to be sufficient for the analysis of experimental data.

In order to obtain an expression for the TED function we first calculate the integral in (1), having in mind all the preceding discussion. Thus the integral is

$$\int_0^{E_F+\varepsilon} \exp\left\{-\left[\frac{W-E_F}{\varphi f_0}\right]^2 + \left[\frac{W-E_F}{\varphi d_0}\right] - c\right\} \mathrm{d}W =$$

$$= \varphi f_0 \left(\frac{\pi}{4}\right)^{1/2} \exp\left(\frac{f_0^2}{4d_0^2} - c\right)\left[\operatorname{erf}\left(\frac{\varepsilon}{\varphi f_0} - \frac{f_0}{2d_0}\right) + \operatorname{erf}\left(\frac{E_F}{\varphi f_0} + \frac{f_0}{2d_0}\right)\right]. \tag{14}$$

Inserting numerical values into the integral (14) one can obtain some idea about the order of magnitude of the function arguments in the last set of square brackets. Thus, we assume $E_F = 10$ eV (which can be accepted as a correct value for tungsten, for instance, but extensive discussion of the inner potential role in the TED has been developed by Duke and Fauchier [15]); and bearing in mind the typical field-emission conditions (13), and the parameters occurring in the integral (14), we obtain approximate values of the error-function arguments

$$\begin{aligned} &\frac{\varepsilon}{\varphi f_0} \ll \frac{f_0}{2d_0}, \\ &\frac{\varepsilon}{\varphi f_0} - \frac{f_0}{2d_0} \approx -7\,, \qquad \frac{E_F}{\varphi f_0} + \frac{f_0}{2d_0} \approx 14\,. \end{aligned} \tag{15}$$

Thus, one can easily verify that the second error function in integral (14) may be very well replaced by unity. Finally, multiplying (14) by the supply function (cf. (1)) we get the following TED equation:

$$j'_{TF}(\varepsilon) = \left(\frac{2\pi^{3/2} me}{h^3}\right) \varphi f_0\, f(\varepsilon) \exp\left[\left(\frac{f_0}{2d_0}\right)^2 - c\right] \operatorname{erfc}\left(\frac{f_0}{2d_0} - \frac{\varepsilon}{\varphi f_0}\right). \tag{16}$$

This is somewhat inconvenient in computations since it contains an internal product of the small quantity by a large one. A convenient form of (16) may be simply obtained by introducing into it the asymptotic expansion for the erfc (x) function (c.f. formula (7.1.23) in [16]). The argument of this function varies between such limits (c.f. the relations (15)) that a good accuracy is secured by the asymptotic approximation. Taking the above into account, (16) may be developed into the form

$$j'_{TF}(\varepsilon) = \left(\frac{2\pi^{1/2} me}{h^3}\right) \varphi f_0\, \mathrm{e}^{-c} f(\varepsilon) \exp\left[\left(\frac{\varepsilon}{\varphi d_0}\right) - \left(\frac{\varepsilon}{\varphi f_0}\right)^2\right] \zeta^{-1} \sum_{n=0}^{n_0} \Gamma\left(n + \frac{1}{2}\right) (2\zeta^2)^{-n}, \tag{17}$$

where $\zeta = (f_0/2d_0) - (\varepsilon/\varphi f_0)$ and $\Gamma(x)$ is the Euler gamma function, and n_0 denotes any (finite) number of expansion terms. The sum of the asymptotic series in (17) does not differ significantly from $\pi^{1/2}$. Once again we note that this equation applies to the typical field-emission conditions provided that $|\varepsilon| \leqq 1.5$ eV. When $f_0 \to \infty$ i.e. when the quadratic term in the exponential of (17) is vanishing, then this equation goes over to

$$j'_{\mathrm{TF}}(\varepsilon)^* = \left(\frac{J^*_{\mathrm{0F}}}{\varphi d_0}\right) \mathrm{e}^{\varepsilon/\varphi d_0} (1 + \mathrm{e}^{\varepsilon/kT})^{-1} . \tag{18}$$

Equation (18) is known as the Young equation [17], with

$$J^*_{\mathrm{0F}} = \left[\frac{4\pi me(\varphi d_0)^2}{h^3}\right] \mathrm{e}^{-c} = \left(\frac{e^3}{8\pi h}\right)\left(\frac{F^2}{\varphi}\right) [t(y)]^{-2}\, \mathrm{e}^{-c} =$$

$$= 1.5414 \times 10^8 \left(\frac{F^2}{\varphi}\right) [t(y)]^{-2}\, \mathrm{e}^{-c}\ \mathrm{A\ cm^{-2}} , \tag{19}$$

being the first order (uncorrected) Fowler-Nordheim (FN) current density, at zero temperature. (N. B.* denotes here and henceforth the uncorrected quantity.)

3. Total Emission Current

Integrating the expression for the TED over the whole energy range, one obtains the total temperature-dependent field-emission-current density, i.e. the thermal field emission (TFE) current density. To find the relevant equation we use the expression (16) in a slightly different form. Therefore, we can write the relation

$$J_{\mathrm{TF}} = \left(\frac{2\pi^{3/2}\, me}{h^3}\right) \varphi f_0 \exp\left[\left(\frac{f_0}{2d_0}\right)^2 - c\right] \int\limits_{-\infty}^{\infty} \left[1 + \operatorname{erf}\left(\frac{\varepsilon}{\varphi f_0} - \frac{f_0}{2d_0}\right)\right] f(\varepsilon)\, \mathrm{d}\varepsilon \tag{20}$$

in which J_{TF} is the second-order TFE current density at a given temperature T and electric field strength F. The tacit assumption in (20), that integration can be extended beyond the allowed energy range for ε, introduces a small error. This means that there exists an upper limit to the temperature, for which (20) is still exact within the assumed approximation for the tunneling function; this limit is considered in Section 5.

The integral in (20) may be evaluated in special functions using the Sommerfeld method of integration. We write only the final formula useful for us; details of its derivation can be found in the handbook [18]. In consequence we have

$$\int\limits_{-\infty}^{\infty} \left[1 + \operatorname{erf}\left(\frac{\varepsilon}{\varphi f_0} - \frac{f_0}{2d_0}\right)\right] f(\varepsilon)\, \mathrm{d}\varepsilon = \int\limits_{-\infty}^{0} \left[1 + \operatorname{erf}\left(\frac{\varepsilon}{\varphi f_0} - \frac{f_0}{2d_0}\right)\right] \mathrm{d}\varepsilon +$$

$$+ \sum_{n=1}^{\infty} \lambda_n (kT)^{2n} \left\{\frac{\mathrm{d}^{2n-1}}{\mathrm{d}\varepsilon^{2n-1}} \left[1 + \operatorname{erf}\left(\frac{\varepsilon}{\varphi f_0} - \frac{f_0}{2d_0}\right)\right]\right\}_{\varepsilon=0} \tag{21}$$

where λ_n is a dimensionless numerical constant given by the following formula:

$$\lambda_n = (2^{2n} - 2)\left[\frac{\pi^{2n}}{(2n)!}\right] |B_{2n}| , \qquad n = 1, 2, \ldots \tag{22}$$

with B_{2n} being the Bernoulli number. For $n \geqq 5$ the following relation is useful:

$$\lambda_n = 2\left(1 - \frac{1}{2^{2n}} + \frac{1}{3^{2n}} - \frac{1}{4^{2n}} + \ldots\right).$$

It is easy to notice that the first term in the right-hand side of (21) determines the zero temperature emission current J_{OF}. The calculation of the integral does not raise any difficulty:

$$\int_{-\infty}^{0}\left[1+\operatorname{erf}\left(\frac{\varepsilon}{\varphi f_0}-\frac{f_0}{2d_0}\right)\right]\mathrm{d}\varepsilon=\frac{\varphi f_0}{\pi^{1/2}}\exp\left[-\left(\frac{f_0}{2d_0}\right)^2\right]-\frac{\varphi f_0^2}{2d_0}\operatorname{erfc}\left(\frac{f_0}{2d_0}\right). \tag{23}$$

Multiplying this result by the factor standing before the integral in (20), and replacing the error function by its asymptotic expansion, we get after some elementary calculations

$$J_{OF}=J_{OF}^{*}s_1\,, \tag{24}$$

$$s_1=1-6\xi^2+60\xi^4-840\xi^6+15120\xi^8-\ldots\,, \tag{25a}$$

where ξ is given by

$$\xi=\frac{d_0}{f_0}=(12C)^{-1/2}\left[\frac{v(y)}{1-y^2}\right]^{1/2}[t(y)]^{-1}=0.11045\left(\frac{F^{1/2}}{\varphi^{3/4}}\right)\left[\frac{v(y)}{(1-y^2)}\right]^{1/2}[t(y)]^{-1}. \tag{25b}$$

This J_{OF} is the corrected FN current density, at zero temperature.

As far as the second term in (21) is concerned, we express the derivatives of the error function through the Hermite polynomials in the following manner:

$$\left[\frac{\mathrm{d}^{2n-1}}{\mathrm{d}\varepsilon^{2n-1}}\operatorname{erf}\left(\frac{\varepsilon}{\varphi f_0}-\frac{f_0}{2d_0}\right)\right]_{\varepsilon=0}=\frac{2}{\pi^{1/2}}\left(\frac{1}{\varphi f_0}\right)^{2n-1}\times$$
$$\times\,\mathrm{H}_{2n-2}\left(\frac{f_0}{2d_0}\right)\exp\left[-\left(\frac{f_0}{2d_0}\right)^2\right],\qquad n=1,2,\ldots \tag{26}$$

where $\mathrm{H}_n(x)$ is the n-th degree Hermite polynomial (c.f. formula (7.1.19) in [16]).

When the derived results ((21), (23), (24), and (26)) are combined with (20), the result for the TFE current density is

$$J_{TF}=J_{OF}\left[1+s_1^{-1}\left(\frac{f}{d}\right)^2\sum_{n=1}^{\infty}\lambda_n\left(\frac{kT}{f}\right)^{2n}\mathrm{H}_{2n-2}\left(\frac{f}{2d}\right)\right]. \tag{27}$$

Here the parameters d_0 and f_0 were replaced by $d=\varphi d_0$, $f=\varphi f_0$ in accordance with traditional notation; s_1 denotes the sum defined in (25a). A slightly more convenient expression for computation can be obtained by replacing the Hermite polynomials by the Laguerre ones. Thus

$$J_{TF}=J_{OF}\left\{1+s_1^{-1}\left(\frac{f}{d}\right)^2\sum_{n=1}^{\infty}\Lambda_n\left(\frac{kT}{f}\right)^{2n}\mathrm{L}_{n-1}^{-1/2}\left[\left(\frac{f}{2d}\right)^2\right]\right\}, \tag{28}$$

where

$$\Lambda_n=(-1)^{n-1}\,2^{2(n-1)}\,(n-1)!\,\lambda_n \tag{29}$$

and $\mathrm{L}_n^{-1/2}(x)$ is the n-th degree generalized Laguerre polynomial. In both above cases one has to use the recurrence relations for numerical evaluation of the corresponding polynomials (c.f. formulae (22.7.12) and (22.7.13) in [16]).

When $f\to\infty$ (27) goes over, as did (17) to the well-known Murphy and Good formula [6]

$$J_{TF}^{*}=J_{OF}^{*}\left[1+\sum_{n=1}^{\infty}\frac{(2^{2n}-2)\,|\mathrm{B}_{2n}|\,\pi^{2n}}{(2n)!}\left(\frac{kT}{d}\right)^{2n}\right]=J_{OF}^{*}\frac{\pi p}{\sin\pi p} \tag{30}$$

with p defined as kT/d (c.f. formula (1.411.11) in [19]). In Fig. 1 the formulae (28) and (30) are compared. Note that the relation (30) breaks down at $p\geqq 1$.

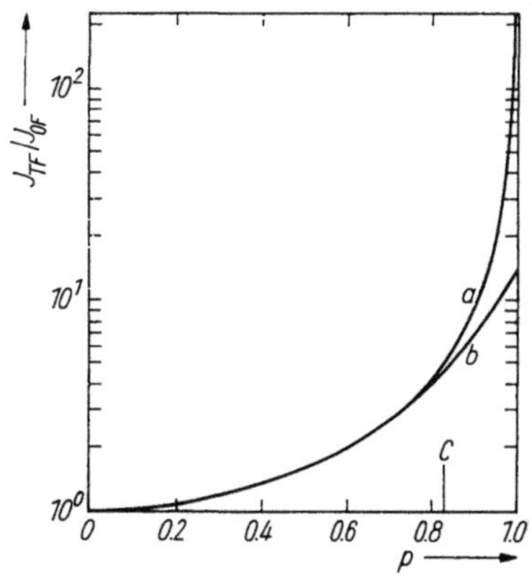

Fig. 1. Relative change of thermal field-emission current plotted as a function of the dimensionless variable $p = kT/d$. (a) According to the Murphy and Good formula (30); (b) according to formula (28), for typical field-emission conditions; (c) point on the abscissa corresponding to the temperature $T = (5/3)\ T_i$

We now use, once again returning to the TED equation, the expression (24) to rewrite (17) in final form

$$j'_{TF}(\varepsilon) = \left(\frac{J_{OF}}{d}\right)\left(\frac{f}{2d}\right)\exp\left[\left(\frac{\varepsilon}{d}\right) - \left(\frac{\varepsilon}{f}\right)^2\right][(1 + e^{\varepsilon/pd})\,\zeta]^{-1}\left(\frac{s_2}{s_1}\right), \tag{31}$$

where

$$s_2 = 1 - (2\zeta^2)^{-1} + 3(2\zeta^2)^{-2} - 15\,(2\zeta^2)^{-3} + 105(2\zeta^2)^{-4} - \dots$$

$$\zeta = \frac{f}{2d} - \frac{\varepsilon}{f}$$

and s_1 is given by the relation (25a). The plot of (31) is shown in Fig. 2. The plot of the retarding energy distribution (RED) is presented in Fig. 3; at the same time the RED function is defined by

$$j_{TF}(\varepsilon) = \int_{\varepsilon}^{\infty} j'_{TF}(t)\,dt\,, \qquad j_{TF}(-\infty) = J_{TF} \tag{32}$$

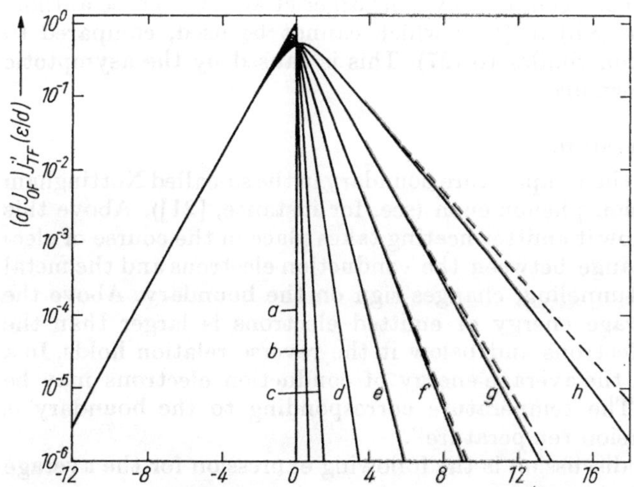

Fig. 2. Total energy distributions of field emitted electrons plotted as a function of the relative electron energy ε/d at various values of the parameter $p = kT/d$. (a) 0, (b) 0.05, (c) 0.1, (d) 0.2, (e) 0.3, (f) 0.4, (g) 0.5, (h) 0.6. TED curves were determined on the basis of (31) under typical field-emission conditions; for comparison, the leading edges represented by the dashed lines were plotted according to Young's formula (18) for the three values of p corresponding to adjacent TED curves

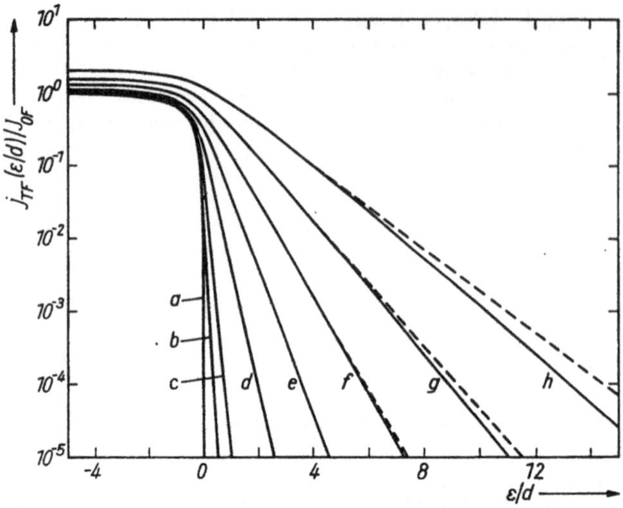

Fig. 3. Retarding energy distributions of field emitted electrons plotted as a function of the relative electron energy ε/d at various values of the parameter $p = kT/d$. (a) 0, (b) 0.05, (c) 0.1, (d) 0.2, (e) 0.3, (f) 0.4, (g) 0.5, (h) 0.6. RED curves were determined on the basis of (32) under typical field-emission conditions; for comparison, the leading edges represented by the dashed lines were plotted according to the first-order RED equation for the three values of p corresponding to adjacent RED curves

(see [5]). The plot of the $j_{TF}(\varepsilon)$ function was obtained using numerical integration of the integral (32); the limitations regarding (31) hold also in this case. It is clearly seen in both figures that the second-order correction causes a bend of the TED and RED curve tails toward lower values of the ordinate.

Equation (31) requires additional comment. There exists another way to obtain an equivalent TED function; it has been used by Vorburger et al. [20]. These authors have proposed the formula (see (8b) in [20]) which cannot be used, compared to (16), for derivation of an equation similar to (27). This is caused by the asymptotic nature of the Vorburger et al. formula.

4. The Inversion Temperature

It is well known that there exists the temperature boundary in the so-called Nottingham effect that occurs in field emission phenomenon (see, for instance, [21]). Above this boundary emitter cooling and below it emitter heating takes place in the course of electron emission. The energy exchange between the conduction electrons and the metal lattice, that accompanies the tunneling, changes sign on the boundary. Above the temperature boundary the average energy of emitted electrons is larger than the average energy of conduction electrons and below it the reverse relation holds. In a free-electron-gas approximation, the average energy of conduction electrons may be equated to the Fermi energy. The temperature corresponding to the boundary is usually referred to as the "inversion temperature".

The starting point for further discussion is the following expression for the average energy of emitted electrons:

$$\langle\varepsilon\rangle = \frac{\int\limits_{-\infty}^{\infty} \varepsilon j'_{TF}(\varepsilon)\,d\varepsilon}{\int\limits_{-\infty}^{\infty} j'_{TF}(\varepsilon)\,d\varepsilon}. \tag{33}$$

The denominator in expression (33) is always larger than zero during the emission process. Thus, at the inversion temperature expression (33) reduces to the simple condition

$$\int_{-\infty}^{\infty} \varepsilon j'_{\mathrm{TF}}(\varepsilon)\, \mathrm{d}\varepsilon = 0\,. \tag{34}$$

To evaluate the integral in condition (34) we use the method already used for (20). Equation (34) then reduces to an algebraic equation involving the inversion temperature T_{i}.

By virtue of straightforward transformations and some elementary calculations (see Appendix), one may derive from condition (34) the equation

$$-a_0 + \sum_{n=1}^{7} a_n x^n = 0\,, \tag{35}$$

where the variable

$$x = \left(\frac{\pi k T_{\mathrm{i}}}{d}\right)^2 \tag{36}$$

includes the inversion temperature being sought. The numerical coefficients $a_0, \ldots, a_n$ bring in the barrier parameters (see the quotient (25b)) into (35), as follows:

$$a_0 = s_3 = 1 - 12\xi^2 + 180\xi^4 - 3360\xi^6 + 75600\xi^8 - \ldots \tag{37}$$

$$a_1 = \frac{s_4}{6} = \frac{1 - 2\xi^2 + 12\xi^4 - 120\xi^6 + 1680\xi^8 - \ldots.}{6} \tag{38}$$

$$a_2 = \frac{21}{360} \tag{39}$$

$$a_3 = \frac{31}{3024}\,(1 - 6\xi^2) \tag{40}$$

$$a_4 = \frac{889}{604800}\,(1 - 20\xi^2 + 60\xi^4) \tag{41}$$

$$a_5 = 1.9202 \times 10^{-4}\,(1 - 42\xi^2 + 420\xi^4 - 880\xi^6) \tag{42}$$

$$a_6 = 2.3797 \times 10^{-5}\,(1 - 72\xi^2 + 1512\xi^4 - 10080\xi^6 + \ldots) \tag{43}$$

$$a_7 = 2.85 \times 10^{-6}\,(1 - 110\xi^2 + 3960\xi^4 - 55440\xi^6 + \ldots)\,, \tag{44}$$

where $\xi = d/f$; the symbols s_3 and s_4 are explained in the Appendix. Since the coefficients (37) to (44) are real and positive (35) has only one positive root (Descartes criterion). It is this case that is valid for us because of (36). In general, the value of the algebraic equation root, for realistic values of the barrier parameters, does not significantly differ from $(\pi/2)^2$. Thus, the exact value of the root may be easily found by iteration. When $f \to \infty$ (i.e. $T_{\mathrm{i}} \to T_{\mathrm{i}}^*$) the solution of (35) gives the value $kT_{\mathrm{i}}^*/d = 0.500082$ which differs from the exact value ($kT_{\mathrm{i}}^*/d = 0.5$) at the fifth decimal place. Clearly, the accuracy of approximation leading from (34) to (35) is close to 0.016%. Note that the first order inversion temperature is obtainable from the relation

$$T_{\mathrm{i}}^* = \frac{d}{2k} = 566.28\left(\frac{F}{\varphi^{1/2}}\right)[t(y)]^{-1} \quad (\mathrm{K})\,. \tag{45}$$

5. Conclusions

This paper has presented formulae which permit a more detailed analysis of field-emission experimental data. The formulae derived are more accurate than those commonly used and may be applied in a wider energy range.

We may usefully summarize the properties of the derived formulae, as follows.

The central role is played by expression (31) for the TED function, because it is this quantity that is measured in systems with the geometry of the classical field-electron-emission microscope, as mentioned earlier. The expression for the TED function was derived by retaining the quadratic term in a WKB phase integral expansion. Owing to this, the energy range in which expression (31) applies, under the typical field-emission conditions (13), was extended up to 1.5 eV above and below the Fermi energy. In principle, including terms of order higher than quadratic in a phase-integral expansion seems to be possible. This would lead, of course, to further raising the accuracy of equations. Nevertheless, the accuracy of the expression for the TED function cannot be more strictly specified, since only the first-order form (see expression (2)) of the WKB tunneling probability approximation is known at present.

As far as the formulae derived from (16) are concerned, their accuracy is determined by the accuracy of the original expression for the TED function. In the literature it is assumed that the Murphy and Good formula (30) is valid for temperatures not higher than $(5/3)T_i$ [22]. The expression (28) for the TFE current density is, obviously, more accurate than formula (30) within the assumed WKB approximation. The new expression gives current density values about 10% lower than the Murphy and Good formula on the upper limit of the useful temperature range. Also, the inversion temperature calculated from (35) is less by approximately 0.5% than that calculated by means of the first-order relation (45).

Acknowledgement

The Polish Academy of Sciences is gratefully acknowledged for partial financial support of this work.

Appendix

Derivation of the $a_0, \dots, a_n$ coefficients in (35).

By analogy to the integral (21) we have

$$\int_{-\infty}^{\infty} Q(\varepsilon)\, f(\varepsilon)\, \mathrm{d}\varepsilon = \int_{-\infty}^{0} Q(\varepsilon)\, \mathrm{d}\varepsilon + \sum_{n=1}^{\infty} \lambda_n (kT_i)^{2n} \left[\frac{\mathrm{d}^{2n-1}}{\mathrm{d}\varepsilon^{2n-1}} Q(\varepsilon)\right]_{\varepsilon=0} = 0\,, \qquad \text{(A1)}$$

where

$$Q(\varepsilon) = \varepsilon \left[1 + \operatorname{erf}\left(\frac{\varepsilon}{f} - \frac{f}{2d}\right)\right]$$

and $d = \varphi d_0$, $f = \varphi f_0$. λ_n is given by the relation (22). To simplify the notation we introduce auxiliary quantities

$$a = 1/f\,, \qquad b = f/2d$$

and then the integral in the right-hand side of (A1) may be represented by

$$\int_{-\infty}^{0} Q(\varepsilon)\, \mathrm{d}\varepsilon = -\,(2a^2)^{-1}\left[b^2 \operatorname{erfc}(b) - \frac{b}{\pi^{1/2}}\, \mathrm{e}^{-b^2} + \frac{1}{2} \operatorname{erfc}(b)\right] =$$

$$= -(4\pi^{1/2} a^2 b^3)^{-1}\, \mathrm{e}^{-b^2}\, s_3\,. \qquad \text{(A2)}$$

Here s_3 is the sum of the asymptotic series which appeared from elementary transformations of the expression in square brackets of (A2) (c.f. coefficient (37)).

The derivatives in (A1) are given by

$$\left[\frac{\mathrm{d}}{\mathrm{d}\varepsilon}\,Q(\varepsilon)\right]_{\varepsilon=0} = \operatorname{erfc}(b) = (b\pi^{1/2})^{-1}\,\mathrm{e}^{-b^2}\,s_4\,, \tag{A3}$$

$$\left[\frac{\mathrm{d}^{2n-1}}{\mathrm{d}\varepsilon^{2n-1}}\,Q(\varepsilon)\right]_{\varepsilon=0} = (2\pi^{-1/2})\,(2n-1)\,a^{2n-2}\,\mathrm{H}_{2n-3}(b)\,\mathrm{e}^{-b^2}\,, \qquad n = 2, 3, \ldots \tag{A4}$$

with $\mathrm{H}_n(x)$ being the Hermite polynomial. The sum s_4 arises from the asymptotic expansion of the erfc (b) function (c.f. coefficient (38)).

Further calculations are as follows: (i) we retain only seven expansion terms in (A1); (ii) substituting (A2), (A3), and (A4) into (A1) and multiplying the obtained result by $4\pi^{1/2}a^2b^3\,\mathrm{e}^{b^2}$ we obtain (35); (iii) the last step in our calculations consists in developing Hermite polynomials into explicit form; this immediately leads to the coefficients (39) to (44).

References

[1] P. H. Cutler and R. H. Good, Jr., Phys. Rev. **104**, 308 (1956).
[2] J. W. Gadzuk and E. W. Plummer, Rev. mod. Phys. **45**, 487 (1973).
[3] E. Burstein and S. Lundqvist (Ed.), Tunneling Phenomena in Solids, Plenum Press, New York 1969.
[4] T. Wolfram (Ed.), Inelastic Electron Tunneling Spectroscopy, Springer-Verlag, 1978.
[5] C. J. Workowski, J. Phys. D **16**, 445 (1983).
[6] E. L. Murphy and R. H. Good, Jr., Phys. Rev. **102**, 1464 (1956).
[7] D. W. Juenker, Phys. Rev. **99**, 1155 (1955).
[8] P. H. Cutler and J. C. Davis, Surface Sci. **1**, 194 (1964).
[9] P. H. Cutler and D. Nagy, Surface Sci. **3**, 71 (1965).
[10] M. N. Read and P. J. Jennings, Surface Sci. **74**, 54 (1978).
[11] A. Modinos and G. Oxinos, J. Phys. C **14**, 1373 (1981).
[12] J. M. Baribeau and J. D. Carette, Surface Sci. **112**, 241 (1981).
[13] W. Ekardt, Phys. Rev. B **26**, 536 (1982).
[14] R. Stratton, Phys. Rev. **135**, A794 (1964).
[15] C. B. Duke and J. Fauchier, Surface Sci. **32**, 175 (1972).
[16] M. Abramowitz and I. A. Stegun (Ed.), Handbook of Mathematical Functions, NBS, Washington 1964.
[17] R. D. Young, Phys. Rev. **113**, 110 (1959).
[18] N. W. Ashcroft and N. D. Mermin, Solid State Physics, Appendix C, Holt, Rinehart, and Winston, New York 1976.
[19] I. S. Gradshteyn and I. M. Ryzhik, Table of Integrals, Series and Products, Fiz.-Mat.-Lit., Moscow 1963 (in Russian).
[20] T. V. Vorburger, D. Penn, and E. W. Plummer, Surface Sci. **48**, 417 (1975).
[21] L. W. Swanson, L. C. Crouser, and F. M. Charbonnier, Phys. Rev. **151**, 327 (1966).
[22] S. G. Christov, Surface Sci. **70**, 32 (1978).

(Received May 31, 1985)

phys. stat. sol. (b) 133, 195 (1986)

Subject classification: 71.10 78.70; S9.11

Institute of Physics, Rostov State University[1])

Multiple Scattering Approach to the XANES Theory of Alkali Halide Crystals

I. Crystalline Potential in the X-Ray Absorption Spectra Problem

By

L. A. BUGAEV, I. I. GEGUSIN, V. N. DATSYUK, A. A. NOVAKOVICH, and R. V. VEDRINSKII

Two schemes of the crystalline potential for the slow electron in the conduction band of alkali halides are proposed. The first one (semiempirical) is based on the assumption that in the case of an ion in the crystal one can use the same energy dependence of logarithmic derivatives (of radial solutions at the atomic sphere radius) as in the case of the isolated ion. The modifications required consist in some potential corrections, that are due to the Madelung field as well as to the relaxation and polarization response for both, hole and electron. The second model (ab initio) suggests the straightforward solution of Poisson's equation. Both approaches are compared, so it is found that they lead to satisfactory agreement for a wide energy interval in the continuum.

Предложены две схемы построения кристаллического потенциала, описывающего движение медленного электрона в зоне проводимости щелочно-галоидных кристаллов. Первая из них (полуэмпирическая) основывается на предположении, что для иона, находящегося в кристалле, можно использовать ту же самую энергетическую зависимость логарифмических производных (радиальных решений на радиусе атомной сферы), что и для изолированного иона. Необходимые изменения состоят в некоторых поправках к потенциалу, обусловленных как маделунговским полем, так и релаксационным и поляризационным откликом на дырку и электрон. Вторая модель основана на непосредственном решении уравнении Пуассона. Сопоставлены оба подхода и обнаружено, что они приводят к удовлетворительному согласию для широкого интервала энергий в непрерывном спектре.

1. Introduction

The X-ray absorption spectra of crystals can be divided into two intervals. The X-ray absorption near edge structure (XANES) covers the region about 40 eV beyond the threshold, followed by the extended X-ray absorption fine structure (EXAFS), which goes typically 1000 eV further. It was shown [1, 2] that the single-scattering approximation is often sufficient to describe EXAFS spectra, so the main problem is to construct the appropriate crystalline or molecular potential. On the other hand, the XANES problem is not solved yet. Different approaches have been used to explain XANES of polyatomic systems: band structure [3], MO LCAO for small clusters [4], SCF SW Xα method [5, 6], the effective one-center potential method [7], and the multiple scattering technique, based on the Green's function (GF) formalism [8, 9]. The band structure calculations were carried out for a number of metals [3]. The other methods were applied to particular spectra only.

Indeed the band structure approach provides reasonable results for metals, where the core hole is almost completely screened by the conduction electrons. However, severe problems arise for insulators, where the core hole potential is markedly stronger.

[1]) Prospekt Stachky 194, 344090 Rostov on Don, USSR.

From the point of view of XANES interpretation two important shortcomings are connected with the LCAO approaches. Firstly, because of the fixed shape of the atomic orbital the energy dependence of the radial wave functions is ignored, which is essential in the continuum. Secondly, owing to the limited basis set normally used, some principal features of the spectra cannot be explained [10]. For example, KCl XANES spectra display not only the peaks formed by the K and Cl d-bands, which could be in principle interpreted in terms of LCAO states, but also the peaks formed by the interaction of d-bands with the nearly free electron bands, as it was shown previously [11].

More rigorous investigations are required in the case of the effective one-center potential method [7]. It seems necessary to compare the spectra resulting from a multi-center potential with the corresponding ones resulting from the appropriate single-center potential, which is obtained when the former is averaged over angles. The SCF SW Xα method as well as multiple scattering method allow to overcome the drawbacks mentioned above. The last technique was successfully applied to explain XANES spectra of Cu [12] and of K and Cl in KCl [13, 14].

In the following series of papers a systematic application of the multiple scattering technique to the XANES calculation of alkali halides (AH) is presented. These crystals were chosen because of several reasons: i) the alkali halide XANES spectra appear to have good contrast, and a number of peaks is concentrated in the near-edge region, ii) there exists a lot of experimental investigations [15], and iii) the spectra of a given atom that belongs to different crystals distinguish essentially (for example, the Na K-spectrum in NaCl and NaF, or the Cl K-spectrum in KCl and NaCl).

In the first paper two schemes of constructing the crystalline potential are discussed: the semiempirical (the "isolated ions" model) and the ab initio (the "crystalline potential") one. To confirm the reliability of the developed schemes the binding energies of the core levels in alkali halides are evaluated. The Xα and Hartree-Fock (HF) approximations are analyzed when treating the exchange interaction in the continuum. The results of both approximations are compared in terms of the logarithmic derivatives of the radial equation solutions.

In the second paper the band structure approach and the cluster approximation are considered. The spectra resulting from the band structure approach are compared with the experimental ones, so the conclusion of the necessity to go beyond the frames of the band structure approach is drawn. Further the size effects in the cluster approximations are analyzed. It is shown how the "cluster" spectra converge to the "band" ones when the cluster size increases. The formation of a band gap in the finite size cluster is found.

The third paper gives a critical survey of the main concepts, that were earlier put forward in connection with AH XANES. The results of K-spectra calculations for six crystals: LiF, NaF, KF, LiCl, NaCl, and KCl, are presented. The sensitivity of the results to some potential effects is treated in the case of the Na K-spectrum in NaCl taken as an example. All the calculated spectra are compared with the experimental ones. Finally, the advantages, drawbacks, and possible perspectives are discussed.

2. Semiempirical Potential ("Isolated Ions") Model

The numerical methods of the self-consistent crystal potential calculations are rather cumbersome, and the results obtained are not transparent. It is especially true in the case of an impurity center. Hence it seems to be more advantageous to construct a model potential. In so doing one must be guided by reasonable physical assumptions.

In the case of AH crystals it is naturally to suppose the crystal electron density to be equal to the superposition of the isolated ion densities.

Two different approaches to the problem of the potential construction are proposed in this paper. Both approaches employ the "muffin-tin" (MT) approximation, which is suitable for AH crystals [16]. The first one is based on the assumption that in the case of an ion in the crystal one can use the same energy dependence of logarithmic derivatives (of radial solutions at the atomic sphere radius) as in the case of the isolated ion. The only modification consists in the energy shift ΔV_n, caused by the crystalline environment. (Here n denotes the number of the ion in the lattice.) This so-called "isolated ions" model was previously described [14].

If one treats the ideal crystal potential, the shift ΔV_n is equal to the sum of the Madelung potential $E_{n,\mathrm{M}}$ and some terms, which are due to the polarization of the environment by an electron in the conduction band:

$$\Delta V_n^0 = \pm E_{n,\mathrm{M}} + \Delta V_{n,\mathrm{relax}} + \Delta V_{n,\mathrm{corr}} + \Delta E_{n,\mathrm{ML}} \,. \tag{1}$$

Here the terms $\Delta V_{n,\mathrm{relax}}$ and $\Delta V_{n,\mathrm{corr}}$ are the atomic relaxation and correlation corrections, which are due to the polarization of the n-th ion by the conduction electron. The extra-atomic relaxation correction $\Delta E_{n,\mathrm{ML}}$ may be obtained within the Mott-Littleton approach [17].

If an ion in the crystal contains a core hole, then, firstly, the corresponding logarithmic derivatives are to be recalculated by means of the SCF HF approach, thus accounting for the atomic relaxation effect. Secondly, additional terms should be included in the expression for the potential corrections:

$$\Delta V_n^{\mathrm{c}} = \begin{cases} \Delta V_n^0 - 2\,\Delta E_{n,\mathrm{ML}}\,, & \text{if} \quad n = 1\,, \qquad (2\mathrm{a}) \\ \Delta V_n^0 - 2/\varepsilon_\infty R_n\,, & \text{if} \quad n \neq 1\,. \qquad (2\mathrm{b}) \end{cases}$$

Here $n = 1$ is the number of the excited ion with a core hole, R_n is the distance from the n-th ion to the excited one, and ε_∞ is the optical dielectric constant (its AH values are taken from [18]).

The required logarithmic derivatives in the continuum energy region result from the solution of a single Hartree-Fock equation. The corresponding potentials are the self-consistent potentials of the isolated ions with or without the core hole.

To check the model proposed the binding energies of the occupied core levels were calculated. Analogously to (1, 2a) the ionization energy of the i-th core level of the n-th ion may be evaluated as follows:

$$E_{ni} = E_{ni}^0 + \Delta E_{ni,\mathrm{relat}} + \Delta E_{ni,\mathrm{corr}} \pm \Delta E_{n,\mathrm{M}} + \Delta E_{n,\mathrm{ML}} \,. \tag{3}$$

The term E_{ni}^0 is the atomic binding energy evaluated as the difference of HF total energies of two states, the ground one and the excited one with the core hole. The atomic relativistic correction $\Delta E_{ni,\mathrm{relat}}$ is taken into account for 1s levels only [19]. The terms $\Delta E_{n,\mathrm{M}}$ and $\Delta E_{n,\mathrm{ML}}$ represent Madelung and Mott-Littleton corrections, respectively. The atomic correlation correction $\Delta E_{ni,\mathrm{corr}}$ is estimated by comparison of the theoretical and experimental core binding energies of the noble gas atoms, neighbouring to the alkali and halide ions in the periodic system. The estimated correlation corrections are displayed in Table 1.

The absolute values of experimental core binding energies were evaluated in the following manner. The energy differences between the core level and the valence band midpoint were calculated by means of XPS data [20], whereas the absolute positions of valence band midpoints were taken from the review article [21]. The comparison of the theoretical binding energies with the experimental ones (Table 2) confirms the high accuracy of the "isolated ion" model. Note that good agreement is

Table 1

The correlation corrections (eV) resulting from the total energies of free Ne and Ar atoms

atom	level i	E_i^0	$E_{i,\,\mathrm{relat}}$	$E_{i,\,\mathrm{exp}}$	$E_{i,\,\mathrm{corr}}$
Ne	1s	868.5	1.0	870.1	0.6
	2p	19.8	0	21.5	1.7
Ar	1s	3195.4	12.3	3206.1	−1.6
	2p	248.7	0	249.2	0.5
	3p	15.0	0	15.8	0.8

Table 2

The various contributions (eV) to ionization energies of the occupied levels in KCl, NaCl, LiCl, KF, NaF, and LiF crystals

system	ion	level i	$E_i^0 \pm E_\mathrm{M}$	$\Delta E_{i,\,\mathrm{relat}} + \Delta E_{i,\,\mathrm{corr}}$	ΔE_ML	E_i^theor	E_i^exper
KCl	K^+	1s	3599.3	13.8	−1.9	3611.2	3611.5
$E_\mathrm{M} = 8.0$		2p	300.1	0.5	−1.9	298.7	298.7
		3p	22.6	0.8	−1.9	21.5	22.2
	Cl^-	1s	2818.3	8.8	−1.3	2825.8	2825.2
		2p	204.7	0.5	−1.3	203.9	203.5
		3p	10.5	0.8	−1.3	10.0	10.2
NaCl	Na^+	1s	1077.6	1.6	−2.5	1076.7	1076.6
$E_\mathrm{M} = 8.9$		2p	36.8	1.7	−2.5	36.0	35.8
	Cl^-	1s	2819.2	8.8	−1.6	2826.4	—
		2p	205.6	0.5	−1.6	204.5	204.2
		3p	11.4	0.8	−1.6	10.6	10.9
LiCl	Li^+	1s	64.7	1.0	−3.2	62.5	62.7
$E_\mathrm{M} = 9.8$	Cl^-	2p	206.5	0.5	−1.8	205.2	205.7
		3p	12.3	0.8	−1.8	11.3	11.7
KF	K^+	1s	3598.1	13.8	−1.7	3610.2	—
$E_\mathrm{M} = 9.4$		2p	298.7	0.5	−1.7	297.5	297.3
		3p	21.2	0.8	−1.7	20.3	—
	F^-	1s	687.7	1.6	−1.8	687.5	687.0
		2p	10.7	1.7	−1.8	10.6	10.5
NaF	Na^+	1s	1075.7	1.6	−2.1	1075.2	1076.0
$E_\mathrm{M} = 10.9$		2p	34.9	1.7	−2.1	34.5	34.6
	F^-	1s	689.2	1.6	−1.6	689.2	688.7
		2p	12.2	1.7	−1.6	12.3	11.8
LiF	Li^+	1s	62.0	1.0	−2.9	60.0	59.9
$E_\mathrm{M} = 12.5$	F^-	1s	690.8	1.6	−1.8	690.6	689.0
		2p	13.8	1.7	−1.8	13.7	12.8

achieved without taking into account the additional terms Σ_2 and E_R, suggested earlier by Mahan [17].

The final problem to be solved is the determination of the interstitial potential V_{MT} (MT zero). It is proposed to treat this quantity as a parameter, chosen in such a way that the theoretical gap width would be equal to the experimental one. When the X-ray absorption problem is considered, an additional contribution to the MT zero of the finite size cluster is given by the screened core hole potential $-2/(\varepsilon_\infty r)$.

3. Ab initio Potential ("Crystalline Potential") Model

Another approach to the potential problem is based on the representation of the charge density in a crystal $\varrho_{\mathrm{cryst}}(\boldsymbol{r})$ as superposition of the free ion densities $\varrho_n(\boldsymbol{r} - \boldsymbol{R}_n)$. In the case of an ideal crystal ground state it follows:

$$\varrho^0_{\mathrm{cryst}}(\boldsymbol{r}) = \sum_n \varrho^0_n(\boldsymbol{r} - \boldsymbol{R}_n) \,. \tag{4}$$

The Coulomb contribution to the crystal MT potential may be found as solution of the Poisson equation with subsidiary periodic boundary conditions on the unit cell surface [22]. The exchange part of the potential will be treated later by means of Xα or HF approaches. In the following such a model will be called "ab initio" or "crystalline potential" one.

As to the X-ray absorption problem, it involves an excited state of the crystal, containing the localized core hole. It is suggested that the excited state spectrum may be adequately described if the atomic relaxation effect is introduced. It means that for any point $\boldsymbol{r}$ in the crystal the electron density should be expressed in terms of the ground state density and of the excited atom density:

$$\varrho^i_{\mathrm{cryst}}(\boldsymbol{r}) = \varrho^0_{\mathrm{cryst}}(\boldsymbol{r}) + \varrho^i_1(\boldsymbol{r} - \boldsymbol{R}_1) - \varrho^0_1(\boldsymbol{r} - \boldsymbol{R}_1) \,. \tag{5}$$

Here $\varrho^i_1(\boldsymbol{r} - \boldsymbol{R}_1)$ denotes the density of the excited ion with the hole in the core level i. The required densities are extracted from the atomic calculations based on HF as well as Xα approaches. Thus the Coulomb potential for the excited state may be also calculated.

As long as the total density (5) is determined, the exchange Xα potential is allowed to be easily evaluated. Further it will be shown that the traditional Xα potential with $\alpha = 1$ is too strong for the purpose of adequate XANES description. Better results are expected if one solves a single HF equation for the electron wave function in the continuum energy region. The difficult problem of HF equation solution for a many-atomic system is simplified in the case of AH crystals. It is because of the strong localization of all occupied states inside the atomic spheres. Hence the non-local character of exchange potential is to be taken into account inside the atomic spheres only, whereas in the interstitial space the crystal potential is treated as local one (V_{MT}).

Inside the atomic spheres the electron wave function may be expanded in partial wave orbitals. Thus, every radial orbital $P_{\varepsilon l}(r)$ should satisfy the following HF equation [23]:

$$\left[-\frac{\mathrm{d}^2}{\mathrm{d}r_n^2} + \frac{l(l+1)}{r_n^2} + V^{\mathrm{coul}}_{\mathrm{cryst}}(r_n)\right] P_{\varepsilon l}(r_n) -$$

$$- \frac{2}{r_n} \sum_{p\lambda} P_{p\lambda}(r_n) \sum_k \Gamma^k_{\lambda,l} Y^k_{p\lambda,\varepsilon l}(r_n) = \varepsilon P_{\varepsilon l}(r_n) \,, \tag{6}$$

$$\boldsymbol{r}_n = \boldsymbol{r} - \boldsymbol{R}_n \,, \qquad r_n \leqq b_n \,.$$

Here ε is the electron energy in continuum, l is the orbital moment of the partial wave, $V_{\text{cryst}}^{\text{coul}}(r_n)$ is the Coulomb part of the crystal potential, $p\lambda$ are the quantum numbers of the occupied states, $\Gamma_{\lambda,l}^k$ are Slater coefficients (where $k = |\lambda - 1|$, $|\lambda - l + 2|, \ldots, \lambda + l$), b_n is the radius of the atomic sphere considered, and

$$Y_{p\lambda,\varepsilon l}^k(r_n) = \int_0^{r_n} P_{p\lambda}(r')\, P_{\varepsilon l}(r')\, (r'/r_n)^k\, dr' + \int_{r_n}^{b_n} P_{p\lambda}(r')\, P_{\varepsilon l}(r')\, (r_n/r')^{k+1}\, dr'. \quad (7)$$

The occupied state orbitals $P_{p\lambda}$ are taken in the "frozen core" approximation.

In order to investigate the effects of the exchange potential on XANES spectra two examples were chosen, namely K- and L-absorption spectra of potassium in KCl crystal. Both spectra were treated using $X\alpha$ ($\alpha = 1$, 0.75, 0.5) and HF approaches. The calculations were carried out for the case of a 27-atom cluster of the ideal crystal. The method employed was described earlier [13].

As one can see (Fig. 1) the results of $X\alpha$ and HF calculations tend to be similar when α is approximately 0.4. This conclusion confirms the $X\alpha$ results, concerning f-states [24]. Increase of the parameter α makes the $X\alpha$ potential to be too strong when compared with the HF one. On the other hand, the value $\alpha \approx 0.4$ seems to be too small, if one tries to obtain the energies of the occupied states. For example, if

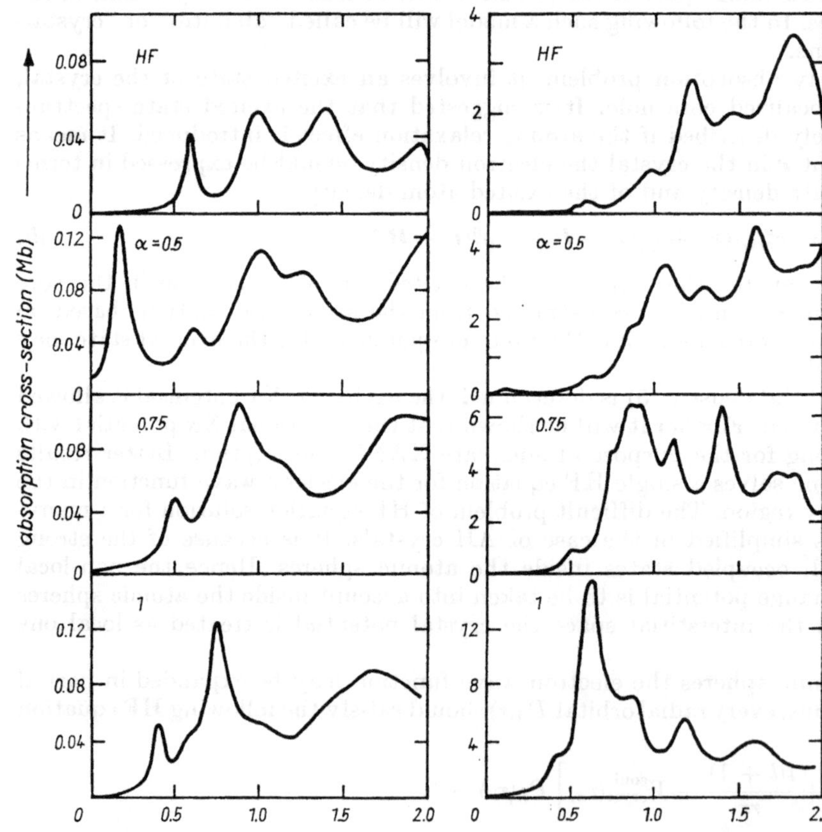

Fig. 1. XANES cluster calculations of K-spectrum (left) and L-spectrum (right) in KCl crystal. The continuum state orbitals are results from $X\alpha$ ($\alpha = 1$, $\alpha = 0.75$, and $\alpha = 0.5$) as well as from HF approaches

the $X\alpha$ potential weakens, the valence band states emerge above the MT zero. These lead to improper absorption, which must be forbidden by the Pauli exclusion principle (see Fig. 1, the low-energy peaks in the case $\alpha = 0.5$). These results prove the impossibility of an adequate theoretical description of both occupied and empty electron states of a crystal by means of a unique $X\alpha$ potential. Thus, in the following the HF exchange potential is employed, though one might attempt to use $X\alpha$ potentials with different values of the parameter α: $\alpha = 0.4$ for the case of empty states and $\alpha = 1$ for the case of occupied ones.

The proposed model non-self-consistent potential should be tested concerning the question: which are the atomic (ionic) densities that enter (4), (5)? Bearing this problem in mind the electronic structure of the ideal NaCl crystal was studied. The electronic atomic densities (Na^0 and Cl^0) as well as the ionic ones (Na^+ and Cl^-) were used. These densities were evaluated within both $X\alpha$ ($\alpha = 1$) and HF approaches. As a result four various versions of the MT potential were established. The corresponding energies of some levels are presented in Table 3. In the case of valence and conduction bands the APW method was used. The core orbitals were determined as single-site solutions, and the proper core energies are found from the requirement of exponential decrease of the radial wave functions.

Table 3

The core and band eigenenergies (Ryd, relative to MT zeros) in NaCl crystal for four various versions of the "crystalline potential"

energy level	potential version			
	I	II	III	IV
Cl 1s	−204.53	−204.65	−206.05	−206.09
Na 1s	− 77.45	− 77.36	− 78.40	− 78.34
Cl 2p	− 14.48	− 14.59	− 15.35	− 15.38
$1\Gamma_1$(Cl 3s)	− 1.11	− 1.18	− 1.35	− 1.38
$1\Gamma_{15}$(Na 2p)	− 2.04	− 1.95	− 2.35	− 2.30
$2\Gamma_{15}$(Cl 3p)	− 0.18	− 0.23	− 0.40	− 0.42
$2\Gamma_1$	0.27	0.27	0.22	0.22
$1\Gamma_{2'5}$	0.73	0.72	0.66	0.66
$1\Gamma_{2'}$	0.99	0.99	0.98	0.98
$3\Gamma_{15}$	1.18	1.19	1.13	1.14
$3\Gamma_1$	1.63	1.65	1.59	1.61

I: Na^+ and Cl^-; $X\alpha$ ($\alpha = 1$) [25].
II: Na^0 and Cl^0; $X\alpha$ ($\alpha = 1$) [25].
III: Na^+ and Cl^-; Hartree-Fock.
IV: Na^0 and Cl^0; Hartree-Fock.

One can see that the energy variations in the core, valence and conduction band regions are quite different. They imply that the charge redistributions considered cause only small potential modifications (≈ 0.1 Ryd) outside core regions, while in the deep core regions the potential changes significantly (≈ 1 to 2 Ryd). Thus, the calculations of the core binding energies turn out to be unreliable until the self-consistent approach is applied. On the other hand, the agreement between the theoretical and experimental core binding energies that is achieved within the "isolated ion" model, implies that the latter should be really close to the self-consistent one. Nevertheless, the small sensitivity of the conduction band structure to the details of the charge distribution (Table 3) allows to accept" the crystalline potential" model

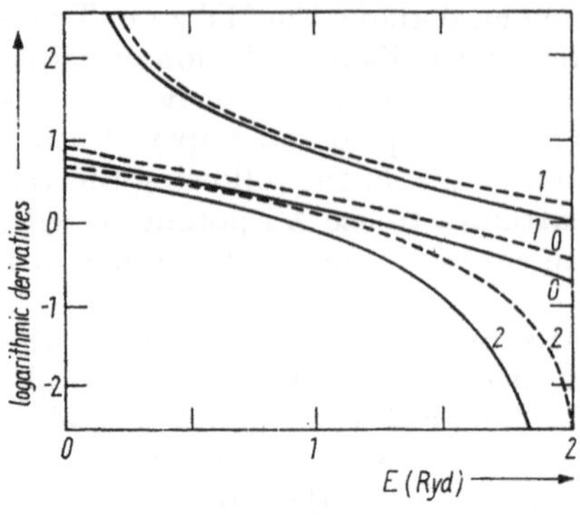

Fig. 2. The logarithmic derivatives of K radial orbitals, which are the solutions of a single HF equation. The ground state potentials of KCl crystal are employed: semiempirical (full curves) and ab initio (dashed). The values of the moment are shown above the corresponding curves. MT zeros of both models are brought together

for the purpose of XANES calculations. Moreover, the ab initio model seems to be preferable because its applicability is not restricted to some narrow class of compounds.

In order to compare the two potential models proposed the logarithmic derivatives of s-, p-, and d-radial wave functions at the MT sphere radius were calculated in the case of KCl crystal (Fig. 2). One can see that the respective logarithmic derivatives resulting from both potentials are in close agreement within a rather extended energy interval. The agreement may be improved if the average interstitial potential V_{MT} is supplied by appropriate relative shifts.

References

[1] P. A. Lee, P. H. Citrin, P. Eisenberger, and B. M. Kincaid, Rev. mod. Phys. **53**, 769 (1981).
[2] R. V. Vedrinskii and L. A. Bugaev, Fiz. tverd. Tela **27**, 280 (1985).
[3] J. E. Mueller, O. Jepsen, and J. W. Wilkins, Solid State Commun. **42**, 365 (1982).
[4] L. Glenn and C. G. Dodd, J. appl. Phys. **39**, 5372 (1968).
[5] J. L. Dehmer and D. Dill, J. chem. Phys. **61**, 693 (1973).
[6] V. P. Sachenko, E. V. Polozhentsev, Yu. F. Migal, R. V. Vedrinskii, and V. V. Kolesnikov, Phys. Letters A **48**, 169 (1974).
[7] A. A. Pavlychev, Fiz. tverd. Tela **26**, 1558 (1984).
[8] R. V. Vedrinskii and A. A. Novakovich, Fiz. Metallov i Metallovedenie **39**, 7 (1975).
[9] P. J. Durham, J. B. Pendry, and C. H. Hodges, Solid State Commun. **38**, 159 (1981).
[10] L. A. Grunes, Phys. Rev. B **27**, 2111 (1983).
[11] I. I. Gegusin, V. N. Datsyuk, and R. V. Vedrinskii, phys. stat. sol. (b) **109**, 563 (1982).
[12] G. N. Greaves, P. J. Durham, G. Diakun, and P. Quinn, Nature **294**, No. 5837, 139 (1982).
[13] R. V. Vedrinskii, I. I. Gegusin, V. N. Datsyuk, A. A. Novakovich, and V. L. Kraizman, phys. stat. sol. (b) **111**, 433 (1982).
[14] R. V. Vedrinskii et al., Solid State Commun. **44**, 1401 (1982).
[15] S. T. Pantelides, Phys. Rev. B **11**, 2391 (1975).
[16] P. D. de Cicco, Phys. Rev. **153**, 93 (1967).
[17] G. D. Mahan, Phys. Rev. B **21**, 4791 (1980).
[18] V. Kumar Srivastava, phys. stat. sol. (b) **114**, 667 (1982).
[19] K. Siegbahn et al., Elektronnaya Spektroskopiya, Mir, Moskva 1971.
[20] V. V. Nemoshkalenko and V. G. Aleshin, Elektronnaya Spektroskopiya Kristallov, Naukova Dumka, Kiev 1976.
[21] R. T. Poole, J. G. Jenkin, J. Liesegang, and R. C. Leckey, Phys. Rev. B **11**, 5179 (1975).
[22] V. V. Dyakin, R. F. Egorov, and V. P. Shirokovskii, phys. stat. sol. **36**, 447 (1969); S. Asano and J. Yamashita, J. Phys. Soc. Japan **30**, 667 (1971).
[23] F. Perrot, phys. stat. sol. (b) **52**, 163 (1972).
[24] M. S. Woolfson and B. W. Holland, Solid State Commun. **38**, 961 (1981).
[25] F. Herman and S. Skillman, Atomic Structure Calculations, Prentice Hall Inc., Englewood Cliffs (N.Y.) 1963.

(Received July 10, 1985)

phys. stat. sol. (b) 133, 203 (1986)

Subject classification: 79.20; S4

Department of Electronics, College of Engineering, University of Osaka Prefecture[1])

Monte-Carlo Studies of Energy Straggling of Electrons in Solids

By

M. KOTERA[2]), K. MURATA, and K. NAGAMI

An energy straggling distribution is incorporated in the Monte-Carlo simulation of keV-electron scattering in solids. This distribution is obtained by modifying the original Blunck and Leisegang distribution based on the Landau theory. The average energy loss of the distribution is adjusted to fit the stopping power of the Rao Sahib and Wittry equation (the modified Bethe equation). Since an electron trajectory is simulated by a stepwise fashion, the energy straggling of an electron even in a bulk specimen can be calculated. By comparing with an experimental energy distribution of transmitted electrons from a thin film, the validity of the present model is discussed. The electron path length distribution in a bulk specimen due to the energy straggling is also discussed.

In die Monte-Carlo-Simulation von keV-Elektronen-Streuung in Festkörpern wird eine Energieausbreitungsverteilung eingeführt. Diese Verteilung wird durch Modifizierung der ursprünglichen Verteilung von Blunck und Leisegang auf der Grundlage der Landautheorie erhalten. Der mittlere Energieverlust der Verteilung wird an die Bremsleistung der Rao Sahib-Wittry-Gleichung (die modifizierte Bethe-Gleichung) angepaßt. Da eine Elektronenbahn stufenweise simuliert wird, kann die Energieausbreitung eines Elektrons auch in einer Volumenprobe berechnet werden. Durch Vergleich mit einer experimentellen Energieverteilung der transmittierten Elektronen aus einer dünnen Schicht wird die Gültigkeit des angegebenen Modells diskutiert. Die Verteilung der Elektronenweglänge in einer Volumenprobe auf Grund der Energieausbreitung wird ebenfalls diskutiert.

1. Introduction

It is generally known that Monte-Carlo simulation is the most flexible and accurate approach to study keV electron interactions with solids. Many kinds of models of simulation are proposed and utilized in various fields with considerable success [1 to 13]. However, conventional Monte-Carlo simulation based on the continuous slowing down approximation (CSDA) model, using the stopping power equation (e.g. the Bethe equation) is not sufficient to describe an energy distribution of electrons. For example, a calculated energy distribution of transmitted electrons from a thin film target using the Bethe energy loss law shows a much sharper peak than the experimental result [7, 9, 10]. This is mainly because that in the CSDA model all possible discrete energy loss processes of an electron are averaged and the electron energy straggling effect is ignored. Several attempts have already been proposed by some authors to describe the energy straggling effect using the Monte-Carlo technique up to now [7 to 16]. Since most of them used differential cross-sections for several types of collision which are generally rare to happen, they have to take much CPU time

---

[1]) Mozu, Umemachi, Sakai, Osaka 591, Japan.

[2]) Present address: Department of Electronic Engineering, Osaka Institute of Technology, Omiya, Asahi-ku, Osaka, Japan.

to obtain a statistical result. The energy straggling theories have already been developed in the field of high-energy physics by Williams [17], Landau [18], and Blunck and Leisegang [19], etc. By using these straggling theories in Monte-Carlo simulation of electron scattering, it is expected that a more simplified simulation can be made. This paper provides a new simple Monte-Carlo simulation model to describe an electron energy loss distribution in solids based on these theories.

It is known that these theories give quite reasonable values for the energy straggling at high-energy electrons (≈ 1 MeV). However, for example, the Landau distribution includes the possibility of an infinite energy loss, and therefore a conception of the average energy loss does not exist. Besides, neither angular scattering of an electron nor electron absorption in a specimen is taken into account in these theories. In a present model these problems are numerically solved in a Monte-Carlo simulation of electrons. This simulation can be applied to obtain not only the energy distribution of electrons but also every kind of physical quantity which is reported with conventional simulations.

In the single scattering model [1, 2] an electron trajectory is simulated by an assembly of straight steps of the elastic scattering mean free path. This step length can be set to correspond to a distance used in a straggling theory, whereas there are two important points to consider for the distance. One is that the straggling theory is basically a multiple scattering theory and this distance should not be very short. Another is that in order to consider an energy dependence of an energy loss distribution the distance cannot be very long. However, in practical use of the theory a large discrepancy may not be found in many cases in spite of these facts, and some kinds of usage of the theory have been reported [20, 21]. As a numerical technique in a present model, the step length used for the straggling is taken tentatively to make an average energy loss of the energy distribution in one step fit the Rao Sahib and Wittry equation [22] (the modified Bethe equation). The step length in a present calculation is not very long or short, and the calculation did not give unreasonable values for many quantities in a wide energy range ($E_0 = 5$ to 20 keV). Then, we adopt this method in all energy range in the simulation from 20 down to 0.1 keV. In order to consider an electron energy in a step, a linear interpolation of energy in a step is adopted using energies at the beginning and the end of the step. With this approximation electron absorption in a solid can be expressed assuming that the electron stops when its energy is below a certain threshold energy.The angular scattering of electrons which has an energy dependence can also be considered in the step. In the present paper the angular scattering is not considered for simplicity.

2. Theories and Simulation Model

Blunck and Leisegang [19] approximated the Landau distribution by a sum of four Gaussian distributions as follows and gave the values of C_ν, λ_ν, and γ_ν for each ν:

$$(\Phi(\lambda))_{\text{Landau}} = \sum_{\nu}^{4} C_\nu \exp\left[-\frac{(\lambda-\lambda_\nu)^2}{\gamma_\nu^2}\right], \tag{1}$$

where

$$\lambda = \frac{\Delta}{\xi} - \left[1 - C + \ln\left(\frac{4E}{I^2}\,\xi\right)\right], \tag{2}$$

$$\xi = \frac{\pi N_A e^4 \varrho Z}{EA}\,x\,, \tag{3}$$

under non-relativistic condition. C is the Euler constant, N_A the Avogadro number, and I the average ionization energy of a specimen. The probability of losing energy between Δ and $\Delta + d\Delta$ from the initial energy E in a step length x for an electron is

$$f(\Delta) = \frac{1}{\xi}\,\Phi(\lambda)\,. \tag{4}$$

They took into account the second moment of the resonance term to the Landau theory and obtained the following equation:

$$(\Phi(\lambda))_{BL} = \sum_{\nu}^{4} \frac{C_\nu \gamma_\nu}{\sqrt{\gamma_\nu^2 + b^2}} \exp\left[-\frac{(\lambda - \lambda_\nu)^2}{\gamma_\nu^2 + b^2}\right], \tag{5}$$

where

$$b^2 = \frac{2\overline{K}_r^2}{\xi^2}\,x\,, \tag{6}$$

$$\overline{K}_r^2 = 1.5 \sum_i I_i \overline{K}_{ri}\,, \tag{7}$$

$$\sum_i \overline{K}_{ri} = \frac{\xi}{x} \sum_i \frac{n_i}{Z} \ln\left(\frac{2E}{I_i}\right). \tag{8}$$

$\overline{K}_r^2$ is calculated using the data of the binding energy I_i and the number of electrons n_i at each binding energy for all core electrons in an atom. For aluminum the following equation is obtained with the table of the binding energy [23]:

$$b^2 = \frac{3}{\xi}\,(0.2241 + 0.2922 \ln E)\,, \tag{9}$$

where E is in keV units. When b^2 is zero, $(\Phi(\lambda))_{BL}$ agrees with $(\Phi(\lambda))_{Landau}$. The solid line in Fig. 1 shows the original distribution of the Landau theory which is obtained by a numerical calculation in the way Hénoc and Maurice [20] have reported. The dashed line shows the distribution of $(\Phi(\lambda))_{Landau}$ calculated with (1). There is a clear difference between these two lines if λ is larger than 15. In order to study precisely the energy straggling e.g. for 20 keV incident electrons in an aluminum target with the present model, it is necessary to modify the original Blunck and Leisegang distribution to fit the Landau distribution more exactly up to $\lambda \approx 50$. Then, in the present study we introduced the fifth parameters ($\nu = 5$) to the original Blunck and Leisegang distribution. We tentatively determined those as $C_5 = 0.002$, $\lambda_5 = 20.0$, and $\gamma_5 = 20.0$. This modified Blunck and Leisegang distribution $(\Phi(\lambda))_{MBL}$ is plotted by a dash-dotted line in Fig. 1 and hereafter it is called as the MBL distribution. This fits the original Landau distribution fairly well for $\lambda \lesssim 50$.

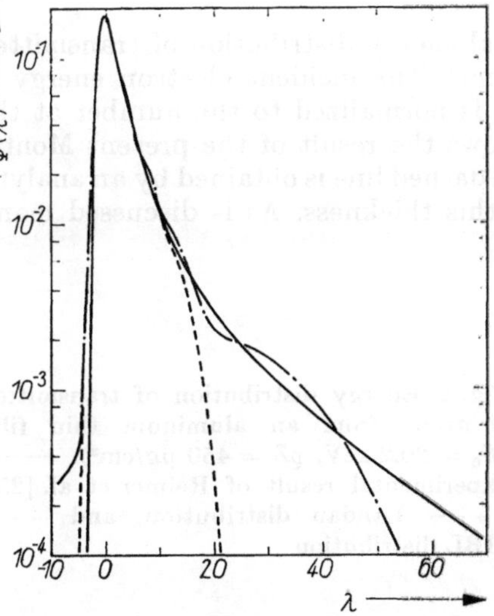

Fig. 1. Energy straggling distributions, $\Phi(\lambda)$, calculated with the original Landau theory [18] (———), the original Blunck and Leisegang equation [19] (– – – –), and the MBL equation (— · —)

Chen and Warshaw [24] discussed a relationship between ξ and I (in (2)) to study a criterion where the Landau theory is applicable in terms of the specimen thickness. In the present simulation in order to make the average value of the MBL distribution agree with the Rao Sahib and Wittry stopping power, the following relation is introduced:

$$\xi = QI \,. \tag{10}$$

After several trials the average energy loss fits the Rao Sahib and Wittry value within 5% error from 20 to 0.1 keV when Q has the following energy dependence:

$$Q(E) = 0.55E^{0.45} \,. \tag{11}$$

At the end of one step in the simulation the electron energy distributes from zero to the initial energy, and the larger energy loss probability is ignored.

An electron energy loss in one step is calculated using a uniform random number r distributed from 0 to 1, that is, the value of λ is selected by the following equation:

$$r = \frac{\int\limits_{\lambda_0}^{\lambda} \Phi(\lambda')\,\mathrm{d}\lambda'}{\int\limits_{\lambda_0}^{\lambda_E} \Phi(\lambda')\,\mathrm{d}\lambda'} \,, \tag{12}$$

where λ_0 and λ_E are values of λ when an energy loss of an electron in the step is zero and the initial energy, respectively. The selected value of λ is converted to an energy loss Δ with (2). The Monte-Carlo simulation of electron trajectories is continued until the electron energy is less than 0.1 keV.

3. Results and Discussions

The energy distributions of transmitted electrons from a thin film have been investigated experimentally by many authors [7, 21 to 24]. In a present study we refer to the experimental data of Reimer et al. [25]. By using a small detecting aperture ($\alpha = 0.75 \times 10^{-2}$ rad), they detected only electrons which could be thought to have passed straight through the target.

The solid line in Fig. 2 shows the experimental energy distribution of transmitted electrons from an aluminum target of 450 μg/cm². The incident electron energy is 20.2 keV. The number of transmitted electrons is normalized to the number at the most probable energy. The dash-dotted line shows the result of the present Monte-Carlo calculation with the MBL distribution. The dashed line is obtained by an analytical calculation of the Landau distribution for this thickness. As is discussed many

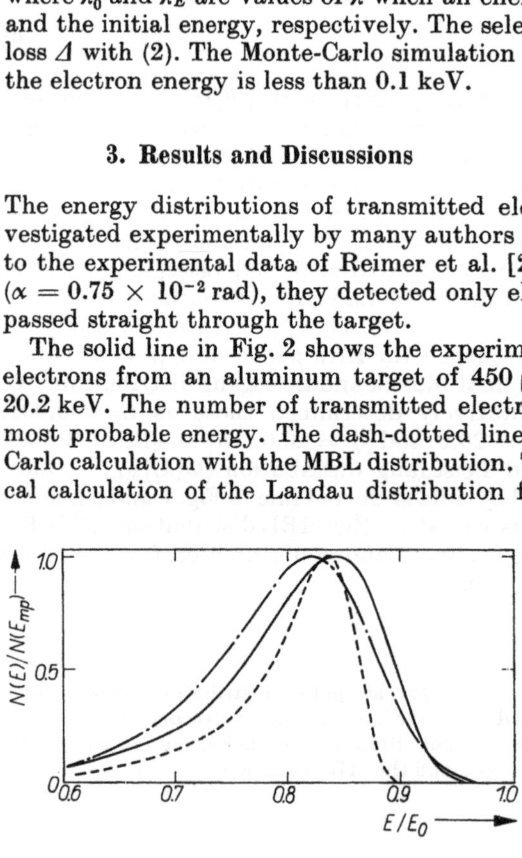

Fig. 2. Energy distribution of transmitted electrons from an aluminum thin film ($E_0 = 20.2$ keV, $\varrho Z = 450$ μg/cm²). ——— experimental result of Reimer et al. [25], ---- Landau distribution, and —·— MBL distribution

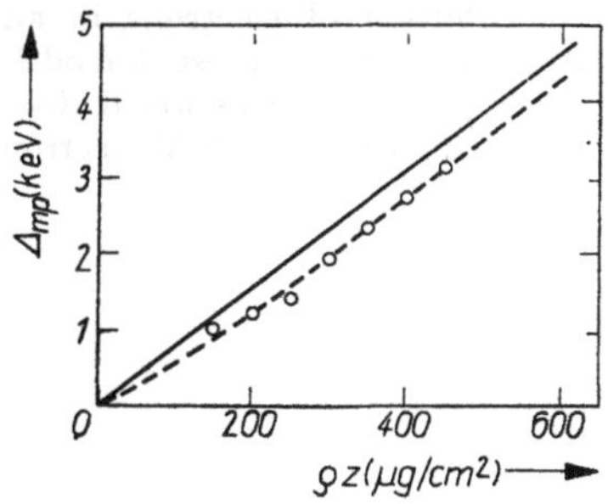

Fig. 3

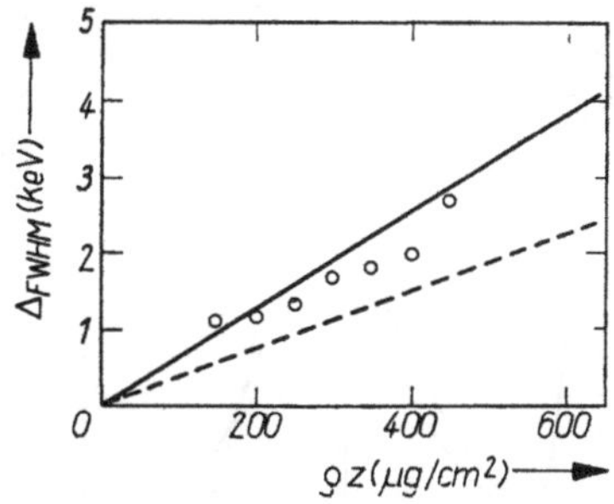

Fig. 4

Fig. 3. Film mass thickness (ϱZ) dependence of the most probable energy (Δ_{mp}) of the transmitted electron in Al ($E_0 = 20.2$ keV). ○ experimental results of Reimer et al. [25], ——— present calculation, - - - - theoretical value after Landau [18]

Fig. 4. Film mass thickness (ϱZ) dependence of the full width at half maximum of the transmitted electron energy distribution (Δ_{FWHM}) in Al ($E_0 = 20.2$ keV). ○ experimental results of Reimer et al. [25], ——— present calculation, - - - - theoretical value after Landau [18]

times for the Landau distribution, the calculated most probable energy is close to the experiment and the width is underestimated. For the MBL distribution, the calculated most probable energy is underestimated but the width is closer to the experiment. The differences in the most probable energy and the width between these calculated distributions are mainly attributed to their b^2 dependence of the MBL distribution. As discussed by Birkhoff [26], the Blunck and Leisegang distribution (also the MBL distribution) is getting wider and the most probable energy is getting low with increasing value of b^2.

Fig. 3 shows a most probable energy loss Δ_{mp} of transmitted electrons as a function of film thickness of aluminum at 20.2 keV electron incidence. The data of Reimer et al. are shown by circles. The dashed line shows a value analytically calculated with the Landau theory using $\lambda_{mp} = -0.225$ [27] for each aluminum thickness. The curve has a little bend because of the logarithmic term in (2). The dashed line fits quite well with all experimental points. The solid line shows the result of simulation with the MBL distribution. The first step length of this simulation corresponding to (3) is 370 μg/cm². Since we took a linear interpolation of an energy loss in one step, only the value at this thickness of the solid line agrees exactly with the analytically calculated value, and the line is straight up to 370 μg/cm². Even if the present simulation overestimates the most probable energy, the difference is slight, and remembering the average energy loss is adjusted to fit the Rao Sahib and Wittry equation, this difference will not give a big influence on the simulation of the electron behavior in solids.

Fig. 4 shows the full width at half maximum of the energy distribution Δ_{FWHM} of transmitted electrons as a function of film thickness of aluminum at 20.2 keV electron incidence. Analytically calculated results with $\lambda_{FWHM} = 4.02$ [27] in the Landau theory are shown by the dashed line. The curve obtained by the Landau distribution is generally lower than experimental points. On the contrary, the simulated result with the MBL distribution shown by the solid line is larger than the experiment. This solid line is also straight up to 370 μg/cm². If we take a smaller thickness than 370 μg/cm² as the first step length, b^2 in (9) becomes large and then, λ_{FWHM} gets larger than the present value. For example, if we take the first step length as 100 μg/cm², Δ_{FWHM} at this thickness becomes 1.2 keV and it is too large. When the film thickness

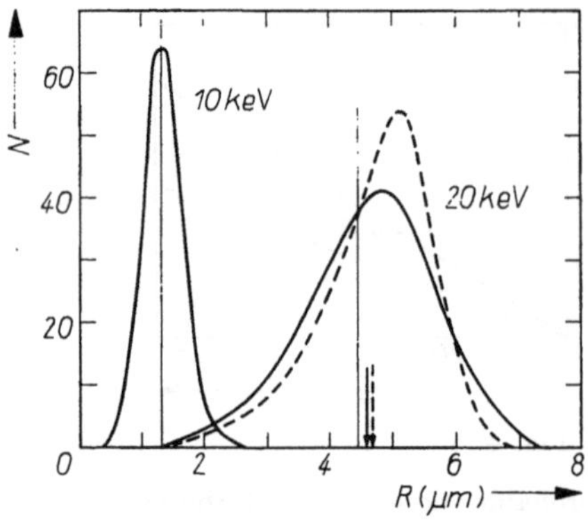

Fig. 5. Path length distribution of electrons in an aluminum target calculated with the present model. Number of electron stops (N) in each assumed division of the path length (R) for 10 and 20 keV electron incidence

is very small, the Blunck and Leisegang theory will give a much wider energy distribution than the experiment. It is necessary to avoid such a condition with a reasonable extrapolation or interpolation as we adopted here. It might give a better result in the Δ_{FWHM} using the distribution that recently Perez et al. [28] or Liljequist [16] proposed, taking into account all orders of the resonance term with a reasonable approximation.

The energy straggling effect gives rise to a distribution of the electron mean free path length in a solid. Assuming that an electron stops effectively when its energy is below 0.1 keV, the path length distribution is easily calculated. Fig. 5 shows the electron path length distribution in a bulk aluminum target at 10 and 20 keV electron incidence. The number at the vertical axis is obtained by counting the number of electron stops in every 50 nm division in the electron path for 10 keV, and in every 100 nm division for 20 keV when 1000 electron trajectories are calculated. The solid line and the dashed line show results with the MBL distribution and with the Landau distribution, respectively. The same procedure has been applied to calculate the Landau distribution. Because the Landau energy loss distribution is narrower than the MBL distribution, the distribution of the dashed line is narrower than that of the solid line. The vertical fine lines in the figure are electron ranges which are derived from integrations of the reciprocal of the Rao Sahib and Wittry stopping power. The calculated result for the MBL distribution is widely spread around them. Since these distributions are obtained by the energy straggling distribution, the shapes are very similar to the straggling distribution. The minimum path length of an electron agrees with the step length at the initial energy. The arrows of the solid and dashed line in the figure show average depths of each distribution, and they are very close to the vertical line at 20 keV. Those three values at 10 keV almost agree with each other in this figure. The difference between the average path length for the MBL distribution and the Rao Sahib and Wittry electron range is below 5%. The incident electron energy dependence of a full width at half maximum of the path length distribution using the MBL distribution is calculated for 5, 10, 15, and 20 keV and we obtained the following equation:

$$R_{FWHM} = 1.85 \times 10^{-2} E^{1.60}, \tag{13}$$

where R_{FWHM} is in μm and E in keV units. For example, the value of R_{FWHM} calculated with (13) in aluminum at $E = 20$ keV is 2.23 μm. On the other hand, R_{FWHM} under the same condition obtained by Salvat and Parellada [14] with their Monte-Carlo simulation model is about 2.78 μm, and there is about 20% difference between the results.

4. Conclusion

Although the Bethe theory or the Rao Sahib and Wittry equation expresses the average energy loss of an electron very well and it is widely used, it should be noted that it does not give any information on the energy loss distribution. On the other hand, though both theories of Landau, and Blunck and Leisegang express the energy loss distribution quite well even in the low-energy region as we discussed here, neither theory produces a theoretical value of the average energy loss. In this paper these three theories are combined in Monte-Carlo simulation, and the calculated result shows reasonable values for the energy distribution of transmitted electrons from a thin film. In future the present model should be extended to take into account the angular scattering effects of electrons in a specimen. Then, the energy distribution arising from transmitted electrons, the thickness dependence of the transmission or the backscattering coefficient, etc. can be obtained.

References

[1] K. Murata, T. Matsukawa, and R. Shimizu, Japan. J. appl. Phys. **10**, 678, 1290 (1971).
[2] D. F. Kyser and K. Murata, IBM J. Res. Developm. **18**, 352 (1974).
[3] G. Love, M. G. Cox, and V. D. Scott, J. Phys. D **11**, 7 (1978).
[4] M. Kotera, K. Murata, and K. Nagami, J. appl. Phys. **52**, 997 (1981).
[5] M. Kotera, K. Murata, and K. Nagami, J. appl. Phys. **52**, 7403 (1981).
[6] K. Murata, M. Kotera, and K. Nagami, J. appl. Phys. **54**, 1110 (1983).
[7] R. Shimizu, Y. Kataoka, T. Ikuta, T. Koshikawa, and H. Hashimoto, J. Phys. D **9**, 101 (1976).
[8] K. Murata, M. Kotera, and K. Nagami, Proc. 8th Internat. Conf. X-Ray Optics and Microanalysis, Ed. D. R. Beaman, R. E. Ogilvie, and D. B. Wittry, Pendell Publ. Co., Michigan 1980 (p. 434).
[9] K. Murata, D. F. Kyser, and C. H. Ting, J. appl. Phys. **52**, 4396 (1981).
[10] I. Adesida, R. Shimizu, and T. E. Everhart, J. appl. Phys. **51**, 5962 (1980).
[11] E. R. Krefting and L. Reimer, Quantitative Analysis with Electron Microprobes and Secondary Ion Mass Spectroscopy, Ed. E. Preuss, Zentralbibliothek der KFA, Jülich 1973 (p. 114).
[12] S. Ichimura, M. Aratama, and R. Shimizu, J. appl. Phys. **51**, 2853 (1980).
[13] D. Liljequist, J. Phys. D **11**, 839 (1978).
[14] F. Salvat and J. Parellada, J. Phys. D **17**, 185 (1984).
[15] A. Desalvo, A. Parisini, and R. Rosa, J. Phys. D **17**, 2455 (1984).
[16] D. Liljequist, J. appl. Phys. **57**, 657 (1985).
[17] E. J. Williams, Proc. Roy. Soc. **A125**, 420 (1929).
[18] L. Landau, J. Phys. (USSR) **8**, 201 (1944).
[19] O. Blunck and S. Leisegang, Z. Phys. **128**, 500 (1950).
[20] J. Hénoc and F. Maurice, Nat. Bur. Standards spec. Publ. **460**, 61 (1976).
[21] M. Chung and K. Tai, Proc. 8th Internat. Conf. Electron and Ion Beam Science and Technology, Electrochemical Society, Princeton (N.J.) 1978 (p. 242).
[22] T. S. Rao Sahib and D. B. Wittry, J. appl. Phys. **45**, 5060 (1974).
[23] M. Someno and I. Yasumori, Hyomenbunseki (Surface Analysis), Kodansha 1976 (p. 327) (in Japanese).
[24] J. J. Chen and S. D. Warshaw, Phys. Rev. **84**, 355 (1951).
[25] L. Reimer, K. Brockmann, and U. Rhein, J. Phys. D **11**, 2151 (1978).
[26] R. D. Birkhoff, Hdb. Phys., Vol. **34**, Springer-Verlag, Berlin 1958 (p. 53).
[27] H. D. Maccabee and D. G. Papworth, Phys. Letters A **30**, 241 (1969).
[28] J. Ph. Perez, J. Sevely, and B. Jouffrey, Phys. Rev. A **16**, 1061 (1977).

(Received May 28, 1985)

phys. stat. sol. (b) 133, 211 (1986)

Subject classification: 61.80 and 78.70; S5; S5.11

Institute of Nuclear Physics, Academy of Sciences of the USSR, Novosibirsk[1])

Nondipole Radiation of Electrons at Planar Channeling

By

V. N. Baier, V. M. Katkov, and V. M. Strakhovenko

The nondipole radiation of electrons is considered for any shape of the interplane potential. A concrete analysis and comparison with the radiation of positrons are carried out for a potential of the "inverse parabola" type. It is shown that for the description of the radiation of particles with energy $\varepsilon \lessapprox 10$ GeV in the region of the maximum of the spectral distribution it is possible to take into account the few first harmonics of the motion only. This substantially simplifies the calculation, especially for the potentials with complex shape. The diagrams which illustrate the radiation spectrum as a function of the angular width and of the direction of beam incidence in a thin crystal, are depicted. A detailed comparison of the results with available experimental data is made.

Рассмотрено недипольное излучение электронов для любого вида межплоскостного потенциала. Конкретный анализ и сравнение с излучением позитронов проведен в потенциале типа ,,перевернутая парабола''. Показано, что для описания излучения частиц с энергией $\varepsilon \lessapprox 10$ GeV в районе максимума спектрального распределения оказывается возможным ограничиться учетом нескольких первых гармоник движения, что существенно упрощает расчет, особенно для потенциалов сложной формы. Построены графики, иллюстрирующие зависимость спектра излучения от угловой ширины и направления влета пучка падающих частиц в тонком кристалле. Проведено детальное сравнение результатов с имеющимися экспериментальными данными.

1. Introduction

The present paper deals with the nondipole radiation of electrons at planar channeling. If the transverse momentum of a radiating particle is $p_\perp \ll m$, then one can use the dipole approximation; but if $p_\perp \gtrsim m$, then the radiation becomes significantly nondipole. Making allowance for the fact that $p_\perp \sim \varepsilon\vartheta_c$ ($\varepsilon(m)$ is the energy (mass) of a particle, ϑ_c is the Lindhard angle: $\vartheta_c = \sqrt{2U_0/\varepsilon}$, U_0 is the depth of the potential well), we have $p_\perp/m \sim \sqrt{\varrho_0}$, where ϱ_0 is the parameter introduced in [1],

$$\varrho_0 = \frac{2\varepsilon U_0}{m^2}\,. \tag{1.1}$$

Thus, at $\varrho_0 \gtrsim 1$ ($\varepsilon \gtrsim \varepsilon_{nd} = m^2/2U_0$) the radiation gets nondipole. For the (110) plane, $\varepsilon_{nd} \approx 5.2$ GeV in Si(d) ($U_0 = 25$ eV) and $\varepsilon_{nd} \approx 0.93$ GeV in W ($U_0 = 140$ eV). The planar-channeling radiation in the nondipole region has been discussed in [1 to 6].

The effect under study is, in fact, a manifestation of the general mechanism of radiation at quasi-periodic motion (periodic in a frame moving with the particle's average velocity). It is this mechanism which determines radiation in undulators and in the field of a monochromatic plane wave. A theory of radiation during quasi-periodic motion, including the quantum region where the radiation recoil should be taken into consideration, has been developed by the authors in [1, 7].

[1]) Prospekt Lavrenteva 11, 630090 Novosibirsk 90, USSR.

In application to the channeling radiation problem, the radiation characteristics, calculated in [1, 7], are to be averaged over the states with different transverse energy $\varepsilon_\perp$:

$$\varepsilon_\perp = \frac{p_\perp^2}{2\varepsilon} + U(x)\,, \tag{1.2}$$

where $U(x)$ is the inter-plane potential, taking into account the dependence of the motion parameters on $\varepsilon_\perp$.

The motion of electrons in any adequate potential of the crystal planes is a non-harmonic one. This circumstance significantly complicates the solution of the radiation problem in comparison with the case of positrons for which the oscillator potential can be employed in the nondipole region (this question has been discussed in [5]). The latter enables the analytical stage in calculations to be substantially advanced. The radiation from positrons in the nondipole region has been analysed in the authors' paper [5] (see [4, 6] as well) which contains also a comparison with experiments. For electrons, a detailed consideration is made in the present paper.

2. Description of the Radiation for Arbitrary Potential at a Given Transverse Energy

With an arbitrary potential of planar channeling, for the radiation spectrum at a given value of $\varepsilon_\perp$ we have

$$\frac{\mathrm{d}I}{\mathrm{d}\omega} = \frac{i\alpha\omega}{(2\pi\gamma)^2}\int\limits_{-\infty}^{\infty}\frac{\mathrm{d}t}{t-i0}\int\limits_{-\pi}^{\pi}\mathrm{d}\tau\left[1+\frac{a(\omega)}{2}\,(g(\tau-t)-g(\tau+t))^2\right]\times$$
$$\times\exp\left\{-i\xi\left[2t(1+\langle g^2\rangle)+\int\limits_{-t}^{t}\mathrm{d}\varphi(g^2(t+\tau)-\langle g^2\rangle)\,\frac{1}{2t}\left(\int\limits_{-t}^{t}\mathrm{d}\varphi\, g(\varphi+\tau)\right)^2\right]\right\} \tag{2.1}$$

using formula (4.2) of [7]. In the above formula $\gamma = \varepsilon/m$; $\alpha = 1/137$; ω is the photon frequency, $\xi = (\omega\varepsilon)/[(\varepsilon-\omega)\,2\gamma^2\omega_0]$, $\omega_0(\varepsilon_\perp)$ is the motion frequency, $g(\varphi) = \gamma[v_x(\varphi) - \langle v_x\rangle]$, the x-coordinate is perpendicular to the planes forming a channel, v is the velocity; $\langle\ldots\rangle$ stands for an average value; $a(\omega) = 1 + \omega^2/2\varepsilon(\varepsilon-\omega)$; and the nondipoleness parameter is $\varrho = 2\langle g^2\rangle$. Formula (2.1) describes the radiation at any values of the parameter ϱ_0, including the quantum region when the radiation recoil needs to be taken into account [1]. It should bear in mind that (2.1), as (4.2) in [7], is straightforwardly derived from the results of [8] which presents a quantum theory of the interaction between charged particles and a field of a plane electromagnetic wave. The total radiation probability is expressed via an imaginary part of the mass operator of a particle in the field of a wave (see formula (3.36) in [8]). The frequency-differential radiation probability (spectrum) can be obtained using (3.31) of the cited publication. In this case, we are to have inmind that the variable "u" is related, by virtue of (2.21) in [8], to the fourth momentum of the emitted photon $k: u = (\varkappa k)/(\varkappa p)$, where $\varkappa(p)$ are the four-momenta of the wave (of the particle). The situation under discussion corresponds to that in which a wave propagating towards an ultrarelativistic particle whose average velocity is directed along the z-axis:

$$u = \frac{\omega + k_z}{\varepsilon + p_z} \approx \frac{\omega}{\varepsilon}\,. \tag{2.2}$$

Omitting the integral over u in (3.31), we have the radiation probability, differential with respect to ω, and multiplying the result by the ω, we obtain the radiation spectrum in the wave field.[2]) The radiation under quasi-periodic motion coincides with quasiclassical accuracy (the terms $\omega_0/\varepsilon_\perp$ are neglected, ω_0 is the frequency of motion), with that in a field of the corresponding wave. The vector potential of the latter is $e\boldsymbol{A}(\varkappa x) = \boldsymbol{p}(\varkappa x) - \langle \boldsymbol{p} \rangle$, $eA^0 = 0$, e is the particle charge.

If we base ourselves on (3.10) in [7], describing also the polarization properties of the radiation, then we obtain for the radiation spectrum an expression in the form of a sum over harmonics,

$$\frac{\mathrm{d}I}{\mathrm{d}\omega} = \frac{\alpha\omega}{(2\pi\gamma)^2} \sum_{n=1}^{\infty} \vartheta\left(n - \xi\left(1 + \frac{\varrho}{2}\right)\right) \times$$
$$\times \int_{-\pi}^{\pi} \mathrm{d}t_1 \, \mathrm{d}t_2 \, \mathrm{J}_0\left(2 \sqrt{\xi\left(n - \xi\left(1 + \frac{\varrho}{2}\right)\right)} \int_{t_2}^{t_1} \mathrm{d}\varphi \, g(\varphi)\right) \times$$
$$\times \left[1 + \frac{a(\omega)}{2} \left(g(t_2) - g(t_1)\right)^2\right] \exp\left\{i\left(n - \frac{\xi\varrho}{2}\right)(t_1 - t_2) + i\xi \int_{t_2}^{t_1} \mathrm{d}\varphi \, g^2(\varphi)\right\}, \tag{2.3}$$

where J_0 is the Bessel function and

$$\vartheta(x) = \begin{cases} 1, & x > 0 \\ 0, & x < 0 \end{cases};$$

for the rest of the notations see (2.1). Formulae (2.1) and (2.3) are completely equivalent and describe the radiation from both electrons and positrons. To derive explicit expressions, it is necessary to find $v_x(\varphi)$, i.e. to solve the problem of one-dimensional motion. This is possible in any potential.

In what follows, we shall make a comparison between the radiation of electrons and positrons. For the positron case, a parabolic potential has been used in [5], and for electrons it corresponds to an "inverse parabola" potential which describes well the electron motion. For the more suitable potential introduced in [10], the expressions for v_x, derived in [1, 10] should be used.

In the parabolic ("inverse parabola" for electrons) potential we have, from (2.3) for the radiation spectrum at a given energy $\varepsilon_\perp$,

$$\frac{\mathrm{d}I}{\mathrm{d}\omega} = \frac{\alpha}{\pi^2 d} \vartheta_c \xi_0 \left(1 - \frac{\omega}{\varepsilon}\right) \sum_{n=1}^{\infty} \vartheta\left(\frac{n}{s} - \xi_0\left(1 + \frac{\varrho}{2}\right)\right) \times$$
$$\times \int_{-\pi}^{\pi} \mathrm{d}t_1 \, \mathrm{d}t_2 \, \mathrm{J}_0\left(2(q_1 - q_2) \sqrt{\varrho_0 \xi_0 \left(\frac{n}{s} - \xi_0\left(1 + \frac{\varrho}{2}\right)\right)}\right) \times$$
$$\times \left[1 + \varrho_0 \frac{a(\omega)}{2} (\psi_1 - \psi_2)^2\right] \cos\left[n(t_1 - t_2) + \frac{\xi_0 \varrho_0}{2} (r_1 - r_2)\right], \tag{2.4}$$

where $\xi_0 = (\omega\varepsilon)/[(\varepsilon - \omega) \, 2\gamma^2 \omega_{\mathrm{os}}]$, ω_{os} is the $\varepsilon_\perp$-independent frequency of motion in the parabolic potential: $\omega_{\mathrm{os}} = (2/d) \, \vartheta_c$, d is the distance between planes forming the channel, $s = T/T_{\mathrm{os}}$, T is the period of motion, $T_{\mathrm{os}} = 2\pi/\omega_{\mathrm{os}}$; index (1, 2) indicates the

[2]) This procedure is permissible because no integration by parts over the variable u has been made when deriving formula (3.31) in [8].

Table 1
The functions involved in (2.4)

function	particles and type of motion			
	e_c^+	e_{nc}^+	e_c^-	e_{nc}^-
$\psi(t)$	$\sqrt{z}\cos(st)$		$\sqrt{z}\,\mathrm{ch}\,st - \mathrm{sh}\,st$	
$q(t)$	$\sqrt{z}\sin t$	$\sqrt{z}\sin st - t/\pi$	$1 - \mathrm{ch}\,st + \sqrt{z}\,\mathrm{sh}\,st$	$1 - \mathrm{ch}\,st + \sqrt{z}\,\mathrm{sh}\,st - \frac{t}{\pi}$
$r(t)$	$\frac{z}{2}\sin 2t$	$\frac{1}{2}\sin 2st - \frac{4q(t)}{\pi s} - \frac{t}{\pi}\sqrt{z-1}$	$\sqrt{z}(1 - \mathrm{ch}\,2st) - \frac{2t\sqrt{z}}{\pi} + \frac{1+z}{2}\mathrm{sh}\,2st$	$\sqrt{z}(1 - \mathrm{ch}\,2st) + \frac{1+z}{2}\mathrm{sh}\,2st - \frac{t}{\pi}\sqrt{z} - \frac{4}{\pi s}q(t)$
$s(z)$	1	A/π	$2f/\pi$	f/π
$\frac{\varrho(z)}{\varrho_0}$	z	$z + \frac{\sqrt{z-1}}{A} - \frac{2}{A^2}$	$z + \frac{\sqrt{z}}{f} - 1$	$z + \frac{\sqrt{z}}{f} - 1 - \frac{2}{f^2}$

dependence of the function on t_1, or t_2. The explicit expressions for the functions $q(t)$, $\psi(t)$, and $r(t)$ are given in Table 1 for $t > 0$; at $t < 0$ we have $\psi(-t) = \psi(t)$, $q(-t) = -q(t)$, and $r(t) = -r(-t)$; the table also presents $s = s(z)$ and $\varrho(z)$. The variable z is equal to $\varepsilon_\perp/U_0$; index "c" is for the particles in the channel ($z < 1$) and index "nc" is referred to the particles ($z > 1$) above the barrier; the following notations have been introduced:

$$f(z) = \frac{1}{2}\ln\left|\frac{\sqrt{z}+1}{\sqrt{z}-1}\right| \quad \text{and} \quad A(z) = \arcsin\frac{1}{\sqrt{z}}.$$

With increase of z from 0 to 1 the motion period of an electron in the channel increases and above the barrier decreases monotonously, and at $z \gg 1$, $T \sim 1/\sqrt{z}$; in the vicinity of the point $z = 1$ the period $T \sim \ln[1/|1 - z|]$. The parameter $\varrho(z)$ increases in the channel with growing z, attains a maximum, $0.43\varrho_0$ at $z \approx 0.85$, and then reduces to 0. Above the barrier, $\varrho(z)$ grows from 0, takes the value $\varrho_{\max}^{nc} \approx 0.13\varrho_0$ at $z \approx 1.001$, then decreases, and $\varrho/\varrho_0 \approx 2/45z$ at $z \gg 1$. This result coincides with that for the positron case because $\varrho \approx \gamma^2(U^2 - \langle U^2\rangle)/\varepsilon\varepsilon_\perp$ highly above the barrier and is independent of the charge sign of the particle. In general, with $z \gg 1$ the radiation does not depend of the sign of charge (see, e.g., Section 4 in [1]), since in this limit the result coincides with the Born approximation. Recall (see [9]) that the total electron radiation intensity in the potential under consideration is

$$I(z) = I_e\left(1 - z + \frac{\sqrt{z}}{f(z)}\right), \tag{2.5}$$

where $I_e = 16e^2U_0^2\gamma^2/3m^2d^2$, it decreases from $2I_e$ at $z = 0$ to 0 at $z = 1$, and then increases, at $z > 1$, approaching $I_{as} = \frac{2}{3}I_e$.

Formulae (2.1), (2.3), and (2.4) may be used for any ϱ_0. In case of a moderate nondipoleness ($\varrho_0 \approx 1$), the condition $\varrho \ll 1$ is satisfied for the above-barrier electrons, and it is possible to describe their radiation using the dipole formulae which are considerably simpler (cf. [5]). At $\varrho_0 \approx 1$, the description of radiation from electrons in the channel can be made significantly simpler as well. The velocity $v_x(\varphi)$ (just as the function $g(\varphi)$) in (2.1) can be expanded in Fourier series

$$v_x(\varphi) = \sum_{l=1}^{\infty} v_{2l-1} \cos\left((2l-1)\,\varphi\right), \qquad v_{2l-1} = \frac{1}{\pi}\int_0^{2\pi} \mathrm{d}\varphi\, v_x(\varphi) \cos\left((2l-1)\,\varphi\right). \tag{2.6}$$

The values of v_{2l-1} for appropriate potentials are given in [1] and [9]. The quantity v_{2l-1} falls as the motion harmonic l grows, and the corresponding radiation frequency shifts to the right. Thus, the contribution of the higher motion harmonics l to the region of a maximum of the resultant spectrum ($\xi \approx 1$) proves to be suppressed for two reasons: (i) we have found ourselves on the left slope of the curve for a given l and (ii) even in its maximum the harmonic l contributes much less as compared with the first harmonic at $\xi \approx 1$. Correspondingly, at $\xi \approx 1$, for the purpose of describing the radiation from channeled particles one can restrict oneself to a few terms in (2.6). If we retain only the first term $v_1 \cos\varphi$ in (2.6), then substituting an appropriate $g(\varphi)$ into (2.1) we are led to (4.6) from [7], wherein the following substitution should be made: $\varrho \to \varrho_1 = \gamma^2 v_1^2$. We also present here an expression for the spectrum with taking into account also a small addition contributed by the term $v_3 \cos 3\varphi$

$$\frac{\mathrm{d}I}{\mathrm{d}\omega} = \frac{i\alpha\omega}{2\pi\gamma^2}\int_{-\infty}^{\infty}\frac{\mathrm{d}t}{t-i0}\,\mathrm{e}^{-iZ_1}\left[\mathrm{J}_0(Z_0) + a(\omega)\,\varrho_1 \sin^2 t\,(\mathrm{J}_0(Z_0) - i\mathrm{J}_1(Z_0)) + \varrho_1\frac{v_3}{v_1}G(t)\right], \tag{2.7}$$

where

$$\begin{aligned} G(t) &= 2a(\omega)\sin t \sin 3t\,(\mathrm{J}_2(Z_0) + i\mathrm{J}_1(Z_0)) + \xi\left(\sin 2t - \frac{2\sin t\sin 3t}{3t}\right)\times \\ &\quad \times [b\mathrm{J}_1(Z_0) + i(b-1)\,\mathrm{J}_1'(Z_0)] + \xi\left(\sin 2t\cos 2t - \frac{2\sin t \sin 3t}{3t}\right)\times \\ &\quad \times [ib\mathrm{J}_2(Z_0) - (b-1)\,\mathrm{J}_2'(Z_0)], \qquad b = 1 + \varrho_1 a(\omega)\sin^2 t, \\ Z_0 &= \xi\varrho_1\left(\frac{\sin^2 t}{t} - \frac{\sin 2t}{t}\right), \qquad Z_1 = 2t\xi\left(1 + \frac{\varrho_1}{2}\right) - \xi\varrho_1\frac{\sin^2 t}{t}. \end{aligned} \tag{2.8}$$

To make better the convergence of the integral in (2.7), one can use the subtraction procedure described in [7].

The presented approach calls for a knowledge only of the Fourier harmonics of velocity and, hence, is applicable to the description of radiation from both the electron and positrons in any potential. We note that the dependence $T(\varepsilon_\perp)$ is taken into account exactly and the quantity ϱ_1 is close to ϱ. Since for electrons at $z \to 1$ the period $T(z)$ tends to infinity in any adequate potential, the non-harmonicity of motion will be largest at $z \to 1$. Thus, in the "inverse parabola" potential, with $z = (0.3; 0.6$ and $0.9)$ v_3/v_1 is equal, correspondingly, to 0.13; 0.15, and 0.22 and $v_5/v_1 = (0.045,$ 0.056, and $0.089)$. However, because at $z \to 1$ the intensity (2.5) $I(z) \to 0$, the radiation from particles which have undergone a significant non-harmonic motion, may be neglected at a finite population at $z \to 1$. Even in the case when the population in the channel is $\sim f(z)$ (thin crystal, the particles are uniformly distributed over the angle of incidence $\vartheta_0 \leqq \vartheta_c$; very thick crystal) and $I(z)\, f(z) \to I_e$ at $z \to 1$, the fraction of particles in the vicinity of $z = 1$ is small and their radiation lies within low frequencies.

3. Analysis of the Radiation in Thin Crystals

Dechanneling in thin crystals may be neglected by definition and the distribution over $\varepsilon_\perp$ is determined by the initial conditions. If the incident beam is uniformly distributed within the interval ϑ_1^0 to ϑ_2^0, then in the "inverse parabola" potential we have (see, e.g., formula (A.4) in [1])

$$\frac{\mathrm{d}N}{\mathrm{d}z} = \frac{\vartheta(z - y_1^2)\, \vartheta(1 + y_2^2 - z)}{2(y_2 - y_1)} \ln \frac{C}{B}, \tag{3.1}$$

where $y_{1,2} = \vartheta_{1,2}^0/\vartheta_c$, $C = (1 + \sqrt{z})\, \vartheta(y_2^2 - z) + (y_2 + \sqrt{1 + y_2^2 - z})\, \vartheta(z - y_2^2)$, $B = (y_1 + \sqrt{1 + y_1^2 - z})\, \vartheta(1 + y_1^2 - z) + \sqrt{z - 1}\, \vartheta(z - 1 - y_1^2)$. The spectra $\mathrm{d}I/\mathrm{d}\omega$, averaged over $\mathrm{d}N/\mathrm{d}z$ (see formula (3.1)), are presented in Fig. 1 and 2 which demon-

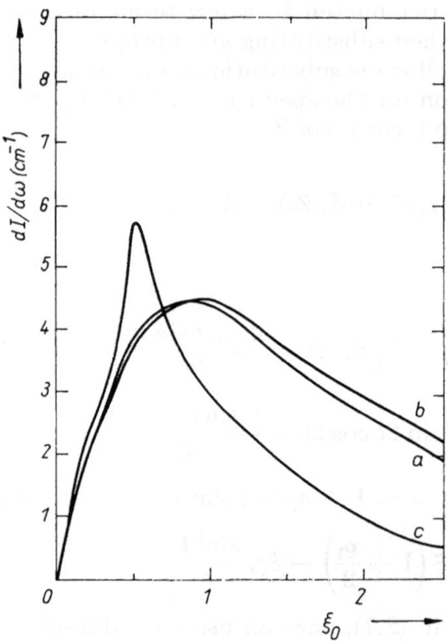

Fig. 1

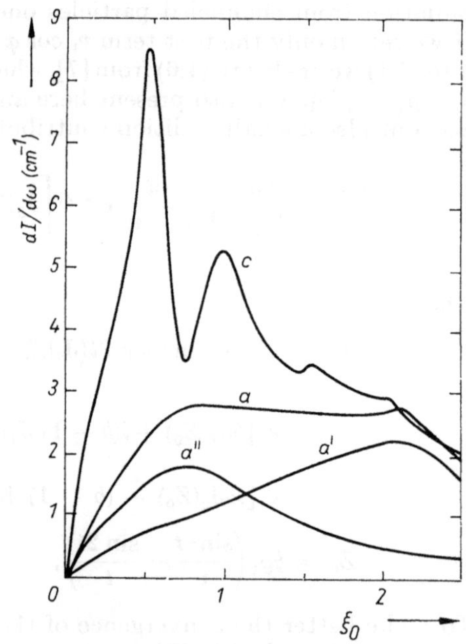

Fig. 2

Fig. 1. Spectral intensity of radiation from electrons with $\varepsilon = 10$ GeV upon their motion along the (110) plane of a Si crystal when the initial beam is uniformly distributed over the angle of incidence in the 0 to 20 μrad range. (a) Representation of calculation by means of (2.7) at $v_3 = 0$, (b) calculation using (2.4) and (c) is for the positron spectrum under the same conditions. The value of $\xi_0 = 1$ corresponds to the energy $\omega = 110$ MeV

Fig. 2. The same as in Fig. 1 but for 40 to 60 μrad angles of incidence. Curve a'' shows the contribution from the channeled electrons, curve a' is the contribution of the above-barrier electrons

strate the spectra in Si ((110) plane at $\varepsilon = 10$ GeV ($\xi_0 = 1$ corresponds to $\omega = 110$ MeV)) for the intervals of incident angles (0 to 20) μrad (Fig. 1), and of (40 to 60) μrad (Fig. 2); the Lindhard angle is $\vartheta_c = 71$ μrad. In Fig. 1, for curve a $dI^{nc}/d\omega$ has been calculated by means of the dipole formulae (see [9]), and $dI^c/d\omega$ has been calculated according to (2.7) (with the term $\sim v_3$ omitted). Curve b is a result of calculations by means of the exact formula (2.4). It is seen that taking into account only one harmonic of motion (at a given $\varepsilon_\perp$) is a good approximation. In addition, the discrepancy between curves a and b is more noticeable on their right slope. For comparison, we also present the radiation spectrum for positrons (curve c) obtained under the same conditions in [5]. Just as in the dipole case, the spectral curve for electrons is wider and has a lower height in its maximum (cf. [1, 9, 10]) when comparing with that for positrons, although the difference in height in the nondipole region becomes not so significant. It is seen from Fig. 2 that the rotation of the narrow beam by the angle $\sim \vartheta_c$ leads to sweeping changes in the spectrum. The channeled (a″) and above-barrier (a′) particles contribute at different frequencies and their contributions are approximately equal in magnitude; as a consequence, the summary spectrum (a) has a wide plateau. The corresponding spectrum for positrons [5] is shown by curve c. Note that when rotating a wide beam by an angle of about ϑ_c a maximum (keeping its position) from the channeled particles is still clearly seen in the radiation spectrum[3]).

Reference [5] has indicated simple rules to find the characteristic points (maxima and minima) of the spectral curve for positrons, which are based on an analysis of the z-dependence of the intensity and of the argument of ϑ-functions in (2.4). For electrons these structures are washed out on account of a large width of the spectral curves which is due to a substantial spread of the motion periods. The estimates become less definite as a result. The spectral curve has, as a rule, one maximum which is due to the first harmonic (term with $n = 1$ in (2.4)) of radiation from channeled particles. It follows from (2.4) that at $n = 1$ the radiation frequencies lie within the interval $0 \leqq \xi_0 \leqq \xi_1, \xi_1 = [s(1 + \varrho/2)]^{-1}$. The quantity $s(1 + \varrho/2)$ grows monotonously with increasing z ($0 < z < 1$), and ξ_1 shifts to the left with increasing z. For estimation, it is convenient to introduce a "weighted intensity": $F(z) = I(z)\, dN/dz$. If $\max F = F(z_1)$, then as z increases from 0 to z_1 the boundary frequency shifts to the left and the peak becomes higher $(\sim F(z))$. At larger z both functions $F(z)$ and $\xi_\perp$ decrease. As a result, the maximum of the resultant distribution $dI_c/d\omega$ lies to the left from the point $\xi_m^c(z_1) = [s(z_1)\,(1 + \varrho(z_1)/2)]^{-1}$. If the function $F(z)$ has a sharp peak at z_1, then the maximum $dI^c/d\omega$ will lie near $\xi_m^c(z_1)$. If at $z > 1$, $\max F(z) = F(z_2)$, then, similarly, the maximum of $dI^{nc}/d\omega$ is to the left from $\xi_m^{nc}(z_2) = [s(z_2)\,(1 + \varrho(z_2)/2)]^{-1}$. For illustration, we consider the case when a narrow beam is incident at ϑ_0 near ϑ_c (Fig. 2). Here the function $F(z)$ has a clear maximum both in the channel ($z_1 \approx 0.7$) and above the barrier ($z_2 \approx 1.4$); the corresponding frequencies are: $\xi_m^c \approx 0.94$ and $\xi_m^{nc} \approx 2.4$. In Fig. 2 the maxima of the spectral curves correspond to $\xi_{max}^c \approx 0.8$ and $\xi_{max}^{nc} \approx 2.1$. We would like also to discuss an important case when $(dN/dz) \sim f(z)$. As shown in [11], such a distribution is established in thick crystals in the channel and above the barrier at not too large $z = \varepsilon_\perp/U_0$. This case is realized also in thin crystals if $0 \leqq \vartheta_0 \leqq \vartheta_c$ (see (3.1)); whereas if $0 \leqq \vartheta_0 \lesssim \vartheta_c$, then the distribution is not strongly different, in configuration, from $f(z)$. The function $F(z) = I(z)\, f(z)$ has a maximum at $z_1 \approx 0.695$, and $\xi_m^c(z_1) \approx 1.3/(1 + \varrho_0/5)$. For the (110) plane, at $\varepsilon = 10$ GeV $\xi_m^c \approx 0.94$ in Si (cf. Fig. 1; curve a).

[3]) The corresponding curves as well as a more comprehensive discussion and comparison with experiment are contained in [19].

4. Radiation in Thick Crystals

For a thick crystal, the thickness is $L \gg l_d$, where l_d is the dechanneling length, introduced in [1],

$$l_d = \frac{\alpha}{2\pi} \varrho_0 L_{rad} . \tag{4.1}$$

Here L_{rad} is the radiation length in an appropriate amorphous medium. If $\varepsilon \lesssim 10^{11}$ eV then multiple scattering will dominate in evolution of the distribution function (DF) over $\varepsilon_\perp$ (see [11]). We shall take this evolution into account just as in [12], assuming that the incident beam is a Gaussian one with respect to the incident angles (its width is Δ_0). Namely we write down the initial distribution in which the substitution $\Delta_0^2 \to \Delta_0^2 + \frac{1}{2} \vartheta_s^2(l)$ is made ($\vartheta_s^2(l)$ is the average squared angle of multiple scattering in an appropriate amorphous medium[4]) at depth l). If the channeling radiation is not collimated, then one can obtain the distribution over z, averaged over the crystal thickness, from formula (4) in [12] where it is necessary to make an elementary integration over v:

$$\left\langle \frac{dN}{dz} \right\rangle = \frac{1}{L} \int_0^L \frac{dN(l)}{dz} dl = \sqrt{\frac{2}{\pi\lambda}} \frac{1}{1-\beta} \left\{ f(z) \left[e^{-z/2\lambda} - \sqrt{\beta}\, e^{-z/2\tilde{\Delta}_0^2} \right] - \right.$$
$$\left. - \frac{1}{\lambda} \int_{\sqrt{z}}^{\sqrt{z/\beta}} du\, e^{-u^2/2\lambda} + \frac{1}{\sqrt{\lambda}} \int_0^{f(z)} du\, g_1(ug_2 - g_1) \left[\frac{1}{\sqrt{\lambda}} e^{-g_2^2/2\lambda} - \frac{1}{\tilde{\Delta}_0} e^{-g_2^2/2\tilde{\Delta}_0^2} \right] \right\}, \tag{4.2}$$

where

$$\lambda = \frac{L}{l_d} + \tilde{\Delta}_0^2 , \qquad \tilde{\Delta}_0 = \frac{\Delta_0}{\vartheta_c}, \qquad \beta = \frac{\tilde{\Delta}_0^2}{\lambda}, \qquad g_2(u) = \frac{dg_1(u)}{du},$$
$$g_1(u) = \sqrt{|1-z|}\, \{\vartheta(1-z) \operatorname{ch} u + \vartheta(z-1) \operatorname{sh} u\} . \tag{4.3}$$

From (4.2), there is no difficulty in deriving an asymptotic distribution at $L \gg l_d$. In this limit, $\beta \to 0$ and $\lambda \to \infty$, and we find, within an accuracy up to terms $\sim 1/\sqrt{\lambda}$,

$$\left\langle \frac{dN}{dz} \right\rangle \approx \sqrt{\frac{2 l_d}{\pi L}}\, f(z)\, e^{-z/2\lambda} \tag{4.4}$$

so that for $z \ll \lambda$ we have $\langle dN/dz \rangle \sim f(z)$. An analysis of the solution of the kinetic equation, which was made in [11], has shown in a general form that at very large depths this distribution $\sim f(z)$ what corresponds to a uniform one in transverse phase space. Thus, at $L \to 0$ the expression (4.2) is exact (for the Gaussian beam) and has a correct asymptotics at $L \to \infty$. This permits one to hope for a satisfactory description of the function $\langle dN/dz \rangle$ at any l.

In the foregoing, we have estimated the position of the maximum of the spectral curve at $L \gg l_d$. Let us now show that these curves prove to be similar at different (large) thicknesses. Indeed, the resultant spectrum is given by an integral

$$\int_0^\infty \left\langle \frac{dN}{dz} \right\rangle \frac{dI}{d\omega} dz . \tag{4.5}$$

[4]) In an adiabatic approximation, the energy losses may be taken into account in $\vartheta_s^2(l)$ [13].

Consider now the case of large L when it is possible to use (4.4). We split the integration region into two parts: 0 to z_0 and z_0 to ∞; with $1 \ll z_0 \ll \lambda$. After that, in the first part one can substitute $\exp(-z/2\lambda) \to 1$ and we obtain a contribution, including L, only as the factor $L^{-1/2}$. It is the region ($z \approx 1$) that contributes to the maximum of the spectrum ($\xi_0 \approx 1$). In the second part ($z \gg 1$) the particles are in the high-lying above-barrier states and the dipole formulae may be used. Making use of formula (7) in [9], we have roughly, at $z \gg 1$: $(\mathrm{d}I/\mathrm{d}\omega) \sim \xi_0/z$ (photons with a frequency up to $\xi_0 \approx 2\sqrt{z}$ are emitted). Although the contribution of these particles breaks the similarity, it is strongly suppressed at $\xi_0 \approx 1$. The region up to $z \approx \lambda$ contributes to the integral $\int_{z_0}^{\infty} \mathrm{d}z \ldots$, so that the frequency boundary of the spectrum corresponds to $\xi_0 \approx 2\sqrt{\lambda}$. Thus, at $\xi_0 \approx 1$ the shape of the spectrum proves to be universal, and the difference in height is determined by the scaling factor[5]) $\sim L^{-1/2}$. The accuracy of this expression is not worse than $\lambda^{-1/2}$ (the order of the terms omitted in going from (4.2) to (4.4)).

We have calculated the radiation from electrons upon channeling in the (110) plane of a diamond single crystal with thickness $L_1 = 0.1$ mm (Fig. 3), $L_2 = 1$ mm and $L_3 = 1.7$ mm at $\varepsilon = 4.5$ GeV; then $\vartheta_c \approx 110$ μrad, $\varrho_0 \approx 0.95$ and at $\omega = 54$ MeV $\xi_0 = 1$; this corresponds to the experimental conditions [14]. The authors of the cited publication estimate the angular width of the incident beam as $\Delta_0 \approx 60$ μrad. The theoretical curve 1 in Fig. 3 is the sum of the contributions from the radiation in a continuous potential of the planes ($\langle \mathrm{d}N/\mathrm{d}z \rangle$ is taken in the form (4.2)) and the bremsstrahlung[6]) (the latter contribution is depicted as the dashed line; as seen, it is negligibly small). The crystals L_2 and L_3 are thick enough; for them; (see (4.2)) λ is: $\lambda_2 = 7.7$, $\lambda_3 = 13$. The expected accuracy of the similarity of the spectral curves will be about $\lambda^{-1/2}$ (30 to 40%) and the asymptotic scale factor is $(L_2/L_3)^{-1/2} \approx 1.3$. For $\xi_0 \lesssim 2$ neglecting bremsstrahlung the calculated curves are similar within an accuracy better than 5% and the scaling factor is 1.2; its difference from $(L_2/L_3)^{-1/2}$ means that the asymptotic regime does not yet occur. Nevertheless, the accuracy of similarity, at $\xi_0 \lesssim 2$, has proved to be much better than the estimate with respect to $\lambda^{-1/2}$.

The accuracy of the dipole approximation is possible to estimate by calculation in this approximation and by means of exact formulae. Consider the radiation in the (100) plane in Si at $\varepsilon = 1.2$ GeV ($\vartheta_c = 150$ μrad, $\varrho_0 \approx 0.13$; $\xi_0 = 1$ corresponds to $\omega = 4.8$ MeV); at $L = 0.24$ mm, $\Delta_0 \approx 100$ μrad that is in accord with the experimental conditions [15]. The calculation with the use of (4.2) is given in Fig. 4 where curve 1 is the contribution from the channeled particles, curve 2 is the contribution given by the above-barrier particles, the dashed line is for the contribution of the bremsstrahlung, and curve 3 shows the sum of all these contributions. In the dipole approximation, the radiation has been calculated using the formulae from [12]. Comparison with curves 1 and 2 in Fig. 4 has shown that the dipole curves lie 5 to 10% higher ($\approx \varrho_0$). We would like to attract the reader's attention to the fact that the shape of the spectrum in Fig. 4 is mainly determined, in the region of maximum, by the contribution of the particles in the channel. The above-barrier particles give a dominant contribution only on the right slope (at $\xi_0 > 2$). This is inconsistent with the assertion in [15] that taking into account the contribution of the above-barrier particles only is a good approximation. Note the good agreement of the theory, free from fitted parameters, with the experiment [15] (unlike the theoretical arguments in [15]).

---

[5]) Upon collimation of the photon beam to an angle $\lesssim \vartheta_c$, a factor $\sim (1/L) \ln (L/l_d)$ arises. Recall that all the formulae incorporate the intensity recalculated per unit length.

[6]) We describe bremsstrahlung using the standard formulae.

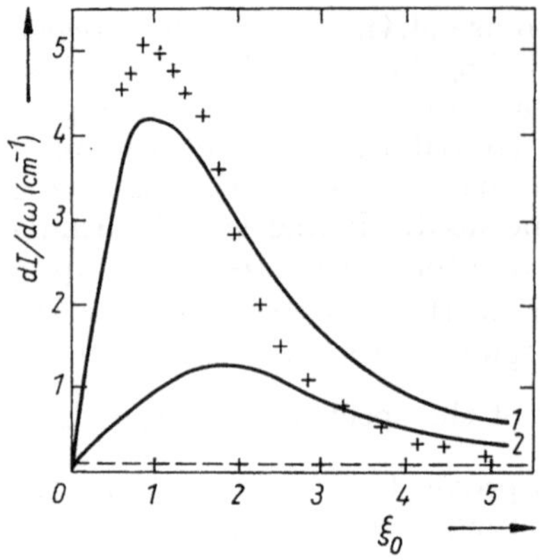

Fig. 3

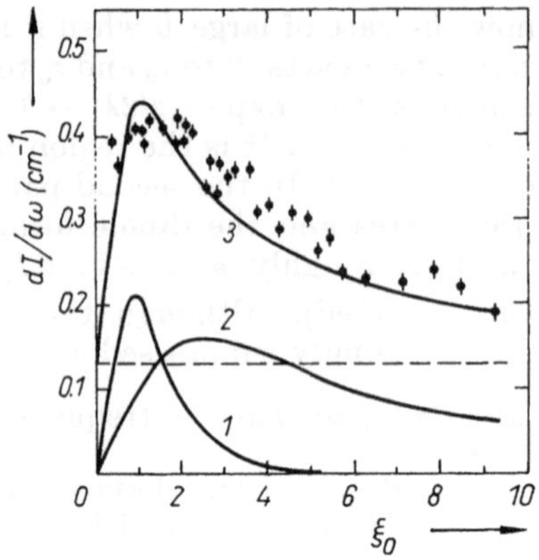

Fig. 4

Fig. 3. Spectral intensity of radiation, per unit thickness, from electrons with $\varepsilon = 4.5$ GeV upon motion along the (110) plane in diamond whose thickness is $L_1 = 0.1$ mm. The value of $\xi_0 = 1$ corresponds to $\omega = 54$ MeV. The dashed line shows the bremsstrahlung contribution; (2) is the contribution of the above-barrier electrons, and (1) is the resultant spectrum; the experimental data (+) are from [14]

Fig. 4. Spectral intensity of radiation, per unit thickness, from electrons with $\varepsilon = 1.2$ GeV upon motion along the (100) plane in a Si crystal of $L = 0.24$ mm. The value of $\xi_0 = 1$ corresponds to $\omega = 4.8$ MeV. The bremsstrahlung contribution is shown by the dashed line. (1) Is the contribution from the channeled electrons, (2) is that from the above-barrier ones, and (3) is the resultant spectrum; the experimental data (●) from [15]

In [16] the radiation from electrons with energy $\varepsilon = 4.3$ GeV has been measured in diamond. The divergence of the incident beam was $\Delta_0 \approx 100$ μrad [16] and the beam of radiated photons was collimated to a solid angle of 10^{-7} ster.[7]) The results in [16] are presented in terms of $\mathrm{d}(I^c + I^{nc})/\mathrm{d}I^{br}$. We give them in Fig. 5 in an absolute normalization using ($\omega \ll \varepsilon$) the value of $\mathrm{d}I^{br}/\mathrm{d}\omega \approx 4k/(3L_{rad})$, the coefficient k takes into account the influence of the collimator for a disoriented crystal. Within the frame of the accepted multiple-scattering model it is easy to find $k \approx 0.7$.

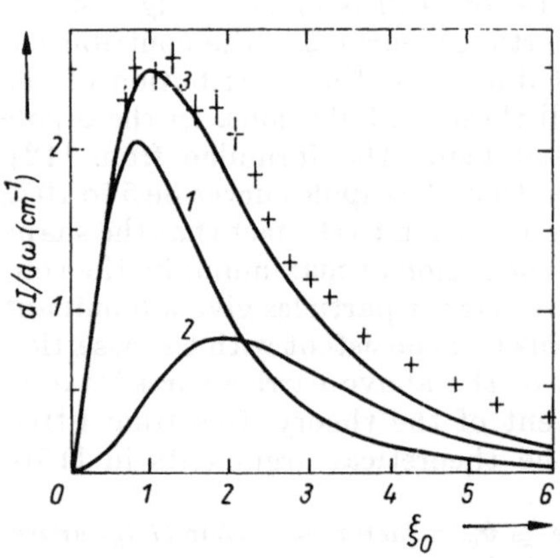

Fig. 5. Spectral intensity of radiation, per unit thickness, from electrons with $\varepsilon = 4.3$ GeV upon their motion along the (011) plane of a diamond crystal with $L_4 = 0.1$ mm. The value of $\xi_0 = 1$ corresponds to $\omega = 50$ MeV, $\vartheta_c = 113$ μrad, $\varrho_0 \approx 0.91$. (1) Is the contribution of the channeled electrons, (2) is that of the above-barrier ones, and (3) is the summed contribution without the bremsstrahlung taken into account; the experimental data have been taken from [16]

[7]) We thank A. O. Aganiants for discussion of the conditions of the experiment.

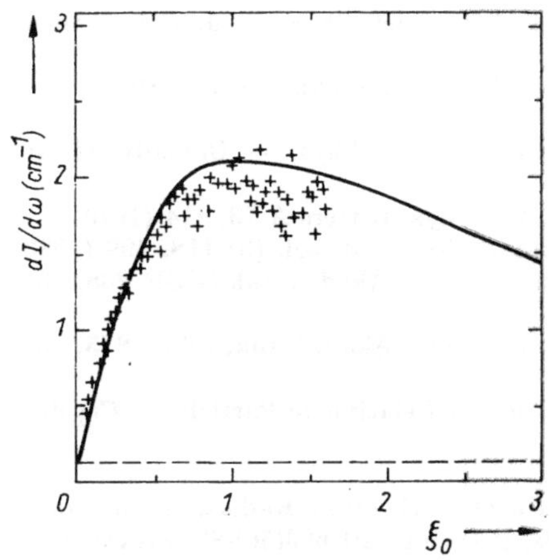

Fig. 6

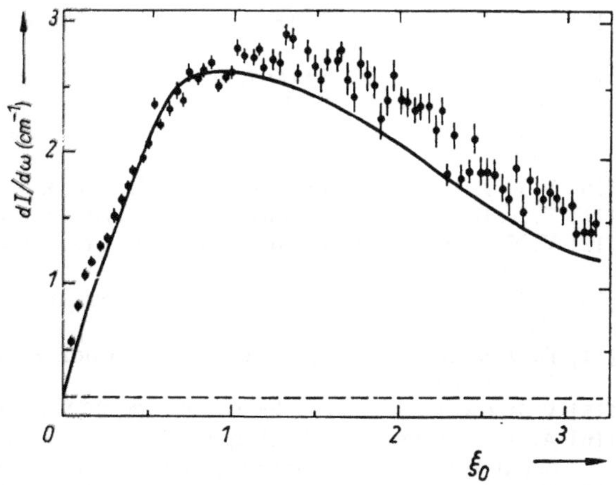

Fig. 7

Fig. 6. Spectral intensity of radiation, per unit thickness, from electrons with $\varepsilon = 10$ GeV upon motion along the (110) plane of a Si crystal of $L_1 = 0.135$ mm; the experimental data have been taken from [17]. The value of $\xi_0 = 1$ corresponds to $\omega = 110$ MeV. The dashed line shows the bremsstrahlung. The solid curve is the summed spectrum

Fig. 7. The same as in Fig. 6 for $\varepsilon = 7$ GeV and crystal thickness $L_2 = 0.1$ mm. The experimental data have been taken from [18]. The value of $\xi_0 = 1$ corresponds to $\omega = 65$ MeV

In [17] and [18] the authors have analysed the radiation of channeling electrons in the (110) plane of a silicon crystal at the energy $\varepsilon = 10$ GeV and $\varepsilon = 7$ GeV, respectively. The crystal thickness was $L_1 = 0.135$ mm [17] and $L_2 = 0.1$ mm [18]; $\vartheta_c^{1(2)} = 71(95)$ μrad, $\varrho_0^{1(2)} \approx 1.9(1.33)$; $\xi_0 = 1$ corresponds to $\omega^{1(2)} = 110(65)$ MeV. In [17] the events with $\vartheta_0 \leqq 50$ μrad have been detected and the accuracy of measuring the angles is also estimated as 50 μrad. Therefore, we put $\Delta_0 = 100$ μrad, bearing in mind the notes of [18] about the experimental set-up in [17], the quantity Δ_0 in [18] was $\Delta_0 \approx 60$ μrad. Fig. 6 and 7 give the calculational results and experimental data for $\varepsilon = 10$ GeV [17] and $\varepsilon = 7$ GeV [18] which we have represented in absolute units.

Thus, we can see that the developed theory of radiation from electrons at planar channeling, not containing fitting parameters, describes quite well the experiments for various media, planes, and thicknesses in a broad interval of energies. It follows from this, in particular, that the used simple dechanneling model is quite adequate.

References

[1] V. N. Baier, V. M. Katkov, and V. M. Strakhovenko, Radiation of Relativistic Particles at Planar Channeling, Preprint INP 80-03, Novosibirsk 1980 (in Russian).
[2] M. A. Kumakhov and Ch. G. Trikalinos, phys. stat. sol. (b) **99**, 449 (1980).
[3] V. A. Bazylev, V. I. Glebov, and N. K. Zhevago, Soviet Phys. — J. exper. theor. Phys. **51**, 31 (1980).
[4] N. F. Shulga et al., Zh. eksper. teor. Fiz. **82**, 50 (1982).
[5] V. N. Baier, V. M. Katkov, and V. M. Strakhovenko, Dokl. Akad. Nauk SSSR **266**, 605 (1982); Preprint INP 81-139, Novosibirsk 1981.
[6] V. A. Bazylev et al., Zh. eksper. teor. Fiz. **80**, 608 (1981).

[7] V. N. BAIER, V. M. KATKOV, and V. M. STRAKHOVENKO, Soviet Phys. — J. exper. theor. Phys. 53, 688 (1981).
[8] V. N. BAIER, V. M. KATKOV, A. I. MILSTEIN, and V. M. STRAKHOVENKO, Soviet Phys. — J. exper. theor. Phys. 42, 400 (1975).
[9] V. N. BAIER, V. M. KATKOV, and V. M. STRAKHOVENKO, Soviet Phys. — Doklady 24, 469 (1979).
[10] V. N. BAIER, V. M. KATKOV, and V. M. STRAKHOVENKO, Phys. Letters A 73, 414 (1979).
[11] V. N. BAIER, V. M. KATKOV, and V. M. STRAKHOVENKO, phys. stat. sol. (b) 118, 499 (1983).
[12] V. N. BAIER, V. M. KATKOV, and V. M. STRAKHOVENKO, Dokl. Akad. Nauk SSSR 268, 1365 (1983).
[13] V. N. BAIER, V. M. KATKOV, and V. M. STRAKHOVENKO, Dokl. Akad. Nauk, SSSR 275, 1369 (1984).
[14] R. O. AVAKYAN et al., 2-nd All Union Conf. Radiation of Relativistic Particles in Crystals, Cheget 1982.
[15] V. B. GANENKO et al., Ukr. fiz. Zh. 28, 302 (1983).
[16] A. O. AGANIANTS et al., Angular Dependence of 4.3 GeV Electron Radiation Spectra at Small Angles of Entry into Diamond Single Crystals, Preprint YPI-666(56)-83, Erevan 1983.
[17] M. ATKINSON et al., Radiation from Planar Channeled 5 to 55 GeV/c Positrons and Electrons, CERN-EP/82-03 (1982).
[18] J. BAK et al., Measurement of Channeling Radiation and Its Polarization and X-ray Excitation together with Deviation from Landau Distribution, CERN/PSCC/82-94 (1982).
[19] V. N. BAIER, V. M. KATKOV, and V. M. STRAKHOVENKO, Nondipole Radiation of Electrons at Planar Channeling, Preprint INP 86—116, Novosibirsk 1984.

(Received October 26, 1984)

phys. stat. sol. (b) **133**, 223 (1986)

Subject classification: 71.45 and 78.70; 53

Department of Physics, University of Ioannina[1]) (a)
and The Institute for Ultrafast Spectroscopy and Lasers,
Department of Physics and Department of Electrical Engineering,
The City College of New York[2]) (b)

Plasmon Bands in X-Ray Spectra of Polycrystalline Beryllium

By

D. K. PAPADEMETRIOU (a, b), D. KATSANOS (a), and A. G. DOUKAS (b)

The plasmon bands are observed in the X-ray spectra of polycrystalline Be, using CuKα_1, α_2, radiation for $k > k_c$. The spectra show four peaks at the long wavelength region of Raman bands, assigned to the $3sp_xp_y$-$3sp_z$ splitting of the 3ps hybridized band.

Nous avons observe des bandes plasmons dans le spectre de rayons-X de Be polycrystaline excite par l'emission de Cu/Kα_1, α_2 pour des valeurs de $k > k_c$. Le spectre a quatre maxima dans la region des longues longueurs d'ondes des bandes Raman. Nous avons assigné ces maxima aux transitions $3sp_xp_y$-$3sp_z$ de la bande hybride 3ps.

1. Introduction

A number of interesting phenomena arise in X-ray inelastic scattering by free electrons in a solid, with decreasing scattering angle φ [1]. These phenomena involve excitation of longitudinal plasma oscillations. Plasma oscillations arise from Coulomb interactions among the electrons. The problem of small-angle inelastic scattering was treated in great detail in the studies of some groups [2 to 5]. The plasmon scattering is a combination process in which the frequency of the scattered quanta is smaller than that of the incident quanta by the frequency of the plasma oscillations. The equation

$$\varepsilon(\boldsymbol{k}, \omega) = 0 \,, \tag{1}$$

where $\varepsilon(\boldsymbol{k}, \omega)$ is the longitudinal dielectric constant, determines the spectrum $\omega(\boldsymbol{k})$ of the longitudinal plasma waves.

According to Nozieres and Pines [3] the differential inelastic-scattering cross section $\sigma(\omega, \vartheta)$ is determined by the Fourier image of the density–density correlation function for the electrons and is expressed in terms of the longitudinal dielectric constant $\varepsilon(\boldsymbol{k}, \omega)$,

$$\sigma(\omega, \vartheta) = \sigma_T S(\boldsymbol{k}, \omega) = \sigma_T \frac{hk^2}{4\pi^2 e^2 n_0} \operatorname{Im} \varepsilon^{-1}(\boldsymbol{k}, \omega) \,, \tag{2}$$

where σ_T is the Thomson scattering cross section, $S(\boldsymbol{k}, \omega)$ the structure factor, $\boldsymbol{k}$ the momentum inparted to the medium, and n_0 the density of the valence electrons.

Experimental study of the plasmon scattering cross section is useful for determining the form of the dielectric constant and also for determining the validity of any

[1]) Ioannina, Greece.

[2]) 138th str. and Convent. Ave., New York, N.Y. 10031, USA.

particular model of the behavior of the electrons in a solid from which ε is being calculated theoretically.

In a recent experiment on polycrystalline Be [6] the X-ray spectra of inelastically scattered CuK_β radiation, showed the existence of plasmon bands approximately 21 eV from the position of the primary K_β band for scattering angles $\varphi > \varphi_c$ where $\varphi_c = 16.0°$. Two small bands (b_1 and b_2) with an energy separation of 6 to 7 eV at the long-wavelength region of Raman bands, were observed. These bands were characterized as bonding-antibonding bands. Under irradiation of CuK_{α_1,α_2} on the same sample Be, we observed [7] two small b_1 and b_2 bands with energy shift ΔE 155 and 161 eV, respectively, from he position of primary K_{α_1} line. These bands were assigned to $3sp_xp_y$–$3sp_z$ splitting of 3ps hybridized band. The assignment is supported by an ab initio unrestricted Hartree-Fock cluster calculation [8].

The present work is a continuation of our previous studies on the inelastic scattering of Cu X-ray radiation by light elements [6, 7]. Our plan is twofold-to verify the existence of plasmon scattering for momentum transfer $k > k_c$ and to verify the existence of bonding-antibonding bands at the long-wavelength side of Raman bands.

2. Methods and Materials

The experimental arrangement and procedure have been previously described in detail [7]. The X-ray tube was operating at 55 kV, 190 mA. The used scatterer was polycrystalline Be, 4 mm in thickness, purity 99.99% and was purchased by Metals Research Division Co. The spectrometer resolution was $E/\Delta E \approx 1050$. A Ge single crystal (220) orientation with $2d = 0.4$ nm was used as an energy analyser because of higher intensity. Small angles were used in order to avoid the overlapping of Compton bands, due to K_{α_1} and K_{α_2} lines, with the observed bands b_1, b_2, b_1', and b_2'. The spectra were obtained using a scintillation counter by step-scanning. An automatic step-scanner of 0.01° was used and the time duration of each step was set to 400 s. The spectra were scanned five times in order to improve the statistical error which is estimated to be 1.5%.

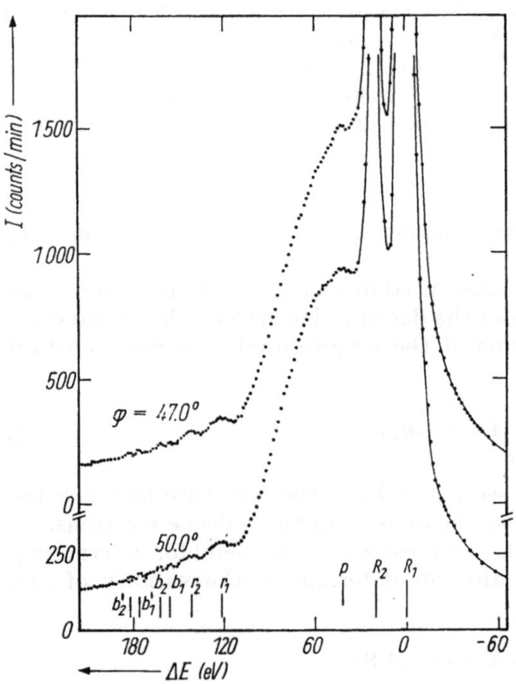

Fig. 1. Spectral distributions of the scattered X-ray radiation versus the energy loss from the primary band K_{α_1} at scattering angles φ of 47.0° and 50.0°. The arrows R_1, R_2, r_1, r_2, b_1, b_2, b_1', and b_2' correspond to the peak positions of K_{α_1}, K_{α_2} Rayleigh lines, Raman, and bonding — antibonding bands, respectively. I is the intensity in counts/min, φ the scattering angle, and the points are points of the experimental curves

3. Results

The obtained X-ray spectra at two different scattering angles of $\varphi = 47.0°$ and 50.0° are displayed in Fig. 1. The

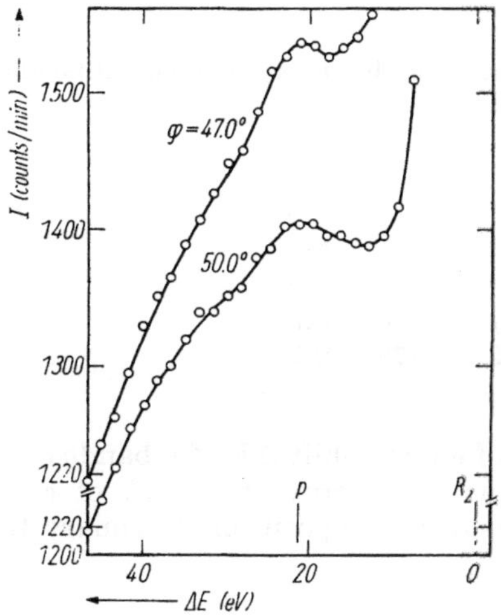

Fig. 2. The spectral region of interest where the plasmon p bands are observed at $\varphi = 47.0°$ and $50.0°$

marks with the letters R_1, R_2 correspond to the peak positions of K_{α_1} and K_{α_2} primary lines, respectively. The letter p corresponds to the plasmon band due to K_{α_2}. The b_1 and b_2 to the bonding-antibonding bands due to K_{α_1}. The b_1' and b_2' are the two new bonding-antibonding bands due to K_{α_2}. Fig. 2 shows the spectral region of interest where the plasmon bands are observed. Fig. 3 a and b show the region where the b_1, b_2, b_1', and b_2' bands are observed. Tables 1 and 2

Table 1

The values of energy shift ΔE of p bands from the position of primary K_{α_2} band

scattering angle φ	energy shift ΔE (eV)
47.0°	21.0
50.0°	21.0

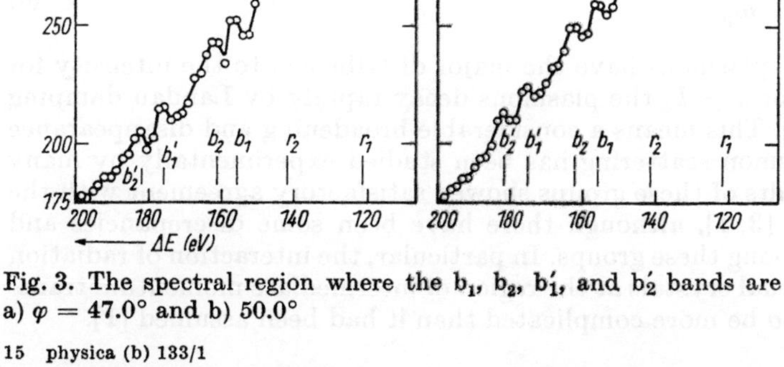

Fig. 3. The spectral region where the b_1, b_2, b_1', and b_2' bands are observed at scattering angle a) $\varphi = 47.0°$ and b) $50.0°$

Table 2
The values of energy shift ΔE of r_1, r_2 Raman and b_1, b_2, b_1' and b_2' bonding-antibonding bands from the position of primary K_{α_1} line

scattering angle φ	energy shift ΔE (eV)					
	r_1	r_2	b_1	b_2	b_1'	b_2'
47.0°	121	142	156	161	176	181
50.0°	122	142	156	162	175	181

summarize our results. Table 1 contains the values of energy shift ΔE of p bands from the position of primary K_{α_2} band. Table 2 contains the energy shifts ΔE of r_1, r_2 Raman and b_1, b_2, b_1', b_2' bonding-antibonding bands from the position of primary K_{α_1} band.

4. Discussion

The spectral distributions of the scattered X-ray radiation versus the energy loss ΔE from the primary K_{α_1} line, at scattering angles of $\varphi = 47.0°$ and $50.0°$ are shown in Fig. 1. In these spectra we see that the plasmon band p is about at the same position where the theory provides for small scattering angles $\varphi > \varphi_c = 16.0°$. Fig. 1 also shows the position of b_1, b_2, b_1', b_2' bonding-antibonding bands due to the primary K_{α_1} and K_{α_2} lines. Fig. 2 shows the regions of the spectra where the plasmon p bands are observed. Fig. 3 a and b show the spectral region of interest where the b_1, b_2, b_1', b_2' bonding-antibonding bands are observed.

The observed band p in both spectra is assigned to the plasmon band produced by K_{α_2}. This band is due to loss energy of the primary radiation creating collective oscillations of the electron gas of the scatterer. The shift ΔE of p band is (21 ± 1) eV from the position of K_{α_2}. The plasmon band produced by K_{α_1} is not observed because this one is overlapping with the intense K_{α_2} band. According to the random phase approximation (RPA) the plasmons do not decay into excitations when the momentum transfer k is smaller than the critical value $k_c = \omega_p/v_F = (1.71\sqrt{r_s})$, where $\omega_p = \sqrt{4\pi n_0 e^2/m}$ the frequency of the plasmon oscillations, $v_F = (4.22/r_s) \times 10^8$ cm/s is the Fermi velocity, and n_0 the density of the valence electrons. In our experiment the values of k are 32 nm^{-1} and 34 nm^{-1} for the two scattering angles of $\varphi = 47.0°$ and $50.0°$, respectively. At small momenta in a degenerate electron gas the dispersion law of the plasmons is [9]

$$\omega(\boldsymbol{k}) = \omega_p + \frac{3}{10}\frac{(\omega_F k)^2}{\omega_p}. \tag{3}$$

According to the RPA the plasmons have the major distribution to the intensity for small scattering angles. For $k > k_c$ the plasmons decay rapidly by Landau damping in electrons and holes [10]. This means a considerable broadening and disappearance of the plasmon band. Plasmon scattering has been studied experimentally by many groups [11 to 22]. The results of these groups showed satisfactory agreement with the conclusions of the theory [3, 4], although there have been some discrepancies and ambiguity of the results among these groups. In particular, the interaction of radiation with the electron gas in actual crystals at the region of intermediate momentum transfer ($k \sim 1 - 2k_c$) proved to be more complicated than it had been assumed [1].

The dispersion law of the plasmons in beryllium for different scattering angles has been found to be linear with respect to k^2 [15]. The linearity of the dispersion law with respect to k^2 has been also observed in other studies [11, 16, 18]. Marinos et al. [13] observed no plasmon dispersion in Be for $k \geqq 0.8k_c$. Eisenberger et al. observed in lithium [20] and graphite [21] a shift of the peak that was linear in k^2 at $k \gtrsim 1.5k_c$ and $k \gtrsim 1.3k_c$, respectively.

Our results suggest that the dispersion of plasmons ceases when we exceed the critical value k_c of momentum transfer to the system corresponding at critical scattering angle $\varphi_c = 16.0°$.

The problem of observing plasmon lines at $\varphi > \varphi_c$ has not yet been solved. Since the conclusion of decay of plasmons at $k > k_c$ was made within the framework of the random phase approximation (RPA) this theory needs further refinement.

The splitting of bonding-antibonding states in Be has been suggested by Inoue and Yamashita [23]. In our case, we used the double primary line K_{α_1,α_2} the observed peaks are four the b_1, b_2 peaks because of K_{α_1} and the b_1', b_2' because of K_{α_2}. Each pair of these bands corresponds to the transitions of K-band electrons to the 3ps hybridized band. The agreement between theory and experiment confirms the existence of observed bands at the determined positions in the spectra.

5. Conclusions

In this work using the CuK_{α_1,α_2} X-ray radiation and as scatterer polycrystalline Be, we have verified,

(i) the existence of plasmon bands for $k > k_c$;

(ii) the existence of two b_1 and b_2 bonding-antibonding bands at the long-wavelength region of Raman bands; and

(iii) the existence two new b_1' and b_2' bands at the long-wavelength region of b_1 and b_2 bands. The last two bands 20 eV about from b_1 and b_2 were predicted in our previous work [7].

Acknowledgements

We thank Prof. Alfano for many helpful discussions. D.K.P. also wishes to thank Prof. Alfano for his extended hospitality and encouragement during his sabbatical at the Institute for Ultrafast Spectroscopy and Lasers. This work is supported by the Department of Physics, University of Ioannina, Greece.

References

[1] V. A. Bushuev and R. N. Kuzmin, Soviet Phys. — Uspekhi **20**, 415 (1980).
[2] A. H. Compton, Bull. mat. Res. Council **4**, No. 20 ,19 (1922); Phys. Rev. **21**, 207; 483 (1923).
[3] P. Nozieres and D. Pines, Phys. Rev. **113**, 1254 (1959).
[4] Y. Ohmura and N. Matsudaira, J. Phys. Soc. Japan **19**, 1355 (1964).
[5] V. M. Agranovich and V. L. Ginzburg, Zh. eksp. teor. Fiz. **40**, 913 (1961).
[6] D. Papademetriou and D. Miliotis, Solid State Commun. **48**, 799 (1983).
[7] D. Papademetriou and D. Miliotis, J. Phys. Soc. Japan **51**, 2966 (1982).
[8] A. D. Zdetsis and D. Miliotis, Solid State Commun. **42**, 277 (1982).
[9] D. Pines, Elementary Excitations in Solids, W. A. Benjamin, Inc., New York 1963.

[10] L. D. LANDAU, Zh. eksp. teor. Fiz. **16**, 574 (1946).
[11] G. PRIFTIS, Phys. Rev. B **2**, 54 (1970).
[12] G. D. PRIFTIS, Phys. Rev. B **8**, 3134 (1973).
[13] D. MARINOS and D. MILIOTIS, phys. stat. sol. (b) **68**, K133 (1975).
[14] C. KOUMELIS, D. LEVENTOURI, and K. ALEXOPOULOS, phys. stat. sol. **46**, K89 (1971).
[15] T. SUZUKI and A. TANOKURA, J. Phys. Soc. Japan **29**, 972 (1970).
[16] A. TANOKURA, N. HIROTA, and T. SUZUKI, J. Phys. Soc. Japan **27**, 515 (1969); **28**, 1382 (1970).
[17] N. G. ALEXANDROPOULOS, J. Phys. Soc. Japan **31**, 1790 (1971).
[18] P. EISENBERGER, P. M. PLATZMAN, and K. C. PANDY, Phys. Rev. Letters **31**, 311 (1973).
[19] P. M. PLATZMAN and P. EISENBERGER, Phys. Rev. Letters **33**, 152 (1974).
[20] P. EISENBERGER, P. M. PLATZMAN, and P. SCHMIDT, Phys. Rev. Letters **34**, 18 (1975).
[21] P. EISENBERGER and P. M. PLATZMAN, Phys. Rev. B **13**, 934 (1976).
[22] W. SCHÜLKE and W. LAUTNER, phys. stat. sol. (b) **66**, 211 (1974).
[23] S. T. INOUE and J. YAMASHITA, J. Phys. Soc. Japan **35**, 677 (1973).

(Received August 16, 1985)

phys. stat. sol. (b) **133**, 229 (1986)

Subject classification: 71.38 and 73.40; S7.12; S7.15

Department of Physics, University of Antwerp[1])

Screening of the Electron–Phonon Interaction in GaAs Heterostructures

By

WU XIAOGUANG, F. M. PEETERS, and J. T. DEVREESE[2])

The effect of the screening of the electron–phonon interaction on the polaron binding energy (E) and the polaron mass (m^*) is investigated within second order perturbation theory and for different approximations to the two-dimensional electron gas response (random-phase approximation, Hartree-Fock and Thomas-Fermi approximation). For the electron–phonon contribution to the polaron ground state energy a dynamical screening is applied which takes into account the dielectric response at all frequencies. The influence of the finite width of the electron layer on E and m^* is investigated. The results are applied to GaAs–$Al_xGa_{1-x}As$ heterostructures and it is found that the mass renormalization for $n_e = 10^{11}$ cm^{-2} is 0.4% which is a factor of eight smaller than the effective mass of the unscreened ideal 2D polaron.

Mit Störungsrechnung zweiter Ordnung und verschiedenen Näherungen des zweidimensionalen Elektronengas-Response (Random-Phase-Näherung, Hartree-Fock- und Thomas-Fermi-Näherung) wird der Einfluß der Abschirmung der Elektron–Phonon-Wechselwirkung auf die Polaronen-Bindungsenergien (E) und Polaronenmasse (m^*) untersucht. Für den Elektron–Phonon-Beitrag zur Polaronen-Grundzustandsenergie wird eine dynamische Abschirmung angewendet, die die dielektrische Response bei allen Frequenzen berücksichtigt. Der Einfluß der endlichen Werte der Elektronenschicht auf E und m^* wird untersucht. Die Ergebnisse werden auf GaAs–$Al_xGa_{1-x}As$-Heterostrukturen angewendet und es wird gefunden, daß die Massenrenormierung für $n_e = 10^{11}$ cm^{-2} 0,4% beträgt, was um den Faktor acht kleiner ist als die effektive Masse des unabgeschirmten idealen 2D-Polarons.

1. Introduction

In quasi-two-dimensional electron systems formed at the interface of two materials [1] (e.g. Si-space charge layers, GaAs–$Al_xGa_{1-x}As$ structure, InSb inversion layer, . . .), the electron density is such that many-particle properties like screening, Fermi-Dirac statistics, . . . of the electron system are expected to be important. These many-particle characteristics of the electron system will modify the electron–phonon interaction and in particular the polaron binding energy and the polaron mass renormalization. In several treatments of the polaron effects [2 to 5] in these quasi-two-dimensional electron systems the following approximations are made: (i) a single-particle approach is used and (ii) an infinitely thin two-dimensional (2D) electron layer at zero temperature is considered. One then finds that the polaron effective mass (m^*) becomes $m^*/m_b = 1 + (\pi/8)\,\alpha$ with m_b the electron band mass and α the electron–phonon coupling constant. For GaAs heterostructures this leads to a mass renormalization

[1]) Universiteitsplein 1, B-2610 Wilrijk-Antwerpen, Belgium.

[2]) Also at: University of Antwerp (RUCA, Belgium) and the Eindhoven University of Technology, Eindhoven, The Netherlands.

correction of almost 3% which compares with a correction of 1% in the case of an electron in bulk GaAs.

Cyclotron resonance experiments on GaAs heterostructures, on the other hand, indicate [6 to 7] that the polaron mass renormalization at small magnetic fields is considerably smaller than 1%. This means that a decrease of polaron effects was observed in this quasi-2D electron system in comparison with the 3D system. This is contrary to the 2D single polaron picture which predicted an enhancement of the polaron mass correction with a factor 2.66 compared to the 3D result. The inclusion of the finite width of the electron layer only, cannot explain the experimental results because a polaron correction between that for the ideal 2D and the ideal 3D result is then expected.

Recently Das Sarma [2, 5] investigated the effect of screening of the electron–phonon interaction and found that it can reduce the polaron effects sufficiently to explain the experimental findings. In [2] Das Sarma used Thomas-Fermi screening in the electron–phonon interaction. Such an approximation has the remarkable effect that, in the ideal 2D case, the electron–phonon contribution to the polaron binding energy E and to the effective mass m^* is independent of the electron density n_e of the electron layer. This unexpected behaviour is a consequence of the fact that in the ideal 2D-limit the Thomas-Fermi screening length k_{TF} ($k_{TF} = 2m_b e^2/\hbar\varepsilon_\infty$, with ε_∞ the high-frequency dielectric constant of the material) is independent of the electron density n_e.

In the present paper we will question the Thomas-Fermi approach to the screening of the electron–phonon interaction on two levels: (i) we notice that for static screening the Thomas-Fermi approach is valid (see e.g. [8] for the 3D case) for wave vectors $k \ll k_F$ (k_F is the Fermi wave vector). For the electron–optical phonon interaction, wave vectors such that $k \sim k_{LO}$ are relevant ($k_{LO} = (2m_b\omega_{LO}/\hbar)^{1/2}$ where ω_{LO} is the optical-phonon frequency). Consequently, the Thomas-Fermi static screening of the electron–phonon interaction is expected to be applicable if $k_{LO} \ll k_F$. For GaAs heterostructures this implies the condition $n_e > 10^{12}\ \text{cm}^{-2}$ on the electron density. This is a much higher density than the electron densities in the experiments of [6] and [7] (typically $n_e = 5 \times 10^{11}\ \text{cm}^{-2}$). (ii) For the polaron binding energy E and the polaron mass m^* Das Sarma used the one-particle expression in which he replaced the electron–phonon interaction by a Thomas-Fermi [2] and a RPA (random phase approximation) screened electron–phonon interaction [5].

In the present paper we re-investigate the influence of screening on the polaron binding energy and the polaron mass renormalization within different approximations for the screening while the electron–phonon interaction is treated within second-order perturbation theory. In Section 2 the unitary transformation method of [9] is applied to calculate the contribution of the electron–phonon interaction to the many-polaron ground state energy E and to the electron–electron interaction. In this approximation dynamical effects of the screening are incorporated. For the ideal 2D case the energy E reduces to the one-polaron result in the limit of low electron density. In the opposite limit $n_e \to \infty$ we obtain $E \to 0$.

In Section 3 the electron–phonon contribution to the electron self-energy is calculated in order to obtain the polaron mass renormalization. The definition of the electron effective mass is taken from Falicov and Garcia [10] who defined a classical cyclotron electron mass for a many-particle system. In Section 4 numerical results for E and m^* are presented and discussed for the different approximations: (i) for the ideal 2D limit and (ii) for the realistic case of a finite width of the electron layer. Parameters describing GaAs–$Al_xGa_{1-x}As$ heterostructures with $10^{10}\ \text{cm}^{-2} < n_e < 10^{12}\ \text{cm}^{-2}$ are considered. Our conclusions are presented in Section 5 and are qualitatively in accordance with those of Das Sarma's paper (reduction of E and m^* due to the screening of the electron–phonon interaction) but are quantitatively different.

2. The Electron-Phonon Correction to the Ground State Energy

The Hamiltonian of the system is given by

$$H = \sum_{j=1}^{N_s} \frac{p_j^2}{2m_b} + \sum_{k} \hbar\omega_{LO} a_k^+ a_k + \sum_{i=j}^{N_s} V(r_i - r_j) +$$
$$+ \sum_{j=1}^{N_s} \sum_{k} (V_k a_k \, e^{ikr_j} + V_k^* a_k^+ \, e^{-ikr_j}) , \quad (1)$$

where $p(r)$ is the momentum (position) operator of an electron (N_s is the total number of electrons). $a_k^+(a_k)$ is the creation (annihilation) operator of a phonon with wave vector k and energy $\hbar\omega_{LO}$. In (1)

$$V_k = \hbar\omega_{LO} \left(\frac{4\pi\alpha}{Vk^2}\right)^{1/2} \left(\frac{\hbar}{2m_b\omega_{LO}}\right) \langle\psi_0| \, e^{ik_z z} \, |\psi_0\rangle \, .$$

Here $\psi_0(z) = (b^3/2)^{1/2} \, z \, e^{-bz/2}$ is the variational wave function of the electron in the z-direction which is normal to the electron layer [1]. We will not discuss the variational wave function here. We just point out that the parameter b is a known function of the electron density, of the depletion charge density of the system and of the effective mass of the electron, namely $b = (48\pi m_b Ne^2/\hbar^2\varepsilon_\infty)^{1/3}$, where $N = n_d + (11/32) \, n_e$ with n_d and n_e the depletion and the carrier densities, respectively [1]. In the following, we will not only consider the quasi-two-dimensional system but also the exact 2D system which can be easily obtained by taking the limit $b \to \infty$.

The quasi-two-dimensional characteristics of the system enter the problem through the modifications of the bare electron–electron interaction and electron–phonon interaction [11, 12]. $V(r - r')$ represents the Coulomb interaction between two electrons in the quasi-two-dimensional electron layer. It is obtained by averaging over the electron wave function in the z-direction. The Fourier transform of $V(r)$ is $V(k_\parallel) = 2\pi e^2/k_\parallel\varepsilon_\infty \times \times f(k_\parallel, b)$, ($k_\parallel$ is the component of k parallel to the 2D electron layer) with $f(k, b) = = (8b^3 + 9kb^2 + 3k^2b)/8(k + b)^3$. This result can be understood [11] as the replacement of the normal dielectric constant ε_∞ by a wave vector dependent dielectric function $\varepsilon_\infty(k) = \varepsilon_\infty/f(k, b)$. In the exact 2D limit (i.e. $b \to \infty$) we have $f(k, \infty) = 1$.

For simplicity, we take only the first sub-band into account. This simplification is justified by a study of Rahman and Mills [13] (see also [12]) who showed that the effects of higher sub-bands are not so important in the study of the electron–phonon interaction in the inversion layer. The reason is that the intersub-band transition rate induced by the electron–phonon interaction is quite small. In the present paper, the limitation to the first sub-band is introduced for convenience because it will severely reduce our numerical work.

To study the ground state properties of the polaron gas, we employ two approaches: a Lee-Low-Pines unitary transformation approximation (LLP) as introduced in [9] for the 3D polaron gas and a first-order expansion of the electron–phonon interaction correction to the electron self-energy as developed in [2 and 5].

2.1 Dynamical screening of the electron–phonon interaction

The unitary transformation of Lee-Low-Pines has been generalized by Lemmens et al. [9] in a study of the ground state energy of a 3D polaron gas in degenerate semiconductors. The approximation is perturbative (in electron–phonon coupling) in nature. Here we will only give a brief outline of the calculation. For details we refer to [9].

First, an unitary transformation U is performed on the Hamiltonian. The unitary transformation is taken to be

$$U = \exp\left(\sum_{j=1}^{N_j} s(\boldsymbol{r}_j)\right),$$

where

$$s(\boldsymbol{r}) = \sum_{\boldsymbol{k}} (f_{\boldsymbol{k}} a_{\boldsymbol{k}} \, \mathrm{e}^{i\boldsymbol{k}\boldsymbol{r}} - f_{\boldsymbol{k}}^* a_{\boldsymbol{k}}^+ \, \mathrm{e}^{-i\boldsymbol{k}\boldsymbol{r}})$$

which results in a displacement of the phonon coordinates. Consequently the expectation value of the Hamiltonian is calculated over the state with zero real phonons and where the electron gas is in a state described by a wave function ψ. The $f_{\boldsymbol{k}}$ are determined by minimizing the ground state energy which leads to

$$f_{\boldsymbol{k}} = \frac{V_{\boldsymbol{k}}}{\hbar\omega_{\mathrm{LO}} + (\hbar^2/2m_\mathrm{b})\, k_{||}^2/S(k_{||})}.$$

After some algebraic work we obtain for the electron–phonon interaction correction to the ground state energy the following result:

$$E = -\sum_{\boldsymbol{k}} \frac{|V_{\boldsymbol{k}}|^2 \, S(k_{||})}{\hbar\omega_{\mathrm{LO}} + (\hbar^2/2m_\mathrm{b})\, k^2/S(k_{||})}$$

$$= -\alpha\hbar\omega_{\mathrm{LO}} k_{\mathrm{LO}}^{-1} \int_0^\infty \mathrm{d}k \, \frac{S^2(k)}{S(k) + k^2 k_{\mathrm{LO}}^{-2}} \, f(k, b)\,, \tag{2}$$

where k_{LO} is a wave vector scale introduced by the LO-phonons. The $S(k)$ is the static structure factor for the 2D electron system in the absence of electron–phonon interaction. This function has been calculated before in different approximations. In the present work we will consider $S(k)$ within the random phase approximation [14] (RPA) and within the Hartree-Fock approximation (HF) [15]. Note that $S(k)$ is defined by

$$S(k) = \int_0^\infty \mathrm{d}\omega \; S(k, \omega)$$

with

$$S(k, \omega) = \frac{1}{\pi V(k)} \, \mathrm{Im} \, \frac{-1}{\varepsilon(k, \omega)}$$

the dynamical structure factor and $\varepsilon(k, \omega)$ the dielectric function. Note that $S(k)$ is different from $S(k, 0)$, the zero-frequency (or static) limit of the dynamical structure factor. $S(k)$ contains all frequencies of $\varepsilon(k, \omega)$ explicitly. This is the underlying reason why we call this approach "dynamical screening". We point out that the effect of the finite width of the electron layer has been taken into account in the calculation of the dielectric function, since $\varepsilon(k, \omega) = 1 - V(k)\, \chi(k, \omega)$ ($\chi(k, \omega)$ is the polarization of the electron gas) where $V(k)$ contains the effect of the finite width by $V(k) = (2\pi e^2/k\varepsilon_\infty) \times$ $\times f(k, b)$.

It is interesting to point out that after the unitary transformation is performed, there results an effective electron–electron interaction [9],

$$\sum_{i<}^{N} V_{\mathrm{eff}}(\boldsymbol{r}_i - \boldsymbol{r}_j) \tag{3}$$

in the Hamiltonian, where now

$$V_{\text{eff}}(\boldsymbol{r} - \boldsymbol{r}') = V(\boldsymbol{r} - \boldsymbol{r}') - 2 \sum_{\boldsymbol{k}} (V_{\boldsymbol{k}} f_{\boldsymbol{k}}^* + V_{\boldsymbol{k}}^* f_{\boldsymbol{k}} - \hbar\omega_{\text{LO}} f_{\boldsymbol{k}} f_{\boldsymbol{k}}^*) \, e^{i\boldsymbol{k}(\boldsymbol{r}-\boldsymbol{r}')}$$
$$= V(\boldsymbol{r} - \boldsymbol{r}') - 2 \sum_{\boldsymbol{k}} \frac{|V_{\boldsymbol{k}}{}^2|}{[\hbar\omega_{\text{LO}} + (\hbar^2/2m_{\text{a}}) \, k_{||}^2/S(k_{||})]^2} \times$$
$$\times \left[\hbar\omega_{\text{LO}} + \frac{\hbar}{m_{\text{b}}} \, k_{||}^2/S(k_{||})\right] e^{i\boldsymbol{k}(\boldsymbol{r}-\boldsymbol{r}')} . \tag{4}$$

The last term on the right-hand side of (4) can be understood as a modification of the electron–electron interaction due to the electron–phonon interaction. In Section 4, we will discuss this electron–phonon interaction correction to the direct Coulomb electron–electron interaction.

Considering the ideal 2D case, it can be verified that, as the electron density in the 2D electron layer goes to infinity, the static structure factors in the RPA and in the HF approximation coincide, and give an electron–phonon interaction correction to the energy E which tends to zero (i.e. $|E| \sim 1/(n_{\text{e}})^{1/2}$). In the opposite limit of low density, we obtain the familiar one-polaron correction, i.e. E reduces to $-(\pi/2) \, \alpha\hbar\omega_{\text{LO}}$ for the ideal 2D system (i.e. $b = \infty$). The numerical results will be presented in Section 4.

2.2 Static screening of the electron–phonon interaction

In a recent work, Das Sarma [2, 5] considered the influence of the electron–phonon screening and the finite width of the 2D electron layer on the polaron effects in semiconductor heterostructures by using a simple perturbation scheme. His approach in fact implies that the ground state energy and the polaron mass are calculated within a one-polaron picture but where the screening is taken into account by replacing the electron–phonon interaction coefficients $|V_{\boldsymbol{k}}|^2$ by $|V_{\boldsymbol{k}}|^2 \varepsilon^{-2}(k, 0)$ with $\varepsilon(k, 0)$ the static dielectric function. This results in

$$E = -\alpha\hbar\omega_{\text{LO}} k_{\text{LO}}^{-1} \int\limits_0^\infty \mathrm{d}k \, \frac{1}{\varepsilon^2(k, 0)} \, \frac{f(k, b)}{(1 + k^2 k_{\text{LO}}^{-2})} \tag{5}$$

and

$$m^* = m_{\text{b}} \left[1 + 2\alpha k_{\text{LO}}^{-1} \int\limits_0^\infty \mathrm{d}k \, \frac{k^2 k_{\text{LO}}^{-2}}{(1 + k^2 k_{\text{LO}}^{-2})^3} \, \frac{f(k, b)}{\varepsilon^2(k, 0)}\right]. \tag{6}$$

Ultimately, in [2] Das Sarma considered the dielectric function in the Thomas-Fermi (TF) approximation, i.e. $\varepsilon(k, 0) = (1 + k_{\text{TF}}/k)^{-1}$ where k_{TF} is the Thomas-Fermi wave vector.

In order to make a comparison with our calculation we also take $\varepsilon(k, 0) = \varepsilon_{\text{RPA}}(k, 0)$ in (5) and (6), and we will call it the *static RPA approximation* which was studied by Das Sarma and Mason [5] in a very recent publication. It must be noticed that only for $k < 2k_{\text{F}}$ (k_{F} is the Fermi wave vector) $\varepsilon_{\text{RPA}}(k, 0)$ equals the Thomas-Fermi result for the dielectric function.

Focusing on the exact 2D limit case, it is simple to verify that with an infinite electron density, the results of [2] will be the same as the results of the static RPA approximation of [5]. However, in the low electron density limit, the results of [2] fail to give the usual single-polaron energy and effective mass correction, while the static RPA approximation is able to give the correct result.

The numerical results for (5) and (6) in the TF approximation and in the static RPA will be presented in Section 4.

3. The Electron–Phonon Correction to the Electron Effective Mass

At present we have not developed an equivalent unitary transformation approach to calculate the electron effective mass as in Section 2.1. Therefore, we have to rely on an approach based upon a static screening treatment of the electron–phonon interaction. First we will calculate the electron–phonon interaction contribution to the electron self-energy $\Sigma(k) = \Sigma(k, E(k) - E(0))$ which is obtained by solving the equation $E(k) = \hbar^2k^2/2m_\text{b} + \Sigma(k, E(k) - E(0))$ for each value of the electron wave vector $\boldsymbol{k}$. Consequently we use the Falicov-Garcia [10] definition for the effective electron mass $m^* = (\hbar^2/2\pi)\ \partial A(E)/\partial E|_{E=E_\text{F}}$ with $A(E) = 2\pi \int_0^{E(k)} \text{d}k\ k$ which in our notation becomes (we take the Fermi surface at $k = k_\text{F}$)

$$\frac{m^*}{m_\text{b}} = \left(1 + \frac{m_\text{b}}{\hbar^2 k_\text{F}} \frac{\partial}{\partial k} \Sigma(k)|_{k=k_\text{F}}\right)^{-1} \tag{7}$$

with k_F the Fermi wave vector. Equation (7) corresponds to the classical electron cyclotron resonance mass.

The first-order Feynman diagram contribution to the electron self-energy due to the electron–phonon interaction is given by [6]

$$\Sigma(\boldsymbol{k}, \omega) = \frac{i}{\hbar} \int \frac{\text{d}\boldsymbol{q}}{(2\pi)^4}\, \text{d}\omega'\, \frac{|V_{\boldsymbol{q}}|^2}{\varepsilon^2(q_{\|}, \omega')}\, D(q_{\|}, \omega')\, G_0(\boldsymbol{q}_{\|} + \boldsymbol{k}, \omega + \omega')\,, \tag{8}$$

where $G_0(k, \omega)$ is the unperturbed electron Green's function. The phonon Green's function $D(q, \omega)$ is approximated by its unperturbed expression $D^0(q, \omega)$. For the dielectric function we take the static RPA result $\varepsilon_\text{RPA}(k, 0)$. In making these simplifications we neglect the frequency dependence of the dielectric function and we assume that the LO-phonon frequency is unchanged by the electron gas which is reasonable for $\omega_\text{p} \ll \omega_\text{LO}$ with ω_p the wave vector dependent plasmon frequency. The condition $\omega_\text{p} \ll \omega_\text{LO}$ is valid for not too high electron densities, i.e. $n_\text{e} < 10^{12}\ \text{cm}^{-2}$ for GaAs heterostructures.

In the limit of low electron density, (8) will only reduce to the expression for the one-polaron self-energy if we take $k = 0$ and $\omega = 0$. The usual one polaron energy $E = -(\pi/2)\, \alpha\hbar\omega_\text{LO}$ and effective mass $m^*/m_\text{b} = 1 + (\pi/8)\, \alpha$ for the ideal 2D system (the effective mass is defined in a similar way as in Section 2.2) are then obtained.

The function $\Sigma(k)$ which is the electron–phonon interaction correction to the electron energy as a function of the wave vector, has been calculated for different values of the electron densities n_e for (i) the exact 2D limit and (ii) with incorporation of the effects of the finite width of the electron layer. The depletion charge density n_d is chosen [17] to be $8 \times 10^{10}\ \text{cm}^{-2}$. The numerical results will be presented in the next section.

4. Numerical Results and Discussion

In Fig. 1 the electron–phonon interaction correction to the ground state energy of a polaron in a 2D polaron gas is presented for the different approximations discussed in Section 2. We have plotted the dynamical screening result with $S(k)$ in the RPA (thick solid curve) and in the HF (dash-dotted) and the static screening result with $\varepsilon(q, 0)$ in the RPA (thin solid curve) which is the result of [5] and in the TF (dashed curve) approximation which is the result of [2]. Physical parameters are taken which correspond to the GaAs–$\text{Al}_x\text{Ga}_{1-x}$As heterostructures, i.e. $\hbar\omega_\text{LO} = 36.77$ meV, dielectric constant $\varepsilon_\infty = 10.9$, band mass $m_\text{b} = 0.0657 m_\text{e}$ and electron–phonon coupling constant $\alpha = 0.068$.

Except for the Thomas-Fermi screening result all other approaches studied here give a decreasing $|E|$ with increasing electron density. Consequently there is an enhanced screening of the electron–phonon interaction with increasing electron density of the 2D electron layer. For all densities ($n_e < 10^{12}$ cm^{-2}) the dynamical screening approach gives a larger binding energy $|E|$ than the static screening approach. The static screening result within the RPA approaches the TF result of Das Sarma [2] in the high electron density limit. In the low density limit ($n_e \to 0$) we obtain (with the dynamical screening in the RPA and in the HF) $E/\hbar\omega_{LO} = -(\pi/2)\,\alpha$, while for the large electron density ($n_e \to \infty$ and thus $k_F \to \infty$, since $k_F = (2\pi n_e)^{1/2}$) $E/\hbar\omega_{LO} = -1.215\alpha(k_{LO}/k_F)$. From Fig. 1 it is apparent that the dynamical screened 2D electron–phonon binding energy is still larger than the single-particle 3D result when $n_e < 2 \times 10^{11}$ cm^{-2}.

The result for the energy in the TF approximation does not vary with the electron density. This is a consequence of the fact that in 2D the Thomas-Fermi wave vector is density independent. As a consequence Das Sarma's approach of [2] is unable to recover the correct 2D single polaron limit $E = -(\pi/2)\,\alpha\hbar\omega_{LO}$ when $n_e \to 0$.

In Fig. 2 we are concerned with the effect of the finite width of the electron–phonon interaction to the electron ground state energy. The same notations are used as in Fig. 1. When the density of the electron layer increases, the width of the electron layer decreases (b increases) and we would expect that the energy correction approaches the 2D limit when $n_e \to \infty$. However, the higher the electron density the stronger the screening effects on the electron–electron interaction and the electron–phonon interaction and in the limit $n_e \to \infty$ we expect that the electron–phonon interaction would be screened out and therefore $E = 0$. This behaviour is indeed found within the present dynamical screening approach. We observe from Fig. 2 that the energy correction $|E|$ due to the electron–phonon interaction decreases as a function of the electron den-

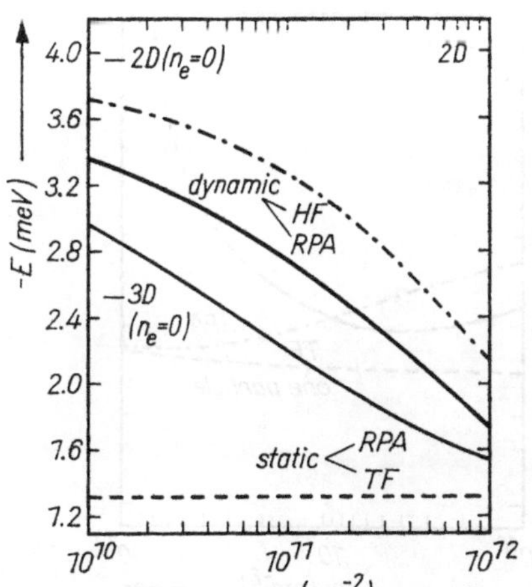

Fig. 1

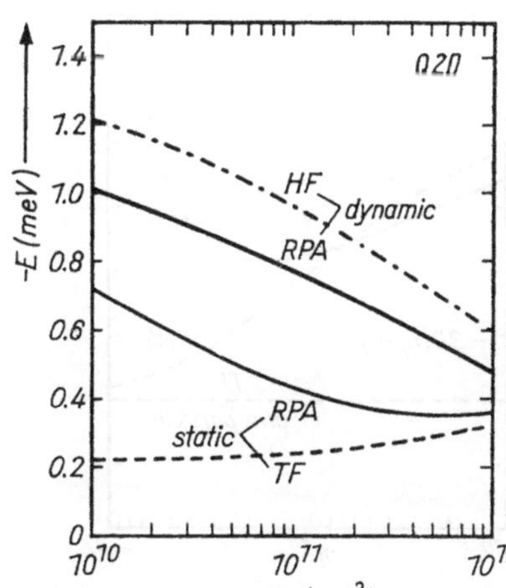

Fig. 2

Fig. 1. The contribution of the electron–phonon interaction to the ground state energy as a function of the electron density. The results for (i) the LLP approximation (static structure factor taken from the RPA (thick solid curve) and the Hartree-Fock (dash-dotted curve) approximation), (ii) the static RPA approximation (thin solid curve) and (iii) the static TF approximation (dashed curve) are shown. The energy correction is calculated for the ideal 2D system

Fig. 2. Same as Fig. 1 but now including the effect of the finite width of the electron layer

sity in the generalized LLP approximation (dynamical screening), while the result with the TF screening has the opposite behaviour, it increases with increasing electron density. For the static RPA approximation $|E|$ decreases first at low density ($n_e < 5 \times 10^{11}\ cm^{-2}$) with increase of n_e while at high n_e it starts to increase and approaches the TF result.

The reason why the result with TF screening increases with n_e can be easily understood. In the TF approach there is no effect of the electron density on the screening. But on increase of n_e, the electron layer width decreases which increases the polaron effects and as a result $|E|$ increases. In a more correct treatment we notice from Fig. 2 that the increase in polaron effects due to a decrease of the layer width when n_e increases, is smaller than the screening effects which as a result diminishes the polaron effects with increasing electron density n_e.

For 3D systems the Thomas-Fermi wave vector is electron density dependent in contrast to the 2D case. Our present study suggests that the Thomas-Fermi approximation to the screening effect in studying the electron–phonon interaction effect is not valid in 2D, although it may be valid for 3D systems.

In a recent work, Stern and Das Sarma [17] calculated the sub-band structure of GaAs heterostructures. Their calculated sub-band energy is of the order of 50 meV. We observe that the correction to the sub-band energy due to the electron–phonon interaction is more than a factor 50 smaller for experimental electron densities and thus of little consequence experimentally.

In Fig. 3 and 4, we show the mass renormalization due to the electron–phonon interaction calculated in different approximation schemes for the exact 2D and the quasi-2D (Q2D) electron system, respectively. In determining the effective mass from the

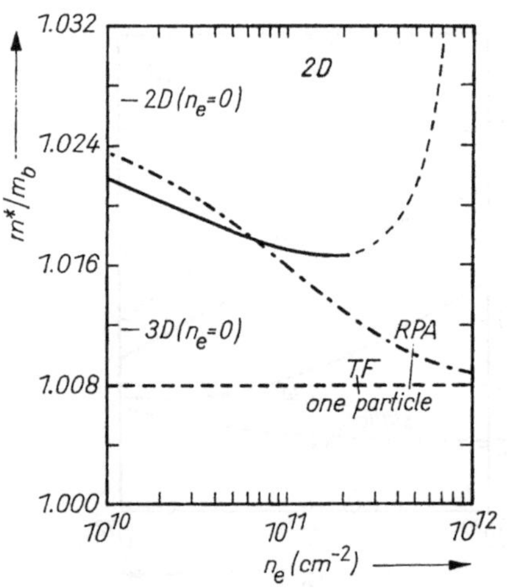

Fig.3

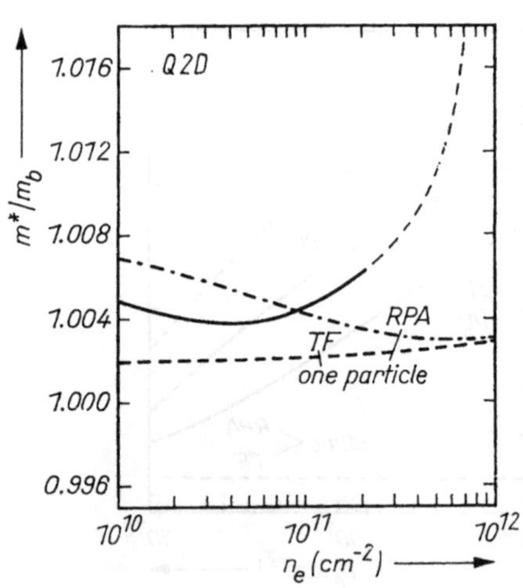

Fig. 4

Fig. 3. The effective mass calculated within different approximation schemes for the ideal 2D system. The m_F^* (calculated from polaron energy dispersion relation, solid curve) and m_D^* (in approximation of TF, dashed curve) and m_R^* (in static RPA approximation, dash-dotted curve) are shown as functions of the electron density for the ideal 2D system. The result for m_F^* is not valid for large electron densities and is therefore indicated by a thin dashed curve

Fig. 4. Same as Fig. 3, but including the effect of the finite width of the electron layer

energy dispersion relation obtained in Section 2.2, we use the definition

$$\frac{m_F^*}{m_b} = \left(1 + \frac{m_b}{\hbar^2 k_F} \frac{\partial}{\partial k} \Sigma(k)|_{k=k_F}\right)^{-1}$$

as was mentioned before. In the exact 2D limit (see Fig. 3), the Das Sarma (m_D^*) definition for the effective mass (dashed curve) with TF screening leads to a constant, which is a consequence of the Thomas-Fermi approximation as was discussed above. The effective mass, calculated in the static RPA approximation (m_R^*, dash-dotted curve) using (6), decreases as the density of the electron layer increases. The effective mass derived from the first-order diagram expansion approximation (m_F, solid curve) decreases first and then increases as the density increases further. In the low density limit m_R^* and m_F^* are nearly the same. As the electron density increases m_R^* approaches m_D^*. In Fig. 4, the effect of the finite width of the electron layer is included. We observe that m_D^* increases with an increase of n_e. m_R^* and m_F^* decrease first and then increase when the electron density increases. The electron density dependence of m_R^* and m_D in Fig. 3 and 4 is similar to the n_e dependence of $|E|$ calculated within the same approximation.

The behaviour of m_F^* can be understood as follows: As the density increases, the effective electron–phonon interaction decreases because of an increased screening effect and this would result in a smaller mass renormalization. This is also found for small electron densities. On the other hand, the energy dispersion relation for the electron at $k = k_F$ becomes more non-parabolic with increase of n_e and this can be viewed as an enhancement of the polaron mass renormalization. We have thus two competing mechanisms which have opposite effects on the effective mass. It is then reasonable to have the behaviour shown in Fig. 3 and 4 for the effective mass as a function of n_e: decreasing first and then increasing. The increase in m_F^* with increasing electron density is more dramatic when k_F approaches k_{LO} which occurs for electron densities around $n_e = 10^{12}$ cm^{-2}. This effect will be discussed in more detail in the following paragraphs.

A simple calculation of the electron energy dispersion relation as modified due to the electron–phonon interaction can be obtained by calculating $\Sigma(k, \hbar^2k^2/2m_b)$. But at high electron density and at large wave vectors such a calculation breaks down, since at the wave vector $k = k_{LO}$ the real part of the self-energy diverges [4]. The reason is that for an electron wave vector such that $E(k) \to \hbar\omega_{LO}$ degenerate perturbation theory should be used in order to calculate the electron self-energy. Our method for the calculation of the electron energy dispersion relation overcomes this difficulty. By taking a self-consistent perturbation procedure, i.e. to solve the equation $E(k) = \hbar^2k^2/2m_b \times + \Sigma(k, E(k) - E(0))$, which is known as "improved Wigner-Brillouin perturbation theory" (see [18] and [3], F. M. Peeters and J. T. Devreese, [19]), the divergency is removed.

In Fig. 5 and 6 the real part of the electron–phonon interaction correction to the electron self-energy is shown for different values of the electron density in the 2D limit (Fig. 5) and in the case of a finite width for the electron layer (Fig. 6). At $k = 0$, $\Sigma(0)$ increases ($|\Sigma(0)|$ decreases) as the electron density increases. This is consistent with our previous calculation of the ground state energy correction (see Fig. 1). Notice that $\Sigma(0) \neq E$ because in the ground state the different electrons have different k-vectors which is a consequence of the Fermi-Dirac statistics. Also note that at low electron density the electron self-energy does not depend strongly on k. However at higher electron densities $\Sigma(k)$ does change more dramatically with k because of the strong non-parabolic effect induced by the electron–phonon interaction. Consequently the derivative $\partial\Sigma(k)/\partial k$ taken at $k = k_F$ becomes more negative with increasing n_e and there-

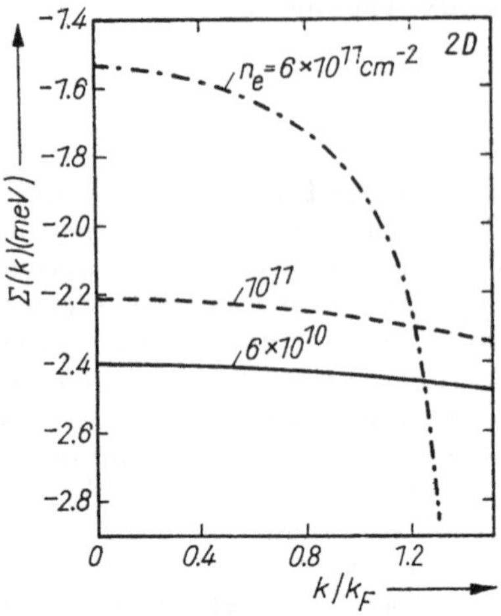

Fig. 5

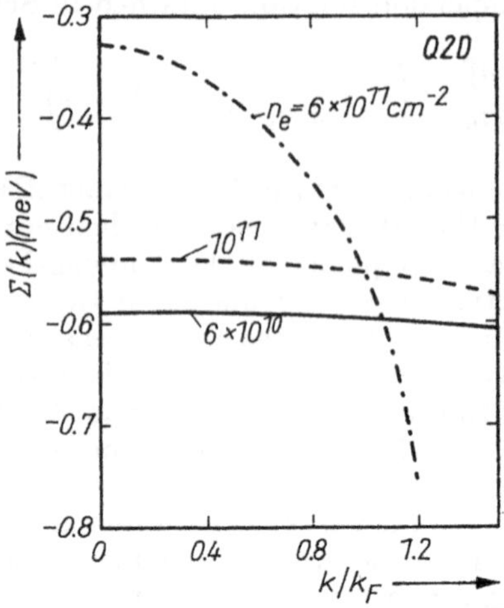

Fig. 6

Fig. 5. The electron–phonon interaction correction to the electron self-energy for different electron densities, i.e. $n_e = 6 \times 10^{10}\,\text{cm}^{-2}$ (solid curve), $n_e = 10^{11}\,\text{cm}^{-2}$ (dashed curve), and $n_e = 6 \times 10^{11}\,\text{cm}^{-2}$ (dash-dotted curve) as a function of the wave vector k for the ideal 2D system

Fig. 6. Same as Fig. 5 but including the effect of the finite width of the electron layer

fore the mass renormalization increases with increasing electron density as is apparent from Fig. 3 and 4 for $n_e > 10^{11}\,\text{cm}^{-2}$.

Next we discuss the origin of the sharp increases in the effective mass m_F for $n_e > 6 \times 10^{11}\,\text{cm}^{-2}$ (Fig. 3 and 4). For $k > k_{LO}$ the polaron energy momentum curve $E(k)$ calculated in Section 3 *asymptotically* approaches the LO-phonon energy as $k \to \infty$. This is different from the 3D situation where $E(k)$ is supposed to equal $\hbar\omega_{LO}$ at a certain finite k_{crit} [19]. As a consequence in 2D, $E(k) \to \hbar\omega_{LO}$ with increasing k when $k > k_{LO}$ and therefore $\Sigma(k) \to \hbar\omega_{LO} + E(0) - \hbar^2k^2/2m_b$. This limiting behaviour is apparent in Fig. 5 and 6 for the high electron density curve. (In Fig. 5 and 6 the wave vector scale is given by $k_F \sim n_e^{1/2}$ while the scale for the electron–phonon interaction is k_{LO}.) In the above large wave vector limit the derivative becomes $\partial\Sigma(k)/\partial k \sim -\hbar^2k/m_b$ and as a result of the Falicov-Garcia definition for the effective mass we obtain $m_F \to \infty$ for $k_F \gg k_{LO}$ which is the case for large electron densities.

However, in our calculation of the electron energy dispersion relation, the contribution of the direct electron–electron interaction to the electron self-energy ($\Sigma_{e,e}$) is not included. We should calculate $\Sigma = \Sigma_{e,e} + \Sigma_{e,ph}$ and consider $\partial\Sigma/\partial k = \partial\Sigma_{e,e}/\partial k + \partial\Sigma_{e,ph}/\partial k$ at $k = k_F$. In the present paper we only calculated $\Sigma_{e,ph}$ and neglected $\Sigma_{e,e}$. Inclusion of the electron–electron interaction contribution to the electron self-energy will modify the effective mass in the high electron density region. The present results for m_F^* breaks down for large electron densities (we estimated $n_e \approx 2 \times 10^{11}\,\text{cm}^{-2}$).

As mentioned in Section 2.1, the electron–phonon interaction leads to a modification of the Coulomb interaction (see (4)). In Fig. 7, we show the electron–phonon interaction correction to the Coulomb electron–electron interaction. This correction is given by

$$eV(r) = -4\alpha\omega_{LO}k_{LO}^{-1} \int_0^\infty dk\, J_0(kr)\, S(k)\, \frac{(S(k) + 2k^2k_{LO}^{-2})}{(S(k) + k^2k_{LO}^{-2})^2}, \tag{9}$$

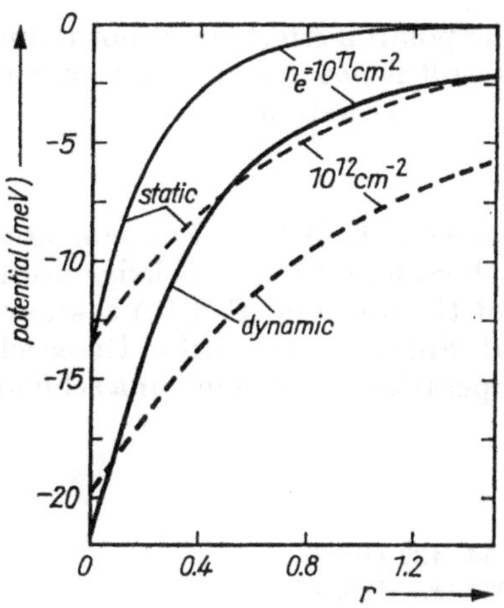

Fig. 7. The contribution of the phonon-mediated electron–electron interaction to the direct Coulomb potential is shown as a function of r (in units of $1/k_F$) (in real space) for different values of the electron density, i.e. $n_e = 10^{11}\ \mathrm{cm}^{-2}$ (solid curves) and $n_e = 10^{12}\ \mathrm{cm}^{-2}$ (dashed curves). The potential is calculated within the dynamical RPA screening (thick curves) approximation and within the static TF screening (thin curves) approximation for the ideal 2D system

where $J_0(x)$ is the Bessel function of the first kind of index zero. Following the one particle approach of Das Sarma we defined another potential in analogy with (9),

$$eV(r) = -4\alpha\omega_{LO}k_{LO}^{-1} \int_0^\infty dk\, J_0(kr)\, \frac{1}{\varepsilon(k,0)}\, \frac{(1 + 2k^2k_{LO}^{-2})}{(1 + k^2k_{LO}^{-2})^2}, \tag{10}$$

where for the static dielectric function we will use $\varepsilon(k, 0) = \varepsilon_{TF}(k, 0)$ the Thomas-Fermi approximation. An important feature of the correction to the electron–electron potential induced by the electron–phonon interaction is that the potential (equations (9) and (10)) is finite at $r = 0$.

5. Conclusions

We have calculated the electron–phonon interaction correction to the ground state energy of the interacting 2D polaron gas within a dynamical screening scheme. The screening effects due to the electron–electron interaction (without phonon mediated electron–electron interaction) are included into our treatment of the electron–phonon interaction exactly within the RPA approximation. Comparing our results with the results from TF screening, we found that both results show a reduction of the electron–phonon interaction due to the screening effects, but our results are quantiatively different from these of [2]. A dynamical treatment of the screening effect on the electron–phonon interaction results in larger polaron effects for electron densities $n_e < 10^{12}\ \mathrm{cm}^{-2}$ in comparison with a static screening approach. The electron–phonon correction to the ground state energy per polaron was also calculated within a Hartree-Fock approximation and within the static RPA approximation. Our results indicate that for electron densities $n_e < 10^{12}\ \mathrm{cm}^{-2}$ the screening effects are less severe than predicted by Das Sarma in [2] and [5].

The electron–phonon interaction contribution to the electron self-energy and to the polaron mass renormalization was also compared with the static screening results of [5]. Here we found a larger mass renormalization than predicted from TF screening over the whole electron density range studied, i.e. $10^{10}\ \mathrm{cm}^{-2} < n_e < 10^{12}\ \mathrm{cm}^{-2}$.

In the present paper, we studied the ground state energy and the polaron mass renormalization in the absence of a magnetic field. In cyclotron resonance experiments [20] on InSb inversion layers it was found that at the resonance condition (i.e.

$\omega_c \approx \omega_{LO}$, ω_c is the cyclotron resonance frequency) the polaron effects were not reduced by the many-particle effects (see also [21]). This result is not in contradiction with the present results because we treated the non-resonance situation.

Acknowledgements

This work is partially sponsored by F.K.F.O. (Fonds voor Kollektief Fundamenteel Onderzoek, Belgium), project No. 2.0072.80. The authors wish to acknowledge useful discussions with F. Brosens and L. Lemmens. One of the authors (F.M.P.) acknowledges financial support from the Belgian National Fund for Scientific Research. Wu X. wishes to thank the International Culture Co-operation of Belgium for a scholarship.

References

[1] T. Ando, B. Fowler, and F. Stern, Rev. mod. Phys. **54**, 437 (1982).
[2] S. Das Sarma, Phys. Rev. B **27**, 2590 (1983); B **37**, 4034 (E) (1985).
[3] D. M. Larsen, Phys. Rev. B **30**, 4595 (1984).
S. Das Sarma, Phys. Rev. Letters **52**, 859, 1580(E) (1984).
F. M. Peeters and J. T. Devreese, Phys. Rev. B **31**, 3689 (1985).
Wu Xiaoguang, F. M. Peeters, and J. T. Devreese, Phys. Rev. B **31**, 3420 (1985).
[4] S. Das Sarma and A. B. Mason, Ann. Phys. (USA), to be published.
S. Das Sarma, Surface Sci. **142**, 341 (1984).
[5] S. Das Sarma and A. B. Mason, Phys. Rev. B **31**, 5536 (1985); B **32**, 2656(E) (1985).
[6] D. C. Tsui, Th. Englert, A. Y. Cho, and A. C. Gossard, Phys. Rev. Letters **44**, 341 (1980).
[7] W. Seidenbusch, G. Lindemann, R. Lassnig, T. Ellinger, and E. Gornik, Surface Sci. **142**, 375 (1984).
[8] A. L. Fetter and J. D. Walecka, Quantum Theory of Many-Particle Systems, McGraw-Hill Publ. Co., New York 1971 (p. 175).
[9] L. F. Lemmens, J. T. Devreese, and F. Brosens, phys. stat. sol. (b) **82**, 439 (1977).
[10] L. M. Falicov and N. Garcia, Solid State Commun. **17**, 437 (1975).
[11] B. Vinter, Phys. Rev. B **13**, 4447 (1976).
[12] G. Kawamoto, R. Kalla, and J. J. Quinn, Surface Sci. **98**, 589 (1980).
G. Kawamoto, J. J. Quinn, and W. L. Bloss, Phys. Rev. B **23**, 1875 (1981).
[13] T. S. Rahman and D. L. Mills, Phys. Rev. B **23**, 4081 (1981).
[14] F. Stern, Phys. Rev. Letters **18**, 546 (1967).
[15] A. Czachor, A. Holas, S. R. Sharma, and S. Singwi, Phys. Rev. B **25**, 2144 (1982).
[16] G. D. Mahan, in: Polarons in Ionic Crystals and Polar Semiconductors, Ed. J. T. Devreese, North-Holland Publ. Co., Amsterdam 1972 (p. 553).
[17] F. Stern and S. Das Sarma, Phys. Rev. B **30**, 840 (1984).
[18] G. Lindemann, R. Lassnig, W. Seidenbusch, and E. Gornik, Phys. Rev. B **28**, 4693 (1983).
[19] D. M. Larsen, Phys. Rev. **144**, 697 (1966).
See also [16] (p. 237).
[20] M. Horst, U. Merkt, and J. D. Kotthaus, Phys. Rev. Letters **50**, 754 (1983).
[21] R. Lassnig and W. Zawadzki, Surface Sci. **142**, 388 (1984).

(Received September 10, 1985)

phys. stat. sol. (b) 133, 241 (1986)

Subject classification: 71.25 and 72.15, 75.50, S1.2

Department of Physics, Monash University, Clayton[1])

The Band Structure of Co_3Ga and Co_2MnGa and the Electrical Conductivity of Related Alloys

By

G. C. FLETCHER

Band structure calculations are carried out for hypothetical nonmagnetic alloys Co_3Ga and Co_2MnGa and for the ferromagnetic alloy Co_2MnGa, all with the $L2_1$ structure. The results are used to investigate the electrical conductivity of (Co, Ga) and Co(Ga, TM) alloys (where TM is a transition metal), particularly as this involves hopping conduction. Explanation of the latter by the presence of antistructure Co or TM atoms on the Ga subslattice seems to be confirmed although they do not affect normal conduction appreciably. Mn assists hopping conduction less than antistructure Co since it provides fewer good hopping states, albeit better ones.

Es werden Bandstrukturberechnungen für die hypothetischen nichtmagnetischen Legierungen Co_3Ga and Co_2MnGa sowie für die ferromagnetische Legierung Co_2MnGa mit $L2_1$-Struktur durchgeführt. Die Ergebnisse werden benutzt, um die elektrische Leitfähigkeit von (Co, Ga)- und Co(Ga, TM)-Legierungen (wobei TM ein Übergangsmetall ist) zu untersuchen, insbesondere da diese eine Hoppingleitung einschließt. Eine Erklärung der letzteren mit Antistruktur-Co- oder TM-Atomen auf Ga-Untergitterplätzen scheint bestätigt, obwohl sie die normale Leitung nicht wesentlich beeinflussen. Mn unterstützt die Leitfähigkeit weniger als Antistruktur-Co, da es weniger gute, obgleich bessere, Hoppingzustände liefert.

1. Introduction

Alloys of transition metals with nontransition ones generally show interesting structural, magnetic, and electrical properties. In this respect considerable work has been carried out on alloys of Fe, Co, Ni with Ga, Al which adopt the B2 or CsCl structure and ternary (generally Heusler) alloys, where Ga, Al is partly replaced by a different transition metal and which usually have the $L2_1$ structure. For a number of years Booth and co-workers [1] studied the (Co, Ga) and (Co, Ga, TM) series of alloys (where TM denotes another transition metal) and their results have been summarized in a recent paper by Whittle, et al. [2]. In particular their explanation of the high resistivity and its temperature dependence for these alloys, involving hopping conduction at low temperatures, has been discussed in the light of a band calculation for CoGa by Whittle et al. [3]. Three other band calculations for Heusler alloys are of added interest in this regard and a brief presentation of them is given here.

2. Method and Parameters of Calculation

Two of the band calculations were made for hypothetical alloys, namely Co_3Ga, in which every alternate Ga atom in CoGa has been replaced by Co, and nonmagnetic Co_2MnGa, where the replacement is by Mn, both alloys having the $L2_1$ structure. The third calculation was for Co_2MnGa with the same structure but in its experimentally observed ferromagnetic state. The results for Co_3Ga can be used to discuss antistruc-

[1]) Clayton, Victoria 3168 Australia.

ture (AS) Co atoms in (Co, Ga) alloys, i.e. ones on the Ga sublattice, which are relevant to superparamagnetic clusters in those alloys and hence ferromagnetism. The other calculations, apart from being for Heusler alloys per se, were intended to give insight into the effects of replacing Ga atoms by a transition metal other than Co, nonmagnetic or magnetic.

All calculations were carried out by the Hubbard method [4] as generalized by Wakoh [5] and relevant parameters used are given in Table 1. In all cases muffin-tin potentials were constructed in the usual way (Loucks [6]) with a muffin-tin sphere radius of 2.340 at. units for all atoms, using free atom wave functions from Clementi [7] and full Slater exchange. l_{max} was chosen as four and six preferred reciprocal lattice vectors (000), $(\bar{1}\bar{1}1)$, $(\bar{1}\bar{1}\bar{1})$, $(\bar{2}00)$, $(\bar{1}1\bar{1})$ and $(\bar{1}11)$ in units of $2\pi/a$ were used, a being the face-centred cubic Bravais lattice constant. With four atoms per unit cell, of which three had an $l = 2$ resonance at energies near the Fermi level ε_F, the Hubbard method then involved solving a 21×21 secular equation. As arranged in the present calculations the first five coefficients of each eigenvector referred approximately to five linearly independent d-type tight-binding functions centred on one set of Co atoms. The next five referred to similar functions on the second set of Co atoms on the original Co sublattice, the next five to ones on the Co or Mn atoms on the original Ga sublattice and the last six approximately to orthogonalized plane waves (OPW) with the preferred wave vectors given above. It was therefore quite simple to find the symmetry of wave functions and in particular their t_{2g} or e_g character around different atoms.

3. Calculation for Co_3Ga

No stoichiometric ordered alloy Co_3Ga exists but, as mentioned already, a band calculation was carried out for an alloy, in which every alternate atom of Ga in CoGa was replaced by Co, in an effort to gain further insight into the effects of AS Co atoms on the conductivity of (Co, Ga) alloys. Comparison is therefore made with the calculation for CoGa described by Whittle et al. [3]; the relevant results of that calculation are as follows.

(i) Flat (Co) d-band dispersion curves across the Brillouin zone (BZ) are separated from higher energy, nearly free electron curves in several directions in the BZ by distinct gaps. This produces a very high density of states peak followed by a deep minimum with ε_F in this trough.

(ii) This situation corresponds to one discussed by Mott and Davis [8] which leads to localization of electrons with energy $\sim \varepsilon_F$ and to conduction at low temperatures mainly by thermally activated hopping between atoms. Put another way the relaxation time (τ) approximation for normal electrical conductivity when there is cubic symmetry

$$\sigma \sim \iint_{FS} \tau \, |\nabla_{\boldsymbol{k}}\varepsilon| \, dS$$

will be small because $|\nabla_{\boldsymbol{k}}\varepsilon|$ is small over a small Fermi surface (FS) and hopping conduction becomes comparable, if not dominant.

(iii) Since nearest neighbour atoms on the Co sublattice are in $\langle 100 \rangle$ directions, states of e_g symmetry round each atom should overlap most and be most conducive to hopping conduction. Nearly all occupied states within 0.01 Ryd of ε_F were of this symmetry in the calculation but, of course, the Co–Co separation is quite large.

(iv) An AS Co atom links atoms on the Co sublattice in $\langle 111 \rangle$ directions at a smaller separation and should improve hopping conduction if occupied states of t_{2g} symmetry are involved.

Table 1

Physical data and parameters used for the band calculations. The average potential V_0 is that calculated between the muffin-tin spheres. The parameters are the energy ε_0 and strength Γ of each $l = 2$ resonance and d-band energy ε_1. For ease of comparison of the various calculations ε_0, ε_1, ε_F and the energy scales on the figures are absolute, i.e. without subtraction of V_0. For Co_3Ga CoII is the antistructure atom

alloy	state	assumed configuration		lattice constant	average potential	Hubbard parameters				Fermi level
		Co	Mn	a (at. units)	V_0 (Ryd)		ε_0 (Ryd)	Γ (Ryd)	ε_1 (Ryd)	ε_F (Ryd)
Co_3Ga	non-magnetic	$3d^{7.8}\,4s^{1.2}$	—	10.909	−1.476	Co I	−0.747	0.323	−0.946	−0.849
		$3d^8 4s$				Co II	−0.734	0.322	−0.932	
Co_2MnGa	non-magnetic	$3d^8 4s$	$3d^6 4s$	10.928	−1.368	Co	−0.488	0.346	−0.708	−0.559
						Mn	−0.282	0.464	−0.581	
Co_2MnGa	ferro-magnetic	$3d_{\uparrow}^{4.26} 3d_{\downarrow}^{3.74} 4s_{\uparrow}^{.5} 4s_{\downarrow}^{.5}$	$3d_{\uparrow}^{4.505} 3d_{\downarrow}^{1.495} 4s_{\uparrow}^{.5} 4s_{\downarrow}^{.5}$	10.928	−1.389	↑ Co	−0.453	0.448	−0.737	
						↑ Mn	−0.421	0.500	−0.738	−0.610
					−1.346	↓ Co	−0.370	0.481	−0.678	
						↓ Mn	−0.047	0.550	−0.399	

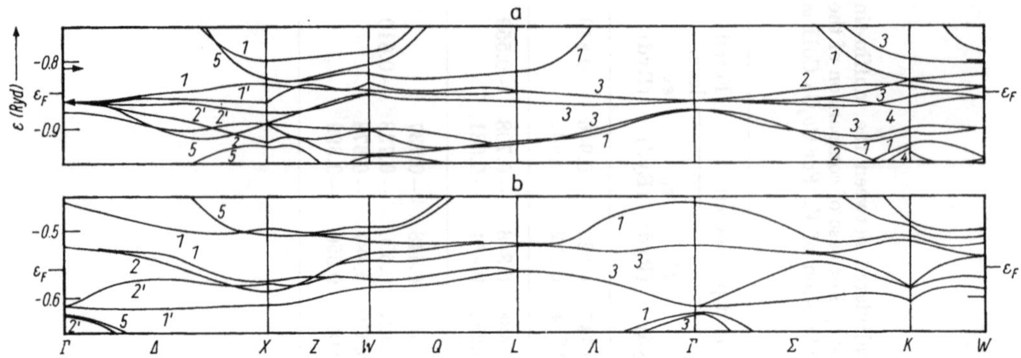

Fig. 1. One-electron energies $\varepsilon(\boldsymbol{k})$ near the Fermi level ε_F calculated for nonmagnetic a) Co_3Ga and b) Co_2MnGa for $\boldsymbol{k}$ along symmetry directions of the Brillouin zone. Numbers on curves refer to irreducible representations of the relevant symmetry group of $\boldsymbol{k}$. The Fermi level is denoted by ε_F while in part a) arrows indicate its position relative to the curves for up- and down-spin in a ferromagnetic Co_3Ga

The results of the calculation for Co_3Ga, which may be regarded as local bands around a nonmagnetic AS Co atom in a (Co, Ga) alloy, are shown in Fig. 1 and 2. A noticeable difference from CoGa in the density of states curve is the subband split off completely at low energies, which accommodates just two electrons per unit cell, is nearly free electron in character and is a feature of calculations for Heusler alloys, both here and by other authors. The simple explanation lies in the change from B2 to $L2_1$ symmetry with a doubling of the unit cell. If the B2 BZ is folded into the $L2_1$ one and the dispersion curves (of CoGa, say) are perturbed by the corresponding change of crystal potential, especially at the new zone boundaries, the appearance of the energy gap follows naturally. The Co d-bands of CoGa are broadened and lowered in Co_3Ga relative to the nearly free electron bands, changes traceable to the presence of a greater fraction of Co atoms in two different environments.

With regard to normal conduction Fig. 1a shows that $|\nabla_{\boldsymbol{k}}\varepsilon|$ is generally very small near ε_F but the Fermi surface proves to be rather larger in area than that for CoGa (folded into the $L2_1$ BZ); the relevant states near ε_F have almost no OPW component so overall the AS Co atoms seem unlikely to inprove the normal conduction very much. With regard to the symmetry of occupied states within 0.01 Ryd of ε_F a good proportion are strongly t_{2g} around AS Co and e_g around the other Co atoms, a combination

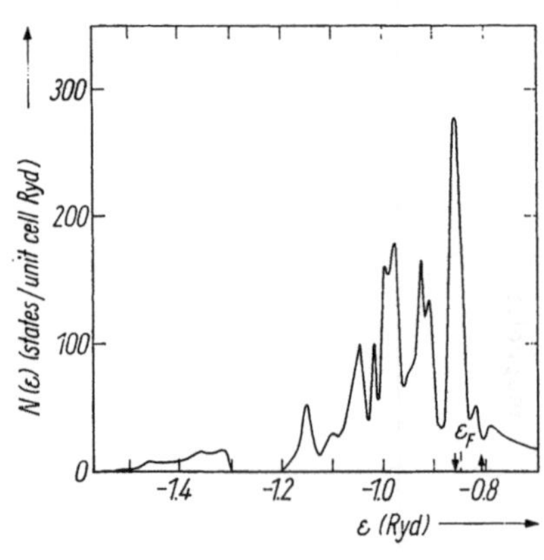

Fig. 2. Density of states curve for nonmagnetic Co_3Ga. The Fermi level and its position for a ferromagnetic alloy are indicated as in Fig. 1a

which should be nearly as good for hopping conduction as t_{2g} around both. Adopting a qualitative scale 0 to 3 of suitability for this, an average figure of about 1.5 was calculated for these states and it does seem likely that hopping conduction is largely due to links through AS Co atoms.

It may be remarked that Co_xGa_{100-x} alloys up to $x = 65$ have been investigated by Whittle et al. [3] and the results for the most Co-rich may be compared with the band calculation here. These display a conductivity larger by a factor of five than that of the equiatomic alloy and are ferromagnetic. Adjusting the band calculation approximately to allow for a moment of $\approx 2\mu_B$ per Co atom (Amamou and Gautier [9]) by rigidly splitting the bands puts the Fermi level for up-spin amongst the OPW bands (for which $|\nabla_k \varepsilon|$ is much larger) while moving it very little for down-spin. This is shown in Fig. 1a and 2 and it suggests that the increased conductivity arises as a consequence of the Co-rich alloys going ferromagnetic and not the presence of AS Co atoms; the conduction is normal and largely involves electrons of majority spin.

4. Calculation for Nonmagnetic Co_2MnGa

Again no nonmagnetic stoichiometric ordered alloy Co_2MnGa exists at low temperatures but only a ferromagnetic one. However, a band calculation was performed for such an alloy with $L2_1$ structure as a first step to investigating the replacement of half the Ga atoms in CoGa by a transition metal other than Co. In this respect there have been a number of experimental investigations of the effect of such replacement on structural, magnetic, and electric properties, e.g. in $Co_2Fe_xGa_{2-x}$ by Whittle et al. [10]. The results of the present calculation are shown in Fig. 1b and 3, which may be compared with those for Co_3Ga. Looking at the density of states curve, apart from the small band split off at the lowest energies for the same reason as discussed before, the next band up to the gap at ≈ -0.6 Ryd is almost pure Co 3d in character with some OPW hybridization but not involving the Mn atoms. On the other hand, the band containing ε_F above the gap is very much a mixed one with d character on both Co and Mn atoms. The overall d-bands are much broader than for Co_3Ga and, although ε_F falls within them again, $|\nabla_k \varepsilon|$ at ε_F is not as small; the Fermi surface is more complicated but normal conduction will be discussed later for the ferromagnetic calculation. Compared with Co_3Ga there are only about one-third as many occupied states within 0.01 Ryd of ε_F but they are rather more conducive to hopping conduction (an average "suitability" figure calculated as before has a value of 1.8), again most having e_g symmetry around Co atoms,

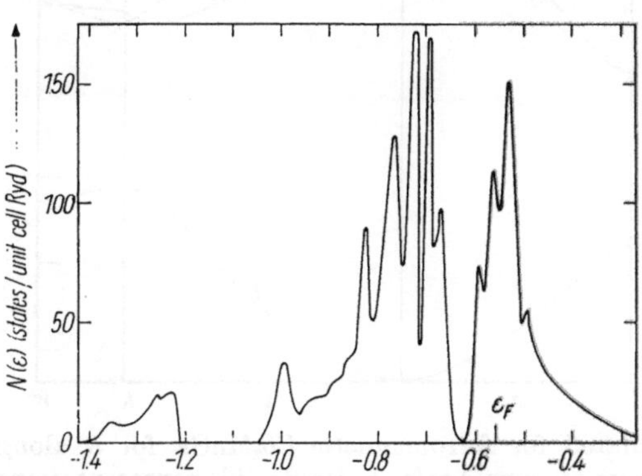

Fig. 3. Density of states curve for nonmagnetic Co_2MnGa. ε_F denotes the Fermi level

t_{2g} around Mn ones. Altogether the hopping conduction would seem to be reduced by using Mn rather than Co to replace Ga, a result similar to that reported for Fe by Whittle et al. in their work on non-ferromagnetic (Co, Ga) and Co(Ga, Fe) alloys [3, 10].

5. Calculation for Ferromagnetic Co_2MnGa

As mentioned already the Heusler alloy Co_2MnGa is in fact ferromagnetic with a Curie temperature of 694 K and a moment of $4.05\mu_B$ per molecule; according to Webster [11] this is made up of moments $0.52\mu_B$ on each Co atom and $3.01\mu_B$ on the Mn atom. A band calculation was carried out for this alloy for comparison with experiment but also, by comparison with the calculation for the nonmagnetic alloy, to investigate the effects of going ferromagnetic. The results of the calculation are shown in Fig. 4 to 8 and are reasonably self-consistent in that they correspond to moments of 0.40 and $3.44\mu_B$ on Co and Mn atoms respectively. Looking at the density of states for up- and down-spin both have a nearly free electron band split off at the bottom for the same reason as discussed before. The predominantly d-bands, compared with those calculated for the nonmagnetic alloy, are narrower for up-spin with non real separation of Co and Mn bands while the down-spin ones are broader with a near-gap very close to ε_F and which separates nearly pure Co bands (below) from nearly pure Mn ones (above). The Fermi surfaces are much simpler than for the nonmagnetic case. Considering first normal conduction $|\nabla_k \varepsilon|$ at ε_F has been reduced considerably for the up-spin bands compared with the nonmagnetic ones and the Fermi surface is of comparable area whereas $|\nabla_k \varepsilon|$ at ε_F is larger for the down-spin bands but the Fermi surface is small. The average

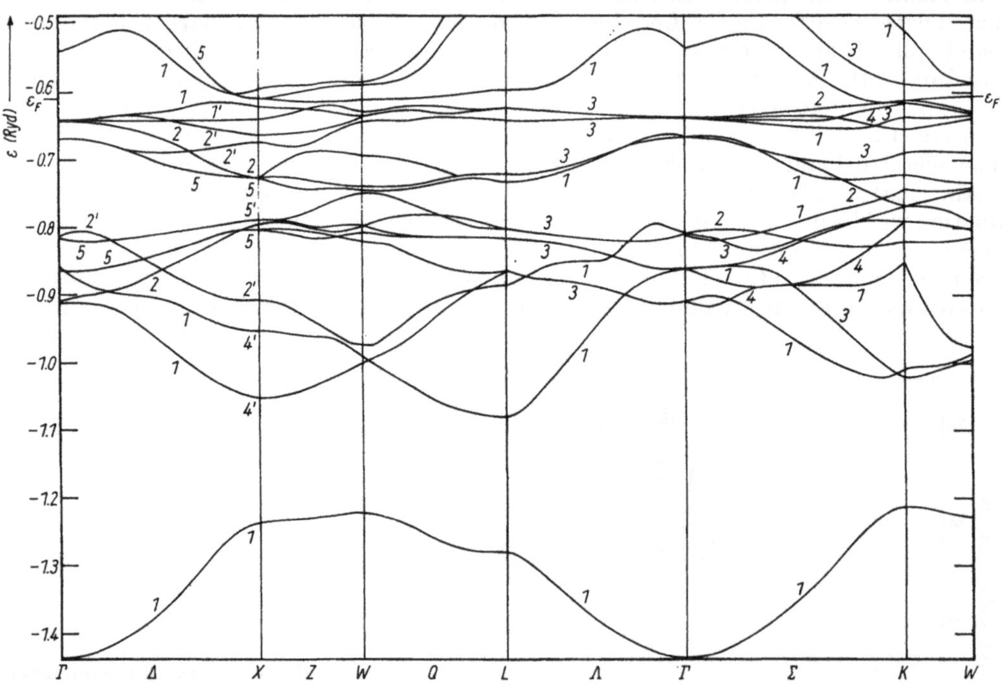

Fig. 4. Up-spin one-electron energies $\varepsilon(\boldsymbol{k})$ calculated for ferromagnetic Co_2MnGa for $\boldsymbol{k}$ along symmetry directions of the Brillouin zone. Numbers on curves refer to irreducible representations of the relevant symmetry group of $\boldsymbol{k}$. The Fermi level is denoted by ε_F

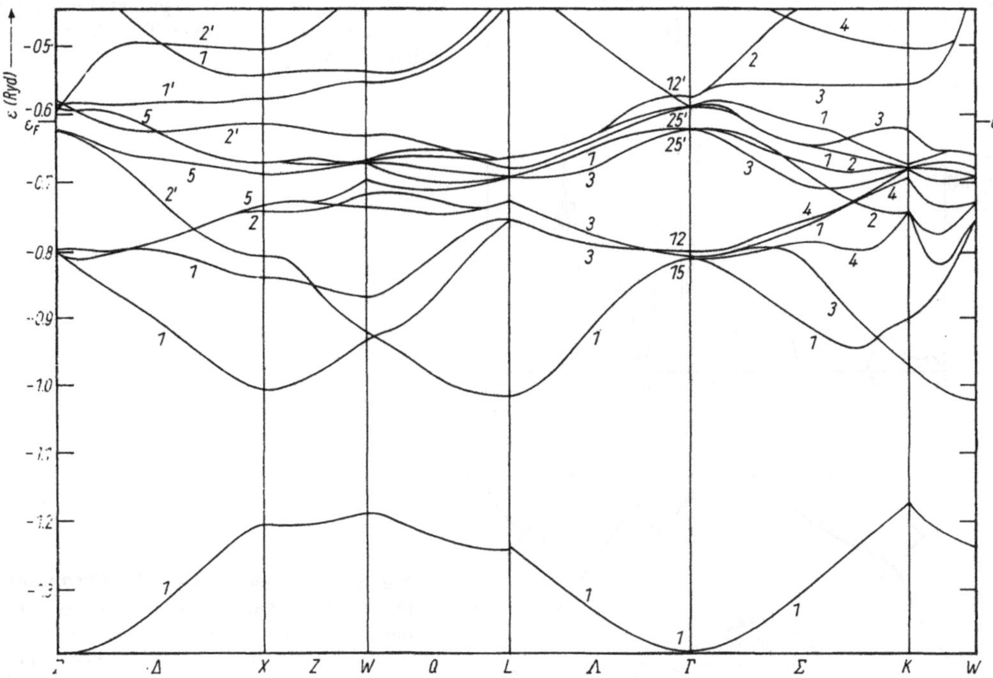

Fig. 5. Down-spin one-electron energies $\varepsilon(\boldsymbol{k})$ calculated for ferromagnetic Co_2MnGa for $\boldsymbol{k}$ along symmetry directions of the Brillouin zone. Numbers on curves refer to irreducible representations of the relevant symmetry group of $\boldsymbol{k}$. The Fermi level is denoted by ε_F

OPW content of the states near ε_F is about the same but rather surprisingly the down-spin states have a much higher average OPW content than the up-spin ones. Overall the effect on normal conduction of going ferromagnetic would not seem to be great; more up-spin than down-spin states are involved but both are approximately equally responsible for the current. With regard to hopping conduction investigation of the symmetry of occupied states within 0.01 Ryd of ε_F shows that down-spin states are virtually not involved as far as Co → Mn → Co hopping is concerned as they have scarce-

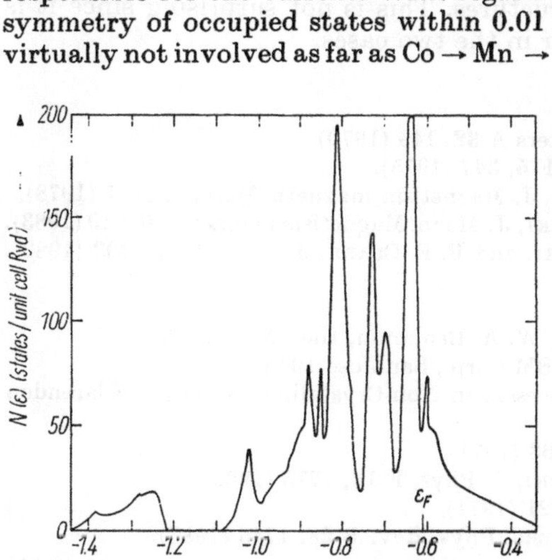

Fig. 6. Density of up-spin states curve for ferromagnetic Co_2MnGa. The Fermi level is denoted by ε_F

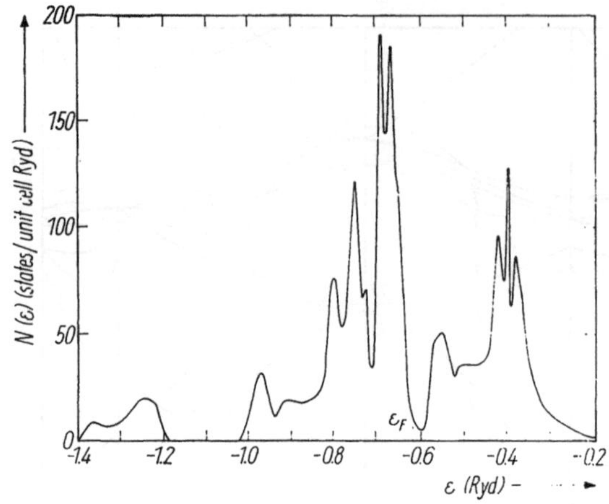

Fig. 7. Density of down-spin states curve for ferromagnetic Co_2MnGa. The Fermi level is denoted by ε_F

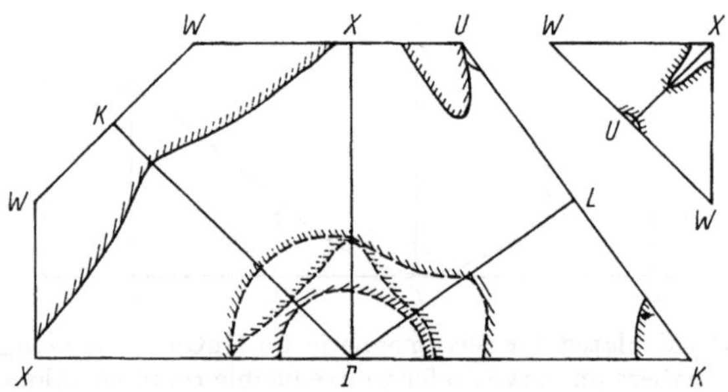

Fig. 8. Cross-sections of the Fermi surface for ferromagnetic Co_2MnGa by symmetry planes of the Brillouin zone. ——— up-spin; - - - - down-spin. Hachuring is on the occupied side of each piece of the surface

ly any component on the Mn atoms. On the other hand, the up-spin states have become more conducive to this hopping with an average "suitability" value of 2.0. The overall effect of becoming ferromagnetic is perhaps a slight reduction in hopping conduction.

One final comment is that the results of the calculation for ferromagnetic Co_2MnGa are very similar to those for Co_2MnAl by Kübler et al. [12], i.e. in the widths of both up- and down-spin density of states curves, positions of peaks and troughs and location of the Fermi level within their structures. This is not surprising since it is only the inner cores of Al and Ga that differ in the two cases.

References

[1] J. G. Booth and J. D. Marshall, Phys. Letters A **32**, 149 (1970).
J. G. Booth and R. G. Pritchard, J. Phys. F **5**, 347 (1975).
J. G. Booth, R. Cywinski, and J. G. Prince, J. Magnetism magnetic Mater. **7**, 127 (1978).
[2] G. L. Whittle, P. E. Clark, and R. Cywinski, J. Magn. Magnetism magnetic **30**, 319 (1983).
[3] G. L. Whittle, G. C. Fletcher, R. Cywinski, and P. E. Clark, J. Phys. F **12**, 303 (1982).
[4] J. Hubbard, J. Phys. C **2**, 1222 (1969).
[5] S. Wakoh, Phys. Letters A **31**, 333 (1970).
[6] T. Loucks, Augmented Plane Wave Method, W. A. Benjamin, Inc., New York 1967.
[7] E. Clementi, Tables of Atomic Functions, IBM Corp., San Jose 1965.
[8] N. F. Mott and E. A. Davis, Electronic Processes in Non-Crystalline Materials, Clarendon Press, Oxford 1971.
[9] A. Amamou and F. Gautier, J. Phys. F **4**, 563 (1974).
[10] G. L. Whittle, P. E. Clark, and R. Cywinski, J. Phys. F **10**, 1775 (1980).
[11] P. J. Webster, J. Phys. Chem. Solids **32**, 1221 (1971).
[12] J. Kübler, A. R. Williams, and C. B. Sommers, Phys. Rev. B **28**, 1745 (1983).

(Received September 10, 1985)

phys. stat. sol. (b) 133, 249 (1986)

Subject classification: 71.25; 76.80; S1.63

Department of Physics, S. V. College, Aligarh[1])

Electronic Properties of Iron Chalcogenides FeS, FeS_2, $FeSe_2$, and $FeTe_2$

By

O. P. SINGH and V. P. GUPTA

Lavin's modifications of Phillips and Van Vechten (PVV) theory are discussed and hence ionic gap (E_c), homopolar gap (E_h), and ionicity (f_i) are investigated for iron chalcogenides (FeS, FeS_2, $FeSe_2$, and $FeTe_2$). A correlation between evaluated values of ionicity and Mössbauer isomeric shift is also discussed.

Für die Eisenchalkogenide (FeS, FeS_2, $FeSe_2$ und $FeTe_2$) werden die Ionenlücke (E_c), die homopolare Bandlücke (E_h) und die Ionizität (f_i) untersucht und mit der von Lavin vorgenommenen Modifikation der Theorie von Phillips und Van Vechten diskutiert. Die Wechselbeziehung zwischen den berechneten Werten der Ionizität und der Mößbauer-Isomerieverschiebung wird diskutiert.

1. Introduction

Crystal ionicity (f_i) is one of the very important parameters concerning semiconducting materials and can be used in tackling problems related to various fields such as crystal structure, band structure, elastic constants, effective charges, cohesive energies, heat of formation, etc. [1]. The values of average gap (E_g), ionic gap (E_c), and hence Phillips ionicity ($f_i = E_c^2/E_p^2$) can be derived using basic crystallographic parameters and following the Phillips and Van Vechten (PVV) theory [2, 3]. This theory has been successfully applied to investigate the bond properties of different varieties of crystals [4 to 6].

Pyrite is one of the most common sulfide minerals and a well-known semiconductor. Its semiconducting nature has induced many workers to study its electronic and semiconducting properties [7 to 9] with the objective of its possible use in devices. Significantly its band gap is about 1.0 eV which brings it very near to silicon ($E_p = 1.1$ eV). It has halite structure (NaCl type) in which anion sites are taken by S_2 molecules. Fe atoms and molecular centers are on a face-centered cubic lattice. Each Fe atom is coordinated by six sulfur atoms, while a sulfur atom is bonded to three iron and one sulfur atoms. Triolite (FeS) or iron monosulfide has got NiAs-type structure. In this structure the cation and anion lie, respectively, on interpenetrating simple hexagonal and close-packed hexagonal lattices. At present triolite (commonly known as pyrrhotite) has attracted considerable interest of geophysists, minealogists, and physicists. The reason for the interest is the strongly anisotropic ferrimagnetism, the high electrical conductivity, and its occurrence in sulfide ore deposits of economic interest. In this work we have investigated the Phillips ionicity of pyrrhotite, pyrite, and some other iron chalcogenides [FeS_2, (marcasite), $FeSe_2$, $FeTe_2$].

2. Theory

PVV theory was proposed for elemental and binary compound (AB) semiconductors having tetrahedral coordinations. Later on it was extended to octahedral compounds, too. According to PVV theory,

$$E_p^2 = E_h^2 + E_c^2 ,$$

[1]) Aligarh 202001, India.

where

$$E_h = \frac{40.468}{r_{AB}^{2.5}},$$

$$E_c = be^2 \left(\frac{Z_A}{r_A} - \frac{Z_B}{r_B}\right) e^{-K_s r_{AB}/2},$$

where r_A, r_B, and r_{AB} are covalent radii of atoms A, B and the interatomic distance between atoms A and B, respectively. Z_A, Z_B are numbers of valence electrons of atoms A and B, respectively, b is an adjustable parameter, K_s is the Thomas Fermi screening parameter. In order to generalize PVV theory Lavin [4] proposed the following modifications in the original PVV concept:

1. r_0 ($r_0 = r_{AB}/2$) is introduced in place of r_A and r_B;

2. the value of b is defined in terms of average coordination number $\overline{N}_c$ ($b = 0.089\overline{N}_c^2$);

3. for complex systems like AB_n the valence electrons will spend more time on atom B, therefore they will experience an n times greater core change (nZ_B) than in a similar AB compound;

4. in case of transition and noble metal compounds it was proposed [4] to introduce all the d electrons in bond formation. Furthermore a variable (σ) is introduced in order to match the theoretical results with experimental data.

The final relation for transition and noble metal materials was presented in the following form:

$$E_c = be^2 \left(\frac{Z_A^*}{r_0} - n\frac{Z_B}{r_0}\right) e^{-K_s r_0},$$

where $Z_A^* = \sigma Z_A$.

The modifications 1, 2 and part of modification 4 are based on sound theoretical grounds and they appear to be justified and essential. We feel that the modification 3 and the latter part of the modification 4 are not justified. Our main objection against these are summarized below:

It is very clear from PVV theory and Lavin's modifications that the value of E_h depends only on the interatomic distances while E_c is a function of interatomic distances, valence number of atoms taking part in the bond formation, and the coordination number in the crystal. Furthermore the contribution to E_c comes only from heteropolar bonds, while to E_h both (heteropolar and homopolar bonds) contribute [5]. In case of elemental and binary compounds (AB), we have a single type of bonds and the coordination around cation and anion is the same. On the other hand, the compounds of complex structure AB_n (such as FeS_2), possess two types of bonds and different coordinations around cation and anion. In case of FeS (which in binary analog of FeS_2) we have only one type of bond Fe–S and the coordination around Fe and S is six. Compared to this in FeS_2 the cations (Fe) are still coordinated by six sulfur atoms (other sulfur atoms are relatively farther and therefore the interaction with them will be negligible), but S atoms are coordinated by four atoms (three Fe and one S atoms). In addition to three heteropolar bonds (Fe–S) there is an extra S–S bond in FeS_2. Since this extra bond is homopolar in nature and therefore it should not affect E_c. Thus whatever changes will occur, they will be only because of changes in coordination number which are taken care of through the value of b. Therefore, one can work without modification 3 proposed by Levine. Moreover, as long as the surroundings of the cation is not disturbed, the valence electron distribution around cation and anion will not be very different, as is obtained in case of the binary analog.

Table 1

compound	isomeric shift (mm/s)		Pauling resonating ionicity
FeF_2	1.4	[16]	0.9
$FeCl_2$	1.19	[16]	0.77
$FeBr_2$	1.13	[16]	0.74
FeI_2	1.04	[16]	0.7
$Fe_{1-x}S$	0.72	[13]	0.7
FeS_2 (marcasite)	0.277	[15]	0.6
FeS_2 (pyrite)	0.32	[13]	0.6
$FeSe_2$	0.395	[14]	0.59
$FeTe_2$	0.471	[14]	0.56
FeP_2	0.09	[14]	0.402
$FeAs_2$	0.39		0.395
$FeSb_2$	0.455		0.39

Regarding modification 4 we agree with the proposition that all the d electrons or part of d electrons are involved in bond formation (as they are loosely bound). But the value of Z_A^*, when it is greater than the total number of electrons (including all d electrons) cannot be justified until it is confirmed experimentally. The problem lies in modification 3, where the number of valence electrons for B is changed n times. Accordingly the value of Z_A has to be increased in order to match the experimental data. Hence, we feel that PVV theory with modifications 1, 2, and part of 4 can very well investigate complex structures like FeS_2, too. Based on the above concept the value of the ionic gap is given as

$$E_c = be^2 \frac{Z_A - Z_B}{r_0} e^{-K_s r_0} .$$

It is a well-known fact [10] that the sharing of d electrons in the bond formation in a material directly depends on ionicity (f_i) or covalency ($f_h = 1 - f_i$). The number of d electrons taking part in bond formation is directly proportional to the covalency of the bond. Moreover there is a correlation between ionicity and Mössbauer isomeric shift [11 to 13]. For ionic compounds there is a direct correlation, while in case of covalent compounds the ionicity and isomeric shift are inversely related, i.e. as the ionicity decreases the isomeric shift increases (see Table 1).

Now pyrite and pyrite-type chalcogenides are covalent in nature [10] and their isomeric shift is increasing from marcasite to $FeTe_2$ and hence their ionicity should be decreasing. It is clear from Table 1 that Pauling ionicity is in decreasing order, therefore, the number of d electrons taking part in bond formation should be increasing from marcasite to $FeTe_2$. On the oth-

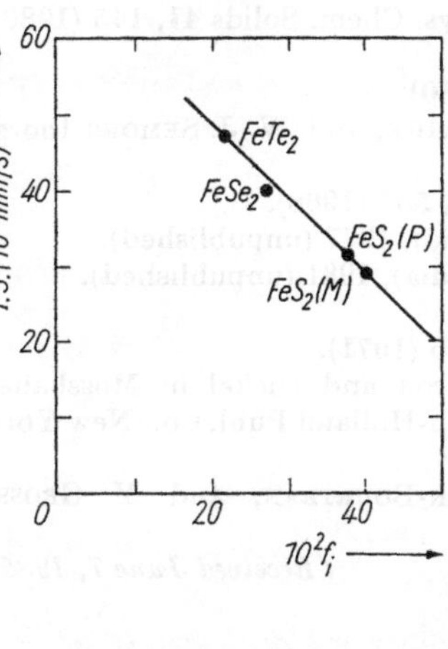

Fig. 1. Isomeric shift as a function of calculated ionicity of pyrite-type materials

Table 2

material	Z_{Fe}	E_h (eV)	E_c (eV)	E_p (eV)	f_i	isomeric shift (mm/s)
FeS	3	4.1	7.18	8.27	0.75	0.73 [13]
FeS_2 (marcasite)	4	5.3648	4.40	6.937	0.402	0.277 [15]
FeS_2 (pyrite)	4	5.27	4.1748	6.723	0.386	0.32 [13]
$FeSe_2$	4	4.6898	2.8645	5.4953	0.272	0.395 [14]
$FeTe_2$	4.5	3.8896	2.03	4.3913	0.215	0.471 [14]

Table 3

material	f_i (Phillips)	f_i (Pauling)
FeS_2 (marcasite)	0.402	0.6
FeS_2 (pyrite)	0.386	0.6
$FeSe_2$	0.272	0.59
$FeTe_2$	0.215	0.56

er hand, FeS is an ionic material [17] and its isomeric shift is 0.72. Hence in this case a smaller number of d electrons should take part in bond formation. Working on this line and using the modified PVV theory we have evaluated the ionic gap, homopolar gap, and hence the ionicity for these materials. The evaluated parameters are presented in Table 2. In Table 3 we present a comparison of the values of Phillips ionicity and Pauling ionicity. We find a similar trend in the values of both ionicities. In Fig. 1 we present a graph between calculated ionicity of pyrite-type materials and their isomeric shift and we find an inverse correlation between them.

3. Conclusion

Based on modified PVV theory the value of average gap, ionic gap, and hence Phillips ionicity have been determined for iron chalcogenides. An inverse correlation between isomeric shift and ionicity is found in pyrite-type dichalcogenides, which reflects their covalent character.

References

[1] B. F. LAVIN, J. chem. Phys. 59, 1463 (1973).
[2] J. C. PHILLIPS, Bonds and Bands in Semiconductors, Academic Press, Inc., London/New York 1973 (Chap. 1, 2, 7).
[3] J. C. PHILLIPS and J. A. VAN VECHTEN, Phys. Rev. 2, 2147 (1970).
[4] B. F. LAVIN, Phys. Rev. B 7, 2591 (1973).
[5] V. P. GUPTA, A. GUPTA, and V. K. SRIVASTAVA, phys. stat. sol. (b) **108**, 323 (1981).
[6] V. P. GUPTA and O. P. SINGH, phys. stat. sol. (b) **125**, K1 (1984).
[7] V. P. GUPTA, N. M. RAVINDRA, and V. K. SRIVASTAVA, J. Phys. Chem. Solids **41**, 145 (1980).
[8] H. HORITA, J. Phys. Soc. Japan **33**, 1723 (1972).
[9] H. HORITA and T. SUJUKI, Japan. J. appl. Phys. 19, 391 (1980).
[10] T. A. BITHER, R. J. BOUCHARD, W. H. CLAUD, P. C. DONOHUE, and W. J. SEMONS Inorg. Chem. 7, 2208 (1968).
[11] G. GERTH, P. KIENLE, and K. LUCHNER, Phys. Letters A 27, 557 (1968).
[12] B. K. JAIN, Ph.D. Thesis, Roorkee University, Roorkee (India), 1977 (unpublished).
[13] V. P. GUPTA, Ph.D. Thesis, Roorkee University, Roorkee (India), 1981 (unpublished).
[14] R. G. ARNOLD, Econ. Geol. 57, 72 (1962).
[15] A. KJEKSUS and D. G. NICHALSON, Acta chem. Scand. 25, 886 (1971).
[16] R. INGALLS, F. VAN DER WOUDE, and G. A. SAWATZKY, Iron and Nickel in Mössbauer Isomeric Shift, Ed. G. K. SHENOY, and F. E. WAGNER, North-Holland Publ. Co., New York 1978 (p. 363).
[17] M. WINTENBERGER, B. SROUR, C. MAYER, F. HARTMANN-BOURTRAN, and Y. GROSS, J. Physique 39, 965 (1978).

(Received June 7, 1985)

phys. stat. sol. (b) 133, 253 (1986)

Subject classification: 71.38; 71.35; S12.1

Georgi Nadjakov Institute of Solid State Physics, Bulgarian Academy of Sciences, Sofia[1])

Solitary Excitations in Molecular Chains in the Presence of Vibronic States

By

KH. I. PUSHKAROV and M. T. PRIMATAROVA

Solitary excitations due to the combination of an electronic excitation with intramolecular (both totally symmetric and nontotally symmetric) vibrations are shown to take place in molecular chains. Two cases are considered: a) when the molecules of the chain are fixed at their equilibrium positions and b) when the chain molecules are oscillating. In the second case solitary bound states of the chain deformation and the exciton-intramolecular excitation can exist and propagate along the chain. The parameters of the solitons and the energy of the system are calculated.

Es wird gezeigt, daß in Molekülketten solitäre Anregungen, verursacht durch die Wechselwirkung der Elektronenanregungen mit intramolekularen (sowohl totalsymmetrischen als auch nicht totalsymmetrischen) Schwingungen, existieren. Zwei Fälle werden untersucht: a) wenn die Kettenmoleküle an ihren Gleichgewichtsstellen fixiert sind und b) wenn die Kettenmoleküle oszillieren. Im zweiten Fall können solitäre gebundene Zustände der Kettendeformation und der Exziton-Intramolekülanregung existieren und sich entlang der Kette fortpflanzen. Die Parameter der Solitonen und die Energie des Systems werden berechnet.

1. Introduction

The problem of the existence of solitary excitations in molecular crystals have been studied intensively [1 to 9]. The most interesting effects have been obtained when the presence of an exciton causes a deformation of a molecular chain and forms a solitary bound state with it. This problem is studied in detail in harmonic and anharmonic approximation with respect to the molecular motion in the chain.

But the molecules forming the crystal are, in general, more complicated systems of electron excitations and intramolecular vibrations. The theory of the so-called vibronic excited states in molecular systems has been developed in a series of papers, reviews and books [10 to 20].

In the present paper we shall study the problem of the existence of solitary excitations in molecular chains when the presence of the vibronic excitations is taken into account. It will be shown that solitary bound states of excitons and intramolecular vibrations can exist and propagate in such media both in the case of totally symmetric vibrations and in the case of non-totally symmetric ones.

2. Hamiltonian and Basic Equations

We consider the simplest model of one-dimensional molecular crystals containing one molecule per elementary cell. When each crystal molecule interacts with its nearest neighbours only, the Hamiltonian corresponding to this situation can be written in

---

[1]) Lenin Boul. 72, 1784 Sofia, Bulgaria.

Heitler-London approximation as follows [21]:

$$H = \mathscr{E}_0 + T + U + \hbar\Omega_0 \sum_n b_n^+ b_n + \sum_n [\Delta E + D_{n+1n}^{(0)} + D_{nn-1}^{(0)} + \hbar\,\Delta\Omega(b_n^+ b_n + \tfrac{1}{2})]\, B_n^+ B_n +$$
$$+ \sum_n [M_{n+1n}^{(0)} V_n^+ V_{n+1} + \text{h.c.}] + \sum_n [M_{n+1n}^{(1)} V_n^+ (b_n^+ b_{n+1} + \text{h.c.})\, V_{n+1} + \text{h.c.}] +$$
$$+ \sum_n \tilde{\mathscr{M}}_{n+1n}(b_n^+ b_{n+1} + \text{h.c.}) + \sum_n D_{n+1n}^{(1)}(b_n^+ b_{n+1} + \text{h.c.})\,(B_n^+ B_n + B_{n+1}^+ B_{n+1})\,. \tag{1}$$

In (1) $\mathscr{E}_0$ is the energy of the molecule ground state including the interaction energy of the chain molecules corresponding to their ground state, ΔE is the minimum difference of the excited and ground electronic terms of the chain molecule, $\Delta\Omega \equiv \Omega_1 - \Omega_0$ is the frequency difference of the nuclear vibrations in the excited and ground electronic states (for non-totally symmetric vibrations usually $\Delta\Omega < 0$), $D_{mn}^{(0)}$ is the change of the interaction energy of the molecule with the others during the transition to the excited electronic state in Condon approximation, $M_{mn}^{(0)}$ is the resonance matrix element of the molecules m and n in the same approximation, $D_{mn}^{(1)}$ and $M_{mn}^{(1)}$ characterize the correction to the Condon approximation taking into account the possibility of existence of non-totally symmetric vibrations, $\tilde{\mathscr{M}}_{mn}$ is the resonance matrix element of the intramolecular phonons. The resonance matrix elements mentioned have the form

$$M_{mn}^{(l)} = \frac{|\boldsymbol{d}^{(l)}|^2}{2^l\,|\boldsymbol{n} - \boldsymbol{m}|^3}\,(1 - 3\cos\theta^{(l)})\,; \qquad l = 0, 1\,, \tag{2a}$$

$$\tilde{\mathscr{M}}_{mn} = \frac{|\boldsymbol{\delta}|^2}{|\boldsymbol{n} - \boldsymbol{m}|^3}\,(1 - 3\cos\alpha)\,, \tag{2b}$$

$$D_{mn}^{(1)} = \frac{1}{2}\left[\left(\sqrt{\frac{\Omega_0}{\Omega_1}} - 1\right)\tilde{\mathscr{M}}_{mn} + \frac{\sqrt{\Omega_0/\Omega_1}}{|\boldsymbol{n} - \boldsymbol{m}|^3}\,(\boldsymbol{\partial}\cdot\boldsymbol{\delta} - 3\,|\boldsymbol{\partial}|\,|\boldsymbol{\delta}|\cos\alpha\cos\gamma)\right], \tag{2c}$$

where $\boldsymbol{d}^{(0)}$ is the dipole moment of the vibronic transition in the molecule in Condon approximation, $\boldsymbol{d}^{(1)}$ is a correction to the Condon approximation in the expression for the dipole moment of the vibronic transition, $\boldsymbol{\delta}$ is the dipole moment of the one-phonon transition in the electronic ground state of the molecule, $\boldsymbol{\partial}$ is the correction to the dipole moment of the one-phonon transition in the first excited electronic state of the molecule determined by the change in the electronic wave function under the influence of the non-totally symmetrical vibrations of the nuclei, $\theta^{(0)}$, $\theta^{(1)}$, α, and γ are the angles which the chain direction makes with $\boldsymbol{d}^{(0)}$, $\boldsymbol{d}^{(1)}$, $\boldsymbol{\delta}$, and $\boldsymbol{\partial}$, respectively. So generally speaking the matrix elements (2) can take positive and negative values depending on these angles. B_n^+, b_n^+, V_n^+ are the creation operators of the electronic excitation, intramolecular phonon and vibronic excitation of the molecule n, respectively, and B_n, b_n, V_n are the corresponding annihilation operators. The operators B_n^+, B_n as well as b_n^+, b_n satisfy Bose commutation relations.

The vibronic operator has the form

$$V_n = B_n \exp\,[\xi(b_n^+ - b_n)]\,\exp\,[\beta(b_n^+ b_n^+ - b_n b_n)]\,, \tag{3}$$

where ξ and β characterize the interconnection of electronic excitations and the intramolecular vibrations,

$$\xi = \frac{1}{2}\,q\Big/\sqrt{\frac{\hbar}{2m\Omega_0}} \tag{3a}$$

represents the ratio of the displacement from the equilibrium position of the normal vibrational coordinate q to the double amplitude of the zero molecule vibration in its ground state,

$$\beta = \frac{1}{4} \ln \frac{\Omega_0}{\Omega_1}. \tag{3b}$$

Finally

$$T = \sum_n p_n^2/2m \tag{4}$$

is the kinetic energy of the molecules in the chain, where p_n and m are momentum and mass of the molecules, respectively, and

$$U = \frac{mv_0^2}{2} \sum_n (u_n - u_{n-1})^2 \tag{5}$$

is the strain energy of the chain in harmonic approximation, u_n and v_0 being the molecule displacements from their equilibrium positions and the sound velocity, respectively. The lattice constant is chosen as unity.

In our considerations we shall use the following representation of the vibronic operator (3) [18, 21] ($\xi = 0$):

$$V_n = (1 - r^2)^{1/4} B_n , \tag{6}$$

where

$$r \equiv \frac{\Omega_0 - \Omega_1}{\Omega_0 + \Omega_1}. \tag{7}$$

Because of the dipole character of the molecule interaction (2) we adopt

$$M_{nn-1}^{(l)} \approx M_0^{(l)} - M_1^{(l)}(u_n - u_{n-1}) , \qquad M_1^{(l)} \equiv \frac{\partial M_{nn-1}^{(l)}}{\partial(u_n - u_{n-1})} ; \qquad l = 0, 1 , \tag{8}$$

$$\tilde{\mathcal{M}}_{nn-1} \approx \tilde{\mathcal{M}}_0 - \tilde{\mathcal{M}}_1(u_n - u_{n-1}) , \qquad \tilde{\mathcal{M}}_1 \equiv \frac{\partial \tilde{\mathcal{M}}_{nn-1}}{\partial(u_n - u_{n-1})} , \tag{9}$$

$$D_{nn-1}^{(0)} \approx D_0 - D_1(u_n - u_{n-1}) , \qquad D_1 \equiv \frac{\partial D_{nn-1}^{(0)}}{\partial(u_n - u_{n-1})} , \tag{10}$$

$$D_{nn-1}^{(1)} \approx \mathcal{D}_0 - \mathcal{D}_1(u_n - u_{n-1}) , \qquad \mathcal{D}_1 \equiv \frac{\partial D_{nn-1}^{(1)}}{\partial(u_n - u_{n-1})} . \tag{11}$$

The substitution of (6) to (11) into (1) leads to the following form of the Hamiltonian:

$$H = \mathcal{E}_0 + H_B + H_b + H_w + H_{int} , \tag{12}$$

where

$$H_B = \tilde{\varepsilon} \sum_n B_n^+ B_n + \mathcal{M}_0 \sum_n B_n^+(B_{n+1} + B_{n-1}) , \tag{12a}$$

$$H_b = \varepsilon_0 \sum_n b_n^+ b_n + \tilde{\mathcal{M}}_0 \sum_n b_n^+(b_{n+1} + b_{n-1}) , \tag{12b}$$

$$H_w = T + U , \tag{12c}$$

$$H_{\text{int}} = \Delta\varepsilon \sum_n b_n^+ b_n B_n^+ B_n + \mathcal{D}_0 \sum_n (b_n^+ b_{n+1} + \text{h.c.}) (B_n^+ B_n + B_{n+1}^+ B_{n+1}) +$$
$$+ R_0 \sum_n (b_n^+ b_{n+1} + \text{h.c.}) (B_n^+ B_{n+1} + \text{h.c.}) - D_1 \sum_n (u_{n+1} - u_{n-1}) B_n^+ B_n -$$
$$- \mathcal{M}_1 \sum_n (u_{n+1} - u_n) (B_n^+ B_{n+1} + \text{h.c.}) - \tilde{\mathcal{M}}_1 \sum_n (u_{n+1} - u_n) (b_n^+ b_{n+1} + \text{h.c.}) -$$
$$- \mathcal{D}_1 \sum_n (u_{n+1} - u_n) (b_n^+ b_{n+1} + \text{h.c.}) (B_n^+ B_n + B_{n+1}^+ B_{n+1}) -$$
$$- R_1 \sum_n (u_{n+1} - u_n) (b_n^+ b_{n+1} + \text{h.c.}) (B_n^+ B_{n+1} + \text{h.c.}) \,. \tag{12d}$$

In (12a) to (12d) the following notations have been made for simplicity:

$$\tilde{\varepsilon} \equiv \Delta E + 2D_0 + \tfrac{1}{2} \Delta\varepsilon \,, \qquad \Delta\varepsilon \equiv \hbar \, \Delta\Omega \,, \qquad \varepsilon_0 \equiv \hbar\Omega_0 \,, \tag{13}$$

$$\mathcal{M}_i \equiv \sqrt{1 - r^2} \, M_i^{(0)} \,; \qquad i = 0, 1 \,, \tag{14}$$

$$R_i \equiv \sqrt{1 - r^2} \, M_i^{(1)} \,; \qquad i = 0, 1 \,. \tag{15}$$

One can see that H_B, H_b, and H_w describe the pure exciton, phonon and lattice vibrations parts of H, respectively, and H_{int} represents the interaction between excitons, phonons, and lattice vibrations.

Using the commutation relations

$$[B_n, B_m^+] = \delta_{nm} \,, \qquad [B_n, B_m] = 0 \,,$$

$$[b_n, b_m^+] = \delta_{nm} \,, \qquad [b_n, b_m] = 0 \,,$$

as well as those of the operators u_n and p_n,

$$[u_n, p_m] = i\hbar\delta_{nm} \,,$$

we obtain the following equations of motion:

$$i\hbar \frac{\partial B_n}{\partial t} = \tilde{\varepsilon} B_n + \mathcal{M}_0(B_{n+1} + B_{n-1}) + \Delta\varepsilon \, b_n^+ b_n B_n + \mathcal{D}_0[b_n^+ (b_{n+1} + b_{n-1}) + \text{h.c.}] \, B_n +$$
$$+ R_0[(b_n^+ b_{n+1} + \text{h.c.}) \, B_{n+1} + (b_n^+ b_{n-1} + \text{h.c.}) \, B_{n-1}] - D_1(u_{n+1} - u_{n-1}) \, B_n -$$
$$- \mathcal{M}_1[(u_{n+1} - u_n) \, B_{n+1} + (u_n - u_{n-1}) \, B_{n-1}] -$$
$$- \mathcal{D}_1[(u_{n+1} - u_n) \, (b_n^+ b_{n+1} + \text{h.c.}) + (u_n - u_{n-1}) \, (b_n^+ b_{n-1} + \text{h.c.})] \, B_n -$$
$$- R_1[(u_{n+1} - u_n) \, (b_n^+ b_{n+1} + \text{h.c.}) \, B_{n+1} + (u_n - u_{n-1}) \, (b_n^+ b_{n-1} + \text{h.c.}) \, B_{n-1}] \,, \tag{16}$$

$$i\hbar \frac{\partial b_n}{\partial t} = \varepsilon_0 b_n + \tilde{\mathcal{M}}_0(b_{n+1} + b_{n-1}) + \Delta\varepsilon \, B_n^+ B_n b_n +$$
$$+ \mathcal{D}_0[B_{n+1}^+ B_{n+1} b_{n+1} + B_{n-1}^+ B_{n-1} b_{n-1} + B_n^+ B_n (b_{n+1} + b_{n-1})] +$$
$$+ R_0[(B_n^+ B_{n+1} + \text{h.c.}) \, b_{n+1} + (B_n^+ B_{n-1} + \text{h.c.}) \, b_{n-1}] -$$
$$- \tilde{\mathcal{M}}_1[(u_{n+1} - u_n) \, b_{n+1} + (u_n - u_{n-1}) \, b_{n-1}] -$$
$$- \mathcal{D}_1[(u_{n+1} - u_n) \, (B_n^+ B_n + B_{n+1}^+ B_{n+1}) \, b_{n+1} +$$
$$+ (u_n - u_{n-1}) \, (B_n^+ B_n + B_{n-1}^+ B_{n-1}) \, b_{n-1}] -$$
$$- R_1[(u_{n+1} - u_n) \, (B_n^+ B_{n+1} + \text{h.c.}) \, b_{n+1} + (u_n - u_{n-1}) \, (B_n^+ B_{n-1} + \text{h.c.}) \, b_{n-1}] \,, \tag{17}$$

$$\frac{\partial u_n}{\partial t} = \frac{p_n}{m}, \tag{18}$$

$$\begin{aligned}\frac{\partial p_n}{\partial t} &= mv_0^2(u_{n+1} + u_{n-1} - 2u_n) - D_1(B_{n+1}^+ B_{n+1} - B_{n-1}^+ B_{n-1}) - \\ &- \mathcal{M}_1[B_n^+ (B_{n+1} - B_{n-1}) + \text{h.c.}] - \tilde{\mathcal{M}}_1[b_n^+ (b_{n+1} - b_{n-1}) + \text{h.c.}] - \\ &- \mathcal{D}_1[(b_n^+ b_{n+1} + \text{h.c.}) (B_n^+ B_n + B_{n+1}^+ B_{n+1}) - \\ &- (b_{n-1}^+ b_n + \text{h.c.}) (B_n^+ B_n + B_{n-1}^+ B_{n-1})] - \\ &- R_1[(b_n^+ b_{n+1} + \text{h.c.}) (B_n^+ B_{n+1} + \text{h.c.}) - (b_{n-1}^+ b_n + \text{h.c.}) (B_n^+ B_{n-1} + \text{h.c.})] .\end{aligned} \tag{19}$$

Eliminating p_n from (18) and (19) we obtain

$$\begin{aligned}m \frac{\partial^2 u_n}{\partial t^2} &= mv_0^2(u_{n+1} + u_{n-1} - 2u_n) - D_1(B_{n+1}^+ B_{n+1} - B_{n-1}^+ B_{n-1}) - \\ &- \mathcal{M}_1[B_n^+ (B_{n+1} - B_{n-1}) + \text{h.c.}] - \tilde{\mathcal{M}}_1[b_n^+ (b_{n+1} - b_{n-1}) + \text{h.c.}] - \\ &- \mathcal{D}_1[(b_n^+ b_{n+1} + \text{h.c.}) (B_n^+ B_n + B_{n+1}^+ B_{n+1}) - \\ &- (b_{n-1}^+ b_n + \text{h.c.}) (B_n^+ B_n + B_{n-1}^+ B_{n-1})] - \\ &- R_1[(b_n^+ b_{n+1} + \text{h.c.}) (B_n^+ B_{n+1} + \text{h.c.}) - (b_{n-1}^+ b_n + \text{h.c.}) (B_n^+ B_{n-1} + \text{h.c.})] .\end{aligned} \tag{20}$$

Equations (16), (17), and (20) are the basic equations of the problem. We shall solve them in continuum limit treating the quantities B, b, and u as classical ones. So we are interested in the case when the deformation region caused by the presence of electronic and phonon excitations has a size $\mathcal{L}$ much larger than the lattice constant ($\mathcal{L} \gg 1$). Then, treating B_n^+, B_n, b_n^+, b_n, and u_n as smooth functions of the position and going over to a continuum approximation,

$$f_n(t) \to f(x, t) , \qquad f = B^+, B, b^+, b, u , \tag{21}$$

we can write (16), (17), and (20) in the form

$$\begin{aligned}i\hbar \frac{\partial B}{\partial t} &= \left[\bar{\varepsilon} + 2\mathcal{M}_0 - 2(D_1 + \mathcal{M}_1) \frac{\partial u}{\partial x}\right] B + \mathcal{M}_0 \frac{\partial^2 B}{\partial x^2} + \\ &+ \left\{\Delta\varepsilon + 4 \left[\mathcal{D}_0 + R_0 - (\mathcal{D}_1 + R_1) \frac{\partial u}{\partial x}\right]\right\} |b|^2 B ,\end{aligned} \tag{22}$$

$$\begin{aligned}i\hbar \frac{\partial b}{\partial t} &= \left[\varepsilon_0 + 2\tilde{\mathcal{M}}_0 - 2\tilde{\mathcal{M}}_1 \frac{\partial u}{\partial x}\right] b + \tilde{\mathcal{M}}_0 \frac{\partial^2 b}{\partial x^2} + \\ &+ \left\{\Delta\varepsilon + 4 \left[\mathcal{D}_0 + R_0 - (\mathcal{D}_1 + R_1) \frac{\partial u}{\partial x}\right]\right\} |B|^2 b ,\end{aligned} \tag{23}$$

$$m \frac{\partial^2 u}{\partial t^2} = mv_0^2 \frac{\partial^2 u}{\partial x^2} - 2 \frac{\partial}{\partial x} [(D_1 + \mathcal{M}_1) |B|^2 + \tilde{\mathcal{M}}_1 |b|^2 + 2(\mathcal{D}_1 + R_1) |b|^2 |B|^2] . \tag{24}$$

Now, we look for a solution of (22) to (24) in the form of a wave propagating along the chain with constant velocity v,

$$f(x, t) = f(x - vt)\ ; \quad \text{(see (21))}.$$

Then, integrating (24) we obtain

$$\frac{\partial u}{\partial x} = 2\,\frac{(D_1 + \mathcal{M}_1)\,|B|^2 + \tilde{\mathcal{M}}_1\,|b|^2 + 2(\mathcal{D}_1 + R_1)\,|b|^2\,|B|^2}{mv_0^2(1 - s^2)}, \qquad s = v/v_0\,. \tag{25}$$

The substitution of (25) into (22) and (23) leads to the following set of nonlinear differential equations:

$$i\hbar\,\frac{\partial B}{\partial t} = (\tilde{\varepsilon} + 2\mathcal{M}_0)\,B + \mathcal{M}_0\,\frac{\partial^2 B}{\partial x^2} - [a(D_1 + \mathcal{M}_1)\,|B|^2 - c\,|b|^2]\,B - {}$$
$$- 2(\mathcal{D}_1 + R_1)\,[2a\,|B|^2 + \tilde{a}\,|b|^2]\,|b|^2\,B\,, \tag{26}$$

$$i\hbar\,\frac{\partial b}{\partial t} = (\varepsilon_0 + 2\tilde{\mathcal{M}}_0)\,b + \tilde{\mathcal{M}}_0\,\frac{\partial^2 b}{\partial x^2} + [c\,|B|^2 - \tilde{a}\tilde{\mathcal{M}}_1\,|b|^2]\,b - {}$$
$$- 2(\mathcal{D}_1 + R_1)\,[a\,|B|^2 + 2\tilde{a}\,|b|^2]\,|B|^2\,b\,, \tag{27}$$

where

$$a \equiv \frac{4(D_1 + \mathcal{M}_1)}{mv_0^2(1 - s^2)}, \quad \tilde{a} \equiv \frac{4\tilde{\mathcal{M}}_1}{mv_0^2(1 - s^2)}, \quad c \equiv \Delta\varepsilon + 4(\mathcal{D}_0 + R_0) - a\tilde{\mathcal{M}}_1\,. \tag{28}$$

In the case when the motion of the chain molecules is not taken into account, i.e $H_\mathrm{w} = 0$ in (12) and H_int has the form

$$H_\mathrm{int} = \Delta\varepsilon \sum_n b_n^+ b_n B_n^+ B_n + \mathcal{D}_0 \sum_n (b_n^+ b_{n+1} + \text{h.c.})\,(B_n^+ B_n + B_{n+1}^+ B_{n+1}) + {}$$
$$+ R_0 \sum_n (b_n^+ b_{n+1} + \text{h.c.})\,(B_n^+ B_{n+1} + \text{h.c.})\,, \tag{29}$$

the procedure described above leads to the following set of nonlinear differential equations:

$$i\hbar\,\frac{\partial B}{\partial t} = (\tilde{\varepsilon} + 2\mathcal{M}_0)\,B + \mathcal{M}_0\,\frac{\partial^2 B}{\partial x^2} + [\Delta\varepsilon + 4(\mathcal{D}_0 + R_0)]\,|b|^2\,B\,, \tag{30}$$

$$i\hbar\,\frac{\partial b}{\partial t} = (\varepsilon_0 + 2\tilde{\mathcal{M}}_0)\,b + \tilde{\mathcal{M}}_0\,\frac{\partial^2 b}{\partial x^2} + [\Delta\varepsilon + 4(\mathcal{D}_0 + R_0)]\,|B|^2\,b\,. \tag{31}$$

3. Solitary Bound Formations

It is convenient to treat the case when the chain molecule motion is taken into account separately from the case when this motion is absent.

3.1 Case of fixed chain molecules

In this case (30) and (31) have the following soliton solutions:

$$B(x, t) = \left(\frac{1}{\mathcal{L}}\,\frac{\tilde{\mathcal{M}}_0}{\mathcal{M}_0 + \tilde{\mathcal{M}}_0}\right)^{1/2} \mathrm{e}^{i(k_1 x - \omega_1 t + \varphi_1)}\,\mathrm{sech}\,\frac{x - x_0 - vt}{\mathcal{L}}, \tag{32}$$

$$b(x, t) = \left(\frac{1}{\mathcal{L}}\,\frac{\mathcal{M}_0}{\mathcal{M}_0 + \tilde{\mathcal{M}}_0}\right)^{1/2} \mathrm{e}^{i(k_2 x - \omega_2 t + \varphi_2)}\,\mathrm{sech}\,\frac{x - x_0 - vt}{\mathcal{L}}, \tag{33}$$

where

$$k_1 = -\frac{\hbar v}{2\mathcal{M}_0}, \qquad k_2 = -\frac{\hbar v}{2\tilde{\mathcal{M}}_0}, \tag{34}$$

$$\hbar\omega_1 = \tilde{\varepsilon} + 2\mathcal{M}_0 - \frac{\hbar^2 v^2}{4\mathcal{M}_0} + \frac{\mathcal{M}_0}{\mathcal{L}^2}, \tag{35a}$$

$$\hbar\omega_2 = \varepsilon_0 + 2\tilde{\mathcal{M}}_0 - \frac{\hbar^2 v^2}{4\tilde{\mathcal{M}}_0} + \frac{\tilde{\mathcal{M}}_0}{\mathcal{L}^2}, \tag{35b}$$

$$\mathcal{L} = \frac{2(\mathcal{M}_0 + \tilde{\mathcal{M}}_0)}{\Delta\varepsilon + 4(\mathcal{D}_0 + R_0)}. \tag{36}$$

In the formulae above $\mathcal{L}$, v, x_0, $\varphi_{1,2}$ are length, velocity, centre, and initial phases of the solitons, respectively. We also consider the situation when the condition

$$\sum_n B_n^+ B_n = 1 \tag{37a}$$

is realized, i.e. only one exciton is created in the chain. If we also assume the presence of one phonon excitation, i.e.

$$\sum_n b_n^+ b_n = 1\,, \tag{37b}$$

the following normalization condition has to be used in continuum approximation:

$$\int_{-\infty}^{\infty} (|B|^2 + |b|^2)\,\mathrm{d}x = 2\,. \tag{38}$$

It follows from (32) and (33) that the whole solitary formation propagates with constant velocity v having a length $\mathcal{L}$ given by (36).

The analysis of (32) to (36) shows that for the low-lying states, when $\mathcal{M}_0$, $\tilde{\mathcal{M}}_0$, $\mathcal{D}_0$, R_0 are negative (as it was mentioned in Section 2, $\Delta\varepsilon < 0$ for non-totally symmetric vibrations), the excitation energies $\hbar\omega_1$ and $\hbar\omega_2$ have minimum values which are smaller than the case of free excitations of an amount of $\mathcal{M}_0/\mathcal{L}^2$ and $\tilde{\mathcal{M}}_0/\mathcal{L}^2$, respectively.

For the validity of the continuum approximation used the inequality

$$\mathcal{L} \gg 1 \tag{39}$$

must be satisfied which can be reached for realistic values of $\Delta\varepsilon$, $\mathcal{M}_0$, $\tilde{\mathcal{M}}_0$, $\mathcal{D}_0$, and R_0.

Taking into account the continuum form of the Hamiltonian (12 a, b, d) where the dependence of u is neglected ($\partial u/\partial x = 0$), i.e.

$$H_B = \int_{-\infty}^{\infty} \left[(\tilde{\varepsilon} + 2\mathcal{M}_0)\,|B|^2 - \mathcal{M}_0 \left|\frac{\partial B}{\partial x}\right|^2 \right] \mathrm{d}x =$$

$$= 2\left(\tilde{\varepsilon} + 2\mathcal{M}_0 - \frac{\hbar^2 v^2}{4\mathcal{M}_0} - \frac{\mathcal{M}_0}{3\mathcal{L}^2}\right) \frac{\tilde{\mathcal{M}}_0}{\mathcal{M}_0 + \tilde{\mathcal{M}}_0}, \tag{40a}$$

$$H_b = \int_{-\infty}^{\infty} \left[(\varepsilon_0 + 2\tilde{\mathcal{M}}_0)\,|b|^2 - \tilde{\mathcal{M}}_0 \left|\frac{\partial b}{\partial x}\right|^2 \right] \mathrm{d}x =$$

$$= 2\left(\varepsilon_0 + 2\tilde{\mathcal{M}}_0 - \frac{\hbar^2 v^2}{4\tilde{\mathcal{M}}_0} - \frac{\tilde{\mathcal{M}}_0}{3\mathcal{L}^2}\right) \frac{\mathcal{M}_0}{\mathcal{M}_0 + \tilde{\mathcal{M}}_0}, \tag{40b}$$

$$H_{\text{int}} = [\Delta\varepsilon + 4(\mathcal{D}_0 + R_0)] \int_{-\infty}^{\infty} |b|^2 |B|^2 \, dx = \frac{4[\Delta\varepsilon + 4(\mathcal{D}_0 + R_0)]}{3\mathcal{L}} \frac{\mathcal{M}_0\tilde{\mathcal{M}}_0}{(\mathcal{M}_0 + \tilde{\mathcal{M}}_0)^2} \tag{40c}$$

we find the total energy of the system

$$E_{\text{tot}} = \mathcal{E}_0 + 2\left[\left(\tilde{\varepsilon} + 2\mathcal{M}_0 - \frac{\hbar^2 v^2}{4\mathcal{M}_0}\right)\frac{\tilde{\mathcal{M}}_0}{\mathcal{M}_0 + \tilde{\mathcal{M}}_0} + \left(\varepsilon_0 + 2\tilde{\mathcal{M}}_0 - \frac{\hbar^2 v^2}{4\tilde{\mathcal{M}}_0}\right)\frac{\mathcal{M}_0}{\mathcal{M}_0 + \tilde{\mathcal{M}}_0}\right] + $$
$$+ \frac{[\Delta\varepsilon + 4(\mathcal{D}_0 + R_0)]^2 \mathcal{M}_0\tilde{\mathcal{M}}_0}{3(\mathcal{M}_0 + \tilde{\mathcal{M}}_0)^3}.$$

3.2 *Case of oscillating molecules*

Now, we shall consider the case when the chain molecules oscillate about their equilibrium positions. In this case the nonlinear equations (26) and (27) have the following soliton solutions:

$$B(x, t) = \left(\frac{1}{(1 + \zeta)\mathcal{L}} \frac{\text{th}\,\eta}{\eta}\right)^{1/2} e^{i(k_1 x - \omega_1 t + \varphi_1)} \left(1 + \text{sech}\,\eta\,\text{ch}\,\frac{x - x_0 - vt}{\mathcal{L}}\right)^{-1/2}, \tag{41a}$$

$$b(x, t) = \left(\frac{\zeta}{(1 + \zeta)\mathcal{L}} \frac{\text{th}\,\eta}{\eta}\right)^{1/2} e^{i(k_2 x - \omega_2 t + \varphi_2)} \left(1 + \text{sech}\,\eta\,\text{ch}\,\frac{x - x_0 - vt}{\mathcal{L}}\right)^{-1/2}. \tag{41b}$$

Here k_1 and k_2 are given by (34),

$$\hbar\omega_1 = \tilde{\varepsilon} + 2\mathcal{M}_0 - \frac{\hbar^2 v^2}{4\mathcal{M}_0} + \frac{\mathcal{M}_0}{4\mathcal{L}^2}, \tag{42a}$$

$$\hbar\omega_2 = \varepsilon_0 + 2\tilde{\mathcal{M}}_0 - \frac{\hbar^2 v^2}{4\tilde{\mathcal{M}}_0} + \frac{\tilde{\mathcal{M}}_0}{4\mathcal{L}^2}, \tag{42b}$$

$$\mathcal{L} = \frac{c(\mathcal{M}_0 + \tilde{\mathcal{M}}_0) + a(\mathcal{D}_1 + \mathcal{M}_1)\tilde{\mathcal{M}}_0 + \tilde{a}\tilde{\mathcal{M}}_1\mathcal{M}_0}{c^2 - a\tilde{a}\mathcal{M}_1\tilde{\mathcal{M}}_1} \frac{\eta}{\text{th}\,\eta}, \tag{43}$$

$$\eta = \frac{2\sqrt{G}}{1 + \zeta}, \qquad G = \frac{2\zeta(\mathcal{D}_1 + R_1)(2a + \zeta\tilde{a})}{3\mathcal{M}_0}, \tag{44}$$

$$\zeta \equiv \frac{c\mathcal{M}_0 + a(\mathcal{D}_1 + \mathcal{M}_1)\tilde{\mathcal{M}}_0}{c\tilde{\mathcal{M}}_0 + \tilde{a}\tilde{\mathcal{M}}_1\mathcal{M}_0}. \tag{45}$$

For the existence of the solitary formation (41a), (41b), because of the continuum approximation used, the inequality (39) must be satisfied. As the dipole moments $\boldsymbol{d}^{(0)}$, $\boldsymbol{d}^{(1)}$, $\boldsymbol{\delta}$, and $\boldsymbol{\partial}$ of the molecules can make different angles $\theta^{(0)}$, $\theta^{(1)}$, α, and γ with the chain direction ($\Delta\varepsilon$ can be positive or negative) the condition (39) may be satisfied.

Equations (26) and (27) are self-consistent to the order of $\mathcal{L}^{-5/2}$ while in the absence of the acoustic branch (the molecules are fixed at their equilibrium positions) the nonlinear terms in (30) and (31) are of the order of $\mathcal{L}^{-3/2}$. In both cases, besides (39), the whole amplitudes B and b must be small. This leads to the condition

$$\frac{\text{th}^2\,\eta}{\eta} \ll \left(1 + \begin{Bmatrix}\zeta \\ \zeta^{-1}\end{Bmatrix}\right) \frac{c(\mathcal{M}_0 + \tilde{\mathcal{M}}_0) + a(\mathcal{D}_1 + \mathcal{M}_1)\tilde{\mathcal{M}}_0 + \tilde{a}\tilde{\mathcal{M}}_1\mathcal{M}_0}{c^2 - a\tilde{a}(\mathcal{D}_1 + \mathcal{M}_1)\tilde{\mathcal{M}}_1}, \tag{46}$$

where ζ and in ζ^{-1} in the brackets correspond to B and b, respectively.

Other soliton solutions of (26) and (27) have the form

$$B(x, t) = \left(\frac{1}{(1+\zeta)\,\mathscr{L}}\,\frac{\operatorname{tg}\eta}{\eta}\right)^{1/2} e^{i(k_1 x - \omega_1 t + \varphi_1)} \left(1 + \sec\eta \operatorname{ch}\frac{x - x_0 - vt}{\mathscr{L}}\right)^{-1/2}, \tag{47a}$$

$$b(x, t) = \left(\frac{\zeta}{(1+\zeta)\,\mathscr{L}}\,\frac{\operatorname{tg}\eta}{\eta}\right)^{1/2} e^{i(k_2 x - \omega_2 t + \varphi_2)} \left(1 + \sec\eta \operatorname{ch}\frac{x - x_0 - vt}{\mathscr{L}}\right)^{-1/2}. \tag{47b}$$

In (47 a, b) k_1 and k_2, and $\hbar\omega_1$ and $\hbar\omega_2$ are given by (34) and (42 a, b), respectively. The soliton length $\mathscr{L}$ has the form (43) where th η is changed by tg η, η being determined according to (44) with $-G$ instead of G (the radicant must be positive!).

As for the previous case the conditions (39) and (46) (where th η is changed by tg η) must be satisfied.

When $\mathscr{M}_0$ and $\tilde{\mathscr{M}}_0$ have negative values both (41a, b) and (47a, b) are energetically more favourable: one can see that in this case the expressions for the energies (35a, b) and (42a, b) contain negative terms $\sim \mathscr{L}^{-2}$.

Using (25) and (41a, b) or (47a, b) the deformation caused by the presence of solitary excitations can be determined. Integrating (25) we find that the total length change of the molecular chain due to the solitary formation has the value

$$\Delta = \int\limits_{-\infty}^{\infty} \frac{\partial u}{\partial x}\,\mathrm{d}x =$$

$$= \begin{cases} \dfrac{4}{(1+\zeta)\,mv_0^2(1-s^2)}\left[D_1 + \mathscr{M}_1 + \zeta\tilde{\mathscr{M}}_1 + \dfrac{2(\mathscr{D}_1 + R_1)\,\zeta}{(1+\zeta)\,\mathscr{L}}\,\dfrac{\eta\coth\eta - 1}{\eta^2}\right] & \text{for (41 a, b)}\,, \\[2ex] \dfrac{4}{(1+\zeta)\,mv_0^2(1-s^2)}\left[D_1 + \mathscr{M}_1 + \zeta\tilde{\mathscr{M}}_1 + \dfrac{2(\mathscr{D}_1 + R_1)\,\zeta}{(1+\zeta)\,\mathscr{L}}\,\dfrac{1 - \eta\operatorname{cotg}\eta}{\eta^2}\right] & \text{for (47 a, b)}\,. \end{cases}$$

Depending on the signs of the matrix elements Δ can be positive or negative, i.e. there is a possibility of elongation or striction of the molecular chain.

Finally, we shall calculate the total energy of the system. Using the Hamiltonian (12a to d) in continuum approximation we find

$$\begin{aligned} E_{\text{tot}} = \frac{2}{1+\zeta}&\left\{\tilde{\varepsilon} + 2\mathscr{M}_0 - \frac{\hbar^2v^2}{4\mathscr{M}_0} - \frac{\mathscr{M}_0}{8\mathscr{L}^2}\,\frac{\operatorname{sh}\eta}{\eta}\left(\frac{1}{1+\operatorname{ch}\eta} + \frac{\operatorname{sh}\eta - \eta}{\operatorname{sh}^3\eta}\right) + \right. \\ &+ \left.\zeta\left[\varepsilon_0 + 2\tilde{\mathscr{M}}_0 - \frac{\hbar^2v^2}{4\tilde{\mathscr{M}}_0} - \frac{\tilde{\mathscr{M}}_0}{8\mathscr{L}^2}\,\frac{\operatorname{sh}\eta}{\eta}\left(\frac{1}{1+\operatorname{ch}\eta} + \frac{\operatorname{sh}\eta - \eta}{\operatorname{sh}^3\eta}\right)\right]\right\} + \\ &+ \frac{2}{(1+\zeta)^2\,\mathscr{L}}\left(\left\{[\Delta\varepsilon + 4(\mathscr{D}_0 + R_0)]\,\zeta - \right.\right. \\ &- \left.\frac{(1-3s^2)\,[a(D_1 + \mathscr{M}_1) + \zeta^2\tilde{a}\tilde{\mathscr{M}}_1] + 4\zeta a\tilde{\mathscr{M}}_1 s^2}{2(1-s^2)}\right\}\frac{\eta\coth\eta - 1}{\eta^2} - \\ &- \left.\frac{(\mathscr{D}_1 + R_1)\,\zeta(a + \zeta\tilde{a})}{(1+\zeta)\,\mathscr{L}\eta^2}\,\frac{1-3s^2}{1-s^2}\left[3\coth\eta\left(\coth\eta - \frac{1}{\eta}\right) - 1\right]\right) \end{aligned}$$

for (41 a, b), and

$$E_{\text{tot}} = \frac{2}{1+\zeta}\left\{\tilde{\varepsilon} + 2\mathcal{M}_0 - \frac{\hbar^2 v^2}{4\mathcal{M}_0} - \frac{\mathcal{M}_0}{8\mathcal{L}^2}\frac{\sin\eta}{\eta}\left(\frac{1}{1+\cos\eta} + \frac{\eta - \sin\eta}{\sin^3\eta}\right) + \right.$$

$$+\zeta\left[\varepsilon_0 + 2\tilde{\mathcal{M}}_0 - \frac{\hbar^2 v^2}{4\tilde{\mathcal{M}}_0} - \frac{\tilde{\mathcal{M}}_0}{8\mathcal{L}^2}\frac{\sin\eta}{\eta}\left(\frac{1}{1+\cos\eta} + \frac{\eta - \sin\eta}{\sin^3\eta}\right)\right]\Bigg\} +$$

$$+\frac{2}{(1+\zeta)^2\,\mathcal{L}}\left(\left\{[\Delta\varepsilon + 4(\mathcal{D}_0 + R_0)]\,\zeta - \right.\right.$$

$$-\left.\frac{(1-3s^2)\,[a(D_1+\mathcal{M}_1) + \zeta^2\tilde{a}\tilde{\mathcal{M}}_1] + 4\zeta a\tilde{\mathcal{M}}_1 s^2}{2(1-s^2)}\right\}\frac{1-\eta\operatorname{cotg}\eta}{\eta^2} -$$

$$\left.-\frac{(\mathcal{D}_1 + R_1)\,\zeta(a+\zeta\tilde{a})}{(1+\zeta)\,\mathcal{L}\eta^2}\,\frac{1-3s^2}{1-s^2}\left[3\operatorname{cotg}\eta\left(\operatorname{cotg}\eta - \frac{1}{\eta}\right) + 1\right]\right)$$

for (47 a, b).

4. Solitary Excitations in the Case of Totally Symmetric Vibrations

In this section we shall consider the situation when totally symmetric intramolecular vibrations take place. In this connection two cases will be investigated: A) when the displacements of the nuclei from their equilibrium positions are small ($\xi^2 \ll 1$) and the frequency of the intramolecular vibrations remains unchanged and B) when $\xi^2 \ll 1$ as in the upper case but there is a frequency change ($\Delta\Omega \neq 0$).

Case A: $\xi^2 \ll 1$, $\Delta\Omega = 0$

For this case the vibronic operator V_n can be presented in the form

$$V_n = B_n[1 + \xi(b_n^+ - b_n)]\,, \tag{48}$$

where the linear terms with respect to ξ are taken into account. Then, the Hamiltonian turns into

$$H = \mathscr{E}_0 + T + U + \varepsilon_0 \sum_n b_n^+ b_n + (\Delta E + 2D_0) \sum_n B_n^+ B_n + $$
$$+ \sum_n M_{n+1n}^{(0)}(B_n^+ B_{n+1} + \text{h.c.})\,, \tag{49}$$

where T and U are given by (4) and (5), respectively, and $M_{n+1n}^{(0)}$ has the form (2a). We adopt that (8) takes place and $D^{(0)} = D_0$. Using the same procedure as in Section 2, in continuum approximation we find the following expression for the deformation shape,

$$\frac{\partial u}{\partial x} = \frac{2M_1^{(0)}}{mv_0^2(1-s^2)}\,|B|^2\,. \tag{50}$$

Taking into account (50) we obtain the following equations for B and b:

$$i\hbar\,\frac{\partial B}{\partial t} = (\Delta E + 2D_0 + 2M_0^{(0)})\,B + M_0^{(0)}\,\frac{\partial^2 B}{\partial x^2} - \frac{4M_1^{(0)2}}{mv_0^2(1-s^2)}\,|B|^2\,B\,, \tag{51}$$

$$i\hbar\,\frac{\partial b}{\partial t} = \varepsilon_0 b\,. \tag{52}$$

For $M_0^{(0)} < 0$ the nonlinear Schrödinger equation (51) has the soliton solution of the form

$$B(x, t) = \frac{1}{\sqrt{2\mathcal{L}}}\, e^{i(kx - \omega t + \varphi)} \operatorname{sech} \frac{x - x_0 - vt}{\mathcal{L}}, \tag{53}$$

where

$$k = -\frac{\hbar v}{2M_0^{(0)}}, \tag{54}$$

$$\hbar\omega = \Delta E + 2D_0 + 2M_0^{(0)} - \frac{\hbar^2 v^2}{4M_0^{(0)}} + \frac{M_0^{(0)}}{\mathcal{L}^2}, \tag{55}$$

$$\mathcal{L} = \frac{|M_0^{(0)}|\, m v_0^2(1 - s^2)}{M_1^{(0)2}}, \tag{56}$$

and the normalization condition

$$\int_{-\infty}^{\infty} |B|^2 \,dx = 1 \tag{57}$$

has been used.

Equation (52) does not influence the solution (53). So in such a system Davydov-type solitons can propagate. The energy of the system read from H_b is then

$$\Delta W \equiv E_{\text{tot}} - H_b - \mathcal{E}_0 = $$
$$= \Delta E + 2D_0 + 2M_0^{(0)} - \frac{\hbar^2 v^2}{4M_0^{(0)}} + \frac{M_1^{(0)4}(1 - 5s^2)}{3M_0^{(0)} m^2 v_0^4 (1 - s^2)^3}. \tag{58}$$

Case B: $\xi^2 \ll 1$, $\Delta\Omega \neq 0$

In this case the vibronic operator V_n has the form

$$V_n = B_n[1 + \xi(b_n^+ - b_n)]\,(1 - r^2)^{\frac{1}{2}\left(b_n^+ b_n + \frac{1}{2}\right)},$$

which for small ξ leads to the following Hamiltonian of the system:

$$H = \varepsilon_0 \sum_n b_n^+ b_n + (\Delta\tilde{E} + \tfrac{1}{2}\Delta\varepsilon) \sum_n B_n^+ B_n + \Delta\varepsilon \sum_n b_n^+ b_n B_n^+ B_n +$$
$$+ \sum_n \mathcal{M}_{n+1n}(B_n^+ B_{n+1} + \text{h.c.})\,, \tag{59}$$

where $\Delta\tilde{E} = \Delta E + 2D_0$. (We do not include the site dependence of $D^{(0)}$ and take $D^{(0)} = D_0$),

$$\mathcal{M}_{n+1n} = \sqrt{1 - r^2}\, M_{n+1n}^{(0)}\,,$$

r being given by (7).

With the help of (8) and (14) in continuum approximation, we obtain the following set of nonlinear differential equations:

$$i\hbar \frac{\partial B}{\partial t} = (\Delta\tilde{E} + 2\mathcal{M}_0 + \Delta\varepsilon\, |b|^2)\, B + \mathcal{M}_0 \frac{\partial^2 B}{\partial x^2} - \frac{4\mathcal{M}_1^2}{m v_0^2 (1 - s^2)} |B|^2 B\,, \tag{60}$$

$$i\hbar \frac{\partial b}{\partial t} = (\varepsilon_0 + \Delta\varepsilon\, |B|^2)\, b\,, \tag{61}$$

where the form of the deformation region is given by (50) ($M_1^{(0)}$ being changed by $\mathcal{M}_1$).

Multiplying (61) and its complex conjugate by b^+ and b, respectively, after addition we obtain

$$|b|^2 = \text{const} = b_0 \,. \tag{62}$$

The substitution of (62) into (60) leads to the following nonlinear Schrödinger equation

$$i\hbar \frac{\partial B}{\partial t} = (\Delta\tilde{E} + 2\mathcal{M}_0 + \Delta\varepsilon b_0^2)\, B + \mathcal{M}_0 \frac{\partial^2 B}{\partial x^2} - \frac{4\mathcal{M}_1^2}{mv_0^2(1 - s^2)} |B|^2 B \,. \tag{63}$$

Equation (63) has a normalized solution of the form (53), where k, $\hbar\omega$, and $\mathcal{L}$ have the form (54) to (56), respectively, but with $M_0^{(0)}$ and $M_1^{(0)}$ replaced by $\mathcal{M}_0$ and $\mathcal{M}_1$, respectively ($\mathcal{M}_0$ and $\mathcal{M}_1$ being negative!).

The substitution of the soliton solution for B into (61) leads to the equation

$$i\hbar \frac{\partial b}{\partial t} = \left[\varepsilon_0 + \frac{\Delta\varepsilon}{2\mathcal{L}} \operatorname{sech}^2 \frac{x - x_0 - vt}{\mathcal{L}}\right] b \tag{64}$$

with the solution

$$b(x, t) = b_0 \exp\left\{i\left[\frac{\varepsilon_0}{\hbar v}(x - x_0 - vt) + \frac{\Delta\varepsilon}{\hbar v} \operatorname{th} \frac{x - x_0 - vt}{\mathcal{L}}\right]\right\}. \tag{65}$$

The energy of the system read from H_b is then

$$\begin{aligned}\Delta W &\equiv E_{\text{tot}} - H_b - \mathcal{E}_0 \\ &= \Delta\tilde{E} + \Delta\varepsilon b_0^2 + 2\tilde{\mathcal{M}}_0 - \frac{\hbar^2 v^2}{4\mathcal{M}_0} + \frac{\mathcal{M}_1^4(1 - 5s^2)}{3\mathcal{M}_0 m^2 v_0^4 (1 - s^2)^3}\,.\end{aligned}$$

Integrating $\partial u/\partial x$ we obtain that the total length change of the molecular chain due to the soliton formation is $\Delta = 2\mathcal{M}_1/\big(mv_0^2(1 - s^2)\big)$, i.e. the phonon plane waves do not contribute.

If we do not take into account the motion of the molecules in the chain we have the equations

$$i\hbar \frac{\partial B}{\partial t} = (\Delta\tilde{E} + 2\mathcal{M}_0)\, B + \mathcal{M}_0 \frac{\partial^2 B}{\partial x^2} + \Delta\varepsilon\, |b|^2 B \,, \tag{66}$$

$$i\hbar \frac{\partial b}{\partial t} = (\varepsilon_0 + \Delta\varepsilon\, |B|^2)\, b \,, \tag{67}$$

instead of (60) and (61). Making the same steps as above we obtain for b (62) and the solutions of (66) and (67) in the form

$$B(x, t) = B_0\, \mathrm{e}^{i(kx - \omega t + \varphi)} \,,$$

$$b(x, t) = b_0\, \mathrm{e}^{i(\tilde{k}x - \tilde{\omega}t + \tilde{\varphi})} \,,$$

where

$$k = -\hbar v/\mathcal{M}_0 \,, \qquad \hbar\omega = \Delta\tilde{E} + 2\mathcal{M}_0 + \Delta\varepsilon b_0^2 - \mathcal{M}_0 k^2 \,,$$

$$\tilde{k} = \frac{\varepsilon_0 + \Delta\varepsilon B_0^2}{\hbar v} \,, \qquad \hbar\tilde{\omega} = \varepsilon_0 + \Delta\varepsilon B_0^2 \,.$$

One can see that from the last two expressions the well-known phonon dispersion law $\tilde{\omega} = \tilde{k}v$ follows.

References

[1] A. S. Davydov and N. I. Kislukha, phys. stat. sol. (b) **59**, 465 (1973).
[2] W. Weidlich and W. Heudorfer, Z. Phys. **268**, 133 (1974).
[3] A. S. Davydov and N. I. Kislukha, phys. stat. sol. (b) **75**, 735 (1976).
[4] A. S. Davydov and N. I. Kislukha, Zh. eksper. teor. Fiz. **71**, 1090 (1976).
[5] N. I. Kislukha, Ukr. fiz. Zh. **21**, 829 (1976).
[6] A. S. Davydov and A. A. Eremko, Ukr. fiz. Zh. **22**, 881 (1977).
[7] M. Mabuchi, J. Phys. Soc. Japan **41**, 735 (1976).
[8] A. S. Davydov, Uspekhi fiz. Nauk **138**, 603 (1982).
[9] Kh. I. Pushkarov and M. T. Primatarova, Preprint IC/84/31, ICTP, Trieste 1984.
[10] A. S. Davydov, Theory of Light Absorption in Molecular Crystals, Izd. Akad. Nauk Ukr. SSR, Kiev 1951 (in Russian).
[11] W. A. Bingel, Canad. J. Phys. **37**, 680 (1959).
[12] E. G. Ray, Austr. J. Chem. **14**, 354 (1960).
[13] E. G. Ray, Austr. J. Chem. **16**, 295; 315 (1963).
[14] E. I. Rashba, Zh. eksper. teor. Fiz. **50**, 1064 (1966).
[15] E. I. Rashba, Zh. eksper. teor. Fiz. **54**, 542 (1968).
[16] A. S. Davydov and A. A. Serikov, phys. stat. sol. **42**, 603 (1970).
[17] A. S. Davydov and A. A. Serikov, phys. stat. sol. (b) **44**, 127 (1971).
[18] A. A. Serikov, phys. stat. sol. (b) **44**, 733 (1971).
[19] H. Sponer and E. Teller, Rev. mod. Phys. **13**, 76 (1941).
[20] M. A. Eliashevich, Atomic and Molecular Spectroscopy, Izd. Fizmatgiz, Moscow 1962 (in Russian).
[21] A. S. Davydov, Theory of Solid State, Izd. Nauka, Moscow 1976 (in Russian).

(Received November 30, 1984)

phys. stat. sol. (b) 133, 267 (1986)
Subject classification: 79.20; S2

Department of Physics, Waseda University, Tokyo[1])

Stopping Power of Atoms near Metal Surfaces at Grazing Incidence Condition

By
M. Kato and Y. H. Ohtsuki

The energy loss of slow atoms is studied from the viewpoint of the indirect excitation of low-energy electron–hole pairs, i.e., the excitations of electron–hole pairs through the electron exchange process from the metal to the projectile's atomic states, and vice versa. For the slow atom, $v \ll v_F$ (v and v_F are the velocity of the projectile and the Fermi velocity, respectively), electrons have enough velocity for response, thus the atomic states are not well defined due to the screening effect and the electron exchange processes. By constructing the model Hamiltonian which describes such effects from first principles the stopping power at grazing incidence condition, especially for the Cs surface, is calculated.

Der Energieverlust langsamer Atome wird vom Standpunkt einer indirekten Anregung von niederenergetischen Elektronen–Loch-Paaren untersucht, d. h. die Anregungen von Elektronen–Loch-Paaren durch den Elektronenaustauschprozeß vom Metall zu den atomaren Zuständen des Projektils und umgekehrt. Für langsame Atome, $v \ll v_F$ (v und v_F sind die Geschwindigkeiten des Projektils bzw. die Fermigeschwindigkeit) haben die Elektronen genügend Energie für Response, somit sind die atomaren Zustände infolge des Abschirmeffekts und der Elektronenaustauschprozesse nicht gut definiert. Durch Konstruktion eines Modellhamiltonians, der solche Effekte von den ersten Prinzipien her beschreibt, wird die Bremsleistung bei streifendem Einfall, speziell für die Cs-Oberfläche, berechnet.

1. Introduction

The stopping power in the surface region is not well defined generally. However, when the velocity of the projectile atom is much less than the velocity of metal electrons, especially at grazing incidence condition, the meaning of this revives, since the projectile atom spends much time near the surface. At this condition, the stopping power becomes important, for example, with respect to the skipping motion [1] and hydrogen atoms reflected from the first wall of a thermo-nuclear reactor.

In this circumstance, the electronic stopping power is proportional to the velocity of the projectile. Several authors concretely calculated such a stopping power within a definite model [2, 3]. They assumed that the charge of the projectile does not change during scattering. However, most of the hydrogen atoms reflected from the metal surface are neutralized [4]. This means that when the atom approaches the metal surface, there exist frequent electron transfer processes between the metal and the travelling atom.

We investigate the excitation of low-energy electron–hole pairs by these processes. From this point of view, several authors treated the problem somewhat qualitatively [5, 6]. In this paper we apply the Anderson-Newns model Hamiltonian [8] to description of this energy loss mechanism. Since we usually treat the hopping terms in this Hamiltonian as parameters, it seems to have disadvantages in a concrete calculation.

[1]) Okubo 3-4-1, Shinjuku, Tokyo 160, Japan.

However, we approximately construct this model Hamiltonian from first principles and concretely calculate the stopping power at grazing incidence condition.

2. Expression for Stopping Power

At each position of the slow atom, electrons quickly adjust their states to the new equilibrium states, thus the electron state is very near to its equilibrium state. According to the theories [6, 7], on such a situation, the expression for the electronic stopping power is

$$S_e(t) = -\frac{2}{\pi\hbar v} \operatorname{Re}\left[\int_0^\infty \frac{d\omega}{\omega} \int dr_1 \int dr_2 \int_{-\infty}^{t} dt' \, e^{-i\omega(t-t')} \dot{V}(r_1, t) \, \dot{V}(r_2, t') \times \right.$$

$$\left. \times \operatorname{Im}\left[P^t(r_1, r_2, \omega)\right]\right] - \frac{1}{\pi\hbar v} \int_0^\infty \frac{d\omega}{\omega} \int dr_1 \int dr_2 \int_{-\infty}^{t} dt_1 \int_{-\infty}^{t} dt_2 \times$$

$$\times e^{-i\omega(t_1-t_2)} \dot{V}(r_1, t_1) \, \dot{V}(r_2, t_2) \frac{\partial}{\partial t} \operatorname{Im}\left[P^t(r_1, r_2, \omega)\right]; \tag{1}$$

$$\operatorname{Im}\left[P^t(r_1, r_2, \omega)\right] = -\pi\hbar \sum_m \langle 0_t | \hat{n}(r_1) | m_t\rangle \langle m_t | \hat{n}(r_2) | 0_t\rangle \delta\left(\omega - \frac{1}{\hbar} (\varepsilon_m^t - \varepsilon_0^t)\right),$$

where $V(r, t)$ is the scattering potential, v the velocity of the atom, $P^t(r, r', \omega)$ the polarization propagator, $\hat{n}(r)$ the induced charge density, $|m_t\rangle$ the state at time t adiabatically evolved from $|m_{-\infty}\rangle$, and ε_m^t the corresponding energy. If we apply the linear-response theory to the calculation of the polarization part in (1), including a two-body interaction between metal electrons, then we obtain the well-known expression

$$S_e(s) = \frac{(Ze)^2}{\pi} \int \frac{d^2\varkappa}{\varkappa} \int_0^\infty \frac{d\omega}{\omega} \frac{\varkappa v_\perp}{(\omega - \boldsymbol{\varkappa} \cdot \boldsymbol{v}_{||})^2 + (\varkappa v_\perp)^2} \left[(\boldsymbol{\varkappa} \cdot \boldsymbol{v}_{||})^2 + (\varkappa v_\perp)^2\right] \times$$

$$\times e^{-2\varkappa s} \operatorname{Im}\left[\frac{kI(\boldsymbol{\varkappa}, \omega) - \pi}{kI(\boldsymbol{\varkappa}, \omega) + \pi}\right], \tag{2}$$

where

$$I(\boldsymbol{\varkappa}, \omega) = \int dq \, \frac{1}{\varkappa^2 + q^2} \frac{1}{\varepsilon_{\mathrm{RPA}}(\boldsymbol{\varkappa}, q; \omega)}; \quad \varkappa = |\boldsymbol{\varkappa}|,$$

where Ze is the charge of the ion, $\varepsilon_{\mathrm{RPA}}(\boldsymbol{\varkappa}, q; \omega)$ the usual RPA dielectric function, s the distance from the surface, $v_\perp$ and $\boldsymbol{v}_{||}$ are the normal and the parallel velocity components to the metal surface, respectively. Equation (2) is for the incoming case. We have a similar expression for the outgoing case. It is noteworthy that this formula is not restricted to the slow ion case. Ferrell, et al. [2], and Núñez, et al. [3] have evaluated (2) with the approximated RPA dielectric function. However, in obtaining (2), a metal potential is approximated by the infinite barrier potential and, moreover, we assume that the external ion does not change its charge. Hence, (2) breaks down when electron capture and loss processes are frequent, for example, as is the case of slow hydrogen or alkali metal atoms. In the latter case we apply to (1) the strong coupling Hamiltonian, i.e., the Anderson-Newns Hamiltonian [8],

$$\hat{H}(t) = \sum_{k} \varepsilon_k C_k^+ C_k + \varepsilon_a C_a^+ C_a + \sum_{k} [V_k(t)\, C_k^+ C_a + \text{h.c.}], \tag{3}$$

where C_k^+ and C_k, etc., are the creation and destruction operators of the state $|\boldsymbol{k}\rangle$, respectively. The third term of (3) represents the electron hopping between the metal states $|\boldsymbol{k}\rangle$ and the atomic state $|\mathrm{a}\rangle$. In (3), we neglected the intra-atomic Coulomb repulsion for simplicity, but we discuss about it in Section 4. The time dependence of the Hamiltonian is originated from the trajectory of the atom which is assumed to be classical. We consider metal electrons as the free electron gas confined in the semi-infinite metal with a finite surface potential with wave functions

$$\varphi_k(r) = e^{ik_x x}\, e^{ik_y y}\, \Phi_{k_z}(z), \tag{4}$$

where the z-axis is the normal to the metal surface. Due to the translational invariance parallel to the surface we have the following Hamiltonian, which was introduced by van Wunnik et al. [9] in order to investigate the effect of the parallel velocity on the negative hydrogen formation,

$$\hat{H}(t) = \sum_{k} \varepsilon_k C_k^+ C_k + \varepsilon_a C_a^+ C_a + \sum_{k} [V_{k-Q}(s)\, e^{\frac{i}{\hbar}(\varepsilon_{k-Q}-\varepsilon_k)t}\, C_k^+ C_a + \text{h.c.}],$$

$$\boldsymbol{Q} = \frac{m\boldsymbol{v}_{\|}}{\hbar}, \tag{5}$$

where m is the electron mass and s the distance from the surface to the vacuum. Thus s is represented by the normal velocity, $v_{\perp}$, and time t. But at grazing incidence condition ($v_{\|} \gg v_{\perp}$) we may neglect the effect of the normal velocity in first approximation. Hence, s serves as the impact parameter.

We calculate the stopping power of (1) by making use of (5). The first-order contribution to the stopping power, which is proportional to the parallel velocity, $v_{\|}$, is

$$S_e(s) = mv_{\|} \frac{2\pi}{\hbar} \int\limits_{|k|=k_F} ds_k\, k_F \cos^2\theta\, |V_k(s)|^2\, \varrho_a^s(\varepsilon_F), \tag{6}$$

where k_F is the Fermi wave number and θ the angle between $\boldsymbol{k}$ and $\boldsymbol{v}_{\|}$, and

$$\varrho_a^s(\varepsilon) = \frac{1}{\pi} \frac{\Delta^s(\varepsilon)}{(\varepsilon - \varepsilon_a - \Lambda^s(\varepsilon))^2 + (\Delta^s(\varepsilon))^2}; \quad \Delta^s(\varepsilon) = \pi \sum_{k} |V_k(s)|^2\, \delta(\varepsilon - \varepsilon_k), \quad \Lambda^s(\varepsilon) = P\, \frac{1}{\pi} \int d\varepsilon'\, \frac{\Delta^s(\varepsilon')}{\varepsilon - \varepsilon'}. \tag{7}$$

If the k-dependence of the matrix elements is weak, then we have

$$S_e(s) = mv_{\|} \frac{2\pi}{\hbar} N \langle |V_k(s)|^2 \rangle\, \varrho_a^s(\varepsilon_F), \tag{8}$$

where N is the number of electrons in the system and $\langle\ \rangle$ means the average at the Fermi energy. Using the Anderson phase shift, (8) is rewritten in more familiar form

$$S_e(s) = \frac{8}{3\pi\hbar}\, \varepsilon_F \sin^2 \delta^s(\varepsilon_F),$$

$$\sin \delta^s(\varepsilon) = \frac{\Delta^s(\varepsilon)}{\sqrt{(\varepsilon - \varepsilon_a - \Lambda^s(\varepsilon))^2 + (\Delta^s(\varepsilon))^2}}, \tag{9}$$

where the phase shift is evaluated at the Fermi energy (ε_F) and calculated from the Hamiltonian

$$\hat{H}(s) = \sum_k \varepsilon_k C_k^+ C_k + \varepsilon_a^+ C_a C_a + \sum_k [V_k(s) C_k^+ C_a + \text{h.c.}] . \tag{10}$$

Equation (9) has similarity to the electric resistance. As is well known, the electronic stopping power of slow atoms in metals is determined by the quantities only at the Fermi energy due to Fermi statistics. Equation (10) is the same as (3), if we take a static limit of (3). Therefore, the stopping power of slow atoms is given by the adsorbate-induced electronic structure.

3. Hamiltonian

In this section, in order to clarify the meanings of the coupling coefficient V_k we derive the approximated Hamiltonian, which has just the same form as (5), based on the transfer Hamiltonian formalism [10], which we sometimes encounter on the theoretical treatment of inelastic electron tunneling spectroscopy (IETS). There exists the vacuum between the metal surface and the atom which approaches the surface. We consider it as a barrier corresponding to the barrier in the case of IETS.

In principle, we know both electron states of the metal and the atom if they are isolated from each other. But, when the atom approaches the metal we have not enough knowledge of it due to a many-body effect. However, what we can say at least about the potential is as follows: near the atom it is very similar to the potential of the isolated atom, on the other hand, in or near the metal, to that of the isolated metal. Hence we divide the whole space into two half spaces by a certain surface, which we call the transfer surface, between the metal and the atom (Fig. 1). We label it A(B)-space which contains the metal (atom). Thus we assume the scattering potential as follows:

$$V(r) = \begin{cases} V_m(r) & \text{in A-space} \\ V_a(r) & \text{in B-space}, \end{cases} \tag{11}$$

where $V_m(r)$ and $V_a(r)$ are the unperturbed potentials of the metal and the atom, respectively.

Starting from the single-electron Hamiltonian,

$$H(r) = -\frac{\hbar^2}{2m} \nabla^2 + V(r), \tag{12}$$

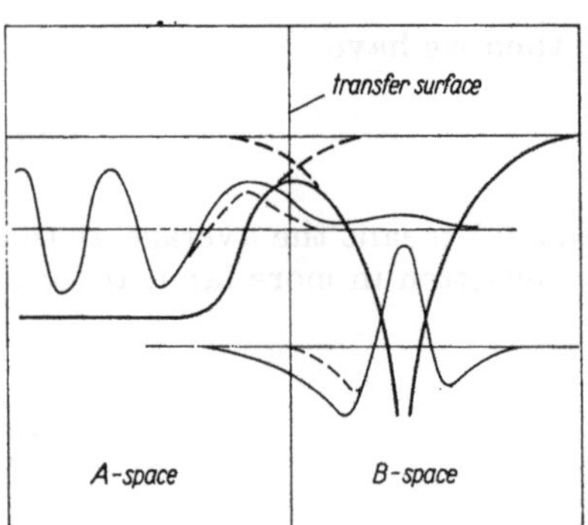

Fig. 1. Illustration of the potential, the wave functions, and the transfer surface. The dashed lines are the unperturbed potentials and wave functions

where $V(r)$ is the scattering potential, we evaluate the expectation value of (12) using the approximated forms of (11). In the second quantized form, we obtain

$$\hat{H} = \int \mathrm{d}r\, \hat{\psi}^+(r)\, H(r)\, \hat{\psi}(r)\,, \tag{13}$$

where $\hat{\psi}^+(r)$ and $\hat{\psi}(r)$ are the creation and destruction field operators, respectively. The field operators can be expanded in terms of metal electron states and atomic electron states. But, for simplicity, we consider mono-valent atoms like hydrogen and alkali metal atoms, so we take only one state for the atomic states. Then we have

$$\hat{\psi}(r) = \sum_k (C_k - S_{ka}C_a)\, \varphi_k(r) + C_a\varphi_a(r)\,,$$

where $\varphi_a(r)$ and $\varphi_k(r)$ are the atomic and the metal electron states, respectively, and C_a and C_k's are destruction operators, and S_{ka} means the overlap integral due to non-orthogonality between the atomic and the metal electron states. We retain the first-order contribution of S_{ka}'s in (13), then we obtain

$$\begin{aligned}\hat{H} &= \sum_{k,k'} \int \mathrm{d}r\, \varphi_k^*(r)\, H(r)\, \varphi_k(r)\, [C_k^+C_k - S_{ka}C_k^+C_a - S_{ak}C_a^+C_{k'}] + \\ &\quad + \int \mathrm{d}r\, \varphi_a^*(r)\, H(r)\, \varphi_a(r)\, C_a^+C_a + \\ &\quad + \sum_k [\int \mathrm{d}r\, \varphi_k^*(r)\, H(r)\, \varphi_a(r)\, C_k^+C_a + \text{h.c.}]\,.\end{aligned}$$

For $\langle \boldsymbol{k}|\, H\, |\boldsymbol{k}'\rangle$ and $\langle a|\, H\, |a\rangle$, we find

$$\begin{aligned}\int \mathrm{d}r\, \varphi_k^*(r)\, H(r)\, \varphi_{k'}(r) &= \int_{A+B} \mathrm{d}r\, \varphi_k^*(r) \left[-\frac{\hbar^2}{2m}\nabla^2 + V_m(r)\right] \varphi_{k'}(r) + \\ &\quad + \int_B \mathrm{d}r\, \varphi_k^*(r)\, [V_a(r) - V_m(r)]\, \varphi_{k'}(r)\end{aligned} \tag{14}$$

and

$$\begin{aligned}\int \mathrm{d}r\, \varphi_a^*(r)\, H(r)\, \varphi_a(r) &= \varepsilon_a S_{aa}(B) + \int_A \mathrm{d}r\, \varphi_a(r)\, H(r)\, \varphi_a(r) = \\ &= \varepsilon_a S_{aa}(B) + \int_A \mathrm{d}r\, \varphi_a^*(r) \left[-\frac{\hbar^2}{2m}\nabla^2 + V_m(r)\right] \varphi_a(r)\,,\end{aligned} \tag{15}$$

where

$$S_{aa}(B) = \int_B \mathrm{d}r\, \varphi_a^*(r)\, \varphi_a(r)\,.$$

The last term of (14) is negligible when the atom is comparably far from the metal. Thus we approximate (14) as

$$\int \mathrm{d}r\, \varphi_k^*(r)\, H(r)\, \varphi_k(r) = \varepsilon_k \delta_{k'k}\,. \tag{14'}$$

From (13), (14'), and (15), we have

$$\begin{aligned}&\sum_{k,k'} \int \mathrm{d}r\, \varphi_k^*(r)\, H(r)\, \varphi_{k'}(r)\, [C_k^+C_{k'} - S_{ka}C_k^+C_a - S_{ak}C_a^+C_{k'}] + \\ &+ \int \mathrm{d}r\, \varphi_a^*(r)\, H(r)\, \varphi_a(r)\, C_a^+C_a \\ &\approx \sum_k \varepsilon_k C_k^+C_k + \varepsilon_a(s)\, C_a^+C_a - \sum_k \varepsilon_k (S_{ak}C_a^+C_k + S_{ka}C_k^+C_a)\,,\end{aligned} \tag{16}$$

where

$$\varepsilon_a(s) = \varepsilon_a S_{aa}(B) + \int\limits_A dr\, \varphi_a^*(r) \left[-\frac{\hbar^2}{2m} \nabla^2 + V_m(r)\right] \varphi_a(r) .$$

Next, for $\langle \boldsymbol{k}| H |a\rangle$ we have

$$\int dr\, \varphi_{\boldsymbol{k}}^*(r)\, H(r)\, \varphi_a(r) = \int\limits_A dr\, \varphi_{\boldsymbol{k}}(r)\, \vec{\boldsymbol{H}}(r)\, \varphi_a(r) + \varepsilon_a S_{\boldsymbol{k}a}(B) , \tag{17}$$

where

$$S_{\boldsymbol{k}a}(B) = \int\limits_B dr\, \varphi_{\boldsymbol{k}}^*(r)\, \varphi_a(r) .$$

Here we can obtain the more symmetric form, as follows.

In the same-order approximation we have

$$\int\limits_A dr\, \varphi_{\boldsymbol{k}}^*(r)\, \overleftarrow{H}(r)\, \varphi_a(r) - \varepsilon_{\boldsymbol{k}} S_{\boldsymbol{k}a}(A) = 0 , \tag{18}$$

where

$$S_{\boldsymbol{k}a}(A) = \int\limits_A dr\, \varphi_{\boldsymbol{k}}^*(r)\, \varphi_a(r) .$$

In (17) and (18), the arrow indicates that the differential operator operates in the corresponding direction.

Subtracting (18) from (17), then we obtain

$$\int dr\, \varphi_{\boldsymbol{k}}^*(r)\, H(r)\, \varphi_a(r) = -\frac{\hbar^2}{2m} \int\limits_S dS\, [\varphi_{\boldsymbol{k}}^*(r)\, \nabla\varphi_a(r) - \varphi_a(r)\, \nabla\varphi_{\boldsymbol{k}}^*(r)]\, \boldsymbol{n} + \\ + \varepsilon_a S_{\boldsymbol{k}a}(B) + \varepsilon_{\boldsymbol{k}} S_{\boldsymbol{k}a}(A) , \tag{19}$$

where S indicates the transfer surface which is the domain of surface integration and $\boldsymbol{n}$ is the normal vector of S directed to B-space. In (19) we use Green's theorem with the suitable boundary condition. The integrand of the first term in (19) is the off-diagonal matrix element of the current density operator. Similarly, we obtain expressions for $\langle a| H |\boldsymbol{k}\rangle$. This is the hermite conjugate of (19).

Therefore, from (16), (19), and the hermite conjugate of (19), we obtain the following strong coupling Hamiltonian:

$$\hat{H} = \sum_{\boldsymbol{k}} \varepsilon_{\boldsymbol{k}} C_{\boldsymbol{k}}^+ C_{\boldsymbol{k}} + \varepsilon_a C_a^+ C_a + \sum_{\boldsymbol{k}} [V_{\boldsymbol{k}} C_{\boldsymbol{k}}^+ C_a + \text{h.c.}] , \tag{20}$$

where

$$V_{\boldsymbol{k}} = -i\hbar \int\limits_S dS\, \hat{J}_{\boldsymbol{k}a} n + (\varepsilon_a - \varepsilon_{\boldsymbol{k}})\, S_{\boldsymbol{k}a}(B) , \tag{21}$$

$$\hat{J}_{\boldsymbol{k}a} = -i\, \frac{\hbar}{2m}\, [\varphi_{\boldsymbol{k}}^*(r)\, \nabla\varphi_a(r) - \varphi_a(r)\, \nabla\varphi_{\boldsymbol{k}}^*(r)] .$$

Accordingly, it is clear that off-diagonal matrix elements describe electron transfer through the transfer surface between the metal and the atom. Moreover, this Hamiltonian is hermite at least when $S_{\boldsymbol{k}a} = S_{\boldsymbol{k}a}(A) + S_{\boldsymbol{k}a}(B) = 0$, that is, $|a\rangle$ is orthogonal to $|\boldsymbol{k}\rangle$. But, the second term of (21) is notable. This is the energy difference times the overlap integral in the half space between two states. When the physical process is the exact resonance, this term has no contribution [10]. In the present case, however, it is not negligible. It should be noted that the appearance of this term in (21) is consistent with the other derivation of the Anderson Hamiltonian [15].

Assuming metal electrons as the free electron gas confined in the semi infinite metal, we obtain

$$\hat{H}(t) = \sum_k \varepsilon_k C_k^+ C_k + \varepsilon_a C_a^+ C_a + \sum_k [V_{k-Q}(s)\, e^{\frac{i}{\hbar}(\varepsilon_{k-Q}-\varepsilon_k)t}\, C_k^- C_a + \text{h.c.}] \tag{22}$$

which is the desired expression (see (5)). However, (22) is different from (5) due to non-orthogonality, on account of, for example, the anti-commutation relation $[C_k, C_a^+] = S_{ka}$. But we find that (5) and (22) are the same in first approximation. Another difficulty in this treatment is that we cannot treat the scattering states of the system, which are highly excited states but are needed in order to calculate $\Lambda^s(\omega)$. However, when the atom is comparably far from the metal, $\Lambda^s(\omega)$ is much smaller than other physical quantities, so we neglect it.

4. Result and Discussion[2])

We approximate metal electrons by the free electron gas confined in the semi infinite metal with the finite step potential, of which depth is the sum of ε_F and Φ_w (work function). Thus, the wave functions of metal electrons are represented by

$$\varphi_k(r) = \begin{cases} \dfrac{1}{\sqrt{\Omega}}\, e^{i(k_x x + k_y y)} \sin(k_z z + \eta)\,; & z < 0 \\ \dfrac{1}{\sqrt{\Omega}}\, e^{i(k_x x + k_y y)} \dfrac{k_z}{k_0}\, e^{-\delta z}\,; & z > 0\,, \end{cases} \tag{23}$$

where $\sin \eta = k_z/k_0$, $\delta = \sqrt{k_0^2 - k_z^2}$, $k_0 = \sqrt{2(\varepsilon_F + \Phi_w)}$ and Ω is the volume of the box for normalization and the region of $z > 0$ is the vacuum. We calculate for the metals with $\varepsilon_F = 1.5$ eV, $\Phi_w = 2.0$ eV and with $\varepsilon_F = 5.5$ eV, $\Phi_w = 4.5$ eV. The former corresponds to the clean Cs surface, the latter corresponds to Au, if we take the free electron gas model. But this simple model is too crude for Au. We calculate the case of hydrogen and lithium beams. As the atomic states, we use the s-type states,

$$\varphi_{1s}(r) = \frac{a^{3/2}}{\sqrt{\pi}}\, e^{-aR}\,,$$

$$\varphi_{2s}(r) = \frac{a^{3/2}}{\sqrt{\pi}}\,(1 - aR)\, e^{-aR}, \tag{24}$$

$$R = |r - s|\,.$$

It is natural to take 2s-state for Li with $a = 0.66$ in (24). For hydrogen, we calculate for 1s-state with $a = 1$ and for 2s-states with $a = 0.5$ in (24), independently. We choose the metal surface as the transfer surface, where the metal surface means the position of the step, since this choice is the simplest and the most realistic one [11]. Matrix elements are easy to calculate and, in addition to this, only the values at the Fermi level are necessary, thus the calculation is rather simple.

We calculate (8) or (9), using the Hamiltonian (22) with the hopping matrix elements (21). The calculated results are shown in Fig. 2a and Fig. 3a. These results are comparable to other theories by Ferrell et al. [2], Núñez, et al. [3], and von Gemmingen [12] (Fig. 2b, 3b). In general, the value of the stopping power is sensitive to the relative position between the atomic level and the Fermi level (see (9)). The energy levels of

---

[2]) In this section we use the atomic units.

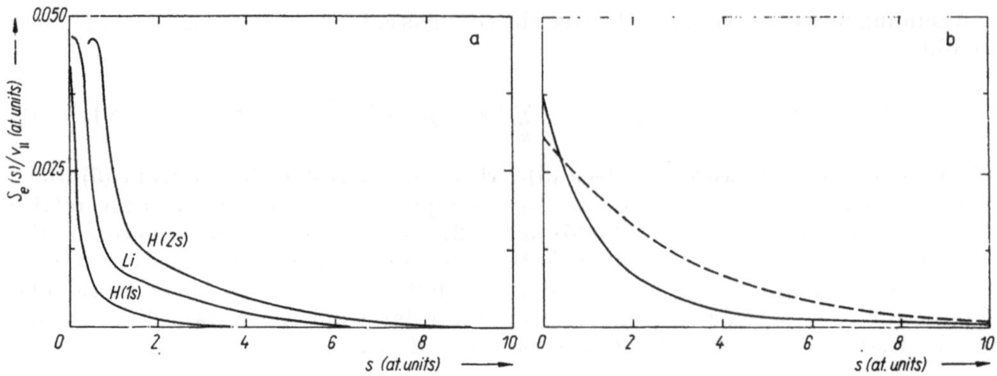

Fig. 2. a) $S_e(s)/v_{\parallel}$ as a function of s, where $S_e(s)$ is the stopping power, $v_{\parallel}$ the parallel velocity component, and s the distance from the metal surface to the vacuum. H and Li scattered by the metal with $\varepsilon_F = 1.5$ eV and $\Phi_w = 2.0$ eV, which corresponds to the Cs metal with clean surface. b) $S_e(s)/v_{\parallel}$ as a function of s, from other theories. Full line the result from Ferrell et al. [2], dashed line the result from von Gemmingen [12]. $rs = 5.9$, corresponding to the Cs metal

1s-state and 2s-state of a hydrogen atom are -13.6 eV and -3.4 eV from the vacuum level, respectively, but the Fermi energy levels are a few eV below the vacuum level. Hence the contribution from the 1s-state of hydrogen is small. Fig. 3a shows that if the Li approaches closer to the metal surface than 2.5 at. units, the stopping power becomes reduced. In this region the direct scattering of metal electrons by the potential of the atom becomes important, though we neglect it in (14).

In this paper we neglected the image interaction and the intra-atomic Coulomb repulsion. The image interaction is important especially for alkali metal atoms [16, 17]. The intra-atomic Coulomb repulsion becomes important for hydrogen, especially for hydrogen scattered by Cs. Moreover, the negative hydrogen yield is enhanced at low velocity [4], thus we have to incorporate this charge state, which will be one of the next steps to consider.

It should be noted that the present energy loss mechanism is different from the electron–hole pair excitation mechanism pointed out by Nørskov et al. [18]. The latter mechanism works while ions reach their adiabatic charge state. Hence, the energy

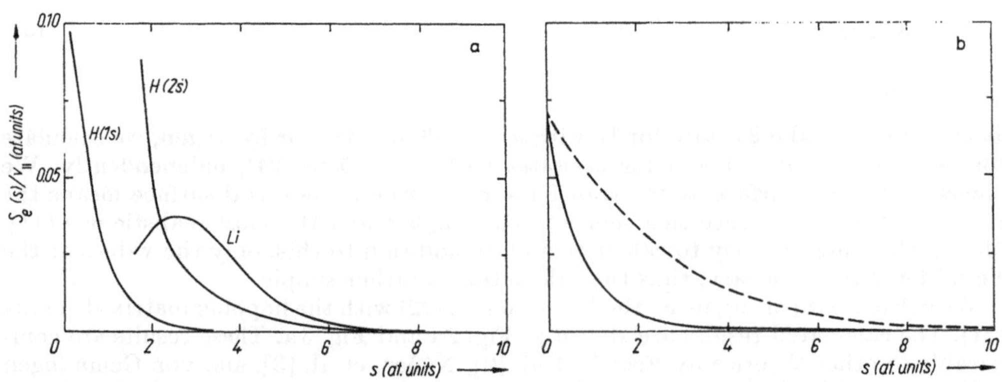

Fig. 3. a) $S_e(s)/v_{\parallel}$ as a function of s. s is the distance from the metal surface to the vacuum. H and Li scattered by the metal with $\varepsilon_F = 5.5$ eV and $\Phi_w = 4.5$ eV. b) $S_e(s)/v_{\parallel}$ as a function of s, from other theories. Full line the result from Ferrell, et al. [2], dashed line the result from von Gemmingen [12]. $rs = 3.0$, corresponding to $\varepsilon_F = 5.5$ eV

loss by this process is dependent on the initial state of the ion. However, this process has only a little contribution to the present calculation of the stopping power, so we neglected it.

One of the authors (Y.H.O) has proposed the semi-classical formula [13] for the electronic stopping power under the same condition as in the present case, originating from an idea of Firsov [14]. This formula is represented by

$$S_e(s) = mv_{\|}P(s)\,, \tag{25}$$

where $mv_{\|}$ is the momentum transfer and $P(s)$ the phenomenological neutralization probability per unit time. We can clarify the meaning of (25) as follows from first principles. When a slow atom like hydrogen approaches the metal surface at grazing incidence, firstly the electron capture and the loss processes are out of the balance depending on the initial state, but they quickly reach the equilibrium. Hence there exists the constant two-way current of electrons between the atom and the metal, which corresponds to $P(s)$. These two-way currents are accompanied by the momentum transfer $mv_{\|}$ resulting in the stopping force of the atom, since the atom is moving at velocity $v_{\|}$. From the classical viewpoint, there is momentum transfer in electron capture processes; on the contrary, in electron loss processes there is no momentum transfer. Therefore, $P(s)$ is estimated by the neutralization probability per unit time [13]. However, it should be noted that according to (6) to (9), $P(s)$ is not exactly the neutralization probability per unit time, due to Fermi statistics, though $P(s)$ is quantitatively nearly the same as the neutralization probability per unit time.

Acknowledgements

The authors wish to thank Dr. U. von Gemmingen, Dr. R. Kawai, Dr. Y. Yamashita, and Dr. A. Isii for their useful discussions.

References

[1] Y. H. Ohtsuki, K. Koyama, and Y. Yamamura, Phys. Rev. B 18, 5044 (1979).
[2] T. L. Ferrell, P. M. Echenique, and R. H. Ritchie, Solid State Commun. 32, 419 (1979).
[3] R. Núñez, P. M. Echenique, and R. H. Ritchie, J. Phys. C 13, 4229 (1980).
[4] W. Eckstein, Inelastic Particle-Surface Collisions, Ed. E. Taglauer and W. Heiland, Springer-Verlag, 1981 (p. 157).
[5] E. G. d'Agliano, P. Kumar, W. Schaich, and H. Suhl, Phys. Rev. B 11, 2122 (1975).
[6] A. Nourtier, J. Physique 38, 479 (1977).
[7] P. Minnhagen, J. Phys. C 15, 2293 (1982).
[8] R. Brako and D. M. Newns, Surface Sci. 108, 253 (1981).
[9] J. N. M. van Wunnik, R. Brako, K. Makoshi, and D. M. Newns, Surface Sci. 126, 618 (1983).
[10] J. Bardeen, Phys. Rev. Letters 6, 57 (1961).
[11] N. D. Lang and A. R. Williams, Phys. Rev. B 18, 616 (1978).
[12] U. von Gemmingen, thesis, University of Munich (1981).
[13] Y. H. Ohtsuki, Nuclear Instrum. and Methods B 2, 280 (1984).
[14] O. V. Firsov, Zh. eksper. teor. Fiz. 36, 1517 (1959); Soviet Phys. — J. exper. theor. Phys. 9, 1076 (1959).
[15] J. Kanamori, K. Terakura, and K. Yamada, Progr. theor. Phys. (Kyoto) 41, 1426 (1969).
[16] M. Remy, J. chem. Phys. 53, 2487 (1970).
[17] Ming L. Yu and N. D. Lang, Phys. Rev. Letters 50, 127 (1983).
[18] J. K. Nørskov and B. I. Lundqvist, Surface Sci. 89, 251 (1979).

(Received July 24, 1985)

phys. stat. sol. (b) 133, 277 (1986)

Subject classification: 71.20; S2; S3; S4

Departamento de Fisica Teorica, Universidad de Valladolid[1])

Density Functional Pseudopotential Calculation of the Cohesive Properties of Ordered (CsCl-Type) Compounds of Nontransition Metals

By

D. J. González

The density functional formalism is used to calculate the electron density, equilibrium volume, and heat of formation of ordered (CsCl-type) alloys of nontransition metals. A cellular method is employed to separate the alloy in atomic cells with appropriate matching conditions at the cell boundaries. The ion cores are substituted by empty core pseudopotentials. The results are compared with other theoretical results and with the experiment.

Der Dichtefunktionalformalismus wird benutzt, um Elektronendichte, Gleichgewichtsvolumen und Bildungswärme geordneter (CsCl-Typ)-Legierungen von Nicht-Übergangsmetallen zu berechnen. Eine Zellenmethode wird benutzt, um die Legierung in atomare Zellen mit geeigneten Anpaßbedingungen an den Zellengrenzen zu separieren. Die Ionenrümpfe werden substituiert durch leere Rumpf-Pseudopotentiale. Die Ergebnisse werden mit anderen theoretischen Ergebnissen und dem Experiment verglichen.

1. Introduction

In previous papers we have used the density functional formalism to study the cohesive properties of homovalent disordered binary solid alloys of nontransition metals with valence $Z = 1$ or 2, employing a cellular method [1 to 4]. In this paper we extend the method to ordered alloys. In this method the ionic cores of the component atoms are characterized by weak pseudopotentials and the alloy is considered as composed of two types of atomic cells with appropriate matching conditions imposed at the cell boundaries. The calculations have been performed for several heterovalent alloys, assuming an ordered CsCl structure. Among the alloys studied in this work (LiMg, MgAl, LiAl, LiTl, NaMg) only two of them (LiMg, LiTl) actually present an ordered phase with the CsCl structure in their experimental equilibrium phase diagram. The others do not have such an ordered CsCl phase, but we think that the model calculations reported here for them are, nevertheless, informative.

2. Density Functional Formalism for Alloys. Computation of the Electron Density

According to the density functional formalism [5], the ground state energy of a system (metal or alloy in our case) formed by a static lattice of local ionic potentials plus a gas of interacting conduction electrons with density $n(\boldsymbol{r})$ can be written in the following way (Hartree at. units are used throughout in this paper):

$$E = F[n] + \int n(\boldsymbol{r})\, V_{\rm i}(\boldsymbol{r})\, {\rm d}\boldsymbol{r} + \tfrac{1}{2} \int n(\boldsymbol{r})\, V_{\rm e}(\boldsymbol{r})\, {\rm d}\boldsymbol{r} + U_{\rm ii}\,, \tag{1}$$

[1]) Valladolid, Spain.

where $F[n]$ is the sum of the kinetic, exchange, and correlation energies of the electrons. $V_i(\boldsymbol{r})$ is the total ionic potential at $\boldsymbol{r}$,

$$V_i(\boldsymbol{r}) = \sum_{\boldsymbol{s}} v_s(\boldsymbol{r} - \boldsymbol{s}) , \tag{2}$$

v_s being the potential due to the ion core located at the lattice site $\boldsymbol{s}$. $V_e(\boldsymbol{r})$ is the Coulomb potential of the conduction electrons,

$$V_e(\boldsymbol{r}) = \int \frac{n(\boldsymbol{r})}{|\boldsymbol{r} - \boldsymbol{r}'|} \, d\boldsymbol{r}' , \tag{3}$$

and U_{ii} is the Coulomb ion–ion interaction. The integrals in (1) and (3) are extended over the total volume of the system. v_s is now approximated by an empty core pseudopotential [6]

$$v(\boldsymbol{r}) = \begin{cases} 0 ; & r < r_c \\ -\dfrac{Z}{r} ; & r > r_c , \end{cases} \tag{4}$$

where r_c is the empty core radius and Z the ionic charge. The ground state energy and the ground state electron density are obtained by minimization of E with respect to $n(\boldsymbol{r})$ subject to the normalization of $n(\boldsymbol{r})$. The ionic positions are held fixed in the process. The corresponding Euler equation becomes

$$\frac{\delta F[n]}{\delta n(\boldsymbol{r})} + V(\boldsymbol{r}) = \mu , \tag{5}$$

where $V(\boldsymbol{r}) = V_i(\boldsymbol{r}) + V_e(\boldsymbol{r})$ and μ is the chemical potential. $F[n]$ is now approximated by

$$F[n] = \int f_r[n] \, d\boldsymbol{r} ; \tag{6}$$

$$f_r[n] = c_k n(\boldsymbol{r})^{5/3} + \frac{\lambda}{8} \left[\frac{(\nabla n(\boldsymbol{r}))^2}{n(\boldsymbol{r})} \right] + c_x n(\boldsymbol{r})^{4/3} + \\ + [c_1 \ln (3/4\pi n(\boldsymbol{r}))^{1/3} + c_2] \, n(\boldsymbol{r}) , \tag{7}$$

where $c_k = \frac{3}{10} (3\pi^2)^{2/3}$, $c_x = -\frac{3}{4} (3/\pi)^{2/3}$, $c_1 = 0.0155$, $c_2 = -0.0575$ and the value of the parameter λ will be discussed later. The first and second terms in (7) give the kinetic energy density, the third term is the exchange energy density, and the last term is the correlation energy density of the conduction electrons. Using (6) and (7) in (5) one obtains

$$\frac{5}{3} c_k n(\boldsymbol{r})^{2/3} + \frac{\lambda}{8} \left(\frac{\nabla n(\boldsymbol{r})}{n(\boldsymbol{r})} \right)^2 - \frac{\lambda}{4} \left(\frac{\nabla^2 n(\boldsymbol{r})}{n(\boldsymbol{r})} \right) + \frac{4}{3} c_x n(\boldsymbol{r})^{1/3} + \\ + c_1 \ln (3/4\pi n(\boldsymbol{r}))^{1/3} + c_2 - \frac{1}{3} c_1 + V(\boldsymbol{r}) = \mu . \tag{8}$$

This differential equation must be integrated coupled with Poisson's equation

$$\nabla^2 V(\boldsymbol{r}) = -4\pi n(\boldsymbol{r}) . \tag{9}$$

The boundary conditions appropriate to an AB alloy are

$$(\nabla n(\boldsymbol{r}))_{O_A} = (\nabla V(\boldsymbol{r}))_{O_A} = (\nabla n(\boldsymbol{r}))_{O_B} = (\nabla V(\boldsymbol{r}))_{O_B} = 0 , \tag{10}$$

where O_A and O_B are the centers of the A- and B-type lattice sites, respectively (evidently, in the case of a pure metal there is only one type of site). A cellular method is now used to integrate (8) and (9). The volume of the equiatomic alloy is first decomposed into atomic cells: N_A equal atomic cells of type A and N_B equal atomic cells of type B ($N_A = N_B = N/2$, where N is the total number of atoms in the alloy). The atomic cell volumes Ω_A and Ω_B must satisfy the equation

$$\Omega_A + \Omega_B = 2\Omega_{AB}, \tag{11}$$

where Ω_{AB} is the average volume per atom in the AB alloy. Equations (8) and (9) must be integrated inside each cell of the alloy with appropriate matching conditions at the cell surfaces. In order to make the calculations feasible, the atomic cells are taken to be spherical (the volumes Ω_A and Ω_B again satisfying (11); these volumes will be precisely characterized below) and $n(\boldsymbol{r})$ is assumed to be spherically symmetric in each cell. Two natural matching conditions are

$$n(R_A) = n(R_B), \tag{12}$$

$$\left(\frac{dn}{dr}\right)_{R_A} = -\left(\frac{dn}{dr}\right)_{R_B}, \tag{13}$$

where R_A and R_B are the atomic cell radii. A matching condition can also be imposed on $V(\boldsymbol{r})$ [7]. The electrostatic potential at the surface of an i-cell (i = A, B) is the sum of two terms

$$V(R_i) = V_{in}(R_i) + V_{out}(R_i). \tag{14}$$

$V_{in}(R_i)$ is due to the net electric charge inside the cell,

$$V_{in}(R_i) = -\frac{Q_i}{R_i}, \tag{15}$$

where $Q_i = Z_i - q_i$ (Z_i and q_i are the positive charge of the ionic core and the number of conduction electrons inside the cell, respectively). On the other hand, $V_{out}(R_i)$ is due to the rest of the electrons and ions of the crystal. This is a Madelung potential which depends on the lattice structure. Since the alloy is equiatomic and completely ordered,

$$V_{out}(R_i) = \alpha \frac{Q_i}{l}, \tag{16}$$

where l is the lattice constant and α the Madelung constant. If matching of the electrostatic potential is imposed,

$$V(R_A) = V(R_B), \tag{17}$$

then, from (15) and (16) plus the charge neutrality condition,

$$Q_A = -Q_B = Q, \tag{18}$$

one obtains

$$\frac{1}{R_A} + \frac{\alpha}{l} = \frac{1}{R_B} - \frac{\alpha}{l}. \tag{19}$$

l can now be written in terms of R_A and R_B. This relation depends on the lattice structure. If the ordered alloy has the CsCl structure, then

$$l^3 = \tfrac{4}{3}\pi(R_A^3 + R_B^3), \tag{20}$$

where l is the smallest distance between two A atoms or between two B atoms. Then, from (19) and (20) (with $\alpha(\text{CsCl}) = 2.03536$) one obtains [7]

$$\frac{R_A}{R_B} = 0.936448. \tag{21}$$

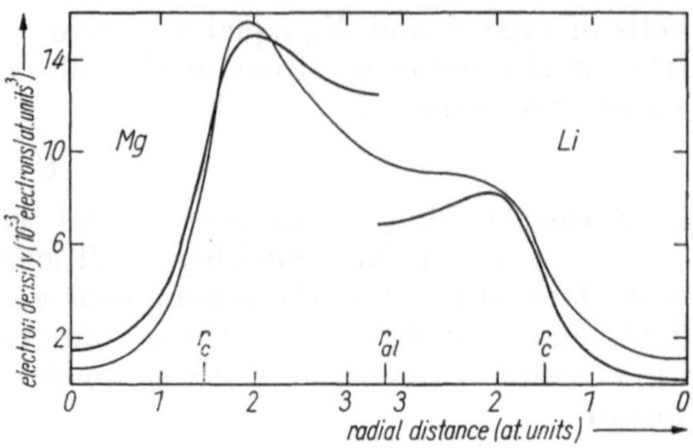

Fig. 1. Electron density in an atomic cell of Li and in an atomic cell of Mg in the ordered alloy LiMg (thick continuous line). The electron density in the pure metals (thin continuous lines) is given for comparison. r_{al} marks the boundary between the cells in the alloy

This result is easily generalized to other ordered structures. In conclusion, the ratio R_A/R_B is fixed by the type of ordered structure if we impose (17). Finally, the fourth matching condition is

$$\mu_A = \mu_B , \tag{22}$$

that is, the statement that the chemical potential is the same constant in all the cells. From (8), (12), (13), and (17) it is clear that (22) is equivalent to

$$\left(\nabla^2 n(\boldsymbol{r})\right)_{R_A} = \left(\nabla^2 n(\boldsymbol{r})\right)_{R_B} . \tag{23}$$

In conclusion, the electron density in the ordered CsCl-type compound is computed by integrating (8) and (9) inside an A-cell and inside a B-cell, subject to the matching conditions (12), (13), (17), and (23). The practical method followed to integrate these equations is:

(i) Fix first l. This immediately fixes R_A and R_B through (20) and (21).

(ii) Start the numerical integration at O_A with initial values $(dn/dr)_{O_A} = (dV/dr)_{O_B} = 0$ and arbitrary values of $n(O_A)$ and $V(O_A)$ and integrate out until $r = R_A$.

(iii) Start now the integration at O_B with initial values $(dn/dr)_{O_B} = (dV/dr)_{O_B} = 0$ and arbitrary values of $n(O_B)$ and $V(O_B)$ and integrate out until $r = R_B$. By changing $n(O_B)$ and $V(O_B)$, two of the matching conditions can be fulfilled, for instance, $n(R_A) = n(R_B)$ and $V(R_A) = V(R_B)$.

(iv) Finally the whole process is repeated for new values of $n(O_A)$ and $V(O_A)$ until the four matching conditions are satisfied.

An example of calculated self-consistent electron densities is given in Fig. 1 for the compound LiMg at the theoretical equilibrium volume. The electron densities for the pure metals are also given for comparison.

3. Calculation of Total Energies. Heat of Formation

If $E_a(A)$, $E_a(B)$, and $E_a(AB)$ are the energies per atom of the pure metals and of the equiatomic alloy, respectively, then the heat of formation per atom is defined as

$$\Delta H = E_a(AB) - \tfrac{1}{2}\left(E_a(A) + E_a(B)\right) . \tag{24}$$

The cell partitioning introduced in Section 2 induces a convenient cell partitioning of the energies in (24). $F[n]$ can be decomposed into a sum of integrals over atomic cells,

$$F[n] = \int_{\Omega_s} f_r[n] \, d\boldsymbol{r} , \tag{25}$$

where the integral is extended over the volume of the atomic cell centered at the lattice site $\mathbf{s}$. The three Coulomb terms of (1) (their sum is here labelled E_c) can be combined and also written in cell form [8],

$$E_c = \sum_s \left[\frac{1}{2} \int_{\Omega_s} \int_{\Omega_s} \frac{n(\mathbf{r})\, n(\mathbf{r}')}{|\mathbf{r} - \mathbf{r}'|} \,\mathrm{d}\mathbf{r}\, \mathrm{d}\mathbf{r}' + \int_{\Omega_s} n(\mathbf{r})\, v_s(\mathbf{r})\, \mathrm{d}\mathbf{r}\right] + E_M . \tag{26}$$

E_M is a Madelung energy which gives the electrostatic interaction between the cells. Since $n(\mathbf{r})$ has spherical symmetry inside each cell,

$$E_M = \frac{1}{2} \sum_{s \neq s'} \frac{Q_s Q_{s'}}{|\mathbf{s} - \mathbf{s}'|}, \tag{27}$$

where Q_s is the net electric charge inside the $\mathbf{s}$ cell. Finally, using (25) to (27)

$$E_a(\mathrm{AB}) = \frac{1}{N} \left[\sum_s I^{\mathrm{all}}(\mathbf{s}) + E_M\right], \tag{28}$$

$$I^{\mathrm{all}}(\mathbf{s}) = \int_{\Omega_s} f_r[n]\, \mathrm{d}\mathbf{r} + \frac{1}{2} \int_{\Omega_s} \int_{\Omega_s} \frac{n(\mathbf{r})\, n(\mathbf{r}')}{|\mathbf{r} - \mathbf{r}'|} \,\mathrm{d}\mathbf{r}\, \mathrm{d}\mathbf{r}' + \int_{\Omega_s} n(\mathbf{r})\, v_s(\mathbf{r})\, \mathrm{d}\mathbf{r} . \tag{29}$$

Since all the terms $I^{\mathrm{all}}(\mathbf{s})$ corresponding to A cells are equal (we denote each of them by $I^{\mathrm{all}}(\mathrm{A})$) and the same happens for the terms associated to B cells, (28) becomes

$$E_a(\mathrm{AB}) = \frac{1}{N} \left\{\frac{N}{2} \left[I^{\mathrm{all}}(\mathrm{A}) + I^{\mathrm{all}}(\mathrm{B})\right] + E_M\right\}. \tag{30}$$

Using (18), E_M can be written as

$$E_M = -\frac{1}{2} \frac{N\alpha Q^2}{l}. \tag{31}$$

Then, the energy per atom in the alloy AB is

$$E_a(\mathrm{AB}) = \frac{1}{2} \left[I^{\mathrm{all}}(\mathrm{A}) + I^{\mathrm{all}}(\mathrm{B}) - \frac{\alpha Q^2}{l}\right]. \tag{32}$$

Using a similar notation, the energies per atom of the pure metals are

$$E_a(\mathrm{A}) = I^{\mathrm{met}}(\mathrm{A}) , \quad E_a(\mathrm{B}) = I^{\mathrm{met}}(\mathrm{B}) . \tag{33}$$

4. Results

The preceding formalism has been applied to the study of five heterovalent nontransition alloys, assuming an ordered CsCl structure. A parameter, λ, appears in (7) which affects the gradient correction to the kinetic energy. In our previous work on disordered homovalent alloys with $Z = 1$ (alkaline alloys) or $Z = 2$ (alkaline-earth alloys), a value $\lambda = 1$ was found adequate both for pure metals and alloys. This value of λ corresponds to the gradient correction introduced by Weizsäcker [9]. Our preliminary attempts to use the same prescription for heterovalent compounds yielded unrealistic values for the heat of formation. This is understandable for the following reason: the formalism presented above and applied to pure metals, gives accurate values for the compressibilities of most simple metals with $Z = 1$ or 2 [10 to 12] but the results are less accurate for metals with $Z = 3$ and larger [10]. Since one of the important contributions to the heat of formation of alloys can be described as an elastic compression or expansion of atomic cells, it is clear that the pure metal compressibilities play an important role [1]. In order to account for this contribution as accurately as possible we have followed the prescription of fixing λ and the empty core radius r_c, for each pure metal, by requiring the formalism to reproduce the experimental volume per atom and compressibility. In fact a similar philosophy was followed by Chelikowsky in his study of the cohesive properties of pure metals [13, 14]. He assigned a different value of λ to each element, although the method followed by Chelikowsky to calculate λ was different from the one we follow here. This author fixed λ by fitting to a free atom property. The core radii and the values of λ obtained in our work are given in Table 1.

Table 1

Empty core radius (at. units) and coefficient of the gradient correction to the kinetic energy (λ in (7)) obtained by fitting the theoretical equilibrium volume and the theoretical bulk modulus to experiment

metal	r_c	λ
Li	1.508	1/5.908
Na	1.830	1
Mg	1.443	1/1.978
Al	1.145	1/2.875
Tl	1.655	1/4.954

Concerning the ordered alloy, still remains the problem of fixing a value of λ to be used in the computations. Hodges and Stott [15] have suggested that the gradient correction introduced by Kirzhnits [16] (that is, $\lambda = 1/9$ in (7)) should be more appropriate for an ordered alloy. Nevertheless, preliminary computations carried out by us have also yielded unrealistic heats of formation. These attempts suggested that a parameter intermediate between $\lambda = 1$ and 1/9 should be used for the ordered alloy. The results presented in this paper have been obtained according to the following criterion: given an ordered alloy AB we have characterized each metal by a core radius (r_c^i) and a parameter (λ_i) obtained as explained above, but the parameter used for the alloy (λ_{AB}) has been chosen as the arithmetic mean of both ($\lambda_{AB} = \frac{1}{2} (\lambda_A + \lambda_B)$).

Among the alloys studied in this work, only two of them (LiMg and LiTl) actually show the existence of an ordered phase with the CsCl structure in their experimental equilibrium phase diagram. LiAl shows an ordered phase with the B32 structure. MgAl and NaMg do not show any ordered phase.

The equilibrium lattice constant, l_0, obtained by minimizing the energy of the alloy as a function of the lattice constant, is given in Table 2. Theoretical results obtained by other authors are also given. A similar comparison is provided for the relative deviation (β) of the alloy volume with respect to the average of the pure metal volumes.

For the LiMg alloy there are experimental values of β corresponding to the ordered alloy [17], disordered alloy [18], and liquid alloy [19], which are rather similar. Our calculated value for the ordered alloy is in good agreement with the experimental one. This alloy has also been studied by Hafner [20] and Zaremba [21]. Hafner used standard pseudopotential theory, that is, second-order perturbation theory plus linear dielectric screening. Zaremba used the density functional formalism. In his computations Zaremba partitioned the alloy into cells defined by surfaces on which

Table 2

Equilibrium lattice constant l_0 (smallest distance between two A atoms or between two B atoms) in at. units, and relative deviation β (%) of the volume per atom of the alloy with respect to the average of the pure metal volumes

alloy	this work		other works		experiment
	l_0	β	l_0	β	β
LiMg	6.69	−1.1	6.43 [20, 23]	−12.0 [20, 23]	−3.0 [17], 0**) [19] −5.9***) [18]
			6.18 [21]	−12.0 [21]	
MgAl	6.39	−3.1		−3.6*) [23]	−3.0****) [22]
LiAl	6.33	−1.9		−26.6 [20, 23]	−15.5 [24]
LiTl	6.95	−0.9			
NaMg	7.22	−8.1			

*) Theoretical result of Hafner for the disordered solid alloy.
**) Data for the liquid alloy.
***) Data for the disordered solid alloy.
****) Data for the liquid alloy.

the gradient of the electron density vanishes. Their results for β are greater than the experimental one. In the MgAl alloy our theoretical result is very similar to both the experimental result for the liquid alloy [22] and the theoretical result of Hafner and Sommer [23] corresponding to the disordered solid alloy. Our calculated value for the LiAl ordered alloy is much smaller than the experimental result [24]. This system has also been studied by Hafner and Sommer [23] using standard pseudopotential theory. These authors obtained a result greater than the experimental one. In fact Table 2 shows that our method tends to give values for β which are smaller than the experimental ones while Hafner's method gives results which are greater.

The calculated heats of formation of the CsCl compounds are given in Table 3. For all the alloys, there is qualitative agreement with the experiment. Our results for LiMg, LiAl, and NaMg are in semiquantitative agreement with other theoretical calculations [21, 23, 25]. Although LiAl has an ordered phase with B32 structure our calculation has been performed assuming a CsCl structure. Nevertheless, we think that our ΔH is similar to the one we would obtain using the right structure because, as some authors have pointed out [7], changes in binding energies with lattice structure are generally much smaller than the binding energies themselves.

Table 3

Calculated and measured heats of formation (cal/g-atom)

alloy	this work	others works	experiment
LiMg	−1001	−2286 [20, 23]	compounds
		−1500 [21]	
MgAl	7487	670 [17]	
LiAl	−4821	−1900 [20, 23]	compounds
LiTl	−1003		−6400 ± 500*) [19]
NaMg	8742	3612 [17]	

*) Data for the liquid alloy.

The only quantitative experimental information on ΔH for the ordered alloys studied in this paper is $\Delta H(\mathrm{LiTl}) = -(6400 \pm 500)$ cal/g-atom [27] which agrees in sign with our calculated value. Although MgAl and NaMg do not show any ordered phase in their experimental phase diagrams, there is a computation of ΔH for these alloys, carried out by Pettifor and Gelatt [25], assuming an ordered CsCl structure. There is agreement in sign between both sets of results, although our calculated ΔH are greater.

In conclusion, a density functional pseudopotential method has been used to compute the cohesive properties of ordered CsCl-type compounds formed by nontransition metals. The results are in good agreement with the experimental features of the phase diagrams of these alloys. Nevertheless, we think that a more exhaustive study about the coefficient of the gradient correction to the kinetic energy must be done in order to improve the results.

Acknowledgement

I am very grateful to Dr. J. A. Alonso for valuable discussions while this work was being carried out.

References

[1] M. P. Iñiguez, J. A. Alonso, D. J. González, and M. L. Zorita, J. Phys. F **12**, 1907 (1982).
[2] D. J. González, J. A. Alonso, and M. P. Iñiguez, phys. stat. sol. (b) **114**, 495 (1982).
[3] D. J. González, M. L. Zorita, J. A. Alonso, and M. P. Iñiguez, phys. stat. sol. (b) **119**, 589 (1983).
[4] D. J. González, C. M. Zapata, and J. A. Alonso, phys. stat. sol. (b) **129**, 483 (1985).
[5] P. Hohenberg and W. Kohn, Phys. Rev. **136**, B864 (1964).
[6] N. W. Ashcroft, Phys. Letters (Netherlands) **23**, 48 (1966).
[7] A. R. Williams, J. Kübler, and C. D. Gelatt, Phys. Rev. B **19**, 6094 (1979).
[8] H. Brooks, Nuovo Cimento, Suppl. 7, 165 (1958).
[9] C. F. Weizsäcker, Z. Phys. **96**, 431 (1935).
[10] A. Meyer, I. H. Umar, and W. H. Young, Phys. Rev. B **4**, 3287 (1971).
[11] J. M. López and J. A. Alonso, phys. stat. sol. (b) **104**, 307 (1981).
[12] D. J. González and J. A. Alonso, Physica (Utrecht) **112B**, 67 (1982).
[13] J. R. Chelikowsky, Solid State Commun. **33**, 1201 (1980).
[14] J. R. Chelikowsky, Phys. Rev. B **21**, 3074 (1980).
[15] C. Hodges and M. J. Stott, Phil. Mag. **26**, 375 (1972).
[16] D. A. Kirzhnits, Soviet Phys. — J. exper. theor. Phys. **5**, 64 (1957).
[17] F. H. Herbstein and L. Averbach, Acta metall. **4**, 407 (1956).
[18] L. D. Firth, N. H. A. Nowaira, and N. Scott, J. Phys. F **4**, L200 (1974).
[19] H. Ruppersberg, J. Saar, W. Speicher, and P. Heitjans, J. Physique **41**, 68 (1980).
[20] J. Hafner, J. Phys. F **6**, 1243 (1976).
[21] E. Zaremba, Solid State Commun. **44**, 931 (1982).
[22] S. Steev and S. Woerner, Z. Metallk. **56**, 771 (1965).
[23] J. Hafner and F. Sommer, CALPHAD **1**, 325 (1977).
[24] K. Kuriyama and N. Masaki, Acta cryst. B **31**, 1793 (1975).
[25] D. G. Pettifor, Physical Metallurgy, Ed. R. W. Cahn and P. Haasen, Elsevier, Amsterdam 1983 (p. 73).
[26] O. Kubachewsky and E. Ll. Evans, Metallurgical Thermochemistry, 3. Edition, Pergamon Press, London/Oxford 1958.

(Received July 22, 1985)

phys. stat. sol. (b) 133, 285 (1986)

Subject classification: 71.25; S5; S5.11; S5.12

Odessa State University[1]) *(a),*
and International Centre for Theoretical Physics, Trieste
and Faculty of Physics, University of Sofia[2]) *(b)*

Bond Charge Approximation for Valence Electron Density in Elemental Semiconductors

By

V. K. BASHENOV (a), V. E. GORBACHEV (a), and D. I. MARVAKOV (b)

The spatial valence electron distribution in silicon and diamond is calculated in adiabatic bond charge approximation at zero temperature when bond charges have Gaussian shape and their tensor character is taken into account. Agreement between theory and experiment is achieved. For this purpose Xia's ionic pseudopotentials and Schulze-Unger's dielectric function are used. By two additional parameters A_B and Z'_B the spatial extent is described of the bond charge and local-field corrections, respectively. The parameter Z'_B accounts for the ratio between the Coulomb and exchange correlation interactions of the valence electrons. Its silicon and diamond values have different signs.

Пространственное распределение плотности валентных электронов в Si и алмазе вычислено при 0 K в приближении адиабатических зарядов на связях гауссовской формы с учетом их тензорной природы. Согласие теории и опыта достигнуто применением ионных псевдопотенциалов Ксиа и диэлектрических функций Шульца-Унгера, которые не включают эффекты локального поля, а также двух параметров A_B и Z'_B. Параметр A_B характеризует пространственную протяженность заряда на связи, а параметр локального поля Z'_B учитывает соотношение кулоновского и обменно-корреляционного взаимодействия валентных электронов. Величины Z'_B отличаются знаком в Si и алмазе.

1. Introduction

Many properties of elemental silicon-type semiconductors depend on the spatial valence electron distribution. X-ray diffraction investigations in these crystals show that the electron density distribution cannot be approximated by a simple superposition of spherical distributions centred on the lattice sites. A considerable part of the electron density is concentrated into so-called bond charges. The first theory of these quantities has been developed by Phillips [1]. However, he neglects the local-field effects in the valence electron response to the ionic pseudopotentials. The further investigations have suggested that such effects are very important in elemental silicon-type semiconductors [2].

In semiconductors the valence electron distribution $\varrho(\boldsymbol{r})$ is usually determined by a sum of $|\psi_n(\boldsymbol{k}, \boldsymbol{r})|^2$ over all valence bands labeled by n and over all mean-value points $\boldsymbol{k}$ of the Brillouin zone [3]. The mentioned theoretical calculations are very labour-consuming. One must determine the Bloch-wave function $\psi_n(\boldsymbol{k}, \boldsymbol{r})$ as a self-consistent solution of the one-electron Schrödinger equation where exchange and correlation valence electron effects are included. Unfortunately, these effects are insufficiently studied. Another kind of calculations of $\varrho(\boldsymbol{r})$ are based on different model considera-

[1]) Odessa 270000, USSR.
[2]) Sofia 1126, Bulgaria.

tions [4 to 7]. As a rule they give a bond charge compressed along the line connecting neighbouring atoms while the experiment [8] indicates its extent.

In our paper the valence electron distribution of silicon-type semiconductors is calculated in the framework of the microscopical dielectric screening theory [9]. The preliminary results have been given in [10]. The adiabatic bond charge model for the valence electron density of covalent crystals is described in the next section. In Section 3 we select the dielectric function with vanishing local-field effects and the form factor of a suitable ionic pseudopotential. As known both functions are very important because they determine the main contributions to the valence electron density which come from the diagonal screened bare ions of the crystal lattice. In Section 4 we discuss the rest part of the valence electron distribution. It is represented by the Gaussian bond charges and determined by two parameters. The first of them, A_B, characterizes the spatial extent of the bond charges and the other accounts for the local-field effects. At last, in Section 5 we present our conclusions and possibilities for further improvements of the bond charge approximation in semiconducting crystals. This approximation can especially be perspective in investigations of the vibrational and electronic spectra of covalent crystals and in calculations of their total energy.

2. Adiabatic Bond Charges

In covalent semiconductors the attendance of the bond charges can be explained by different approaches. Phillips [1] has presented that if one considers the dielectric screening of the bare ions a part of the valence electron charges is concentrated on the bonds between neighbouring atoms. Bertoni et al. [4] have noted that this part of the valence electron charge is referred to the off-diagonal elements of the dielectric matrix. By the Brillouin-Wigner perturbation theory Makhnovetskii and Krasko [5] have shown that the bond charges arise from the coherent scattering of electron waves on some Bragg planes. But we regard [10] that the most direct way for introducing the bond charges consists in a consideration of the dynamical electron density variations when atoms accomplish harmonical displacements from their lattice sites.

In fact, the linear variations of the electron density $\delta\varrho(\boldsymbol{Q})$ are related to the Fourier components $\varrho(\boldsymbol{G})$ of the static electron density by

$$i\boldsymbol{w} \cdot \boldsymbol{G}\varrho(\boldsymbol{G}) = \lim_{\boldsymbol{q}\to 0} \delta\varrho(\boldsymbol{Q}) , \tag{1}$$

where $\boldsymbol{w}$ means an arbitrary translation of the crystal as a whole, $\boldsymbol{G}$ denotes a reciprocal lattice vector, $\boldsymbol{Q} = \boldsymbol{q} + \boldsymbol{G}$, and $\boldsymbol{q}$ is a vector reduced to the first Brillouin zone. This condition determines the invariance of $\varrho(\boldsymbol{r})$ when a rigid translation on a vector $\boldsymbol{w}$ is made. In this case all the Fourier transforms $u_\alpha^k(\boldsymbol{q})$ of the displacements are equal to $w_\alpha\, \delta(\boldsymbol{q})$ ($\delta(\boldsymbol{q})$ is the Kronecker symbol). Here α and k denote the rectangular components and the k-th bare ion of the Wigner-Seitz cell, respectively. After extracting the bond charges from $\delta\varrho(\boldsymbol{G})$ we can determine, following (1), the bond charge contributions to the static valence electron density. Then the problem consists in developing a suitable model for $\delta\varrho(\boldsymbol{Q})$ which can be used in the case of elemental semiconductors.

In the microscopic theory [9] the linear variations of the electron density are given by

$$\delta\varrho(\boldsymbol{Q}) = \sum_{\boldsymbol{G}'} \chi(\boldsymbol{Q}, \boldsymbol{Q}')\, \delta V(\boldsymbol{Q}') , \tag{2}$$

where $\chi(\boldsymbol{Q}, \boldsymbol{Q}')$ is the static charge density response function of the valence electrons. Let us introduce the linear variations of the electron–ion potential by

$$\delta V(\boldsymbol{Q}) = i \sum_{\alpha\, k} Q_\alpha u_\alpha^k(\boldsymbol{q})\, v_k(\boldsymbol{Q})\, e^{i\boldsymbol{G}\cdot \boldsymbol{r}_k} . \tag{3}$$

Here $v_k(\boldsymbol{Q})$ is the local pseudopotential of the k-th bare ion placed at $\boldsymbol{r}_k$ in the Wigner-Seitz cell. Without any loss of generality $\chi(\boldsymbol{Q}, \boldsymbol{Q}')$ can be written in the following form:

$$\chi(\boldsymbol{Q}, \boldsymbol{Q}') = \bar{\chi}(Q)\, \delta_{G,G'} + \tilde{\chi}(\boldsymbol{Q}, \boldsymbol{Q}'), \tag{4}$$

where $\chi(Q)$ describes the homogeneous and isotropic part of the static charge correlation function. By analogy to (4) the longitudinal dielectric matrix $\varepsilon(\boldsymbol{Q}, \boldsymbol{Q}')$ and the polarization operator $\pi(\boldsymbol{Q}, \boldsymbol{Q}')$ of valence electrons can be represented in a similar way.

Following (4) we divide $\delta\varrho(\boldsymbol{Q})$ in two parts. The first one $\delta\varrho_i(\boldsymbol{Q})$ corresponds to the homogeneous and isotropic response of the valence electron density and it reads

$$\delta\varrho_i(\boldsymbol{Q}) = i \left[\frac{1}{\bar{\varepsilon}(Q)} - 1\right] \sum_{\alpha,k} Q_\alpha u_\alpha^k(\boldsymbol{q})\, f_k(\boldsymbol{Q})\, e^{i\boldsymbol{G}\cdot\boldsymbol{r}_k}, \tag{5}$$

where $\bar{\varepsilon}(Q)$ is the dielectric valence electron function with vanishing local-field corrections. The form factor of the k-th bare ion is $f_k(\boldsymbol{Q}) = v_k(\boldsymbol{Q})/v(Q)$ and $v(Q)$ denotes the Coulomb potential. The second part of the valence electron density representation $\delta\varrho_B(\boldsymbol{Q})$ corresponds to $\chi(\boldsymbol{Q}, \boldsymbol{Q}')$, whose exact form is unknown. In our previous paper [10] the following approximative expression for $\delta\varrho_B(\boldsymbol{Q})$ has been suggested:

$$\delta\varrho_B(\boldsymbol{Q}) = i \left[\frac{1}{\bar{\varepsilon}(Q)}\right] \sum_{\alpha,s} Q_\alpha u_\alpha^s(\boldsymbol{q})\, f_s(\boldsymbol{Q})\, e^{i\boldsymbol{G}\cdot\boldsymbol{r}_s}, \tag{6}$$

where s labels the bond charge at the $\boldsymbol{r}_s$ point of the Wigner-Seitz cell and its form factor is interpolated by $f_s(\boldsymbol{Q}) = \exp(-A_B Q^2)$. This representation of $\delta\varrho_B(\boldsymbol{Q})$ has been made as suggested by the dipole approximation in the lattice dynamics of covalent crystals [9].

The adiabatic bond charge model is introduced when we minimize the electronic part of the crystal potential with respect to $\delta\varrho_B(\boldsymbol{Q})$,

$$\frac{\delta V(\boldsymbol{Q})}{\bar{\varepsilon}(Q)} + \sum_{G'} \left[\frac{v(Q)}{\bar{\varepsilon}(Q)}\delta_{G,G'} - \tilde{\pi}^{-1}(\boldsymbol{Q}, \boldsymbol{Q}')\right] \bar{\varepsilon}(Q')\, \delta\varrho_B(\boldsymbol{Q}') = 0, \tag{7}$$

where $\tilde{\pi}(\boldsymbol{Q}, \boldsymbol{Q}')$ is the inhomogeneous and anisotropic part of the valence electron polarization operator. After taking into account (3) and (6), the latter equality reads

$$\sum_{\beta,s'} [P_{\alpha\beta}^{ss'}(\boldsymbol{q}) - V_{\alpha\beta}^{ss'}(\boldsymbol{q})]\, u_\beta^{s'}(\boldsymbol{q}) = \sum_{\gamma,k} E_{\alpha\gamma}^{sk}(\boldsymbol{q})\, u_\gamma^k(\boldsymbol{q}). \tag{8}$$

Here the following abbreviations are used:

$$P_{\alpha\beta}^{ss'}(\boldsymbol{q}) = \sum_{G,G'} Q_\alpha f_s(\boldsymbol{Q})\, \tilde{\pi}^{-1}(\boldsymbol{Q}, \boldsymbol{Q}')\, f_{s'}(\boldsymbol{Q}')\, Q'_\beta\, e^{-i(\boldsymbol{G}\cdot\boldsymbol{r}_s - \boldsymbol{G}'\cdot\boldsymbol{r}_{s'})}, \tag{9}$$

$$V_{\alpha\beta}^{ss'}(\boldsymbol{q}) = \sum_{G} Q_\alpha f_s(\boldsymbol{Q})\, \frac{v(Q)}{\bar{\varepsilon}(Q)}\, f_{s'}(\boldsymbol{Q})\, Q_\beta\, e^{-i\boldsymbol{G}\cdot(\boldsymbol{r}_s - \boldsymbol{r}_{s'})}, \tag{10}$$

$$E_{\alpha\gamma}^{sk}(\boldsymbol{q}) = \sum_{G} Q_\alpha f_s(\boldsymbol{Q})\, \frac{v(Q)}{\bar{\varepsilon}(Q)}\, f_k(\boldsymbol{Q})\, Q_\gamma\, e^{-i\boldsymbol{G}\cdot(\boldsymbol{r}_s - \boldsymbol{r}_k)}. \tag{11}$$

Let us introduce the matrix $S(\boldsymbol{q})$ with elements $S_{\alpha\beta}^{ss'}(\boldsymbol{q})$ which satisfy the set of equations

$$\sum_{\beta,s'} S_{\alpha\beta}^{ss'}(\boldsymbol{q})\, [P_{\beta\gamma}^{s's''}(\boldsymbol{q}) - V_{\beta\gamma}^{s's''}(\boldsymbol{q})] = \delta_{ss''}\delta_{\alpha\gamma}. \tag{12}$$

After determining the bond charge displacements from (8) we have

$$u_\alpha^s(\boldsymbol{q}) = \sum_{\beta,\gamma,k,s'} S_{\alpha\beta}^{ss'}(\boldsymbol{q})\, E_{\beta\gamma}^{s'k}(\boldsymbol{q})\, u_\gamma^k(\boldsymbol{q}) \tag{13}$$

and then $\delta\varrho_B(Q)$ becomes

$$\delta\varrho_B(Q) = i\left[\frac{1}{\bar{\varepsilon}(Q)}\right] \sum_{\alpha,\beta,\gamma,s,s'} Q_\alpha f_s(Q)\, u_\gamma^k(q)\, e^{iG\cdot r_s}\, S_{\alpha\beta}^{ss'}(q)\, E_{\beta\gamma}^{s'k}(q)\,. \tag{14}$$

Here the linear valence electron variations are referred to the inhomogeneous and anisotropic response in the adiabatic bond charge model. It is not difficult to show that in such an approximation $\tilde{\chi}(Q, Q')$ can be written in the following form:

$$\tilde{\chi}(Q, Q') = \sum_{\alpha,\beta,s,s'} Q_\alpha \frac{f_s(Q)}{\bar{\varepsilon}(Q)}\, e^{iG\cdot r_s}\, S_{\alpha\beta}^{ss'}(q)\, e^{-iG\cdot r_{s'}} \frac{f_{s'}(Q')}{\bar{\varepsilon}(Q')} Q'_\beta\,. \tag{15}$$

The static valence electron distribution can be calculated by (1) if we use (5) and (14). At that, when $q \to 0$ the singular part of $\delta\varrho(Q)$ vanishes because the acoustic sum rule [9] holds. Then the regular part gives

$$\varrho(G) = \left[\frac{1}{\bar{\varepsilon}(G)} - 1\right] \sum_k f_k(G)\, e^{G\cdot r_k} + \left[\frac{1}{\bar{\varepsilon}(G)}\right] \sum_s f_s(G) e^{iG\cdot r_s} \sum_{\alpha,\gamma} Z_{\alpha\gamma}^s \frac{G_\alpha G_\gamma}{G^2}, \tag{16}$$

where the components of the effective bond charge tensor read

$$Z_{\alpha\gamma}^s = \lim_{q\to 0} \sum_{\beta,k,s'} \bar{S}_{\alpha\beta}^{ss'}(q)\, E_{\beta\gamma}^{s'k}(q)\,. \tag{17}$$

Here $\bar{S}_{\alpha\beta}^{ss'}(q)$ satisfy the set of equations (12) with $V_{\alpha\beta}^{ss'}(q)$ replaced by $\bar{V}_{\alpha\beta}^{ss'}(q)$. We note that $\bar{V}_{\alpha\beta}^{ss'}(q)$, $\bar{E}_{\beta\gamma}^{sk}(q)$ are determined by (10) and (11) when terms with $G = 0$ are dropped. Our consideration indicates a nontrivial dependence of the effective bond charge tensor on the interaction parameters.

Phillips' theory can be reproduced in a point bond charge approximation when A_B is equal to zero and $Z_{\alpha\beta}^s = Z_B\delta_{\alpha\beta}$. Also one must replace $\bar{\varepsilon}(Q)$ by the total dielectric function of valence electrons $\varepsilon(Q)$. In this case (16) takes the form

$$\varrho(G) = \left[\frac{1}{\varepsilon(G)} - 1\right] \sum_k f_k(G)\, e^{iG\cdot r_k} + \left[\frac{1}{\varepsilon(G)}\right] \sum_s Z_s\, e^{iG\cdot r_s}\,. \tag{18}$$

Then the average valence electron density is described by

$$\varrho(0) = \left[\frac{1}{\varepsilon(0)} - 1\right] \sum_k (-Z_k) + \left[\frac{1}{\varepsilon(0)}\right] \sum_k Z_s\,, \tag{19}$$

where Z_k denotes the valence of the k-th bare ion. From the crystal electroneutrality $\varrho(0)$ must be equal to $\sum_k Z_k$ and therefore $\sum_s Z_s = \sum_k Z_k$. In a silicon-type unit cell four bond charges with $Z_s = 2$ must persist because there are two bare ions with $\sum_k Z_k = 8$.

Moreover, in our model the tensor character of the bond charge can be accounted for. The symmetry considerations give in silicon-type semiconductors $Z_{\alpha\beta}^s$ in the following form:

$$Z_{\alpha\beta}^s = Z_s\delta_{\alpha\beta} + Z'_B(1 - \delta_{\alpha\beta})\, \nu_\alpha^s \nu_\beta^s\,. \tag{20}$$

Here ν_α^s are equal to ± 1 in accordance with bond charge positions $r_s = \pm(a/8)\,(\nu_1^s, \nu_2^s, \nu_3^s)$ and a denotes the lattice constant. The electroneutrality condition leads to $Z_s = 2$ while the ratio between the Coulomb and exchange correlation interactions of valence electrons can change the magnitude and sign of Z'_B.

3. Dielectric Function and Bare Ion Form Factor

According to (16) the spatial valence electron distribution in covalent crystals is very sensitive to the form of the dielectric function $\bar{\varepsilon}(Q)$. In our adiabatic bond charge model this function does not include local-field corrections until Phillips' theory [1] operates with a dielectric function with the Coulomb and exchange correlation effects taken into account. The difference between $\bar{\varepsilon}(Q)$ and $\varepsilon(Q)$ is quite pronounced in the small-Q limit: Then $\bar{\varepsilon}(Q \to 0) = \bar{\varepsilon}_0$, $\varepsilon(Q \to 0) = \varepsilon_0$, where ε_0 denotes the dielectric crystal constant. In order to estimate $\bar{\varepsilon}_0$ in silicon-type semiconductors we will consider their optical properties in terms of the optical permittivity $\varepsilon(\boldsymbol{q}, \omega)$ which is wave vector and frequency dependent.

The theory of the optical permittivity of a covalent semiconductor with allowance for the effect of static exchange correlation interaction of valence electrons has been developed in [11]. It has been shown that in these semiconductors the Coulomb interaction effects lead to a shift of the characteristic optical bands toward higher frequencies but such a shift can be compensated by the exchange correlation interaction. To perform the calculation of $\varepsilon(\boldsymbol{q}, \omega)$ one has to know the charge density correlation function of valence electrons $\chi(\boldsymbol{q} + \boldsymbol{G}, \boldsymbol{q} + \boldsymbol{G}'; \omega)$ at different $\boldsymbol{G}$ and $\boldsymbol{G}'$ reciprocal lattice vectors. This correlation function is determined in mean-field approximation [12] by

$$\chi^{-1}(\boldsymbol{Q}, \boldsymbol{Q}'; \omega) = \pi^{-1}(\boldsymbol{Q}, \boldsymbol{Q}'; \omega) - \delta_{\boldsymbol{G}\boldsymbol{G}'}[v(Q) + w(\boldsymbol{Q}, \omega)], \tag{21}$$

where $\pi^{-1}(\boldsymbol{Q}, \boldsymbol{Q}'; \omega)$ denotes the inverse polarization matrix for valence electrons in plane wave representation, $v(Q)$ and $w(\boldsymbol{Q}, \omega)$ are the Coulomb and exchange correlation potentials, respectively. Then $\varepsilon(\boldsymbol{q}, \omega)$ can easily be calculated for a model semiconductor with two nondegenerate flat bands v and c separated by a gap $E_g = \hbar\Delta$. The Wannier functions $|\mathrm{v}, l\rangle$ and $|\mathrm{c}, l\rangle$ are strongly localized in the l-th Wigner-Seitz cell. Then we have

$$\pi(\boldsymbol{Q}, \boldsymbol{Q}'; \omega) = \delta_{\boldsymbol{G}\boldsymbol{G}'} \frac{4\Delta^2}{(\omega + i0^+)^2 - \Delta^2} \langle \mathrm{v}, 0| \, \mathrm{e}^{-i\boldsymbol{Q}\cdot\boldsymbol{r}} \, |\mathrm{c}, 0\rangle \,. \tag{22}$$

Here $|\mathrm{v}, 0\rangle$ and $|\mathrm{c}, 0\rangle$ are referred to the central cell for valence v- and conduction c-bands and the interaction is adiabatically turned on.

Following [11] one can obtain

$$\varepsilon(\boldsymbol{q}, \omega) = 1 - \frac{v(q)\,\pi(\boldsymbol{q}, \boldsymbol{q}; \omega)}{1 + v(q)\,[1 - F(\boldsymbol{q}, \omega)]\,\pi(\boldsymbol{q}, \boldsymbol{q}; \omega)}, \tag{23}$$

where (21) and (22) are taken into account and

$$F(\boldsymbol{q}, \omega) = \frac{\sum\limits_{\boldsymbol{G}} [v(Q) + w(\boldsymbol{Q}, \omega)]\,\pi(Q, Q; \omega)}{v(q)\,\pi(\boldsymbol{q}, \boldsymbol{q}; \omega)}. \tag{24}$$

We have $F(\boldsymbol{q}, \omega) = 1$ when local-field effects are neglected. It means $\pi(\boldsymbol{Q}, \boldsymbol{Q}; \omega) = 0$ for $\boldsymbol{G} \neq \boldsymbol{0}$ and $w(\boldsymbol{Q}, \omega) = 0$. Then $\varepsilon(\boldsymbol{q}, \omega)$ coincides with $\bar{\varepsilon}(\boldsymbol{q}, \omega)$ from [11]. It is not difficult to establish the long-wavelength behaviour of the polarization operator $\pi(\boldsymbol{q}, \boldsymbol{q}; \omega)$ because $\bar{\varepsilon}(\boldsymbol{q}, \omega)$ satisfies the Kramers-Kronig relations and the sum rule for $\bar{\varepsilon}_2(\omega) = \lim\limits_{\boldsymbol{q} \to \boldsymbol{0}} \operatorname{Im} \bar{\varepsilon}(\boldsymbol{q}, \omega)$ holds.

As follows from (22) $\bar{\varepsilon}_2(\omega)$ has only one pole at $\omega = \Delta$. But the optical spectra of elemental semiconductors show two typical maxima at ω_1 and ω_2 [2, 13]. If we use (23) the characteristic frequencies are determined by the following equation:

$$1 + \frac{\omega_\mathrm{p}^2}{\omega^2 - \Delta^2}\,[1 - F(\omega)] = 0\,, \tag{25}$$

where $\omega_p = (4\pi e^2 \varrho_0/m)^{1/2}$ is the plasmon frequency of valence electrons with density ϱ_0 and $F(\omega) = \lim\limits_{q \to 0} F(\boldsymbol{q}, \omega)$. Next, let us account for the fact that at high frequencies all electrons can be considered as free-like and they form a homogeneous electron gas. Then we approximate $F(\omega)$ by

$$F(\omega) = 1 + \frac{\Omega^2 \Delta^2 (D - 1)}{D \omega_p^2 (\omega^2 - \Omega^2)}, \tag{26}$$

where Ω, Δ, and D have to be extracted from the optical spectra of elemental semiconductors.

In order to estimate the above parameters we use data for ε_0, $E_1 = \hbar\omega_1$, and $E_2 = \hbar\omega_2$ in the case of silicon-type semiconductors given in [13]. Further, one can easily solve (25) and then Ω and Δ read

$$\Omega = \frac{\omega_1 \omega_2 (\varepsilon_0 - 1)^{1/2}}{\omega_p}, \qquad \Delta = (\omega_1^2 + \omega_2^2 - \Omega^2)^{1/2}. \tag{27}$$

The optical dielectric constant of the crystal is

$$\varepsilon_0 = 1 + D \frac{\omega_p^2}{\Delta^2}, \tag{28}$$

where the local-field effects are taken into account by D. The empirical values of $\hbar\omega_p$, E_1, E_2, and ε_0 for silicon-type semiconductors [13] and the calculated values of $\hbar\Omega$, $\hbar\Delta$, and D are presented in Table 1. Here one can also find the values of

$$\bar{\varepsilon}_0 = 1 + \frac{\omega_p^2}{\Delta^2}, \tag{29}$$

Table 1

Energies of the optical gaps (in eV) and dielectric constants of silicon-type crystals

crystal	$\hbar\omega$	E_1	E_2	ε_0	$\hbar\Omega$	$\hbar\Delta$	D	$\bar{\varepsilon}_0$
diamond	31.2	9.8	12.5	5.7	8.5	13.4	0.87	6.4
Si	16.6	3.4	4.3	11.7	2.8	4.6	0.84	13.7
Ge	15.6	2.1	4.4	16.0	2.3	4.3	1.14	14.2
α-Sn	12.7	1.4	3.1	24.0	1.6	3.0	1.28	19.0

which exceeds ε_0 in diamond and silicon as has been expected in [14].

At last, in elemental semiconductors we take the dielectric function in the form [15]

$$\bar{\varepsilon}(Q) = 1 + \frac{\bar{\varepsilon}_0 - 1}{[1 + L^2(\bar{\varepsilon}_0 - 1) Q^2] (1 + fQ^2)^2} \tag{30}$$

with $L^2 = \pi/4k_F$ and $f = 3/4k_F^2$. As has been shown [16] this dielectric function leads to the best results in theoretical investigations of the band structure in silicon-type semiconductors. The form factors of bare ions will be approximated by

$$f_k(Q) = -Z_k \, e^{-\frac{Q^2}{4}} + \frac{Q^2}{4\pi} \sum_{i=1}^{5} C_{k,i} \left(\frac{\pi}{\alpha_{k,i}}\right) e^{-\frac{Q^2}{4\alpha_{k,i}}}, \tag{31}$$

where the ionic parameters $C_{k,i}$ and $\alpha_{k,i}$ can be found in Xia's paper [17]. In Fig. 1 we compare $f_k(Q)$ functions for the Si^{4+} ion when different ionic pseudopotentials are used

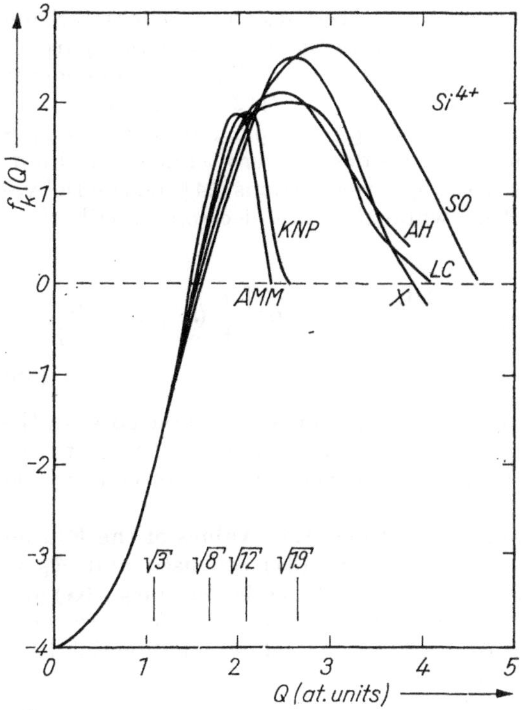

Fig. 1. Form factors of Si^{4+} ion when the ionic pseudopotentials of X [17], AH [18], LC [19], SO [20], KNP [21], and AMM [22] are used

[17 to 22]. Obviously, for $\varrho(G)$ calculations X, AH, LC, and SO functions are more adapted if non-truncated local ionic pseudopotentials [17 to 20] are applied. We found that Xia's function (31) was more adjustable in our case. Moreover, the ionic pseudopotentials [21, 22] can only be used for band structure calculations of elemental semiconductors in terms of the non-spherical local pseudopotential method as has been shown in [23].

4. Valence Electron Distribution in Silicon and Diamond

After substituting (20) into (16) the following expression for the Fourier components of the valence electron density is established:

$$\varrho(\boldsymbol{G}) = \varrho_i(\boldsymbol{G}) + \varrho_B(\boldsymbol{G}) + \varrho'_B(\boldsymbol{G})\,, \tag{32}$$

where at an arbitrary $\boldsymbol{G} = (2\pi/a)\,(h, k, l)$ in silicon-type semiconductors we have

$$\varrho_i(\boldsymbol{G}) = 2f_i(G)\left[\frac{1}{\varepsilon(G)} - 1\right]\cos\left[\frac{\pi}{4}\,(h + k + l)\right], \tag{33}$$

$$\varrho_B(\boldsymbol{G}) = 2\,\frac{f_B(G)}{\varepsilon(G)}\left[1 + \cos\left[\frac{\pi}{2}\,(h + k)\right] + \cos\left[\frac{\pi}{2}\,(k + l)\right] + \cos\left[\frac{\pi}{2}\,(l + h)\right]\right], \tag{34}$$

$$\varrho'_B(\boldsymbol{G}) = 2Z'_B\,\frac{f_B(G)}{\varepsilon(G)}\times$$

$$\times\frac{\left[\begin{array}{c}(hk + kl + lh) - (lh - hk + kl)\cos\dfrac{\pi}{2}\,(h+k) - (hk - kl+lh)\cos\dfrac{\pi}{2}\,(k + l) - \\ - (kl - lh + hk)\cos\dfrac{\pi}{2}\,(l + h)\end{array}\right]}{h^2 + k^2 + l^2}. \tag{35}$$

Here $f_i(G)$ is the form factor of the bare ion given by (31), $f_B(G) = \exp\,(-A_BG^2)$ the bond charge form factor, and Z'_B the off-diagonal components of the effective bond charge tensor. The reference point has been fixed on the bond charge.

In further consideration we have to determine parameters A_B and Z'_B characterizing the bond charge spatial extent and local-field effects, respectively. For this purpose we minimize the mean-square deviation of the calculated Fourier components $\varrho_{theor}(\boldsymbol{G})$ from the measured one $\varrho_{exp}(\boldsymbol{G})$. Unfortunately, $\varrho_{total}(\boldsymbol{G})$ [24] is experimentally determined where $\varrho_{core}(\boldsymbol{G})$ also persists. But we need $\varrho_{total} - \varrho_{core}$. Moreover, ϱ_{total} is usually available at room temperature while we consider the zero-temperature limit. For Si we use form factor data from X-ray scattering investigations [24] and the Debye-Waller factor value equal to 0.46×10^{-20} m^2 while the core electron contributions have been approximated by

$$\varrho_{core}(\boldsymbol{G}) = 2\,\frac{(A_k - Z_k) + (A_k + Z_k + 2B_k)\,(GR_k)^2}{[1 + (GR_k)^2]^2}\cos\left[\frac{\pi}{4}\,(h + k + l)\right], \tag{36}$$

where A_k is the nuclear charge, B_k and R_k denote parameters of the ionic core. In the case of silicon we determine $B_k = 2.169$ and $R_k = 0.289$ at. units by fitting to the measured form factors for large-angle scattering because then the core electrons yield the main contribution.

Table 2 contains $\varrho_{exp}(\boldsymbol{G})$ for silicon as well as the theoretical values of the Fourier electron density components calculated when (30) and (31) are used and $A_B = 0.7$ at. units, $Z'_B = 0.05$. There is a satisfactory coincidence between $\varrho_{theor}(\boldsymbol{G})$ and $\varrho_{exp}(\boldsymbol{G})$ but at $\boldsymbol{G} = (2\pi/a)\,(1, 1, 1)$ a large difference persists. To make agreement between theory and experiment improved we use Tyuterev's dielectric function [25],

$$\bar{\varepsilon}(Q) = \frac{fQ^4 + Q^2 + \xi^2}{fQ^4 + Q^2 + \xi^2\left[\left(\frac{1}{\bar{\varepsilon}_0} - D\right)\frac{\sin QR}{QR} + D\cos QR\right]}, \tag{37}$$

where $f = 3/4k_F^2$, $\xi^2 = 4k_F/\pi$, $D = -0.041$ and $R = 4.529$ at. units have been determined when $\bar{\varepsilon}_0 = 13.74$. Unfortunately, $\varrho_{theor}(1, 1, 1) = -2.104$ while all other components remain unchanged. Also, the difference arises when one puts $f = 0$ because our

Table 2

The Fourier components of the valence electron density in silicon (in electrons per Wigner-Seitz cell)

hkl	$\varrho_{exp}(\boldsymbol{G})$	$\varrho_i(\boldsymbol{G})$	$\varrho_B(\boldsymbol{G})$	$\varrho'_B(\boldsymbol{G})$	$\varrho_{theor}(\boldsymbol{G})$
000	8	7.416	0.584	0	8
111	−1.780	−0.935	−1.198	0.060	−2.073
220	0.084	−0.073	0	0.020	0.093
311	0.338	0.122	0.209	0.006	0.337
222	0.380	0	0.325	0	0.325
400	0.364	0.155	0.116	0	0.271
331	−0.052	−0.084	0.027	−0.008	−0.065
422	−0.126	−0.068	0	0.002	−0.066
333	−0.097	−0.031	−0.003	0.005	−0.029
511	−0.076	−0.031	−0.003	0.001	−0.033
440	−0.080	−0.016	0.002	0	−0.014
531		−0.005	0.000	0.000	−0.005
442		0	0	0.000	0.000

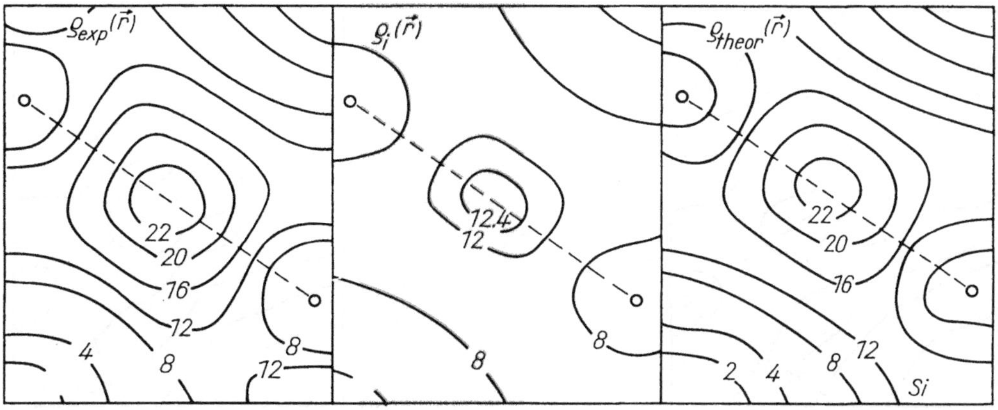

Fig. 2. Valence electron density distribution on {110} plane in silicon. Figures indicate the number of electrons in the Wigner-Seitz cell

consideration becomes identical to the Thomas-Fermi approximation for the dielectric function.

The contours of the electronic density represented by $\varrho_{\text{exp}}(\boldsymbol{r})$, $\varrho_{\text{i}}(\boldsymbol{r})$, and $\varrho_{\text{theor}}(\boldsymbol{r})$ on {110} crystallographic plane in silicon case are mapped in Fig. 2. One may see a good agreement between the theoretical and experimental data although the difference between the Fourier component with $\boldsymbol{G} = (2\pi/a)\ (1, 1, 1)$ has been noted. In silicon the bond charge has a nearly spherical shape. Also, our $\varrho_{\text{theor}}(\boldsymbol{r})$ data are very close to the results given in [3] where non-local pseudopotential calculations have been performed. But our consideration has been based on the local pseudopotential approach. We suppose the crystal pseudopotential is a local one but its atomic-like ingredients lose the spherical symmetry from the bond charge contributions [23, 26].

By analogy to silicon $\varrho_{\text{exp}}(\boldsymbol{G})$ of diamond has been determined from the measured form factors in X-ray experiments [27]. The Debye-Waller factor has been adopted equal to $0.15 \times 10^{-20}\ \text{m}^2$ [28] and the fitted parameters of the ionic core are $B_k = 0.741$ and $R_k = 0.204$ at. units. Also, we suggest that some anomalous dispersion contributions have been taken into account because the ionic core parameters B_k and R_k have been fitted to empirical data at large scattering angles. Moreover, we well understand that our $\varrho_{\text{exp}}(\boldsymbol{G})$ results are based on a very rough approximation for the core electron contributions and neglect anharmonic temperature corrections expected to be essential in silicon.

In Table 3 we present the results for diamond with $A_{\text{B}} = 0.4$ at. units and $Z_{\text{B}}' = -0.3$. The agreement between theory and experiment is not so good as in the case of silicon. In Fig. 3 the contours of the equivalent electronic density on the {110} plane in diamond are plotted. Let us note that the spatial valence electron density distribution has been acceptably described in our adiabatic bond charge model. Particularly, this model gives a correct explanation of the bond charge extent along the line connecting neighbouring atoms. Also, in covalent semiconductors the bond charge shape is a very sensitive function of the local-field parameter Z_{B}'. In Fig. 4 the electron density distribution $\varrho_{\text{theor}}(\boldsymbol{r}) - \varrho_{\text{i}}(\boldsymbol{r})$ is shown on the {110} plane of diamond when Z_{B}' takes different values. Our results indicate that the spherical bond charge distribution can be obtained for $Z_{\text{B}}' > 0$ while for $Z_{\text{B}}' < 0$ the bond charge becomes longer and the decoupling of two equal parts is possible. These features of the spatial valence electron distribution in covalent semiconductors have not been taken into account in [1].

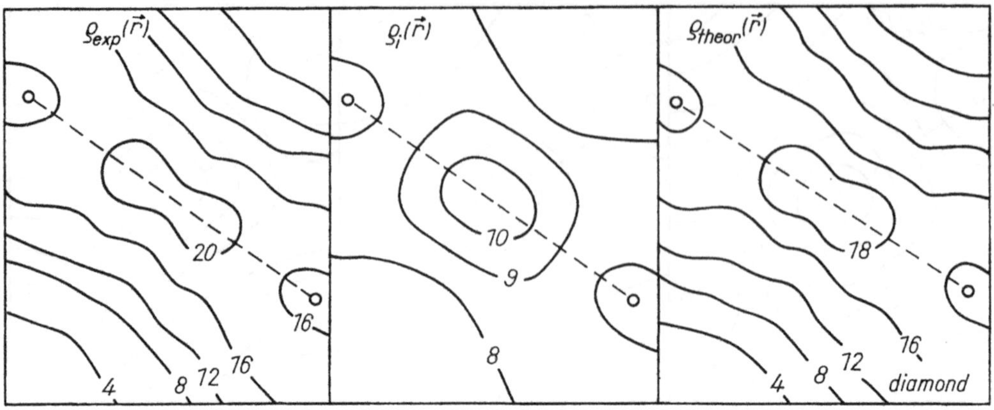

Fig. 3. Valence electron density distribution on {110} plane in diamond

Table 3

The Fourier components of the valence electron density in diamond (in electron per Wigner-Seitz cell)

hkl	$\varrho_{exp}(G)$	$\varrho_i(G)$	$\varrho_B(G)$	$\varrho'_B(G)$	$\varrho_{theor}(G)$
000	8	6.750	1.250	0	8
111	−1.946	−0.645	−1.061	−0.318	−2.024
220	−0.295	−0.026	0	−0.069	−0.094
311	0.124	0.056	0.083	−0.011	0.128
222	0.298	0	0.119	0	0.119
400	0.310	0.098	0.003	0	0.101
331	0.009	−0.051	0.005	0.001	−0.045
422	−0.060	−0.029	0	−0.001	−0.030
333	−0.002	−0.009	0.000	0.000	−0.009
511	−0.002	−0.009	0.000	0.000	−0.009
440	−0.097	0.001	0.000	0	0.002
531	−0.031	0.004	0.000	0.000	0.004
442	0.000	0	0	0.000	0

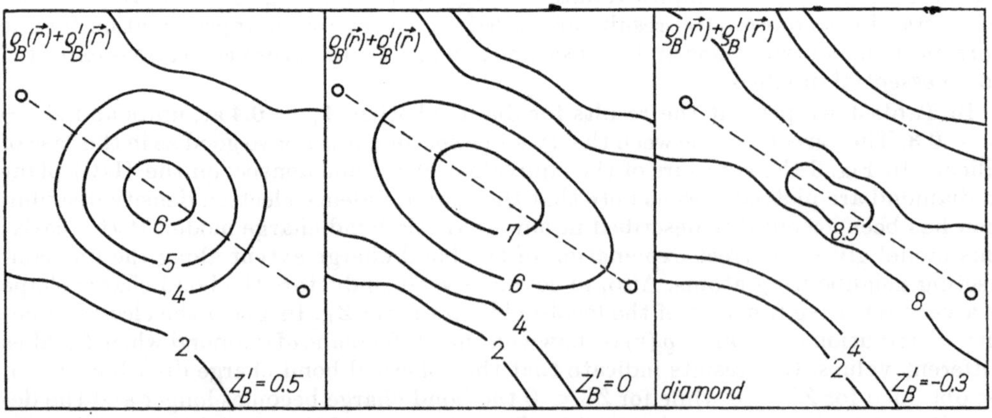

Fig. 4. Electron densities on {110} plane in diamond for different values of the local-field parameter Z'_B

5. Discussion

The microscopic adiabatic bond charge model of the spatial valence electron distribution in covalent crystals has been presented. The model has been based on the theory of dielectric screening [9] and on a rigorous sum rule (1) in which the linear electron density variation is bound to the static electron distribution in the crystal. In distinction from Phillips' theory [1] our model gives a correct proportion between the Coulomb and exchange correlation interactions of valence electrons and leads to acceptable results in silicon and diamond. However, the theoretical results are very sensitive to the selection of the dielectric function $\bar{\varepsilon}(Q)$, form factor of the bare ion $f_i(\mathbf{Q})$, bond charge form factor $f_B(\mathbf{Q})$, and local-field parameter Z'_B. In our consideration Schulze-Unger's dielectric function [15] and Xia's ionic pseudopotentials [17] have been used. The bond charge form factor has been approximated by the Gaussian function $\exp(-A_B Q^2)$. Then two parameters A_B and Z'_B have been found by a fitting procedure.

Certainly, a right selection of the empirical values for the Fourier components of the valence electron density has been an important point in our calculations. A very simple procedure has been used to take into account the temperature and core electron corrections. The anharmonic effects which are more important in silicon [29, 30] than in diamond have been neglected and the role of core electrons has been accounted for in a simple manner accordingly (36). It is possible that in silicon the divergence between theory and experiment for $\varrho(1, 1, 1)$ is a consequence of the non-realistic empirical value. Nevertheless, our results for the spatial electron distribution in silicon and diamond are better than other model considerations [4 to 7].

Maybe to consider another kind of $f_B(Q)$ is a useful task and we hope to make it in the near future. When we use the Gaussian function for $f_B(\mathbf{Q})$ a new feature of diamond in comparison with silicon has been discovered. It consists in the negative local-field parameter Z'_B while in silicon it is nearly zero. Also, in diamond the parameter A_B is smaller than the same one in silicon. It means that in the first case bond charges are more localized in space. The adiabatic bond charge model can also applied to Ge and α-Sn for finding the natural trends of A_B and Z'_B parameters in elemental semiconductors. Another useful spreading of the bond charge model concerns the zincblende crystals. These investigations are in progress and the results will be published in a forthcoming paper.

Acknowledgements

One of the authors (D.I.M.) would like to thank Prof. Abdus Salam, the International Atomic Energy Agency, and UNESCO for hospitality at the International Centre for Theoretical Physics, Trieste.

References

[1] J. C. Phillips, Phys. Rev. **166**, 832 (1968).
[2] G. Strinati, H. J. Mattausch, and W. Hanke, Phys. Rev. B **25**, 2867 (1982).
[3] J. R. Chelikowsky and M. L. Cohen, Phys. Rev. B **10**, 5095 (1974).
[4] C. M. Bertoni, V. Bartolani, C. Calandra, and F. Nizzoli, J. Phys. C **6**, 3612 (1973).
[5] A. B. Makhnovetskii and G. L. Krasko, phys. stat. sol. (b) 81, 263 (1977).
[6] A. Baldereschi, K. Maschke, A. Milchev, R. Pickenhain, and K. Unger, phys. stat. sol. (b) **108**, 511 (1981).
[7] U. Pietsch, phys. stat. sol. (b) **126**, 595 (1984).
[8] Y. M. Yang and P. Coppens, Solid State Commun. **15**, 1555 (1974).
[9] S. K. Sinha, R. P. Gupta, and D. L. Price, Phys. Rev. B **9**, 2564; 2573 (1974).
[10] V. K. Bashenov, V. E. Gorbachev, and D. I. Marvakov, phys. stat. sol. (b) **128**, K59 (1985).

[11] V. Ya. Altshul, V. K. Bashenov, and A. G. Petukhov, Fiz. tverd. Tela **24**, 2925 (1982); Soviet Phys. – Solid State **24**, 1657 (1982).
[12] P. U. Giaquinta, M. Parrinello, E. Tosatti, and M. P. Tosi, J. Phys. C **9**, 2031 (1976).
[13] A. N. Pikhtin and A. D. Yaskov, Fiz. Tekh. Poluprov. **12**, 1047 (1978); Soviet Phys. – Semicond. **12**, 622 (1978).
[14] V. K. Bashenov, R. A. Alarashi, and M. G. Foigel, phys. stat. sol. (b) **63**, 403 (1974).
[15] K.-R. Schulze and K. Unger, phys. stat. sol. (b) **66**, 491 (1974).
[16] V. K. Bashenov, I. K. Doicho, and A. G. Petukhov, phys. stat. sol. (b) **93**, K131 (1979).
[17] J. Xia, Chin. J. Semicond. **5**, 121 (1984).
[18] J. A. Appelbaum and D. R. Hamann, Phys. Rev. B **8**, 1777 (1973).
[19] S. G. Louie and M. L. Cohen, Phys. Rev. B **13**, 2461 (1976).
[20] K. Shindo and I. Ohkoschi, J. Phys. Soc. Japan **42**, 1423 (1977).
[21] O. V. Konstantinov, S. K. Nasibullaev, and M. M. Panakhov, Fiz. Tekh. Poluprov. **11**, 881 (1977).
[22] F. Aymerich, F. Meloni, and G. Mula, Phys. Rev. B **15**, 3980 (1977).
[23] V. K. Bashenov, Y. K. Markolenko, and A. G. Petukhov, phys. stat. sol. (b) **96**, K91 (1979).
[24] P. J. E. Aldred and M. Hart, Proc. Roy. Soc. **A332**, 223 (1973).
[25] V. G. Tyuterev, Izv. vuzov Fiz. **11**, 99 (1981).
[26] V. K. Bashenov and Y. K. Markolenko, phys. stat. sol. (b) **97**, K145 (1980).
[27] P. F. Price and E. N. Maslen, Acta cryst. **A34**, 173 (1978).
[28] P. F. Price, E. N. Maslen, and F. H. Moore, Acta cryst. **A34**, 171 (1978).
[29] J. B. Roberto, B. W. Batterman, and D. T. Keating, Phys. Rev. B **9**, 2590 (1974).
[30] P. F. Price, E. N. Maslen, and S. L. Mair, Acta cryst. **A34**, 183 (1978).

(Received July 26, 1985)

phys. stat. sol. (b) 133, 297 (1986)

Subject classification: 13.4 and 19; 22.8

Department of Physics, M.D. University, Rohtak[1])

Solution of the Spin-Hamiltonian with Orthorhombic HF and g-Tensors ($S = 1/2$)

II. Application to VO^{2+} Centres in Oxide Glasses[2])

By

V. P. SETH, A. YADAV, and R. S. BANSAL

A theory developed previously is used in the analysis of ESR spectra of VO^{2+} in ternary borate glasses. The theoretical optical basicity of these glasses is calculated. It is found that $|P|$ and $|A'_{||}|$ are larger and K is smaller in those glasses for which the optical basicity is large.

Eine früher entwickelte Theorie wird für die Analyse von VO^{2+}-ESR-Spektren in ternären Boratgläsern entwickelt. Die theoretische optische Basizität dieser Gläser wird berechnet. Es wird gefunden, daß in den Gläsern, für die die optische Basizität groß ist, $|P|$ und $|A'_{||}|$ größer und K kleiner ist.

1. Introduction

In a previous paper [1] we developed a theory appropriate for analyzing ESR spectra characterized by a spin-Hamiltonian $\mathcal{H} = \beta \boldsymbol{H} \cdot \mathbf{g} \cdot \boldsymbol{S} + \boldsymbol{S} \cdot \mathbf{A} \cdot \boldsymbol{I}$ for $S = 1/2$ where the symbols have their usual meaning. In the present paper the theory is applied to VO^{2+} in borate glasses. ESR parameters have been determined using the present theory and the second-order perturbation theory given by Bleaney et al. [2]. The theoretical optical basicity of the glasses has also been calculated.

2. Experimental

The starting materials used in the present study were analar grade reagents of Li_2CO_3, $CdCO_3$, $CaCO_3$, $MgCO_3$, $BaCO_3$, and H_3BO_3. 1.0 mol% V_2O_5 was added to each batch. Each batch was first sintered [3] for about 6 h at (600 ± 50) °C and then melted in silica crucibles in an electric furnace at (1150 ± 50) °C. The melts remained at the firing temperature for 45 min and during firing, which was done in air, they were stirred until they appeared to be homogeneous and bubble free. The melts were poured on a carbon plate and pressed with another one. No annealing was done. Table 1 gives the composition in mol% of the glasses studied in the present paper.

Table 1

The compositions (in mol%) of all the borate glasses

glass number	B_2O_3	Li_2O	CdO	MgO	CaO	BaO
CdV	72	10	18	—	—	—
MV	70	10	—	20	—	—
CaV	70	10	—	—	20	—
BV	70	10	—	—	—	20

[1]) Rohtak (Maryana), India.

[2]) Part I see phys. stat. sol. (b) **132**, 255 (1985).

The glasses were crushed into pieces for ESR measurements. The ESR spectra were recorded at 290 and 90 K on a JEOL-FE3X spectrometer operating in the X-band. A magnetic field modulation of 100 kHz was applied. DPPH with $g = 2.0036$ was used as a standard for the determination of the spin-Hamiltonian parameters.

3. Results and Discussion

Fig. 1 shows the ESR spectra of VO^{2+} in these glasses. The spectra show a structure which is characteristic of the hyperfine interaction of a single unpaired electron with ^{51}V (99.76% abundant, nuclear spin $I = 7/2$). The spectrum can be described by an axial spin-Hamiltonian

$$\mathcal{H} = g_{\parallel}\beta S_z H_z + g_{\perp}\beta(S_x H_x + S_y H_y) + A_{\parallel} S_z I_z + A_{\perp}(S_x I_x + S_y I_y)\,, \qquad (1)$$

where the symbols have their usual meaning and quadrupole and nuclear Zeeman interaction terms are neglected [1, 4].

The principal results of the above spin-Hamiltonian are that the eight hyperfine components of the electron–nuclear spin interaction have envelopes with divergencies at the magnetic field positions Δ given by (21) of [1] and shoulders at positions Δ given by (23) of [1] where ν is the spectrometer frequency. $h\nu$ in these equations has been substituted in mT by using the relations $h\nu = g\beta H$. The spin-Hamiltonian parameters were calculated by an iterative procedure giving the best fit to (21) and (23) of [1]. First-order perturbation theory solution was used as the first trial solution in the iteration. The values of $g_{\parallel}$, $g_{\perp}$, $A_{\parallel}$, and $A_{\perp}$ were also calculated using the first-order perturbation theory [5] and second-order perturbation theory [6] and are given in Table 2 for comparison. The uncertainty in the values of g and A is ± 0.0006 and ± 0.5, respectively, for the second-order perturbation theory calculation and the present theory calculation, whereas in the first-order perturbation theory calculation the uncertainty is quite large. It is clear from these values that the neglect of the second-order perturbation term resulted in the overestimation of the spin-Hamiltonian parameters especially of the values of $g_{\perp}$ and $A_{\perp}$. The present theory gives the values of these parameters slightly different from the values obtained using the second-order perturbation theory. The method proposed by Muncaster and Parke [7] shows that $A_{\parallel}$ and $A_{\perp}$ are negative. On lowering the temperature to liquid nitrogen temperature the spectra were virtually identical.

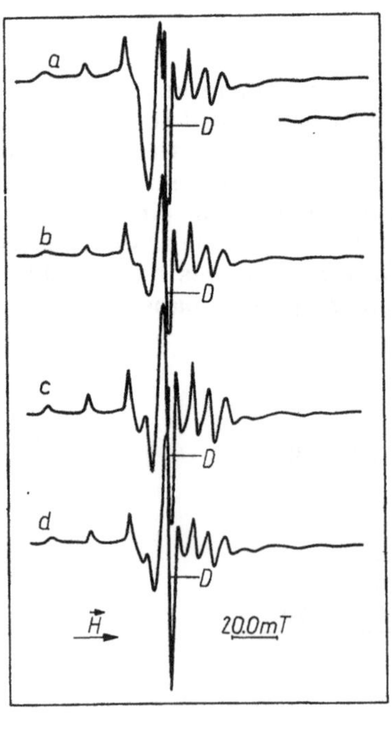

Fig. 1 The ESR spectra of VO^{2+} ions in $Li_2O \cdot MeO \cdot B_2O_3$ (Me = Cd, Mg, Ca and Ba) at X-band ($\nu = 9.445$ GHz) and 290 K, (a) CdV, (b) MV, (c) CaV, and (d) BV glasses, D represents the DPPH line (0,3375 T), Spectrum below (a) is for glass CdV with a ten times increase in grain

Table 2

ESR parameters of the VO^{2+} in glasses obtained by using the first-order and second-order perturbation theory and the present theory

glass number	theory	$g_{\|\|}$	$g_{\perp}$	$A_{\|\|}$ (10^{-4} cm^{-1})	$A_{\perp}$ (10^{-4} cm^{-1})
CdV	first-order	1.9410	2.0077	168.6	73.8
	second-order	1.9376	1.9729	167.8	61.7
	this work	1.9372	1.9734	168.3	61.9
MV	first-order	1.9450	2.0077	168.2	71.5
	second-order	1.9431	1.9738	167.4	59.5
	this work	1.9424	1.9738	167.9	59.9
CaV	first-order	1.9403	1.9987	169.1	71.7
	second-order	1.9370	1.9703	168.2	59.8
	this work	1.9373	1.9702	168.9	60.2
BV	first-order	1.9433	2.0019	165.9	67.5
	second-order	1.9404	1.9692	165.8	56.1
	this work	1.9414	1.9703	165.7	56.7

Kivelson and Lee [8] gave the following relations:

$$A_{||} = -P[K + \tfrac{4}{7} - \Delta g_{||} - \tfrac{3}{7}\,\Delta g_{\perp}]\,, \tag{2}$$

$$A_{\perp} = -P[K - \tfrac{2}{7} - \tfrac{11}{14}\,\Delta g_{\perp}]\,, \tag{3}$$

where $P = 2\gamma\beta\beta_N\langle r^{-3}\rangle$ is the dipolar hyperfine coupling parameter and K is proportional to the amount of the isotropic Fermi contact interaction, $\Delta g_{||} = g_{||} - g_e$ and $\Delta g_{\perp} = g_{\perp} - g_e$. Using (2) and (3) the values of P and K were calculated for the $A_{||}$ and $A_{\perp}$ values of the present theory and are given in Table 3. From the molecular orbital

Table 3

ESR parameters and theoretical values of optical basicity of the glasses

glass	$\|P\|$	K	$\|A'_{\|\|}\|$	$\|A'_{\perp}\|$	$\Delta g_{\|\|}/\Delta g_{\perp}$	Λ_{th}
CdV	116.7	0.79	75.7	30.7	2.25	0.460
MV	119.1	0.77	76.6	31.4	2.10	0.478
CaV	119.4	0.76	77.6	31.1	2.03	0.496
BV	120.2	0.73	77.7	31.3	1.90	0.508

theory, it can be shown that the components $A_{||}$ and $A_{\perp}$ consist of the contributions $A'_{||}$ and $A'_{\perp}$ of the $3d_{xy}$ electron to the hyperfine structure and the PK term arising due to the anomalous contribution of the s-electrons. Equations (2) and (3) can be split into their component parts,

$$\begin{aligned} A_{||} &= -P[\tfrac{4}{7} - \Delta g_{||} - \tfrac{3}{7}\,\Delta g_{\perp}] - PK \\ &= A'_{||} - PK\,, \end{aligned} \tag{4}$$

$$\begin{aligned} A_{\perp} &= +P[\tfrac{2}{7} + \tfrac{11}{14}\,\Delta g_{\perp}] - PK \\ &= A'_{\perp} - PK\,. \end{aligned} \tag{5}$$

The values $A'_{||}$ and $A'_{\perp}$ were calculated using the above equations and are given in Table 3. It is clear that $|A'_{||}|$ increases from CdO to BO glasses whereas PK decreases.

The values of $\Delta g_{\parallel}/\Delta g_{\perp}$ were also calculated and are given in Table 3. The lowering of the values of this ratio from CdV to BV shows the improvement [9] of the octahedral symmetry of the V^{4+} site.

It is possible to calculate the theoretical optical basicity [10] by

$$\Lambda_{\text{th}} = \sum_i \frac{Z_i r_i}{2\gamma_i}, \tag{6}$$

where Z_i is the oxidation number of the cation i, r_i the ratio of the cation i to the total number of oxides, and γ_i the basicity moderating parameter. The parameter for the cation is given by

$$\gamma_i = 1.36(x_i - 0.26)\,, \tag{7}$$

where x_i is the Pauling electronegativity [11] of this cation. Theoretical values of the optical basicity were calculated and are included in Table 3. It may be noted that $|P|$ and $|A'_{\parallel}|$ are larger and K is smaller in those glasses for which the optical basicity is large.

Acknowledgement

This work was supported by CSIR, New Delhi.

References

[1] V. P. Seth, A. Yadav, and R. S. Bansal, phys. stat. sol. (b) **131**, 255 (1985).
[2] B. Bleaney, K. D. Bowers, and M. H. L. Pryce, Proc. Roy. Soc. **A228**, 147 (1955).
[3] L. Ya. Mazelev, Borate Glasses, Consultants Bureau, New York 1960 (p. 11).
[4] V. P. Seth, V. K. Jain, and R. K. Malhotra, J. non-crystall. Solids **57**, 199 (1983).
[5] H. G. Hecht and T. S. Johnston, J. chem. Phys. **46**, 23 (1967).
[6] G. Hochstrasser, Phys. Chem. Glasses 7, 178 (1966).
[7] R. Muncaster and S. Parke, J. non-crystall. Solids **24**, 399 (1977).
[8] D. Kivelson and S. Lee, J. chem. Phys. **41**, 1896 (1964).
[9] A. K. Bandyopadhyay, J. Mater. Sci. **16**, 189 (1981).
[10] J. A. Duffy and M. D. Ingram, J. inorg. nuclear Chem. **37**, 1203 (1975).
[11] L. Pauling, The Nature of the Chemical Bond, 3rd edition, Cornell University Press, New York 1960 (p. 93).

(Received June 19, 1985)

phys. stat. sol. (b) 133, 301 (1986)
Subject classification: 72.40; 71.55; 59

Research Laboratories, Eastman Kodak Company, Rochester[1])

Photocarrier Kinetics in $IrBr_6^{3-}$-Doped AgBr

By

J. P. SPOONHOWER and R. J. DERI

The kinetics of photogenerated carriers in AgBr crystals doped with trivalent iridium are investigated by photoconductivity measurements. The temperature dependence of the photoelectron lifetime is highly dependent on the level of iridium doping. Experiments show that photoelectron–iridium reaction cross-sections have a weak dependence on the number of vacancies bound to the iridium site. Estimates of the cross sections for electron capture at $IrBr_6^{3-}$ centers with one and two associated vacancies are deduced.

Die Kinetik der photogenerierten Träger in mit dreiwertigem Iridium dotierten AgBr-Kristallen wird mittels Photoleitungsmessungen untersucht. Die Temperaturabhängigkeit der Photoelektronenlebensdauer hängt stark vom Niveau der Iridiumdotierung ab. Experimente zeigen, daß Photoelektron–Iridium-Reaktionsquerschnitte eine schwache Abhängigkeit von der Zahl der Leerstellen, die an Iridiumplätzen gebunden sind, besitzen. Berechnungen der Querschnitte für Elektroneneinfang an $IrBr_6^{3-}$-Zentren mit einer oder zwei verknüpften Leerstellen werden durchgeführt.

1. Introduction

Evidence from optical absorption [1], vibrational (Raman and infrared) [1], and electron paramagnetic resonance (EPR) [2] spectroscopies shows that iridium ions replace lattice silver ions substitutionally in AgBr and that the ions are incorporated in the trivalent state. That is, the iridium species is the $IrBr_6^{3-}$ complex, on the basis of bond-length data and thermodynamic arguments [1]. Many experiments show that these centers interact with photocarriers in single crystals of AgBr: (i) Observation of $IrBr_6^{4-}$ formation upon photoexcitation at low temperatures (≈ 180 K) by EPR is evidence for electron trapping by Ir^{3+} centers [2]. Analysis of thermal detrapping rates suggests a trap depth of 0.44 eV for this center. (ii) Observation of a greatly reduced photoconductive decay time suggests that Ir^{3+} doping enhances photoelectron decay [2, 3]. (iii) Interaction of photocarriers with Ir^{3+} has been inferred from iridium-induced quenching of visible photoluminescence [4]. (iv) Low-temperature infrared luminescence associated with iridium doping suggests that the iridium site functions primarily as a radiative recombination center below 140 K [4].

The variety of spectroscopic techniques available for characterization of iridium centers in bulk AgBr crystals permits a reasonably thorough determination of the number and type of Ir^{3+} species in these materials. Previous experiments have characterized the iridium centers in several iridium-doped AgBr crystals [2 to 4]; we present here the kinetic information deduced from observation of photoconductive transients in these materials.

---

[1]) Rochester, New York 14650, USA.

2. Experimental

The materials used in these experiments were AgBr crystals grown from the melt under Br_2 by the Bridgman technique. Most of the crystals studied were grown from the same pure silver bromide crystal, which was prepared from 6N silver shot and distilled acids by a procedure described elsewhere [2]. A portion of this pure crystal was melted and regrown with a small quantity of $IrBr_3$ added as dopant; for lower dopant levels, this doped material was diluted with portions of the original pure crystal. In this manner the level of background impurities was kept constant in samples of various iridium concentrations. As a check on the experimental results, a crystal containing 89 ppm iridium (i.e. 89 μmol of iridium per mole of AgBr) was prepared from different starting materials (different silver shot and acid batches). The behavior of this crystal was essentially identical to that of a sample containing 85 ppm Ir prepared by dilution. Crystal samples were cut from the boules, polished, and etched with sodium cyanide solution to prepare quality optical surfaces. Other details of sample preparation may be found elsewhere [4].

The level of iridium doping was determined by neutron activation analysis, which is accurate to $\pm 5\%$. Analysis of aliquots obtained from adjacent portions of the same crystal boule showed that the iridium is uniformly distributed on a coarse scale (≈ 4 mm). The spatial distribution on a finer scale (≈ 35 μm) was investigated by observation of the spatial variation of the intensity of $IrBr_6^{3-}$ vibrational Raman lines relative to the intrinsic AgBr two-phonon band; this experiment showed a uniform distribution of dopant on this finer scale as well [1]. The levels of background impurities were determined by emission spectroscopy and spark-source mass spectroscopy. The concentration of common cationic impurities such as Fe and Cu was less than 1 ppm, as was the level of iodine.

Photoconductive transients were excited by light pulses of photon energy sufficient for excitation of band-to-band electronic transitions in AgBr. A Quanta Ray DCR-1A pulsed Nd:YAG laser provided optical power. The laser output was frequency doubled to 530 nm and either Raman shifted in a H_2 cell to produce 435.6 nm light or used to pump a Rhodamine 6G dye laser, which was Raman shifted to produce 440 to 470 nm light. The pulsewidth (FWHM), as measured with a fast silicon photodiode, was 8 ns at 420 nm and 3.3 ns at 460 nm [5]. The absorption coefficient of 435 to 470 nm light in AgBr ranges from 10 to 1000 cm^{-1} at room temperature [6]; this shows that the observed photoconductivity is characteristic of bulk exposures.

Photoconductive transients were measured by using a Q-band (36 GHz) microwave instrument at room temperature and an X-band (10.5 GHz) microwave apparatus for lower-temperature measurements (90 to 298 K). Both instruments have TE_{103} reflection cavities constructed from rectangular brass waveguide, a tunable short, and variable coupling irises; these instruments have been described elsewhere [3 ,6, 7]. Under the small-signal conditions used in this work, the observed photosignals are linearly proportional to sample photoconductivity [8]. The instrument time constants were measured by using a semi-insulating GaAs:Cr wafer as the photoconductive sample [3, 6]; the minimum achievable instrument time constants were 3 to 5 ns at Q-band and 30 ns at X-band. These time constants represent the fastest exponential decay processes that could be resolved and are due to cavity bandwidth-limited response [9]. Of course, longer time constants could be attained by varying the iris coupling to alter the cavity bandwidth (loaded quality factor).

3. General Results. Origin of Dopant Effects and Sign of Photocarriers

The photoconductive responses of iridium-doped AgBr and a pure AgBr control at room temperature are shown in Fig. 1. The pure material exhibits a multicomponent

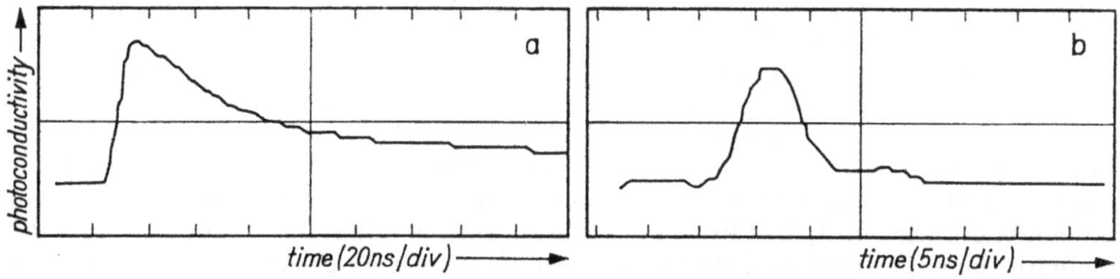

Fig. 1. Photoconductive response of AgBr samples at room temperature. a) Pure AgBr; b) 10.6 ppm Ir^{3+}-doped AgBr. $\lambda = 460$ nm, pulsewidth = 3.5 ns FWHM, $\approx 5 \times 10^{13}$ photons/cm². Some ringing due to finite cavity bandwidth appears in b)

decay (i.e. not a simple exponential) that extends for ≈ 300 ns; photoconductive decay processes extending over 100 to 500 ns are typical for pure AgBr crystals [10, 11]. The photodecay of Ir^{3+}-doped AgBr was always faster than the instrument resolution for iridium levels from 3.9 to ≈ 176 ppm at all temperatures from 90 to 298 K and under a variety of exposure conditions ($\lambda = 337$, 420, 440 to 470 nm; intensity from 10^{12} to 10^{16} photons/cm²). The room-temperature peak photosignal for iridium-doped samples was 1/20 or less of that for pure AgBr. Since photogeneration processes are expected to be identical in pure and doped materials for band-to-band excitation, this change in signal magnitudes shows a greatly enhanced photoconductivity decay rate in doped samples, which can compete with photogeneration on time scales shorter than the light pulsewidth. The photoconductivity in pure AgBr may be interpreted as a measure of photoelectron density due to the greater mobility of electrons [10] than holes [12] in AgBr and the comparable electron and hole densities under bandgap excitation.

The large decrease in photoelectron lifetime and peak density in Ir^{3+}-doped crystals is clearly the result of iridium doping. The constant level of background impurities in the doped and pure control samples (Section 2) and the correlation of increasing photoconductivity with decreasing iridium concentration (Section 5 and [3]) demonstrate that the observed changes are directly induced by addition of iridium and not by changes in extraneous impurity levels. Thus the defect centers responsible for electron loss could be either the $IrBr_6^{3-}$ centers identified by spectroscopy (see below) or the charge-compensating silver ion vacancies introduced by substitutional multivalent cation dopants. The contribution of free vacancies to the observed phenomena can be ruled out, we believe, for several reasons: (i) The large binding energies (≈ 0.15 eV) expected for vacancy–Ir^{3+} complexes (from measured binding energies of other vacancy–cation species in silver halides) [13] suggest that a large number of the vacancies should be bound at room temperature. At lower temperatures, all vacancies should be bound; the fact that the Ir^{3+}-enhanced decay mechanism persists down to 90 K suggests that the process involves ($IrBr_6^{3-} \cdot 2V$) rather than free vacancies. (ii) The measured ionic conductivity of our samples shows that ≈ 4 ppm of free vacancies exist in the pure control at room temperature.[2]) Addition of 3.9 ppm of Ir^{3+} cannot greatly change the vacancy population, although it dramatically affects the photoconductivity.

Therefore, we conclude that ($IrBr^{3-} \cdot 2$ V) centers provide an efficient photoelectron decay mechanism in AgBr crystals. Similar conclusions have been reached

[2]) Ionic conductivity measurements were provided by D. Jeanmaire and were deduced from the dielectric relaxation of ≈ 1 cm disks of AgBr ≈ 200 μm thick.

on the basis of spectroscopic (i.e. EPR [2]) and photographic evidence [14] for a variety of iridium-doped silver halides.

Although the photoconductivity in pure AgBr reflects only photoelectron behavior, it is not initially clear that the photoconductivity in Ir^{3+}-doped samples is representative of electrons as well. The observed photoconductivity in doped materials at room temperature was 1/20 to 1/250 of that of pure AgBr [3], showing that the peak photoelectron density is reduced by at least as much. The ratio of room-temperature electron and hole mobilities in AgBr [10 to 12] suggests that residual hole conductivity would dominate the experimental signal if the hole density were greater than 70 times the electron density. That is, the maximum hole photosignal is 1/70 of the peak photoconductivity of the pure control.

Several arguments suggest that the photosignal observed for doped materials did not represent hole conductivity. (i) If any hole decay occurs during photogeneration, the maximum hole conductivity will be less than 1/70 of the peak photoconductivity of pure AgBr. The photoconductive decay time in doped samples shows a hole lifetime <4 ns, assuming a measurable hole conductivity in these materials. Therefore, substantial hole decay occurs during illumination (light pulsewidth ≈ 3 ns FWHM); this implies that the peak hole density is at most 60% of the total amount of photogenerated holes, so that the maximum contribution from photoproduced holes is 1/120 of the peak photoconductivity of pure AgBr. Since the photosignal of AgBr containing 3.9 or 10.6 ppm Ir^{3+} is greater than 1/40 of the photosignal of the pure control [3], the data obtained for these samples are primarily ($>75\%$) due to photoelectrons. (ii) The temperature dependence of the observed photosignals (Section 6) is indicative of electron conduction. Transient hole conduction is thermally quenched below ≈ 240 K, owing to multiple trapping, which reduces the drift mobility [11]; in contrast, the photoconductivity of the Ir^{3+}-doped samples increases with decreasing temperature. Arguments analogous to (i) suggest that the photoconductivity is primarily ($>70\%$) due to photoelectrons for all the samples studied below 220 K. Therefore, we believe the phototransients observed in Ir^{3+}-doped AgBr represent primarily photoelectrons.

4. Analysis of Kinetics

The large decrease in peak photoconductivity due to ($IrBr_6^{3-} \cdot 2\,V$) and the instrument-limited decay transients in Ir^{3+}-doped samples suggest that the photoelectron decay rate is much faster than the time evolution of the light pulse. In contrast, the photoelectron decay rate in pure materials is slow compared to the time scale of the light pulsewidth, so that the peak photosignal is the time integral of the photocarrier generation rate. The large decrease in peak photosignal in doped materials is due to iridium-induced decay processes that complete with photogeneration during illumination. This suggests that the general rate equation involving first- and second-order photoelectron decay processes,

$$\frac{\mathrm{d}n}{\mathrm{d}t} = \varepsilon\Phi - \left[\frac{n}{\tau} + An^2\right], \tag{1}$$

may be simplified to the quasi-equilibrium case to analyze the phototransients of the Ir^{3+}-doped samples,

$$\varepsilon\Phi = \frac{n}{\tau} + An^2 \quad \text{for all times}\,. \tag{2}$$

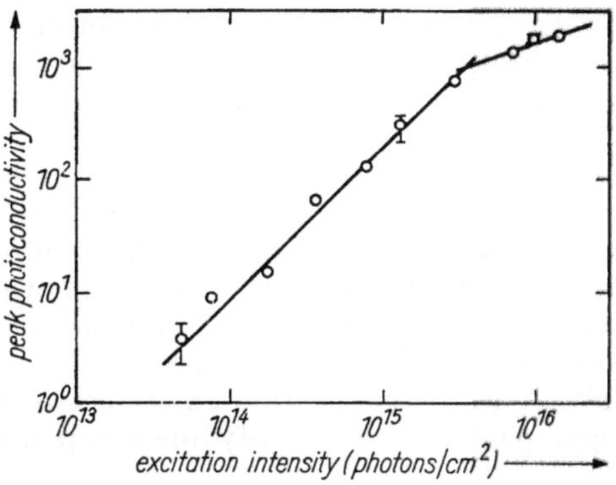

Fig. 2. Intensity dependence of peak photoconductivity in Ir^{3+}-doped AgBr. $\lambda = 450$ nm, 3.0 ppm dopant; typical-size error bars are shown. At low intensity, photosignal is linear in intensity ($\sigma \approx \varphi^m$, $m = 1.08 \pm 0.10$); at high intensity, signal varies with square root of intensity ($m = 0.5 \pm 0.1$)

Here n is the photoelectron density, Φ is the light flux (photons/cm^2 s), τ is the photoelectron lifetime (time constant) for the first-order decay process, and A is the constant for the second-order mechanism. ε is a generalized photoelectron generation efficiency. At the low excitation intensities used for most of our experiments, trap saturation may be ignored.

Experimentally, the maximum photosignal depended linearly on excitation intensity (slope of log–log plot is unity) for pure and doped crystals (Fig. 2) below a critical intensity, which was wavelength dependent. This suggests that, in the linear region, the kinetics are first order in photoelectron density. Equation (2) then reduces to

$$\varepsilon\Phi = \frac{n}{\tau} \quad \text{for all times} \tag{3}$$

which is the usual adiabatic-following result [15], which may be derived from (1) when $(1/\Phi)\,(\mathrm{d}\Phi/\mathrm{d}t) \ll 1/\tau$ and recombination is ignored. Under experimental conditions where the light pulsewidth is long relative to the response time of the microwave apparatus (so that the instrument response does not distort the observed photosignal), the photoconductivity signal evolves in time identically as the light pulse envelope; this fact supports the adiabatic-following model.

The linear dependence of the photoelectron density on excitation intensity in the Ir^{3+}-doped crystals can be explained by electron capture at centers that exist prior to exposure. Electron capture may be followed by subsequent hole capture at the filled electron trap (i.e. recombination), but the data cannot be explained by prior hole kinetics that create the centers that capture electrons (see Appendix A). This fact, in conjunction with earlier arguments that identified the kinetically active center (Section 3), suggests that the observed photoelectron decay kinetics are best described by the reaction

$$e^- + (IrBr_6)^{3-} \rightarrow (IrBr_6)^{4-} \,. \tag{4}$$

This interpretation is consistent with the observation by EPR of electron trapping by $IrBr_6^{3-}$ at low temperatures [2].

5. Room-Temperature Lifetime and Cross Section for Electron Capture by IrB_6^{3-}

Although instrumental limitations prevented direct observation of the photoelectron kinetics in Ir^{3+}-doped AgBr, it is possible to obtain some knowledge of the decay rate from the peak photosignal, which measures the relative peak photoelectron density (Section 3). In particular, it is possible to deduce the photoelectron lifetime parameter τ, which appears in (1) to (3). τ is operationally defined in (3), and the following discussion does not rely on the assumption of simple exponential decay kinetics in the iridium-doped crystals.

The photoelectron lifetime τ has been deduced by a calibrated peak-height technique; the photoresponse of a pure AgBr control was used to determine τ in doped crystals [3]. This approach utilizes the adiabatic-following behavior of the photoresponse of the doped samples and the integrating behavior of the initial photoresponse of pure crystals. Details of the method are described in [3]. The results of such a procedure show a room-temperature photoelectron lifetime ranging from 13 to 150 ps in crystals containing 3.9 to 89 ppm Ir^{3+}. A strong correlation exists between decreasing photoelectron lifetime and increasing dopant level; a roughly linear dependence of $1/\tau$ and N was observed,

$$\frac{1}{\tau} = \sigma v N\,. \tag{5}$$

The cross section for electron capture by $IrBr_6^{3-}$, $\sigma = (26 \pm 5) \times 10^{-2}\,nm^2$, was deduced using the polaron velocity v. This cross section represents an average over all the different vacancy-compensated $IrBr_6^{-3}$ species present in the material.

6. Temperature Dependence

We studied the temperature dependence of the peak photoconductivity to obtain further information pertaining to electron capture by Ir^{3+} centers in AgBr. From the kinetic analysis of Section 4, the peak photosignal S_{max} will be given by

$$\begin{aligned} S_{max,\,pure} &= C\varepsilon \int \Phi\, dt\,, \\ S_{max,\,doped} &= C\varepsilon\tau\Phi_{max}\,, \end{aligned} \tag{6}$$

where C is a constant that is linearly proportional to the carrier mobility μ. Since the index of refraction is weakly temperature dependent, ε should be temperature independent as long as the reciprocal absorption coefficient remains smaller than the sample thickness. This was true for our sample dimensions in the range 90 to 298 K for $\lambda \leqq 450$ nm [16]. Therefore, the temperature dependence of the photosignals should vary as

$$\begin{aligned} S_{max,\,pure} &\sim \mu(T)\,, \\ S_{max,\,doped} &\sim \mu(T)\,\tau(T)\,. \end{aligned} \tag{7}$$

The peak photosignal measured for a pure AgBr crystal as a function of temperature (corrected for finite instrument response by the methods of Appendix B) is shown in Fig. 3; the peak height follows the electron mobility [10] from 90 to 198 K. This

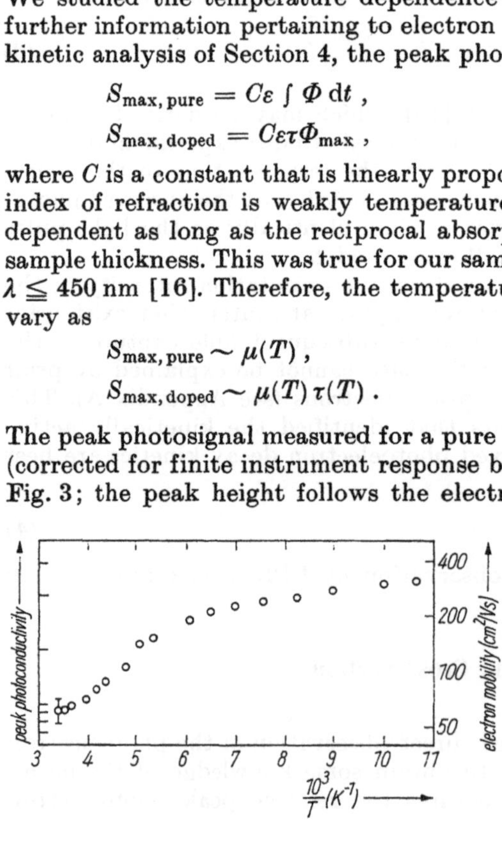

Fig. 3. Temperature dependence of peak photoconductivity of pure AgBr. $\lambda = 447.7$ nm, $\approx 10^{12}$ photons/cm². Electron mobility from [10]

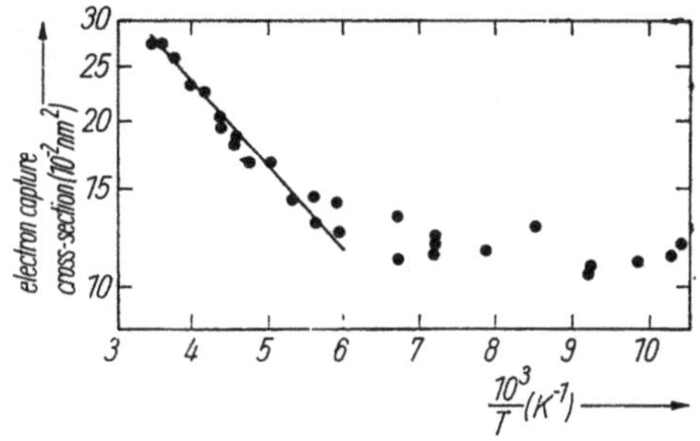

Fig. 4. Temperature dependence of electron capture cross-section of $IrBr_6^{3-}$ in AgBr. Relative cross-section was deduced from peak photoconductivity measurements and normalized to the measured room-temperature cross section of $26 \times \times 10^{-2}$ nm² [3]. Data pertain to crystal containing 89 ppm Ir^{3+}. Straight line represents fit of high-temperature data to $\sigma \approx \sigma_0 \exp(-E_{act}/kT)$ ($E_{act} = 44$ meV)

result confirms the kinetic analysis in this work. In particular, it shows that: (i) The pure AgBr photoconductivity is due to electrons. (ii) No unresolved fast decay processes occur in pure AgBr. If this were not the case, the temperature dependence of the fast processes would result in a different behavior. This assumption was implicit in the lifetime and cross-section measurements of Section 5 and [3]. (iii) Shallow trapping effects that reduce carrier mobility and effective velocity are negligible in this temperature range. This assumption led to the choice of the thermal velocity for v in (5), from which the cross section was deduced. (iv) ε is at best weakly temperature dependent.

The temperature dependence of the photoconductivity of doped material may be used to deduce the temperature dependence of the photoelectron–$IrBr_6^{3-}$ cross section. From (5) and (7) it is clear that

$$\sigma(T) \sim \mu(T)/N\, v(T)\, S_{max,\,doped}(T) \sim \mu(T)/\sqrt{T} S_{max,\,doped}(T)\,. \tag{8}$$

The thermal velocity and the phonon scattering-limited mobility are used in (8) for the reasons discussed in the preceding section. The concentration of kinetically active iridium centers is taken to be the total (temperature-independent) iridium level, so that $\sigma(T)$ represents the average cross section over all the $IrBr_6^{3-}$ centers present in the crystal. Experimental results for $\sigma(T)$ are presented in Fig. 4. The average cross section decreased with decreasing temperature above ≈ 160 K; at lower temperatures the cross section varied only slightly with temperature. Above ≈ 160 K, the temperature dependence of the cross section could be described by an Arrhenius law,

$$\sigma(T) = \sigma_0\, e^{-E/kT}\,, \tag{9}$$

where the activation energy $E = (44 \pm 14)$ meV, and the prefactor deduced from the results of Section 5 is $\sigma_0 \approx 1$ nm². The ratio of room temperature to 160 K cross sections was 2.2 ± 0.4.

7. Discussion. Effect of Vacancy Association and the Nature of the Electron-Capture Process

Doping with Ir^{3+} ions, which occupy substitutional silver ion sites in the AgBr lattice [1], will also introduce silver ion vacancies into the lattice to preserve the electrical neutrality of the crystal. Two vacancies, which are negatively charged with respect to the lattice, may be bound to the iridium ions ("associated") or unassociated and thus capable of motion in the lattice ("free"). These statements are substantiated by room-temperature ionic conductivity measurements on our crystals (see Footnote²)) and similar samples [17]; the ionic conductivity in iridium-doped samples is greater than in pure AgBr. The activation energy of the ionic conductivity is characteristic

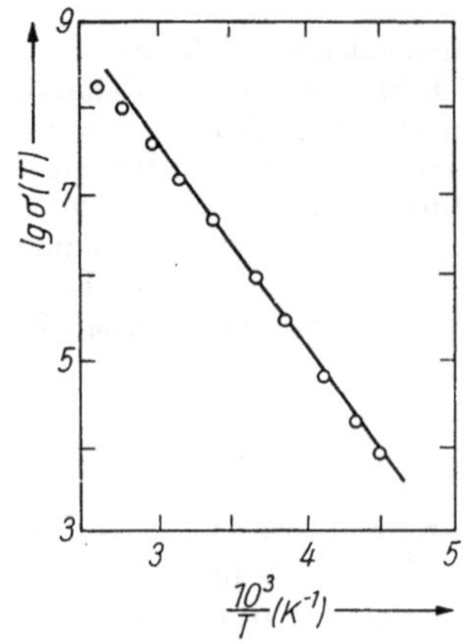

Fig. 5. Ionic conductivity of 85 ppm Ir^{3+}-doped AgBr (courtesy D. Jeanmaire). Straight line shows fit to $\sigma(T) = \exp(-E_{act}/kT)$ ($E_{act} = 0.44$ eV)

Table 1

Ionic conductivity measurements

$[Ir^{3+}]$ (ppm)	σ_{ionic} at 294 K (Ω^{-1} cm^{-1})	E_{act} (eV)	E_k (eV)
0	4×10^{-8}	0.39 ± 0.02	0.14 ± 0.03
85	1.3×10^{-7}	0.44 ± 0.02	0.24 ± 0.05

of vacancies rather than interstitial silver ions, so that these results imply that iridium doping introduces a number of free vacancies into the AgBr lattice. Results obtained for two of the crystals used in this work are presented in Table 1 and Fig. 5.

In the case of divalent, substitutional cationic impurities, the number of free vacancies V is equal to the number of uncompensated defects and is given by a mass-action law involving the total number of impurities N and the equilibrium constant for vacancy compensation K [13],

$$\frac{V}{N} = \frac{K}{2N}\left[\left(1 + \frac{4N}{K}\right)^{1/2} - 1\right]. \tag{10}$$

The constant K varies with temperature as $\exp(-E_k/kT)$, where E_k is an energy of association. At high temperatures, the number of free vacancies is thermally activated, with activation energy $E_k/2$. Therefore, the activation energy for ionic conductivity is

$$E_{act} = E_M + \frac{E_k}{2}, \tag{11}$$

where $E_M = (0.32 \pm 0.02)$ eV is the activation energy for vacancy motion. The association energy E_k deduced for our pure AgBr sample (Table 1) is close to the values deduced by Müller and by Ulrici for divalent impurities (0.16 eV) [13]. The same analysis used for divalent impurities above may also be applied to trivalent cations under the assumption that each ion binds at least one of the two charge-compensating vacancies [13, 17]. This assumption is reasonable, considering the activation energies observed for trivalent ions (Table 1 and [13]) and the enhanced binding expected because of the greater ionic charge. The resultant association energy deduced from

(11) and E_{act} for an iridium-doped sample (Table 1) is significantly greater than that for divalent impurities; similar behavior has been observed for other trivalent defects in AgBr and is ascribed to imperfect shielding of the additional cationic charge by the additional vacancy [13].

The ionic conductivity data show that a substantial number of the Ir^{3+} complexes do not have two vacancies associated at room temperature (e.g. 30% of the Ir^{3+} for a dopant level of 85 ppm). Since the number of associated vacancies will determine the effective charge of the $IrBr_6^{3-}$ complex, it is conceivable that iridium centers with only a single bound vacancy may have a greater electron capture cross section than those with two associated vacancies. Models involving either cascade [18] or multiphonon emission [19] mechanisms of carrier capture suggest that the cross section should depend on the charge of the complex. In this case, the capture rate should depend on the capture cross section of fully compensated (σ_2) and partially compensated centers (σ_1) as

$$\frac{\tau^{-1}}{Nv} = \frac{N_1}{N} (\sigma_1 - \sigma_2) + \sigma_2 , \tag{12}$$

where N_1/N is the fraction of Ir^{3+} centers that are not fully compensated. N_1/N will vary with both temperature and the total iridium concentration N.

Since the decay rate $1/\tau$ varied linearly with N (Section 5 and [3]), vacancy compensation effects did not influence the capture rate to a first approximation. The deviations from strict linearity that occur, however, are consistent with effects expected due to various degrees of charge compensation. Therefore we attempted to fit the room-temperature lifetime data to (12); the uncertainties in $1/\tau$ and N_1/N precluded a clear determination of whether $\sigma_1 = \sigma_2$.

A dependence of cross section on vacancy association is also expected to influence the temperature dependence. Both theoretical and experimental results show that cross sections do not increase rapidly with decreasing temperature [18 to 20]. Thus, if the capture process were primarily due to partially compensated Ir^{3+} centers, an activation energy of at least $\approx E_k/2 \approx 0.12$ eV for the average cross section should be observed. Since the data of Section 6 show a much weaker temperature dependence, the capture process is not primarily due to Ir^{3+} centers that are not fully compensated. If it is assumed that the observed temperature dependence is due to changes in N_1/N, and both σ_1 and σ_2 are temperature independent, then a ratio $\sigma_1/\sigma_2 \approx 2.5$ is deduced; if σ_1 or σ_2 decreases with temperature, σ_1/σ_2 must be smaller. Since the temperature dependence of our samples was essentially independent of dopant level, it appears more appropriate to attribute the change in the average cross section to changes in σ_1, σ_2 rather than N_1/N.

Our results are readily understood in terms of a two-step capture process. Electrons are first captured into a relatively shallow, excited state of the $IrBr_6^{3-}$ complex and then decay into a deeper level to form $IrBr_6^{4-}$. The rate-limiting step in such a capture process at room temperature is the electron decay from the shallow to the deep state, because the observed capture cross section is much smaller than that expected for capture into a hydrogenic state on the basis of theoretical considerations [18, 22] and cross sections observed for hydrogenic centers in other materials [18]. For such a situation,[3]) the capture cross section is given by [20, 21]

$$\sigma = \frac{\nu}{N_c v g^{-1}} \exp (E_1/kT) , \tag{13}$$

---

[3]) Our use of (12) is justified because the emission rate from a shallow hydrogenic level, as calculated from detailed balance and the expected cross section, is ≈ 10 times greater than ν [21].

where $\nu \approx 2 \times 10^{10}\,s^{-1}$ at 294 K is the shallow-to-deep state transition frequency, E_1 is the depth and g the degeneracy of the shallow level, and N_c is the density of states in the conduction band calculated from the electron band mass found by Tamura and Masumi [23]. We expect that the frequency ν will be independent of the degree of vacancy association of the $IrBr_6^{3-}$, so that σ_1/σ_2 will depend only on the ratio of Boltzmann factors involving the shallow-state energies. If the depth of the shallow state arising from fully compensated centers is ≈ 10 meV, equation (12) predicts $\sigma_1/\sigma_2 \approx 2$. A ratio σ_1/σ_2 of such magnitude is not inconsistent with the data, as discussed above; it also suggests that the temperature-dependence data for the more heavily doped samples are characteristic of ν and E_1 rather than N_1/N.

A thermally activated decay frequency ν is characteristic of a process involving multiple phonon emission; it is not expected for radiative, Auger, or cascade capture processes [19]. A cascade decay process is also improbable, owing to the large energy difference [2] between the ground state of $IrBr_6^{4-}$ and the conduction band [19]. Capture by multiphonon emission is an attractive hypothesis for several additional reasons: (i) The magnitude of σ and its activation energy are comparable to those for defects in GaAs and GaP, which are believed to exhibit multiphonon emission carrier capture. (ii) The molecular vibrations of the $IrBr_6^{3-}$ are expected to facilitate such a process. Localized defect vibrations enhance the density of vibrational states in the vicinity of the impurity and thus enhance the transfer of electronic to vibrational energy. In addition, the $T_{1u}(\nu_4)$ mode of the molecular ion (140 cm^{-1}) [24] is resonant with the lattice LO phonon at T (139 cm^{-1}) [25]; such a resonance may facilitate the transfer of localized vibrational energy to the lattice.

8. Conclusions

The cross section for electron capture by $IrBr_6^{3-}$ at room temperature is 0.26 nm^2. It is only weakly dependent on the number of vacancies bound to the Ir^{3+} center; that is, the cross section is not strongly dependent on the net charge of the electron trap. (Carrier capture at charged centers is expected to depend on the square of the effective charge; capture at neutral centers is typically orders of magnitude less than that for charged centers.) This behavior may be understood in terms of a two-step capture process involving a shallow level and a deeper, molecular-like state.

Acknowledgements

We thank R. S. Eachus and R. H. Young for many fruitful discussions, D. Jeanmaire for ionic conductivity measurements, J. L. Pietruszewski for neutron activation analyses, M. Burberry for his gracious loan of the nitrogen laser, and C. Hamer for preparing several of the samples.

Appendix A

Quasi-equilibrium carrier densities for recombination centers

Since the Fermi level in AgBr is deep in the gap prior to exposure, the Shockley-Read-Hall mechanism [26] yields particularly simple results. Consider sequential trapping of two carriers A, B of opposite sign at an iridium site,

$$Ir^{3+} + A \rightarrow Ir^{3+} \cdot A\,, \qquad \text{(A1a)}$$

$$Ir^{3+} + A + B \rightarrow Ir^{3+}\,. \qquad \text{(A1b)}$$

This system is governed by the rate equations

$$\frac{dA}{dt} = \varepsilon\Phi - N\sigma v A\,,$$
$$\frac{dB}{dt} = \varepsilon\Phi - N'\sigma' v' B\,, \tag{A2}$$

and a conservation-of-charge equation. Here N is the density of Ir^{3+} centers and N' is the density of $Ir^{3+} \cdot A$ centers. It is assumed that no saturation of process (A1a) occurs.

The solution of (A2) in the quasi-equilibrium case is

$$A = \frac{\varepsilon\Phi}{N\sigma v}\,, \tag{A3a}$$

$$B = \frac{\varepsilon\Phi}{2N\sigma v}\left(1 + \left[1 + \frac{(2N\sigma v)^2}{\sigma' v' \varepsilon\Phi}\right]^{1/2}\right). \tag{A3b}$$

These solutions have the following properties: (i) The carrier that is captured first (A) has a density linear in excitation intensity. (ii) The density of the carrier captured second (B) is linear in excitation intensity only at high intensities when

$$\sigma' v' \gtrapprox \frac{(N\sigma v)^2}{e\Phi}\,. \tag{A4}$$

At low intensities, the density is proportional to the square root of excitation intensity. (iii) The density of the carrier captured first is always less than or equal to the density of the other carrier: $A/B \leqq 1$. The densities are equal only in the high-intensity regime discussed in (ii).

Experimentally, a signal proportional to photoelectron density (Section 3) varied linearly with exposure intensity (Section 4). This could be consistent with either electrons being captured first or prior hole capture at high intensities. In either case, the electron density is given by

$$n = \varepsilon\Phi\tau\,, \tag{A5}$$

where $\tau = 1/n\sigma v$ is the lifetime measured experimentally (Section 5 and [3]). Note that the measured lifetime refers to the first capture process (A1a) in this model.

If prior hole capture occurs, the electron density is given by (A3b), which reduces to (A5) only for

$$\sigma' \geqq \frac{1}{\varepsilon\Phi\tau^2 v'}\,. \tag{A6}$$

Equation (A6) indicates the lower limit on the electron capture cross section σ' by the trapped hole–iridium (Ir^{4+}) complex. Our estimates show that $\sigma' \geqq 3 \times 10^5\ nm^2$, an unreasonably large capture cross section. This suggests that formation of electron-trapping complexes by prior hole capture does not explain our experimental results.

Appendix B

Calibrated peak-height method

From the kinetic analysis of Section 4, the peak photosignals observed in pure and doped AgBr crystals are given by

$$S_{max,\,pure} = C\varepsilon \int \Phi\, dt\,,$$
$$S_{max,\,doped} = C\varepsilon\Phi_{max}\tau\,. \tag{A7}$$

Here, S_{max} is the observed photosignal peak height and C is a constant that relates the instrumental signal to the number of photoelectrons present in the sample. C depends on the electron mobility, the geometric arrangement of the sample in the microwave cavity, and various instrumental parameters [8]. The assumption that no photoelectron decays during photogeneration in pure AgBr leads to the expression for $S_{max, pure}$ in (A7); this is confirmed by the temperature dependence of the peak photosignal of pure AgBr as discussed in Section 6.

It is assumed that doping does not alter the electron mobility or the photogeneration efficiency ε. The microwave power incident on and reflected from the microwave cavity was constant, as was the iris coupling parameter; this is sufficient to maintain a constant instrumental sensitivity to cavity Q-spoiling [8]. The various samples used for this study were cut to the same size and shape and were positioned in the cavity by using a translational stage with three mutually perpendicular degrees of freedom. By such considerations, variations in geometric sensitivity factors were minimized. Thus, $C\varepsilon$ may be taken to be sample independent, so that (A7) shows that

$$R \equiv \frac{S_{max, doped}}{S_{max, pure}} = \frac{\tau \Phi_{max}}{\int \Phi \, dt}. \tag{A8}$$

The quantity $\int \Phi \, dt/\Phi_{max}$ is readily obtained from numerical integration of the light pulse time evolution, so the peak height ratio is a measure of τ.

Equation (A8) is a correct description of the ideal case where the sample/microwave apparatus behaves as either a perfect follower or a perfect integrator. The finite response time of the microwave instrument, however, introduces deviations from the ideal behavior. The finite response time reduces the peak height from its ideal value. To evaluate correction factors for this effect, we modelled the response of the microwave system by

$$\frac{dS}{dt} + \frac{S}{T} = n\,, \tag{A9}$$

where T is the apparatus time constant, S is the photosignal, and n is the photoelectron density obtained from (1). Equation (A9) can be derived from a perturbation of the (exact) transmission line equations for the microwave apparatus [9] and ignoring coherence terms ("ringing"). Alternatively, (A9) can be obtained from a perturbation of the power-balance equation for the cavity, which inherently ignores coherence effects [9]. The cavity time constant T was obtained from the observed phototransients for each sample, and the light pulse envelope [$\Phi(t)$ in (1)] was taken to be a single half-wave of – a sinusoid a reasonably close fit, as shown in Fig. 6. The ratio of real to ideal signal peak heights γ $(0 \leqq \gamma \leqq 1)$ (γ_D) is a function of apparatus time constant T for doped crystals and a function of T and the photoelectron lifetime for pure crystals (γ_P).

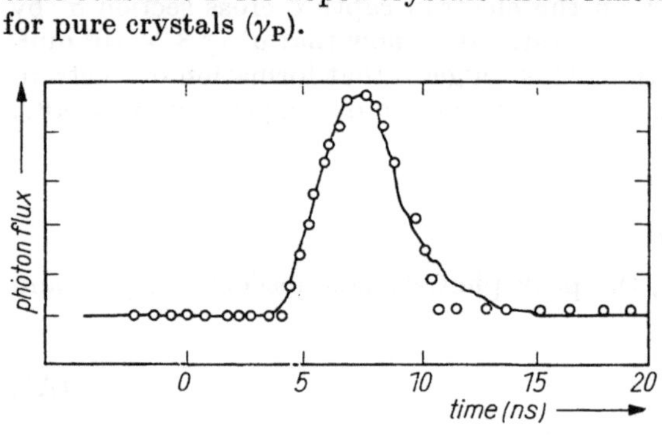

Fig. 6. Experimental light pulse profile. Time evolution of light pulse at $\lambda =$ = 468.3 nm is shown by solid line; fit to sinusoid is shown by circles. Deviation from sinusoidal behavior at long times does not affect calculations of γ_D, since the photoconductivity attains its maximum well before the deviations become significant

Experimental peak height ratios R' were used to deduce R in (A8) from the relation

$$R = \frac{\gamma_P}{\gamma_D} R' . \tag{A10}$$

R was measured at a number of different cavity time constants T (T was varied by varying iris coupling) in an effort to reduce uncertainties introduced by system modelling. The microwave cavity was operated near critical coupling to minimize ringing [9].

References

[1] R. J. Deri and J. P. Spoonhower, J. chem. Phys. 80, 5423 (1984).
[2] R. S. Eachus and R. E. Graves, J. chem. Phys. 61, 2860 (1974).
[3] R. J. Deri and J. P. Spoonhower, Appl. Phys. Letters 43, 65 (1983).
[4] J. P. Spoonhower and C. A. Hamer, J. Lum. 28, 191 (1983).
[5] G. H. McCall, Rev. sci. Instrum. 43, 865 (1972).
[6] R. J. Deri and J. P. Spoonhower, Photogr. sci. Engng. 28, 92 (1984).
[7] R. J. Deri and J. P. Spoonhower, Phys. Rev. B 25, 2821 (1982).
[8] G. Feher, Bell Syst. tech. J. 47, 449 (1957).
R. J. Deri and J. P. Spoonhower, Photogr. sci. Engng. 25, 89 (1981).
[9] R. J. Deri and J. P. Spoonhower, Rev. sci. Instrum. 55, 1343 (1984).
[10] R. K. Ahrenkiel, Phys. Rev. 180, 859 (1969).
[11] V. I. Saunders, R. W. Tyler, and W. West, Photogr. sci. Engng. 16, 87 (1972).
[12] R. K. Ahrenkiel and R. S. van Heyningen, Phys. Rev. 144, 576 (1966).
W. Platikanova and J. Malinowski, phys. stat. sol. 14, 205 (1966).
J. Malinowski, Photogr. sci. Engng. 14, 112 (1970).
[13] P. Müller, phys. stat. sol. 12, 775 (1965).
R. J. Lieb, Thesis, Univ. of North Carolina (Chapel Hill), 1976.
W. Ulrici, phys. stat. sol. 27, 333, 489 (1968).
[14] W. Bahnmueller, Photogr. Korresp. 104, 169, 195 (1968).
J. W. Mitchell, Photogr. sci. Engng. 23, 1 (1979).
B. H. Carroll, Photogr. sci. Engng. 24, 265 (1980).
[15] L. Allen and J. H. Eberly, Optical Resonance and Two-Level Atoms, John Wiley & Sons, New York 1975 (p. 72).
[16] F. Moser and F. Urbach, Phys. Rev. 102, 1519 (1956).
S. Tutihasi, Phys. Rev. 105, 882 (1957).
[17] J. E. Keevert, private communication.
[18] M. Lax, Phys. Rev. 119, 1502 (1960).
V. N. Abakumov, V. I. Perel, and I. N. Yassievich, Soviet Phys. — Semicond. 12, 1 (1978).
[19] C. H. Henry and D. V. Lang, Phys. Rev. B 15, 989 (1977).
[20] G. J. Rees, H. G. Grimmeiss, E. Janzen, and B. Skurstam, J. Phys. C 13, 6157 (1980).
[21] R. M. Gibb, G. J. Rees, B. W. Thomas, B. L. H. Wilson, B. Hamilton, D. R. Wight, and N. F. Mott, Phil. Mag. 36, 1021 (1977).
[22] C. R. Berry, Photogr. sci. Engng. 21, 202 (1973).
[23] H. Tamura and T. Masumi, Solid State Commun. 12, 1183 (1973).
[24] G. L. Bottger and A. E. Salwin, Spectrochim. Acta A 28, 925 (1972).
[25] Y. Fujii, S. Hoshino, S. Sakuragi, and H. Kanzuki, Phys. Rev. B 15, 358 (1977).
[26] R. N. Hall, Phys. Rev. 87, 387 (1952).
W. Shockley and W. R. Read, Phys. Rev. 87, 835 (1952).

(Received July 11, 1985)

phys. stat. sol. (b) **133**, 315 (1986)

Subject classification: 71.70; S10; S10.1

Institute for Physical Research,
Academy of Sciences of the Armenian SSR, Ashtarak[1])

Zero-Field Splitting and g-Values of d^8 Ions in a Trigonal Crystal Field

By

A. K. PETROSYAN and A. A. MIRZAKHANYAN

On the basis of the crystal-field theory the zero-field splitting and g-values of the $^3A_2$ ground state for d^8 impurity ions in a trigonally distorted octahedral field are calculated. Using perturbation theory the analytical expressions connecting the spin-Hamiltonian parameters with the crystal-field parameters are obtained. It is shown that experimental data for Cu^{3+} in Al_2O_3 and Ni^{2+} in Al_2O_3, α-$LiIO_3$, $LiNbO_3$ can be explained taking into account the influence of all the excited levels on the ground state.

На основе теории кристаллического поля проведён расчёт расщепления в нулевом поле и g-факторов основного состояния $^3A_2$ для примесных d^8 ионов в октаэдрическом поле с тригональным искажением. Методом теории возмущений получены аналитические выражения, связывающие параметры спин-гамильтониана с параметрами кристаллического поля. Показано, что экспериментальные данные для Cu^{3+} в Al_2O_3 и Ni^{2+} в Al_2O_3, α-$LiIO_3$ и $LiNbO_3$ можно объяснить при учёте всех возбуждённых уровней, взаимодействующих с основным состоянием.

1. Introduction

Impurity ions of $3d^8$ electron configuration (Ni^{2+}, Cu^{3+}) are intensively investigated by EPR and optical spectroscopy [1, 2]. In a crystal field of octahedral symmetry the ground state of these ions is the $^3A_2$ orbital singlet triply spin-degenerated and considerably remoted from the excited levels (Fig. 1). If the crystal field has an axial component (in particular, a trigonal distortion) the ground state is split into the singlet $|0\rangle$ and the doublet $|\pm 1\rangle$. In the absence of an external magnetic field the zero-field splitting (ZFS) between these levels is equal to the axial parameter D of the spin-Hamiltonian and can reach a magnitude of a few cm^{-1}. As a rule, a trigonal field also leads to a considerable anisotropy of g-values. However, the attempts to explain the magnitudes of ZFS and g-value anisotropy for Ni^{2+} and Cu^{3+} in some crystals by taking into account the interaction with only the nearest excited $^3T_2$ level were unsuccessful, since a wrong sign has been obtained for both D and $\Delta g = g_{\parallel} - g_{\perp}$ [2 to 5].

In this paper the ZFS and g-values for d^8 ions in trigonally distorted octahedral complexes are calculated within the framework of the crystal-field theory. It is shown that the experimental data for the Ni^{2+} and Cu^{3+} ions can be interpreted only by considering all the excited levels, which interact with the ground state.

2. Method of Calculation

In order to obtain analytical expressions the calculations were carried out by perturbation theory using the wave functions of a strong octahedral field. The Hamiltonian

[1]) 378410 Ashtarak 2, USSR.

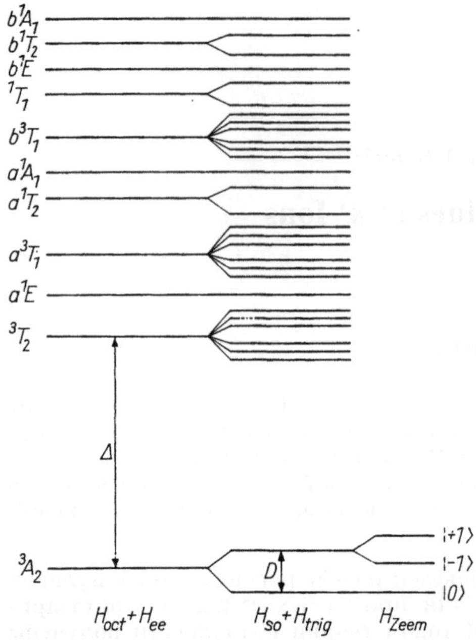

Fig. 1. Energy level scheme of d^8 ions in an octahedral crystal field with a trigonal distortion (splittings not to scale)

for an impurity ion in a trigonally distorted octahedral coordination can be written as

$$\mathcal{H} = H_{\text{oct}}(\Delta) + H_{\text{ee}}(B, C) + H_{\text{so}}(\xi) + + H_{\text{trig}}(v, v') + H_{\text{Zeeman}} . \quad (1)$$

Here Δ is the splitting in octahedral field, B, C are the Racah parameters characterizing the electrostatic interaction between d-electrons, ξ is the one-electron spin–orbit coupling parameter, v and v' are the diagonal and off-diagonal trigonal–field parameters, respectively [2]. It should be noted that covalency is taken into account by introducing the orbital reduction factor k (its departure from unity is a measure of the amount of covalency) and by reducing the values of B, C, ξ below their free-ion values [1, 6]. The explicit form of the Hamiltonian operators and the methods of matrix element calculation are presented in [6, 7].

For the ions of d^8 configuration the splitting in octahedral field and the Coulomb interaction between d-electrons are generally of the same order of magnitude. Therefore, the "strong-field" approach ($H_{\text{oct}} \gg H_{\text{ee}} \gg H_{\text{so}} + H_{\text{trig}}$) used in [8] is not satisfactory in this case. In order to improve the accuracy of calculations we used the "quasi-intermediate-field" approach, where the unperturbed Hamiltonian includes both H_{oct} and the diagonal part of the Coulomb interaction [9]. Hence, the off-diagonal part of H_{ee}, the trigonal field, the spin–orbit coupling, and the Zeeman interaction were simultaneously considered as a perturbation.

3. Results and Discussion

In the case of d^8 ions the ground state ZFS and the g-value anisotropy arise only in third order of perturbation theory as a result of the joint action of trigonal field and spin–orbit coupling. We calculated all the third-order and fourth-order ZFS contributions caused by the interaction of the $^3A_2$ ground state with the excited levels of the $t_2^5e^3(^3T_2, a^3T_1, a^1T_2)$ and $t_2^4e^4(b^3T_1, b^1T_2)$ configurations. The analysis shows that there are four terms linear in the Coulomb interaction, which are the largest among the fourth-order contributions to D. These terms result from the configuration mixing of a^3T_1 with b^3T_1, and their contribution to the ZFS is appreciably larger than in the case of d^3 ions [9]. Other fourth-order terms are much smaller in magnitude and may be ignored, since they do not lead to a noticeable change in D. Thus, in the "quasi-intermediate-field" approach the perturbation expression for the axial parameter is

$$D = \frac{\xi^2 v}{2W_1^2} - \frac{\xi^2 v}{2W_2^2} - \frac{3\sqrt{2}\,\xi^2 v'}{2W_1W_3} + \frac{3\sqrt{2}\,\xi^2 v'}{2W_2W_3} - \frac{18\sqrt{2}\,B\xi^2 v'}{W_1W_3W_4} - - \frac{18\sqrt{2}\,B\xi^2 v'}{W_2W_3W_4} + \frac{6\sqrt{2}\,B\xi^2 v'}{W_2W_3W_5} + \frac{6\sqrt{2}\,B\xi^2 v'}{W_2^2W_5}, \quad (2)$$

where W_i denotes the energy difference between the $^3A_2$ ground state and the corresponding excited level in this approximation (adopting zero energy for the ground state),

$$W_1 = W(^3T_2) = \Delta\,,$$
$$W_2 = W(a^1T_2) = \Delta + 2C + 8B\,,$$
$$W_3 = W(a^3T_1) = \Delta + 12B\,,$$
$$W_4 = W(b^3T_1) = 2\Delta + 3B\,,$$
$$W_5 = W(b^1T_2) = 2\Delta + 2C + 2B\,.$$

Analogous calculations have also been performed for the g-values of d^8 ions. To first order, g is isotropic and spin only. The second-order contribution is orbital, isotropic, and about an order of magnitude lower than the first-order one. In third and fourth order, there are two types of contributions: the first produces small isotropic corrections to g, and the second gives rise to anisotropy in the g-value. Summing all contributions, we derive the following analytical expressions:

$$g_{\|} = g_e + \frac{4k\xi}{W_1} - \frac{(g_e + k/2)\,\xi^2 + \frac{4}{3}\,k\xi v}{W_1^2} - \frac{(g_e - k/2)\,\xi^2}{W_2^2} - \frac{k\xi^2}{W_1 W_2} + \\ + \frac{4\sqrt{2}k\xi v'}{W_1 W_3} + \frac{48\sqrt{2}Bk\xi v'}{W_1 W_3 W_4} - \frac{12Bk\xi^2}{W_1 W_2 W_5} - \frac{6Bk\xi^2}{W_2^2 W_5}\,, \tag{3}$$

$$\Delta g = g_{\|} - g_{\perp} = -\frac{2k\xi v}{W_1^2} + \frac{6\sqrt{2}k\xi v'}{W_1 W_3} + \frac{72\sqrt{2}Bk\xi v'}{W_1 W_3 W_4}\,, \tag{4}$$

where g_Ω is the free-electron g-value of 2.0023.

The obtained expressions have been used by us for the interpretation of experimental data related to the Ni^{2+} and Cu^{3+} ions in some crystals, investigated in detail by EPR and optical spectroscopy [4, 5, 10 to 13]. The calculated values of the spin-Hamiltonian paramotors and the magnitudes of parameters entering in the Hamilonian (1) which are used in calculations are listed in Table 1. In the case of α-$LiIO_3$ and

Table 1

The parameters of d^8 ions in a trigonal crystal field. Experimental data for Al_2O_3 were taken from [1, 4, 5]. Spin-Hamiltonian parameters for α-$LiIO_3$ and $LiNbO_3$ were measured by us, and crystal-field parameters were calculated using optical absorption data from [10, 12, 13]. Except for k, $g_{\|}$, and Δg, all the values are in cm^{-1}. $T = 300$ K

	Al_2O_3:Cu^{3+}	Al_2O_3:Ni^{2+}	α-$LiIO_3$:Ni^{2+}	$LiNbO_3$:Ni^{2+}
Δ	21000	10000	7750	7920
B	1030	800	850	816
C	4850	3400	3600	3224
k	0.68	0.87	0.92	0.83
ξ	600	565	595	540
v	400	600	− 550	− 950
v'	350	500	200	600
D_{calc}	−0.195	−1.14	−2.85	−5.24
D_{exp}	−0.188	−1.38	−3.21	−5.31
$g_{\|, calc}$	2.0789	2.1958	2.281	2.239
$g_{\|, exp}$	2.0788	2.1957	2.280	2.24
Δg_{calc}	0.0015	0.0094	0.021	0.038
Δg_{exp}	0.0016	0.0098	0.023	0.04

$LiNbO_3$ we have determined the trigonal-field parameters v and v' from the fine structure of the optical absorption spectra in polarized light. The values of the orbital reduction factor k and the spin–orbit coupling parameter in crystals ($\xi = k\xi_0$) were calculated taking into account the departure of the $g_{\parallel}$-value from g_e. It should be noted that because of the large ZFS of Ni^{2+} in $LiNbO_3$ and α-$LiIO_3$ [11] the g-values in these crystals were measured by us using the angular dependence of EPR spectra at high frequencies, and the sign of D was determined from the temperature dependence of EPR transition intensities in the millimeter wavelength range.

As can be seen from Table 1, there is a sufficiently good agreement between experimental and calculated data for all the crystals under consideration. In order to check the accuracy of the expressions obtained for D and Δg we have also performed similar calculations using the "intermediate-field" approach (i.e., $H_{oct} + H_{ee}$ were diagonalized simultaneously, and the perturbation was the sum of H_{trig}, H_{so}, and H_{Zeeman}). The difference between the result of both approximations does not exceed 10% and, consequently, the "quasi-intermediate-field" approach is quite satisfactory. Its advantage is in the relatively simple form of the analytical expressions.

It is well known that for a given ion the values of parameters Δ, B, C, k, ξ have only small variations from one crystal to another, if the coordination and the type of ligand are not changed [1, 2]. Therefore, using average values of these parameters for Ni^{2+} and Cu^{3+} in sixfold oxygen coordination, one can obtain an approximate form of the ZFS and Δg dependences on the trigonal-field parameters from (2) and (4) (D, v, v' in units of cm^{-1}),

$$Ni^{2+}: \quad D \approx 10^{-3}\,(2.2v - 6.5v')\,, \qquad \Delta g \approx 10^{-5}\,(-1.6v + 4.8v')\,,$$

$$Cu^{3+}: \quad D \approx 10^{-4}\,(2.9v - 8.9v')\,, \qquad \Delta g \approx 10^{-6}\,(-1.9v + 6.3v')\,.$$

These relations are valid over a reasonably wide range of parameter values and can be used for the evaluation of the spin-Hamiltonian parameters in cases where no experimental data are yet available. As it has been shown by Macfarlane [9, 14], for d^3 ions, having the orbital singlet $^4A_2$ ground state, the ZFS and Δg values effectively depend only on v'. That is why there is no direct correlation between the magnitudes of the spin-Hamiltonian parameters for d^3 and d^8 ions in the same crystals.

4. Conclusions

On the basis of the results discussed in the previous section we conclude:

1. In the case of d^8 ions an appreciable contribution to the ZFS and g-value anisotropy is caused by terms proportional to the off-diagonal trigonal-field parameter v'. Therefore, the consideration of only the diagonal parameter v leads to discrepancy with experimental data [4, 5]. From (2) to (4) it is obvious that the inclusion of interactions with only nearby excited levels is not correct, since the matrix element with higher levels is often large enough to compensate for the energy denominator.

2. The larger values of D and Δg for Ni^{2+} as compared with Cu^{3+} in the same crystal are caused mainly by the difference in magnitude of the splitting in octahedral field: $\Delta(Cu^{3+}) \approx 2\Delta(Ni^{2+})$.

3. The largest ZFS and Δg values for d^8 ions are observed in crystals having trigonal-field parameters of opposite signs (as in α-$LiIO_3$ and $LiNbO_3$). In other cases, where v and v' are of the same sign, D and Δg may be close to zero in spite of the considerable trigonal distortion.

4. Comparing (2) and (4) one can conclude that for d^8 ions in a trigonal crystal field the approximate relation $D \approx -(1/4)\,\xi\,(g_{\parallel} - g_{\perp})$ is valid. This expression may be used for estimations of the spin-Hamiltonian parameters.

Acknowledgement

We are grateful to Dr. S. N. Lukin from the Physico-Technical Institute, Donetsk for providing the possibility to make EPR measurements at high frequencies.

References

[1] A. ABRAGAM and B. BLEANEY, Electron Paramagnetic Resonance of Transition Ions, Clarendon Press, Oxford 1970.
[2] C. J. BALLHAUSEN, Introduction to Ligand Field Theory, McGraw-Hill Publ. Co., New York 1962.
[3] H. KAMIMURA, Phys. Rev. **128**, 1077 (1962).
[4] S. GESCHWIND and J. P. REMEIKA, J. appl. Phys. **33**, 370 (1963).
[5] W. E. BLUMBERG, J. EISINGER, and S. GESCHWIND, Phys. Rev. **130**, 900 (1963).
[6] S. SUGANO, Y. TANABE, and H. KAMIMURA, Multiplets of Transition-Metal Ions in Crystals, Academic Press, New York/London 1970.
[7] J. S. GRIFFITH, The Theory of Transition Metal Ions, Cambridge University Press, Cambridge 1961.
[8] K. ŽĎÁNSKÝ, Phys. Rev. **159**, 201 (1967).
[9] R. M. MACFARLANE, J. chem. Phys. **47**, 2066 (1967).
[10] L. M. BELYAEV, B. N. GRECHUSHNIKOV, G. F. DOBRZHANSKII, N. N. DIMENKO, Y. N. MARTISHEV, Z. B. PEREKALINA, and M. S. SMORODINA, Kristallografiya **22**, 650 (1977).
[11] A. A. MIRZAKHANYAN, Fiz. tverd. Tela **23**, 2452 (1981).
[12] L. ARIZMENDI, J. M. CABRERA, and F. AGULLO-LOPEZ, Ferroelectrics **26**, 823 (1980).
[13] L. ARIZMENDI, J. M. CABRERA, and F. AGULLO-LOPEZ, Ferroelectrics **56**, 79 (1984).
[14] R. M. MACFARLANE, Phys. Rev. B **1**, 989 (1970).

(Received May 29, 1985)

phys. stat. sol. (b) 133, 321 (1986)

Subject classification: 13.4 and 19; 22.9

Institute of Molecular Physics, Polish Academy of Sciences, Poznań[1])

Low-Symmetry Doped Cu(II) Complexes in $(gly)_2CaCl_2 \cdot 4\,H_2O$ Single Crystal: EPR Studies

By

S K. HOFFMANN, J. GOSLAR, and L. S. SZCZEPANIAK

EPR results for copper(II) ions doped in $(gly)_2CaCl_2 \cdot 4\,H_2O$ single crystals are reported. The copper(II) complex formed in the chromophore CuN_2Cl_4 displays rhombic crystal field symmetry with an unusual spin-Hamiltonian parameter sequence: $g_x = 2.115$, $g_y = 2.034$, $g_z = 2.308$, and $A_x = 40.1$, $A_y = 113.4$, $A_z = -73.0 \times 10^{-4}\ \text{cm}^{-1}$. The g-factors are related to the ground state wave function $\alpha\,|a d_{x^2-y^2} - b d_{z^2}\rangle$ with $\alpha^2 = 0.86$, $a = 0.9835$, and $b = 0.181$, and the untypical hyperfine splitting parameters result from the very low value of the Fermi contact parameter of $\varkappa = 0.087$.

Mit Cu(II)-Ionen dotierte Einkristalle von $(gly)_2CaCl_2 \cdot 4\,H_2O$ werden mittels EPR untersucht. Die entstandenen Komplexe des Chromophors CuN_2Cl_4 besitzen rhombische Kristallfeldsymmetrie mit einer außergewöhnlichen Sequenz der Parameter des Spin-Hamiltonoperators: $g_x = 2{,}115$; $g_y = 2{,}034$; $g_z = 2{,}308$ und $A_x = 40{,}1$; $A_y = 113{,}4$; $A_z = -73{,}0 \times 10^{-4}\ \text{cm}^{-1}$. Die g-Faktoren werden dem Grundzustand $\alpha\,|a d_{x^2-y^2} - b d_{z^2}\rangle$ und den Werten $\alpha^2 = 0{,}86$; $a = 0{,}9835$ und $b = 0{,}181$ zugeordnet. Die untypischen Werte der Hyperfeinstrukturaufspaltungs-Parameter werden durch einen sehr geringen Wert des Fermi-Kontaktterms $\varkappa = 0{,}087$ hervorgerufen.

1. Introduction

EPR studies of paramagnetic centres in diamagnetic crystals are successful in determining both the position of a centre in a crystal unit cell and the details of the electronic structure of the centre. In crystals containing glycine molecules doped Cu(II) ions are generally trapped interstitially between two glycine molecules which coordinate to the central ion in trans-positions. The planar stereochemistry of the CuO_2N_2 chromophore results in a nearly axial crystal field symmetry as has been found from EPR and ENDOR studies [1 to 7].

In $(gly)_2CaCl_2 \cdot 4\,H_2O$ we have found, however, a strongly non-axial Cu(II) EPR spectrum with uncommon sequence of the spin-Hamiltonian parameters. It suggests a different kind of Cu(II) coordination in the crystal or a strong distortion of the $Cu(gly)_2$ stereochemistry. Detailed studies of the problem are presented in this paper.

Bis(glycine)calcium(II)dichloride tetrahydrate, $(NH_2CH_2COOH)_2CaCl_2 \cdot 4\,H_2O$, crystallises in a tetramolecular monoclinic unit cell of dimensions $a = 1.3067$, $b = 0.6818$, $c = 1.5536$ nm and $\beta = 91.40°$, in space group $P2_1/c$ [8]. There are two non-equivalent glycine molecules in the structure with a positional disorder of the nitrogen atom of one of the two glycines. The disorder, uncommon for glycine compounds, exists in the crystal because the structure is not a densely packed one and there is enough space for a glycine molecule to exist in two orientations. Calcium is coordinated to eight oxygen atoms, five from the glycine molecules and the rest from the water molecules, forming a distorted dodecahedron.

[1]) Smoluchowskiego 17, PL-60-179 Poznań, Poland.

There exists a possibility to substitute copper(II) ions on the Ca(II) site. The situation has been found in $(gly)_2BaCl_2 \cdot H_2O$ where doped Cu(II) ions were detected by EPR on barium(II) sites [9]. In our crystal, however, only spectra with ligand shfs were observed. Thus the possibility of Cu(II) coordination in CuO_8 dodecahedra can be ruled out.

2. Experimental

Single crystals of diglycine calcium(II) chloride tetrahydrate were grown by slow evaporation of a saturated aqueous solution containing stoichiometric proportions of glycine and calcium chloride with 10% of copper(II) carbonate as dopand. The crystals grow at room temperature as platelets elongated in [010] direction with well shaped {101} faces. The crystal morphology is presented in Fig. 1.

The EPR measurements were carried out using a RADIOPAN SE/X-2543 spectrometer (X-band) with TE_{102} rectangular microwave cavity, and 100 kHz modulation. Magnetic field was monitored with on automatically tracked RADIOPAN JTM-41 nuclear magnetometer, and DPPH was used as internal g-factor marker.

3. EPR Data Analysis

EPR spectra were recorded in ab, bc, and ca crystal planes. The spectrum consists of three hyperfine structure quartets with one of them of double integral intensity. The spectrum observed along the principal z-axis of the I_a–Cu(II) complex is presented in Fig. 2. The spectrum can be described in terms of a spin-Hamiltonian including Zeeman and hyperfine interactions

$$H = \beta \sum_{i=x,y,z} g_i B_i S_i + \sum_{i=x,y,z} A_i S_i I_i \tag{1}$$

in the x, y, z principal axes system which has been found to be identical for g-factor and hyperfine tensors in our crystal. Superhyperfine structure terms are omitted in (1) although a shfs splitting is observed for some crystal orientations, as can be seen on high-field hfs lines in Fig. 2. The shfs was not well resolved, however, in most orientations.

To obtain the principal values of g-factor and A-splitting the angular dependence of the spectrum was analyzed in two steps. In the first step the angular dependence was analyzed using the first-order method of Waller and Rogers [10] for rotations about three orthogonal axes. The g-values were estimated from the center of each quartet of lines while copper(II) A-splittings were determined from mean hyperfine spacings. Principal values of g were obtained as square roots of the diagonal elements

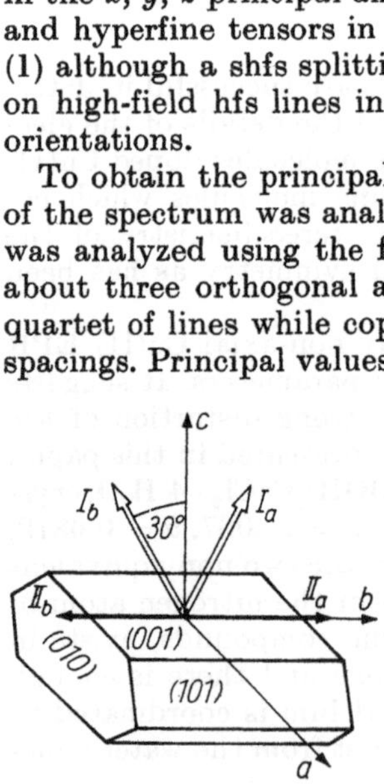

Fig. 1. The morphology of the $(gly)_2CaCl_2 \cdot 4\,H_2O$ single crystal. The z-axis directions of doped Cu(II) complexes are marked by I_a, I_b, and II_a, II_b

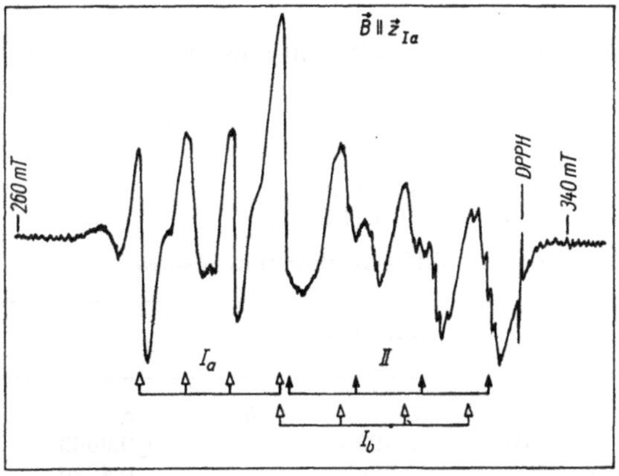

Fig. 2. X-band EPR spectrum of $(gly)_2 \cdot CaCl_2 \cdot 4\ H_2O:Cu^{2+}$ at 298 K. The arrows mark positions of hyperfine lines of Cu(II) complexes

of the g^2-tensor. The principal values of the copper hyperfine interaction were obtained by diagonalization of the tensor $\tilde{\mathbf{A}}^2 = \mathbf{g}^{-1}\mathbf{g}\mathbf{A}^2\mathbf{g}\mathbf{g}^{-1}$. The data so obtained were then used as initial parameters for computer fitting of the angular dependences to Sakaguchi et al.'s second-order equations [11]. The fitting confirmed a coincidence between principal axes of g^2-tensor and A^2-tensor and is presented by solid lines in Fig. 3.

All hfs quartets denoted by I_a, I_b, and II in Fig. 3 display the same spin-Hamiltonian parameters and differ in direction cosines only. The parameters and direction cosines are summarized in Table 1.

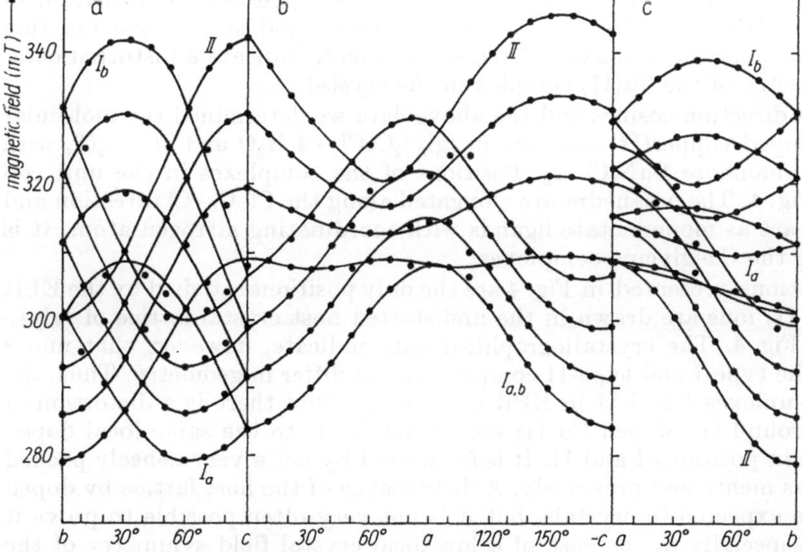

Fig. 3. Angular dependence of the EPR of $(gly)_2CaCl_2 \cdot 4\ H_2O:Cu^{2+}$ spectrum in bc, ca, and ab crystal planes

Table 1

Spin Hamiltonian parameters and direction cosines of Cu^{2+} centers in $(gly)_2CaCl_2 \cdot 4\,H_2O$ crystals

$g_z = 2.308\,(2)$	$A_z = -73.0\,(2)$*)
$g_y = 2.034\,(2)$	$A_y = 113.4\,(1.5)$
$g_x = 2.115\,(2)$	$A_x = 40.1\,(1.5)$

Centers I and II have the same parameters, with different direction cosines

	center I**)			center II**)		
	a	b	c	a	b	c
x	0.5092	±0.7669	0.3907	±0.4226	0	±0.9063
y	±0.8607	−0.45369	±0.2312	0.9063	0	0.4226
z	0	±0.4540	−0.8910	0	±1	0

*) in $10^{-4}\,cm^{-1}$

**) upper signs for I_a and II_a, and lower signs for I_b and II_b

4. Molecular Structure of Cu(II) Centers

The EPR data indicate that there exist three identical but magnetically non-equivalent copper(II) centers denoted by I_a, I_b, and II in Table 1, and in Fig. 2 and 3. The double intensity of the EPR spectrum of center II indicates that two magnetically equivalent type-II centers exist in the crystal unit cell. Experimentally determined directions of the z-axes of the centers are marked in Fig. 1. The ligand shfs splitting observed for the all centers near perpendicular complex orientations amounts to about $18(2) \times 10^{-4}\,cm^{-1}$, and the number of shfs lines suggests a coordination of Cu(II) by nitrogen ^{14}N and chlorine ^{35}Cl atoms. The ground state wave function $d_{x^2-y^2}$ of Cu(II) ions and the non-axial g^2-tensor symmetry indicate a distorted octahedral stereochemistry of the Cu(II) complex in the crystal.

Using the z-axis direction cosines and the above data we determined the molecular structure of the doped copper(II) complex in $(gly)_2CaCl_2 \cdot 4\,H_2O$ as $Cu(gly)_2Cl_4$ with an octahedral chromophore $CuN_2Cl_2Cl_2$. Positions of the complexes in the unit cell are presented in Fig. 4. The octahedra are elongated along the Cl–Cu–Cl direction and glycine molecules act as monodentate ligands with coordinating nitrogen atom. It is uncommon among the Cu–glycine complexes.

The Cu(II) positions presented in Fig. 4 are the only positions satisfied by the EPR data. The copper(II) ions are drawn in the undistorted host crystal lattice of $(gly)_2 \cdot CaCl_2 \cdot 4\,H_2O$ in Fig. 4. The crystallographical data indicate, however, that under these conditions the type-I and type-II complex should differ in geometry. Thus, the identity of the complexes I and II in EPR data proves that there is a distortion of the host lattice around the doped Cu(II) ions which leads to the same local doped complex structure in positions I and II. It is facilitated by not a very densely packed crystal structure as mentioned previously. A disturbance of the host lattice by doped Cu(II) ions can be expected in crystals, but it is not very often possible to prove it from EPR data, especially in the case of a low local crystal field symmetry of the doped complex. The distortion has been observed by EPR, however, in aquatris (L-glutamato) cadmium monohydrate [12], and in bis(glycinate)cadmium monohydrate [13] where Cu(II) ions are substituted on Cd(II) sites.

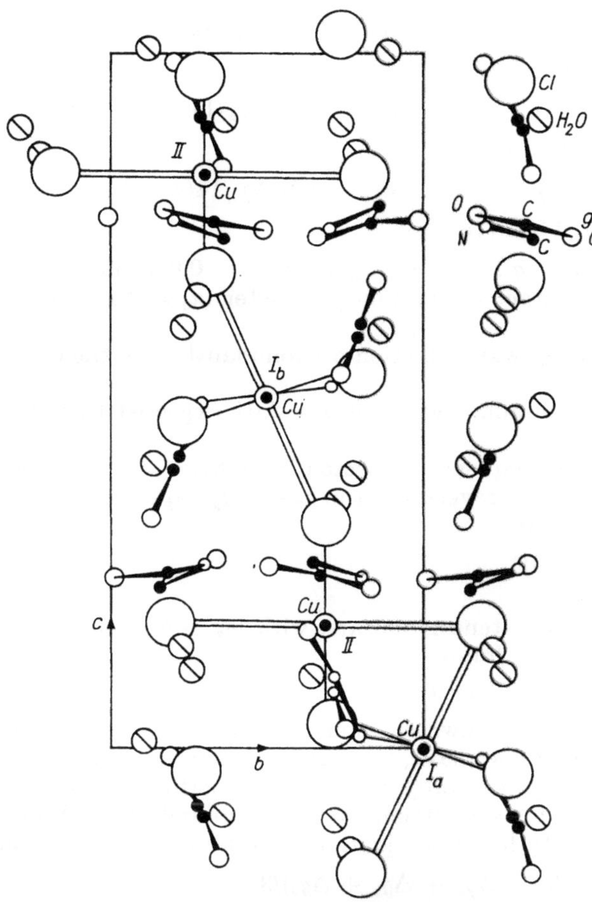

Fig. 4. The unit cell projection on bc crystal plane with positions of doped Cu(II) ions in undisturbed molecular structure. The double-solid lines represent the EPR determined *z*-axis directions

5. Spin-Hamiltonian Parameters

EPR data indicate non-axial crystal field symmetry of the Cu(II) complexes. It is impossible, however, to decide what is the exact point symmetry of the crystal field about Cu(II). We can suggest that there is a rhombic symmetry D_{2h} since the most common indication of a lower-symmetry Cu(II) site, i.e. monoclinic or triclinic, is a non-coincidence between the extrema of the hyperfine splitting and the principal *g*-tensor axes [14].

The ground state wave function of Cu(II) in D_{2h} symmetry can be written in the form

$$\psi = \alpha[a d_{x^2-y^2} - b d_{z^2}]\,, \tag{2}$$

where $a^2 + b^2 = 1$, and α^2 is a covalency parameter describing the density of unpaired etectrons on the Cu(II) ion. Following McGarvey's procedure [15] one can find a relationship between the principal values of the spin-Hamiltonian tensors (1) and the ground state wave function parameters:

$$\begin{aligned} g_x &= g_e - 2\alpha^2\lambda(a + b\sqrt{3})^2/E_{yz}\,, \\ g_y &= g_e - 2\alpha^2\lambda(a - b\sqrt{3})^2/E_{xz}\,, \\ g_z &= g_e - 8\alpha^2\lambda a^2/E_{xy}\,, \end{aligned} \tag{3a}$$

$$A_x/P_0 = -\alpha^2\varkappa + \tfrac{2}{7}\alpha^2(a^2 - b^2) + \Delta g_x - (3a + b\sqrt{3})\,\Delta g_y/14(a - b\sqrt{3}) + + b\sqrt{3}\,\Delta g_z/14a - 4\sqrt{3}\,\alpha^2 ab/7\,,$$
$$A_y/P_0 = -\alpha^2\varkappa + \tfrac{2}{7}\alpha^2(a^2 - b^2) + \Delta g_y - (3a - b\sqrt{3})\,\Delta g_x/14(a + b\sqrt{3}) - - b\sqrt{3}\,\Delta g_z/14a + 4\sqrt{3}\,\alpha^2 ab/7\,, \quad (3\text{b})$$
$$A_z/P_0 = -\alpha^2\varkappa - \tfrac{4}{7}\alpha^2(a^2 - b^2) + \Delta g_z + (3a + b\sqrt{3})\,\Delta g_y/14(a - b\sqrt{3}) + + (3a - b\sqrt{3})\,\Delta g_x/14(a + b\sqrt{3})\,,$$

where $g_e = 2.0023$, and $\Delta g_i = g_i - g_e$. $P_0 = g_e\beta_e g_N\beta_N\langle r^{-3}\rangle = 0.036\ \text{cm}^{-1}$, $\lambda = -829\ \text{cm}^{-1}$, and $\varkappa$ is the Fermi contact interaction parameter ($\varkappa_0 = 0.43$ for the free Cu(II) ion).

A symmetry allowed mixing of the d_{z^2} wave function predominantly into the $d_{x^2-y^2}$ ground state results in three effects.

(i) The g_y-value is lower than the minimal value of $g \approx 2.04$ expected from the orbital energy E_{yz}.

(ii) A reversal of the normal $g_y > g_x$ sequence, expected from $E_{xz} < E_{yz}$, appears. As a result we have $g_z > g_x > g_y$, and the reversal of the g_x and g_y sequence appears for a very small mixing coefficient, when

$$E_{yz}/E_{xz} = (a + b\sqrt{3})^2/(a - b\sqrt{3})^2\,.$$

It is practically fulfilled for $b \approx 0.01$.

(iii) A significant non-axiality of the g-tensor appears, with $(g_x - g_y)$ determined mainly by the mixing coefficient value as follows:

$$(g_x - g_y)/(\Delta g_x + \Delta g_y) = 2\sqrt{3}\,ab/(1 + 2b^2)\,.$$

The orbital splittings E_{ij} in (3a) are unknown and must be treated as parameters in calculations. Together with α^2, a, and $\varkappa$ we have, thus, the six unknowns in (3). We do not know, moreover, the signs of experimentally determined hyperfine parameters, and the only $A_z < 0$ can be assumed for a predominant $d_{x^2-y^2}$ ground state. Equations (3) have been solved as follows. Addition of equations (3b) leads to the relation

$$\alpha^2\varkappa = -(A_x + A_y + A_z)/3P_0 + (\Delta g_x + \Delta g_y + \Delta g_z)/3\,.$$

Thus $\varkappa$ can be eliminated from (3b), and an iterative procedure leads to the α^2 and a^2 values. The calculations have been performed for all possible sign combinations of A_x, A_y and A_z, but the only $A_z = -73.0$, $A_y = +113.4$, and $A_z = +40.1 \times 10^{-4}\ \text{cm}^{-1}$ gave real values and reasonable orbital splittings calculated from g_x, g_y, and g_z. The resulting electronic structure parameters are summarized in Table 2.

Table 2

Electronic structure parameters for Cu(II) complexes in $(gly)_2CaCl_2 \cdot 4\,H_2O$

a	b	α^2	$\varkappa$	E_{xy}	E_{xz}	E_{yz}
0.9835	0.1810	0.86	0.087	18070	20060	21270

The g-factors are determined by 3d-orbital mixing in the ground state, the hyperfine splitting parameters, however, are affected mainly by the Fermi contact interaction parameter $\varkappa$. The low value of $\varkappa = 0.087$ (Table 2) in our crystal results in an unusual sequence of A_i-parameters, with $|A_z| < |A_y|$.

As a rule, the $\varkappa$-values are lower in crystals compared to the free ion value $\varkappa = 0.43$. The $\varkappa$-parameter values in some crystals are given in Table 3. The theoretical background of the phenomenon is still under discussion [26], although a spin polarization

Table 3

Fermi contact hyperfine splitting parameters

compound	$\varkappa$	ref.
free copper(II) ion	0.43	—
amine copper(II) complexes	0.3 to 0.4	[16]
aqua copper(II) complexes	0.25	[17]
copper(II) Tutton salts	0.2 to 0.4	[18]
α-glycine: Cu^{2+}	0.375	[19]
L-alanine: Cu^{2+}	0.360	[19]
ZnF_2: Cu^{2+}	0.37	[19, 20]
TiO_2: Cu^{2+}	0.352	[19, 21]
$(gly)_2BaCl_2 \cdot H_2O$: Cu^{2+}	0.27	[9]
$(gly)_2CaCl_2 \cdot 4\,H_2O$: Cu^{2+}	0.087	this paper
L-histidine hydrochloride $\cdot H_2O$: Cu^{2+}	0.071	[19, 22]
ND_4Cl: Cu^{2+}-centre II	0.055	[23]
NH_4Cl: Cu^{2+}	0.03	[24]
$LiAlSiO_4$: Cu^{2+}	−0.102	[25]

is generally accepted as responsible for the isotropic hyperfine splitting. The observed $\varkappa$-value is due to a small difference between large competing spin polarizations of core s-shells and the outer 4s shell of the Cu(II) ion. As a consequence the $\varkappa$-value is very sensitive to changes in the radial distribution of atomic orbitals [27]. Especially sensitive to the variation of the ligand field are radial distributions of the 3s and 4s orbitals [26]. In a low crystal field symmetry, an additional 4s admixture to the ground state can be induced as allowed by symmetry conditions. The 4s admixture — similar to the spin polarization of the 4s orbital — gives a negative contribution to the Fermi term, leading to decrease in the net $\varkappa$-value. This mechanism seems to be responsible for the very low $\varkappa$-value in our crystal.

Acknowledgements

The paper was supported by the Polish Academy of Sciences under Project MR.I.9 The authors wish to thank Prof. J. Stankowski for his interest and encouragement.

References

[1] A. Lösche and W. Windsch, phys. stat. sol. **11**, K55 (1965).
[2] W. Windsch and M. Welter, Z. Naturf. **22a**, 1 (1967).
[3] J. Stankowski, Acta phys. Polon. **33**, 387 (1968).
[4] J. Stankowski, A. Wieckowski, and S. Hedewy, J. magnetic Resonance **15**, 498 (1974).
[5] W. Windsch, M. Welter, and W. Driesel, Ferroelectrics **8**, 551 (1974).
[6] R. Böttcher, D. Heinhold, S. Wartewig, and W. Windsch, J. molecular Structure **46**, 363 (1978).
[7] R. Böttcher, D. Heinhold, and W. Windsch, Chem. Phys. Letters **65**, 452 (1979).
[8] S. Natarajan and J. K. Mohana Rao, Z. Krist. **152**, 179 (1980).
[9] V. G. Krishnan, S. G. Sathyanarayan, and G. Sivarama Sastry, J. chem. Phys. **66**, 1715 (1977).
[10] W. G. Waller and M. T. Rogers, J. magnetic Resonance **9**, 92 (1973).
[11] U. Sakaguchi, Y. Arata, and S. Fijiwara, J. magnetic Resonance **9**, 118 (1973).
[12] R. P. Bonomo, J. R. Pilbrow, and G. R. Sinclair, J. Chem. Soc. Dalton Trans. 489 (1983).
[13] J. Goslar, L. Szczepaniak, and A. Dezor, Acta Phys. Polon. **A63**, 671 (1983).
[14] J. R. Pilbrow and M. R. Lowrey, Rep. Progr. Phys. **43**, 433 (1980).

[15] B. R. McGarvey, in: Transition Metal Chemistry, Vol. 3, Ed. R. L. Carlin, Dekhor, New York 1966 (p. 160).
[16] Teh-Ching Chiang, J. chem. Phys. **48**, 1814 (1968).
[17] A. Abragam and M. H. L. Pryce, Proc. Roy. Soc. **A205**, 135 (1951).
[18] T. Bhaskar Rao and M. Narayana, phys. stat. sol. (b) **106**, 601 (1981).
[19] B. N. Misra and R. Kripal, Chem. Phys. **19**, 17 (1977).
[20] J. D. Swallen, B. Johnson, and H. M. Gladney, J. chem. Phys. **52**, 4078 (1970).
[21] T. C. Ensign, Te-Tse Chang, and A. H. Kahn, Phys. Rev. **188**, 703 (1969).
[22] R. Hirasawa and H. Kon, J. chem. Phys. **56**, 4467 (1972).
[23] S. H. Hagen and N. J. Trappeniers, Physica (Utrecht) **66**, 166 (1973).
[24] N. Kuroda and A. Kawamori, J. Phys. Chem. Solids **32**, 1233 (1971).
[25] J. L. Berchot, D. Vivien, D. Gourier, J. Thery, and R. Collongues, J. Solid State Chem. **34**, 199 (1980).
[26] A. Rockenbauer, J. magnetic Resonance **35**, 429 (1979).
[27] R. E. Watson and A. J. Freeman, Phys. Rev. **120**, 1134 (1960); **123**, 2027 (1961).

(Received June 21, 1985)

phys. stat. sol. (b) **133**, 329 (1986)

Subject classification: 71.45 and 73.20; S2

National Centre for Scientific Research of Vietnam, Ho-Chi-Minh-City[1])

Static Polarizability and Dipolar Surface Plasmon of Small Metal Particles

II. Including the Lattice Structure[2])

By

TRAN THOAI DUY BAO

The frequency of the dipolar surface of small particles is calculated using the sum rule techniques due to Lushnikov et al. The effect of surface diffuseness of electron density is included. The screening of the electrons is described within the Thomas-Fermi approximation. The lattice structure and the lattice relaxation through the introduction of pseudopotentials are taken into account. The ground state charge density is obtained from a variational estimate of the total energy of the clusters based on the local density functional formalism. The surface plasmon frequency is red-shifted from its classical value.

Die Frequenz von Dipol-Oberflächenplasmonen kleiner Metallcluster wird unter Verwendung der Summenregel von Lushnikov et al. berechnet. Der Einfluß der Verbreiterung der Elektronendichte an der Oberfläche wird berücksichtigt. Die Abschirmung der Elektronen wird im Rahmen der Thomas-Fermi-Näherung beschrieben. Der Einfluß der Gitterstruktur und der Gitterrelaxation wird durch Einführung von Pseudopotentialen berücksichtigt. Die Ladungsdichte im Grundzustand wird durch die Variation der Gesamtenergie der Cluster im Formalismus des lokalen Dichtefunktionals bestimmt. Es wird eine Rotverschiebung der Frequenz des Oberflächenplasmons erhalten.

1. Introduction

The study of collective excitations in metal clusters has a rather long history. Works have been focused on the frequency and the lifetime of the dipolar surface plasmon. Theoretical results of the lifetime [1 to 3] showed some disagreements with experimental data for silver by Schulze and Abe [4]. In a recent work [5] we found that the corrected RPA result of the lifetime by Lushnikov et al. [2] is in quantitative agreement with data by Schulze and Abe.

However, the frequency of the dipolar surface plasmon is still a subject of controversial discussion. Using the quantum-mechanic dielectric function in the infinite barrier model, many authors predict a blue shift [3, 6] of the frequency of the surface plasmon compared to its classical value $\omega_s = \omega_p/\sqrt{3}$ with $\omega_p^2 = 4\pi e^2 n/m$. The blue shift is also obtained from the semi-classical calculations using the Thomas-Fermi approximation with step-density [7] or the specular reflection model [8]. Taking into account roughly the surface diffuseness, other authors have obtained a red shift [9, 10] by increasing the diffuseness parameter. What we need is a detailed knowledge of the electronic density profile near the surface of the cluster. For this point the infinite barrier model is oversimplifying with respect to the properties of a real surface. Using a simple model based on the local density formalism, Cini [11] has calculated

---

[1]) 1 Mac Dinh Chi, Ho-Chi-Minh-City, Vietnam.

[2]) Part I see phys. stat. sol. (b) **128**, 731 (1985).

the ionization potential and the electron affinity of small metal clusters in strict analogy to Smith's work on the theory of work functions of plane metal surfaces [12]. In this type of calculation one avoids automatically a super-artificial surface barrier because the electron density profile follows from minimizing the surface energy of the particle. The density, in turn, determines a reasonable approximation to the surface barrier confining the electrons within the cluster. Work functions determined in this way are in reasonable agreement with experimental data [11], and we believe that this type of charge density might be useful for the problem under discussion. A similar procedure has also been used [13] for determining the ionization potential and electron affinity.

Very recently, Ekardt [14] investigated the self-consistently determined absorption spectrum of small metal particles within the local density formalism and obtained a red shift of the surface plasmon frequency. A red shift is also obtained in the local Thomas-Fermi approximation by using the variational method for determining the electron density of small clusters with jellium background (see [15]).

In this work, we use the variational spherical model [15, 16] in the Thomas-Fermi approximation for studying the frequency shift by taking into account the effects of the lattice structure and of the relaxation of the lattice. We apply this model to Na clusters having b.c.c. structure and density parameter $r_s = 4$ with $1/\bar{n} = (4\pi/3)\, r_s^3$ ($\bar{n}$ average density).

2. Static Polarizability

The starting point of our method is a variational determination of the electronic charge density $n(r)$ in the ground state. In strict analogy to the original work by Smith [12] we adopt for this procedure the following family of trial densities:

$$n(r) = \frac{\bar{n}}{1 + 6/(\beta R)^2 + 3/(\beta R)^3\, e^{-\beta R}} \{[1 - 0.5\, e^{\beta(r-R)}]\, r\, \theta(R - r) + + 0.5\, e^{-\beta(r-R)}\, \theta(r - R)\}\,. \tag{1}$$

Here, R is the cluster radius, $\bar{n}$ the average density, $\theta(x)$ the unit step, and β is to be determined by minimizing the total energy $E(n)$. The prefactor in (1) assures correct normalization. We use for the exchange and correlation energy the RPA expression [17]. In [16] we obtained almost the same results for the polarizability and the surface plasmon frequency by performing the variational calculation with the Wigner and the RPA expressions.

In this model the positive ions are represented by Ashcroft pseudopotentials

$$V_{ps}(r) = \begin{cases} \dfrac{-Z}{r} & \text{for} \quad r > r_c\,, \\ 0 & \text{for} \quad r < r_c \end{cases}$$

where r_c is the core radius and Z is the valency of the ion core. For Na clusters we take $r_c = 1.6$ at. units given by Ashcroft [18].

Supposing that the clusters have a central atom we treat the pseudopotentials of the neighboring atoms with the procedure of Loucks [22], i.e. the pseudopotentials are expanded in spherical harmonics about the origin of the central atom. An approximation equivalent to this procedure was made by Lang and John for plane surface [19].

By adding the electrostatic interaction energy between the ions, and between the ions and the electronic charges to the electronic energy, we get the total energy of the cluster. We minimize this energy by varying both the lattice constant a and the density parameter β.

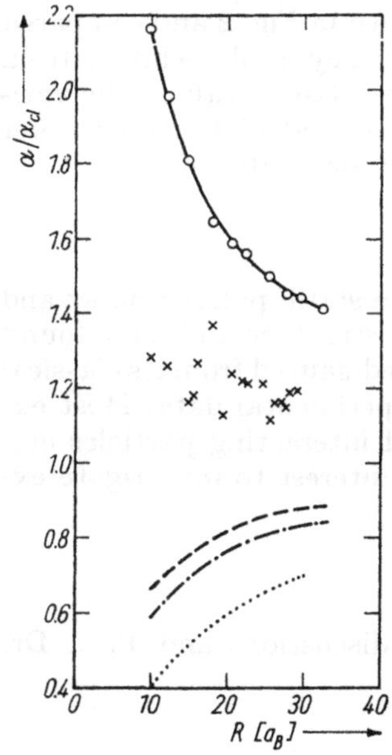

Fig. 1. Static polarizability of Na clusters ($r_s = 4$) in units of their classical value R^3. Crosses results in pseudopotentials; continuous line results in jellium model; dashed line step-density Thomas-Fermi; dash-dotted line specular reflection model [8]; dotted line quantum infinite barrier model [6]

Having obtained a reasonable ground state density, we solve next the Poisson equation in the Thomas-Fermi approximation:

$$\Delta V = k_{\mathrm{TF}}^2[n(r)]\, V\,, \tag{2}$$

$$k_{\mathrm{TF}}^2[n(r)] = 4\left[\frac{3n(r)}{\pi}\right]^{1/3}. \tag{3}$$

From the asymptotic solution of this equation for the dipole fields

$$V(r \to \infty) \to \cos\theta \left[r - R^3\alpha_{\mathrm{D}}(R, r_s)\,\frac{1}{r^2}\right] \tag{4}$$

the dipolar polarizability

$$\alpha = \alpha_{\mathrm{D}}(R, r_s)\, R^3 \tag{5}$$

is obtained. Here, $\alpha_{\mathrm{D}}(R, r_s)$ gives the correction factor in the Thomas-Fermi approximation with surface diffuseness to the polarizability R^3 of a classical metal particle.

The results for Na clusters with $r_s = 4$ are shown in Fig. 1 and compared with various calculations [6, 8]. There are two competing effects which determine the behavior of the static polarizability. First, the quantum size effect due to reduction of cluster radii also causes a reduction of the static polarizability. Second, the surface diffuseness gives rise to an enlarged total dipole moment of the particle. In the jellium model, our previous calculations [15] have shown a strong enhancement of the static polarizability for small particles (see also continuous line in Fig. 1). Taking into account the lattice structure and the lattice relaxation, the surface diffuseness will be reduced and results in a slight enhancement of the static polarizability for small particles. Because of shell structure we have got small oscillations of the static polarizability versus the particle radius.

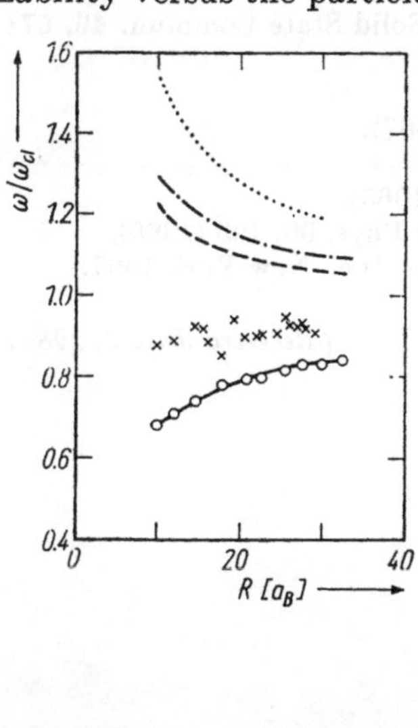

3. Surface Plasmon Energy

Using the sum rule techniques due to Lushnikov et al. [2, 20, 21] we obtained the following expression [16] for the collective frequency of the surface plasmon of the particles:

$$\omega_s = \frac{\omega_p/\sqrt{3}}{\sqrt{\alpha_{\mathrm{D}}(R, r_s)}}\,, \tag{6}$$

Fig. 2. Surface plasmon frequency in units of its classical value. The surface plasmon frequency is obtained by using (6). See also Fig. 1

where $\alpha_D(R, r_s)$ was defined in Section 2. The results are shown in Fig. 2 and compared with various calculations [6, 8]. The surface plasmon frequency is blue-shifted from its value in the jellium model [15], but red-shifted from its classical value. The variation of the surface frequency versus the particle radius in our calculations is weaker than in other calculations. This variation is still small even for small radii.

4. Conclusion

We have studied the effects of the lattice structure on the static polarizability and the surface frequency in the case of Na clusters with b.c.c. structure and have found that the average trend of the surface plasmon frequency is red-shifted from its classical value. This result allows a direct comparison with the experimental data. Most experiments showing a blue shift are referring to samples of interacting particles embedded in a matrix [3, 4]. Therefore it would be of great interest to investigate experimentally clusters of the light metals in the gas phase.

Acknowledgements

The author wishes to thank Dr. W. Ekardt for fruitful discussions and Prof. Dr. E. Zeitler for his continuing interest and support.

References

[1] A. KAWABATA and R. KUBO, J. Phys. Soc. Japan **21**, 1765 (1966).
[2] A. A. LUSHNIKOV and A. J. SIMONOV, Z. Phys. **270**, 17 (1974).
[3] L. GENZEL, T. P. MARTIN, and U. KREIBIG, Z. Phys. **B21**, 339 (1975).
[4] W. SCHULZE and H. ABE, Faraday Disc. Royal Soc. Chem. **14**, 87 (1980).
[5] D. B. TRAN THOAI and W. EKARDT, Solid State Commun. **41**, 687 (1982).
[6] M. J. RICE, W. R. SCHNEIDER, and S. STRAESSER, Phys. Rev. B **8**, 473 (1973).
[7] M. CINI, Surface Sci. **62**, 148 (1977).
[8] B. B. DASGUPTA and R. FUCHS, Phys. Rev. B **24**, 554 (1981).
[9] P. ASCARELLI and M. CINI, Solid State Commun. **18**, 385 (1967).
[10] R. RUPPIN, J. Opt. Soc. Amer. **55**, 449 (1976).
[11] M. CINI, J. Catalysis **37**, 187 (1975).
[12] J. R. SMITH, Phys. Rev. **181**, 522 (1969).
[13] J. L. MARTINS, R. CAR, and J. BUTTET, Surface Sci. **106**, 265 (1981).
[14] W. EKARDT, Phys. Rev. Letters **52**, 1925 (1984).
[15] TRAN THOAI DUYBAO, phys. stat. sol. (b) **128**, 731 (1985).
[16] W. EKARDT, D. B. TRAN THOAI, F. FRANK, and W. SCHULZE, Solid State Commun. **46**, 571 (1983).
[17] N. D. LANG, Solid State Phys. **28**, 225 (1973).
[18] N. W. ASHCROFT and D. C. LANGRETH, Phys. Rev. **155**, 682 (1967).
[19] N. D. LANG and W. KOHN, Phys. Rev. B **1**, 4555 (1970).
[20] A. A. LUSHNIKOV and D. F. ZARETSKY, Nuclear Phys. **66**, 35 (1965).
[21] A. B. MIGDAL, A. A. LUSHNIKOV, and D. F. ZARETSKY, Nuclear Phys. **66**, 193 (1965).
[22] T. L. LOUKS, Augmented Plane Wave Method, W. A. Benjamin, Inc., New York 1967.

(Received July 2, 1985)

phys. stat. sol. (b) **133**, 333 (1986)

Subject classification: 78.55; 71.55; 78.50; S8

Institut de la Science Electronique, Université des Sciences d'Okayama[1]) *(a) et Institut de Physique, Faculté des Sciences, Université d'Okayama*[2]) *(b)*

Luminescence et transition non radiative $^3T_{1u} \rightarrow {}^3A_{1u}$ dans le luminophore BaS:Bi^{3+}

Par

S. ASANO (a), Y. NAKAO (b), N. YAMASHITA (b) et I. MATSUYAMA (b)

Emission and excitation spectra of BaS:Bi^{3+} phosphors are measured at 300, 80, and 6 K. In the excitation spectra the A and C excitation bands are observed. In the emission spectra only one band is observed which is assigned to the transition $^3A_{1u} \rightarrow {}^1A_{1g}$ on the basis of the measurement of the time-resolved emission spectra and the decay constant of the phosphorescence. The diffused reflection spectra and the magnetic effect on the phonon structure of the emission band are also measured at 80 and 4.2 K, respectively. The former is compared with the excitation spectra. The reason why the emission band $^3T_{1u} \rightarrow {}^1A_{1g}$ cannot be observed at low temperatures is explained by the fact that the probability of the tunnel effect $^3T_{1u} \rightarrow {}^3A_{1u}$ is larger than that of the luminescent transition $^3T_{1u} \rightarrow {}^1A_{1g}$.

Измерены спектры излучения и возбуждения люминофоров BaS:Bi^{3+} при 300, 80 и 6 K. На спектре возбуждения наблюдаются полосы возбуждения A и C. На спектре излучения наблюдается только одна полоса, которая отожествляется с переходом $^3A_{1u} \rightarrow {}^1A_{1g}$ на основе измерения спектра фосфоресценции по временному разложению и постоянной затухания фосфоресценции. Помимо того, спектр диффузного отражения и магнитный эффект на фононную структуру полосы излучения измеряются при 80 и 4,2 K соответственно. Первый сравнен со спектром возбуждения. Та причина, по которой полоса излучения $^3T_{1u}$ — $^3A_{1u}$ не может наблюдаться при низких температурах, объясняется выводом, что вероятность туннельного эффекта $^3T_{1u}$ — $^3A_{1u}$ больше чем та же луминесцирующего перехода $^3T_{1u}$ — $^1A_{1g}$.

1. Introduction

Les chalcogénures alcalino-terreux activés à l'ion Bi^{3+} ou Pb^{2+} montrent les trois bandes d'excitation A, B et C pareillement aux halogénures alcalins activés à l'ion du type Tl^+. Ils montrent d'ailleurs les deux bandes d'émission I et II qu'on peut observer séparément et qui correspondent respectivement aux transitions électroniques $^3T_{1u} \rightarrow$ $\rightarrow {}^1A_{1g}$ et $^3A_{1u} \rightarrow {}^1A_{1g}$ et il apparaît sur elles la structure vibrationnelle à basse température [1 à 4]. Cela les distingue des halogénures alcalins activés à l'ion du type Tl^+, dont les bandes d'émission I et II se chevauchent entièrement l'une sur l'autre et ne montrent aucune structure vibrationnelle même à basse température.

On a déjà étudié la luminescence des luminophores MgS, CaS et SrS activés à l'ion Bi^{3+} et l'on s'est aperçu que le rapport de l'intensité de la bande I à celle de la bande II diminue graduellement dans le passage de MgS à SrS [1, 4]. Par suite il est intéressant d'étudier la luminescence du luminophore BaS:Bi^{3+} dont on ne peut trouver qu'un petit nombre d'investigations publiées.

On traite dans cet article des spectres d'émission, d'excitation et d'absorption diffuse du luminophore BaS:Bi^{3+}, de la structure vibrationnelle sur ses bandes d'émission et

[1]) Ridai-cho, Okayama 700, Japan.

[2]) Tsushima, Okayama 700, Japan.

d'excitation, de l'effet magnétique sur la bande II et de la constante de déclin de la phosphorescence de la bande II.

En plus, en se basant sur la forme calculée des surfaces d'énergie potentielle adiabatique des états électroniques $^3T_{1u}$ et $^3A_{1u}$ et les résultats d'expérience obtenus, on discutera le mécanisme de la transition non radiative $^3T_{1u} \rightarrow {}^3A_{1u}$ qui exerce une grande influence sur les intensités relatives des bandes d'émission I et II.

2. Préparation des échantillons et appareils d'expérience

On a préparé le cristal hôte en poudre BaS en utilisant de l'eau triplement distillée et des produits chimiques avec la pureté du réactif garantie.

On a d'abord précipité du carbonate de barium en mélangeant des solutions aqueuses de NH_4HCO_3 et de nitrate de barium recristallisé. On a fait sécher à 200 °C le dépôt obtenu après le décantage répété à plusieurs reprises. On a sulfuré du carbonate de barium ainsi obtenu en chauffant à 800 °C pendant 1 h dans un courant de H_2S,

$$BaCO_3 + H_2S \rightarrow BaS + CO_2 + H_2O\,. \tag{2.1}$$

Le poids du produit a été réduit justement à 85,85% de la matière partante en accord avec la diminution de poids correspondant à la conversion de $BaCO_3$ en BaS.

Pour préparer le luminophore BaS:Bi^{3+} on a calciné à 1000 °C pendant 1 h dans un courant de $N_2 + H_2S$ un mélange de cristal hôte décrit plus haut et de Bi_2S_3 en très faible teneur. La diffraction aux rayons X a montré que le luminophore ainsi achevé se cristallisait bien en structure cubique du type sel gemme et qu'il n'y avait pas de trace décelable d'oxyde et de carbonate de barium.

Les appareils spectroscopiques utilisés pour la mesure sous l'excitation continue ont été décrits dans un article précédent [3]. Pour mesurer la constante de déclin de la phosphorescence et le spectre de phosphorescence décomposée au ralenti nous nous sommes servis d'un intégrateur boxcar digital (Modèle BX-531, NF (Japon)) et de trois sortes de sources lumineuses fournissant du rayonnement d'excitation discontinu. On en verra la description détaillée dans un article précédent [4].

3. Spectres d'émission et d'excitation

La Fig. 1 présente le spectre d'émission qu'on obtient à 300 et à 80 K en excitant la bande d'excitation A du luminophore BaS:Bi^{3+} et le spectre d'excitation correspondant au pic du spectre d'émission. Le pic d'émission (561 nm) à 300 K se déplace légèrement vers de courtes longueurs d'onde avec l'abaissement de température. Sur le spectre d'excitation on voit de façon manifeste les bandes d'excitation A ($^1A_{1g} \rightarrow {}^3T_{1u}$) et C ($^1A_{1g} \rightarrow {}^1T_{1u}$) qui sont situées respectivement à 448 et à 370 nm. Mais on ne peut pas nettement observer la bande d'excitation B ($^1A_{1g} \rightarrow {}^3E_u$, $^3T_{2u}$), car elle est située tout près de la bande C à cause de l'interaction spin–orbite assez importante de l'ion Bi^{3+} [5] et elle est par conséquent cachée sous l'aile à énergie élevée de la bande C beaucoup plus intense. On obtient un spectre d'émission presque similaire à celui montré dans la Fig. 1, si on excite la bande C.

Il est à noter qu'on ne peut observer à 80 K aucune structure vibrationnelle sur les bandes d'émission et d'excitation, ce qui contraste avec les luminophores MgS, CaS et SrS activés à l'ion Bi^{3+}, qui montrent des structures vibrationnelles assez nettes à 80 K [1, 4]. On aperçoit d'ailleurs l'allure un peu raboteuse du spectre d'excitation dans le domaine à énergie plus élevée que la bande C.

Pour élucider l'allure raboteuse mentionnée ci-dessus on mesure à 80 K le spectre d'absorption diffuse (SAD) du luminophore BaS ne contenant pas d'activateur. La

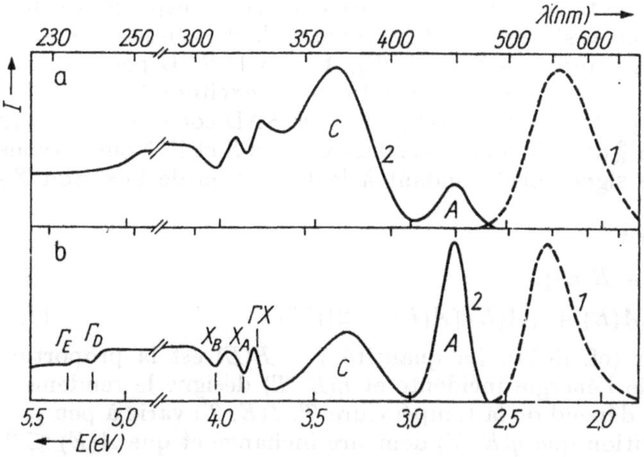

Fig. 1. Spectres d'émission (1) et d'excitation (2) a) à 300 et b) à 80 K pour le luminophore BaS:Bi_2S_3 (0,1 mol%) ne contenant pas de fondant. Longueur d'onde de l'excitation 450 nm (bande A), longueur d'onde de l'observation 545 nm

Fig. 2 montre le spectre d'absorption diffuse

$$A(E) = \frac{[1 - R(E)]^2}{2R(E)} \tag{3.1}$$

en fonction de l'énergie des photons E, $R(E)$ désignant la réflectivité spectrale d'une couche épaisse de cristaux en poudre BaS, mesurée à l'énergie E, relativement au MgO en poudre étalon, au moyen de l'appareillage décrit dans [6]. A l'aide de la formule d'Antonov-Romanovskii [7] nous avons déjà attesté que les pics et les échancrures sur le SAD dans le domaine de l'absorption fondamentale correspondent respectivement aux pics du coefficient d'absorption et aux pics de la réflectivité et que le soulèvement ΓX qui se produit aux environs de la limite d'absorption n'est pas dû à l'absorption par un défaut de réseau ou une impureté mais à la queue de l'absorption fondamentale [8]. Le spectre d'absorption mesuré sur BaS en couche mince [9] (cf. la courbe 2) et le spectre de réflexion mesuré sur BaS monocristallin [10] (cf. la courbe 3)

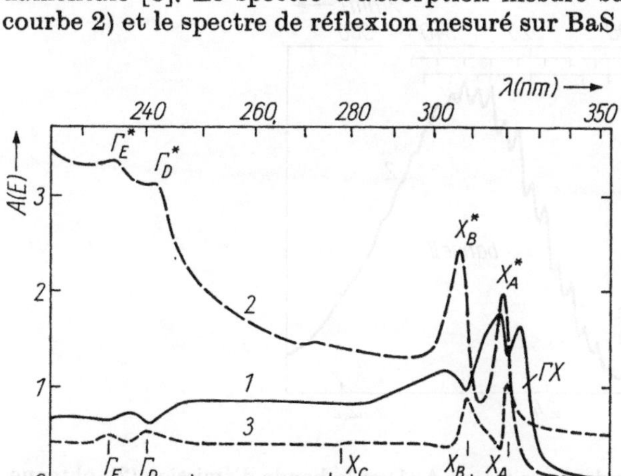

Fig. 2. (1) Spectre d'absorption diffuse $A(E)$ du cristal BaS en poudre à 80 K; (2) le spectre d'absorption $k(E)$ du cristal BaS en couche mince à 113 K [9] et (3) le spectre de réflexion $r(E)$ du monocristal BaS à 2 K [10] sont également reproduits à titre de comparaison. Les ordonnées des courbes (2) et (3) sont portées sur une échelle linéaire à unité arbitraire

sont aussi reproduits sur la même figure à titre d'indication. La correspondance susdite entre les trois sortes de spectres s'avère valide, comme on le voit ci-dessous, aussi pour le cristal BaS. Les quatre échancrures X_A, X_B; Γ_D, Γ_E sur le SAD peuvent être attribuées aux pics de réflexion se référant aux formations des excitons X ($X_5' \to X_3$) et des excitons Γ ($\Gamma_{15} \to \Gamma_1$) [8]. D'autre part, les pics sur le SAD correspondent aux pics d'absorption X_A^*, X_B^*; Γ_D^*, Γ_E^* dus à la formation des excitons X et Γ. Nous n'avons pas pu trouver sur le SAD de signe correspondant à la formation de l'exciton X_C ($X_4' \to X_3$).

Le spectre d'excitation s'éctrit

$$I(E, T) = \eta(E, T)\,[1 - R(E)]$$
$$= \eta(E, T)\,[-A(E) + \{A(E)\,[A(E) + 2]\}^{1/2}] \qquad (3.2)$$

pour un échantillon particulier (cf. (3.1)). La quantité $1 - R(E)$ est la proportion absorbée par le luminophore de l'énergie incidente et $\eta(E, T)$ désigne le rendement quantique de luminescence qui dépend de la température T. $I(E, T)$ varie à peu près linéairement avec $A(E)$ à condition que $\eta(E, T)$ demeure inchangé et que $A(E) \lesssim 3$. En comparant le spectre d'excitation de la Fig. 1 avec le SAD de la Fig. 2, on peut présumer que les échancrures, X_A, X_B; Γ_D, Γ_E sur le spectre d'excitation sont imputables à la formation des excitons X et Γ. Le soulèvement ΓX sur le spectre d'excitation correspond au soulèvement ΓX sur le SAD qui est dû à la queue de l'absorption fondamentale. La queue est partiellement attribuée, comme on le sait, à la transition indirecte de bande à bande $\Gamma_{15} \to X_3$ [8]. Le fait que les spectres d'excitation et d'absorption diffuse ont des allures semblables dans le domaine de l'absorption fondamentale veut dire que le rendement quantique de luminescence est à peu près invariant dans ce domaine et que l'énergie d'excitation est efficacement transférée jusqu'à l'ion Bi^{3+} non excité par la migration de porteurs ou la résonance (cf. (3.2)).

4. Structure vibrationnelle et son effet magnétique

On n'observe la structure vibrationnelle sur la bande d'excitation A et la bande d'émission que quand on refroidit l'échantillon jusqu'à 6 K. La Fig. 3 présente la structure vibrationnelle obtenue à 6 K sur la bande A et la bande d'émission. On prend la raie

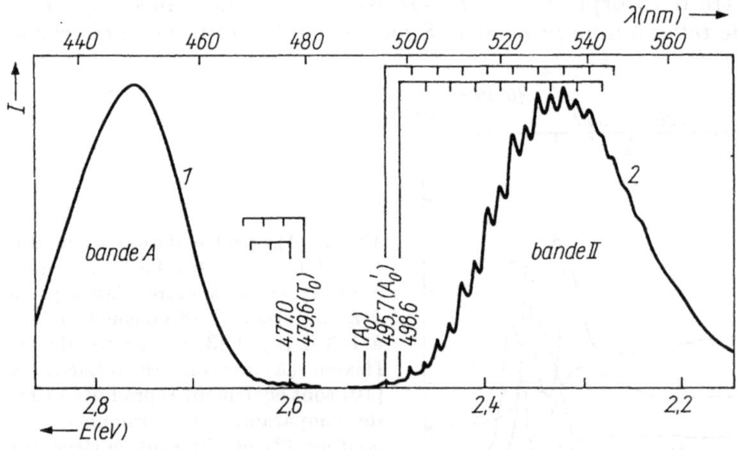

Fig. 3. Structure vibrationnelle sur la bande d'excitation A (1) et la bande d'émission (2), obtenue à 6 K pour le luminophore $BaS{:}Bi_2S_3$ (0,08 mol%). Longueur d'onde d'observation 545 nm, longueur d'onde d'excitation 450 nm (bande A)

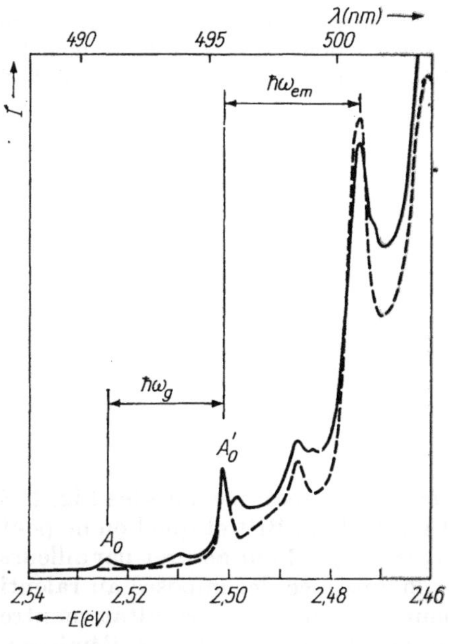

Fig. 4. La queue à énergie élevée de la bande d'émission en l'absence (courbe discontinue) et en présence (courbe continue) de champ magnétique externe ($H = 0.98$ T) à 4,2 K pour le luminophore BaS:Bi_2S_3 (0,08 mol%). Longueur d'onde d'excitation 450 nm (bande A)

vibronique 479,6 nm sur la bande d'excitation A pours a raie à zéro phonon T_0, car un dépouillement plus serré décèle qu'il n'y a pas de raie vibronique ayant l'intervalle énergétique 167 cm^{-1} au-dessus de 479,6 nm. Pour le luminophore BaS:Bi^{3+} la structure vibrationnelle se montre remarquablement affaiblie même à 6 K. Par suite on n'a pas pu observer la raie à zéro-phonon A_0 de la bande d'émission II en l'absence du champ magnétique externe, qui doit être provoquée par l'interaction hyperfine entre le spin des électrons de valence et le spin nucléaire [11, 12]. Pour observer la raie A_0 nous avons utilisé le fait qu'elle croît rapidement avec l'augmentation du champ magnétique externe H [2, 4].

La Fig. 4 présente la queue à énergie élevée de la bande d'émission en l'absence et en présence du champ magnétique externe. On voit que l'intensité de la raie vibronique 491,2 nm est sensiblement renforcée par le champ magnétique. Par suite on choisit cette raie comme raie à zéro-phonon A_0, et la raie vibronique 495,7 nm comparativement intense et qui apparaît aussi dans la Fig. 3 comme raie à un phonon A_0'. Celle-ci provient du champ magnétique induit par la quasi-rotation du complexe octaédrique $Bi^{3+}(S^{2-})_6$ et son intensité est insensible au champ magnétique externe [13].

La Fig. 5 présente la dépendance de l'intensité de la raie A_0 en fonction du champ magnétique externe H pour BaS:Bi^{3+} ainsi que pour les autres chalcogénures alcalino-terreux activés à l'ion Bi^{3+} [4, 14]. On a [2, 14]

$$\frac{I(A_0) - I_0(A_0)}{I(A_0')} = \eta H^2 , \tag{4.1}$$

$I(A_0)$ et $I_0(A_0)$ désignant respectivement les intensités de la raie A_0 en présence et en l'absence du champ magnétique externe H et $I(A_0')$ l'intensité de la raie A_0'. Ici on a prit en pratique pour intensité d'une raie vibronique la hauteur de la raie sur le fond continu de la bande d'émission. En se référant à la Fig. 5 on voit que la formule (4.1) est aussi valable pour le luminophore BaS:Bi^{3+}.

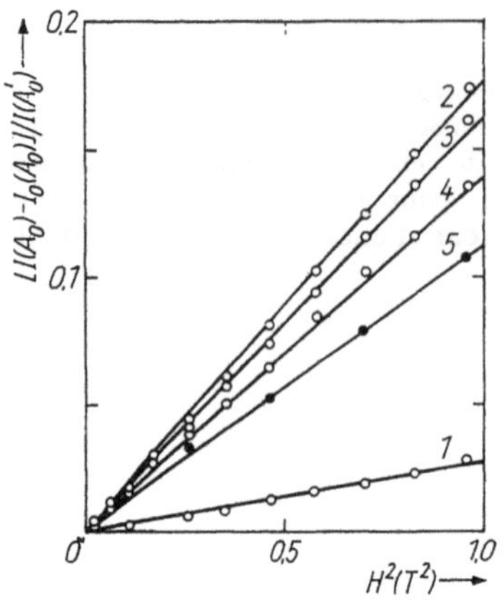

Fig. 5. Variation du rapport $[I(A_0) - I_0(A_0)]/I(A_0')$ avec l'intensité du champ magnétique externe à 4,2 K pour les luminophores (1) CaO, (2) MgS, (3) CaS, (4) SrS et (5) BaS activés à l'ion Bi^{3+}

Ce qui est dit plus haut suggère que la bande d'émission présentée dans les Fig. 1, 3 et 4 correspond à la bande d'émission II ($^3A_{1u} \rightarrow {}^1A_{1g}$) de l'ion Bi^{3+} et que l'on ne peut absolument pas observer la bande d'émission I ($^3T_{1u} \rightarrow {}^1A_{1g}$) . Pour assurer par ailleurs cette conclusion on a mesuré les spectres de phosphorescence décomposée au ralenti correspondant à différentes durées de retardement 2 à 16 ms. Le résultat montre qu'ils sont entièrement similaires l'un à l'autre, ce qui veut dire que la bande d'émission présentée dans les Fig. 1, 3 et 4 correspond à une seule transition électronique.

Nous avons de plus mesuré la courbe de déclin de la bande d'émission en l'absence et en présence du champ magnétique externe H. La courbe a montré un déclin exponentiel. La constante de déclin déterminée à partir de la courbe de déclin à 4,2 K est de 4,4 ms pour $H = 0$ et de 4,1 ms pour $H = 0{,}98$ T. Puisque les états $^3T_{1u}$ et $^1T_{1u}$ se mélangent l'un avec l'autre par l'intermédiaire de l'interaction spin–orbite et que la bande d'émission I correspondant à la transition $^3T_{1u} \rightarrow {}^1A_{1g}$ a une constante de déclin de l'ordre de 10^{-8} s, la bande d'émission observée pour le luminophore $BaS{:}Bi^{3+}$ doit s'identifier avec la bande d'émission II correspondant à la transition $^3A_{1u} \rightarrow {}^1A_{1g}$. La diminution légère de la constante de déclin en présence du champ magnétique externe est due à ce que le mélange entre les états $^3T_{1u}$ et $^3A_{1u}$ augmente à cause du champ magnétique externe et que cela fait augmenter la probabilité de la transition $^3A_{1u} \rightarrow {}^1A_{1g}$ [13].

5. Formes des surfaces d'énergie potentielle adiabatique des états $^3T_{1u}$ et $^3A_{1u}$

Pour élucider le mécanisme de la transition $^3T_{1u} \rightarrow {}^3A_{1u}$ qui agit sur l'intensité de la bande II nous allons étudier les formes de leurs surfaces d'énergie potentielle adiabatique (SEPA). On sait qu'il existe sur les SEPA des états $^3T_{1u}$ et $^3A_{1u}$ dans le sous-espace $\varepsilon_g(Q_2, Q_3)$, trois minima respectifs X et eT correspondant aux trois distorsions tétragonales du complexe octaédrique $Bi^{3+}(S^{2-})_6$ le long des directions [100], [010] et [001]. Il existe, d'autre part, sur les SEPA susdites dans le sous-espace $\tau_{2g}(Q_4, Q_5, Q_6)$ quatre minima respectifs X et eT correspondant aux quatre distorsions trigonales le long des directions [111], [$\bar{1}1\bar{1}$], [$1\bar{1}\bar{1}$] et [$\bar{1}\bar{1}1$]. De plus, dans le cas du couplage intermédiaire les trois ou quatre minima métastables T correspondant respectivement aux distorsions tétragonales ou trigonales peuvent se réaliser sur la SEPA du sous-état supérieur de l'état $^3T_{1u}$ [15, 16].

Pour une des trois distorsions tétragonales ($Q_2 = 0$, $Q_3 = Q$), les équations séculaires se référant à la matrice d'énergie de la configuration $n$sp ont pour expression [15]

$$[y^2 + (2x + \tfrac{1}{2})\, y - 8x^2 + 2x - \tfrac{1}{2}]\,(y - 2x - \tfrac{1}{2}) = 0\,,$$
$$y^3 - 2gy - (12x^2 + 4gx + \tfrac{3}{4})\, y + (16x^2 + \tfrac{1}{2})\, g + 16x^3 + \tfrac{1}{4} = 0 \quad \text{(doublet)}\,, \tag{5.1}$$
$$[y^2 + (2x - 2g + \tfrac{1}{2})\, y + (4x - 1)\, g - 8x^2 + 2x - \tfrac{1}{2}]\,(y - 2x - \tfrac{1}{2}) = 0\,,$$

où

$$y = \frac{E'}{\zeta}\,, \qquad E' = E - Q^2\,, \qquad x = -\frac{bQ}{2\sqrt{3}\zeta} \quad \text{et} \quad g = \frac{G}{\zeta}\,. \tag{5.2}$$

On désigne par E l'énergie potentielle adiabatique mesurée par rapport au centre de gravité des trois états triplets (^3E_u $^3T_{2u}$) ($^3T_{1u}$) ($^3A_{1u}$), ζ le paramètre de l'interaction spin–orbite, G l'énergie de l'interaction d'échange et b la constante de couplage pour le mode vibrationnel ε_g. La quantité $(E - E')/\zeta$ qui exprime le rapport de l'énergie potentielle vibrationnelle au paramètre ζ s'écrit

$$\frac{E - E'}{\zeta} = \frac{(1 - \beta)\, Q^2}{\zeta} = \frac{12\zeta(1 - \beta)\, x^2}{b^2}\,, \tag{5.3}$$

$\beta = (\omega_{em}^2 - \omega_{ex}^2)/\omega_{em}^2$ désignant la différence de la courbure entre les SEPA des états excités $^3T_{1u}$, $^3A_{1u}$ et celle de l'état fondamental $^1A_{1g}$. On peut tirer de (5.2) et de (5.3)

$$\frac{E}{\zeta} = y + Ax^2; \qquad A = \frac{12\zeta(1 - \beta)}{b^2}\,. \tag{5.4}$$

On résout d'abord (5.1) par rapport à y pour des valeurs fixées de g, A et diverses valeurs de x. On peut alors calculer la valeur de E/ζ en fonction de x pour les valeurs fixées de g et de A en portant les valeurs obtenues de y dans (5.4). D'entre les neuf racines des équations (5.4) la moindre racine correspond à la SEPA de l'état $^3A_{1u}$ et les deux racines qui se trouvent immédiatement au-dessus d'elle et qui se composent d'un singlet et d'un doublet, correspondent aux SEPA de l'état $^3T_{1u}$. La Fig. 6 présente l'allure de E/ζ pour les états $^3T_{1u}$ et $^3A_{1u}$ en fonction de x, c'est-à-dire la section droite $Q_2 = 0$ des SEPA dans le sous-espace $\varepsilon_g(Q_2, Q_3)$. Pour $A \gtrsim 6$ il ne se produit pas de minimum métastable T sur la SEPA du sous-état supérieur de l'état $^3T_{1u}$ [17].

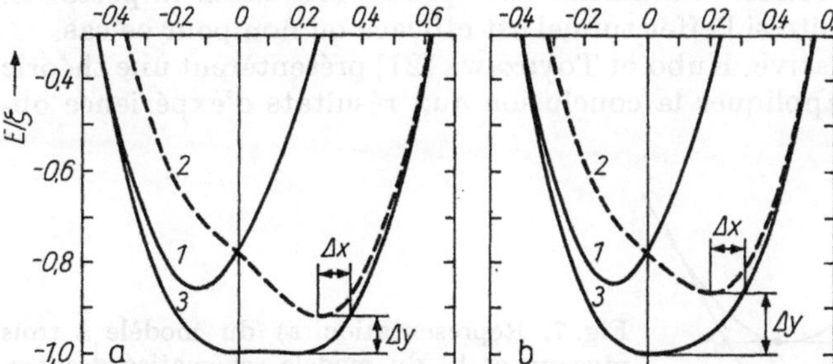

Fig. 6. Sections droites ($Q_2 = 0$) des SEPA calculées des sous-états (1) $^3T_{1u}^z$ et (2) $^3T_{1u}^{x,y}$ et de l'état (3) $^3A_{1u}$ pour a) $A = 6$, $g = 0{,}5$ et b) $A = 7$, $g = 0{,}5$. Le plus profond des deux points stationnaires qui se réalisent respectivement sur les courbes 1 et 2 est un minimum et l'autre est un point de selle ou une arête. Pour A $\geqq$ 6 la section droite de la SEPA de l'état $^3A_{1u}$ s'approche d'une courbe ayant un seul minimum totalement symétrique à $x = 0$

D'après des calculs des auteurs, on obtient les mêmes équations séculaires que les équations (5.1) pour une des quatre distorsions trigonales ($Q_4 = Q_5 = Q_6 = Q\sqrt{3}$), sauf que b dans (5.2) doit être remplacé par la constante de couplage c pour le mode vibrationnel τ_{2g}. Autrement dit, on peut aussi considérer la Fig. 6 comme représentant la section droite $Q_4 = Q_5 = Q_6$ des SEPA des états $^3T_{1u}$ et $^3A_{1u}$ dans le sous-espace $\tau_{2g}(Q_4, Q_5, Q_6)$. Alors les courbes 1, 2 et 3 correspondent respectivement aux sous-états $^3T_{1u}^{z'}$, $^3T_{1u}^{x',y'}$ et à l'état $^3A_{1u}$, où les fonctions d'onde non perturbées en base trigonale $|^3T_{1u}^i\rangle$ ($i = x', y', z'$) se relient à celles en base tétragonale $|^3T_{1u}^j\rangle$ ($j = x, y, z$) [13] par la transformation unitaire

$$\begin{bmatrix} |^3T_{1u}^{x'}\rangle \\ |^3T_{1u}^{y'}\rangle \\ |^3T_{1u}^{z'}\rangle \end{bmatrix} = \begin{pmatrix} \frac{-1}{\sqrt{2}} & \frac{1}{\sqrt{2}} & 0 \\ \frac{-1}{\sqrt{6}} & \frac{-1}{\sqrt{6}} & \sqrt{\frac{2}{3}} \\ \frac{1}{\sqrt{3}} & \frac{1}{\sqrt{3}} & \frac{1}{\sqrt{3}} \end{pmatrix} \begin{bmatrix} |^3T_{1u}^{x}\rangle \\ |^3T_{1u}^{y}\rangle \\ |^3T_{1u}^{z}\rangle \end{bmatrix}, \tag{5.5}$$

6. Choix du modèle pour la transition nonradiative $^3T_{1u} \to {}^3A_{1u}$

Il est à noter que, pour l'excitation de la bande A, la bande d'émission I s'amortit graduellement en comparaison de la bande d'émission II à basse température dans le passage de MgS à SrS [1, 4], et que la bande I en arrive à disparaître pour BaS:Bi^{3+}, comme on le voit Fig. 1 et 3. Puisque la bande II se montre autant ou plus intense que la bande I à basse température, la probabilité de la transition non-radiative $^3T_{1u} \to {}^3A_{1u}$ doit être comparable ou supérieure à celle de la transition luminescente $^3T_{1u} \to {}^1A_{1g}$, qui est partiellement permise à cause du mélange des états $^1T_{1u}$ et $^3T_{1u}$ par l'intermédiaire de l'interaction spin–orbite. Quelques chercheurs ont essayé d'expliquer la décroissance de la bande II avec la température au moyen du modèle à trois niveaux montré dans la Fig. 7a [18 à 20]. On doit alors supposer que la probabilité de la transition nonradiative $^3T_{1u} \to {}^3A_{1u}$ ne dépend pas de la température et qu'elle est de l'ordre de 10^8 s^{-1} pour l'ion Bi^{3+} incorporé dans des oxydes alcalino-terreux. Comme mécanismes de la transition $^3T_{1u} \to {}^3A_{1u}$ qui est très active à basse température, on peut considérer la transition dipolaire magnétique et l'effet tunnel. Toutefois on ne peut pas attribuer au premier mécanisme une probabilité assez importante. Donc on va examiner dans la suite si l'effet tunnel est efficace ou non pour ce cas.

Quant à la transition nonradiative, Kubo et Toyozawa [21] présentèrent une théorie dont on peut immédiatement appliquer la conclusion aux résultats d'expérience ob-

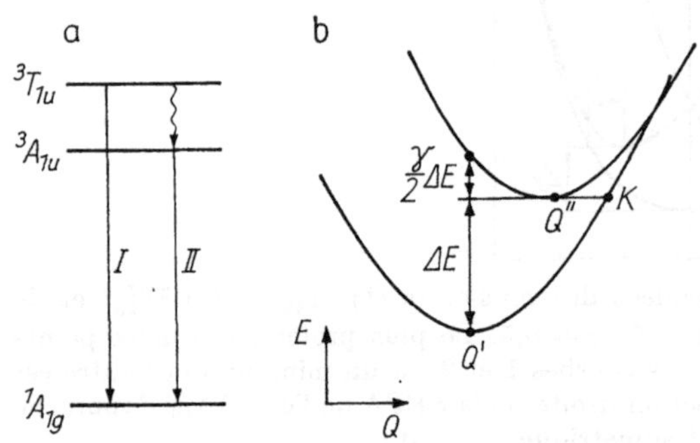

Fig. 7. Représentation a) du modèle à trois niveaux et b) du modèle schématisé des sections droites des SEPA montrées dans la Fig. 6. L'abscisse Q signifie la coordonnée de mode d'interaction en cause, de sorte que l'énergie potentielle vibrationnelle s'exprime par Q^2

tenus sur des cristaux actuels. Après cette théorie quelques travaux théoriques relatifs à la transition nonradiative dans la molécule furent publiés [22, 23]. Ils sont pourtant difficiles à combiner avec les résultats d'expérience et ne tiennent d'ailleurs pas compte de l'influence de l'environnement cristallin. Après s'être assuré qu'on peut aussi appliquer la théorie de Kubo-Toyozawa au système à deux électrons, on la choisit de préférence à toute autre en dépit d'une tendance à la simplification excessive.

Le modèle schématisé des SEPA montrées dans la Fig. 6 est présenté sur la Fig. 7 b, où la SEPA du sous-état inférieur de l'état $^3T_{1u}$ et celle de l'état $^3A_{1u}$ sont remplacées de façon approchée par les deux paraboles similaires. Alors la probabilité pour l'effet tunnel Q'' → K entre deux états électroniques d'un ion d'impureté incorporé dans le cristal polaire s'écrit [21]

$$W = \sqrt{\pi/8}\,\omega_{ex}\,PD\,, \tag{6.1}$$

où

$$P = \left(\frac{1}{N}\right)^{1/2}\left[\left(\frac{\gamma}{2}\right)\exp\left(1-\frac{\gamma}{2}\right)\right]^N, \tag{6.2}$$

$$D = \left(\frac{2}{\gamma}-1\right)^2\gamma''^2 + \frac{4}{\gamma^2}(\gamma\gamma' - \gamma''^2)\left(\frac{1}{N}\right), \tag{6.3}$$

$$N = \frac{\Delta E}{\hbar\omega_{ex}} = \frac{E(T_0) - E(A_0)}{\hbar\omega_{ex}} \tag{6.4}$$

et

$$\gamma = \frac{1}{4\pi\,\Delta E}\left(\frac{1}{\varkappa_0}-\frac{1}{\varkappa}\right)\int[\boldsymbol{E}(^3A_{1u}) - \boldsymbol{E}(^3T_{1u})]^2\,d\boldsymbol{r}\,. \tag{6.5}$$

Les symboles $\varkappa_0$ et $\varkappa$ dans (6.5) désignent les constantes diélectriques optique et statique du cristal, et les paramètres γ' et γ'' sont définis par les expressions correspondantes dans lesquelles l'intégrand est remplacé respectivement par $\boldsymbol{E}(^3T_{1u}{}^3A_{1u})^2$ et $\boldsymbol{E}(^3T_{1u}{}^3A_{1u}) \cdot [\boldsymbol{E}(^3A_{1u}) - \boldsymbol{E}(^3T_{1u})]$. $\boldsymbol{E}(^3T_{1u})$, $\boldsymbol{E}(^3A_{1u})$ et $\boldsymbol{E}(^3T_{1u}{}^3A_{1u})$ sont respectivement les champs électriques dus aux charges $-[e\varphi(^3T_{1u})]^2$, $-e[\varphi(^3A_{1u})]^2$ et $-e\varphi(^3T_{1u}) \times \times \varphi(^3A_{1u})$ spacialement distribuées; $\varphi(^3T_{1u})$ et $\varphi(^3A_{1u})$ désignent respectivement les fonctions d'onde à deux électrons, perturbées par la distorsion tétragonale ou trigonale, des états $^3T_{1u}$ et $^3A_{1u}$. Le champ électrique $\boldsymbol{E}(^3\Gamma_i{}^3\Gamma_j)$ dû à la charge $-e\varphi(^3\Gamma_i)\,\varphi(^3\Gamma_j)$ s'exprime alors par

$$\boldsymbol{E}(^3\Gamma_i{}^3\Gamma_j) = -e\iint\left[\sum_{k=1,2}\frac{\boldsymbol{r}-\boldsymbol{r}_k}{|\boldsymbol{r}-\boldsymbol{r}_k|^3}\right]\varphi(^3\Gamma_i)\,\varphi(^3\Gamma_j)\,d\boldsymbol{r}_1\,d\boldsymbol{r}_2\,, \tag{6.6}$$

compte tenu de l'orthonormali tédes fonctions de spin [13], $\boldsymbol{r}_k$ ($k = 1, 2$) désignant les coordonnées des deux électrons de valence. Pour obtenir le champ électrique $\boldsymbol{E}(^3\Gamma_i)$ dû à la charge $-e[\varphi(^3\Gamma_i)]^2$ on pose simplement $j = i$.

Dans (6.1) le facteur P est très sensible à la valeur de γ, car l'expression mise entre crochets dans (6.2) est toujours inférieure à l'unité et d'habitude $N \gg 1$. Puisque le paramètre γ se relie au facteur de Huang-Rhys S par $\gamma N/2 = S$ [21], la distance de pénétration s'écrit

$$Q''K = \Delta Q = \sqrt{\Delta E}(1 - \sqrt{\gamma/2})\,. \tag{6.7}$$

L'égalité $\gamma = 2$ veut dire que $\Delta Q = 0$, c'est-à-dire Q'' coïncide avec K dans la Fig. 7 b. Ainsi la probabilité pour l'effet tunnel sensiblement dépend de la distance de pénétration ΔQ, à travers laquelle la coordonnée de réseau doit percer. Il s'agit ici d'estimer la valeur de γ. Comme γ est très difficile à évaluer au moyen de (6.5), il faut estimer sa valeur par un autre moyen. Il en est de même des paramètres γ' et γ''. On en verra la description détaillée dans la Section 7.2.

7. Discussion

7.1 Comparaison des constantes spectroscopiques variées

Les valeurs de quelques constantes spectroscopiques et de l'efficacité η pour l'effet magnétique, obtenues expérimentalement dans les Sections 3 et 4, sont disposées sur le Tableau 1, dans lequel les valeurs pour les autres chalcogénures alcalino-terreux activés à l'ion Bi^{3+} sont aussi indiquées à titre d'indication. On voit que la différence d'énergie ΔE diminue régulièrement dans le passage de MgS à BaS. Le fait que la fréquence ω_g demeure sans changement notable en fonction du cation constituant du cristal hôte suggère que le complexe octaédrique $Bi^{3+}(S^{2-})_6$ effectue une vibration rotationnelle harmonique comme s'il était un rotateur rigide [3, 13, 14].

Tableau 1
Positions des raies vibroniques T_0, A_0, A_0', différences d'énergie $\Delta E = E(T_0) - E(A_0)$, $\hbar\omega_g = E(A_0) - E(A_0')$, $\hbar\omega_{ex}$, $\hbar\omega_{em}$ et coefficient η pour l'ion Bi^{3+} incorporé dans les chalcogénures alcalino-terreux

	T_0 (nm)	A_0 (nm)	A_0' (nm)	ΔE (eV)	$\hbar\omega_g$ (cm^{-1})	$\hbar\omega_{ex}$*) (cm^{-1})	$\hbar\omega_{em}$*) (cm^{-1})	η (T^{-2})	bibliographie
CaO:Bi^{3+}	367,0	383,8	388,4	0,148	306	434	493	0,026	[1, 14]
MgS:Bi^{3+}	420,4	437,1	440,5	0,113	177	$\approx$230	$\approx$290	0,177	[1, 14]
CaS:Bi^{3+}	422,1	438,1	442,1	0,107	207	227	276	0,162	[1, 14]
SrS:Bi^{3+}	445,1	460,7	465,0	0,093	200	200	244	0,137	[4]
BaS:Bi^{3+}	479,6	491,2	495,7	0,061	184	167	210	0,112	article présent

*) $\hbar\omega_{ex}$ et $\hbar\omega_{em}$ désignent respectivement l'intervalle entre les raies vibroniques sur les bandes d'excitation et d'émission.

7.2 Estimation de la valeur de la constante b et évaluation des quantités ΔQ, γ, P et D

Pour l'ion Bi^{3+} incorporé dans CaO, MgS et CaS on présente dans le Tableau 2 les valeurs des paramètres G, ζ [1, 24] et b^2 qui figurent dans (5.2). On a convenablement déterminé la valeur de b^2 en se référant à celles des ions du type Tl^+ incorporés dans les halogénures alcalins [25] et, par ailleurs, en comparant entre elles les formes expérimentale et théorique de la bande d'excitation C [26]. On va examiner dans la suite si les valeurs de b^2 ainsi déterminées sont raisonnables ou non. On y dispose aussi les valeurs de $\beta = (\omega_{em}^2 - \omega_{ex}^2)/\omega_{em}^2$ et de $N = \Delta E/\hbar\omega_{ex}$, déterminées à partir du Tableau 1. On peut évaluer le paramètre A en utilisant les valeurs indiquées des paramètres ζ, b^2, β et (5.4) (cf. le Tableau 2). Pour $6 \leqq A < 8$ et $g \approx 0{,}5$ la différence d'énergie réduite, $\Delta y = \Delta E/\zeta$, entre les minima des états $^3T_{1u}$ et $^3A_{1u}$ se trouve dans le domaine 0,1 à 0,13 (cf. la Fig. 6), ce qui est à peu près en accord avec les valeurs expérimentales $\Delta E/\zeta = 0{,}2$ à 0,26 (cf. les Tableaux 1 et 2). En outre, pour $A \gtrsim 6$ la SEPA calculée du sous-état supérieur de l'état $^3T_{1u}$ n'a pas de minimum métastable T (cf. la Fig. 6), ce qui s'accorde bien avec le fait expérimental que l'on n'observe que l'émission A_x, c'est-à-dire la bande I pour la transition $^3T_{1u} \rightarrow {}^1A_{1g}$ chez l'ion Bi^{3+} dans les chalcogénures alcalino-terreux [1, 4]. D'après ce qui est dit plus haut les

Tableau 2
Evaluation des paramètres A et γ à partir des valeurs expérimentalement déterminées des paramètres G, ζ, b^2, β et N.*) On suppose ici la distance de pénétration réduite $\Delta x = (b/2\sqrt{3}\zeta)\,\Delta Q$ égale à 0,1

	G (eV)	ζ (eV)	b^2 (eV)	β	N	A	$\Delta Q/\sqrt{\Delta E}$	γ	$10^4 P$
CaO:Bi^{3+}	0,47	0,75	0,93	0,23	2,8	7,45	0,70	0,18	≈ 60
MgS:Bi^{3+}	0,19	0,44	0,60	0,37	4,0	5,63	0,77	0,11	≈ 2
CaS:Bi^{3+}	0,25	0,44	0,60	0,32	3,8	5,98	0,79	0,09	≈ 1

*) On n'a pas pu nettement observer la bande d'excitation B et obtenir les valeurs fiables des paramètres G et ζ pour les luminophores SrS et BaS activés à l'ion Bi^{3+}. Par suite on n'a plus qu'à les rayer de ce tableau.

valeurs de A, par conséquent celles de b^2, disposées dans le Tableau 2 s'avèrent raisonnables.

Dans la Fig. 6 on voit que pour $6 \leqq A < 8$ la distance de pénétration réduite, Δx, est à peu près égale à 0,1. En portant dans l'équation $\Delta x = (b/2\sqrt{3}\,\zeta)\,\Delta Q$ provenant de la troisième des équations (5.2) $\Delta x = 0{,}1$ et les valeurs des paramètres ζ et b indiquées dans le Tableau 2 on obtient $\Delta Q = 0{,}28$ à $0{,}27\ \mathrm{eV}^{1/2}$. Il en découle les valeurs de $\Delta Q/\sqrt{\Delta E}$ disposées dans le Tableau 2. On peut alors estimer la valeur du paramètre γ au moyen de (6.7) et puis celle du facteur P en utilisant (6.2), qui exerce une grande influence sur la probabilité W (cf. le Tableau 2).

Pour ce qui concerne le dernier facteur D dans (6.1) on a

$$D = \nu^2(2-\gamma)^2 + 4(\mu - \nu^2)\,\frac{1}{N} \tag{7.1}$$

en portant $\gamma' = \mu\gamma$ et $|\gamma''| = \nu\gamma$ dans (6.3). Puisque $\gamma \ll 2$ et $N \approx 4$ (cf. le Tableau 2), on obtient

$$D \approx 3\nu^2 + \mu\,. \tag{7.2}$$

Les formes effectives des fonctions d'onde non perturbées provenant de la configuration nsp nous enseignent que le recouvrement $\varphi(^3T_{1u})\,\varphi(^3A_{1u})$ s'annule partout dans l'espace par suite de l'orthonormalité des fonctions de spin, à moins qu'il n'y ait la distorsion tétragonale ou trigonale, c'est-à-dire dans le cas où $Q_2 = Q_3 = 0$ ou $Q_4 = Q_5 = Q_6 = 0$ [13]. Le recouvrement croît rapidement avec les mélanges $^3A_{1u}$–^3E_u; $^3T_{1u}$–$^3T_{2u}$ ou $^3A_{1u}$–$^3T_{2u}$; $^3T_{1u}$–^3E_u, $^3T_{2u}$, qui se produisent respectivement par suite de la distorsion tétragonale ou trigonale [15]. En se référant à la définition des paramètres γ, γ' et γ'' (cf. (6.5) et (6.6)), on voit que $\mu \lesseqgtr \nu < 1$ et que les paramètres μ et ν sont d'autant plus grands que le recouvrement est plus important. En admettant un fort changement $10^{-2} \lesseqgtr \mu < 1$ et en tenant compte de (7.2) on a $10^{-2} \lesseqgtr D < 1$.

7.3 Estimation de la valeur de la probabilité W pour l'effet tunnel $^3T_{1u} \to {}^3A_{1u}$

Nous allons estimer la valeur de la probabilité W pour l'effet tunnel $^3T_{1u} \to {}^3A_{1u}$ en utilisant les valeurs déjà obtenues des quantités P et D. En portant dans (6.1) les valeurs estimées $P \approx 10^{-4}$ à 10^{-3}, $\omega_{ex} \approx 4$ à $8 \times 10^{13}\ \mathrm{s}^{-1}$ et $10^{-2} \lesseqgtr D < 1$, on peut finalement présumer que la probabilité W pour l'effet tunnel $^3T_{1u} \to {}^3A_{1u}$ est de l'ordre de 10^8 à $10^{10}\ \mathrm{s}^{-1}$, bien qu'il soit difficile de déterminer individuellement la valeur de W pour chacun des luminophores. Cela veut dire que la probabilité W se trouve comparable ou supérieure à celle ($\approx 10^8\ s^{-1}$) de la transition $^3T_{1u} \rightarrow {}^3A_{1u}$ partiellement permise. C'est pourquoi la bande d'émission I se montre autant ou moins intense que la bande d'émission II et la bande d'émission I disparaît parfois entièrement, comme on le voit dans le cas du luminophore BaS:Bi^{3+}.

Ainsi on peut considérer que l'effet tunnel, remarquablement efficace même à basse température, est le mécanisme de la transition nonradiative $^3T_{1u} \rightarrow {}^3A_{1u}$. Le fait que le nombre de phonons effectif $N = \Delta E/\hbar\omega_{ex}$ n'est pas si grand rend l'effet tunnel notablement efficace pour l'ion Bi^{3+} dans les chalcogénures alcalino-terreux.

Bibliographie

[1] S. ASANO et N. YAMASHITA, phys. stat. sol. (b) **105**, 305 (1981).
[2] S. ASANO et N. YAMASHITA, phys. stat. sol. (b) **108**, 549 (1981).
[3] N. YAMASHITA, T. OHIRA, H. MIZUOCHI et S. ASANO, J. Phys. Soc. Japan **53**, 419 (1984).
[4] N. YAMASHITA, S. IWASAKI, S. ASANO, M. OHISHI et K. OHMORI, J. Phys. Soc. Japan **53**, 4425 (1984).
[5] S. SUGANO, J. chem. Phys. **36**, 122 (1962).
[6] M. N. ALENTSEV et A. A. CHEREPNEV, Zh. eksper. teor. Fiz. **26**, 473 (1954).
[7] V. V. ANTONOV-ROMANOVSKII, Zh. eksper. teor. Fiz. **26**, 459 (1954).
[8] S. ASANO, N. YAMASHITA et Y. OGAWA, phys. stat. sol. (b) **118**, 89 (1983).
[9] R. J. ZOLLWEG, Phys. Rev. **111**, 113 (1958).
[10] Y. KANEKO, Thèse, Université de Tokyo, 1984 (communication privée).
[11] A. F. ELLERVEE, A. I. LAISAAR et A. M. A. OPER, Zh. eksper. teor. Fiz., Pisma **33**, 24 (1981); Soviet Phys. — J. exper. theor. Phys., Letters **33**, 21 (1981).
[12] N. YAMASHITA et S. ASANO, Phys. Letters A **96**, 375 (1983).
[13] S. ASANO, phys. stat. sol. (b) **105**, 613 (1981).
[14] S. ASANO et N. YAMASHITA, Phys. Letters A **86**, 191 (1981).
[15] A. FUKUDA, Phys. Rev. B **1**, 4161 (1970).
[16] A. RANFAGNI, Phys. Rev. Letters **28**, 743 (1972).
[17] A. RANFAGNI et G. VILIANI, J. Phys. Chem. Solids **35**, 25 (1974).
[18] A. E. HUGHES et G. P. PELLS, phys. stat. sol. (b) **71**, 707 (1975).
[19] G. BOULON, J. Physique **32**, 333 (1971).
[20] D. DE SI DANG, R. ROMESTAIN, D. SIMKIN et A. FUKUDA, Phys. Rev. B **18**, 2989 (1978).
[21] R. KUBO et Y. TOYOZAWA, Progr. theor. Phys. (Kyoto) **13**, 160 (1955).
[22] R. ENGLMAN et J. JORTNER, Mol. Phys. **18**, 145 (1970).
[23] S. FISCHER, J. chem. Phys. **53**, 3195 (1970).
[24] N. YAMASHITA et S. ASANO, J. Phys. Soc. Japan **40**, 144 (1976).
[25] A. FUKUDA, K. CHO et H. J. PAUS, Luminescence of Crystals, Molecules, and Solutions, Ed. F. WILLIAMS, Plenum Press, New York 1973 (p. 478).
[26] K. CHO, J. Phys. Soc. Japan **25**, 1372 (1968).

(Received July 3, 1985)

phys. stat. sol. (b) 133, 345 (1985)

Subject classification: 18 and 20; 21.4

Laboratoire des Eléments de Transition dans Les Solides, E.R. 210 du C.N.R.S., Meudon Bellevue[1])

Optical Properties of EuF_3 and GdF_3 and Paramagnetic Susceptibility of EuF_3

By

C. K. JAYASANKAR, E. ANTIC-FIDANCEV, M. LEMAITRE BLAISE, and P. PORCHER

The optical properties of EuF_3 and GdF_3 are reinvestigated. Crystal field parameters derived from the data by using the $D_{3h} \rightarrow C_{2v}$ descending symmetry procedure for EuF_3 are (in cm^{-1}): $B_0^2 = 465$, $B_2^2 = 375$, $B_0^4 = -232$, $B_2^4 = -96$, $B_4^4 = -114$, $B_0^6 = -490$, $B_2^6 = 83$, $B_4^6 = -518$, $B_6^6 = -162$. The temperature dependent paramagnetic susceptibility of EuF_3 is calculated by using the simulated wave functions. The calculated and experimental paramagnetic susceptibilities are in very good agreement.

Les propriétés optiques de EuF_3 et GdF_3 ont été réexaminées. A partir des niveaux observés, les paramètres de champ cristallin ont été déterminés par la méthode de symétrie descendante $D_{3h} \rightarrow C_{2v}$ pour EuF_3, les paramètres sont: $B_0^2 = 465$, $B_2^2 = 375$, $B_0^4 = -232$, $B_2^4 = -96$, $B_4^4 = -114$, $B_0^6 = -490$, $B_2^6 = 83$, $B_4^6 = -518$, $B_6^6 = -162$ (en cm^{-1}). En outre, l'évolution en fonction de la température de la susceptibilité paramagnétique de EuF_3 a été calculée grâce aux fonctions d'onde déduites de la simulation. L'accord entre l'expérience et le calcul est excellent.

1. Introduction

With the recent increasing interest in the magnetic and optical properties of the trivalent rare earth ions in different host lattices [1 to 4], we have thought it valuable to reinvestigate in detail the isostructural EuF_3 and GdF_3 compounds. From emission and absorption spectra of EuF_3, a detailed energy level scheme has been established, although this compound has been studied in the past [5]. Moreover, the absorption spectrum of the isostructural GdF_3 is observed for the first time.

For lanthanides, the crystal field effect is comparatively smaller than the spin–orbit interaction. Therefore, the selection rules for the crystal field spectra are nothing but those of the free atom where it changes according to the local crystal field at the site of the ion. According to the Laporte selection rule for transitions between two levels, electric dipole transitions are forbidden for transitions occurring within the same configuration. However, forced electric dipole transitions are allowed by mixing of states from configurations of opposite parity.

Here we derived the sets of crystal field parameters simulating the energy level scheme of EuF_3 and GdF_3. For EuF_3, the wave functions generated from the simulation were used to calculate the paramagnetic susceptibility as a function of temperature and compared with the reported experimental and theoretical values of Kern et al. [6].

2. Analysis of the Spectra

Caspers et al. [5] analysed the absorption (900 to 5500 cm^{-1}), fluorescence (19000 to 13000 cm^{-1}), and diffuse reflection (17000 to 22000 cm^{-1}) spectra of EuF_3 powder at

[1]) Place A. Briand, F-92190 Meudon Bellevue, France.

different temperatures. They were able to identify most of the Stark components belonging to the ground multiplet ^7F_J ($J = 0, 1, 2, 3, 4, 5, 6$) and the three lowest excited states of ^5D_J ($J = 0, 1, 2$).

Transitions $^7F_0 \rightarrow {}^7F_{2,3,4,5,6}$ have been directly measured by infrared absorption as well as fluorescence studies. The three components of the $^7F_1$ state lie in the region of strong lattice vibrations and also the fluorescence spectra around 17000 cm^{-1} are very complex in which case strongest transitions are taken as $^5D_0 \rightarrow {}^7F_1$. Finally these were confirmed by far infrared absorption and diffuse reflection experiments. Hence Caspers et al. [5] have successfully established the sequence of Stark levels of ^7F and ^5D levels without any ambiguity.

In this work we also reinvestigated the fluorescence spectra of EuF_3 powder by selective excitation of the $^5D_0$ level but by using a tunable dye laser accorded on the $^5D_0 \rightarrow {}^7F_0$ wavelength (578.3 nm) at 77 K. Here we were able to observe the transitions of $^5D_0 \rightarrow {}^7F_{1,2,3,4}$. With this we established Stark levels of $^7F_1$, $^7F_2$, $^7F_3$, and $^7F_4$. We have not studied $^7F_5$ and $^7F_6$ Stark levels because of experimental restrictions. The observed levels of ours agree quite well with those reported by Caspers et al. [5]. Moreover, they reported most of the Stark levels of the ^7F multiplet. The analysed energies of the Stark levels of ^7F_J are presented in Table 1. As seen from Table 1 three lines

Table 1

Experimental and calculated energy levels (in cm^{-1}) for EuF_3. The $D_{3h} \rightarrow C_{2v}$ case

level	exp. [5]	calculated					
		D_{3h}	irr. rep.	set I C_{2v}	irr. rep.	set II C_{2v}	irr. rep.
$^7F_0$	0	0	A_1'	0	A_1	0	A_1
$^7F_1$	258	250	A_2'	247	B_2	253	B_2
	420	440	E''	418	B_1	413	B_1
	454	440		464	A_2	467	A_2
$^7F_2$	997	1005	E''	1002	A_2	991	A_2
	1013	1005		1016	B_1	1024	B_1
	1045	1050	A_1'	1033	A_1	1041	A_1
	1098	1118	E'	1104	B_2	1102	B_2
	1144	1118		1138	A_1	1138	A_1
$^7F_3$	1865	1894	A_2''/A_1''	1891	B_1	1883	B_1
	1881	1899	E''	1897	A_2	1895	B_2
	1898	1899		1905	B_2	1911	B_1
	1917	1907	A_2'	1916	A_2	1925	A_2
	1959	1959	E'	1952	B_2	1951	B_2
	1990	1959		1961	A_1	1961	A_1
	1995	1983	A_1''/A_2''	1981	B_1	1976	A_2
$^7F_4$	2775	2804	E'	2770	A_1	2784	A_1
	2810	2804		2806	B_2	2799	B_1
	2819	2811	A_1'	2811	A_2	2819	B_2
	2851	2817	A_2''/A_1''	2862	A_2	2859	A_1
	2862	2872	E''	2870	A_1	2881	A_2
	2904	2872		2889	B_1	2891	B_1
	2923	2943	A_1''/A_2''	2930	B_1	2921	B_2
	2964	2988	E'	2967	B_2	2960	A_1
	2996	2988		2996	A_1	2990	A_2

Table 1 (continued)

level	exp. [5]	calculated D_{3h}	irr. rep.	set I C_{2v}	irr. rep.	set II C_{2v}	irr. rep.
$^7F_5$	3779	3783	E″	3778	B_1	3785	A_2
	3807	3783		3786	A_2	3787	B_1
	3864	3864	A_2''/A_1''	3869	A_2	3879	A_1
	3916	3920	E′	3905	B_2	3908	B_1
	3928	3920		3937	A_1	3928	B_2
		3974	A_1''/A_2''	3963	B_1	3956	A_2
	3983	4008	E′	3989	B_2	3969	B_2
	4007	4008		4004	A_1	3983	B_1
	4018	4023	E″	4016	A_2	4028	B_2
	4037	4023		4043	B_1	4050	A_1
	4065	4051	A_2'	4057	B_2	4066	A_2
$^7F_6$	4827	4857	A_1'	4842	A_1	4845	A_1
		4857	A_2'	4843	B_2	4845	B_2
	4897	4883	E′	4890	A_2	4895	A_2
		4883		4890	B_1	4895	B_1
		5008	E′	5013	B_2	5005	A_1
	5029	5008		5019	A_1	5008	B_2
	5084	5095	A_2''/A_1''	5094	A_2	5085	B_1
	5122	5113	A_1''/A_2''	5113	B_1	5093	A_2
	5136	5161	E′	5143	A_1	5131	A_1
		5161		5157	B_2	5149	B_2
		5189	E″	5165	B_1	5156	B_1
		5189		5222	A_1	5240	A_1
	5229	5189	A_1'	5222	A_2	5240	A_2
residue		13365 ±		4587 ±		6862 ±	
σ		±19		±12		±14	

for $^5D_0 \rightarrow {}^7F_1$ (magnetic dipole transition), five lines for $^5D_0 \rightarrow {}^7F_2$, seven lines for $^5D_0 \rightarrow {}^7F_3$, and nine lines for $^5D_0 \rightarrow {}^7F_4$ (electric dipole transitions) do not permit a point site symmetry higher than C_2.

Of course we missed to observe one level for $^5D_0 \rightarrow {}^7F_5$ and six levels for $^5D_0 \rightarrow {}^7F_6$ which is not at all a major factor because all the expected numbers of splittings are observed for $J = 1, 2, 3, 4$ for a symmetry lower than C_2.

The absorption spectrum of GdF_3 has been recorded at room temperature (32000 to 36800 cm^{-1}), which is sufficient since the $^8S_{7/2}$ is not split by the crystal field. The rare earth fluoride matrices are transparent in the UV region. Almost all possible Stark components of ^6P_J ($J = 5/2, 7/2$) and ^6I_J ($J = 7/2, 9/2, 11/2, 13/2, 15/2$, and $17/2$) levels have been recorded (Table 2). Although the barycentre positions observed here are not far from those of $LaF_3{:}Gd^{3+}$ [1] some difference appears in the energy sequence of certain levels. For example the total splitting of $^6P_{7/2}$ is 122 cm^{-1} for GdF_3 and 52 cm^{-1} for $LaF_3{:}Gd^{3+}$. On the contrary ^6I levels having a higher J value do not exhibit a great difference. Of course, structures are different, but this feature suggests a particular importance of the second-order crystal field parameters.

Table 2
Energy levels of Gd^{3+} in GdF_3

$^{2s+1}L_J$ level	energy (cm^{-1})			
	experimental		calculated (from Table 3)	
$^6P_{7/2}$	32120	32175	32120	32174
	32151		32150	
	32188		32186	
	32242		32242	
$^6P_{5/2}$	32724	32766	32734	32765
	32760		32762	
	32813		32801	
$^6P_{3/2}$	not observed			
$^6I_{7/2}$	35883	35921	35913	35925
	35912		35923	
	35930		35932	
	35959		35934	
$^6I_{9/2}$	36225	36264	36244	36263
	36253		36256	
	36261		36265	
	36283		36275	
	36296		36276	
$^6I_{17/2}$	36305			
	36310			
	36316			
	36322			
	36331			
	36342			
	36355			
$^6I_{11/2}$	36499	36542	36522	36543
	36524		36532	
	36532		36540	
	36545		36550	
	36567		36558	
	36585		36559	
$^6I_{(13/2,15/2)}$	36622			
	36637			
	36647			
	36656			
	36664			
	36677			
	36689			
	36694			
	36703			
	36706			
	36729			
	36740			

2.1 Point site symmetries considered for simulation

The crystal structure determination of YF_3 powder has been carried out in [7] by the powder X-ray diffraction method. The structures of EuF_3 and GdF_3 are isomorphous with that of YF_3, its space group being $D_{2h}^{16}(P_{nma})$. The unit cell contains four molecules, cell parameters (in nm) are: $a = 0.6622(0.6570)$, $b = 0.7019(0.6984)$, $c = 0.4396(0.4393)$ for $EuF_3(GdF_3)$. The europium (or gadolinium) site symmetry C_s can be observed more readily through the projection along the c-axis of the orthorhombic structure. The coordination number for Eu (or Gd) is nine where six fluorine neighbours are at the corners of a regular trigonal prism with an europium (or Gd) in the centre, the remaining three are adjacent to the lateral faces of this prism. Although there are substantial deviations from threefold symmetry due to antiparallelism of the triangular basis of the coordination polyhedron, we can assume that the point symmetry is in fact as derived from D_{3h} which will give the idea to make simulation for crystal field calculation.

Therefore both spectroscopic (Table 1 and 2) and crystallographic results show that the symmetry of the point site occupied by the rare earth ion is not higher than C_2. The spectroscopic results clearly show that the C_{2v} (treated as in the simulation) symmetry of the point site is very close to a higher symmetry. We can apply the descending symmetry method, by considering first a relatively high symmetry for the point site, D_{3h}. Initially a few crystal field parameters are considered then adding extra parameters in order to apply the procedure of descending symmetries [8]. For instance according to the group transformation properties the simplest expansion of a D_{3h} potential is to describe it along the C_3 axis yielding only four nonzero crystal field parameters. But it is always possible to express the same crystal field potential by referring to a C_2 axis, orthogonal to C_3, which also exists in the point group. In this case the crystal field potential is nothing but C_{2v} consisting of nine nonzero parameters of which only four are independent. The respective crystal field Hamiltonian are as follows:

$$H_{D_{3h}} = B_0^2C_0^2 + B_0^4C_0^4 + B_0^6C_0^6 + B_6^6(C_{-6}^6 + C_6^6)\,,$$

$$\begin{aligned}H_{C_{2v}} = {} & B_0^2C_0^2 + B_2^2(C_{-2}^2 + C_2^2) + B_0^4C_0^4 + B_2^4(C_{-2}^4 + C_2^4) + \\ & + B_4^4(C_{-4}^4 + C_4^4) + B_0^6C_0^6 + B_2^6(C_{-2}^6 + C_2^6) + \\ & + B_4^6(C_{-4}^6 + C_4^6) + B_6^6(C_{-6}^6 + C_6^6)\,.\end{aligned}$$

2.2 Simulation of the EuF_3 energy level scheme

For the $4f^6$ configuration of Eu^{3+}, it is well known that the crystal field calculation can be performed accurately on a strongly reduced $|SLJM_J\rangle$ basis, since only the ^7F ground term can be considered [9] which means the diagonalization of a 49×49 matrix for a low symmetry. For D_{3h} the order of magnitude of crystal field parameters is obtained rapidly since the $^7F_1$ splitting mainly depends on B_0^2, the $^7F_2$ splitting on B_0^2 and B_0^4 and $^7F_{3,4,5,6}$ splittings on B_0^2, B_0^4, B_0^6, and B_6^6 parameters. From these starting values the precise calculation is performed by minimizing the root-mean-square (r.m.s.) deviation $\sigma = (\sum_i \sigma_i^2/(N - P))^{1/2}$ where σ_i is the individual error of the i-th level, $\sum_i \sigma_i^2$ is the residue, N the number of levels, and P the number of parameters involved. The experimental and calculated values for different site symmetries are presented in Table 1 and the respective crystal field parameters are collected in Table 3. As is seen from Table 1, the r.m.s. deviation is not good (around 19 cm^{-1}) for D_{3h}. Moreover the sign of B_6^6 is not determined. Change in sign does not modify the energy positions but permutes the label of the A_1'' and A_2'' irreducible representations which is the reason to consider two starting sets of parameters for C_{2v}. When C_{2v} symmetry is considered, the simulation is also carried out rapidly. Finally, the crystal field parameters are close to expected values of the D_{3h} point group described along the C_2 axis. We obtained less residue for set I (positive B_6^6 starting value) than

Table 3
Crystal field parameters for EuF_3 towards descending symmetry

parameter	along C_3 axis	equivalent values along C_2 axis sign +	sign −	set I from sign +	set II from sign −
B_0^2	−680 ± 12	340		465 ± 15	477 ± 18
B_2^2		416		375 ± 10	352 ± 13
B_0^4	−395 ± 26	−148		−232 ± 33	−242 ± 34
B_2^4		−156		− 96 ± 25	− 82 ± 27
B_4^4		−207		−114 ± 24	−206 ± 29
B_0^6	633 ± 36	−656	261	−490 ± 35	581 ± 38
B_2^6		−539	133	83 ± 39	− 66 ± 32
B_4^6		−345	− 99	−518 ± 23	3 ± 45
B_6^6	483 ± 25	−286	−316	−162 ± 24	−422 ± 31
residue	13365 ±			4587 ±	6862 ±
σ	±19			±12	±14

the set II (negative B_6^6 starting value) of Table 3. Moreover as seen from Table 1 it is clear that if we do not consider the $^7F_3$ multiplet, the residue is reduced by 50% with the same crystal field parameters.

When the crystal field parameters were determined, we performed a diagonalization of a larger matrix consisting of the levels ^7F, $^5D(1)$, $^5D(2)$, $^5D(3)$, $^5F(1)$, $^5F(2)$, $^5G(1)$, $^5G(2)$, $^5G(3)$, $^3P(1)$, $^3P(2)$, $^3P(3)$, $^3P(4)$, $^3P(5)$, $^3P(6)$, $^1S(1)$, $^1S(2)$, $^1S(3)$, $^1S(4)$, corresponding to the basis of 671 $|SLJM_J\rangle$ kets which is sufficient to simulate correctly ^7F and ^5D levels [10]. For the C_{2v} symmetry this 671 $|SLJM_J\rangle$ basis has been split into two submatrices with dimensions of 345 $|SLJM_J\rangle$ ($\mu = 0$) and 326 $|SLJM_J\rangle$ ($\mu = +1$). The considered free ion parameters are chosen in order to obtain best simulation for ^7F and ^5D levels by varying slightly their values around the classical value of europium [11], i.e., $E^0 = 2917.0$, $E^1 = 5544.0$, $E^2 = 24.83$, $E^3 = 585.0$, $\alpha = 20.0$, $\beta = -640.0$, $\gamma = 1750.0$, $\xi = 1295.0$ (units in cm^{-1}).

2.3 Simulation of the GdF_3 energy level scheme

The simulation of the energy levels of GdF_3 is rather difficult to perform. Levels having a multiplicity of six which we have observed are mixed with other multiplicities and yield a large $|SLJM_J\rangle$ basis to be taken into account. Nevertheless, one attempt can be made by considering the main components of the free ion wave function [12] according to the method described in [13]. Of course, the J-mixing effect is not taken into account which can create perturbation on the ^6I levels since levels having different J value stand in the same region. If the crystallographic structures are the same, the Gd^{3+} crystal field parameters are expected to be only slightly different from those of Eu^{3+}. In fact, previously determined second rank crystal field parameters for GdF_3 [15] are very close to those of EuF_3. When the above determined crystal field parameters are input in the matrice diagonalization, the calculated energy levels of GdF_3 are in good agreement for the ^6P levels (Table 2), for which the second-order crystal field parameters are the most important. On the contrary the simulation of the ^6I levels is rather bad, since it gives approximately a scale factor of two between experimental and calculated values. Finally this type of calculation underlines that Gd^{3+} is also a good probe for determining the second rank crystal field parameters [14, 15].

2.4 Paramagnetic susceptibility of EuF_3

For rare earth compounds the difference between experimental paramagnetic susceptibility and the Curie-Weiss law is generally due to crystal field effects. This law is not applicable for Eu^{3+} because of its ground level $^7F_0$ which is nonmagnetic. This ultimately indicates only the second-order term in perturbation theory is responsible for the paramagnetic susceptibility at low temperatures. When temperature increases, $^7F_1$ then $^7F_2$ start to be populated and these levels also contribute to the magnetic susceptibility along with $^7F_0$ level [16].

In an attempt to see how well the recent refinement of EuF_3 energy level scheme could be used to simulate the magnetic susceptibility of this ionic compound, if one considers up to second-order interaction terms in the perturbation equation, the paramagnetic susceptibility, according to Van Vleck, will be written as [10]

$$\chi(T) = N \sum_i \left\{ \frac{(E_i^{(1)})^2}{kT} - 2E_i^{(2)} \right\} B_i \qquad (1)$$

with

$$E_i^{(1)} = \beta \langle \Psi_i | \, L + 2S | \Psi_i \rangle \,,$$

$$E_i^{(2)} = \beta^2 \sum_{j \neq i} \frac{\langle \Psi_i | \, L + 2S | \Psi_j \rangle \, \langle \Psi_j | \, L + 2S | \Psi_i \rangle}{E_i^{(0)} - E_j^{(0)}} \,.$$

Here N is the Avogadro number, k the Boltzmann constant, β the Bohr magneton, B_i the thermal partition law coefficient, Ψ_i and Ψ_j are the eigenfunctions of the unperturbed Hamiltonian corresponding to the $E_i^{(0)}$ and $E_j^{(0)}$ eigenvalues which are described on the $|SLJM_J\rangle$ basis, E^k corresponding to the k-th order perturbed state, and $L + 2S$ denotes magnetic interaction associated to a tensorial operator of rank one. The reason of change of isotropy noted in the free ion as well as the ion in a cubic symmetry is due to different values (or combinations) of tensor components. The 0, ± 1 components of the tensor are denoted by $\chi_\perp$ and $\chi_\parallel$, respectively, which are anisotropiccomponents. In the case of powder, the mean paramagnetic susceptibility χ_{av} is

$$\chi_{av} = \frac{2\chi_\perp + \chi_\parallel}{3} = \frac{\chi_x + \chi_y + \chi_z}{3} \,.$$

Except for Eu^{3+}, the diagonal part involving $E_i^{(1)}$ in (1) is the most important contribution for the paramagnetic susceptibility. It is corresponding to the quantum expression of the Curie-Weiss law, in fact. For Eu^{3+}, due to the nonmagnetic $^7F_0$ ground level, the diagonal part vanishes at low temperatures when the next higher state $^7F_1$ is not populated. Then the nondiagonal interaction is only responsible for the paramagnetic susceptibility which is independent of temperature and yields a plateau.

The wave functions are generated by using the free ion parameters (see the simulation section) where the spin–orbit coupling constant has been chosen as 1295 cm^{-1} which affects the value of the paramagnetic susceptibility. The program computes automatically the magnetic susceptibility for the principle axis at the site of C_{2v}. The diagonal and nondiagonal matrix elements of the $(L + 2S)$ operator are computed between all the states involved in the wave vectors up to eigenvalues of 5000 cm^{-1} and fed into the Van Vleck formula which also uses the Hamiltonian eigenvalues for

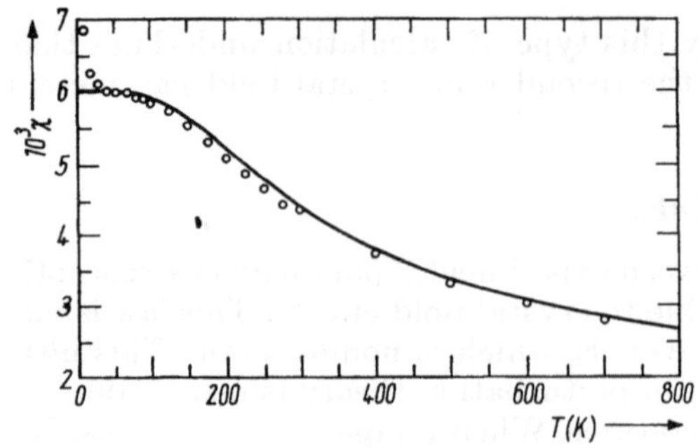

Fig. 1. The selected experimentally measured values (circles) with respect to the computed mean (full line) paramagnetic susceptibility of EuF_3 powder at different temperatures (0 to 800 K)

the energies of the levels. Because of the 671 $|SLJM_J\rangle$, extended basis wave vectors are rather complete, including for example all J-mixing effects and spin–orbit mixing effects (mostly ^7F–^5D) which are of course important for the magnetic calculations. Here the crystal field was assumed to be temperature independent. The selected computed as well as experimental paramagnetic susceptibilities are given in Table 4 and shown in Fig. 1. As seen from Table 4 as well as Fig. 1 the simulation is in good

Table 4

Experimental and calculated temperature dependent paramagnetic susceptibilities (in 10^3 e.m.u. units) of EuF_3

T (K)	calculated				χ_{meas}
	χ_x	χ_y	χ_z	χ_{av}	
1	4.44	9.98	3.69	6.04	—
10	4.44	9.98	3.69	6.04	6.69
20	4.44	9.98	3.69	6.04	6.28
30	4.44	9.98	3.69	6.04	6.13
40	4.44	9.97	3.69	6.04	6.02
50	4.44	9.96	3.70	6.04	6.03
60	4.45	9.94	3.71	6.03	6.03
70	4.46	9.88	3.72	6.02	6.00
80	4.47	9.79	3.75	6.01	5.94
90	4.49	9.67	3.78	5.98	5.92
100	4.50	9.52	3.82	5.95	5.87
125	4.52	9.03	3.91	5.82	5.74
150	4.51	8.45	3.97	5.65	5.53
175	4.47	7.85	4.00	5.44	5.30
200	4.39	7.27	3.99	5.22	5.10
225	4.30	6.74	3.95	5.00	4.86
250	4.19	6.27	3.89	4.78	4.66
275	4.09	5.85	3.82	4.59	4.43
300	3.98	5.48	3.75	4.40	4.35
400	3.59	4.39	3.46	3.81	3.73
500	3.29	3.71	3.21	3.40	3.33
600	3.05	3.26	3.01	3.11	3.04
700	2.86	2.95	2.84	2.88	2.81
800	2.70	2.72	2.70	2.71	2.63
900	2.56	2.53	2.57	2.56	—
1000	2.44	2.38	2.45	2.43	—

agreement with reported data of Kern et al. [6]. But the increasing paramagnetic susceptibility at low temperature is not simulated. The discrepancy between theory and experiment at low temperature is probably due to the small divalent Eu impurity ion which has strong paramagnetic susceptibility at low temperature [6]. Moreover in an attempt to reproduce the calculation by using the wave functions of Kern et al. [6], we derived the position of the plateau as 5.66×10^{-3} instead of 6.00×10^{-3} e.m.u. units.

3. Conclusion

In our present paper we reinvestigated the optical properties along with the paramagnetic susceptibility of EuF_3. The derived sets of crystal field parameters simulate satisfactorily the spectra within a C_{2v} point symmetry. Of course our refinement is not so good because the $^7F_3$ level fits badly which contributes 50% residue.

Here the thermal evolution of the paramagnetic susceptibility was also simulated with an excellent precision from the derived wave functions of the extended basis up to a 671 $|SLJM_J\rangle$ basis by including J mixing. Our theoretical simulation is comparatively better than that reported by Kern et al. [6] due to inclusion of J-mixing effects.

The crystal field parameters derived from EuF_3 simulate the GdF_3 energy level scheme satisfactorily. Moreover the experimental and calculated barycentres of our GdF_3 are quite similar to predicted barycentres of $Gd^{3+}:Cs_2NaGdCl_6$ [17] which indicates that the smoothing procedure of crystal field parameters for the unknown rare earth ion in the same matrix is quite interesting.

Acknowledgement

One of us (C.K.J.) is grateful to the French Government for a fellowship under Indo–French exchange program.

References

[1] C. A. Morrison and R. P. Leavitt, J. chem. Phys. 71, 2366 (1979).
[2] J. J. Dallara, M. F. Reid, and F. S. Richardson, J. phys. Chem. 88, 3587 (1984).
[3] M. Asano and J. A. Koningstein, Chem. Phys. 42, 369 (1979).
[4] R. P. Leavitt, J. B. Gruber, N. C. Chang, and C. A. Morrison, J. chem. Phys. 76, 4775 (1982).
[5] H. H. Caspers, H. E. Rast, and J. L. Fry, J. chem. Phys. 47, 4505 (1967).
[6] S. Kern, P. M. Raccah, and A. Tveten, J. Phys. Chem. Solids 31, 2639 (1970).
[7] A. Zalkin and D. H. Templeton, J. Amer. Chem. Soc. 75, 2453 (1953).
[8] P. Porcher, D. R. Svoronos, M. Leskela, and J. Holsa, J. Solid State Chem. 46, 101 (1983).
[9] P. Porcher and P. Caro, J. chem. Phys. 65, 89 (1976).
[10] J. Huang, J. Loriers, P. Porcher, G. Teste de Sagey, P. Caro, and C. Levy Clement, J. chem. Phys. 80, 6204 (1984).
[11] W. T. Carnall, H. Crosswhite, and H. M. Crosswhite, Argonne National Laboratory Report, 1977, unpublished.
[12] B. G. Wybourne, Phys. Rev. 148, 317 (1966).
[13] E. Antic Fidancev, M. Lemaitre Blaise, and P. Caro, J. chem. Phys. 76, 2906 (1982).
[14] E. Antic Fidancev, M. Lemaitre Blaise, and P. Caro, J. less-common Metals. 93, 137 (1983).
[15] E. Antic Fidancev, M. Lemaitre Blaise, J. Derouet, B. Latourette, and P. Caro, C. R. Acad. Sci. (France) 294, 1077 (1982).
[16] P. Caro and P. Porcher, J. Magnetism magnetic Mater., in the press.
[17] C. A. Morrison, R. P. Leavitt, and D. E. Wortman, J. chem. Phys. 73, 2580 (1980).

(Received July 31, 1985)

phys. stat. sol. (b) **133**, 355 (1986)

Subject classification: 76.30; 71.70; S11.1

Institute of Molecular Physics, Polish Academy of Sciences, Poznań[1]) *(a) and Institute of Physical Research, Academy of Sciences of the Armenian SSR, Ashtarak*[2]) *(b)*

Effect of Pressure and Temperature on the EPR Spectrum of Cr^{3+} Ions in α-$LiIO_3$ Single Crystals

By

M. KRUPSKI (a), A. A. MIRZAKHANYAN (b), W. HILCZER (a), and E. G. SHAROYAN (b)

Results concerning the influence of hydrostatic pressure, uniaxial stress, and temperature on the spin-Hamiltonian parameters of Cr^{3+} in α-$LiIO_3$ single crystals are reported. Assuming that the local compressibility and thermal expansion of the nearest surroundings of the Cr^{3+} ion is the same as of α-$LiIO_3$ crystal lattice it is shown that the temperature and pressure variations on the D-parameter are mainly due to thermal expansion and compressibility of the lattice, whereas the contribution from spin–phonon interactions is several times smaller. The deformation of the oxygen octahedron close to the Cr^{3+} ion is determined.

Исследовано влияние гидростатического давления и осегого сжатия на параметры спин-гамильтониана ионов Cr^{3+} в монокристалах α-$LiIO_3$. Принимая что локальная сжимаемость и термические расширение блиского окружения иона Cr^{3+} такие же самые как решетки α-$LiIO_3$ показано, что температурное изменение аксиального параметра D в основном обусловено тепловым расшерением и сжимаемостью решетки. Вклад спин-фононного взаимодействия в несколько раз меньше. Определена тоже величина искажения ближайшего кислородного октаедра вблизи ионов хрома.

1. Introduction

Crystals of the hexagonal modification of the lithium iodate α-$LiIO_3$ doped with transition metals are of great interest because of the photorefractive effect observed in them recently [1, 2]. The Cr^{3+} ion in α-$LiIO_3$ has been investigated by EPR [3, 4] and has been found to replace a Li^+ ion on the hexagonal axis; charge compensation occurs by way of vacancies of the two closest neighbouring lithium ions lying on the same axis. The dynamics of these vacancies has been studied in other papers [5 to 7] on the basis of measurements of the populations of high- and low-temperature Cr^{3+} centers; the hydrostatic pressure dependence of their concentration has also been investigated [7].

The ground state of Cr^{3+} in lithium iodate is characterized by high splitting ($|2D| = 36$ GHz) and therefore can be expected to be sensitive to pressure-induced changes in the internal crystal field. In this paper we report results bearing on the effect of hydrostatic pressure, uniaxial compressive stress, and temperature on the spin-Hamiltonian parameters of Cr^{3+} in α-$LiIO_3$ single crystal.

---

[1]) Smoluchowskiego 17/19, 60-179 Poznań, Poland.

[2]) 378410 Ashtarak-2, USSR.

2. Experimental

Chromium-doped α-$LiIO_3$ single crystals were grown from $LiIO_3$ aqueous solution doped with $Cr(NO_3)_3$ at pH ≈ 2 by evaporation using VRK-2 equipment constructed in the Institute of Physical Research of the Academy of Sciences of the Armenian SSR. The concentration of chromium ions incorporated into the crystal was determined by EPR as ranging from 0.05 to 0.20 at%.

EPR spectra were measured under hydrostatic pressure up to 500 MPa in the temperature range from 100 to 360 K using an X-band Radiopan SE/X spectrometer. In the pressure equipment constructed in the Institute of Molecular Physics of the Polish Academy of Sciences [8, 9] a corundum resonator was used and the sample located therein was fixed with its C_6 axis parallel to the external magnetic field ($\theta = 0°$). A uniaxial stress equipment similar to that described in [10] was used. Uniaxial compression up to 100 MPa was applied along the C_6 axis to which the external magnetic field was perpendicular ($\theta = 90°$).

3. Results

α-$LiIO_3$ crystals belong to the hexagonal symmetry class. The chromium Cr^{3+} ions are surrounded by an oxygen octahedron, slightly deformed along the trigonal C_3 axis. The C_3 axis of the octahedron and the C_6 axis of the crystal are parallel. Accordingly, the EPR spectrum of α-$LiIO_3$:Cr^{3+} is described by an axially symmetric spin-Hamiltonian of the form

$$\hat{\mathcal{H}} = g_{\parallel}\beta B_z\hat{S}_z + g_{\perp}\beta(B_x\hat{S}_x + B_y\hat{S}_y) + D\left(\hat{S}_z^2 - \tfrac{1}{3}S(S+1)\right); \quad S = \tfrac{3}{2} \qquad (1)$$

with the following parameters [3]: $g_{\parallel} = 1.965$, $g_{\perp} = 1.971$, $D = -18.3$ GHz at room temperature. The negative sign of the axial parameter D was determined from the intensity of the EPR signal at low temperature.

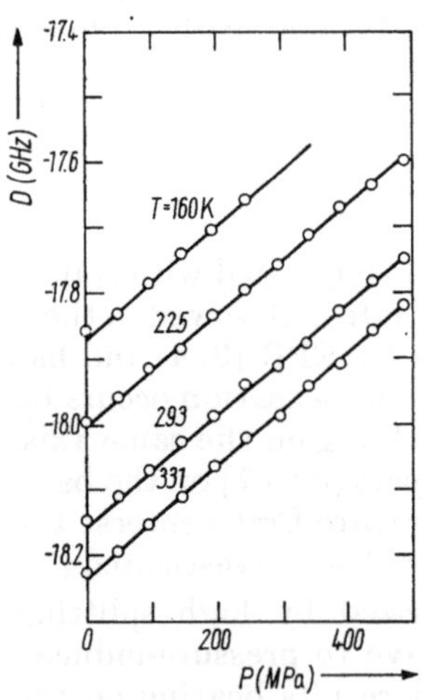

Fig. 1

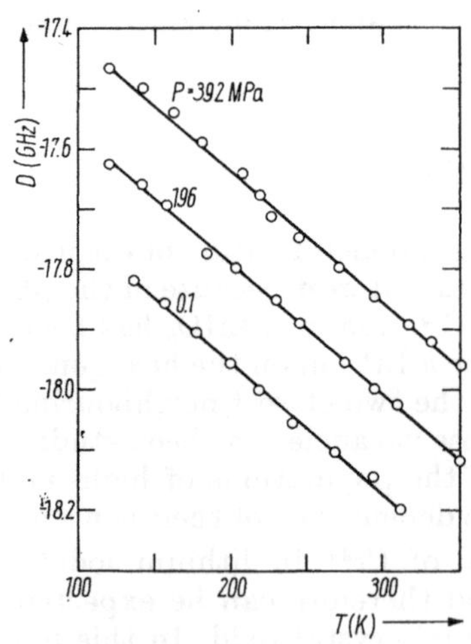

Fig. 2

Fig. 1. Hydrostatic pressure dependence of the axial parameter D of the spin-Hamiltonian for α-$LiIO_3$:Cr^{3+} at different temperatures

Fig. 2. Temperature dependence of the axial parameter D of the spin-Hamiltonian for α-$LiIO_3$:Cr^{3+} at different pressures

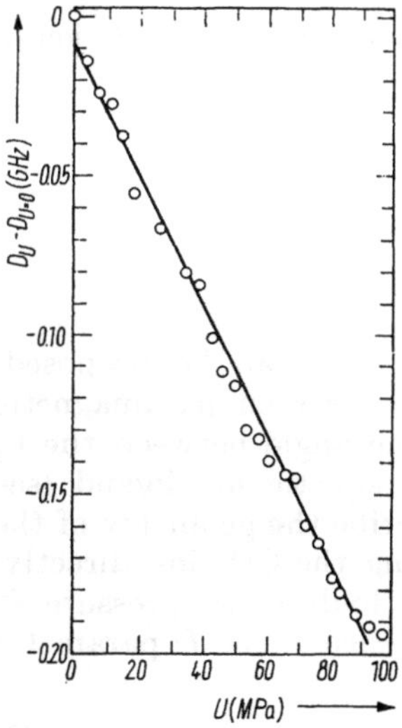

Fig. 3. Uniaxial compressive stress dependence of the axial parameter D of the spin-Hamiltonian for α-$LiIO_3$:Cr^{3+} at room temperature

Our experiments show that $g_{||}$ does not change in the above-mentioned range of temperature and pressure, whereas the absolute value of the axial parameter D markedly decreases with increasing pressure. In Fig. 1, the hydrostatic pressure dependence of D is shown for four values of temperature; the straight lines were obtained by the least-squares method, their slope was found to be the same to within experimental error and amounted to

$$\left(\frac{\partial D}{\partial P}\right)_{T,U} = (0.823 \pm 0.038) \ \text{MHz/MPa} .$$

Fig. 2 shows the temperature dependence of D for three values of pressure. Here, too, the experimental points are described by a linear dependence,

$$\left(\frac{\partial D}{\partial T}\right)_{P,U} = -(2.158 \pm 0.028) \ \text{MHz/K} .$$

The dependence of the axial parameter on uniaxial stress applied along the C_6 axis is shown in Fig. 3. The sample remained undamaged under pressure up to 100 MPa. The experimental points in Fig. 3 were obtained from the dependence of the resonance field on uniaxial compression using the strict solution of the spin-Hamiltonian (1) at $\theta = 90°$. From the linear dependence $D(U)$ the slope was determined as

$$\left(\frac{\partial D}{\partial U}\right)_{P,T} = -(2.037 \pm 0.038) \ \text{MHz/MPa} .$$

4. Discussion

It is known that in general the temperature dependence of the axial parameter D is related both with the changes in the geometry of the nearest surroundings of the doping ion caused by thermal expansion of the lattice and with the lattice vibration contribution due to spin–phonon interactions [11]. The local geometry of the paramagnetic center cannot be determined experimentally. We assume that the changes in the local geometry caused by temperature or pressure reflect directly the changes in the external dimensions of the α-$LiIO_3$ lattice. Thus, we assumed that the D-parameter in the case of axial crystal can be treated as a function of two parameters X and Y. We choose three sets of X and Y:

1. dimensions of the unit cell in the direction parallel to the C_6 axis: $Y = c$ and in the direction perpendicular to this axis $X = a$;

2. the cell volume $X = V = Ga^2c$ and the "shape" $Y = \sigma = G'c/a$, where G and G' are the scaling factors [11];

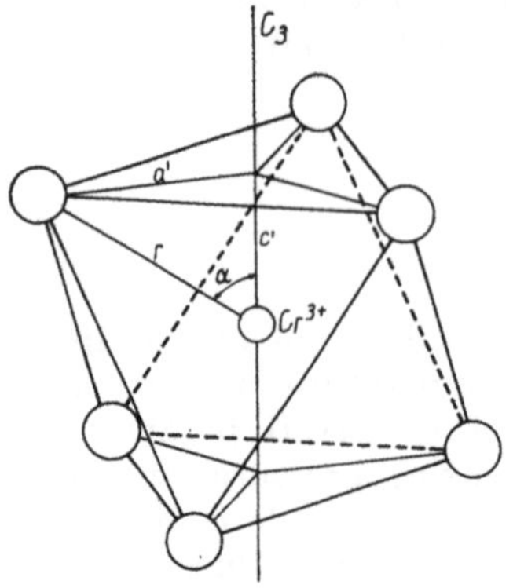

Fig. 4. The oxygen octahedron surrounding the Cr^{3+} ion in α-$LiIO_3$:Cr^{3+}

3. also the set $X = r$, $Y = \alpha$ can be proposed, where r denotes the distance between paramagnetic ion and ligands, α denotes the angle between the C_3 axis and the direction paramagnetic ion–ligand (see Fig. 4). These parameters describe the geometry of the oxygen octahedron surrounding the Cr^{3+} ion directly.

The parameters X and Y are functions of temperature T, hydrostatic pressure P, and uniaxial stress U. The temperature and pressure dependence of the D-parameter can be written as follows:

$$D(T, P, U) = D(X(T, P, U), Y(T, P, U), T) . \tag{2}$$

Differentiating this expression we obtain

$$\left(\frac{\partial D}{\partial T}\right)_{P,U} = \left(\frac{\partial D}{\partial \ln X}\right)_{Y,T} \left(\frac{\partial \ln X}{\partial T}\right)_{P,U} + \left(\frac{\partial D}{\partial \ln Y}\right)_{X,T} \left(\frac{\partial \ln Y}{\partial T}\right)_{P,U} + \left(\frac{\partial D}{\partial T}\right)_{X,Y}, \tag{3}$$

$$\left(\frac{\partial D}{\partial P}\right)_{T,U} = \left(\frac{\partial D}{\partial \ln X}\right)_{Y,T} \left(\frac{\partial \ln X}{\partial P}\right)_{T,U} + \left(\frac{\partial D}{\partial \ln Y}\right)_{X,T} \left(\frac{\partial \ln Y}{\partial P}\right)_{T,U}, \tag{4}$$

$$\left(\frac{\partial D}{\partial U}\right)_{T,P} = \left(\frac{\partial D}{\partial \ln X}\right)_{Y,T} \left(\frac{\partial \ln X}{\partial U}\right)_{P,T} + \left(\frac{\partial D}{\partial \ln Y}\right)_{X,T} \left(\frac{\partial \ln Y}{\partial U}\right)_{P,T}. \tag{5}$$

The first two addends on the right-hand side of (3) are related with the thermal expansion, whereas $(\partial D/\partial T)_{X,Y}$ denotes the contribution of spin–phonon interactions. In order to determine the values of $(\partial D/\partial \ln X)_{Y,T}$, $(\partial D/\partial \ln Y)_{X,T}$, and $(\partial D/\partial T)_{X,Y}$ it is

Table 1

Comparison of the thermal expansion and compressibility coefficients for α-$LiIO_3$:Cr^{3+} at 295 K. Changes in the D-parameter of the spin-Hamiltonian for different pairs of X and Y. $D = -18.16$ GHz

$\{X, Y\} \rightarrow$		$\{a, c\}$	$\{V, \sigma\}$	$\{r, \alpha\}$
$(\partial \ln X/\partial T)_{P,U}$	(10^{-6} K^{-1})	25.3	96.7	32.23
$(\partial \ln X/\partial P)_{U,T}$	(10^{-6} MPa^{-1})	−7.52	−31.35	−10.45
$(\partial \ln X/\partial U)_{P,T}$	(10^{-6} MPa^{-1})	1.59	−16.3	−5.43
$(\partial \ln Y/\partial T)_{P,U}$	(10^{-6} K^{-1})	46.1	20.8	−10.26
$(\partial \ln Y/\partial P)_{U,T}$	(10^{-6} MPa^{-1})	−16.31	−8.79	4.34
$(\partial \ln Y/\partial U)_{P,T}$	(10^{-5} MPa^{-1})	−19.48	−21.07	10.40
$(\partial D/\partial \ln X)_{Y,T}$	(GPa)	−285.7	−68.1	−204.4
$(\partial D/\partial \ln Y)_{X,T}$	(GPa)	81.3	149.4	−302.7
$(\partial D/\partial T)_{X,Y}$	(MHz/K)	1.324	1.324	1.324

necessary to know the temperature and pressure dependence of X and Y. The values of the thermal expansion coefficients are given in [12]. Using the values of elastic constants of lithium iodate [13] and applying usual formulae we obtain values of the compressibility coefficients [14]. Thermal expansion coefficients and the coefficients of compressibility are given in Table 1. Replacing the derivatives in (3) to (5) by the values from Table 1 and our experimental results for the temperature, pressure, and stress dependence, we obtain the values of $(\partial D/\partial \ln a)_{c,T}$, $(\partial D/\partial \ln c)_{a,T}$, and $(\partial D/\partial T)_{a,c}$ (see Table 1). It can be seen that the main contribution to the temperature dependence of the D-parameter is related with the thermal expansion of the crystal lattice,

$$\left(\frac{\partial D}{\partial T}\right)_{P,U} - \left(\frac{\partial D}{\partial T}\right)_{X,Y} = -3.482 \quad \text{MHz/K}\,,$$

and can be explained on ground of high anisotropy of the thermal expansion coefficients of lithium iodate. The contribution of spin–phonon interactions (e.g. the value of $(\partial D/\partial T)_{a,c}$) is of opposite sign and several times smaller. It also can be stated that the D-parameter of Cr^{3+} in α-$LiIO_3$ is more sensitive on changes in the parameter a (in the direction perpendicular to the C_6 axis).

For $X = V$ and $Y = \sigma$ it is necessary to express the derivatives $(\partial \ln V/\partial F)_{\sigma,F}$ and $(\partial \ln \sigma/\partial F)_{V,F}$ ($F = T, P, U$) by the derivatives of a and c. The relations are

$$\frac{\partial \ln V}{\partial F} = 2\,\frac{\partial \ln a}{\partial F} + \frac{\partial \ln c}{\partial F}\,, \tag{6}$$

$$\frac{\partial \ln \sigma}{\partial F} = \frac{\partial \ln c}{\partial F} - \frac{\partial \ln a}{\partial F}\,. \tag{7}$$

The values of these derivatives are presented in the second column of Table 1; this column contains the solution of the system of equations (3) to (5) for V and σ. The explicit temperature dependence coefficient $(\partial D/\partial T)_{V,\sigma}$ has the same value as $(\partial D/\partial T)_{a,c}$. It is evident, because thermal vibrations cannot depend on a set of geometrical parameters. The D-parameter of Cr^{3+} in α-$LiIO_3$ is highly sensitive both to V and σ, and therefore to the geometry and the strength of the local crystal field.

If we introduce $X = r$ and $Y = \alpha$, the temperature, pressure, and stress dependences can be expressed by the $a' = Sa$ and $c' = S'c$ dependences (S and S' are scaling factors),

$$\frac{\partial \ln r}{\partial F} = \frac{1}{1+\sigma'^2}\left(\frac{\partial \ln a}{\partial F} + \sigma'^2\,\frac{\partial \ln c}{\partial F}\right), \tag{8}$$

$$\frac{\partial \ln \alpha}{\partial F} = \frac{1}{\alpha}\,\frac{\sigma'}{1+\sigma'^2}\left(\frac{\partial \ln a}{\partial F} - \frac{\partial \ln c}{\partial F}\right), \tag{9}$$

where $\sigma' = \operatorname{ctg}\alpha = c'/a'$ (see Fig. 4). It is difficult do determine the angle α. We assumed that $\alpha = 54°\,44'$ (the angle for the regular octahedron). The numerical values obtained for this set of parameters X, Y are presented in the third column of Table 1. These results show that the D-parameter depends strongly on the angle α and the distance r. The negative sign of $(\partial D/\partial \ln \alpha)_{r,T}$ means that the D-parameter decreases with increasing α. Since $D < 0$ for α-$LiIO_3$:Cr^{3+}, the absolute value of D increases with increasing α. This means that the angle α is greater than $54°\,44'$; the oxygen octahedron with chromium inside is flattened along the C_3 axis.

In order to determine the deformation angle $(\alpha - \alpha_0)$, the D-parameter can be expanded to first order in $(\alpha - \alpha_0)$ and $(r - r_0)$,

$$D(r, \alpha) = D(r_0, \alpha_0) + (\alpha - \alpha_0)\left(\frac{\partial D}{\partial \alpha}\right)_{r_0} + (r - r_0)\left(\frac{\partial D}{\partial r}\right)_{\alpha_0}. \tag{10}$$

$D(r_0, \alpha_0) = 0$ for the regular octahedron. On the other hand, the theory shows [11] that D can be written in the form

$$D = A(r)\,(\alpha - \alpha_0)\,, \tag{11}$$

where $A(r)$ is a function which depends only on r. Therefore, one can write $(\partial D/\partial r)_{\alpha_0} = 0$, and from (10)

$$D = (\alpha - \alpha_0)\left(\frac{\partial D}{\partial \alpha}\right)_{r_0}. \tag{12}$$

This equation was also found in [15]. At given values of $(\partial D/\partial \alpha)_{\alpha_0}$ and of the D-parameter, the deformation angle was determined to amount to $(\alpha - \alpha_0) = 3° 17'$. Basing on the structural data of α-$LiIO_3$ [16, 17] we determined the angle α for the octahedron with lithium inside, to amount to $52° 28'$. The difference in α for the oxygen octahedron with Li^+ ion inside and for the lithium ion substituted by Cr^{3+} amounts to $5° 33'$.

The value of $(\partial D/\partial \ln r)_{\alpha, T} = -204.4$ GHz denotes that D strongly depends on the paramagnetic ion–ligand distance. Simple calculation shows that $|D| \sim r^{11}$, while Walsh [11], basing on the rigid-ionic model for the octahedron of water molecules surrounding the Ni^{2+} ion, obtained $|D| \sim r^8$. The exponents should be similar in both cases since Cr^{3+} and Ni^{2+} are in the F-state.

To discuss the lattice vibration effect we deal with the linear dependence of the form [11]

$$D = D_0(1 + BT)\,, \tag{13}$$

where D_0 denotes the crystalline field splitting in the absence of the lattice vibrations. This equation can be transformed to the form

$$\left(\frac{\partial \ln D}{\partial T}\right)_{X, T} = \frac{1}{1 + BT}. \tag{14}$$

The experimental value $(\partial D/\partial T)_{X, Y} = 1.324$ MHz implies $B = -71.3 \times 10^{-6}\ K^{-1}$ and $D_0 = -18.55$ GHz. The changes in D caused by the lattice vibrations are about 2.1%.

In our investigations we dealt with the low-temperature paramagnetic center, e.g. the Cr^{3+} ion was associated with two vacancies: $(V$–Cr^{3+}–$V)$ [5]. In this case the chromium ion strongly interacts with ligands, and basing on it the strong deformation of the oxygen octahedron surrounding the Cr^{3+} ion can be explained. At high temperature $(T > 300\ K)$ the complex $(V$–Cr^{3+}–$V)$ is dissociated, and the lithium ions jump into the position of one vacancy $\big((V$–$Cr^{3+}) + V$ complex$\big)$ or substitute both vacancies. In such a case the interaction between Cr^{3+} and ligands is disturbed by lithium ions, and the value of D decreases [5]. Although no pressure measurements of high temperature centers were made it can be predicted that the pressure dependence of D will be similar.

5. Conclusions

On the basis of our data on the influence of hydrostatic pressure, uniaxial stress, and temperature on the EPR spectrum of Cr^{3+} in α-$LiIO_3$ it is shown that the main contribution to the temperature dependence of the axial parameter D of the spin-Hamiltonian originates from the thermal expansion of the crystal lattice. The results of our uniaxial stress experiments show that the high ground state splitting is related with the strong deformation of the oxygen coctahedron surrounding the chromium ion.

Acknowledgements

The authors are strongly indebted to Prof. J. Stankowski and Dr. S. K. Hoffmann for their interest and discussions.

References

[1] A. R. Pogosyan, E. M. Yukin, A. P. Levanyuk, and G. F. Dobrzhanskii, Fiz. tverd. Tela **23**, 3280 (1981).
[2] A. R. Pogosyan, E. M. Yukin, and G. F. Dobrzhanskii, Fiz. tverd. Tela **24**, 3621 (1982).
[3] W. Karthe and R. Kühmstedt, phys. stat. sol. (b) **63**, K5 (1974).
[4] A. Bräuer and D. M. Daraselia, Fiz. tverd. Tela **19**, 1250 (1977).
[5] V. Z. Harutunyan, K. N. Kocharyan, A. A. Mirzakhanyan, L. P. Tambovtseva, and E. G. Sharoyan, Inst. Phys. Res. Acad. Sci. Armenian SSR, Preprint No. 74, 1978 (p. 1).
[6] E. G. Sharoyan, K. N. Kocharyan, A. A. Mirzakhanyan, and V. Z. Harutunyan, Proc. XX. Congr. Ampère, Tallin 1978 (p. 303).
[7] M. Krupski and E. G. Sharoyan, phys. stat. sol. (b) **106**, K95 (1981).
[8] J. Stankowski, A. Gałęzewski, M. Krupski, S. Waplak, and H. Gierszal, Rev. sci. Instrum. **47**, 128 (1976).
[9] J. Stankowski and M. Krupski, Mater. Res. Soc. Symp. Proc., Vol. 22, High-Pressure Technology, Part I, Ed. C. Homan, R. K. MacCrane, and E. Whalley, North-Holland Publ. Co., New York/Amsterdam/Oxford **1984** (p. 109).
[10] J. Szumowski and K. Falkowski, Rev. sci. Instrum. **47**, 252 (1976).
[11] W. M. Walsh, Phys. Rev. **114**, 1473 (1959).
[12] S. C. Abrahams, R. Liminga, P. Marsh, F. Schrey, J. Albertsson, C. Svensson, and Å. Kvick, J. appl. Cryst. **16**, 453 (1983).
[13] S. Haussühl, Acoustica **23**, 195 (1970).
[14] J. F. Nye, Physical Properties of Crystals, Clarendon Press, Oxford 1957.
[15] V. N. Vasyukov, S. N. Łukin, and G. A. Tsintsadze, Fiz. tverd. Tela **20**, 2260 (1978).
[16] A. Emiraliev, A. G. Kocharov, I. I. Yamzin, and V. A. Luybimtsev, Kristallografiya **18**, 1177 (1973).
[17] A. Rozenzweig and B. Morosin, Acta cryst. **20**, 758 (1966).

(Received July 12, 1985)

phys. stat. sol. (b) 133, 363 (1986)

Subject classification: 76.80

Istituto di Ricerca sulle Onde Elettromagnetiche del C.N.R., Firenze[1])

Mössbauer Lineshape in Presence of Relaxation and Diffusion

A Stochastic Approach

By

P. MORETTI and L. CIANCHI

The shape of the Mössbauer relaxation pattern is derived for a Mössbauer ion performing an arbitrary diffusive motion; the formulae used in the literature follow as particular cases.

Nous dérivons la forme de raie Mössbauer en présence de relaxation et de mouvements diffusives de type général. Les formules employées dans la littérature suivent comme des cas particuliers.

1. Introduction

In recent years many studies have been done which deal with the calculation of the shape of Mössbauer spectra when the Mössbauer ion is assumed to perform bounded diffusive motion within a "cage" [1 to 4]. In addition to the contribution to the broadening of the line due to jump diffusion, the change in the hyperfine interaction from one site to another has also been taken into account in many cases [5 to 9].

The formulae used in these references are a straightforward generalization of the usual formula giving the lineshape for a fixed Mössbauer ion in presence of electronic relaxation, first reported by Blume [10], with the assumption that the jumps occur instantaneously and the ion performs many oscillations in the time between two successive jumps [5, 11, 12]. It is interesting, however, to start with a general formula which accounts not only for relaxation effects, but also for bounded and unbounded diffusion (see [12], where diffusion only is taken into account).

In this paper we present a lineshape calculation which, using the concepts of the stochastic theory [13], clearly shows the differences between bounded and unbounded diffusion. The interactions giving rise to changes in electronic states and to nuclear motions are described, as in the usual stochastic theory, by transition probabilities: these are well defined physical quantities and can be calculated by the usual quantum mechanical methods. We would like to point out that the Clauser-Blume's formalism used in previous works [14, 15] for the lineshape calculation seems to be unnecessary, as it was first introduced [16] in order to consider the entire system (nucleus plus electrons) in a framework of stochastic theory. In our case however the method of transition probabilities can well be applied, as we will show, and the introduction of the parameter λ giving the mean frequency of the "chocs" is unnecessary.

[1]) Via Panciatichi 64, 50127 Firenze, Italia.

2. A General Lineshape Expression

The probability of emission of a γ-ray of wave vector $\boldsymbol{k}$ by a nucleus which is a part of a many-body system is given by [13]

$$I(\omega, \boldsymbol{k}) = \text{Re} \int_0^\infty \mathrm{d}t\, \mathrm{e}^{-pt}\, \text{Tr}\big(\varrho B B^\dagger(t)\big) . \tag{2.1}$$

Here $p = -i\omega + \Gamma/2$, Γ being the natural linewidth of the excited level; B is the Hamiltonian describing the interaction between the radiation field and the nucleus and can be written as [17, 18]

$$B = A\, \mathrm{e}^{i\boldsymbol{kR}} , \tag{2.2}$$

where $\boldsymbol{R}$ is the position vector of the center of mass of the nucleus and A acts only on the internal nuclear quantum state; $B^\dagger(t)$ is given by

$$B^\dagger(t) = \mathrm{e}^{i\mathcal{H}t} A^\dagger\, \mathrm{e}^{-i\boldsymbol{kR}}\, \mathrm{e}^{-i\mathcal{H}t} , \tag{2.3}$$

where $\mathcal{H}$ is the Hamiltonian for the entire system containing the nucleus plus its surroundings; ϱ is the equilibrium statistical operator for the system.

At this point it is necessary to introduce some physical assumptions concerning the nuclear motion. We assume that the many-body system in which the nucleus is embedded is a lattice, so that the motion of the Mössbauer ion includes both oscillations and jumps to neighbouring sites (diffusion). The case of a liquid or a dense gas will be considered later. These two types of motion are assumed to be uncorrelated (the frequency of diffusive jumps is much smaller than the frequency of oscillation) [11]. In this way it is possible to write

$$\boldsymbol{R} = \boldsymbol{R}_\mathrm{d} + \boldsymbol{R}_\mathrm{v} , \tag{2.4}$$

i.e., the nuclear motion is considered to be composed of two independent parts, a diffusional and a vibrational part.

In order to evaluate (2.1) we follow the stochastic approach [10, 13] so that $\mathcal{H}$ becomes a random time-dependent Hamiltonian acting on the nuclear variables, varying with the changes of the electronic state of the ion as well as with the changes of its position due to the diffusion. In the following we consider the motion in a lattice, and the nuclear positions will be labelled by a discrete index; greek letters will be used for position indices, latin letters for stochastic indices. So, for example, $|\alpha, a\rangle$ denotes a state in which the nucleus is at the position $\boldsymbol{R}_\alpha$ and the surroundings in the stochastic state $|a\rangle$. Moreover, we denote by V_a the hyperfine Hamiltonian corresponding to the state $|a\rangle$. If the system is in the state $|\alpha, a\rangle$ for $t' = 0$, the probability $\theta_{\substack{\alpha a\\ \beta b}}(\Delta t)$ of finding it in the state $|\beta, b\rangle$ for $t' = \Delta t$ is given by

$$\theta_{\substack{\alpha a\\ \beta b}}(\Delta t) = \delta_{\alpha\beta}\delta_{ab} + \Omega_{\substack{\alpha a\\ \beta b}}\, \Delta t , \tag{2.5}$$

where $\Omega_{\substack{\alpha a\\ \beta b}}$ is the probability per unit time for the system to perform the transition from the state $|\alpha, a\rangle$ to $|\beta, b\rangle$. We can rewrite (2.5) in the compact form

$$\theta(\Delta t) = I + \Omega\, \Delta t . \tag{2.6}$$

The dimension of the matrices θ, Ω, is $(N \times n)$, with N the number of lattice sites and n the number of stochastic states. A formal simplification is obtained in (2.1) introducing the Liouville operator $\mathcal{H}^\times$ associated to $\mathcal{H}$ [10], so that

$$B^\dagger(t) = e^{i\int_0^t dt'\mathcal{H}(t')} A^\dagger e^{-i\boldsymbol{kR}} e^{-i\int_0^t dt'\mathcal{H}(t')} = e^{i\int_0^t dt'\mathcal{H}^\times(t')} A^\dagger e^{-i\boldsymbol{kR}} . \tag{2.7}$$

It is well known [10, 17] that (2.1) must be averaged over the positional and stochastic degrees of freedom, considering all the possible paths of the ion in the time $(0, t)$,

$$I(\omega, \boldsymbol{k}) = \text{Re} \int_0^\infty dt\, e^{-pt}\, \text{Tr}\big(\varrho_N A G(t)\, A^\dagger\big) , \tag{2.8}$$

where the trace is calculated over the nuclear states, ϱ_N is the corresponding statistical operator and

$$G(t) = \big(e^{i\boldsymbol{kR}(t)}\, e^{-i\boldsymbol{kR}(0)}\, e^{i\int_0^t dt'\mathcal{H}^\times(t')}\big)_{av} . \tag{2.9}$$

The average on the positional degrees of freedom can be factorized into vibrational and diffusive average using (2.4) and one obtains (the quantum features of the diffusive motion are ignored)

$$G(t) = \big(e^{i\boldsymbol{k}(\boldsymbol{R}_d(t)-\boldsymbol{R}_d(0))} \langle e^{i\boldsymbol{kR}_v(t)}\, e^{-i\boldsymbol{kR}_v(0)}\rangle\, e^{i\int_0^t dt'\mathcal{H}^\times(t')}\big)_{av} , \tag{2.10}$$

where $\langle ... \rangle$ denotes an average over the vibrational degrees of freedom, and $(...)_{av}$ an average over the lattice diffusive positions of the nucleus and stochastic states of the surroundings. If we pose $\Delta t = t/s$, if $|\alpha, a\rangle$ is the initial state for $t' = 0$, $|\beta, b\rangle$ the final state for $t' = t$, and $p_{\alpha a}$ the a-priori probability for $|\alpha, a\rangle$, $G(t)$ is given by

$$G(t) = \sum_{\substack{\alpha a\\ \beta b}} p_{\alpha a}\, e^{-i\boldsymbol{k}(\boldsymbol{R}_\alpha - \boldsymbol{R}_\beta)} \langle e^{i\boldsymbol{kR}_\beta^{(v)}}\, e^{-i\boldsymbol{kR}_\alpha^{(v)}}\rangle \lim_{s\to\infty} \Big\{ \sum_{\alpha_1 a_1 \dots \alpha_{s-1} a_{s-1}} e^{iV_a^\times \Delta t} \times$$

$$\times\, (\delta_{\alpha\alpha_1}\delta_{aa_1} + \Omega_{\substack{\alpha a\\ \alpha_1 a_1}} \Delta t)\, e^{iV_{a_1}^\times \Delta t} \dots e^{iV_{a_{s-1}}^\times \Delta t} (\delta_{\alpha_{s-1}\beta}\delta_{a_{s-1}b} + \Omega_{\substack{\alpha_{s-1} a_{s-1}\\ \beta\ b}} \Delta t) \Big\} \tag{2.11}$$

since (2.4) can obviously be written as

$$\boldsymbol{R} = \boldsymbol{R}_\alpha + \boldsymbol{R}_\alpha^{(v)} , \tag{2.12}$$

where $\boldsymbol{R}_\alpha^{(v)}$ describes the vibrational motion around $\boldsymbol{R}_\alpha$.

Let us introduce a set of n matrices such that

$$\langle \gamma, f|\, F_c\, |\delta, g\rangle = \delta_{\gamma\delta}\delta_{fc}\delta_{fg} \tag{2.13}$$

and, moreover, we pose

$$\mathcal{H}_0^\times = \sum_m^{1\dots n} V_m^\times F_m + I\mathcal{H}_N^\times , \tag{2.14}$$

where $\mathcal{H}_N$, the unperturbed nuclear Hamiltonian whose eigenvalues are the energy levels E_1 and E_0 with $E_1 - E_0 = \omega_\gamma = \omega_0$, has been extracted from $\mathcal{H}^\times$. Since the factor $\langle ... \rangle$ in (2.11) gives, following Abragam [18],

$$\langle ... \rangle = e^{-(k^2/2)[\langle (R_\beta^{(v)})^2\rangle + \langle (R_\alpha^{(v)})^2\rangle]} = \sqrt{f_\alpha f_\beta} \equiv f_{\alpha\beta} \tag{2.15}$$

(f_γ Mössbauer factor relative to the γ-th site), (2.11) becomes

$$G(t) = \sum_{\substack{\alpha a\\ \beta b}} p_{\alpha a} f_{\alpha\beta}\, e^{-i\boldsymbol{k}(\boldsymbol{R}_\alpha - \boldsymbol{R}_\beta)} \lim_{s\to\infty} \sum_{\alpha_1 a_1 \dots \alpha_{s-1} a_{s-1}} \langle \alpha, a|\, (1 + i\mathcal{H}_0^\times \Delta t)\, |\alpha, a\rangle\, \times$$

$$\times\, \langle \alpha, a|\, (I + \Omega\,\Delta t)\, |\alpha_1, a_1\rangle\, \langle \alpha_1, a_1|\, (1 + i\mathcal{H}_0^\times \Delta t)\, |\alpha_1, a_1\rangle \dots$$

$$\dots \langle \alpha_{s-1}, a_{s-1}|\, (I + \Omega\,\Delta t)\, |\beta, b\rangle \tag{2.16}$$

and using the completeness relation

$$G(t) = \sum_{\substack{\alpha a\\ \beta b}} p_{\alpha a} f_{\alpha\beta}\, e^{-ik(R_\alpha - R_\beta)} \lim_{s\to\infty} \left\langle \alpha, a \left| \left[\left(1 + i\mathcal{H}_0^\times \frac{t}{s}\right)\left(I + \Omega\frac{t}{s}\right)\right]^s \right| \beta, b \right\rangle =$$

$$= \sum_{\substack{\alpha a\\ \beta b}} p_{\alpha a} f_{\alpha\beta}\, e^{-ik(R_\alpha - R_\beta)} \lim_{s\to\infty} \left\langle \alpha, a \left| \left[I + \left(i\mathcal{H}_0^\times I + \Omega\right)\frac{t}{s}\right]^s \right| \beta, b \right\rangle =$$

$$= \sum_{\substack{\alpha a\\ \beta b}} p_{\alpha a} f_{\alpha\beta}\, e^{-ik(R_\alpha - R_\beta)} \langle \alpha, a|\, e^{(i\mathcal{H}_0^\times I + \Omega)t}\, |\beta, b\rangle\,, \tag{2.17}$$

we obtain

$$I(\omega, k) = \mathrm{Re}\,\mathrm{Tr}\,[\varrho_N A(\sum_{\substack{\alpha a\\ \beta b}} p_{\alpha a} f_{\alpha\beta}\, e^{-ik(R_\alpha - R_\beta)} \langle \alpha, a|\, (pI - \Omega - i\mathcal{H}_0^\times I)^{-1}\, |\beta, b\rangle)\, A^\dagger] \tag{2.18}$$

as the most general expression for the lineshape.

3. The Unbounded Diffusive Motion

Expression (2.18) is a very general one but, if we consider the case of an unbounded diffusion between equivalent sites, a remarkable simplification arises. First, the term $f_{\alpha\beta}$ factorizes to give merely the Mössbauer factor relative to the radioactive ion and to the host lattice considered. Furthermore, (2.16) can be written in the form (obviously $N \to \infty$)

$$\begin{aligned} G(t) = \lim_{s\to\infty} \sum_{\substack{ab\\ \alpha}} p_{\alpha a} \sum_{a_1 \dots a_{s-1}} \{\langle a|\, 1 + i\mathcal{H}_0^\times \Delta t\, |a\rangle\, [\sum_{\alpha_1}^{1\dots N} e^{-ik(R_\alpha - R_{\alpha_1})} \times \\ \times \langle \alpha, a|\, I + \Omega\, \Delta t\, |\alpha_1, a_1\rangle]\, \langle a_1|\, 1 + \mathcal{H}_0^\times \Delta t\, |a_1\rangle \times \\ \times [\sum_{\alpha_2}^{1\dots N} e^{-ik(R_{\alpha_1} - R_{\alpha_2})} \langle \alpha_1 a_1|\, I + \Omega\, \Delta t\, |\alpha_2, a_2\rangle] \times \\ \times \dots \langle a_{s-1}|\, 1 + i\mathcal{H}_0^\times \Delta t\, |a_{s-1}\rangle\, [\sum_{\beta}^{1\dots N} e^{-ik(R_{\alpha_{s-1}} - R_\beta)} \times \\ \times \langle \alpha_{s-1}, a_{s-1}| I + \Omega\, \Delta t\, |\beta, b\rangle]\}\,. \end{aligned} \tag{3.1}$$

Taking into account the translational invariance, so that the s summations do not depend on the first position index, we obtain ($\sum_\alpha p_{\alpha a} = p_a$)

$$G(t) = \sum_{a, b} p_a \langle a|\, e^{(i\mathcal{H}_0^\times I + \Omega(k))t}\, |b\rangle\,, \tag{3.2}$$

where

$$\Omega(k) = \sum_{\alpha}^{1\dots N} e^{ikR_\alpha} \langle 0|\, \Omega\, |\alpha\rangle \tag{3.3}$$

is a matrix in the n-dimensional stochastic space; we obtain

$$I(\omega, k) = \mathrm{Re}\,\mathrm{Tr}\,[\varrho_N\, A(\sum_{ab} p_a \langle a|\, (pI - \Omega(k) - i\mathcal{H}_0^\times I)^{-1}\, |b\rangle)\, A^\dagger]\,. \tag{3.4}$$

We now introduce the simplifying hypothesis that the transitions between stochastic states do not depend on positional variables and vice versa. We therefore write

$$\theta_{\substack{\gamma f\\ \delta g}}(\Delta t) = \tau_{\gamma\delta}(\Delta t) \cdot T_{fg}(\Delta t) \quad \text{or} \quad \theta = \tau \otimes T\,, \tag{3.5}$$

where τ and T are probability matrices for positional and stochastic transitions, respectively, and $\otimes$ has the meaning of direct product. The corresponding transition probabilities per unit time are defined by

$$\tau(\Delta t) = I + \tilde{\omega}\,\Delta t \tag{3.6}$$

which is a $(N \times N)$ matrix ($N \to \infty$ for unbounded diffusion), and

$$T(\Delta t) = I + W\,\Delta t \tag{3.7}$$

which is a $(n \times n)$ matrix. One then obtains

$$\theta_{\gamma f, \delta g}(\Delta t) = \delta_{\gamma\delta}\delta_{\ f} + [\tilde{\omega}_{\gamma\delta}\delta_{fg} + W_{fg}\delta_{\gamma\delta}]\,\Delta t \tag{3.8}$$

so that

$$\Omega = \tilde{\omega} \otimes I(n) + W \otimes I(N)\,. \tag{3.9}$$

$I(n)$ and $I(N)$ are unit matrices in n- and N-dimensional spaces, respectively. (3.3) becomes

$$\Omega(\boldsymbol{k}) = I(n)\,\tilde{\omega}(\boldsymbol{k}) + W \tag{3.10}$$

with

$$\tilde{\omega}(\boldsymbol{k}) = \sum_{\alpha} \tilde{\omega}_{\alpha\beta}\,\mathrm{e}^{-i\boldsymbol{k}(\boldsymbol{R}_\alpha - \boldsymbol{R}_\beta)} = \sum_{\alpha} \mathrm{e}^{i\boldsymbol{k}\boldsymbol{R}_\alpha}\,\tilde{\omega}_{0\alpha}\,. \tag{3.11}$$

Using the normalization condition for probabilities, from (3.6) it follows

$$\tilde{\omega}_{00} = -\sum_{\gamma \neq 0} \tilde{\omega}_{0\gamma} \tag{3.12}$$

so that

$$\tilde{\omega}(\boldsymbol{k}) = -\sum_{\alpha \neq 0} (1 - \mathrm{e}^{i\boldsymbol{k}\boldsymbol{R}_\alpha})\,\tilde{\omega}_{0\alpha}\,. \tag{3.13}$$

4. Some Particular Cases

Let us now consider a Mössbauer ion which undergoes bounded diffusive motion. This is the case, for example, of hyperfine Hamiltonian synchronous changing to the nuclear motion [19], of conformational motions in biological macromolecules [20], of the ^{57}Co atom in dilute AlCo alloys [2], and of the iron-containing core of ferritin [3]. In these models relaxation effects are ignored, but it is often useful to suppose that, due to the change in the ion position, the hyperfine Hamiltonian also changes [5]. In this way, to the N nuclear positions will correspond N forms for the hyperfine Hamiltonian (not necessarily all distinct one from another), and the two spaces of the position vectors and of the stochastic states $|a\rangle$ are coincident. We have

$$\mathcal{H}_0^\times = \sum_{m}^{1\ldots N} V_m^\times F_m + I\mathcal{H}_N^\times \tag{4.1}$$

and (2.18) becomes

$$I(\omega, \boldsymbol{k}) = \mathrm{Re}\,\mathrm{Tr}\,[\varrho_N A(\sum_{\alpha\beta} p_\alpha f_{\alpha\beta}\,\mathrm{e}^{-i\boldsymbol{k}(\boldsymbol{R}_\alpha - \boldsymbol{R}_\beta)}\langle\alpha|\,(pI - \tilde{\omega} - i\mathcal{H}_0^\times I)^{-1}\,|\beta\rangle)\,A^\dagger]\,. \tag{4.2}$$

In some cases the term $f_{\alpha\beta}$, accounting for a different vibrational behavior of the Mössbauer ion from one site to another, can be important [20].

If $\mathcal{H}_0^\times = \mathcal{H}_N^\times$, i.e. without relaxation effects, one easily obtains

$$I(\omega, \boldsymbol{k}) \sim \mathrm{Re} \sum_{\alpha\beta} p_\alpha f_{\alpha\beta}\,\mathrm{e}^{-i\boldsymbol{k}(\boldsymbol{R}_\alpha - \boldsymbol{R}_\beta)}\langle\alpha|\,[(p + i\omega_0)\,I - \tilde{\omega}]^{-1}\,|\beta\rangle\,. \tag{4.3}$$

In all these formulae we can pose $p_\alpha = 1/N$.

Let us now show that (3.4) coincides with equations given in [14] where continuous indices are used for nuclear positions. We start by (2.6) defining Ω. We can also write

$$\theta = I + \Omega/\lambda \tag{4.4}$$

as the time-independent probability matrix corresponding to the mean time between two successive pulses $\overline{\Delta t} = 1/\lambda$ [14, 16]. It follows

$$\Omega(\boldsymbol{k}) = \sum_\alpha \mathrm{e}^{i\boldsymbol{k}\boldsymbol{R}_\alpha}\langle 0|\ \Omega\ |\alpha\rangle = \lambda \sum_\alpha \mathrm{e}^{i\boldsymbol{k}\boldsymbol{R}_\alpha}\langle 0|\ (\theta - I)\ |\alpha\rangle = \lambda[\theta(\boldsymbol{k}) - I(n)]\,. \tag{4.5}$$

If (3.5) holds,

$$\Omega(\boldsymbol{k}) = \lambda \sum_\alpha \mathrm{e}^{i\boldsymbol{k}\boldsymbol{R}_\alpha}\langle 0|\ \tau \otimes T - I\ |\alpha\rangle = \lambda[\tau(\boldsymbol{k})\ T - I(n)]\,, \tag{4.6}$$

where

$$\tau(\boldsymbol{k}) = \sum_\alpha \mathrm{e}^{i\boldsymbol{k}\boldsymbol{R}_\alpha}\ \langle 0|\ \tau\ |\alpha\rangle \tag{4.7}$$

can be interpreted, using a continuous picture, as the Fourier transform of the mean probability $\tau(\boldsymbol{R}_\alpha)$ of a single nuclear jump over a distance $\boldsymbol{R}_\alpha$. (4.6), upon inserting in (3.4), gives formula (28) of [14].

In the same way it is possible to obtain the usual results for uncorrelated jump diffusion [17]. In this case (3.4) must be applied, which becomes, by taking into account (3.10) with $W = 0$,

$$I(\omega, \boldsymbol{k}) \sim \mathrm{Re}\,[p + i\omega_0 - \tilde{\omega}(\boldsymbol{k})]^{-1}\,. \tag{4.8}$$

Starting from (3.13) and considering short jumps ($\boldsymbol{k}\boldsymbol{R}_\alpha \ll 1$) in random directions (i.e., neglecting the crystal structure [21]) one obtains, averaging $\tilde{\omega}(\boldsymbol{k})$ over all directions,

$$\tilde{\omega}(\boldsymbol{k}) = -\tfrac{1}{6} \sum_{\alpha \neq 0} \tilde{\omega}_{0\alpha} R_\alpha^2\ |\boldsymbol{k}|^2\,. \tag{4.9}$$

Since (as (4.4))

$$\tau_{0\alpha} = \delta_{0\alpha} + \frac{1}{\lambda}\ \tilde{\omega}_{0\alpha}\,, \tag{4.10}$$

(4.9) becomes

$$\tilde{\omega}(\boldsymbol{k}) = -\frac{\lambda}{6} \sum_{\alpha \neq 0} \tau_{0\alpha} R_\alpha^2 k^2 = -Dk^2\,, \tag{4.11}$$

where $D = (\lambda/6)\langle R^2\rangle$ is the diffusion coefficient in the uncorrelated jump model [17]. We observe that this model can hold, for example, for liquids. If diffusion in a crystal is considered, the correlation between successive jumps of the ion, caused by the vacancy mechanism, is to be taken into account [11, 22 to 24] as well as the crystal structure [21]. In fact, because of the crystal structure the angle between successive jumps cannot be taken to be random, so that the average over all directions in space must be performed in the final expression $I(\omega, \boldsymbol{k})$.

For a nearest-neighbor random jump model, (3.13) gives, in terms of the mean time $1/\lambda$ between jumps (m is the number of nearest neighbors)

$$\tilde{\omega}(\boldsymbol{k}) = -\lambda \left(1 - \frac{1}{m} \sum_\alpha^{1\ldots m} \overline{\cos\,(\boldsymbol{k}\cdot\boldsymbol{R}_\alpha)}\right) = -\lambda\left(1 - \frac{\sin kr}{kr}\right), \tag{4.12}$$

where $r = |\boldsymbol{R}_\alpha|$.

5. The Limit of Continuous Nuclear Motion

We consider in the following the limiting case, when the ion performs an unbounded continuous motion; in particular, the behavior of a free nucleus and of a nucleus performing diffusive continuous motion will be examined.

The concept of transition probability per unit time has no physical meaning now. The following attempts are made to clarify the meaning of $\tilde{\omega}$ in this case.

Let us indicate by $\psi(\boldsymbol{r}, t)$ the wave function describing the motion of the nucleus. The probability amplitude $g(\boldsymbol{r}_1, \boldsymbol{r}_2; t)$ is introduced, as the probability amplitude of finding the nucleus in $\boldsymbol{r} = \boldsymbol{r}_2$ for $t' = t$, when it is certainly in $\boldsymbol{r} = \boldsymbol{r}_1$ for $t' = 0$. In analogy with (2.5) we try to pose

$$g(\boldsymbol{r}_1, \boldsymbol{r}_2; \Delta t) = \delta(\boldsymbol{r}_2 - \boldsymbol{r}_1) + \tilde{\omega}(\boldsymbol{r}_1, \boldsymbol{r}_2)\, \Delta t\,. \tag{5.1}$$

$g(\boldsymbol{r}_1, \boldsymbol{r}_2; \Delta t)$ and $\tilde{\omega}(\boldsymbol{r}_1, \boldsymbol{r}_2)$ are subjected to the relations

$$\begin{aligned} g(\boldsymbol{r}_1, \boldsymbol{r}_2; \Delta t) &= g(0, \boldsymbol{r}_2 - \boldsymbol{r}_1; \Delta t) \equiv g(\boldsymbol{r}_2 - \boldsymbol{r}_1; \Delta t)\,, \\ \tilde{\omega}(\boldsymbol{r}_1, \boldsymbol{r}_2) &= \tilde{\omega}(0, \boldsymbol{r}_2 - \boldsymbol{r}_1) \equiv \tilde{\omega}(\boldsymbol{r}_2 - \boldsymbol{r}_1)\,, \end{aligned} \tag{5.2}$$

denoting translational invariance. It is therefore possible to write, using (5.1),

$$\begin{aligned} \psi(\boldsymbol{r}, \Delta t) &= \int \mathrm{d}^3\boldsymbol{r}_1\, g(\boldsymbol{r} - \boldsymbol{r}_1, \Delta t)\, \psi(\boldsymbol{r}_1, 0) = \psi(\boldsymbol{r}, 0) + \Delta t \int \mathrm{d}^3\boldsymbol{r}_1 \times \\ &\quad \times \tilde{\omega}(\boldsymbol{r} - \boldsymbol{r}_1)\, \psi(\boldsymbol{r}_1, 0)\,.^{2)} \end{aligned} \tag{5.3}$$

It follows

$$\frac{\partial \psi}{\partial t} = \int \mathrm{d}^3\boldsymbol{r}_1\, \tilde{\omega}(\boldsymbol{r} - \boldsymbol{r}_1)\, \psi(\boldsymbol{r}, t) \tag{5.4}$$

and, from the Schrödinger equation,

$$-i\mathcal{H}\psi(\boldsymbol{r}, t) = \int \mathrm{d}^3\boldsymbol{r}_1\, \tilde{\omega}(\boldsymbol{r} - \boldsymbol{r}_1)\, \psi(\boldsymbol{r}_1, t)\,. \tag{5.5}$$

If the nucleus is free, we have

$$\mathcal{H} = -\frac{1}{2m} \nabla^2 \tag{5.6}$$

and performing the Fourier transform,

$$-\frac{i}{2m} k^2 \psi(\boldsymbol{k}, t) = \tilde{\omega}(\boldsymbol{k})\, \psi(\boldsymbol{k}, t) \tag{5.7}$$

with

$$\tilde{\omega}(\boldsymbol{k}) = \int \mathrm{d}^3\boldsymbol{r}\, \mathrm{e}^{i\boldsymbol{k}\boldsymbol{r}}\, \tilde{\omega}(\boldsymbol{r}) \tag{5.8}$$

so that

$$\tilde{\omega}(\boldsymbol{k}) = -i\, \frac{k^2}{2m} \tag{5.9}$$

is the Fourier transform of $\tilde{\omega}(\boldsymbol{r})$, as given by (3.11), when the limit to continuous motion is performed. (4.8) gives in this case (remembering that $\boldsymbol{k}$ is the γ-ray wave vector)

$$I(\omega, \boldsymbol{k}) \sim \mathrm{Re} \left[\frac{\Gamma}{2} - i\left(\omega - \omega_0 + \frac{k^2}{2m}\right)\right]^{-1} \tag{5.10}$$

showing the usual recoil energy of the nucleus [18].

If the nucleus is subjected to diffusion in a fluid, we can assume, using a classical

---

2) (5.3), written in the general form $\psi(\boldsymbol{r}, t) = \int \mathrm{d}^3\boldsymbol{r}_1\, g(\boldsymbol{r}, \boldsymbol{r}_1; t)\, \psi(\boldsymbol{r}_1, 0)$ is the basis of the path integral formulation of quantum mechanics [25]; $g(\boldsymbol{r}, \boldsymbol{r}_1; t)$ is the kernel of this integral equation.

picture, that $\psi(\boldsymbol{r}, t)$ satisfies an isotropic diffusion equation [15]

$$\frac{\partial \psi}{\partial t} = D\nabla^2\psi \,, \tag{5.11}$$

where D is the diffusion coefficient. From (5.4) we obtain

$$-Dk^2\psi(\boldsymbol{k}, t) = \tilde{\omega}(\boldsymbol{k})\, \psi(\boldsymbol{k}, t) \tag{5.12}$$

so that

$$\tilde{\omega}(\boldsymbol{k}) = -Dk^2 \,, \tag{5.13}$$

giving the well-known diffusional broadening first obtained in [17].

6. Conclusion

In this paper we have reformulated the stochastic lineshape theory for Mössbauer spectra, including diffusive motions of the nucleus. We write a general formula, from which all the particular cases of interest can be derived, and which gives the correct results in the continuous limits. It is interesting to observe that (3.4) and (4.2) follow from the same equation, when the characteristic features of the particular situations are taken into account (i.e., translational invariance and bounded motion). In the previous literature, on the contrary, these formulae appear to be completely uncorrelated and perhaps obtained by different methods [26]. The aim of this work, in conclusion, is to complete and clarify a branch of the Mössbauer spectroscopy that, especially in the last few years, has been the object of several studies and experimental works.

References

[1] M. A. KRIVOGLAZ and S. P. REPETSKII, Soviet Phys. – Solid State 8, 2325 (1967).
[2] G. VOGL, W. MANSEL, and P. H. DEDERICHS, Phys. Rev. Letters 36, 1497 (1976).
[3] S. G. COHEN, E. R. BAUMINGER, I. NOWICK, S. OFER, and J. YARIV, Phys. Rev. Letters 46, 1244 (1981).
[4] A. BLÄSIUS, R. S. PRESTON, and U. GONSER, Z. phys. Chem. N.F., 115, 187 (1979).
[5] S. DATTAGUPTA, Solid State Commun. 24, 19 (1977).
[6] F. J. LITTERST and V. D. GOROBCHENKO, phys. stat. sol. (b) 113, K135 (1982).
[7] F. J. LITTERST, A. M. AFANASEV, P. A. ALEKSANDROV, and V. D. GOROBCHENKO, Solid State Commun. 45, 963 (1983).
[8] V. D. GOROBCHENKO and F. J. LITTERST, phys. stat. sol. (b) 127, 351 (1985).
[9] F. J. LITTERST, A. LERF, O. NUYKEN, and H. ALCALÀ, Hyperfine Interactions 12, 317 (1982).
[10] M. BLUME, Phys Rev. 174, 351 (1968).
[11] M. C. DIBAR URE and P. A. FLINN, Phys. Rev. B 15, 1261 (1977).
[12] W. PETRY and G. VOGL, Z. Phys. B 45, 207 (1982).
[13] M. BLUME and J. A. TJON, Phys. Rev. 165, 446 (1968).
[14] S. DATTAGUPTA, Phys. Rev. B 12, 47 (1975).
[15] S. DATTAGUPTA, Phys. Rev. B 14, 1329 (1976).
[16] M. J. CLAUSER and M. BLUME, Phys. Rev. B 3, 583 (1971).
[17] K. S. SINGWI and A. SJÖLANDER, Phys. Rev. 120, 1093 (1960).
[18] A. ABRAGAM, in: Physique des basses temperatures, Ed. C. DE WITT, B. DREYFUS, and P. G. DE GENNES, Presses Univ. de France, Paris 1961 (p. 479).
[19] E. BILL, N. BLAES, K. FISCHER, U. GONSER, K. H. PAULY, R. PRESTON, F. SEEL, R. STAAB, and A. X. TRAUTWEIN, Z. Naturf. 39b, 333 (1984).
[20] H. KELLER and P. G. DEBRUNNER, Phys. Rev. Letters 45, 68 (1980).
[21] M. C. DIBAR URE and P. A. FLINN, Appl. Phys. Letters 23, 587 (1973).
[22] R. C. KNAUER, Phys. Rev. B 3, 567 (1961).
[23] D. WOLF, Appl. Phys. Letters 30, 617 (1977).
[24] O. BENDER and K. SCHROEDER, Phys. Rev. B 19, 3399 (1979).
[25] R. P. FEYNMAN and A. R. HIBBS, Quantum Mechanics and Path Integrals, Chap. 4, McGraw-Hill, New York 1965.
[26] S. DATTAGUPTA, Proc. Internat. Conf. Appl. Mössbauer Effect, Jaipur (India) 1981 (p. 108).

(Received July 3, 1985)

phys. stat. sol. (b) 133, 371 (1986)
Subject classification: 78.70; 71.70; S10.15

Laboratoire des Composés Non-Stoechiométriuqes, Université Paris-Sud[1])

X-Ray Absorption K-Edge Spectra of Titanium in (Ti, Nb)O_2 Rutile Solid Solutions

By
B. POUMELLEC, F. LAGNEL, J. F. MARUCCO, and B. TOUZELIN

The X-ray absorption K-edge spectrum of titanium in (Ti, Nb)O_2, rutile or pseudo-rutile solid solutions is recorded. The second derivative spectra are used to measure the energy shift of the transitions. A partial transfer of Nb 4d electron onto Ti 3d orbitals is found implying a similar energy of these orbitals. On the other hand a decrease of the crystal field is observed when Nb content increases. This may arise from the increase of the mean number of conduction electrons or the increase of the mean size of the metallic octahedron. Lastly, because the electronic structure of the titanium evolves continuously with the niobium content, niobium is certainly randomly distributed around the titanium.

L'absorption X au seuil du titane a été enregistrée à l'aide du rayonnement synchrotron d'orsay pour des solutions solides rutiles ou pseudo-rutiles (Ti, Nb)O_2. Les spectres des derivées secondes ont été analysés et les déplacements énergétiques des différentes transitions mesurés. On en a déduit que l'électron 4d du niobium est partiellement transféré sur le titane impliquant des niveaux Nb 4d et Ti 3d comparables. De plus, on a remarqué une diminution du champ de ligands quand la concentration de niobium augmente. Ceci pourrait provenir de l'écrantage qui augmente avec le nombre d'électrons de conduction ou de l'augmentation de distance des ligands. Enfin puisque le titane voit sa structure électronique évoluer continuement avec la concentration de niobium, celui-ci est certainement distribué aléatoirement autour du titane.

1. Introduction

The solid solutions (Ti, Nb)O_2 are used in the fabrication of optical wave guides. They have rutile crystal form from TiO_2 to 80 at% of niobium at room temperature [1, 2]. The rutile space group is D_{4h}^{14}, the coordination octahedron of oxygen in TiO_2 is slightly distorted, i.e. four Ti–O distances are 0.193 and two Ti–O distances 0.201 nm [1, 3]. For highest niobium content a phase transition takes place by distortion and size increase of the cell [1]. The space group becomes C_{4h}^6. The octahedra in NbO_2 are distorted and actually divided in two sites. The axial Nb–O distances ("axial" corresponds to the axis perpendicular to D_{2h} mirror) are 0.207 and 0.189 nm for the first octahedron and 0.194 and 0.201 nm for the second one. The four other distances are 0.215, 0.212, 0.205, and 0.211 nm for the first one and 0.203, 0.198, 0.215, and 0.219 nm for the second one. In comparison with TiO_2, in NbO_2 the axial M–O (M = Ti, Nb) distances are shortened and the four others are enlarged.

The stability region of these solid solutions has been previously studied [4] showing a small departure from stoichiometry (O/M $<$ 2.006). From this study, stoichiometric compounds have been prepared with non-stoichiometry lower than 0.00005. The experimental preparation procedure is described in [4].

[1]) Bâtiment 415, 91405 Orsay Cédex, France.

2. Experimental

X-ray absorption spectra were recorded at L.U.R.E., Orsay, on the DCI synchrotron radiation on EXAFS 2 line with 0.25 eV energy step. The monochromator was Si(311) calibrated on the first weak peak of 4 at% Nb spectrum at 4967.75 eV.

The absorption spectra were obtained by computing ln (I_0/I) versus energy. Absorption background was then evaluated by a linear least mean square procedure before the threshold between 4950 and 4960 eV and then deduced from ln (I_0/I). Then every spectrum was normalized to unity between the minimum and the maximum.

Moreover to measure clearly the energy position of the peaks, it was necessary to calculate the second derivative of the absorption. This was impossible without smoothing. This was achieved with a second degree polynomial fitted in a 2 or 3 eV wide window. This procedure returns a smooth absorption spectrum, a first derivative, and a second one (Fig. 2a and b). Then the energy position of the peaks was determined by the position of the minima of the second derivative.

3. Absorption Spectra Analysis

3.1 Absorption process

X-ray absorption of material is due to electronic transitions from a deep core level (1s for K-edges) to empty levels above the Fermi level. The intensity of X-ray absorption is related to the Fermi golden rule as [5]

$$I(E) \sim \sum_{i,f} |\langle \Psi_f| \, \boldsymbol{e} \cdot \boldsymbol{r} \, |\Psi_i\rangle|^2 \, \delta(E - E_f + E_i)$$

with E the energy of the photon, Ψ_i the initial Bloch function of the core electron (1s), E_i the corresponding energy, Ψ_f a final Bloch function above the Fermi level: an unoccupied level, E_f the corresponding energy, and $\boldsymbol{e}$ the electric field. Such transitions involving only the dipolar momentum are named dipolar transitions.

In the case of a one-electron transition model for a band with a given symmetry, we can rewrite the sum above and we have $I(E) \sim P(E)\,N(E)$, where $N(E)$ is the density of empty states for which there is a non-zero electronic density on the titanium, $P(E)$ an energy-dependent matrix element overlap factor.

In the one-electron transition model, only one core electron is excited, the others being assumed to be unaffected, remaining frozen in their original states. Many-body effects such as relaxation of valence electrons or core hole and excited electron interaction are neglected. Moreover, for transitions of an electron from 1s to rather localized states as expected in oxides namely 3d, molecular orbital description may be reasonable. In fact, maxima of the band density of states if they are not to far from the Fermi level (30 eV) correspond to atomic orbitals of the cluster. For higher energy, mixing between atomic orbitals with different orbital momentum cancels the possibility to assign simply the maxima [6].

3.2 Assignment

For the bands closest to the Fermi level the probability of transition follows the same selection rule as in the cluster and so the theoretical discussion on the intensity based on group theory is valid [7].

For octahedral symmetry (O_h) around titanium, cluster calculation [8, 9] shows that the empty symmetrized cluster orbital for TiO_6 is t_{2g}, e_g, a_{1g}, and t_{1u} corresponding to mixing between titanium and oxygen ligand orbitals. Nevertheless those molecular orbitals are largely built from titanium atomic orbitals [10]. Hence the t_{2g} orbitals are almost $3d_{xy}$, $3d_{yz}$, and $3d_{zx}$, the e_g orbitals are almost $3d_{x^2-y^2}$ and $3d_{z^2}$, the a_{1g}

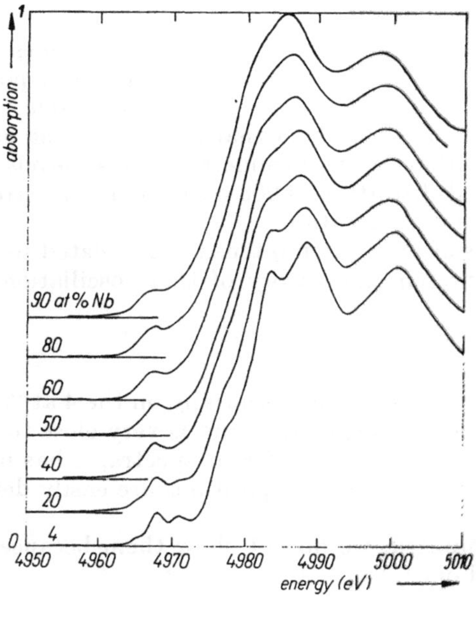

Fig. 1. Normalized absorption on the Ti K-edge vs. the energy from 4 to 90 at% Nb content

orbital is almost 4s, the t_{1u} orbitals are almost $4p_x$, $4p_y$, and $4p_z$. In fact, the point symmetry of Ti in TiO_2 is D_{2h} but the distortion, lowering the O_h symmetry, is weak and not clear in our experiment. Such facts give the possibility to assign the maxima appearing on the 4 at% Nb spectrum (Fig. 1).

We see between 4965 and 4971 eV three peaks, the pre-edge, a weak hump on the edge at about 4977 eV, and a strong peak at about 4983 eV after the edge followed by two other strong peaks, but wider than the previous ones.

3.2.1 Pre-edge

It seems to us that the first shoulder at about 4965 eV is a Frenkel exciton peak previously mentioned by Grunes [7] rather than due to the empty density of states at the top of the valence band which is quite small [9]. This first peak does not appear in any cluster or band calculation so this peak corresponds to a many-body effect. Multiplet splitting of excited states is not possible here because the ground state configuration is $3d^0$, the electron correlation energy amounts to 1 eV and the splitting between the O 2p highest occupied and Ti 3d lowest unoccupied levels is 3 eV. Furthermore the first excited configuration is $1s^1t_{2g}^1$ and the second one is $1s^1e_g^1$. Hence the second feature at 4968 eV arises from the t_{2g} band and the third one from the e_g band. Splitting of these two peaks shows the crystal field splitting which is about 3 eV, consistent with the cluster calculation [8]. On the other hand, transitions from $1s(a_{1g})$ to t_{2g} or to e_g are forbidden because $\boldsymbol{e} \cdot \boldsymbol{r}$ has the t_{1u} symmetry. So $\Psi_f \cdot \boldsymbol{e} \cdot \boldsymbol{r} \cdot 1s$ has no a_{1g} component and subsequently $\langle \Psi_f | \boldsymbol{e} \cdot \boldsymbol{r} | 1s \rangle$ is zero. This result remains true even in D_{2h} symmetry. In fact, the intensity of the 1s-to-t_{2g} (4968 eV) or 1s-to-e_g (4971 eV) transitions (Fig. 1) are not completely zero, this arises from other effects like a dynamic distortion of the site to a lower symmetry or p-quadrupolar interactions.

3.2.2 On the edge

The hump at 4977 eV should be due to the 1s-to-4s forbidden transition. For the same reasons as for the pre-edge peaks, it appears weakly. The splitting between e_g and 4s is about 6 eV in agreement with the cluster calculation [8, 9].

3.2.3 Above the edge

At about 4983 eV and above, strong peaks are present on the spectra. It does not seem that the two latter one (quoted A and B) would be assigned easily. Those peaks are wider than the previous ones and occur at a too large energy to assume that orbital mixing were not present. On the contrary, we think that the first strong peak arises mainly from the three 1s-to-4p transitions. In O_h symmetry these 4p levels are degenerate but not in D_{2h}. In fact the axial distortion (0.008 nm) is not very strong and a splitting is not detectable in regard to our resolution.

For higher energy, the large fluctuations of absorption intensity are related to the distortion of the octahedral site, but at present the theory of these oscillations is not yet well established.

4. Second Derivative Analysis

We have assignated above the different peaks appearing essentially on the 4 at% Nb absorption spectrum. This one is the best resolved spectrum and is very close to the TiO_2 spectrum. In order to quantitatively compare the different spectra, it has been useful to compute the second derivative from which peak positions are easily determined.

Although the first derivatives were computed, we do not show them because all the information is exhibited by the second one.

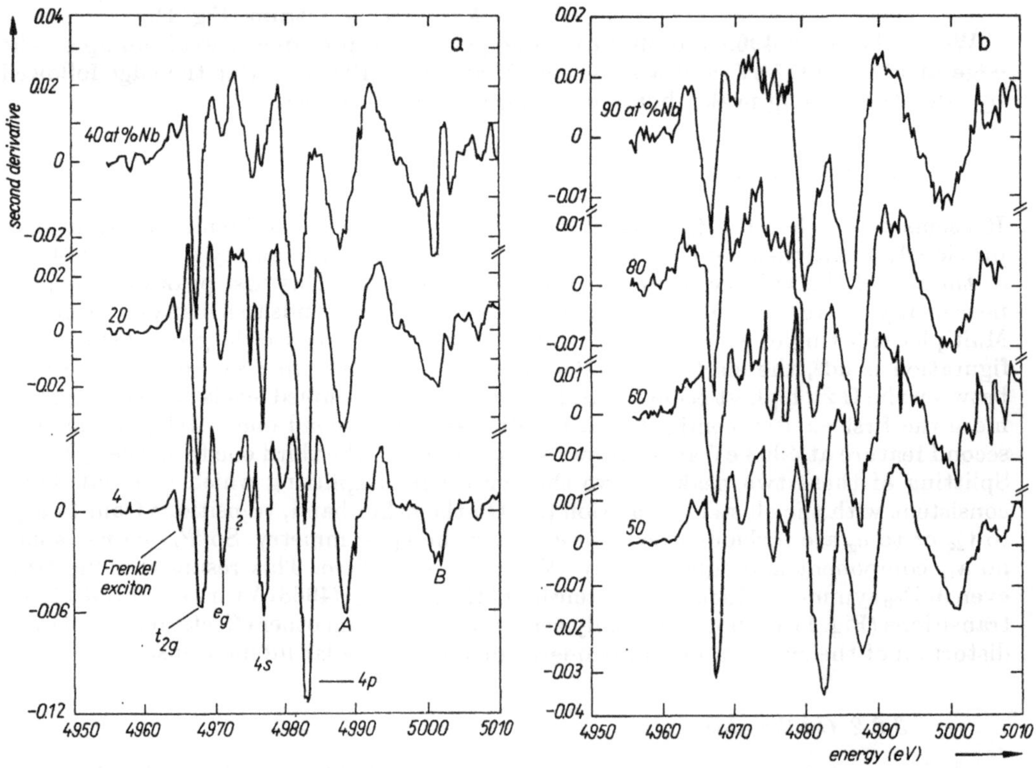

Fig. 2. Second derivatives of the absorption spectra shown in Fig. 1. Assignment of the peaks was made on the best resolved spectrum (4 at%) and energy peak shift was then measured by comparison

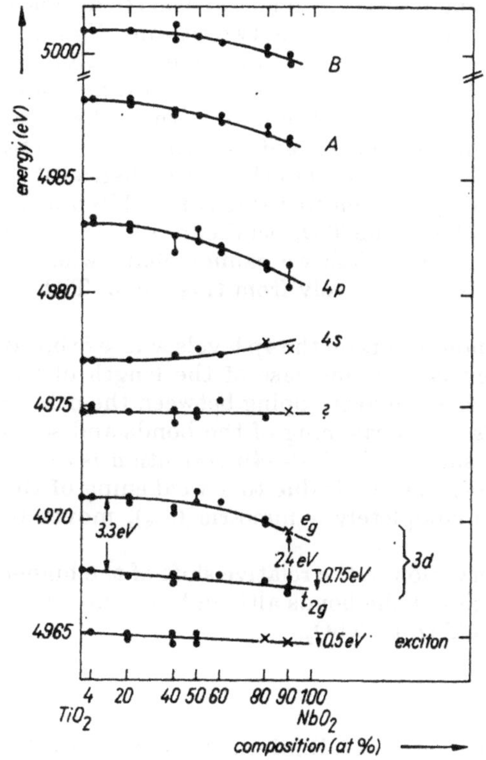

Fig. 3. Energy shift of the different features appearing on the second derivatives. Points are measured and crosses are estimated energies (see text)

Graphs of the second derivatives are shown on Fig. 2a and b. On these spectra we find again the different peaks appearing on absorption spectra but better resolved. Notice that the position of the peaks is often easily identifiable by the energy position of the negative minima. The displacements of the peaks were recorded from 4 to 90 at% Nb and results are collected on Fig. 3 and described below in Section 5.

A new weak feature appears on the Fig. 2 (quoted by interrogation sign), we have no explanation at present of this feature.

For more than 50 at% Nb the exciton peak is no more visible. Nevertheless by comparing with the 4 at% Nb derivative, it does exist till 90 at% Nb and we have estimated its position. It is the same for peaks between 4970 and 4980 eV for which the noise cancels the minima.

From t_{2g} to 4p, peak widths measured by the position of the zeros on 4 at% Nb second derivative are about 1.75 to 2 eV wide. This indicates narrow band of about 1 eV wide because the experimental resolution is about 1 eV. This is consistent with the band calculation made in TiO_2 [11].

For peaks at 4989 and 5001 eV, the width amounts to 4 eV for 4 at% Nb, and to 8 eV for 90 at% Nb where the resolution is worse but unknown.

5. Energy Shift of the Peaks

The energy positions of every peak for all solid solutions are shown in Fig. 3. Absorption spectra shift to the lower energy when niobium content increases. With reference to the exciton or to the t_{2g} peak, it amounts to 0.5 to 0.75 eV. In comparison with the results of Bianconi [12] or Wong et al. [13] which found 1 eV per ionization degree, it can be deduced that a partial electron transfer takes place from Nb $4d^1$ to Ti $3d^0$. This is consistent with the same energy level of the Nb 4d and Ti 3d in (Ti, Nb)O_2 [14].

The crystal field splitting which is the splitting between t_{2g} and e_g, seems to decrease from 3.3 to 2.4 eV as the Nb content increases. This comes from two reasons. First the increase of the mean number of the electrons on titanium and, but weakly, the increase of the conduction electron number of the solid, screens the crystal field and hence diminishes the splitting. The second reason is that the mean size of the octahedron is greater in NbO_2 than in TiO_2. The mean value of M–O distances in the D_{2h} mirror plane is 0.211 nm for one site and 0.209 nm for the other site whereas it is 0.193 nm in TiO_2. Perpendicular to the mirror plane, mean values are 0.198 nm for both of sites whereas it is 0.201 nm in TiO_2. Hence the TiO_6 octahedron is stretched whereas the NbO_6 octahedron is flattened. Nevertheless we cannot claim whether the Ti octahedron in the solid solution evolves continuously from this one in TiO_2 to that one in NbO_2.

The same screening or distortion phenomenon occurs on the 4p levels whose orbitals are oriented as $3d_{x^2-y^2}$ and $3d_{z^2}$ of the e_g levels. The increase of the length of the bonds or the increase of the screening diminishes the overlapping between the orbital of the titanium and of the oxygen, this leads to a weakening of the bonds and so to a lowering of the antibonding levels. As the shift of the 1s-to-4p transition is about 2.5 eV and the chemical shift is about 0.75 eV, the rest is due to a weakening of the bond. At the opposite the 4s level which is completely symmetric (a_{1g}), moves to higher energy.

No large modification is detectable by observation of the relative shift of the higher energy peaks to each other related to the angles of the bonds although we know that the octahedron is significantly distorted from TiO_2 to NbO_2.

6. Conclusion

The X-ray edge absorption spectra bring some very usefull information on the evolution of the electronic and coordination structure of the solid solutions.

The chemical shift of the spectra indicates a partial electronic transfer between Ti and Nb. This implies 3d Ti and 4d Nb levels at the same energy. This is consistent with the fact that neither the band gap nor the flat band potential changes with increasing niobium content [14]. This confirms the hypothesis made to explain the electron transport in (Ti, Nb)O_2 solid solutions [15].

On the other hand, it can be concluded that this solid solution is really disordered as Ti is influenced by the vicinity of Nb. Nevertheless the phase transition which appears for niobium content higher than 80 at% does not yield a discontinuity on the absorption properties. This implies that long-range order is not determinant or not resolved in this spectroscopy and we can conclude that the change of symmetry is continuous from TiO_2 to NbO_2.

Acknowledgements

We would like to acknowledge G. Calas, J. Petiau, A. Manceau, R. Cortes, J. Goulon, and A. Michalovitch for experimental and technical assistance and helpful discussions.

References

[1] K. SAKATA, Acta cryst. **B35**, 2836 (1979).
[2] SH. SHIN, T. H. HALPERN, and P. M. RACCAH, Mater. Res. Bull. **10**, 1061 (1975).
[3] M. G. BLANCHIN, E. VICARIO, and R. A. PLOC, J. appl. Cryst. **10**, 228 (1977).
[4] J. F. MARUCCO, B. POUMELLEC, J. GAUTRON, and P. LEMASSON, J. Phys. Chem. Solids **46**, 709 (1985).

[5] J. E. MULLER, in: EXAFS and Near Edge Structure III, Ed. K. O. HODGSON, B. HEDMAN, and J. E. PENNER-HAHN, Springer-Verlag, Berlin 1984 (p. 7).
[6] N. DAUDE, C. GOUT, and C. JOUANIN, Phys. Rev. B 15, 3226 (1977).
[7] L. A. GRUNES, Phys. Rev. B 27, 2111 (1983).
[8] J. A. TOSSEL, D. J. VAUGHAN, and K. H. JOHNSON, Amer. Mineralogist 59, 319 (1974).
[9] D. W. FISHER, Phys. Rev. B 5, 4219 (1972).
[10] L. A. GRUNES, R. D. LEAPMAN, C. N. WILKER, R. HOFFMANN, and A. B. KUNZ, Phys. Rev. B 25, 7157 (1982).
[11] K. VOS, J. Phys. C 10, 3917 (1977).
[12] A. BIANCONI, Phys. Rev. B 26, 2741 (1982).
[13] J. WONG, F. W. LYTLE, R. P. MESSMER, and D. H. MAYLOTTE, Phys. Rev. B 30, 5596 (1984).
[14] J. GAUTRON, P. LEMASSON, B. POUMELLEC, and J. F. MARUCCO, Solar Energy Mater. 9, 101 (1983).
[15] B. POUMELLEC and J. F. MARUCCO, J. Phys. Chem. Solids 46, 71 (1985).

(Received April 26, 1985)

phys. stat. sol. (b) **133**, 379 (1986)

Subject classification: 78.55; 71.55; S8.16

Dipartimento di Fisica, Università di Firenze[1]) *(a)*
and Dipartimento di Fisica, Università di Parma[2]) *(b),*
Gruppo Nazionale di Struttura della Materia del CNR
e Centro Interuniversitario di Struttura della Materia del MPI

Time-Resolved Luminescence of $CdGaInS_4$ under Pulsed Laser Excitation at 30 K

By

M. Colocci (a), F. Fermi (b), R. Querzoli (a), and A. Vinattieri (a)

A detailed study of the time-resolved luminescence of $CdGaInS_4$ after pulsed laser excitation at $T = 30$ K is presented. Results are given on the time-resolved spectra in the range 50 ns to 10 ms of delay time with respect to excitation, on the time decay of the luminescence intensity at 610 nm, and on the time evolution of the integrated luminescence intensity for different emission wavelengths. Finally a simple model is presented for the recombination processes at low temperatures yielding good agreement with the experimental data.

Nous présentons ici une analyse detaillée de la luminescence, résolue temporellement, de cristaux de $CdGaInS_4$ après excitation laser à 30 K. Nous présentons aussi un simple modèle pour les procès de récombination à basses températures capable de fournir un bon accord avec les données expérimentales.

1. Introduction

$CdGaInS_4$ is a layered crystal compound grown for the first time by Shand [1]; it presents the typical properties of a highly compensated n-type semiconductor and many papers have been published concerning its structural, electrical, and optical properties.

In particular, when excited in the fundamental absorption edge, it shows high photoconductivity and strong photoluminescence (PL). The emission spectrum of $CdGaInS_4$ at 80 K consists of a single wide asymmetrical band peaked at 1.89 eV [1] while two bands have been reported by Irie et al. [2] at lower temperatures with maxima at 2.38 and 1.89 eV, respectively. On the other hand, Irie et al.'s samples were found to be gallium-rich; moreover X-ray and Raman spectra showed additional lines characteristic of $CdGa_2S_4$ [2]. Thus the situation, as far as the PL properties of these crystals are concerned, does not seem to be completely clear.

The fundamental absorption edge of $CdGaInS_4$ has been studied in detail by Abdullaev et al. [3]. They showed the existence of both direct and indirect transitions (the direct gap energy is equal to 2.85 eV at 77 K, while the indirect gap energy is equal to 2.58 eV at 20 K) and the existence of exciton transitions in the indirect absorption region.

Contradictory results have been also reported as far as the space group of $CdGaInS_4$ is concerned. Tentative assignments to C_{3v}^5 [1, 4, 5], C_{3v}^1 [6] and D_{3v}^1 [7] space groups have been proposed; in fact it seems that no definite assignment can be inferred by

[1]) Largo E. Fermi 2, 50125 Firenze, Italy.
[2]) Via M. D'Azeglio 85, 43100 Parma, Italy.

Raman scattering measurements [8] as a consequence of an apparent deep disorder in the internal structure of these crystals.

On the other hand, it seems that, from a general point of view, the main properties of $CdGaInS_4$ can be quite easily understood in the general phenomenological framework in which multinary chalcogenide compounds of the type AB_2X_4 are usually described [1]. Grilli et al. [9] emphasized the fact that the latter systems behave roughly as heavily doped or amorphous semiconductors. In fact different kinds of localized levels seem to be present in the $CdGaInS_4$ energy gap and enter the strong luminescence properties shown by these systems.

In order to achieve a better understanding of these materials, we present in this paper a detailed study of the time-resolved luminescence of $CdGaInS_4$ crystals at low temperature ($T = 30$ K). It is well known, in fact, that radiative recombination of electron-hole pairs is a sensitive indicator for purity and structure of solid-state materials. PL techniques have therefore become widely used as tools for the characterization of semiconductors. In fact details on doping and defects can be inferred from PL spectra; moreover since the time resolution in PL experiments has greatly improved, one has the possibility of obtaining detailed information concerning the initial distribution of the separations of the e–h pairs created by pulsed optical excitation and the subsequent time evolution of this distribution as well as recombination and relaxation rates.

The results on the time-resolved spectra in the range 50 ns to 10 ms of delay time with respect to excitation, the time decay of the luminescence intensity at a given emission wavelength, the time evolution of the integrated luminescence intensity for different emission energies within the 1.89 eV band are discussed and a simple model is presented for the recombination processes yielding a good agreement with the experimental data.

The paper is organized as follows: in Section 2 we describe the experimental apparatus and a presentation is given of the data obtained. In Section 3 we present a discussion of the experimental results based on a simple model for the e–h pair radiative recombination processes. Finally conclusions are given in Section 4.

2. Experimental Apparatus

Samples of $CdGaInS_4$ have been grown at MASPEC-CNR, Parma, by CVD technique in the LTVTP version [10] using iodine as transport agent. The crystal structure has been controlled by X-ray diffraction; the chemical composition has been tested by atomic absorption and compared with solutions having known concentrations of the elements. A satisfactory agreement between the stoichiometric formula and the nominal concentration of the constituents has been found; only a slight sulfur deficiency has been noticed.

The crystal used in the experiment is a yellowish platelet ≈ 100 μm thick with a resistivity of order 10^{-11} Ω^{-1} cm^{-1} at 300 K.

The experimental set-up used is shown in Fig. 1.

The light beam used to excite the PL is a nitrogen pumped dye-laser beam using Coumarin 460. The energy of a single dye-laser pulse is approximately 40 μJ and the time duration is 8 ns. The repetition rate is 5 Hz. The beam wavelength used in most measurements is 467.8 nm, corresponding to a 50% absorption in the sample at 30 K.

The dye-laser beam is focused by lens L1 to a 1 mm spot on the sample kept at 30 K in a cryostat (AIR PRODUCTS-mod. DE202S). The incoming beam makes a 20° angle with the normal to the crystal surface.

The number of absorbed photons per pulse is 5×10^{17} photons/cm^3 corresponding to an average distance of $\approx$10 nm between two absorption sites.

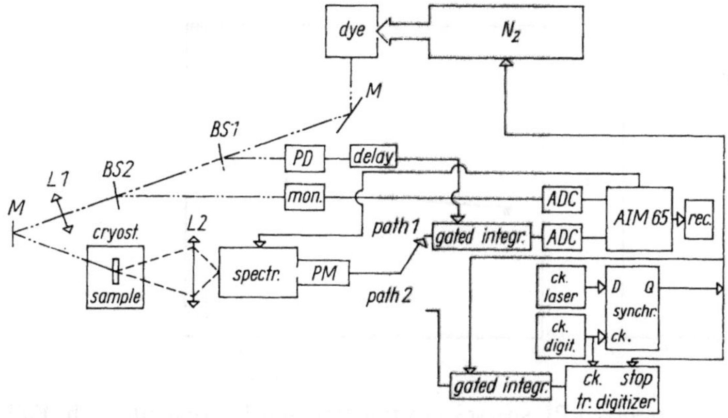

Fig. 1. Experimental apparatus

The PL light is focused by lens L2 (focal length 85 mm, aperture f/1.8) onto the entrance slit of a double monochromator (ORIEL-mod. 7240; focal length 12.5 cm, aperture $f/3.7$; resolution $\Delta\lambda = 3.2$ nm). The light from the exit slit is collected by a PM tube (EMI-mod. 9816 B; cathode spectral response S 20).

Two types of measurements have been performed.

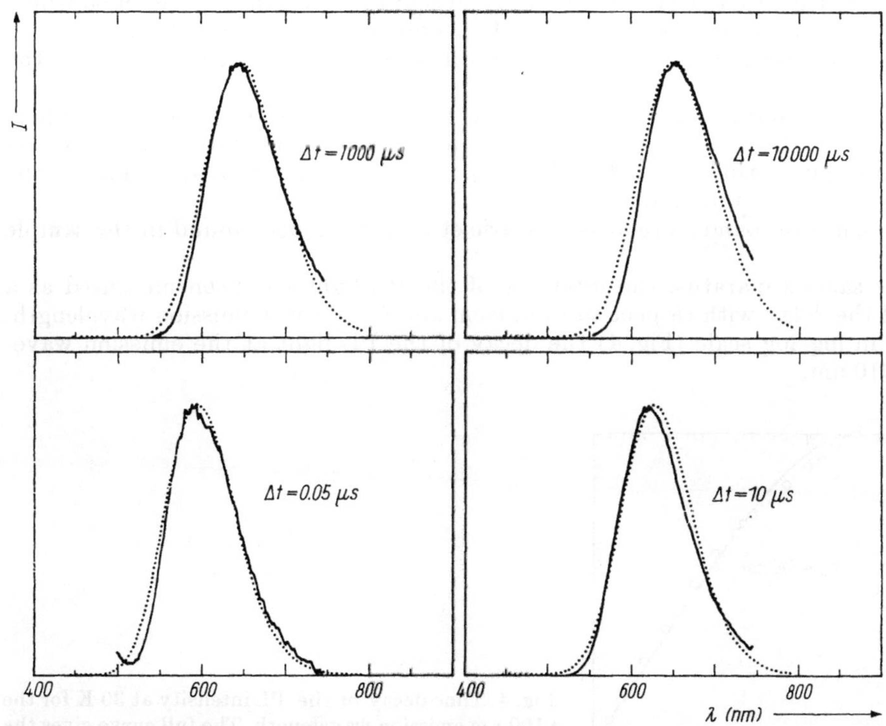

Fig. 2. Time-resolved PL spectra at 30 K for different delay times: $\Delta t = 0.05, 10, 1000, 10000$ μs. The continuous lines give the experimental spectra; the dotted curves are the predictions of the model discussed in Section 3

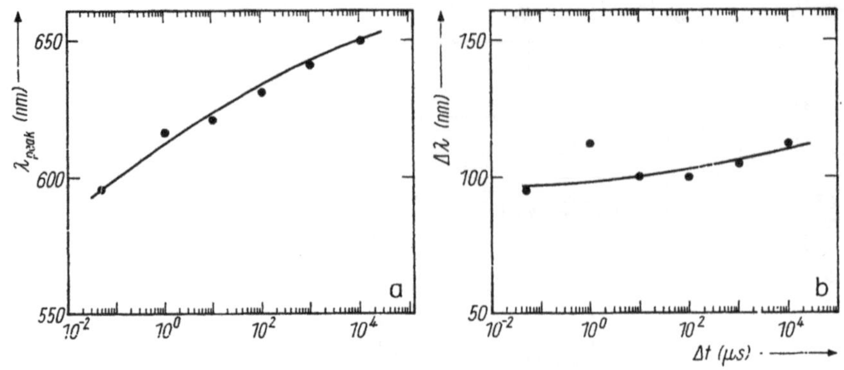

Fig. 3. a) Peak position of the time-resolved PL spectra as a function of delay time (dots). b) Full width at half maximum of the PL spectra as a function of delay time (dots). The continuous lines give the prediction of the model of Section 3

2.1 Time-resolved PL spectra

The signal from the PM tube (path 1) is sent to a gated integrator (STANFORD RES.-mod. SR 250) and the opening of the gate is driven, with a chosen delay with respect to the dye-laser pulse, by the pulse of a fast photodiode PD, shone with a portion of the beam by means of the beam splitter BS1. The signal from the integrator is converted from analogic to digital at each pulse and sent to microprocessor AIM 65 which sums the signals of a prefixed pulse number, normalizes to constant total energy through the monitor signal (mon.) and controls the spectrometer (spectr.) and the recorder (rec.).

Spectra have been taken at delays of 0.05, 1, 10, 100, 1000, and 10000 μs. We report in Fig. 2 some of the spectra after correction for the monochromator efficiency and the photomultiplier response; in Fig. 3 we report the peak position and width of the measured spectra as a function of delay.

No green emission band as reported by Irie et al. [2] has been found in the sample studied.

With the same apparatus, the intensity of the PL light has been measured as a function of the delay with respect to excitation and for a given emission wavelength. We report in log–log scale (Fig. 4) the decay of the PL light at the emission wavelength of 610 nm.

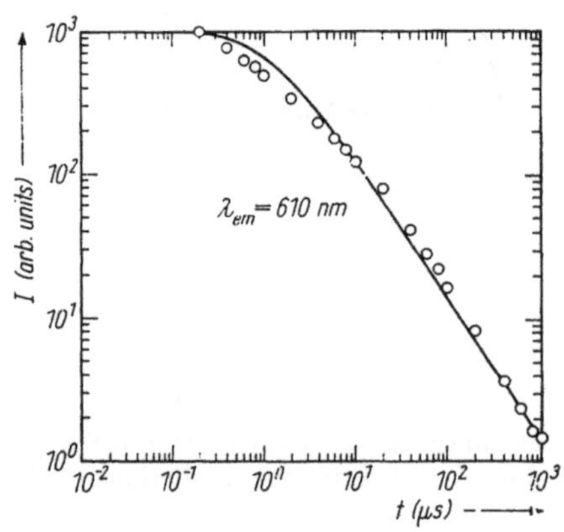

Fig. 4. Time decay of the PL intensity at 30 K for the 6100 nm emission wavelength. The full curve gives the prediction of the model of Section 3

2.2 *Measurement of the integrated PL intensity as a function of delay at a given emission wavelength*

The signal from the PM tube (path 2) is sent to a gated integrator (EVANS-mod. 4130), open by the dye-laser pulse, which integrates the PL light for a time τ, varying in the interval 30 to 3000 μs. The integrated signal is sent to a 100 MHz transient digitizer (LE CROY-mod. 3500 SA) measuring the time behaviour of the integral of the PL intensity. The transient digitizer is driven by an external clock, synchronizing also the firing of the laser, in order to allow for the sum in phase of the signals from a prefixed number of laser pulses.

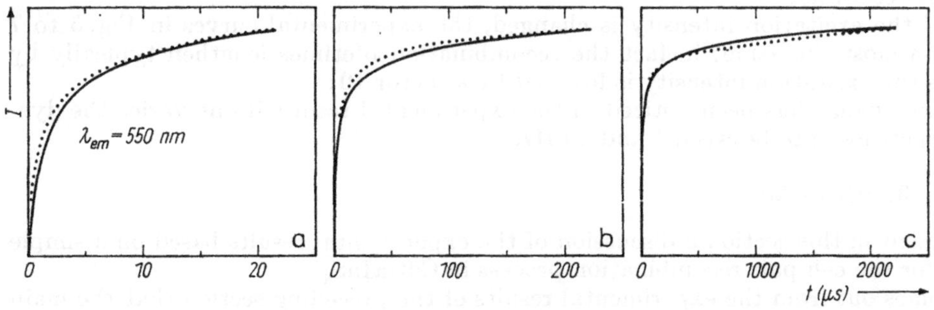

Fig. 5. Integrated PL intensity at λ_{em} = 550 nm (dotted lines) for three different sampling frequencies: a) 10 MHz, b) 1 MHz, c) 100 kHz. The full lines give the predictions of the model

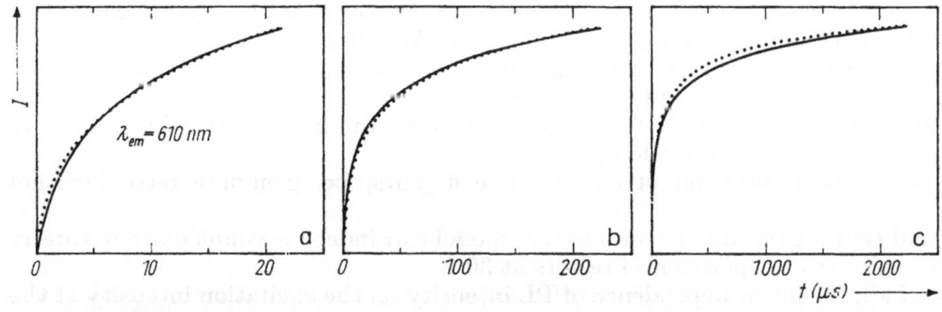

Fig. 6. Integrated PL intensity at λ_{em} = 610 nm (dotted lines) for three different sampling frequencies: a) 10 MHz, b) 1 MHz, c) 100 kHz. The full lines give the predictions of the model

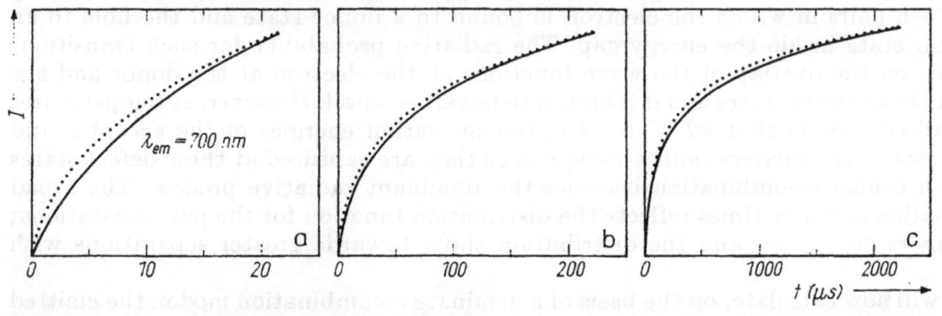

Fig. 7. Integrated PL intensity at λ_{em} = 700 nm (dotted lines) for three different sampling frequencies: a) 10 MHz, b) 1 MHz, c) 100 kHz. The full lines give the predictions of the model

Measurements have been made at the emission wavelengths of 550, 580, 610, 640, 670, and 700 nm; at each wavelength measurements have been performed at five sampling frequencies: 10 MHz, 3 MHz, 1 MHz, 300 kHz, and 100 kHz. Each measurement is composed of 256 points, giving thus the behaviour of the PL integrated intensity in the first 25.6, 76.8, 256, 768, and 2560 μs, respectively. We report in Fig. 5, 6, and 7 the experimental data at three different emission wavelengths.

It can be immediately seen that the experimental curves show a typical pair recombination behaviour.

Moreover we found that:

(i) the PL intensity shows a linear dependence on the excitation intensity at these power levels;

(ii) if the excitation intensity is changed, the experimental curves in Fig. 5 to 7 remain almost the same; in fact the recombination lifetimes lengthen typically by 10% as the excitation intensity is lowered by a factor 10;

(iii) no change has been noticed in the experimental results if one varies the dye-laser repetition rate between 1 and 10 Hz.

3. Discussion

We present in this section a discussion of the experimental results based on a simple model for the e–h pair recombination process in $CdGaInS_4$.

It comes out from the experimental results of the preceding section that the main features of $CdGaInS_4$ PL are very similar indeed to those shown by heavily doped semiconductors such as the ternary compounds AB_2X_4 [9] or amorphous semiconductors such as hydrogenated amorphous silicon or germanium [11 to 13].

Three main models have been introduced in order to explain the experimental data on photoluminescence and photoconductivity in these materials:

(i) recombination of "free" electrons from the conduction band with localized holes trapped in shallow acceptor states [14];

(ii) tunnel recombination of uncorrelated electrons and holes localized at randomly spaced states (distant pair model) [15, 16];

(iii) tunnel recombination of correlated e–h pairs, i.e. geminate recombination [17].

We find that a geminate recombination model can indeed account quite naturally for the whole set of experimental results at 30 K.

First of all, the linear dependence of PL intensity on the excitation intensity at the laser powers used suggests a monomolecular radiative recombination, thus implying a geminate recombination model. Furthermore the PL time decay shifts to longer times as the emission wavelength increases, suggesting radiative transitions involving single e–h pairs in which the electron is bound to a donor state and the hole to an acceptor state inside the energy gap. The radiative probability for such transitions depends on the overlap of the wave functions of the electron at the donor and the hole at the acceptor states and it is known to be rather small. However, at temperatures T small enough so that $kT < E_A, E_D$, the ionization energies of the acceptor and donor states, the carriers cannot escape once they are captured at these defect states and e–h tunnel recombination becomes the dominant radiative process. The broad distribution of decay times reflects the distribution function for the pair separations; close pairs decay fast and the distribution shifts towards greater separations with time.

We will now calculate, on the basis of a geminate recombination model, the emitted PL intensity as a function of both the emission wavelength and time, in order to compare with the experimental data.

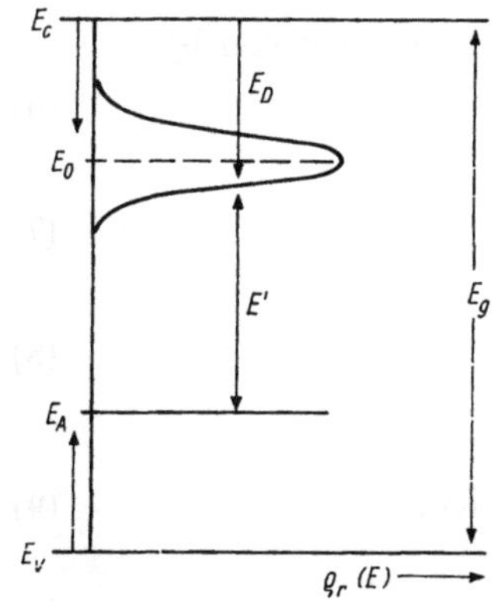

Fig. 8. Joint density of states $\varrho_r(E)$ as assumed in the model. E_g is the band gap energy; $E_A(E_D)$ is the energy of the acceptor (donor) state; σ is the width of the Gaussian distribution peaked at E_0

We assume a simple level scheme for the donor and acceptor defect states within the energy gap in which the defects are simple Gaussian broadened states at energy E_0 and E_A, respectively. The assumption is clearly equivalent to the scheme shown in Fig. 8 where the joint density of recombination states $\varrho_r(E)$ is given as a function of energy inside the energy gap E_g. Of course only the energy difference E' between the donor and acceptor states can be inferred from the PL light.

We assume a simple Coulomb interaction between the ionized defect states, so that the energy of the emitted photon is given by [18]

$$E = h\nu = E_g - (E_A + E_D) + \frac{e^2}{\varepsilon R} \equiv E' + \frac{e^2}{\varepsilon R}, \tag{1}$$

where ε is the static dielectric constant, e the electronic charge, and R the distance between the donor and acceptor states. As in [18] we assume that the probability P of an e–h pair radiative recombination depends on their spatial separation R as

$$P(R) = P_0 \left(\frac{\lambda_{ref}}{\lambda}\right)^3 e^{-\frac{2R}{R_0}}, \tag{2}$$

where $P_0(\lambda_{ref}/\lambda)^3$ is the typical rate of allowed radiative recombinations at wavelength λ (we have assumed for the reference wavelength $\lambda_{ref} \approx 300$ nm) and R_0 is the donor Bohr radius.

As far as the initial spatial distribution of the e–h pairs is concerned, we assume a Gaussian distribution $g(R, 0)$ as it would come out from a pure diffusive model between excitation and capture in the localized states,

$$g(R, 0) = G\, e^{-\left(\frac{R}{R_1}\right)^2}, \tag{3}$$

where R_1 specifies the initial distribution and the capture time is assumed to be negligible compared to the PL lifetimes. No further diffusion is assumed after capture at this temperature.

At time t after excitation, the number dN of donor–acceptor pairs having separation in the interval R, $R + dR$ and relative energy between E' and $E' + dE'$ is

$$dN = 2\pi R\, dR\, g(R, t)\, f(E')\, dE', \tag{4}$$

where

$$f(E') = A\, e^{-\frac{[E' - (E_g - (E_0 + E_A))]^2}{2\sigma^2}}, \tag{5}$$

and an isotropic trap distribution in the crystal layered planes has been assumed.

The number of recombinations in the time interval dt is therefore given by

$$d(dN) = -P(R)\,dN\,dt\,, \tag{6}$$

yielding after integration

$$dN = (dN)_0\,e^{-P(R)t}\,. \tag{7}$$

From (3), (4), and (7) one has

$$g(R, t) = g(R, 0)\,e^{-P(R)t} \tag{8}$$

and the emitted intensity in the energy band dE is

$$I(E, t) \approx \int\limits_{E-e^2/\varepsilon R}^{E-e^2/\varepsilon R+dE} f(E')\,dE' \int\limits_{R_0}^{\infty} 2\pi R\,dR\,g(R, t)\,P(R)\,. \tag{9}$$

One has after integration over E'

$$I(E, t) \approx \int\limits_{R_0}^{\infty} dR\,RP(R)\,g(R, t)\,f(E - e^2/\varepsilon R)\,\frac{dE}{d\lambda}\,d\lambda \approx$$

$$\approx \int\limits_{R_0}^{\infty} dR\,RP_0\left(\frac{\lambda_{ref}}{\lambda}\right)^3 \exp\left\{-\frac{2R}{R_0} - \left(\frac{R}{R_1}\right)^2 - P_0 t\left(\frac{\lambda_{ref}}{\lambda}\right)^3 \times\right.$$

$$\left.\times\, e^{-2R/R_0} - \frac{[E - e^2/\varepsilon R - (E_g - (E_0 + E_A))]^2}{2\sigma^2}\right\}\frac{dE}{d\lambda}\,d\lambda\,. \tag{10}$$

For $t = \underline{t}$ fixed, the integral (10) gives, after numerical integration, the spectrum at time $\underline{t}$ (Fig. 2). For $E = \underline{E}$ fixed, $I(\underline{E}, t)$ gives the decay of PL intensity at the given emission energy $\underline{E}$ (Fig. 4). The integrated PL intensity (Fig. 5, 6, and 7) is obtained by integrating over time $I(E, t)$ at fixed E, namely

$$I_{int} = \int\limits_0^t I(E, t)\,dt \approx$$

$$\approx \int\limits_{R_0}^{\infty} dR\,R\exp\left\{-\left(\frac{R}{R_1}\right)^2 - \frac{[E - e^2/\varepsilon R - (E_g - (E_0 + E_A))]^2}{2\sigma^2}\right\}\times$$

$$\times\left\{1 - \exp\left(-P_0 t\left(\frac{\lambda_{ref}}{\lambda}\right)^3 e^{-2R/R_0}\right)\right\}\frac{dE}{d\lambda}\,d\lambda\,. \tag{11}$$

As is easily seen, the integrated intensity is strongly dependent on R_1 through the exponential factor $\exp\{-(R/R_1)^2\}$ while the dependence of $I(E, t)$ on R_1 is quite weak because of the factor $\exp\{-2R/R_0 - (R/R_1)^2\} \approx \exp(-2R/R_0)$ for $R_1 \gg R_0$ as implied by the experimental data. Both $I(E, t)$ and I_{int} are therefore important in order to determine the free parameters of the model.

It turns out that only four parameters are "a priori" free in our model, namely the energy difference $E_0 - E_A$, σ, R_0, and R_1; ε is then fixed by the value of R_0 through the well-known expression for the "effective" Bohr radius $R_0 = \varepsilon(m_e/\mu^*)\,\hbar^2/m_e e^2$ while P_0 has been assumed equal to $10^8\ s^{-1}$ at the reference wavelength $\lambda_{ref} \approx 300$ nm.

A few comments are now in order concerning the parameters of the model.

(i) The energy difference $E_0 - E_A$ and the width σ of the joint distribution of defect states are fixed by PL spectra at large t. In fact, since close pairs decay fast, at large t only the surviving distant pairs undergo recombination and the influence of the Coulomb term $e^2/\varepsilon R$ becomes negligible. One therefore expects PL spectra to approximately reproduce, at large t, the joint density of recombining states $\varrho_r(E)$. From the spectrum at $t = 10^4$ μs we have taken $E_0 = E_A = 1.86$ eV and $\sigma = 0.14$ eV, corresponding to a full width at half maximum of about 100 nm.

(ii) We have adjusted the parameters R_0 and R_1 to the values $R_0 = 0.54$ nm and $R_1 = 4.0$ nm in order to reproduce the experimental results.

(iii) From the value of R_0 as in (ii) and assuming the effective mass $\mu^* = m_e$, we find for the dielectric constant the value $\varepsilon = 11$.

We have reported in Fig. 2 to 7 the predictions of the model for the values of the parameters given in Table 1.

Table 1

$R_0 = 0.54$ nm	$R_1 = 4.0$ nm
$E_0 - E_A = 1.86$ eV	$\sigma = 0.14$ eV
$\varepsilon = 11$	$P_0 = 10^8\ s^{-1}$

A good overall agreement between the predictions of the model and the complete set of experimental data is immediately seen.

We also notice that the function $P(R)\, g(R, t)$ in (10) is strongly peaked at $R_{peak} \approx \approx R_0 \ln (P_0 t)$. The main contribution to the integral in (10) is thus given by $R = R_{peak}$ and one gets

$$I \approx \frac{\ln (P_0 t)}{(P_0 t)^{1 + R_0/R_1}} \tag{12}$$

in agreement with the results of recent theories on the time-resolved PL of charge carriers in amorphous semiconductors [20]. In fact, if one had chosen a spherical symmetry for the defect distribution rather than a planar one, (4) would read

$$dN = 4\pi R^2\, dR\, g(R, t)\, f(E')\, dE' \tag{13}$$

and one would get

$$I \approx \frac{\ln^2 (P_0 t)}{(P_0 t)^{1 + R_0/R_1}} \tag{14}$$

as in [20].

4. Conclusions

We have presented, in this paper, a detailed study of the time-resolved PL of $CdGaInS_4$ at low temperature.

We have also presented a simple model for the time-resolved luminescence from localized states after pulsed laser excitation able to account for a whole variety of experimental measurements in terms of very few free parameters.

The overall good agreement between the predictions of the model and the experimental data strongly suggests a geminate recombination mechanism in these materials of the same kind as the mechanisms leading to PL in heavily doped or amorphous semiconductors. In particular the shift to lower energies of PL spectra with time seems to be entirely due to Coulomb interaction. Moreover diffusion by hopping seems to be negligible at these temperatures.

On the other hand, the temperature dependence of PL can be easily accomodated in the model; further checks of the model at different T and for different multinary semiconducting compounds will be published in the near future.

Acknowledgements

We thank Drs. L. Zanotti and C. Razzetti for providing us with samples of $CdGaInS_4$ of good quality and Drs. M. Guzzi and L. Tarricone for very enlightening discussions. Finally we deeply thank Mr. M. Neri for help in setting up the experimental apparatus.

References

[1] W. A. SHAND, phys. stat. sol. (a) **3**, K77 (1970).
[2] T. IRIE, H. MIYASHITA, S. ENDO, and H. NAKANISHI, Nuovo Cimento **2D**, 2002 (1983).
[3] G. B. ABDULLAEV, T. G. KERIMOVA, S. S. MAMEDOV, T. R. MECHTIEV, R. K. NANI, and E. Y. SALAEV, phys. stat. sol. (b) **73**, K69 (1976).
[4] K. KAMBAS, phys. stat. sol. (a) **76**, 735 (1983).
[5] G. B. ABDULLAEV, N. V. BAZHOVSKAYA, N. D. YURAEV, D. B. KUSHEV, R. K. NANI, and V. K. SUBASHEEV, phys. stat. sol. (a) **28**, K81 (1975).
[6] K. R. ALLAKHVERDIEV, A. N. ABBASOV, T. R. MEKHTIEV, and R. K. NANI, phys. stat. sol. (b) **94**, K31 (1979).
[7] A. N. ABBASOV, K. R. ALLAKHVERDIEV, T. R. MEKHTIEV, and R. K. NANI, Soviet Phys. — Solid State **23**, 370 (1981).
[8] C. RAZZETTI, P. P. LOTTICI, L. TARRICONE, and L. ZANOTTI, Progr. Crystal Growth Charact. **10**, 353 (1985).
[9] E. GRILLI, M. GUZZI, E. CAMERLENGHI, and F. PIO, to be published.
[10] C. PAORICI, L. ZANOTTI, and G. ZUCCALLI, J. Crystal Growth **43**, 705 (1978).
[11] A. A. ANDREEV, A. V. ZHERZDEV, A. I. KOSAREV, K. V. KOUGIA, and I. S. SHLIMAK, phys. stat. sol. (b) **127**, 273 (1985).
[12] S. KURITA, W. CZAJA, and S. KINMOND, Solid State Commun. **32**, 879 (1979).
[13] J. NOOLANDI, K. M. HONG, and R. A. STREET, Solid State Commun. **34**, 45 (1980).
[14] C. R. WRONSKI and R. E. DANIEL, Phys. Rev. B **23**, 794 (1981).
[15] F. BOULITROP and D. J. DUNSTAN, Solid State Commun. **44**, 841 (1982).
[16] D. J. DUNSTAN, Phil. Mag. B **46**, 579 (1982).
[17] R. A. STREET, Adv. Phys. **30**, 593 (1981).
[18] D. C. THOMAS, J. J. HOPFIELD, and W. M. AUGUSTINYAK, Phys. Rev. **140**, 202 (1965).
[19] J. J. HOPFIELD, Proc. 7th Internat. Conf. Physics of Semiconductors Paris, Academic Press, 1964.
[20] M. SUHRKE and B. ESSER, phys. stat. sol. (b) **126**, 393 (1984).

(Received August 19, 1985)

phys. stat. sol. (b) **133**, 389 (1986)

Subject classification: 73.60

Erevan State University[1])

Calculation of the Charge Carrier Mobility of a Size Quantized Semiconductor Film by the Energy-Loss Method

By

A. L. Vartanian and A. A. Kirakosian

An energy-loss-method formalism applicable to a thin quantized film is developed. Within the frame of a self-consistent field and taking into account the "imaginary" charges in the film surrounding regions the expression for a film dielectric function is derived and, in the long-wavelength limit the analytical form of its imaginary part is found for a simple model of a size quantized film. Under the condition that the electron gas is nondegenerate and only the first subband is filled, the mobility is calculated, when an electron–ionized impurity scattering is taken into account, and compared with the relevant Boltzmann expression.

Развит формализм метода энергетических потерь применительно к тонкой квантованной пленке. В рамках самосогласованного поля выведено выражение для функции диэлектрической проницаемости пленки с учетом ,,мнимых" зарядов в окружающих пленку областях и найден аналитический вид ее мнимой части в длинноволновом приближении для простой модели размерно квантованной пленки. В условиях невырожденности электронного газа и заполнения первой подзоны рассчитана подвижность электрона при рассеянии на заряженной примеси и проведено сравнение с соответствующим больцмановским выражением.

1. Introduction

Lately, together with the traditional method for kinetic coefficient calculations, based on the Boltzmann equation, the energy-loss method (ELM) is also successfully employed. The latter was first used for the mobility calculation of charge carriers (CC) for electron–ionized impurity scattering [1], and later on also for various scattering mechanisms [2 to 6].

In the present work the ELM formalism applicable to a thin semiconductor film under the condition of the quantum size effect (QSE) is developed, when electron–ionized impurity scattering is taken into account.

2. The Dielectric Function for a Semiconductor Film

There are a number of theoretical works devoted to the investigation of the dielectric function (DF) for systems spatially limited in one direction [7 to 10]. In [7] the electrodynamics of an inversion layer on the oxide–semiconductor interface is developed and by taking into account the image forces the expression for the DF is obtained. The analogous two-dimensional (2D) system is also studied in [8]. By use of the known Lindhard expression for the bulk specimen DF the expression of the DF for a thin semiconductor or semimetallic film is derived in [9]. However, the change of the image charges in the film surrounding regions caused by the charge redistribution in the film has been left out of account.

[1]) Mravian Street 1, 375049 Erevan, USSR.

Calculations close to ours are performed in [10] in which, however, the 2D translational invariance of a semiconductor layer is not taken into account, and also the expression for the inverse permeability matrix is not deduced, the knowledge of which is necessary when one investigates screening effects in spatially inhomogeneous systems [11].

Let us proceed to the derivation of the DF expression for a semiconductor film of thickness L and dielectric constant ε_L, placed between two homogeneous media with dielectric constants ε_1 and ε_2, respectively.

We designate by $\psi_{ks}(\mathbf{r}, z)$ and E_{ks} the eigenfunction and the energy spectrum, respectively, of the CC unperturbed Hamiltonian in a medium with 2D symmetry ($\mathbf{k}$ is the 2D wave vector, $s = n, l$; n being the band number, and $l = 1, 2, \ldots$).

We introduce the self-consistent potential V containing both the applied "external" potential V_{ex} and the potential V_i induced by the charge redistribution in the system. Let us represent V in the form of a Fourier expansion [11]:

$$V(\mathbf{r}, z, t) = \sum_{\mathbf{q},\mathbf{g}} V(\mathbf{q} + \mathbf{g}, z, t) \exp\left[i(\mathbf{q} + \mathbf{g})\, \mathbf{r}\right], \tag{1}$$

where $\mathbf{q}$ ranges over values in the first 2D Brillouin zone and $\mathbf{g}$ is the reciprocal lattice vector.

Let the applied "external" field vary with time according to $V_{ex}(\mathbf{r}, z, t) = V_{ex}(\mathbf{r}, z, \omega) \exp\left[(\alpha - i\omega)\, t\right]$, where $\alpha = 0^+$ takes into account the adiabatic switching of the interaction [12], and all the remaining parameters have an analogous time dependence.

Assuming that free CC (conduction band electrons and valence band holes) take part in the screening of an external influence, i.e. $\psi_{ks}(\mathbf{r}, z) = 0$ outside the film ($z < 0$, $z > L$), then the solution of the Poisson equation for the Fourier transform may be represented in the form

$$V_i^j(\mathbf{q} + \mathbf{g}, \omega, z) = G_j(\mathbf{q} + \mathbf{g}, z, z'')\, \chi_\omega^0(\mathbf{q} + \mathbf{g}, \mathbf{q} + \mathbf{g}', z'', z')\, V(\mathbf{q} + \mathbf{g}', \omega, z'), \tag{2}$$

where

$$\chi_\omega^0(\mathbf{q} + \mathbf{g}, \mathbf{q} + \mathbf{g}', z, z') = -\frac{1}{S} \sum_{\mathbf{k}, s, s'} \frac{(f_{\mathbf{k}s} - f_{\mathbf{k}+\mathbf{q}, s'})\, \eta_0(\mathbf{q} + \mathbf{g}, z)\, \eta_0^*(\mathbf{q} + \mathbf{g}', z')}{E_{\mathbf{k}+\mathbf{q}, s'} - E_{\mathbf{k}, s} - \hbar\omega - i\hbar\alpha}, \tag{3}$$

$$\eta_0(\mathbf{q} + \mathbf{g}, z) = \int_S \psi_{\mathbf{k}s}^*(\mathbf{r}, z) \exp\left[-i(\mathbf{q} + \mathbf{g})\, \mathbf{r}\right] \psi_{\mathbf{k}+\mathbf{q}, s'}(\mathbf{r}, z)\, d\mathbf{r}\,. \tag{4}$$

In (2) and henceforth we carry out summation over iterative vectors $\mathbf{g}', \mathbf{g}''$ and integration over iterative coordinates in the transverse direction, extending over the film thickness.

The function $G_j(\mathbf{q} + \mathbf{g}, z, z')$ satisfies the set of equations

$$\left(\frac{\partial^2}{\partial z^2} - q^2\right) G_j(\mathbf{q}, z, z') = -\delta_{jL} \frac{4\pi e^2}{\varepsilon_L} \delta(z - z') \tag{5}$$

and the boundary conditions

$$\begin{aligned} G_1 &= G_L, & \varepsilon_1 G_1' &= \varepsilon_L G_L'; & &\text{when} \quad z = 0 \\ G_L &= G_2, & \varepsilon_L G_L' &= \varepsilon_2 G_2', & &\text{when} \quad z = L. \end{aligned} \tag{6}$$

In (3) to (6) the following designations are introduced: $f_{\mathbf{k}s} \equiv f(E_{\mathbf{k}s})$ is the Fermi-Dirac function, S is the film area, and δ_{jL} is the Kronecker symbol, $j = 1, 2, L$.

Note, that the 2D Fourier transform of two point charges, one of which is at the film point $(0, 0, z')$, satisfies equations and boundary conditions analogous to (5) and (6), respectively [13, 14]. If the second charge is also inside the film, then in the limit $qL \ll 1$ and $\varepsilon_1, \varepsilon_2 \ll \varepsilon_L$ we have for G_L [13, 14]

$$G_L(q, z, z') = \frac{4\pi e^2}{qL\varepsilon_L}\left(q + \frac{\varepsilon_1 + \varepsilon_2}{L\varepsilon_L}\right)^{-1} \equiv G_L(q) . \tag{7}$$

Let us determine the film DF using the relation

$$V_{ex}(\boldsymbol{q} + \boldsymbol{g}, \omega, z) = \varepsilon(\boldsymbol{q} + \boldsymbol{g}, \boldsymbol{q} + \boldsymbol{g}', \omega, z, z')\, V(\boldsymbol{q} + \boldsymbol{g}', \omega, z') . \tag{8}$$

Since $V_i = V - V_{ex}$, from (2) and (8) we obtain

$$\varepsilon(\boldsymbol{q} + \boldsymbol{g}, \boldsymbol{q} + \boldsymbol{g}', \omega, z, z') = \delta_{gg'}\delta(z - z') - G_L(\boldsymbol{q} + \boldsymbol{g}, z, z'') \times \times \chi^0_\omega(\boldsymbol{q} + \boldsymbol{g}, \boldsymbol{q} + \boldsymbol{g}', z'', z') . \tag{9}$$

For the inverse DF $\varepsilon^{-1}(\boldsymbol{q} + \boldsymbol{g}, \boldsymbol{q} + \boldsymbol{g}', \omega, z, z')$ satisfying the condition

$$\varepsilon(\boldsymbol{q} + \boldsymbol{g}, \boldsymbol{q} + \boldsymbol{g}'', \omega, z, z'')\, \varepsilon^{-1}(\boldsymbol{q} + \boldsymbol{g}'', \boldsymbol{q} + \boldsymbol{g}', \omega, z'', z') = \delta_{gg'}\, \delta(z - z') ,$$

one may write the expression

$$\varepsilon^{-1}(\boldsymbol{q} + \boldsymbol{g}, \boldsymbol{q} + \boldsymbol{g}', \omega, z, z') = \delta_{gg'}\, \delta(z - z') + G_L(\boldsymbol{q} + \boldsymbol{g}, z, z'') \times \times \chi_\omega(\boldsymbol{q} + \boldsymbol{g}, \boldsymbol{q} + \boldsymbol{g}', z'', z') , \tag{10}$$

where χ_ω satisfies the integral equation

$$\chi_\omega(\boldsymbol{q} + \boldsymbol{g}, \boldsymbol{q} + \boldsymbol{g}', z, z') = \chi^0_\omega(\boldsymbol{q} + \boldsymbol{g}, \boldsymbol{q} + \boldsymbol{g}', z, z') + + \chi^0_\omega(\boldsymbol{q} + \boldsymbol{g}, \boldsymbol{q} + \boldsymbol{g}'', z, z'')\, G_L(\boldsymbol{q} + \boldsymbol{g}'', z'', z''')\, \chi_\omega(\boldsymbol{q} + \boldsymbol{g}'', \boldsymbol{q} + \boldsymbol{g}', z''', z') . \tag{11}$$

The obtained expressions may significantly be simplified under conditions when $qL \ll 1$. The characteristic length of perturbation variation then is $\lambda = 2\pi/q \gg \gg L \gg a$, where a is the lattice parameter, and, therefore one may ignore the local field effects and set in all expressions $\boldsymbol{g} = \boldsymbol{g}' = \boldsymbol{0}$ [11, 15].

A further simplification may be achieved if using the formula

$$f(\boldsymbol{q}, \omega, z, z') = \sum_{q_3, q_3'} f(\boldsymbol{q}, \omega, q_3, q_3') \exp\,[i(q_3 z + q_3' z')] ,$$

and performing in the expressions $\varepsilon(\boldsymbol{q}, \omega, z, z') \equiv \varepsilon(\boldsymbol{q}, \boldsymbol{q}, \omega, z, z')$ and $\varepsilon^{-1}(\boldsymbol{q}, \omega, z, z') \equiv \varepsilon^{-1}(\boldsymbol{q}, \boldsymbol{q}, \omega, z, z')$ a Fourier expansion over z and z' in the range $[0, L]$. After simple transformations for the dimensionless (unlike $\varepsilon(\boldsymbol{q}, \omega, z, z')$) DF we obtain

$$\varepsilon(\boldsymbol{q}, \omega, q_3, q_3') = \delta_{q_3 q_3'} - \sum_{q_3''} G_L(\boldsymbol{q}, q_3, q_3'')\, \tilde{\chi}^0_\omega(\boldsymbol{q}, q_3'', q_3') , \tag{12}$$

where $\tilde{\chi}^0_\omega(\boldsymbol{q}, q_3, q_3')$ is given by (3) with $\boldsymbol{g} = \boldsymbol{g}' = \boldsymbol{0}$ and replacing in it

$$\eta_0(\boldsymbol{q}, z) \quad \text{by} \quad \eta(\boldsymbol{q}, q_3) = \int_0^L \mathrm{d}z \int_S \mathrm{d}\boldsymbol{r}\, \psi^*_{\boldsymbol{k}s} \exp\,[-i(\boldsymbol{q}\boldsymbol{r} - q_3 z)]\, \psi_{\boldsymbol{k}+\boldsymbol{q},s'}$$

and

$$\eta^*_0(\boldsymbol{q}, z') \quad \text{by} \quad \eta^*(\boldsymbol{q}, -q_3') .$$

Using the diagonality of $G_L(\boldsymbol{q}, q_3, q_3')$ over q_3, q_3' that directly follows from (7), for $\varepsilon(\boldsymbol{q}, \omega, q_3, q_3')$ and $\varepsilon^{-1}(\boldsymbol{q}, \omega, q_3, q_3')$ we obtain

$$\varepsilon(\boldsymbol{q}, \omega, q_3, q_3') = \delta_{q_3, q_3'} - \delta_{q_3, 0} G_L(\boldsymbol{q})\, \tilde{\chi}^0_\omega(\boldsymbol{q}, 0, q_3') ,$$

$$\varepsilon^{-1}(\boldsymbol{q}, \omega, q_3, q_3') = \delta_{q_3, q_3'} + \delta_{q_3, 0} G_L(\boldsymbol{q})\, \frac{\tilde{\chi}^0_\omega(\boldsymbol{q}, 0, q_3')}{1 - G_L(\boldsymbol{q})\, \tilde{\chi}^0_\omega(\boldsymbol{q}, 0, 0)} . \tag{13}$$

For future treatment we need the function $\operatorname{Im} \varepsilon^{-1}(\boldsymbol{q}, \omega, 0, 0) \equiv \operatorname{Im}(1/\varepsilon(\boldsymbol{q}, \omega, 0, 0))$. Let us calculate $\varepsilon(\boldsymbol{q}, \omega, 0, 0) \equiv \varepsilon(\boldsymbol{q}, \omega)$ by assuming that the CC are conduction band electrons. We use the effective mass approximation and the model of an infinitely deep potential well. Then, under the conditions that the electron gas is nondegenerate and only the first subband is filled, we obtain after simple calculations:

$$\begin{aligned} \operatorname{Re} \varepsilon(\boldsymbol{q}, \omega) &= 1 + [\varepsilon(\boldsymbol{q}, 0) - 1] \left(\frac{2mk_{\mathrm{B}}T}{\hbar^2 q^2}\right)^{1/2} [F(\alpha_+) + F(\alpha_-)] , \\ \operatorname{Im} \varepsilon(\boldsymbol{k}, \omega) &= [\varepsilon(\boldsymbol{q}, 0) - 1] \left(\frac{\pi m k_{\mathrm{B}}T}{2\hbar^2 q^2}\right)^{1/2} [\exp(-\alpha_-^2) - \exp(-\alpha_+^2)] , \end{aligned} \tag{14}$$

where $F(x)$ is the Dawson integral [16], $\alpha_\pm = \hbar(2mk_{\mathrm{B}}T)^{-1/2} (q/2 \pm m\omega/\hbar q)$, n is the electron concentration, T is the temperature,

$$\varepsilon(\boldsymbol{q}, 0) = 1 + \frac{\lambda_0^2}{q^2} \left(1 + \frac{\varepsilon_1 + \varepsilon_2}{qL\varepsilon_{\mathrm{L}}}\right)^{-1} \tag{15}$$

is the statistical DF in the long-wavelength limit with the asymptotic behaviour $\varepsilon(q, 0) \sim \lambda_0^2 L q^{-1}$, $\lambda_0 = (4\pi n e^2/\varepsilon_{\mathrm{L}} k_{\mathrm{B}} T)^{1/2}$ being the screening parameter. Note that, using the expression given in [13] for the Fourier transform of an unscreened potential in the film, we obtained, with the help of (15), the expression for the effective potential derived in [13] by linearization of the Poisson equation.

When $\omega \to 0$ and for $q \ll \hbar^{-1}(8mk_{\mathrm{B}}T)^{1/2}$ for the symmetrical case $\varepsilon_1 = \varepsilon_2$ we get from (14)

$$\operatorname{Im} \frac{1}{\varepsilon(\boldsymbol{q}, \omega)} = -\frac{\omega \lambda_0^2}{q^3 \varepsilon(\boldsymbol{q}, 0)} \left(\frac{\pi m}{2k_{\mathrm{B}}T}\right)^{1/2} \left(1 + \frac{2\varepsilon_1}{qL\varepsilon_{\mathrm{L}}}\right)^{-1} \exp\left(-\frac{\hbar^2 q^2}{8mk_{\mathrm{B}}T}\right). \tag{16}$$

3. Calculation of Mobility and Discussion

In order to calculate the mobility it is necessary to have an expression for the energy loss of a charged center moving in a thin film. Using the method developed in [17], after simple transformations we obtain

$$\begin{aligned} -\frac{\mathrm{d}W}{\mathrm{d}t} = \frac{iL\varepsilon_{\mathrm{L}}}{16\pi^3} \sum_{q_3} \int (\boldsymbol{q}\boldsymbol{v}) \, (q^2 + q_3^2) \, V_{\mathrm{ex}}^*(\boldsymbol{q}, q_3) \, \times \\ \times \{\sum_{q_3'} V_{\mathrm{ex}}(\boldsymbol{q}, q_3') \, \varepsilon^{-1}(\boldsymbol{q}, \boldsymbol{q}\boldsymbol{v}, q_3', q_3) - V_{\mathrm{ex}}(\boldsymbol{q}, q_3) \} \, \mathrm{d}\boldsymbol{q} \, , \end{aligned} \tag{17}$$

where $\boldsymbol{v}$ is the electron drift velocity on the film plane. In particular, if the "external" influence does not depend on z, then $V_{\mathrm{ex}}(\boldsymbol{q}, q_3) = V_{\mathrm{ex}}(\boldsymbol{q}) \, \delta_{q_3, 0}$ and instead of (17), we obtain the simple expression

$$\frac{\mathrm{d}W}{\mathrm{d}t} = \frac{L\varepsilon_{\mathrm{L}}}{16\pi^3} \int (\boldsymbol{q}\boldsymbol{v}) \, q^2 \, |V_{\mathrm{ex}}(\boldsymbol{q})|^2 \operatorname{Im} \varepsilon^{-1}(\boldsymbol{q}, \boldsymbol{q}\boldsymbol{v}) \, \mathrm{d}\boldsymbol{q} \, . \tag{18}$$

As is seen, if the DF is known, the problem is reduced to a two-fold multiple integration. Substituting the expression obtained in [13] for the unscreened potential

$$V_{\mathrm{ex}}(q) = \frac{4\pi e}{q^2 L\varepsilon_{\mathrm{L}}} \left(1 + \frac{2\varepsilon_1}{qL\varepsilon_{\mathrm{L}}}\right)^{-1} \tag{19}$$

into (18), under the conditions when $(\lambda_0 L \varepsilon_L \varepsilon_1^{-1})^2 \gg 1$, we obtain

$$-\frac{dW}{dt} = \frac{e^3 n^{1/2} m^{1/2} v^2}{L k_B T \varepsilon_L^{3/2}} f(p)\,, \tag{20}$$

$$f(p) = \frac{\pi^2}{2^{1/2}} \left[\left(p + \frac{1}{2}\right) \exp(p) \operatorname{erfc}(p^{1/2}) - \left(\frac{p}{\pi}\right)^{1/2}\right], \tag{21}$$

$$p = \frac{\pi n e^2 \hbar^2}{2 m \varepsilon_L k_B^2 T^2},$$

erfc (x) is the incomplete probability integral [16]. When the impurity centers are uniformly distributed in a film, then according to [18], the total energy loss is

$$-N \frac{dW}{dt} = \frac{env^2}{\mu}, \tag{22}$$

where N is the charged impurity concentration and μ is the mobility on the film plane. From (20) and (22) we have

$$\mu = \mu_0 f^{-1}(p)\,, \qquad \mu_0 = \frac{L \varepsilon_L^{3/2} k_B T n^{1/2}}{N e^2 m^{1/2}}. \tag{23}$$

Let us compare this expression for μ with that obtained by the Boltzmann method. In [19], under the condition that the first subband of size quantization is filled and that the CC gas is nondegenerate, an expression for the relaxation time is obtained, with the help of which we come to the following result:

$$\mu_B = \mu_0 g(p)\,, \tag{24}$$

$$g(p) = \frac{15}{2^{3/2}\pi} \left[\left(\frac{1}{2} - \frac{p}{5}\right) \operatorname{erfc}(p^{1/2}) \exp(p) + \left(\frac{p}{\pi}\right)^{1/2} \left(1 + \frac{4}{15} p\right)\right]. \tag{25}$$

For a nondegenerate electron gas the parameter $p \ll 1$. Fig. 1 shows the $\Delta\mu/\mu$ versus p dependence for small values of p. For InSb specimen [20] with thickness $L \approx 2.4 \times 10^{-6}$ cm $\varepsilon_L \approx 18$, $m \approx 0.013 m_0$ (m_0 is the free electron mass), $n \approx 10^{16}$ cm^{-3} at $T = 250$ K we have $p \approx 0.016$ and $\mu_B/\mu \approx 2.25$. It follows from the obtained results that the theoretical mobility, obtained within the framework of the ELM, is always less than the relevant Boltzmann expression. This, presumably, should be ascribed to the dynamical character of screening of the "external" perturbation.

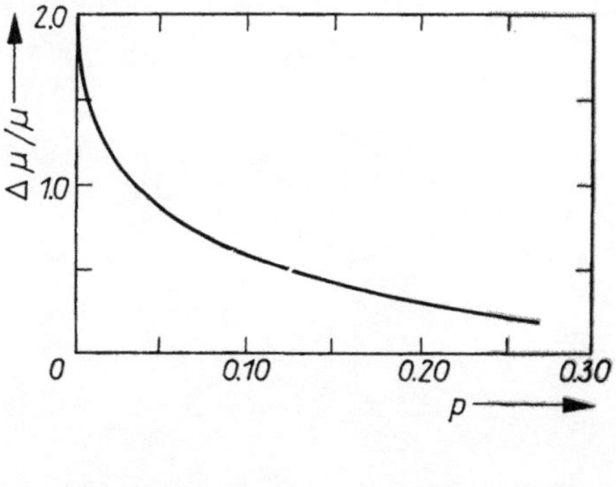

Fig. 1. The dependence of $\Delta\mu/\mu = (\mu_B - \mu)/\mu$ on p

References

[1] E. GERLACH, phys. stat. sol. (b) **61**, K97 (1974).
[2] E. GERLACH and P. GROSSE, Festkörperprobleme **17**, 157 (1977).
[3] J. L. FARVAQUE, J. Physique, Suppl. No. 6, **39**, C2–52 (1978).
[4] E. GERLACH and B. HARBECKE, phys. stat. sol. (b) **100**, 187 (1980).
[5] K. O. KECHECHIAN and A. A. KIRAKOSIAN, Fiz. tverd. Tela **23**, 3186 (1981).
[6] E. GERLACH, phys. stat. sol. (b) **121**, 757 (1984).
[7] D. A. DAHL and L. J. SHAM, Phys. Rev. B **16**, 651 (1977).
[8] C. H. WU and W. HANKE, Solid State Commun. **23**, 829 (1977).
[9] E. A. PASHITZKI and YU. A. ROMANOV, Ukr. fiz. Zh. **15**, 1594 (1970).
[10] J. LEE and H. N. SPECTOR, J. appl. Phys. **54**, 6989 (1983).
[11] S. L. ADLER, Phys. Rev. **126**, 413 (1962).
[12] J. M. ZIMAN, Principles of the Theory of Solids, Izd. Mir, Moscow 1974 (in Russian).
[13] N. S. RYTOVA, Vestnik Moskovsk. Univ., Ser. Fiz. Astron. **3**, 30 (1967).
[14] L. V. KELDYSH, Zh. eksper. teor. Fiz., Pisma **29**, 716 (1979).
[15] O. V. DOLGOV and E. G. MAKSIMOV, Uspekhi fiz. Nauk **135**, 441 (1981).
[16] M. ABRAMOWITZ and I. A. STEGUN (Ed.), Handbook of Mathematical Functions, Izd. Nauka, Moscow 1979 (in Russian).
[17] J. L. FARVAQUE and E. GERLACH, phys. stat. sol. (b) **77**, 651 (1976).
[18] M. G. CALKIN and P. J. NIKOLSON, Rev. mod. Phys. **39**, 361 (1967).
[19] A. YA. SHIK, phys. stat. sol. (b) **34**, 661 (1969).
[20] O. N. FILATOV and I. A. KARPOVICH, Zh. eksper. teor. Fiz., Pisma **10**, 224 (1969).

(Received July 5, 1985)

phys. stat. sol. (b) 133, 395 (1986)

Subject classification: 72.40 and 72.80; 71.25; S12

Department of Physics, University of Roorkee[1])

Carrier Drift Mobilities in Some Organic Solids

A Search for Organic Photovoltaic Materials

I. Charge Transport Calculations in 2-2′ bipyridine and pyridazino [4,5-d] pridazine

By

B. KUMAR

Excess hole and excess electron mobilities are studied in 2-2′ bipyridine and pyridazino [4,5-d] pridazine using tight binding approximation. With suitable molecular orbitals for band structure and using a potential given by Silbey et al. the energy bands are calculated. The applicability of band model and the suitability of these materials for organic photovoltaic devices are discussed in the light of existing theories.

Die Beweglichkeit überschüssiger Löcher und Elektronen in 2-2′-Bipyridin und Pyridazino [4,5-d]-pridazin wird mit der „Tightbinding"-Näherung untersucht. Mit geeigneten Molekularorbitals für die Bandstruktur und unter Verwendung eines von Silbey et al. angegebenen Potentials werden die Energiebänder berechnet. Die Anwendbarkeit des Bandmodells und die Eignung dieser Materialien für organische photovoltaische Bauelemente werden mit existierenden Theorien diskutiert.

1. Introduction

Recently the direct conversion of solar energy to electrical energy using solar cells attracted much attention. Currently various solar cell materials and fabrication techniques are being investigated. However, the reported conversion efficiencies of organic solar cells are very poor because of several factors. One of the main reasons is the high rate of carrier recombination. The carrier recombination is expected to be low in materials showing large differences between electron and hole drift mobilities. We have therefore estimated theoretically excess electron and excess hole drift mobilities in a series of organic semiconductor crystals. Here we report the results of such studies for 2-2′ bipyridine and pyridazino 4,5-d pridazine, which belong to the class of heteroatomic organic semiconductors.

2. Method

For our calculation technique we used a tight binding model. In the tight binding model [1 to 3] the entire wave vector dependence of crystal energy is governed by the equation

$$E_n = \langle \varphi_c(\boldsymbol{r} - \boldsymbol{r}_n)| \; V^{\text{neut}}(\boldsymbol{r} - \boldsymbol{r}_n) \; |\varphi_c(\boldsymbol{r})\rangle \,, \tag{1}$$

where φ_c is the one-electron molecular orbital (MO) appropriate to the excess carrier (hole and electron) and $\boldsymbol{r}_n$ the vector to the centre of molecule n. V^{neut} is the potential experienced by an isolated neutral molecule. The E_n are known as intermolecular integrals (IMI's) and as shown by (1) they depend on the accuracy of MO's and of the potential.

[1]) Roorkee (U. P.), India.

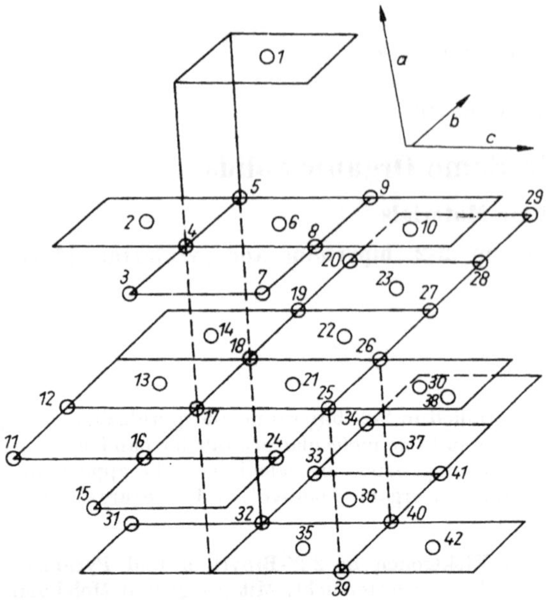

Fig. 1. Schematic representation of the unit cell showing the numbering of molecules

In actual calculations we have constructed MO's [4 to 5] using self-consistent field atomic orbitals (S.C.F.A.O.) and a potential given by Silbey et al. [6]. To evaluate the intermolecular integrals closed form expressions [7] similar to LeBlanc's [3] are utilised. Such expressions are expected to yield more accurate results.

Crystals studied in the present paper belong to space group $P_{2_1/c}$ and have two molecules per unit cell. The numbering of molecules for this space group is given in Fig. 1.

Energy bands for these crystals have been calculated following Mathur and Kumar [8]. According to them, energy eigenvalues are solutions of

$$E = \tfrac{1}{2}(H_{11} + H_{22}) \pm (\{\tfrac{1}{2}(H_{11} - H_{22})\}^2 + |H_{12}|^2)^{1/2}\,, \tag{2}$$

where + and − signs give two branches of the bands corresponding to two molecules per unit cell. The expressions for H_{11} and H_{22} are as follows:

$$\begin{aligned} H_{11} = {} & 2E_3 \cos \boldsymbol{k}\cdot(2\boldsymbol{a} - \boldsymbol{b}) + \\ & + 2E_4 \cos \boldsymbol{k}\cdot 2\boldsymbol{a} + \\ & + 2E_5 \cos \boldsymbol{k}\cdot(2\boldsymbol{a} + \boldsymbol{b}) + \\ & + 2E_7 \cos \boldsymbol{k}\cdot(2\boldsymbol{a} - \boldsymbol{b} + \boldsymbol{c}) + \\ & + 2E_8 \cos \boldsymbol{k}\cdot(2\boldsymbol{a} + \boldsymbol{c}) + \\ & + 2E_9 \cos \boldsymbol{k}\cdot(2\boldsymbol{a} + \boldsymbol{b} + \boldsymbol{c}) + \\ & + 2E_{11} \cos \boldsymbol{k}\cdot(\boldsymbol{a} - \boldsymbol{b} - \boldsymbol{c}) + \\ & + 2E_{12} \cos \boldsymbol{k}\cdot(\boldsymbol{a} - \boldsymbol{c}) + \\ & + 2E_{15} \cos \boldsymbol{k}\cdot(\boldsymbol{a} - 2\boldsymbol{b}) + \\ & + 2E_{16} \cos \boldsymbol{k}\cdot(\boldsymbol{a} - \boldsymbol{b}) + \\ & + 2E_{17} \cos \boldsymbol{k}\cdot\boldsymbol{a} + \\ & + 2E_{18} \cos \boldsymbol{k}\cdot(\boldsymbol{a} + \boldsymbol{b}) + \\ & + 2E_{19} \cos \boldsymbol{k}\cdot(\boldsymbol{a} + 2\boldsymbol{b}) + \\ & + 2E_{20} \cos \boldsymbol{k}\cdot(\boldsymbol{a} + 3\boldsymbol{b}) + \end{aligned}$$

$$
\begin{aligned}
&+ 2E_{24} \cos \boldsymbol{k} \cdot (\boldsymbol{a} - \boldsymbol{b} + \boldsymbol{c}) + \\
&+ 2E_{25} \cos \boldsymbol{k} \cdot (\boldsymbol{a} + \boldsymbol{c}) + \\
&+ 2E_{26} \cos \boldsymbol{k} \cdot (\boldsymbol{a} + \boldsymbol{b} + \boldsymbol{c}) + \\
&+ 2E_{27} \cos \boldsymbol{k} \cdot (\boldsymbol{a} + 2\boldsymbol{b} + \boldsymbol{c}) + \\
&+ 2E_{28} \cos \boldsymbol{k} \cdot (\boldsymbol{a} + 3\boldsymbol{b} + \boldsymbol{c}) + \\
&+ 2E_{29} \cos \boldsymbol{k} \cdot (\boldsymbol{a} + 4\boldsymbol{b} + \boldsymbol{c}) + \\
&+ 2E_{31} \cos \boldsymbol{k} \cdot (\boldsymbol{b} - \boldsymbol{c}) + \\
&+ 2E_{32} \cos \boldsymbol{k} \cdot \boldsymbol{b} + \\
&+ 2E_{33} \cos \boldsymbol{k} \cdot 2\boldsymbol{b} + \\
&+ 2E_{34} \cos \boldsymbol{k} \cdot 3\boldsymbol{b} + \\
&+ 2E_{39} \cos \boldsymbol{k} \cdot \boldsymbol{c} + \\
&+ 2E_{40} \cos \boldsymbol{k} \cdot (\boldsymbol{b} + \boldsymbol{c}) + \\
&+ 2E_{41} \cos \boldsymbol{k} \cdot (2\boldsymbol{b} + \boldsymbol{c}) ,
\end{aligned}
$$

$$
\begin{aligned}
H_{12} = \; & 2E_2 [\cos \boldsymbol{k} \cdot (2\boldsymbol{a} + \tfrac{1}{2}\boldsymbol{b} - \tfrac{1}{2}\boldsymbol{c}) + \cos \boldsymbol{k} \cdot (2\boldsymbol{a} - \tfrac{1}{2}\boldsymbol{b} - \tfrac{1}{2}\boldsymbol{c})] + \\
&+ 2E_6 [\cos \boldsymbol{k} \cdot (2\boldsymbol{a} + \tfrac{1}{2}\boldsymbol{b} + \tfrac{1}{2}\boldsymbol{c}) + \cos \boldsymbol{k} \cdot (2\boldsymbol{a} - \tfrac{1}{2}\boldsymbol{b} + \tfrac{1}{2}\boldsymbol{c})] + \\
&+ 2E_{10} [\cos \boldsymbol{k} \cdot (2\boldsymbol{a} + \tfrac{1}{2}\boldsymbol{b} + \tfrac{3}{2}\boldsymbol{c}) + \cos \boldsymbol{k} \cdot (2\boldsymbol{a} - \tfrac{1}{2}\boldsymbol{b} - \tfrac{1}{2}\boldsymbol{c})] + \\
&+ 2E_{13} [\cos \boldsymbol{k} \cdot (\boldsymbol{a} + \tfrac{1}{2}\boldsymbol{b} - \tfrac{1}{2}\boldsymbol{c}) + \cos \boldsymbol{k} \cdot (\boldsymbol{a} - \tfrac{1}{2}\boldsymbol{b} - \tfrac{1}{2}\boldsymbol{c})] + \\
&+ 2E_{14} [\cos \boldsymbol{k} \cdot (\boldsymbol{a} + \tfrac{3}{2}\boldsymbol{b} - \tfrac{1}{2}\boldsymbol{c}) + \cos \boldsymbol{k} \cdot (\boldsymbol{a} - \tfrac{3}{2}\boldsymbol{b} - \tfrac{1}{2}\boldsymbol{c})] + \\
&+ 2E_{21} [\cos \boldsymbol{k} \cdot (\boldsymbol{a} + \tfrac{1}{2}\boldsymbol{b} + \tfrac{1}{2}\boldsymbol{c}) + \cos \boldsymbol{k} \cdot (\boldsymbol{a} - \tfrac{1}{2}\boldsymbol{b} + \tfrac{1}{2}\boldsymbol{c})] + \\
&+ 2E_{22} [\cos \boldsymbol{k} \cdot (\boldsymbol{a} + \tfrac{3}{2}\boldsymbol{b} + \tfrac{1}{2}\boldsymbol{c}) + \cos \boldsymbol{k} \cdot (\boldsymbol{a} - \tfrac{3}{2}\boldsymbol{b} + \tfrac{1}{2}\boldsymbol{c})] + \\
&+ 2E_{23} [\cos \boldsymbol{k} \cdot (\boldsymbol{a} + \tfrac{5}{2}\boldsymbol{b} + \tfrac{1}{2}\boldsymbol{c}) + \cos \boldsymbol{k} \cdot (\boldsymbol{a} - \tfrac{5}{2}\boldsymbol{b} + \tfrac{1}{2}\boldsymbol{c})] + \\
&+ 2E_{30} [\cos \boldsymbol{k} \cdot (\boldsymbol{a} + \tfrac{1}{2}\boldsymbol{b} + \tfrac{3}{2}\boldsymbol{c}) + \cos \boldsymbol{k} \cdot (\boldsymbol{a} - \tfrac{1}{2}\boldsymbol{b} + \tfrac{3}{2}\boldsymbol{c})] + \\
&+ 2E_{35} [\cos \boldsymbol{k} \cdot (\tfrac{1}{2}\boldsymbol{b} + \tfrac{1}{2}\boldsymbol{c}) + \cos \boldsymbol{k} \cdot (-\tfrac{1}{2}\boldsymbol{b} + \tfrac{1}{2}\boldsymbol{c})] + \\
&+ 2E_{36} [\cos \boldsymbol{k} \cdot (\tfrac{3}{2}\boldsymbol{b} + \tfrac{1}{2}\boldsymbol{c}) + \cos \boldsymbol{k} \cdot (-\tfrac{3}{2}\boldsymbol{b} + \tfrac{1}{2}\boldsymbol{c})] + \\
&+ 2E_{37} [\cos \boldsymbol{k} \cdot (\tfrac{5}{2}\boldsymbol{b} + \tfrac{1}{2}\boldsymbol{c}) + \cos \boldsymbol{k} \cdot (-\tfrac{5}{2}\boldsymbol{b} + \tfrac{1}{2}\boldsymbol{c})] + \\
&+ 2E_{38} [\cos \boldsymbol{k} \cdot (\tfrac{7}{2}\boldsymbol{b} + \tfrac{1}{2}\boldsymbol{c}) + \cos \boldsymbol{k} \cdot (-\tfrac{7}{2}\boldsymbol{b} + \tfrac{1}{2}\boldsymbol{c})] + \\
&+ 2E_{42} [\cos \boldsymbol{k} \cdot (\tfrac{1}{2}\boldsymbol{b} + \tfrac{3}{2}\boldsymbol{c}) + \cos \boldsymbol{k} \cdot (-\tfrac{1}{2}\boldsymbol{b} + \tfrac{3}{2}\boldsymbol{c})] ,
\end{aligned}
$$

where E_3, E_4, ... etc. are the values of IMI's and $\boldsymbol{a}$, $\boldsymbol{b}$, and $\boldsymbol{c}$ are the unit cell dimensions of the crystals.

The expression for H_{22} may be derived from that of H_{11} by just substituting $-\boldsymbol{b}$ for $\boldsymbol{b}$ everywhere. The wave vector dependence of energy can now be found out by substituting the values of H_{11}, H_{22}, and H_{12} in (2).

To have a look on the shape of energy bands we have considered wave vectors $\boldsymbol{k}$ parallel to the reciprocal lattice vectors $\boldsymbol{a}$, $\boldsymbol{b}$, and $\boldsymbol{c}'$ ($\boldsymbol{c}' \perp \boldsymbol{a}, \boldsymbol{b}$). In these three cases the energy expression comes out to be

$$
\begin{aligned}
E(\boldsymbol{k} \parallel \boldsymbol{a}^{-1}) &= [H_{11} \pm H_{12}]\, \boldsymbol{k} \cdot \boldsymbol{b} = \boldsymbol{k} \cdot \boldsymbol{c} = 0 , \\
E(\boldsymbol{k} \parallel \boldsymbol{b}^{-1}) &= [H_{11} \pm H_{12}]\, \boldsymbol{k} \cdot \boldsymbol{c} = \boldsymbol{k} \cdot \boldsymbol{a} = 0 , \\
E(\boldsymbol{k} \parallel \boldsymbol{c}'^{-1}) &= [H_{11} \pm H_{12}]\, \boldsymbol{k} \cdot \boldsymbol{a} = \boldsymbol{k} \cdot \boldsymbol{b} = 0 .
\end{aligned}
$$

The mobilities have been calculated using the constant relaxation time approximation,

$$\mu_{ij} = \frac{e\tau_0}{kT} \langle V_i V_j \rangle .$$

The expression of $\langle V_i V_j \rangle$ for a crystal with two molecules per unit cell may be taken as [2]

$$\langle V_i V_j \rangle = \frac{\int \left\{ \frac{\partial E_+}{\partial k_i} \frac{\partial E_+}{\partial k_j} \exp\left[-\beta E_+(\boldsymbol{k})\right] + \frac{\partial E_-}{\partial k_i} \frac{\partial E_-}{\partial k_j} \exp\left[-\beta E_-(\boldsymbol{k})\right] \right\} \mathrm{d}\boldsymbol{k}}{\hbar^2 \int \left\{\exp\left[-\beta E_+(\boldsymbol{k})\right] + \exp\left[\beta E_-(\boldsymbol{k})\right]\right\} \mathrm{d}\boldsymbol{k}},$$

where

$$\beta = \frac{1}{kT}.$$

The velocity components $\langle V_a^2 \rangle$, $\langle V_b^2 \rangle$, $\langle V_{c'}^2 \rangle$, and $\langle V_a V_{c'} \rangle$ for molecular crystals have been calculated.

So far we discussed the calculations of energy bands and mobilities using a band scheme. However, before attempting such calculations, we must make sure that the charge carrier transport in these crystals can really be discussed by the band model. For doing so, we have looked at criteria given by Singh and Mathur [7] based on Munn and Siebrand's theory [9 to 10], which is the most successful theory of charge transport developed for organic solids. The criterion is based on the numerical values of $|E_n r_n|$, where E_n are IMI's and r_n is the intermolecular distance.

3. Results and Discussion

All the numerical calculations were carried out on an ICL 2960 computer.

Numbering of atomic sites within the molecules are shown in Fig. 2.

The calculated values of intermolecular integrals are collected in Table 1. Table 1 shows that the molecules 4, 32, 21 in 2-2′ bipyridine and molecules 16, 32, 35 in pyridazino 4,5-d pridazine have large IMI's and are expected to dominate charge transport. This is evident from Table 2.

The value of $|E_n r_n|$ and the likely mode of charge carrier transport as determined by Singh and Mathur's [7] criteria in various crystallographic directions in 2-2′ bipyridine and pyridazino 4,s-d pridazine are given in Table 2. Table 2 indicates that in 2-2′ bipyridine band theory is applicable along the ***b***-direction for holes and along ***a***- and ***c***′-directions for electrons. For the other directions a slow phonon hopping is indicated. In pyridazino 4,5-d pridazine, the band theory is expected to explain charge

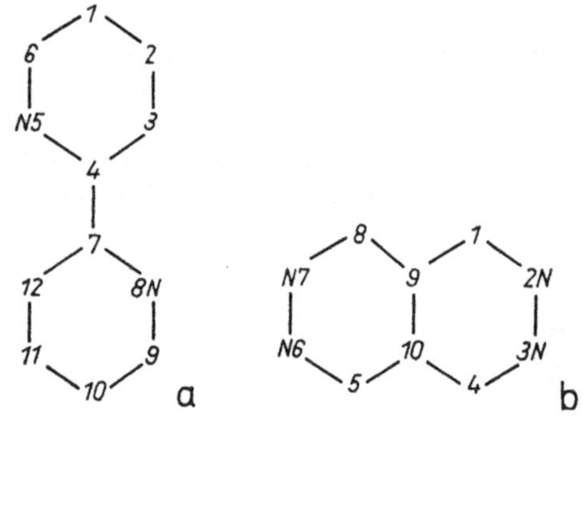

Fig. 2. Numbering of atomic sites in a) 2-2′ bipyridine and b) pyridazino [4,5-d] pridazine

Table 1
Intermolecular integrals (in units of 10^{-4} eV)

molecular number	position			2-2′ bipyridine hole	2-2′ bipyridine electron	pyridazino 4,5-d pridazine hole	pyridazino 4,5-d pridazine electron
0	0	0	0	reference molecule			
1	3	$\frac{1}{2}$	$\frac{1}{2}$	—	—	−0.023	0.014
2	2	$\frac{1}{2}$	$-\frac{1}{2}$	—	—	−0.095	0.101
3	2	−1	0	—	—	0.306	−0.240
4	2	0	0	−99.00	−0.106	1.765	0.941
5	2	1	0	−0.021	0.006	1.116	0.024
6	2	$\frac{1}{2}$	$\frac{1}{2}$	0.134	0.855	−2.925	1.560
8	2	0	1	—	—	−0.009	−0.002
9	2	1	1	0.011	−0.501	—	—
12	1	0	−1	—	—	0.006	0.045
13	1	$\frac{1}{2}$	$-\frac{1}{2}$	0.044	−0.016	−2.266	1.210
14	1	$\frac{3}{2}$	$-\frac{1}{2}$	0.010	0.0	0.015	−0.006
15	1	−2	0	—	—	32.499	2.420
16	1	−1	0	8.508	−9.383	2095.8	470.93
17	1	0	0	−29.994	37.184	—	—
18	1	1	0	−43.594	1.601	88.662	−4.805
19	1	2	0	0.127	0.027	—	—
21	1	$\frac{1}{2}$	$\frac{1}{2}$	120.77	318.96	69.499	−44.00
22	1	$\frac{3}{2}$	$\frac{1}{2}$	0.115	0.994	−0.026	0.005
25	1	0	1	0.604	−0.613	−0.025	0.007
26	1	1	1	0.536	−1.083	0.003	0.0
32	0	1	0	−444.8	137.2	−268.03	−35.989
33	0	2	0	−0.394	−0.282	—	—
35	0	$\frac{1}{2}$	$\frac{1}{2}$	−43.939	37.541	−144.57	130.090
36	0	$\frac{3}{2}$	$\frac{1}{2}$	0.344	1.588	0.0020	−0.004
39	0	0	1	0.004	−0.009	0.060	0.089
40	0	1	1	−0.014	0.019	0.003	−0.006

Table 2
Values of the dominant contribution $|E_n r_n|$ to the mobilities and assignment of transport mechanism

crystals, unit cell dimensions	$\|E_n r_n\|$ (10^{-4} eV, nm) hole			electron		
	$\|a\|$	$\|b\|$	$\|c'\|$ ($\perp ab$)	$\|a\|$	$\|b\|$	$\|c'\|$ ($\perp ab$)
2-2′ bipyridine [11] $a = 0.566$ nm $b = 0.624$ nm $c = 1.346$ nm $\beta = 118°44$	1120(4)*) ph	2775(32) b	712(21) ph	210(17) ph	8600(32) b	1882(21) b
pyridazino (4,5-d) pridazine [12] $a = 0.378$ nm $b = 0.747$ nm $c = 1.039$ nm $\beta = 96°20$	7922(17) b	2002(32) b	747(35) ph	2527(17) b	486(35) ph	672(35) ph

*) Numbers within the parenthesis refer to the molecule to which particular contribution is due
b band theory, ph slow phonon hopping.

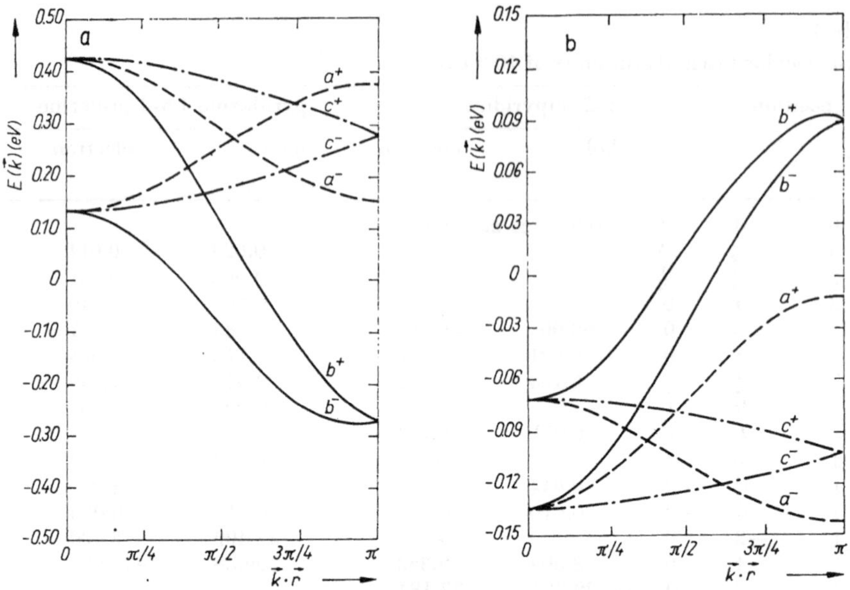

Fig. 3. 2-2′ bipyridine. Structure of a) hole and b) electron band

carrier transport in directions $\boldsymbol{a}$ and $\boldsymbol{b}$ for holes and in $\boldsymbol{a}$-direction for electrons, while a slow phonon hopping mechanism of charge transport seems to be valid in $\boldsymbol{c}'$-direction for holes and in $\boldsymbol{b}$ and $\boldsymbol{c}'$-directions for electrons. Thus having established the general applicability of the band model to the crystals studied here, we feel justified for the band structure calculations. However, the results on mobilities in those directions, where a different transport mechanism is indicated should be accepted with caution.

Energy bands in 2-2′ bipyridine are plotted in Fig. 3 a and b. These figures show that bands are degenerated at $\boldsymbol{k} \cdot \boldsymbol{b}^{-1} = \boldsymbol{k} \cdot \boldsymbol{c}'^{-1} = \pi$. Hole as well as electron bands touch each other at $\boldsymbol{k} \cdot \boldsymbol{r} = 3\pi/5$. The bandwidths are maximum along the $\boldsymbol{c}'^{-1}$-direction for both holes and electrons. The bandwidths given in Table 3 lie between 0.135 and 0.692 eV for holes and between 0.011 and 0.223 eV for electrons. Hole as well as electron bands are highly anisotropic.

Table 3

Bandwidths and splitting (in eV)

	2-2′ bipyridine		pyridazino 4,5-d pridazine	
	hole	electron	hole	electron
a_+^{-1}	0.155	0.083	0.940	0.154
a_-^{-1}	0.135	0.011	0.833	0.222
b_+^{-1}	0.692	0.162	0.091	0.021
b_-^{-1}	0.412	0.223	0.030	0.051
$c_+'^{-1}$	0.143	0.031	0.032	0.036
$c_-'^{-1}$	0.145	0.031	0.033	0.036
$a_{\text{splitting}}$	—	—	0.172	0.140
$c_{\text{splitting}}$	—	—	—	—

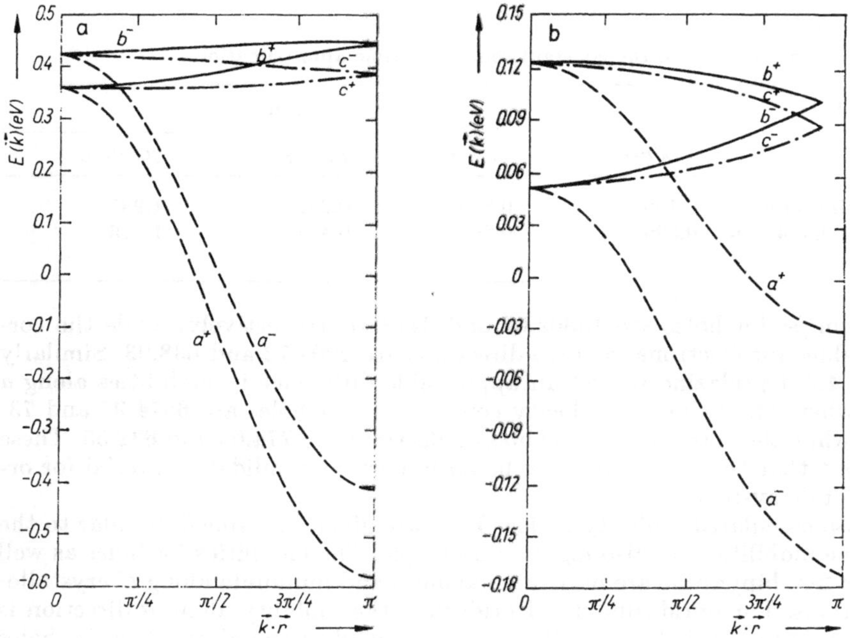

Fig. 4. Pyridazino [4,5-d] pridazine. Structure of a) hole and b) electron band

Fig. 4 a and b show the hole and electron bands in pyridazino 4,5-d pridazine, respectively. The bands in this case are also degenerated at $\boldsymbol{k} \cdot \boldsymbol{b}^{-1} = \boldsymbol{k} \cdot \boldsymbol{c}'^{-1} = \pi$, but they show splitting in $\boldsymbol{a}$-direction. The calculated bandwidths for holes and electrons are also given in Table 3. The calculated bandwidths are maximum along $\boldsymbol{a}^{-1}$ direction and minimum along $\boldsymbol{b}^{-1}$ direction for both holes and electrons. The bandwidth along $\boldsymbol{b}^{-1}$ and $\boldsymbol{c}'^{-1}$ directions for electrons are found to be comparable. The hole bandwidths lie between 0.030 and 0.940 eV, while the electron bandwidths range from 0.021 to 0.223 eV. The bands are highly anisotropic similar to the energy bands in 2-2′ bipyridine.

Table 4 gives the values of the velocity components in the constant isotropic relaxation time approximation. From Table 4 we see that the mobilities for holes and electrons show a quite large difference of their values in all crystallographic directions. In particular along b and c' axes in 2-2′ bipyridine the velocity components in

Table 4
Velocity components (in units of 10^{10} cm^2/s^2)

crystal	$\langle V_a^2\rangle$	$\langle V_b^2\rangle$	$\langle V_{c'}^2\rangle$	$\langle V_a V_{c'}\rangle$
	hole			
2-2′ bipyridine	480.76	10802.02	2448.57	908.90
pyridazino-4,5-d-pridazine	6574.97	507.77	73.13	−41.42
	electron			
2-2′ bipyridine	122.93	2004.72	638.03	208.29
pyridazino-4,5-d-pridazine	744.63	423.64	632.54	49.42

Table 5
Mobility anisotropies in the constant relaxation time approximation

crystal	hole		electron	
	μ_{aa}/μ_{bb}	$\mu_{c'c'}/\mu_{bb}$	μ_{aa}/μ_{bb}	$\mu_{c'c'}/\mu_{bb}$
2-2′ bipyridine	0.169	0.870	0.252	1.981
pyridazino 4,5-d-pridazine	0.899	0.267	0.445	1.846

units of 10^{10} cm²/s² for holes are 10802.02 and 2448.57, respectively, while the corresponding values for electrons in these directions are 2004.72 and 638.03. Similarly in pyridazino 4,5-d pridazine we get an appreciable difference in mobilities along a and c' axes, where the values of velocity components for holes are 6574.97 and 73, respectively, while the corresponding values of electrons are 774.63 and 632.55. These findings suggest that these crystals may be an important candidate material for organic photovoltaic systems.

We have also calculated mobility ratios. They are given in Table 5. Similar to the bandwidths the mobilities are also highly anisotropic. The mobilities for holes as well as electrons in 2-2′ bipyridine are maximum along $\boldsymbol{b}$ and minimum along $\boldsymbol{c}'$ crystallographic directions. For pyridazino 4,5-d pridazine, the mobility along $\boldsymbol{a}$-direction is found to be maximum for holes as well as electrons, while along $\boldsymbol{c}'$-direction for holes and along $\boldsymbol{b}$-direction for electrons, the mobility is minimum. However, no experimental data exist on the mobility anisotropy of these crystals, it is therefore not possible to draw any direct conclusions on the mobility anisotropies. Nevertheless predictions made here about the mobility anisotropies should be very helpful in planning future experiments. Results of such experiments in conjunction with the present predictions should provide a much better understanding of charge transport in these materials.

Acknowledgements

The author wishes to thank Prof. S. C. Mathur, Department of Physics, I.I.T. Delhi for his valuable guidance at various stages of this work. The author would also like to thank Prof. N. C. Varshneya, who as the Head of the Department provided facilities for the work. Financial assistance provided by C.S.I.R., India is also greatly acknowledged.

References

[1] F. Seitz, Modern Theory of Solids, McGraw-Hill Publ. Co., New York 1940 (p. 162).
[2] J. L. Katz, J. Jortner, and S. I. Choi, J. chem. Phys. **39**, 1683 (1963).
[3] O. H. LeBlanc, Jr., J. chem. Phys. **35**, 1275 (1961); **36**, 1082 (1962).
[4] S. C. Mathur and D. C. Singh, Chem. Phys. Letters **5**, 249 (1970).
[5] B. Kumar, Ph.D. Thesis, Delhi 1973.
[6] R. Silbey, J. Jortner, S. A. Rice, and M. T. Vala, Jr., J. chem. Phys. **42**, 733 (1965); **43**, 2925 (1965).
[7] D. C. Singh and S. C. Mathur, Mol. Cryst. liquid Cryst. **27**, 55 (1974).
[8] S. C. Mathur and B. Kumar, Mol. Cryst. liquid Cryst. **23**, 85 (1973).
[9] R. W. Munn and W. J. Siebrand, J. chem. Phys. **52**, 6341 (1970).
[10] R. W. Munn and W. Siebrand, Disc. Faraday Soc. **51**, 17 (1971).
[11] A. Merrit, Jr., L. Lynne, and Edith D. Schoroeder, Acta Cryst. **9**, 801 (1965).
[12] C. Sabelli, P. Tangocci, and P. F. Zanazzi, Acta Cryst. B **25**, 2231 (1969).

(Received June 19, 1985)

phys. stat. sol. (b) 133, 403 (1986)

Subject classification: 73.40; S7.13

Sektion Physik der Humboldt-Universität zu Berlin[1]),
Bereich Tieftemperatur-Festkörperphysik

Quantum Hall Effect in the Inversion Layer of p-Type InSb Bicrystals under High Hydrostatic Pressure

By

W. KRAAK, G. NACHTWEI, and R. HERRMANN

The properties of a two-dimensional electronic system in n-inversion layers in grain boundaries of p-InSb bicrystals are investigated under high hydrostatic pressure (up to 10^3 MPa) in high magnetic fields (up to 14 T). A rapid decrease of the carrier concentration in the inversion layer is observed when hydrostatic pressure is applied. At high pressures and high magnetic fields it is experimentally verified that the Hall resistivity ϱ_{xy} is quantized into integer multiples of h/e^2.

Es werden die Eigenschaften des zweidimensionalen Elektronengases in der n-leitenden Inversionsschicht von Korngrenzen in p-InSb-Bikristallen unter hohem hydrostatischen Druck (bis 10^3 MPa) in hohen Magnetfeldern (bis 14 T) untersucht. Mit wachsendem Druck wird eine schnelle Abnahme der Ladungsträgerkonzentration in der Inversionsschicht beobachtet. Es werden experimentelle Bedingungen realisiert, bei denen der Hall-Widerstand ϱ_{xy} in ganzzahligen Vielfachen von h/e^2 quantisiert ist.

1. Introduction

One of the most interesting problems in quasi-two-dimensional electronic systems is the quantized Hall effect.

In the presence of a strong magnetic field perpendicular to a semiconductor surface/interface the electron motion in such a system is completely quantized due to surface and Landau quantization and the electronic states form two-dimensional subbands e.g. [1]. Recently it was experimentally verified that the Hall conductivity in two-dimensional inversion layers in strong magnetic fields is quantized into integer multiples of e^2/h when the Fermi level lies in the localized states existing between the different levels [2, 3]. The quantum Hall effect manifests itself in the magnetotransport properties as plateaus in the Hall resistivity ϱ_{xy} and as minima in the diagonal resistivity ϱ_{xx}. Recently fractional quantization of the Hall resistivity was observed in the inversion layers of $GaAs/Al_xGa_{1-x}As$ heterostructures in the extreme quantum limit [4, 5]. In previous publications [6 to 9] we have made an experimental study of the magnetotransport properties of grain boundaries at p-InSb bicrystals. Our experimental results can be summarized as follows:

a) In the vicinity of grain boundaries at p-type InSb bicrystals there is a pronounced potential barrier ($\varphi_B \approx 200$ meV) for the carrier flow across the boundary [6].

b) From magnetotransport properties we concluded that a degenerate n-inversion layer exists adjacent to the grain boundary [7].

c) Characteristic features of a quasi-two-dimensional electron gas confined in the symmetrical potential well of the grain boundary have been obtained [8].

d) Two series of quantum oscillations have been observed which correspond to the series of Landau levels with the ground and first higher electronic subbands.

[1]) Invalidenstr. 110, DDR-1040 Berlin, GDR.

e) A rapid decrease of the carrier concentration in the inversion layer is observed when high hydrostatic pressure is applied [9].

In the present paper we report further experiments about the combined effects of hydrostatic pressure (up to 10^3 MPa) and high magnetic fields (up to 14 T) on the transport properties of the degenerate inversion layer of InSb bicrystals. It has been shown that at high pressure and high magnetic fields the Hall resistivity is quantized into integer multiples of h/e^2.

2. Experimental Details

InSb bicrystal samples are grown by the horizontal zone melting method with a double-seed technique. Two [111] p-type InSb crystals are tilted by an angle $\theta = 13°$ along the rotation axis [112]. A symmetrical β-grain boundary in [112] $(1\bar{1}0)$ orientation with a $(1\bar{1}0)$ symmetric plane was obtained. The grown bicrystals are cut by an electroerosive method into bars of $1 \times 1 \times 3.5$ mm³ containing a single grain boundary. After etching the samples in a CP4 solution to remove the damaged surface layer, electrical contacts were applied to the sample. Electrical current (0.1 mm Au) and potential leads (0.025 mm Au) were soldered to the sample using pure indium. The measuring current was always adjusted to levels where an Ohmic behaviour was guaranteed.

The excess acceptor concentration $p = N_A - N_D$ and the carrier mobility μ of the bulk material were determined from the Hall constant measured at 77 K with a magnetic field strength of 1 T ($p_{77K} = N_A - N_D = 5.8 \times 10^{15}$ cm^{-3}; $\mu_{77K} = 4.6 \times 10^3$ cm²/Vs).

For measuring the electrical properties under hydrostatic pressure a bicrystal sample and a monocrystal sample were mounted in a high pressure piston-cylinder apparatus with a pressure channel diameter of 4 mm. All parts of the vessel are made of hardened Be–Cu alloy suited to low temperature applications in high magnetic fields. As a pressure-transmitting medium a 1:1 mixture of transformer oil and kerosene was used. High pressure was generated at room temperature and after that the pressure cell was placed inside the cryostat and slowly cooled down to 4.2 K.

High magnetic field measurements were made to observe quantum properties of the quasi-two-dimensional electrons in the grain boundary region. The high-field experiments ($B > 5$ T) have been performed in a 15 T Bitter-type magnet at the International Laboratory for High Magnetic Fields and Low Temperatures in Wrocław (Poland).

3. Results and Discussion

In the present work we have investigated the magnetotransport properties in a n-inversion layer adjacent to the grain boundary in p-InSb bicrystals at magnetic fields up to 15 T under high hydrostatic pressure up to 10^3 MPa. The experimental results of a previous paper [9] and the results of the present work clearly demonstrate a strong variation of the electronic properties of the inversion layer under hydrostatic pressure. No inversion layer could be detected at high pressures $p > 6 \times 10^2$ MPa. In the quantum oscillations of the magnetoresistivity ϱ_{xx} the oscillations of the first excited subband ($i = 1$) are evident for $p < 3.5 \times 10^2$ MPa, the oscillations of the ground subband ($i = 0$) for $p \leqq 5 \times 10^2$ MPa. From the periodicity of the quantum oscillations the sheet electron densities n_{si} of the two subbands may be evaluated (cf. [8]). The pressure dependence of the electron population of the subbands n_{si} and the total electron density n_s as determined from the low-field Hall coefficient R_H is summarized in Table 1 and Fig. 1. The total occupation of the observed subband n_{si}

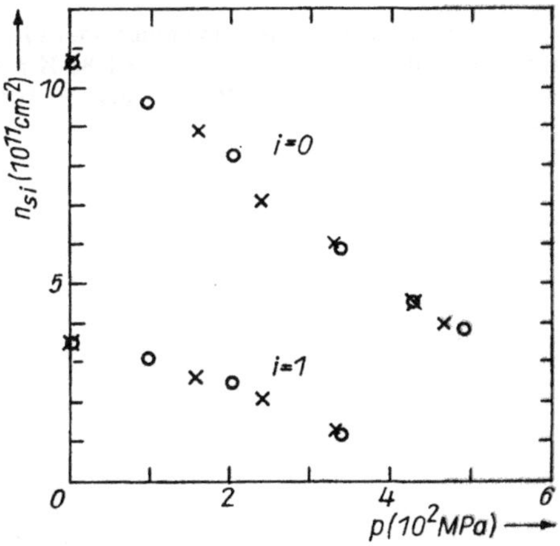

Fig. 1. Pressure dependence of the carrier concentrations n_{si} of the electronic subbands deduced from SdH oscillations. o first experimental series, × second experimental series. The pressure was increased up to 10^3 MPa before the experimental series, then the experimental series were performed with decreasing pressure. No hysteresis effects could be found

agrees well with the total electron density n_s deduced from the Hall coefficient R_H. A rapid decrease of the electron population of the inversion layer and of the carrier concentrations of the electronic subbands n_{si} is observed when pressure is applied.

The strong pressure dependence of the carrier concentration n_{si} is due to a change of the position of the subband edges E_i relative to the Fermi energy E_F under pressure which is primarily determined by the significant change of the direct energy gap E_g between the conduction and valence bands in InSb ($\partial E_g/\partial p \approx 0.15$ meV/MPa [10]).

The weaker pressure dependence of the carrier concentration n_{s1} of the first excited subband (Fig. 1) can be explained by the decreasing electric field at the grain boundary region due to the change of charge balance with increasing pressure. Detailed theoretical calculations will be published in a following paper.

A similar pressure dependence of the carrier concentration has been found in GaAs/GaAlAs heterostructures [11]. This method of monitoring the carrier concentration with hydrostatic pressure can be a useful tool to investigate the quantum properties of 2D structures [9, 12], e.g. for studying the nature of the quantum Hall effect in high magnetic fields and low temperatures.

Table 1

Pressure dependence of the electron populations of the subbands n_{si} (determined from quantum oscillations of ϱ_{xx}) and the total electron density n_s (determined from the low-field Hall coefficient R_H)

p (10^2 MPa)	n_{s0} (10^{11} cm^{-2})	n_{s1} (10^{11} cm^{-2})	$n_{s0}+n_{s1}$ (10^{12} cm^{-2})	n_s (10^{12} cm^{-2})
10^{-3}	10.7	3.5	1.42	1.4*)
1	9.6	3.1	1.27	1.26
1.6	8.9	2.7	1.16	0.98*)
2	8.2	2.5	1.07	1.11
2.4	7.1	2.1	0.92	0.81*)
3.35	6.0	1.3	0.73	0.63*)
3.4	5.9	1.2	0.71	0.77
4.2	4.6	—	0.46	0.42*)
4.65	3.9	—	0.39	0.33*)
4.9	3.8	—	0.38	0.57

*) Values corresponding to one experimental series.

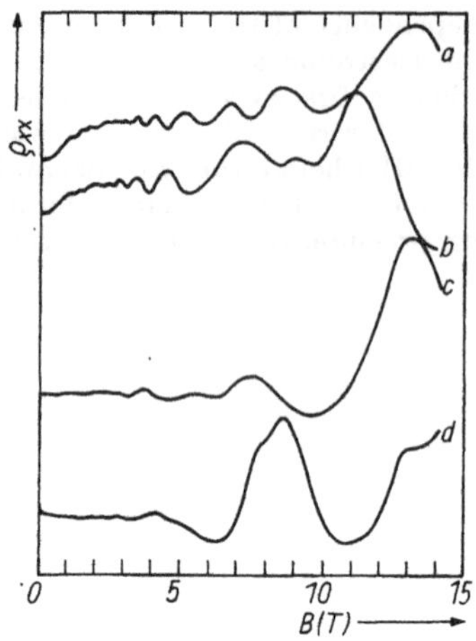

Fig. 2. Quantum oscillations of the resistivity component ϱ_{xx} in transverse magnetic fields (up to 14 T) at $T = 1.85$ K for several pressures p: (a) 0.1, (b) 2.4×10^2, (c) 3.4×10^2, (d) 4.2×10^2 MPa

Fig. 2 and 3 show typical data obtained at 1.9 K for the magnetoresistivity components ϱ_{xx} and ϱ_{xy} as a function of magnetic field B (up to 14 T) for various pressures p. Pronounced quantum oscillations of the diagonal component ϱ_{xx} corresponding to Landau quantization of the electron gas in the inversion layer have been observed in the pressure interval 0.1 MPa $< p <$ 4.2×10^2 MPa. In high magnetic fields ($B >$ $>$ 5 T) pronounced spin splitting is clearly evident for the Landau levels $n \leqq 2$ (Fig. 2b, d). The quantum oscillations reported in Fig. 2 demonstrate clearly a remarkable shift of the peaks of the magnetoresistivity ϱ_{xx} towards lower magnetic fields as the pressure is increased. The pressure dependence of the magnetic field position of the peaks is also a consequence of the change of the position of subband edges E_i relative to the Fermi energy E_F of the bulk material under hydrostatic pressure. This shift of the peaks towards lower magnetic fields is very important for the observation of the quantum Hall effect in rather low magnetic fields even for a structure like the grain boundary in InSb with a relatively high carrier concentration (under zero pressure).

In Fig. 3 the magnetic field dependence of the Hall resistivity ϱ_{xy} is reported for several pressures p. The most striking feature in Fig. 3 is the constant value of the

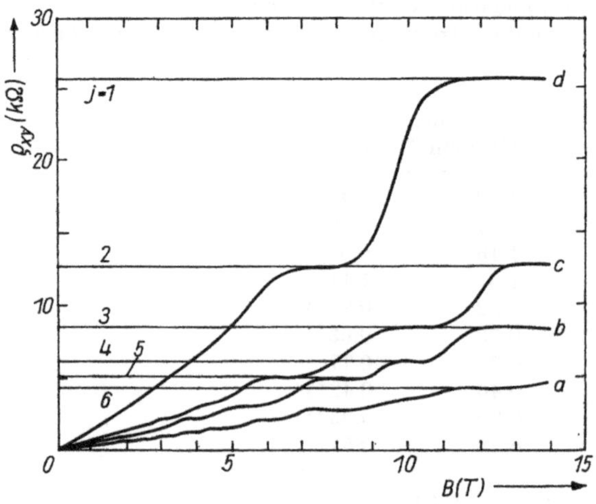

Fig. 3. Hall resistivity ϱ_{xy} vs. magnetic field at $T = 1.85$ K for several pressures p: (a) 0.1, (b) 2.4×10^2, (c) $3.4 \times$ $\times$ 10^2, (d) 4.2×10^2 MPa. Pronounced plateaus for the quantum numbers $j = 1$ to 6 could be found

Table 2

Pressure dependence of the magnetic field position B_{plat} and values ϱ_{xy} of the quantized Hall resistance plateaus

p (10^2 MPa)	j	B_{plat} (T)	ϱ_{xy} (kΩ)	h/je^2 (kΩ)
4.2	1	11.8 to 14	25.850	25.8128
	2	7.1 to 7.8	12.960	12.9064
3.4	2	13.2 to 14	12.902	12.9064
	3	10.3	8.539	8.6043
	5	6.3	5.104	5.1626
2.4	3	12.2 to 14	8.642	8.6043
	4	10.1	6.177	6.4532
	5	8.0	5.110	5.1626
10^{-3}	6	11.4	4.221	4.3021

Hall component ϱ_{xy} (at $\approx$25 kΩ, $\approx$ 12 kΩ, $\approx$6 kΩ etc.) in a finite range of magnetic field corresponding to the situation when the Fermi level E_F lies between the Landau levels. Simultaneously the diagonal resistivity ϱ_{xx} drops to a minimum. A summary of the experimental results is given in Table 2.

The values of ϱ_{xy} obtained in the plateau regions are in good agreement with the predicted values [2, 3]

$$\varrho_{xy}^{th} = \frac{h}{e^2 j}; \qquad j = 1, 2, \dots ,$$

where j is the number of filled Landau levels (without spin and valley degeneracy).

The agreement of the experimental results ϱ_{xy} with the theoretical values is limited likely by the accuracy of the experimental arrangement used. However, it is evident from Table 2 that the width of the plateaus and the accuracy of the Hall resistance increases with increasing pressure p (decreasing n_s) and increasing magnetic field B, i.e. for decreasing Hall resistance quantum number j.

For $p = 4.2 \times 10^2$ MPa the resistance plateaus $j = 1$ ($\varrho_{xy} \approx 25.8$ kΩ) and the plateau $j = 2$ ($\varrho_{xy} \approx 12.9$ kΩ) are evident in the magnetic field intervals 11.8 T $<$ $< B <$ 14 T and 7.1 $< B <$ 7.8 T, respectively. For lower pressures the resistance plateaus with Hall resistance quantum number up to $j = 6$ are identified.

As for the peaks of quantum oscillations ϱ_{xx} a remarkable shift of the Hall resistance plateaus ϱ_{xy} towards lower magnetic fields is observed as the pressure is increased. It is remarkable to note that the Hall resistance quantization under pressure is observable at much lower magnetic fields (Fig. 3) and higher temperatures in comparison with high-mobility Si-inversion layers because of the smaller effective electron mass in InSb. Thus the ultra quantum limit for the grain boundary structure in InSb may be realized under hydrostatic pressure at experimentally attainable magnetic fields (for the investigated bicrystals the ultra quantum limit under atmospheric pressure is realized at $B >$ 40 T).

4. Conclusions

We have found from magnetotransport properties a remarkable change of the electronic properties of the n-inversion layer in InSb bicrystals under high hydrostatic pressure. A rapid decrease of the carrier concentration in the inversion layer is observed when pressure is applied.

The observation of the quantum Hall effect in the inversion layer of InSb-bicrystals under hydrostatic pressure demonstrates once more the universal two-dimensional properties of the electron gas confined in the symmetrical potential well of the grain boundary. The advantages of the method using the effect of hydrostatic pressure for the investigation of the electronic properties of quasi-two-dimensional systems have been demonstrated.

Acknowledgements

The high field experiments have been performed at the International Laboratory for High Magnetic Fields and Low Temperatures in Wrocław (Poland). The authors would like to thank the coworkers of this laboratory for their most valuable technical help and support in the high-field experiments.

References

[1] T. ANDO, A. B. FOWLER, and F. STERN, Rev. mod. Phys. **54**, 437 (1982).
[2] K. v. KLITZING, G. DORDA, and M. PEPPER, Phys. Rev. Letters **45**, 494 (1980).
[3] G. LANDWEHR (Ed.), Proc. Internat. Conf. Application High Magnetic Fields in Semiconductor Physics, Grenoble 1982, Lecture Notes Phys., Vol. 177, Springer-Verlag, 1983.
[4] D. C. TSUI, H. L. STORMER, and A. C. GOSSARD, Phys. Rev. Letters **48**, 1559 (1982).
[5] H. L. STORMER, A. CHANG, D. C. TSUI, J. C. M. HWANG, A. C. GOSSARD, and W. WIEGMANN, Phys. Rev. Letters **50**, 1953 (1983).
[6] R. HERRMANN, W. KRAAK, and G. NACHTWEI, phys. stat. sol. (b) **128**, 337 (1985).
[7] R. HERRMANN, W. KRAAK, G. NACHTWEI, and G. WORM, Solid State Commun. **52**, 843 (1984).
[8] R. HERRMANN, W. KRAAK, G. NACHTWEI, and TH. SCHURIG, phys. stat. sol. (b) **129**, 415 (1985).
[9] W. KRAAK, R. HERRMANN, and G. NACHTWEI, phys. stat. sol. (b) **130**, 781 (1985).
[10] R. KEYES, Phys. Rev. **99**, 490 (1955).
[11] I. M. MERCY, C. BOUSQUET, I. L. ROBERT, A. RAYMOND, G. GREGORIS, J. BEERENS, J. C. PORTAL, P. M. FRIJLINK, P. DELESCLUSE, J. CHEVRIER, and N. T. LINH, Surface Sci. **142**, 298 (1984).
[12] J. L. ROBERT, J. M. MERCY, C. BOUSQUET, A. RAYMOND, I. C. PORTAL, G. GREGORIES, and J. BEERENS, preprint.

(Received September 24, 1985)

phys. stat. sol. (b) 133, 409 (1986)
Subject classification: 72.10 and 72.20; 71.20

Sektion Physik der Humboldt-Universität zu Berlin, Bereich Theoretische Halbleiterphysik[1])

Nonlinear High-Electric-Field Transport in the Presence of Strong Laser Radiation

By

O. ZIEP, A. G. MIRONOV[2]), and R. KEIPER

The Boltzmann equation is investigated in the presence of strong and time-dependent electromagnetic fields taking into account the intra-collisional field effect. The expression obtained for the field-induced tail of the electron–phonon scattering rate leads to the known results for static fields at low frequencies and includes at higher frequencies multiphoton processes.

Die Boltzmann-Gleichung wird bei Anwesenheit starker zeitabhängiger elektromagnetischer Felder unter Einbeziehung des "intra-collisional field-effects" untersucht. Der gewonnene Ausdruck für den feldinduzierten Ausläufer der Stoßzahl infolge Elektron–Phonon-Streuung führt bei tiefen Frequenzen auf die bekannten Resultate für statische Felder und berücksichtigt bei höheren Frequenzen Multiphotonprozesse.

1. Introduction

High-electric-field transport has been studied for many years. Due to the advent of semiconductor devices on the submicrometer scale the investigation of hot-carrier transport is currently arousing greater interest [1]. With regard to the Boltzmann equation an increasing number of papers has been devoted to the problem of nonlinear electric-field quantum transport. Especially, different approaches to non-markovian behaviour and the intra-collisional field effect are still a matter of discussion. Recently, within a correct transport conception the Boltzmann equation has been investigated including static-electric-field dependent collision integrals and scattering rates ν_{ep} due to electron–phonon interaction [2, 3]. As shown there for scattering mechanisms leading to a proportionality of ν_{ep} to the density of final states ϱ in conventional Boltzmann theory the consideration of the intra-collisional field effect consists in replacing ϱ by a static field-dependent density of states ϱ_F expressed by Airy functions. The main emphasis of our work now will be put on the study of transport phenomena which are nonlinear in an external time-dependent field.

In the last few years time-resolved measurements of transport and optical processes in semiconductors have been applied in order to detect nonequilibrium processes on short time scales using laser systems of high excitation power P [4, 5].

It is well known that an intense electromagnetic field changes the energy spectrum in a semiconductor [6] and thus it influences the collision terms also. The aim of the present paper is to calculate the intra-collisional field effect due to a strong electric field $\boldsymbol{F} = \boldsymbol{F}_0 \cos \omega t$ produced e.g. by a laser beam of frequency ω. In accordance to the nonlinear dependence of the scattering rate ν_{ep} on a static electric field $\boldsymbol{F}_0$ we calculate now the corresponding nonlinear relationship between ν_{ep} and the excitation power P.

---

[1]) Invalidenstr. 110, PSF 1297, DDR-1040 Berlin, GDR.
[2]) Permanent address: Faculty of Physics, Moscow State University.

Considering the Franz-Keldysh effect in one-photon electro-optics and the multi-photon absorption coefficient [7] there exists a close analogy to our calculations.

The present paper contains in Section 2 a discussion of the field-induced density of states ϱ_F in an energy region below the band edge where tail states caused by laser radiation become essential. A transport equation with field-dependent collision integrals is obtained in Section 3 for a frequency range $\omega \gg \tau_{rel}^{-1}$ with the relaxation time τ_{rel} supposing a quasistatic behaviour of the electron gas. The obtained generalized Boltzmann equation is appropriate for a numerical treatment within the framework of the traditional Monte Carlo technique. Finally we discuss the field-dependence of the electron–phonon scattering rates ν_{ep} shortly.

2. Local Density of Tail States in the Presence of an ac Field

The local density of states plays an important role in the discussion of quantum kinetic equations [2, 8, 9]. In case of laser radiation the electric-field-dependent wave functions of the electron system are usefully expressed by

$$\Psi_{\boldsymbol{k}}(\boldsymbol{r}, t) = \exp\left(i\boldsymbol{kr} - \frac{i}{\hbar}\int\limits_0^t \mathrm{d}t'\, E(\boldsymbol{k}_{t'})\right), \tag{1}$$

where a parabolic isotropic band structure $E(k) = \hbar^2k^2/2m$ with the effective mass m is regarded. The time-dependent wave vector $\boldsymbol{k}_t = \boldsymbol{k} + (e/\hbar) \int_0^t \boldsymbol{F}(t')\,\mathrm{d}t'$ has been introduced in order to describe the influence of a high-frequency field $\boldsymbol{F} = (0, 0, F_0 \cos \omega t)$.

Due to the nonstationarity of the electron system the z-component of $\boldsymbol{k}_t$ is no longer a good quantum number and the function $E(\boldsymbol{k}_t)$ has to be replaced here by a time-averaged field-dependent quasi-energy W. The density of states $\varrho_F(W)$ for nonstationary states (1) is determined in the sense of a time-averaged value of an observable for the electron system coupled with an electric field.[3])

Expanding (1) in a Fourier series with respect to time according to the treatment of Keldysh [7],

$$\Psi_{\boldsymbol{k}}(\boldsymbol{r}, t) = \exp\left[i\boldsymbol{kr} - \frac{it}{\hbar}\left(E(k) + \frac{1}{2}\,E(\varkappa)\right)\right] \sum_{n=-\infty}^{\infty} \exp\,(in\omega t)\,\varphi_n(k_z)\,, \tag{2}$$

we obtain for the coefficient $\varphi_n(k_z)$ $(n = 0, \pm 1, \ldots)$ the expression

$$\varphi_n(k_z) = \frac{1}{2\pi} \oint\limits_{|v|=1} \frac{\mathrm{d}v}{\sqrt{1 - v^2}} \exp\,[i\chi(n, k_z, v)]$$

with

$$\chi(n, k_z, v) = \frac{\hbar\varkappa}{m\omega} \sqrt{1 - v^2} \left(k_z + \frac{\varkappa v}{4}\right) - n \arcsin v\,, \tag{3}$$

where the parameter $\varkappa = eF_0/\hbar\omega$ has been introduced. In the sense of local spatial homogeneity the electron system is supposed to be inhomogeneous only in time. The field-dependent retarded Green's function $G^{\mathrm{ret}}(\boldsymbol{k}; \Delta t, \tau)$, formulated in terms of the

[3]) We suppose that the observation times are much greater than ω^{-1} and θ^{-1}, θ being the electro-optic frequency.

Wigner variables Δt and τ [2] as

$$G^{\text{ret}}(\boldsymbol{k};\Delta t,\tau)=\frac{1}{i\hbar}\exp\left[-\frac{i}{\hbar}\int\limits_{\tau-\frac{\Delta t}{2}}^{\tau+\frac{\Delta t}{2}}\mathrm{d}t'\,E(k_{t'})\right]\mathrm{H}(\Delta t)\,, \tag{4}$$

reflects the nonstationary electron states (1). $\mathrm{H}(\Delta t)$ denotes the unit step function. In distinction from the static field situation discussed in [2] for the retarded Green's function (4) a representation by the wave vector $\boldsymbol{k}$ has been used. The Fourier transform of function (4) behaves for macroscopic times τ in the energy region W as follows:

$$G^{\text{ret}}(\boldsymbol{k};W,\tau)|_{\tau\to\infty}=\sum_{n=-\infty}^{\infty}\frac{|\varphi_n(k_z)|^2}{W-E(k)-\frac{1}{2}E(\varkappa)+n\hbar\omega+i\delta}\Bigg|_{\delta\to 0^+}. \tag{5}$$

Here the representation (2) has been used.

Function (5) enables us to calculate the field-dependent density of states $\varrho_{\mathrm{F}}(W)$ by means of the expressions

$$\varrho_{\mathrm{F}}(W)=\frac{m}{2\hbar^2\pi^2}\sum_{n=-\infty}^{\infty}\int\limits_{-\infty}^{\infty}\mathrm{d}k_z\,|\varphi_n(k_z)|^2\,\mathrm{H}\left(W-E(k_z)-\frac{1}{2}E(\varkappa)+n\hbar\omega\right), \tag{6}$$

where the integration over the wave vector components k_x and k_y transverse to $\boldsymbol{F}$ has already been performed employing the δ-function in the imaginary part of (5). The interesting field-induced contribution to $\varrho_{\mathrm{F}}(W)$ below the band edge, i.e. at $W<0$ and $E(k_\perp)\approx 0$ is calculated in the following by use of the saddle point method for the integration in (3). The corresponding stationary points v_s in the exponent of (3) which enters the imaginary part of the retarded function (5) at $n\hbar\omega=E(k)+\frac{1}{2}E(\varkappa)-W$ are given by the equation $W-E(k_\perp)-E(k_z+\varkappa v_s)=0$. In the region of tail states an expansion of the function $\chi(n,k_z,v_s)$ up to k_z^2 leads to the approximative form

$$\chi(n,k_z,\gamma)=$$
$$=i\left[n\,\operatorname{arsinh}\gamma-\frac{E(\varkappa)\,\gamma}{2\hbar\omega}\sqrt{1+\gamma^2}+\frac{\hbar\varkappa k_z}{m\omega}\sqrt{1+\gamma^2}-\frac{\gamma E(k_z)}{\hbar\omega\sqrt{1+\gamma^2}}\right], \tag{7}$$

where the real parameter

$$\gamma=\frac{\omega}{eF_0}\sqrt{-2mW}=iv_s|_{k=0}=\sqrt{\frac{-W}{E(\varkappa)}}$$

has been introduced. The squared module of $\varphi_n(k_z)$ is then given by

$$|\varphi_n(k_z)|^2=\left(\frac{m\omega^2}{8\varkappa^2\pi^2(1+\gamma^2)\,(n\hbar\omega-E(k_z)-\frac{1}{2}E(\varkappa))}\right)^{1/2}|\mathrm{e}^{i\chi(n,k_z,\gamma)}|^2\,. \tag{8}$$

Finally, the density of states $\varrho_{\mathrm{F}}(W<0)$ can be arranged as

$$\varrho_{\mathrm{F}}(W<0)=\frac{m^{3/2}\omega^{3/2}\gamma^{1/2}\exp(-c)}{8\pi^3\hbar^{3/2}(-W)\,(1+\gamma^2)^{1/4}}\sum_{n=0}^{\infty}\int\limits_0^{y_n}\frac{\mathrm{d}y}{\sqrt{y}}\sqrt{\frac{b}{b+y_n-y}}\times$$
$$\times\exp(y)\left(\gamma+\sqrt{\gamma^2+1}\right)^{-2(\operatorname{int}(a+1)-a+n)}. \tag{9}$$

Here the abbreviations

$$\begin{aligned} y_n &= \frac{2\gamma}{\sqrt{1+\gamma^2}} \left(\operatorname{int}(a+1) - a + n\right), \\ a &= \frac{\frac{1}{2}E(\varkappa) - W}{\hbar\omega}, \\ b &= \frac{-2W\gamma}{\hbar\omega\sqrt{1+\gamma^2}}, \\ c &= -\frac{2W}{\hbar\omega}\left[\left(1+\frac{1}{2\gamma^2}\right)\operatorname{arsinh}\gamma - \frac{\sqrt{1+\gamma^2}}{2\gamma}\right] \end{aligned} \tag{10}$$

are introduced and int(a) denotes the integer part of a. We remark that (9) holds for the parameter $c \gg 1$ and describes both the low- and the high-frequency cases.

2.1 Static limit of $\varrho_F(W)$

For vanishing frequency ω the parameter $\varkappa$ tends to infinity and, therefore, the limit $\omega \to 0$ cannot be carried out at an arbitrary place in (2), (6), or (9).

In case of quasistatic electric fields we have $\gamma \ll 1$ and the first nonvanishing contribution of the parameter c is given by $c = \frac{4}{3}(-W/\hbar\theta)^{3/2}$ with the electro-optic frequency $\theta = (e^2F_0^2/2m\hbar)^{1/3}$.[4]) Mathematically, the limit $\omega \to 0$ has to be performed after the n-summation and k_z-integration in (6). Changing over to an integration variable $s = n\hbar\omega$ one obtains

$$\varrho_{F_0}(W<0) = \lim_{\omega\to 0} \frac{m^{3/2}\omega \exp\left(-\frac{4}{3}\left(-\frac{W}{\hbar\theta}\right)^{3/2}\right)}{2^{5/2}\hbar^2\pi^3 eF_0} \int_{-\infty}^{\infty} dk_z \int_0^{\infty} ds (s-W)^{-1/2} \times$$
$$\times \exp\left(-2\frac{s}{\hbar\omega}\operatorname{arsinh}\gamma - 2\frac{E(k_z)}{\hbar\omega}\left(\operatorname{arsinh}\gamma - \frac{\gamma}{\sqrt{1+\gamma^2}}\right)\right). \tag{11}$$

Physically the k_z- and s-integrations in (11) are not independent. As was shown in [2] for static fields the k_z-integration corresponds to a time average. In our case the relation $k_z = k_{zt} - (eF_0/\hbar\omega) \sin\omega t$ holds. In order to avoid an undetermined expression which arises in the limit $\omega \to 0$ we suggest the replacement

$$\int_{-\infty}^{\infty} \frac{dk_z}{2\pi} \to \int_{-1}^{+1} \frac{eF_0}{\hbar\omega} dv \quad \text{with} \quad v = \sin\omega t .$$

This difficulty appeared already in [7] where the absorption coefficient for multiphoton processes was calculated. On the basis of our discussion in the limit $\omega \to 0$ one obtains the expression

$$\varrho_{F_0}(W)|_{-W \gg \hbar\theta} \approx \frac{meF_0 \exp\left(-\frac{4}{3}\left(-\frac{W}{\hbar\omega}\right)^{3/2}\right)}{8\pi^2\hbar^2(-W)} \tag{12}$$

which coincides with the known behaviour of the static-field-dependent density of states.

---

[4]) Already at this stage the exponent in (9) coincides with the corresponding result for static fields and the following discussion concerns only the pre-exponential factor.

2.2 High-frequency limit of $\varrho_F(W)$

For electric fields $\boldsymbol{F}$ with high-frequency ω one has $\gamma \gg 1$ and $c \approx -(2W/\hbar\omega) \times \times \ln(2\gamma) \gg 1$. In (9) the term $n = 0$ dominates and describes the contribution of interaction processes of the electron state with $l = \operatorname{int}(-W/\hbar\omega + 1)$ photons to the density of states $\varrho_F(W < 0)$,

$$\varrho_F(W < 0) = \frac{m^{3/2}\omega^{3/2} \exp\left(-\dfrac{W}{\hbar\omega}\right)}{8\pi^3\hbar^{3/2}(-W)\,(4\gamma^2)^l} \int\limits_0^{y_0} \frac{\mathrm{d}y}{\sqrt{y}} \sqrt{\frac{b}{b + y_0 - y}} \exp(y)\,, \tag{13}$$

where $y_0 = 2\big((W/\hbar\omega) + l\big)$, $b = -2W/\hbar\omega$, and $1/\gamma^2 = e^2P/[\varepsilon_0 mcn_r\omega^2(-W)]$ with the vacuum dielectric constant ε_0, the refractive index n_r, and the light velocity c. Thus, the l-photon process is proportional to P^l. In the vicinity of thresholds, i.e. at $W < < 0 < W + l\hbar\omega \ll \hbar\omega$ we get a square-root dependence of the density of states $\varrho_F(W)$,

$$\varrho_F(W < 0) \approx \frac{\sqrt{2}\, m^{3/2-l}(e^2\mu_0 cP)^l \exp\left(-\dfrac{W}{\hbar\theta}\right)\sqrt{W + l\hbar\omega}}{\pi^3\hbar^2 n_r^l \omega^{2l-1}(-4W)^{l+1}}\,. \tag{14}$$

The field-induced tail of the density of states $\varrho_F(W)$ calculated numerically from (9) is represented in Fig. 1 and shows a $P^l\sqrt{W + l\hbar\omega}$ dependence near the thresholds associated with multiphoton processes. With increasing excitation power P the structures in $\varrho_F(W)$ are smeared and the values of $\varrho_F(W)$ at $W < 0$ become comparable with those obtained by

$$\varrho_0(W) = \frac{\sqrt{2}\, m^{3/2} W^{1/2}}{\pi^2\hbar^3}$$

for $W > 0$ above the band edge.

In our treatment the influence of the laser field on $\varrho_F(W)$ has been considered within classical electrodynamics. Consequently, the interaction with the electron system is described by a net absorption process. In this sense $E(k)$ is the energy of band states, $l\hbar\omega$ the energy absorbed by the electron system, and $(1/2)\,E(\varkappa)$ the $\boldsymbol{k}$ and l independent mean oscillation energy in the laser field.

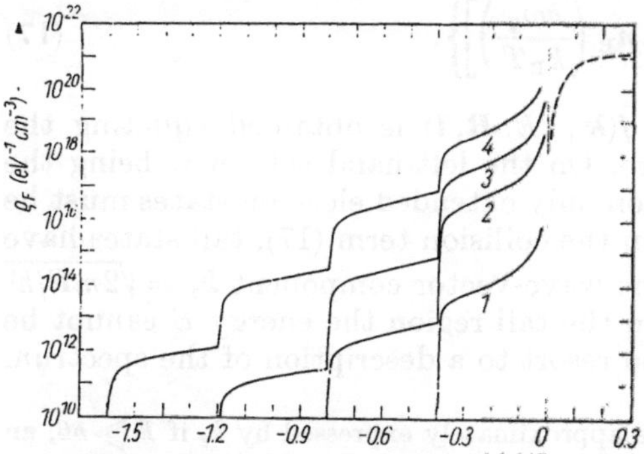

Fig. 1. Density of states $\varrho_F(W)$ in dependence on energy W for various excitation powers (1) $P = 2 \times 10^5$, (2) 10^8, (3) 10^9, and (4) 10^{10} W/cm^2 and $\hbar\omega = 0.4$ eV (InAs laser) compared with the expression $\varrho_0(W) = \sqrt{2}\, m^{3/2}W^{1/2}/\pi^2\hbar^3$ (dashed line) for a parabolic band with $m = 0.43m_0$

3. Transport Equation and Discussion of the Scattering Rate

The local carrier concentration can be represented as an energy integral of the product of the density of states (6) and a time- and position-dependent distribution function $f(\boldsymbol{k}_\perp, E, \boldsymbol{R}, t)$ [2]. Here $E = W - E(k_\perp)$ characterizes the carrier energy of the motion in field direction. For $E(k_\perp) \approx 0$ and negative energies $E < 0$ the function $f(\boldsymbol{k}_\perp, E, \boldsymbol{R}, t)$ describes the occupation of tail states. In the case $E > 0$ it may be approximately characterized by the quasimomentum k_z, which leads to the conventional distribution function $f(\boldsymbol{k}, \boldsymbol{R}, t)$.[5]) Supposing $\omega\tau_{\text{rel}} \gg 1$ with the transport relaxation time τ_{rel} a lower limit for ω is given by infrared frequencies. In this case the distribution function of the carrier system behaves quasistatically. The rigorous representation for the correlation functions $G^{\lessgtr}_{\boldsymbol{R}}(\boldsymbol{k}; W, t)$ is then [2]

$$G^{\lessgtr}_{\boldsymbol{R}}(\boldsymbol{k}; W, t)|_{\omega t \gg 1} = -2 \begin{Bmatrix} f(\boldsymbol{k}_\perp, E, \boldsymbol{R}, t) \\ 1 - f(\boldsymbol{k}_\perp, E, \boldsymbol{R}, t) \end{Bmatrix} \operatorname{Im} G^{\text{ret}}(\boldsymbol{k}; W, t)\,, \tag{15}$$

where the retarded $\boldsymbol{k}$-dependent Greens function (5) is tightly connected with the density of states $\varrho_{\text{F}}(W)$ shown in Fig. 1.

In the following we regard transport under the influence of a weak, perturbing static electric field $\boldsymbol{F}_{\text{s}}(\boldsymbol{r})$ produced e.g. by space charges. Inserting the correlation function into the corresponding equation of motion on the line of [2] one gets

$$\sum_n |\varphi_n(k_z)|^2 \left\{\left(\frac{\partial}{\partial t} + \boldsymbol{v}\frac{\partial}{\partial \boldsymbol{R}} + \frac{e\boldsymbol{F}_{\text{s}}(\boldsymbol{R})}{\hbar}\frac{\partial}{\partial \boldsymbol{k}}\right) f(\boldsymbol{k}_\perp, E, \boldsymbol{R}, t) - I\right\} \times$$
$$\times\, \delta\left(E - E(k_z) - \frac{1}{2}\,E(\varkappa) + n\hbar\omega\right) = 0\,. \tag{16}$$

The collision term I describing one-phonon processes with the coupling parameter $g_{\boldsymbol{q}}$, the frequency $\omega_{\boldsymbol{q}}$, and the Planck function $n_{\text{B}}(x)$ is given by

$$I = \sum_{n', \boldsymbol{k}', \boldsymbol{q}, \eta = \pm 1} \frac{2\pi}{\hbar} |g_{\boldsymbol{q}}|^2 |\varphi_{n'}(k'_z)|^2 \,\delta\Big(E(k_\perp) + E - \eta\hbar\omega_{\boldsymbol{q}} -$$
$$- E(k') - \frac{1}{2}\,E(\varkappa) + n'\hbar\omega\Big)\, \delta_{\boldsymbol{k}', \boldsymbol{k}-\boldsymbol{q}} \Big\{f(\boldsymbol{k}_\perp, E, \boldsymbol{R}, t) \times$$
$$\times [1 - f(\boldsymbol{k}'_\perp, E', \boldsymbol{R}, t)] \left[\frac{1}{2} + \frac{1}{2}\,\eta + n_{\text{B}}\left(\frac{\hbar\omega_{\boldsymbol{q}}}{k_{\text{B}}T}\right)\right] - [1 - f(\boldsymbol{k}_\perp, E, \boldsymbol{R}, t)] \times$$
$$\times f(\boldsymbol{k}'_\perp, E', \boldsymbol{R}, t) \left[\frac{1}{2} - \frac{1}{2}\,\eta + n_{\text{B}}\left(\frac{\hbar\omega_{\boldsymbol{q}}}{k_{\text{B}}T}\right)\right]\Big\}. \tag{17}$$

An equation for the distribution function $f(\boldsymbol{k}_\perp, E, \boldsymbol{R}, t)$ is obtained equating the expression in the paranthesis of (16) to zero. On the left-hand side of it being the field term of the ordinary Boltzmann equation only extended electron states must be considered but on the right-hand side, within the collision term (17), tail states have to be included also. Above the band edge the wave-vector component $k_z = \sqrt{2mE/\hbar^2}$ is used in the sense of local homogeneity. In the tail region the energy E cannot be expressed by a dispersion law and one has to resort to a description of the spectrum

---

[5]) Whereas for static fields the energy E can be approximately expressed by k_z if $E \gg \hbar\theta$, an analogous condition for the high-frequency case ($\omega \gtrsim \theta$) has the form $E \gg \hbar\theta(\theta/\omega)^2$.

by means of the density of states $\varrho_F(W)$. Physically the tail states give only contributions to the carrier concentration and to the scattering rates.

It is noted that the transient case characterized by $\omega\tau_{rel} \approx 1$ and $\gamma \approx 1$ is connected with off-diagonal parts of the distribution function $f_{nn'}(\mathbf{k}_\perp, E, \mathbf{R}, t)$ $(n \neq n')$ with respect to the photon number n. In this case the quasistatic relationship (15) fails. However, under the regarded condition $\gamma \gg 1$ for not too high excitation powers the Boltzmann equation remains valid. But the scattering rates $\nu_{ep}(W)$ are modified near characteristic threshold energies. Particularly, for electron–phonon scattering processes between band and laser-field-induced tail states one gets the following expression for the collision rate $\nu_{ep}(\mathbf{k}, W)$ of the carrier state $\mathbf{k}$:

$$\nu_{ep}(\mathbf{k}, W) = \frac{2\pi}{\hbar} \sum_{\mathbf{k}',\mathbf{q}} \sum_{n'=-\infty}^{\infty} \sum_{\eta=\pm 1} |g_{\mathbf{q}}|^2 \, |\varphi_{n'}(k_z')|^2 \, \delta_{\mathbf{k}',\mathbf{k}-\mathbf{q}} \times \\ \times \left[\frac{1}{2} + \frac{1}{2}\,\eta + n_B\left(\frac{\hbar\omega_{\mathbf{q}}}{k_B T}\right)\right] \times \\ \times \delta\left(W - E(k') - \eta\hbar\omega_{\mathbf{q}} - \frac{1}{2}\,E(\varkappa) + n'\hbar\omega\right). \tag{18}$$

Expression (18) describes transitions of the carriers between the energy levels W, W' due to one-phonon processes assisted by multiphoton absorption ($W' = W - \eta\hbar\omega_{\mathbf{q}} + l\hbar\omega$, $\eta = \pm 1$, $l = 1, 2, \ldots$). In case of transitions between nonequivalent valleys separated by the energy Δ one gets, below a threshold ($W - \Delta - \eta\hbar\omega_{\mathbf{q}_0} < 0$),

$$\nu_{ep}(W) = \frac{2\pi}{\hbar} |g_{\mathbf{q}_0}|^2 \sum_{\eta=\pm 1} \left[\frac{1}{2} + \frac{1}{2}\,\eta + n_B\left(\frac{\hbar\omega_{\mathbf{q}_0}}{k_B T}\right)\right] \varrho_F(W - \Delta - \eta\hbar\omega_{\mathbf{q}_0}), \tag{19}$$

where $g_{\mathbf{q}}$ and $\omega_{\mathbf{q}}$ are approximately taken at the wave-vector distance of nonequivalent valleys $\mathbf{q}_0$ and W is the carrier energy in the lowest valley. Here the laser-field-induced tail of the density of states (9) which has to be numerically evaluated enters (19).

The resulting tail of the scattering rate (19) is represented in Fig. 2a and b for various excitation powers P and laser-field frequencies ω in case of Γ–X intervalley transitions in GaAs. It can be seen that the values of ν_{ep} compete e.g. with acoustical-phonon scattering rates being of the order of $10^{12}\ s^{-1}$ near characteristic threshold energies. At the highest excitation powers P the structures in ν_{ep} due to assisted multiphoton absorption begin to vanish and a more complicated P-dependence of ν_{ep} compared to the power law P^l is observed.

We mention the following relationship between the calculated values ν_{ep} and the intraband absorption coefficient given by $\alpha_{intra} = \hbar\omega R_{intra}/P$, where R_{intra} denotes the total transition rate due to absorption of the laser field $F_0 \cos \omega t$.

For small occupation numbers in the higher valleys the total transition rate due to intraband multiphoton absorption is given by

$$R_{intra} = \sum_{\mathbf{k}} \nu_{ep}(E(\mathbf{k}))\, f(E(\mathbf{k}))\,. \tag{20}$$

Here $f(E(\mathbf{k}))$ denotes the occupation number for band states in the lowest valley and the rate (19) of intervalley transitions has been used. At low temperatures $kT \ll \Delta$ and low carrier concentrations with Fermi energy $E_F \ll \Delta$ we get approximately

$$\alpha_{intra} = \frac{2\pi N\omega\, |g_{\mathbf{q}_0}|^2}{P} \sum_{\eta=\pm 1} \varrho_F(-\Delta - \eta\hbar\omega_{\mathbf{q}_0}) \left[\frac{1}{2} + \frac{1}{2}\,\eta + n_B\left(\frac{\hbar\omega_{\mathbf{q}_0}}{k_B T}\right)\right] \tag{21}$$

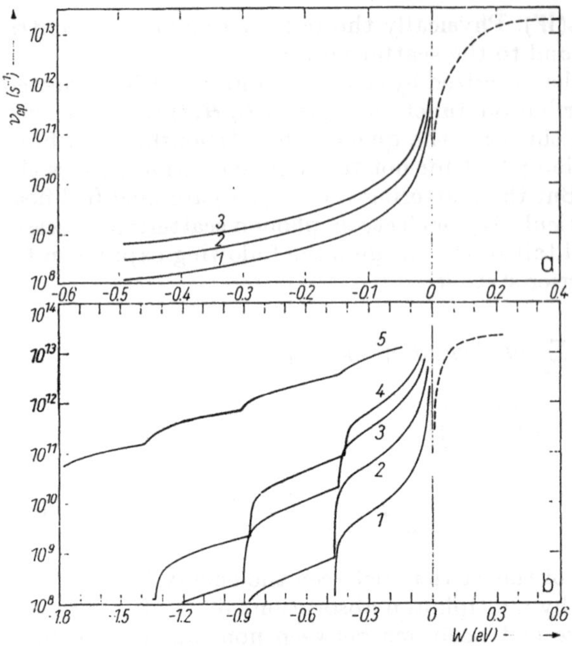

Fig. 2. Scattering rate ν_{ep} of Γ–X intervalley transitions in dependence on the energy distance from the corresponding absorption threshold a) for a Nd:YAG laser with $\hbar\omega = 1.166$ eV and (1) $P = 2 \times 10^8$, (2) 5×10^8, (3) 10^9 W/cm² calculated with the GaAs-like parameters $m = 0.43m_0$, $(2\pi/\hbar)\, g^2_{q_0} = 2 \times 10^{-8}$ eV cm³ s⁻¹; b) for a hydrogen fluoride laser with $\hbar\omega = 0.46$ eV and higher excitation power (1) $P = 10^9$, (2) 10^{10}, (3) 5×10^{10}, (4) 10^{11}, and (5) 10^{12} W/cm²

with the free carrier concentration N. An estimation for GaAs parameters e.g. with $\hbar\omega = 1.65$ eV indicates that this process can only be detected at high excitation powers $P \gtrsim 1$ GW/cm². Thus, a comparison of α_{intra} with the interband absorption coefficient enables us to estimate the rate of carriers excited in upper valleys. As a result of the laser-induced intra-collisional field effect the population of satellite valleys in III–V compounds increases more than predicted by the conventional Boltzmann theory [5].

For one-photon absorption our nonperturbative treatment must coincide with the conventional results for α_{intra}. A discussion of the multiphoton absorption coefficient calculated by Keldysh [7] has been reviewed in [10] where it has been shown that the one-photon absorption limit in direct-gap crystals agrees well with results of conventional perturbation theory.

Acknowledgement

The authors are grateful to Prof. V. L. Bonch-Bruevich for valuable remarks.

References

[1] L. REGGIANI, Proc. 4th Internat. Conf. Hot Electrons, University of Innsbruck 1985 (p. TU1-1)
[2] R. KEIPER and O. ZIEP, phys. stat. sol. (b) **131**, K91 (1985).
[3] O. ZIEP and R. KEIPER, phys. stat. sol. (b) **128**, 779 (1985).
[4] C. L. COLLINS and P. Y. YU, Phys. Rev. B **30**, 4501 (1984).
[5] E. O. GÖBEL, P. H. LIANG, and D. VON DER LINDE, Solid State Commun. **37**, 609 (1981).
[6] I. V. BELOUSOV and V. P. OLEINIK, J. Phys. C **12**, 655 (1979).
[7] L. V. KELDYSH, Zh. eksper. teor. Fiz. **47**, 1945 (1964).
[8] J. R. BARKER, J. Physique **42-C 7**, 245 (1981).
[9] A. P. JAUHO and J. W. WILKINS, Phys. Rev. B **29**, 1919 (1984).
[10] A. VAIDYANATHAN, T. W. WALKER, A. H. GUENTHER, S. S. MITRA, and L. M. NARDUCCI, Phys. Rev. B **20**, 3526 (1979).

(Received November 5, 1985)

phys. stat. sol. **133**, 417 (1986)

Subject classification: 72.30; S7.13

Department of Physics, Government College, Kota[1])

Resonant Excitation of Helicon Waves by Two Microwave Beams in a Solid State Plasma

By

K. P. Maheshwari and G. Tarey

An investigation is made of resonant excitation of helicon waves at the difference frequency of two microwave beams propagating in a magneto-active solid state plasma viz. n-InSb. The two beams propagating obliquely with the magnetostatic field interact in the semiconductor to give rise to a nonlinear ponderomotive force at the difference frequency which resonantly excites helicon waves if the dispersion relation corresponding to helicon waves is satisfied. The results show that microwave beams of power densities 1 MW cm^{-2} can excite helicon waves of power density 0.98 kW cm^{-2} when $\omega_c/\omega = 25$. The power carried by helicon waves depends on the direction of propagation of the pumps relative to magnetostatic field. For the parameters under consideration the power gets optimised at an angle $\theta = 63.4°$. Helicon waves are not excited if the pumps propagate parallel or perpendicular to the magnetic field. The increase in magnetic field increases the power density of the excited helicon wave.

Es wird die Resonanzanregung von Helikonwellen bei der Differenzfrequenz zweier Mikrowellenstrahlen, die sich in einem magnetoaktiven Festkörperplasma, z.B. n-InSb, ausbreiten, untersucht. Die beiden Strahlen breiten sich schräg zum magnetostatischen Feld aus, das im Halbleiter wirkt, und ihre Wechselwirkung gibt Anlaß zu einer nichtlinearen ponderomotorischen Kraft bei der Differenzfrequenz, die eine Resonanzanregung von Helikonwellen ergibt, wenn die den Helikonwellen entsprechende Dispersionsbeziehung erfüllt ist. Die Ergebnisse zeigen, daß Mikrowellenstrahlen mit einer Leistungsdichte von 1 MW cm^{-2} Helikonwellen mit einer Leistungsdichte von 0.98 kW cm^{-2} anregen können, wenn $\omega_c/\omega = 25$ ist. Die von den Helikonwellen getragene Leistung hängt von der Ausbreitungsrichtung der Pumpwellen relativ zum statischen Magnetfeld ab. Für die untersuchten Parameter wird die Leistung bei einem Winkel $\theta = 63.4°$ optimal. Helikonwellen werden nicht angeregt, wenn die Pumpwellen sich parallel oder senkrecht zum Magnetfeld ausbreiten. Die Erhöhung des Magnetfeldes steigert die Leistungsdichte der angeregten Helikonwelle.

1. Introduction

Resonant excitation of waves has been the subject of many theoretical investigations [1 to 3]. Recently we have studied the excitation of magneto-acoustic waves at the difference frequency of two microwave beams propagating in n-InSb [4]. It has been shown that there is a finite probability of excitation of an acoustic wave along with the magneto-acoustic wave. The excited acoustic and magneto-acoustic waves couple at magnetic fields when the thermal pressure is equivalent to the magnetic pressure. In the present study we extend the analysis for the resonant excitation of helicon waves in a solid state plasma, viz. n-InSb. The study is important because helicon waves help in determining the carrier density of solid state plasmas [5]. Although helicon waves have been studied in the past [6, 7] no attempt has been made to study their excitation in a semiconductor by the beating of two microwave beams.

For investigating the resonant excitation of helicon waves we consider two EM waves propagating with their propagation vectors in the xz plane making an angle θ

[1]) Kota 324001, Rajasthan, India.

($0 < \theta < \pi/2$) with the external static magnetic field $\boldsymbol{B}_0 \parallel z$ and the electric fields of the waves along the x-axis. The two waves interact to give rise to a nonlinear ponderomotive force at the difference frequency $\omega = \omega_1 - \omega_2$. The ponderomotive force gives rise to a nonlinear velocity of the electrons which further interact with the external magnetic field and generate a helicon wave if the dispersion relation for helicon waves is satisfied,

$$\omega = \frac{\omega_c c^2 k^2 \cos\theta}{\omega_p^2},$$

where ω_c is the electron cyclotron frequency, c the velocity of light in vacuum, $\boldsymbol{k}$ the difference of propagation vectors of the pump waves, and ω_p the plasma frequency.

In Section 2, the general formulation of the problem has been presented. Writing the basic equations relevant to the problem we have obtained the linear and nonlinear parts of the electron velocity for the geometry under consideration. In Section 3, the interaction of nonlinear velocity with the external magnetostatic field has been studied and the dispersion relation for helicon waves is obtained. In Section 4 the power carried by an excited helicon wave has been obtained. A brief discussion of the results has been presented in Section 5. Our results show that for typical plasma parameters of n-InSb and microwave beams of power densities 1 MW cm^{-2}, the optimum value of power density of the excited helicon wave is 0.98 kW cm^{-2} when $\omega_c/\omega = 25$ and it increases to 3.18 kW cm^{-2} when ω_c/ω increases to 40. The power carried by the helicon waves depends very much on the direction of propagation of pumps relative to the external magnetic field. It is zero (and so the helicon wave is not excited) when pumps propagate parallel or perpendicular to the magnetic field and for parameters under consideration it gets optimised when pumps propagate at an angle $\theta \approx 63.4°$ with the magnetic field.

2. General Formulation: Electron Drift Velocity

We consider two microwave beams of frequencies ω_1 and ω_2 propagating in a homogeneous semiconductor (n-InSb) immersed in an external magnetostatic field $\boldsymbol{B}_0 \parallel z$. The electric vectors of the pumps are considered along the x-axis and their propagation vectors $\boldsymbol{k}_1$ and $\boldsymbol{k}_2$ in the xz-plane both making an angle θ with the z-axis.

$$\boldsymbol{E}_{1,2} = E_{1,2}\hat{x} \exp\left[-i(\omega_{1,2}t - \boldsymbol{k}_{1,2} \cdot \boldsymbol{r})\right] \tag{1}$$

with

$$k_{1,2} = \omega_{1,2}\left[\varepsilon_L - \frac{\omega_p^2}{\omega_{1,2}(\omega_{1,2} - \omega_c)}\right]^{1/2}, \tag{2}$$

$$\omega_p = \left(\frac{4\pi n_e^0 e^2}{m^*}\right)^{1/2}, \tag{3}$$

$$\omega_c = \left|\frac{eB_0}{m^* c}\right|, \tag{4}$$

$$\boldsymbol{B}_{1,2} = -\frac{ic}{\omega_{1,2}}(\nabla \times \boldsymbol{E}_{1,2}), \tag{5}$$

where ω_p is the electron plasma frequency, ω_c the electron cyclotron frequency, $\boldsymbol{B}_{1,2}$ the magnetic fields of the pumps, ε_L the lattice dielectric constant, n_e^0 the equilibrium electron density, and m^* the effective mass of the electron.

The two beams interact in the semiconductor and generate a nonlinear ponderomotive force at the beat frequency $\omega = \omega_1 - \omega_2$. This nonlinear force develops a nonlinear velocity in the electrons which further interacts with the external static magnetic field to excite helicon waves of purely electron origin. In order to investigate the excitation of helicon waves in the semiconductor we consider the following basic equations:

(i) Equation of motion

$$m^* \frac{\partial \boldsymbol{v}}{\partial t} = -e\boldsymbol{E} - \frac{m^* v_{\mathrm{Th}}^2}{n_{\mathrm{e}}^0} \nabla n_{\mathrm{e}} - \frac{e}{c} \boldsymbol{v} \times \boldsymbol{B}_0 - \boldsymbol{F} - m^* \nu \boldsymbol{v} . \tag{6}$$

(ii) Continuity equation

$$\frac{\partial n_{\mathrm{e}}}{\partial t} + \nabla \cdot (n_{\mathrm{e}} \boldsymbol{v}) = 0 . \tag{7}$$

(iii) Poisson's equation

$$\nabla \cdot \boldsymbol{E} = -4\pi e n_{\mathrm{e}} . \tag{8}$$

(iv) Maxwell's equation

$$\nabla \times \boldsymbol{E} = \left(\frac{i\omega}{c}\right) \boldsymbol{B} , \tag{9}$$

$$\nabla \times \boldsymbol{B} = \left(\frac{4\pi}{c}\right) \boldsymbol{J} - \left(\frac{i\omega}{c}\right) \boldsymbol{E} . \tag{10}$$

In (6) to (10), $v_{\mathrm{Th}} \approx (k_{\mathrm{B}} T_{\mathrm{e}}/m^*)^{1/2}$ is the thermal speed of electrons, k_{B} the Boltzmann constant, T_{e} the electron temperature, $\boldsymbol{B}_0$ the external magnetic field, $\boldsymbol{v}$ the electron drift velocity, ν the electron–phonon collision frequency, $\boldsymbol{J}$ the current density, $\boldsymbol{F}$ the ponderomotive force at the beat frequency resulting from the interaction of the two pump waves and given by [8]

$$\boldsymbol{F} = \left[m^* (\boldsymbol{v}_1 \cdot \nabla)\, \boldsymbol{v}_2^* + \frac{e}{c}\, \boldsymbol{v}_1 \times \boldsymbol{B}_{1,2}^* \right] = \frac{e^2}{4 m^* \omega_1 \omega_2} \nabla (\boldsymbol{E}_1 \cdot \boldsymbol{E}_2^*) . \tag{11}$$

Using (11), (7), (8) in (6) the electron drift velocity is obtained as

$$\boldsymbol{v} = \boldsymbol{v}^{\mathrm{l}} + \boldsymbol{v}^{\mathrm{nl}} , \tag{12}$$

where

$$\boldsymbol{v}^{\mathrm{l}} = -\frac{e}{m^*} \left[\frac{1}{\Omega^2 + \omega_{\mathrm{c}}^2} \{(\Omega E_x - \omega_{\mathrm{c}} E_y)\, \hat{\boldsymbol{x}} + (\omega_{\mathrm{c}} E_x + \Omega E_y)\, \hat{\boldsymbol{y}}\} + \frac{E_z \hat{\boldsymbol{z}}}{\Omega} \right], \tag{13}$$

$$\boldsymbol{v}^{\mathrm{nl}} = -\frac{i e^2 E_1 E_2^*}{4 m^* \omega_1 \omega_2} \left[\frac{k_x}{\Omega^2 + \omega_{\mathrm{c}}^2} \{\Omega \hat{\boldsymbol{x}} + \omega_{\mathrm{c}} \hat{\boldsymbol{y}}\} + \frac{k_z}{\Omega} \hat{\boldsymbol{z}} \right], \tag{14}$$

$$\Omega = \left(\nu - i\omega + \frac{i k^2 v_{\mathrm{Th}}^2}{\omega} \right). \tag{15}$$

Thus the electron drift velocity consists of two parts, one linear and the other nonlinear. Linear and nonlinear contributions to the electron drift velocity are given by (13) and (14), respectively.

3. Dispersion Relation of the Excited Wave

The nonlinear velocity (14) further interacts with the external magnetostatic field $B_0 \boldsymbol{z}$ to excite a wave whose dispersion relation is derived below.

Using (9) and (10) the wave equation can be written as

$$-k^2 \boldsymbol{E} + \boldsymbol{k}(\boldsymbol{k} \cdot \boldsymbol{E}) = -\frac{\omega^2}{c^2} \boldsymbol{E} - \frac{4\pi i \omega}{c^2} \boldsymbol{J}^{\mathrm{nl}} . \tag{16}$$

The equation of motion of electrons under the influence of an excited wave would be

$$m^* \frac{\mathrm{d}\boldsymbol{v}^{\mathrm{nl}}}{\mathrm{d}t} = -e\boldsymbol{E} - \frac{e}{c}\,\boldsymbol{v}^{\mathrm{nl}} \times \boldsymbol{B}_0\,. \tag{17}$$

For excitation of low frequencies ($\omega \ll \omega_c$) the first term on the right-hand side of (16) and the left-hand side of (17) can be neglected. Thus, for $\omega \ll \omega_c$ (17) yields.

$$\boldsymbol{E} = \left(\frac{4\pi\omega_c}{\omega_p^2}\right)\boldsymbol{J}^{\mathrm{nl}} \times \hat{\boldsymbol{z}}\,. \tag{18}$$

Using (18) in (16),

$$k^2(\boldsymbol{J}^{\mathrm{nl}} \times \hat{\boldsymbol{z}}) - \boldsymbol{k}[\boldsymbol{k}\cdot(\boldsymbol{J}^{\mathrm{nl}} \times \hat{\boldsymbol{z}})] - \frac{i\omega\omega_p^2}{\omega_c c^2}\boldsymbol{J}^{\mathrm{nl}} = \boldsymbol{0}\,. \tag{19}$$

Since $\boldsymbol{J}_z^{\mathrm{nl}} \times \hat{\boldsymbol{z}}$ vanishes, x and y components of (19) can be written as

$$-\frac{i\omega\omega_p^2}{\omega_c c^2} J_x^{\mathrm{nl}} + (k^2 - k_x^2)\, J_y^{\mathrm{nl}} = 0\,, \qquad k^2 J_x^{\mathrm{nl}} + \frac{i\omega\omega_p^2}{\omega_c c^2} J_y^{\mathrm{nl}} = 0\,.$$

The condition for these homogeneous equations to be compatible is the vanishing of the determinant

$$\begin{vmatrix} -\dfrac{i\omega\omega_p^2}{\omega_c c^2} & (k^2 - k_x^2) \\ k^2 & \dfrac{i\omega\omega_p^2}{\omega_c c^2} \end{vmatrix} = 0\,.$$

Using the fact $k^2 - k_x^2 = k_z^2$ we obtain

$$\omega = \frac{\omega_c c^2 k^2 \cos\theta}{\omega_p^2} \tag{20}$$

which is the dispersion relation of helicon waves [9].

Thus the pumps considered by us under the described geometry are able to excite helicon waves of frequency $\omega = \omega_1 - \omega_2$ and wave vector $\boldsymbol{k} = \boldsymbol{k}_1 - \boldsymbol{k}_2$ in a direction making an angle θ with the external magnetic field $\boldsymbol{B}_0$ if $\omega \ll \omega_c$ holds.

4. Power Density of the Excited Wave

The electric vector of the generated wave is given by

$$\boldsymbol{D}\cdot\boldsymbol{E} = \frac{4\pi i\omega}{c^2}\boldsymbol{J}^{\mathrm{nl}}\,, \tag{21}$$

$$\boldsymbol{J}^{\mathrm{nl}} = -n_e^0 e \boldsymbol{v}^{\mathrm{nl}}\,, \tag{22}$$

$$\boldsymbol{D} = \left[k^2 - \frac{\omega^2}{c^2}\,\varepsilon\right]\boldsymbol{I} - \boldsymbol{k}\boldsymbol{k}\,, \tag{23}$$

$\boldsymbol{I}$ is the unit dyadic,

$$\varepsilon = \begin{pmatrix} \varepsilon_L - \dfrac{\omega_p^2}{\omega^2 - \omega_c^2} & \dfrac{i\omega_p^2\omega_c}{\omega(\omega^2 - \omega_c^2)} & 0 \\ -\dfrac{i\omega_p^2\omega_c}{\omega(\omega^2 - \omega_c^2)} & \varepsilon_L - \dfrac{\omega_p^2}{\omega^2 - \omega_c^2} & 0 \\ 0 & 0 & \varepsilon_L - \dfrac{\omega_p^2}{\omega^2} \end{pmatrix}. \tag{24}$$

Using (14), (22) to (24) in (21) the solution of (21) gives the value of the electric field of the excited helicon wave at the difference frequency as

$$E_x = E' \frac{\omega}{c^2} \left[\frac{\Omega D_{yy} - \omega_c D_{xx}}{D_{xx} D_{yy} - D_{xy} D_{yx}}\right] \sin\theta\,, \tag{25}$$

$$E_y = E' \frac{\omega}{c^2} \left[\frac{\omega_c D_{xx} - \Omega D_{yx}}{D_{xx} D_{yy} - D_{xy} D_{yx}}\right] \sin\theta\,, \tag{26}$$

where

$$E' = -\frac{\omega_p^2 e E_1 E_2^* k}{4 m^* \omega_1 \omega_2 (\Omega^2 + \omega_c^2)}\,, \tag{27}$$

$$D_{xx} = k_z^2 - \frac{\omega^2}{c^2}\left[\varepsilon_L - \frac{\omega_p^2}{\omega^2 - \omega_c^2}\right], \tag{28}$$

$$D_{yy} = k^2 - \frac{\omega^2}{c^2}\left[\varepsilon_L - \frac{\omega_p^2}{\omega^2 - \omega_c^2}\right], \tag{29}$$

$$D_{xy} = -D_{yx} = -\frac{i\omega\omega_c\omega_p^2}{c^2(\omega^2 - \omega_c^2)}\,. \tag{30}$$

Dividing (26) by (25) we obtain ($kc \gg \omega$, $\omega \ll \omega_c$)

$$E_y = i\,|\cos\theta|\, E_x\,, \tag{31}$$

where

$$E_x = iE' \left[\frac{\omega^2\omega_c^2\omega_p^2(\omega^2 - \omega_c^2)}{(k^2c^2\cos^2\theta - \omega^2 S)\,(k^2c^2 - \omega^2 S)\,(\omega^2 - \omega_c^2)^2 - \omega^2\omega_c^2\omega_p^4}\right] \sin\theta\,, \tag{32}$$

$$S = \varepsilon_L - \frac{\omega_p^2}{\omega^2 - \omega_c^2}\,. \tag{33}$$

The power carried by the excited helicon wave is given by [10]

$$P_h = \frac{c_h}{8\pi} \boldsymbol{E} \cdot \boldsymbol{E}^* \frac{\partial}{\partial\omega}(\omega\varepsilon) + \frac{c_h}{8\pi} \boldsymbol{B}_f \cdot \boldsymbol{B}_f^*\,, \tag{34}$$

where c_h is the group velocity of the helicon wave obtained by differentiating (20) with respect to k,

$$c_h = \frac{\partial\omega}{\partial k} = \frac{2kc^2\omega_c}{\omega_p^2}\cos\theta\,. \tag{35}$$

$\boldsymbol{B}_f$ is the fluctuating magnetic field of the wave obtained by using (9),

$$\boldsymbol{B}_f \cdot \boldsymbol{B}_f^* = \frac{2k^2c^2}{\omega^2} E_x E_x^* \cos^2\theta\,. \tag{36}$$

Using (31) to (33), (35) and (36) in (34) we have

$$P_h = P\left[S_1 + \left(S_1 + \frac{2k^2c^2}{\omega^2}\right)\cos^2\theta\right] S_2 \cos\theta \sin^2\theta\,, \tag{37}$$

where

$$P = \frac{2kc^2\omega_c}{\omega_p^2} \frac{\omega_p^4 k^2 c^2 (E_1 E_1^*) (E_2 E_2^*)}{128\pi m^{*2} \omega_1^2 \omega_2^2 \omega^4 S^2}, \tag{38}$$

$$S_1 = \varepsilon_L + \omega_p^2 \frac{\omega^2 + \omega_c^2}{(\omega^2 - \omega_c^2)^2}, \tag{39}$$

$$S_2 = \left[\frac{S\omega_c^2\omega_p^2(\omega^2 - \omega_c^2)\,\omega^4}{(\nu^2 + \omega_c^2)\,\{(k^2c^2\cos^2\theta - \omega^2 S)\,(k^2c^2 - \omega^2 S)\,(\omega^2 - \omega_c^2)^2 - \omega^2\omega_c^2\omega_p^4\}}\right]^2. \tag{40}$$

Equation (37) has been used in computing the power carried by the excited helicon wave.

5. Discussion of Results

The following set of parameters has been used in computing the results:
$\varepsilon_L = 16$, $T_e = 77$ K, $m^* = 0.014 m_e$, m_e free electron mass, $\omega_1 = 2.1 \times 10^{10}\,\text{s}^{-1}$, $\omega_2 = 2.0 \times 10^{10}\,\text{s}^{-1}$, $\omega_p = 10^{10}\,\text{s}^{-1}$, $\nu = 10^{11}\,\text{s}^{-1}$,

$$\frac{cE_1E_1^*}{8\pi} = \frac{cE_2E_2^*}{8\pi} = 1\ \text{MW cm}^{-2}.$$

For the parameters under consideration (37) can be shown to reduce to the simple handy formula

$$P_h = 0.5877 \left(\frac{\omega_c}{\omega}\right)^{5/2} \sin^2\theta \sqrt{\cos\theta} \qquad \text{W cm}^{-2}.$$

The variation of the power density of the excited helicon wave with θ for $\omega_c/\omega = 25$, 30, 35, 40 is shown in Fig. 1. It is found that the magnetic field enhances the power carried by excited helicon waves.

The dependence of power on the inclination of propagation vectors of pump waves towards the magnetic field gives very interesting results. It is found that the low-frequency helicon wave is not excited in the two extreme cases $\theta = 0$ and $\theta = \pi/2$ and for some value of θ lying between 0 and $\pi/2$ the power tends to attain an optimum value. In our case this direction comes out to be $\theta \approx 63.4°$. The explanation for excitation of helicon waves not to occur at $\theta = 0$ and $\pi/2$ is as follows:

(i) At $\theta = 0$, we find from (14) that

$$v_x^{nl} = v_y^{nl} = 0 \quad \text{and} \quad v_z^{nl} \neq 0.$$

As such the last term on the right-hand side of (17) vanishes and the non-vanishing z-component of (17) yields

$$mi\omega v_z^{nl} = eE_z.$$

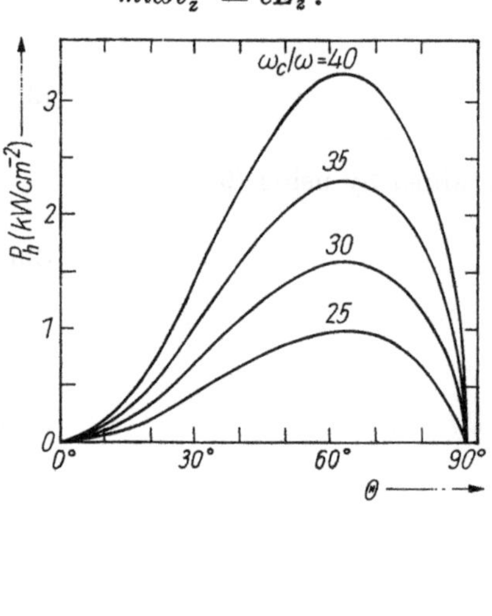

Fig. 1. Variation of the power carried by an excited helicon wave with θ for $\omega_c/\omega = 25$, 30, 35, 40

Using this with the continuity equation (7) and the Poisson equation (8) and following [11], we obtain

$$\omega^2 = \frac{4\pi n_e^0 e^2}{m^*} = \omega_p^2$$

which is the dispersion relation for plasma oscillations. Thus at $\theta = 0$ the helicon wave disappears and ordinary plasma oscillations are excited.

(ii) At $\theta = \pi/2$, we find from (14) that

$$v_x^{\text{nl}} \neq 0\,, \qquad v_y^{\text{nl}} \neq 0\,, \qquad v_z^{\text{nl}} = 0\,.$$

The non-vanishing x and y components of (17) are

$$-m^* i\omega v_x^{\text{nl}} = -eE - \frac{e}{c}\, v_y^{\text{nl}} B_0\,,$$

$$-m^* i\omega v_y^{\text{nl}} = +\frac{e}{c}\, v_x^{\text{nl}} B_0\,.$$

Using these equations with the continuity equation (7) and the Poisson equation (8) and following [11] we obtain

$$\omega^2 = \omega_{\text{p}}^2 + \omega_{\text{c}}^2 = \omega_{\text{h}}^2$$

which are high-frequency hybrid modes. Thus at $\theta = \pi/2$ also the helicon wave disappears. This result is in agreement with the discussion given in [5].

Acknowledgement

Partial financial assistance from the University Grant Commission, New Delhi, is thankfully acknowledged.

References

[1] J. E. Willet and B. Maragechi, Plasma Phys. 19, 785 (1977).
[2] M. S. Sodha, Tarsem Singh, and R. P. Sharma, J. appl. Phys. 51, 2516 (1980).
[3] Md. Salimullah, Phys. Letters A 81, 522 (1981).
[4] K. P. Maheshwari and Ganesh Tarey, phys. stat. sol. (b) 116, 339 (1983).
[5] Martin C. Steele and Bayram Vural, Wave Interactions in Solid State Plasmas, McGraw-Hill Publ. Co., New York 1969 (p. 89).
[6] J. Larson and L. Stenflo, J. geophys. Res. 80, 2325 (1975).
[7] M. S. Sodha and R. R. Sharma, phys. stat. sol. (b) 94, 667 (1979).
[8] M. S. Sodha, A. K. Ghatak, and V. K. Tripathi, Self Focusing of Laser Beams, Tata McGraw-Hill Publ. Co., New Delhi 1974.
[9] E. M. Lifshitz and L. P. Pitaevski, Physical Kinetics, Pergamon Press, New York 1981 (p. 237).
[10] V. L. Ginzburg, Propagation of Electromagnetic Waves in Plasma, Gordon & Breach, New York 1960 (p. 436).
[11] F. F. Chen, Introduction to Plasma Physics, Plenum Press, New York 1974 (p. 72, 91).

(Received April 23, 1985)

phys. stat. sol. (b) 133, 425 (1986)

Subject classification: 77.40

Department of Physics, Kazan State University[1])

The Realization of the Generalized Transfer Equation in a Medium with Fractal Geometry

By

R. R. NIGMATULLIN

It is shown that in a medium representing an example of "Koch's tree"-type fractional structure the diffusion process is described by a generalized transfer equation in partial derivations. Such a structure can serve as a model of a porous medium where the diffusion process takes place. The geometry of an inhomogeneous medium can serve as the dicisive factor in the explanation of the "universal response" phenomenon. A range of frequencies is found where such "superslow" diffusion process can be observed.

Показано, что в среде, представлющей собой пример фрактальной структуры типа "дерева Коха" процесс диффузии описывается обобщенным уравнением переноса в дробных производных. Такая структура может служить моделью пористой среды, где имеет место процесс диффузии. Геометрия неоднородной среды может служить решающим фактором в объяснении явления универсального отклика. Найден интервал частот где наблюдается такой "сверх медленный" диффузионный процесс.

1. Introduction

The detailed analysis of the experimental dielectric loss data shows that the polarization current fading $J(t) = \mathrm{d}P(t)/\mathrm{d}t$ ($P(t)$ is the total polarization) for a certain temporal interval is submitted to the following expression [1]:

$$J(t) \sim t^{-n}; \qquad \omega_M^{-1} \ll t \ll \omega_P^{-1} \tag{1a}$$

that accords with a susceptibility of the type

$$\chi'(\omega) = b_1\omega^{n-1}; \qquad \chi''(\omega) = b_2\omega^{n-1}. \tag{1b}$$

In (1) $0 < n < 1$, n is the characteristic frequency of the loss. Formula (1a) is known as Curie-von Schweidler law [2] and the intensive experimental investigations of recent years [3] show that this law is observed in a wide class of materials irrespective of their physical nature, type of chemical bond, and polarization. These facts give us the right to call such a behaviour universal [3]. It should be noted although that the absorption of electric field energy is non-resonant, mainly occurring in the low-frequency range ($\omega \leqq 10^8$ Hz), and the exponent n practically does not depend on temperature and is connected with heterogeneities of the medium [4].

At present there are three approaches to the explanation of the "universal response" (UR) phenomenon. In [5 to 7] by means of the IR-divergence hypothesis the theory of the UR has been developed. From this theory expressions of the type (1) can be obtained and the parameter n is connected with intra-cluster motion. In details this approach has been given in the monograph [3].

The geometry of a heterogeneous medium has been taken into account in [8 to 10] where the TEISI model was suggested, being based on the ideas of fractal geometry

[1]) ul. Lenina 18, 420008 Kazan, USSR.

[11]. From this model a relaxation equation in fractional derivatives follows and the exponent of the fractional derivatives is connected with the fractal dimension of a medium and the susceptibility is described by (1b).

In [12, 13] for the explanation of the UR a strict quantum-statistical equation have been obtained taking into account the retardation of a system. For a definite distribution of local fields, satisfying the scaling hypothesis, the generalized transfer equation in the fractional derivatives and the relaxation equation have been obtained, and the latter coincides with the results of [8 to 10].

From the results of [12, 13] the expressions (1) can be also obtained.

In this paper without pretending to a complete explanation of the UR we are going to show how the geometry of the medium changes the diffusion equation and it accepts the form of the generalized transfer equation obtained in [12]. In particular results, confirming some facts of the UR phenomenon follow from this consideration.

2. Deduction of the Generalized Transfer Equation

Let us consider the process of concentration smoothing which takes place in a homogeneous channel of length l with a small area of cross section S ($S \ll l^2$). This process obeys the usual diffusion equation

$$\frac{\partial c(x, t)}{\partial t} = D \frac{\partial^2 c(x, t)}{\partial x^2}. \tag{2}$$

Choosing the initial and boundary conditions in the simplest form

$$c(x, 0) = 0\,, \qquad c(0, t) = \bar{c}_0 \tag{3}$$

and using the Laplace transformation

$$C(x, p) = \int_0^\infty c(x, t) \exp(-pt)\, dt \doteqdot c(x, t) \tag{4}$$

it is convenient to present (2) in the form of a pair of the operator equations

$$\frac{\partial J_0(x, p)}{\partial x} = -pSC_0(x, p)\,, \tag{5a}$$

$$\frac{\partial C_0(x, p)}{\partial x} = -(DS)^{-1} J_0(x, p)\,, \tag{5b}$$

where $J_0(x, p)$ is the operator value of a flux in the point x. It is connected with the usual flux $j(x, t)$ by a formula which is analogous to (4). Not giving the solutions of the equations (2) or (5) which are well-known, we write down only the expressions for the propagation coefficient $\gamma_0(p)$

$$\gamma_0(p) = \left(\frac{p}{D}\right)^{1/2} \tag{6}$$

which follows from the equation

$$\frac{\partial^2 C_0(x, p)}{\partial x^2} = \gamma_0^2(p)\, C_0(x, p) \tag{7}$$

and the diffusion conductivity

$$\frac{J_0(x, p)}{C_0(x, p)} \overset{\text{det}}{=} Y_0(x, p) = DS\gamma_0(p) \operatorname{th} [\gamma_0(p)\,(l - x)]\,. \tag{8}$$

When $\gamma_0(p)\, l \gg 1$, i.e. for frequencies ω satisfying the condition

$$\omega \gg \omega_{\min}^{(0)} = \frac{D}{l^2} \tag{9}$$

the flux $J_0(x, p)$ and the density $C_0(x, p)$ are connected by the relationship

$$J_0(x, p) \approx SD^{1/2}p^{1/2}C_0(x, p)\,. \tag{10}$$

One can write (10) using the determination of the fractional derivatives [14] in the form

$$j(x, t) = SD^{1/2}\, \frac{\mathrm{d}^{1/2}c_0(x, t)}{\mathrm{d}t^{1/2}}. \tag{11}$$

The expression (11) connecting the total flux $j_0(x, t)$ and the concentration of particles (ions) in a liquid has been obtained and realized practically in [15].

Now let us change the geometry of the channel where the process of concentration smoothing is realized in the following way. Let us choose an element Δx and a system of the semi-infinite channels with small cross section S_1 ($S_1/S = \varkappa \leqq 1$) and the concentration ξ per length which step away from the element Δx in the perpendicular direction. We shall call this structure "a comb". It is represented on Fig. 1. This structure presents an example of a "Coch tree"-type fractal structure. For the fractal structures the metric dimension (Hausdorff-Besicovitch dimension) exceeds the topological one $d > d_{\mathrm{T}}$. The properties of fractal structures are described in the book [11]. For such a comb the flux $j(x, t)$ is branched over a system of channels of the first order and the operator value for the element $\Delta J_1(x, p)$ over the length Δx of the main channel is written in the following way:

$$\Delta J_1(x, p) = -\left(pS + \bar{\xi}\sqrt{\frac{\varkappa}{S}}\; DS\gamma_0(p)\right)\Delta x\,. \tag{12}$$

Here $\bar{\xi} = \xi/1$ is an average concentration of the "comb teeth" with reference to the concentration of the comb having no blanks between the teeth. This concentration is assumed to be unity. The parameter $\varkappa$ takes into account the difference in the thickness of the channels. When $\bar{\xi}\sqrt{\varkappa S}\, D\gamma_0(p) \gg pS$ which is equivalent to the inequality

$$\omega \ll \omega_{\max} = \frac{\varkappa\bar{\xi}^2 D}{S} \tag{13}$$

the flux $J_1(x, p)$ is distributed over the totality of the first order channels and equations

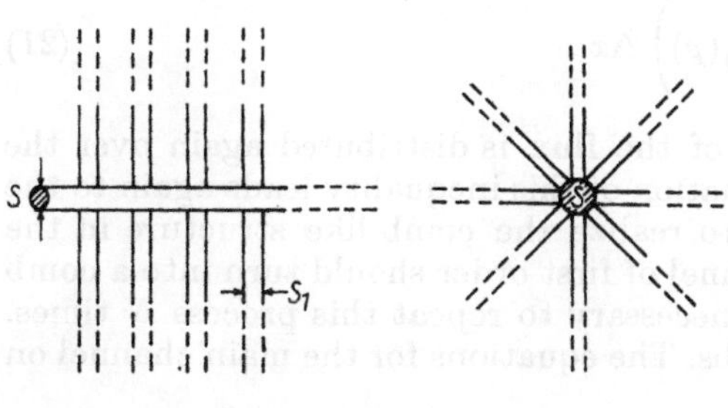

Fig. 1. The comb-like structure of a medium showing an example of the Koch's tree type fractal structure. The teeth of this semiinfinite comb step away in various directions from the main channel being perpendicular to it

describing the diffusion take the form

$$\frac{\partial J_1(x, p)}{\partial x} = -\sqrt{\frac{\varkappa\bar{\xi}^2}{S}}\,\gamma_0(p)\,DSC_1(x, p)\,, \tag{14a}$$

$$\frac{\partial C_1(x, p)}{\partial x} = -(DS)^{-1}\,J_1(x, p)\,. \tag{14b}$$

Excluding $J_1(x, p)$ from the system (14) we get

$$p^{1/2}C_1(x, p) = \mathcal{D}_1\,\frac{\partial^2 C_1(x, p)}{\partial x^2}\,, \tag{15}$$

where $\mathcal{D}_1 = (DS/\varkappa\bar{\xi}^2)^{1/2}$. Using again the determination of the fractional derivatives we get the generalized transfer equation

$$\frac{\partial^{1/2} c_1(x, t)}{\partial t^{1/2}} = \mathcal{D}_1\,\frac{\partial^2 c_1(x, t)}{\partial x^2}\,. \tag{16}$$

For the first time the equation of the form (16) was obtained in [12]. The solution of the system (14) leads to the following expression for the propagation coefficient:

$$\gamma_1(p) = \left(\frac{\varkappa\bar{\xi}^2}{S}\right)^{1/4}\,[\gamma_0(p)]^{1/2} \tag{17}$$

and to the diffusion conductivity for the main channel $(\gamma_1(p)l \gg 1)$

$$\frac{J_1(x, p)}{C_1(x, p)} = Y_1(x, p) = DS\gamma_1(p)\,. \tag{18}$$

From (18) it follows

$$j(x, t) = \left(\varkappa\bar{\xi}^2(DS)^3\right)^{1/4}\,\frac{d^{1/4}c_1(x, t)}{dt^{1/4}}\,. \tag{19}$$

Equation (19) shows that for a semi-infinitive comb structure on Fig. 1 the flux $j_1(x, t)$ and concentration $c_1(x, t)$ along the main channel axis are connected with each other at any point by means of a fractional derivative of order 1/4. It should be noted also that the low frequency estimation following from the condition $\gamma_1(p)\,l \gg 1$ equals

$$\omega \gg \omega_{\min}^{(1)} = \left(\frac{S}{\varkappa\bar{\xi}^2 l^2}\right)\frac{D}{l^2}\,. \tag{20}$$

The low frequency limit decreases $(S/\varkappa\bar{\xi}^2 l^2 < 1)$ as compared with the condition (9). If every vertical channel on Fig. 1 turns into a comb of analogous form, then instead of the condition (12) it is necessary to write the following equality:

$$\Delta J_2(x, p) = -\left(pS + \bar{\xi}\sqrt{\frac{\varkappa}{S}}\,DS\gamma_1(p)\right)\Delta x\,. \tag{21}$$

When $\bar{\xi}(\varkappa S)^{1/2}\,D\gamma_1(p) \gg pS$ the main share of the flux is distributed again over the totality of a semi-infinitive comb. The estimation of this inequality leads again to the condition (13). It becomes clear now how to realize the comb-like structure in the general case. It is necessary that every channel of first order should turn into a comb of the form represented in Fig. 1 and it is necessary to repeat this process N times. Let N denote the total number of such combs. The equations for the main channel on the N-th stage take the form

$$\frac{\partial J_N(x, p)}{\partial x} = -\left(\frac{\varkappa\bar{\xi}^2}{S}\right) \gamma_{N-1}(p)\, DSC_N(x, p)\,, \tag{22a}$$

$$\frac{\partial C_N(x, p)}{\partial x} = -(DS)^{-1}\, J_N(x, p)\,. \tag{22b}$$

From the system (22) one can get the recurrence formula connecting the propagation coefficients

$$\gamma_N(p) = \left(\frac{\varkappa\bar{\xi}^2}{S}\right)^{1/4} [\gamma_{N-1}(p)]^{1/2}\,. \tag{23}$$

From (23) it follows

$$\gamma_N(p) = \left(\frac{\varkappa\bar{\xi}^2}{S}\right)^{1/2} \left(\frac{pS}{\bar{\xi}^2\varkappa D}\right)^{1/2^{N+1}}; \qquad N = 0, 1, 2, \ldots\,. \tag{24}$$

The diffusion conductivity is expressed by the formula

$$\frac{J_N(x, p)}{C_N(x, p)} = Y_N(x, p) = DS\gamma_N(p)\,. \tag{25}$$

The relationship between the total flux $j_k(x, t)$ and the concentration $c_k(x, t)$ at any point of the k-th channel counting from the main one ($k = N$) has the form

$$j_k(x, t) = \varkappa \left(\frac{\varkappa\bar{\xi}^2 D}{S}\right)^{1/2 - 1/2^{k+1}} SD^{1/2}\, \frac{\mathrm{d}^{1/2^{k+1}} c_k(x, t)}{\mathrm{d}t^{1/2^{k+1}}}\,. \tag{26}$$

The generalized transfer equation for the main channel takes the form

$$\frac{\partial^\alpha c_N(x, t)}{\partial t^\alpha} = \mathcal{D}_N\, \frac{\partial^2 c_N(x, t)}{\partial x^2}\,, \tag{27}$$

where $\alpha = 1/2^N$, $\mathcal{D}_N = D^\alpha S^{1-\alpha} (\varkappa\bar{\xi}^2)^{\alpha-1}$. The limits for $\omega_{\min}$ and $\omega_{\max}$ are obtained from the conditions

$$\gamma_N(p)\, l \gg 1\,, \qquad \left(\frac{\varkappa\bar{\xi}^2}{S}\right)^{1/2} \gamma_{N-1}(p)\, DS \gg pS$$

and can be presented by the inequalities

$$\varkappa\bar{\xi}^2\, \frac{D}{S} \left(\frac{S}{\varkappa\bar{\xi}^2 l^2}\right)^{2^N} \ll \omega \ll \varkappa\, \bar{\xi}^2\, \frac{D}{S}\,. \tag{28}$$

Thus we showed that the one-dimensional diffusion equation (2) in the medium obeying the fractal condition $d > d_T$ [11] is changed considerably and the concentration smoothing process in the main channel is described by (27) for the frequency range satisfying the condition (28).

3. Discussion of the Results Obtained

The comb-like structure considered in the previous section can serve as a model of a porous medium, the values being equal to their average ones. It is interesting to compare some results of this model with a number of experimental facts typical of the UR phenomenon. From (26) when $c(0, t) = c_0 = \text{const}$

$$j_k(x, t) = (S^2D)^{1/2} \left(\frac{\varkappa\bar{\xi}^2 D}{S}\right)^{1/2-1/2^{k+1}} \Gamma\left(1 - \frac{1}{2^{k+1}}\right) t^{-1/2^{k+1}} \tag{29}$$

the exponent flux fading follows and the exponent $n = 1/2^{k+1}$ does not depend on temperature.

For a more detailed estimation of the inequality (28) it is necessary to know the values $\varkappa$, ξ, l, D, S. In literature there is a great variety of their values depending on the diffusion process type. Nevertheless one can make preliminary estimations for zeolites for which the above-considered model can be applied. According to the literature data [16] the average value of S lies in the interval (10 to 10^3) $\times$ 10^{-20} m², the ratio $l/\sqrt{S}$ is of the order 10^2 to 10^3. The data taken from kinetic absorption experiments give an average value for the diffusion coefficient $D = 10^{-10}$ to 10^{-14} m²/s. Taking $S \approx 10^{-18}$ m², $D \approx 10^{-11}$ m²/s we get $\omega_{\max} \approx 10^7$ Hz. The lower limit $\omega_{\min}$ depends on the ramified system of subchannels and can lead to small values of the order of tens Hz and less. The geometry of a medium can influence essentially the interpretation of the UR and anomalous low frequency dispersion experiments [3] because it is difficult to expect atomic processes in the frequency range of the order of Hz and less.

Acknowledgements

The author expresses his gratitude to Prof. B. I. Kochelaev for his interest in this paper and wants to note that the idea about the possible realization of fractional derivatives of the order 2^{-N} in such comb-like structure is due to Prof. R. Sh. Nigmatullin.

References

[1] A. K. JONSCHER, Nature (London) **267**, 673 (1977).
[2] E. VON SCHWEILDER, Ann. Phys. (Leipzig) **24**, 711 (1907).
[3] A.K. JONSCHER, Dielectric Relaxation in Solids, Chelsea Dielectric Press, London 1983.
[4] L. A. DISSADO, Physica Scripta **1**, 110 (1982).
[5] L. A. DISSADO and R. M. HILL, Nature (London) **281**, 286 (1979).
[6] R. M. HILL and A. K. JONSCHER, Contemp. Phys. **24**, 75 (1983).
[7] L. A. DISSADO and R. M. HILL, Phil. Mag. **B41**, 625 (1980).
[8] A. L. MEHAUTE, A. GUIBER, M. DELAYE, and C. FILIPPI, C.R. Acad. Sci., (France) **294**, 835 (1982).
[9] A. L. MEHAUTE and G. CREPY, Solid State Ionics **9/10**, 17 (1983).
[10] A. L. MEHAUTE, J. stat. Phys. **36**, 665 (1984).
[11] B. MANDELBROT, Fractals: Form, Chance, and Dimension, W. H. Freeman Co., San Francisco 1977.
[12] R. R. NIGMATULLIN, phys. stat. sol. (b) **123**, 739 (1984).
[13] R. R. NIGMATULLIN, phys. stat. sol. (b) **124**, 389 (1984).
[14] K. OLDHAM and J. SPANIER, The Fractional Calculus, academic Press, New York/London 1974.
[15] R. SH. NIGMATULLIN and M. R. VIYASELEV, J. analyt. Chem. **19**, 545 (1964).
R. SH. NIGMATULLIN and B. A. BELAVIN, Trudy KAI **82**, 58 (1964); **85**, 144 (1964).
[16] D. W. BRECK, Zeolite Molecular Sieves, John Wiley & Sons, NewYork/London 1974.

(Received July 24, 1985)

Erratum

To the paper by V. V. Krasnogolovets and P. M. Tomchuk

The Coulomb Correlations and the Electric Field Local Heterogeneities on the Proton Conductivity of Hydrogen Bonded Chains

phys. stat. sol. (b) **130**, 807 (1985)

the authors request the following rectifications:

On page 809, equations (5) and (6) read correctly

$$i\hbar\,\frac{\partial \varrho_{\mathrm{T}}}{\partial t} = [H_{\mathrm{T}}, \varrho_{\mathrm{T}}]\,, \qquad i\hbar\,\frac{\partial \varrho_E}{\partial t} \approx [H_E, \varrho] + [H_{\mathrm{T}}, \varrho_E]\,, \tag{5}$$

$$H_{\mathrm{T}} = H + H_E\,, \qquad H_E = H_E^{(0)} + H_E^{(1)}\,; \tag{6}$$

instead of

$$i\hbar\,\frac{\partial \varrho_{\mathrm{T}}}{\partial t} = i\hbar\,\frac{\partial \varrho_E}{\partial t} = [H_{\mathrm{T}}, \varrho]\,, \tag{5}$$

$$H_E = H_E^{(1)} + H_E^{(2)}\,. \tag{6}$$

Short Notes

phys. stat. sol. (b) 133, K1 (1986)

Subject classification: 63.20 and 78.30; S7

Laboratoire de Physique des Solides, INSA, Toulouse[1] (a), Institute of Applied Physics, Kishinev[2] (b), and Physikalisches Institut der Universität, Würzburg (c)

Cadmium Antimonide Lattice Excitations

By

M. GOIRAN (a), A. ZWICK (a), R. CARLES (a), J. LEOTIN (a), E. ARUSHANOV (b), and M. VON ORTENBERG (c)

We report in this letter the first infrared transmission measurements in CdSb using the Fourier Transform Spectroscopy (FTS) technique in the wave number region 30 to 300 cm^{-1}. The measurements were performed at 2 K on a pure p-type sample having an acceptor concentration $N_A = 2x10^{21}$ m^{-3} at 77 K. A magnetic field up to 18 T was applied in order to clearly distinguish lattice absorption from electronic excitations. A recording of the CdSb Raman spectrum gives support to our conclusions.

CdSb is a semiconductor with an orthorhombic structure. Its space group is D_{2h}^{15} and the lattice parameters are the following: a = 0.6471 nm; b = 0.8253 nm; c = 0.8526 nm /1/. The CdSb type crystal can be considered as a heavily deformed sphalerite structure in which each atom is surrounded tetrahedrally by one similar atom and three atoms of the other kind. Each unit cell contains 8 molecules Cd-Sb giving 48 vibrational modes at k = 0. The decomposition of the mechanical representation is as follows:

$$\Gamma = 6\,A_g + 6\,B_{1g} + 6\,B_{2g} + 6\,B_{3g} + 6\,A_u + 6\,B_{1u} + 6\,B_{2u} + 6\,B_{3u}\,.$$

The three acoustic modes belong to the B_{1u}, B_{2u}, and B_{3u} representations. Among the remaining 21 odd modes, the 6 A_u modes are infrared inactive and the 15 B_u are infrared active. All the even modes denoted by the index g are Raman active.

A thin wedge-shaped slice was cut from a single crystal ingot perpendicularly to the a axis. The experiment was performed on a rapid scan Fourier trans-

---

1) avenue de Rangueil, 31077 Toulouse. Cedex, France.

2) ul. Grosul 5, 277028 Kishinev, USSR.

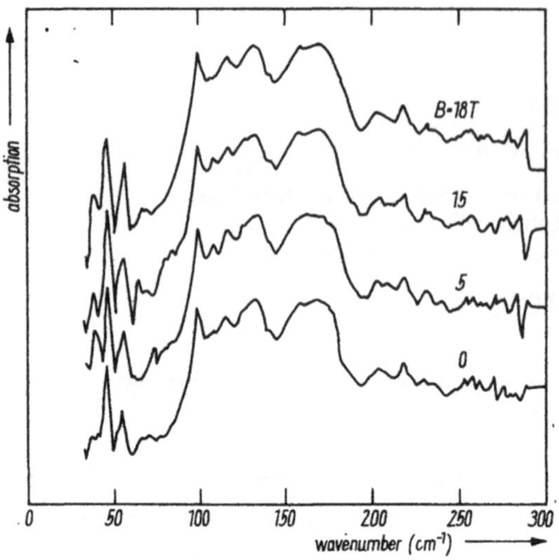

Fig. 1. Absorption versus energy (cm^{-1}) for different values of the magnetic field parallel to the a axis (0, 5, 15, and 18 T). The temperature of the sample is 2 K. The acceptor concentration is $2x10^{21}$ m^{-3}. The experiments were carried out in transmission and the experimental curves correspond to transmission decreasing upwards

form spectrometer Brucker I.F.S. 114 coupled with a data analyser Nicolet 1180 Data system. The detector was a carbon glass bolometer cooled at 2 K. Infrared absorption was measured between 30 and 900 cm^{-1} for different values of the dc magnetic field: 0, 5, 15, and 18 T parallel to the a axis. From 300 up to 900 cm^{-1} no absorption has been found, so we only present the data below 300 cm^{-1}.

Fig. 1 displays the absorption measured at different magnetic field values. This figure clearly shows no shift of the absorption lines when the magnetic field applied to the sample rises up to 18 T. The amplitudes of the weak features around 40 and 70 cm^{-1} and the broadening of the main peak at 100 cm^{-1} exhibit some enhancements when increasing the magnetic field value. Three absorption bands dominate the spectra. The strongest one occurring between 100 and 200 cm^{-1} is broad with some additional structure. At lower wave numbers, between 30 and 70 cm^{-1} a weak absorption band reveals two more pronounced structures. The third band, above 200 cm^{-1}, is nearly flat with poorly resolved peaks.

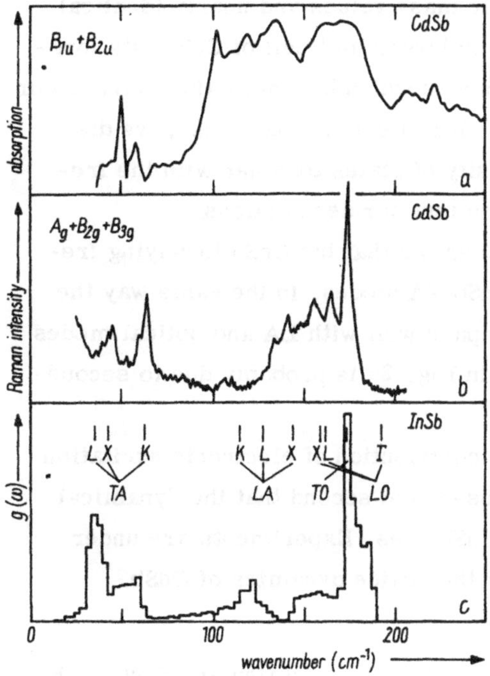

Fig. 2. a) IR absorption spectrum measured on a p-type CdSb, $2x10^{21}$ m^{-3} at zero magnetic field, the temperature is 4 K; b) Raman spectrum of CdSb, the temperature is 77 K, λ = 488 nm; c) InSb phonon density of states g(ω) with phonon energies belonging to high symmetry points Γ, L, X, and K (Kunc et al. /4/)

The same sample has already been investigated in the same temperature and excitation energy ranges under monochromatic excitation and by sweeping the magnetic field /2/. The recorded absorption lines were related to electronic excitations, either cyclotron resonance or impurity transitions. In contrast, the present spectra do not exhibit clearly such electronic excitations probably because of a lower sensitivity of the FTS technique which superimposes to them strong lattice absorption peaks.

In a previous work Stukan et al. /3/ detected a well defined absorption peak at 50 cm^{-1} found also in Fig. 1. As their peak absorption intensity varied when changing the impurities concentration, they concluded that the absorption was due to impurity ionization. However, in our experiment, by using a magnetic field, the data show clearly that the absorption peak is not an impurity one.

In order to confirm that the absorption was mainly due to lattice excitations we have performed a Raman experiment on the sample at 77 K under the 488 nm A^+ laser excitation wavelength. Fig. 2a recalls the infrared absorption measured at zero magnetic field. The theoretical analysis predicts the occurrence of 10 peaks with B_{1u} and B_{2u} symmetries. Fig. 2b gives the Raman spectrum at 77 K containing the three A_g, B_{2g}, and B_{3g} symmetries. We can notice that the Raman active modes lie in the same energy range as the IR active modes.

We tentatively interpret the infrared lattice absorption spectrum by comparing it with the phonon density of states of indium antimonide /4/, a zincblende-type material having similar atomic mass values and nearly identical atomic distances. By taking into account the larger unit cell of CdSb, its long-wavelength modes can be viewed as InSb modes which become either infrared or Raman active due to the resulting zone-folding effect. In that sense, we displayed in Fig. 2c the InSb one-phonon density of states together with the frequency of high symmetry modes in order to test our assumptions.

The comparison between Fig. 2a, b, c shows that the CdSb low-lying frequency peaks are clearly related to the InSb TA modes. In the same way the high frequency bands (up to 200 cm^{-1}) compare well with LA and optical modes of InSb. The third band (above 200 cm^{-1}) in Fig. 2a is probably due to second-order processes.

The measurements show first that the contribution of electronic excitation in the IR and Raman spectra is weak or absent and second that the dynamical properties of CdSb are closely related to InSb ones. Experiments are under way in order to determine more precisely the lattice dynamics of CdSb.

References

/1/ R.W.G. WYCKOFF, Crystal Structures, Vol. 1, J. Wiley and Sons, Ed. New York.

/2/ J. LEOTIN, M. GOIRAN, and S. ASKENAZY, phys. stat. sol. (b) 123, K43 (1984).

J. LEOTIN, M. GOIRAN, S. ASKENAZY, M. VON ORTENBERG, M. SINGH, P.R. WALLACE, and E.K. ARUSHANOV, Proc. 17th Internat. Conf. Phys. Semicond., San Francisco 1984, Springer-Verlag, New York 1984.

/3/ STUKAN , V. TRIFONOV, and V. SHEVCHENKO, Fiz. Tekh. Poluprov. 8, 1664 (1971).

/4/ K. KUNC, M. BALKANSKI, and M.A. NUSIMOVICI, phys. stat. sol. (b) 72, 229 (1975).

(Received November 29, 1985)

phys. stat. sol. (b) 133, K5 (1986)

Subject classification: 75.10

Departamento de Física, Universidade Federal de Minas Gerais, Belo Horizonte[1)]

Soliton Excitations in a One-Dimensional Antiferromagnet with Two Anisotropies[2)]

By

A.S.T. PIRES and S.L. TALIM

The aim of this note is to present a brief study of soliton excitations in a classical one-dimensional antiferromagnet with two anisotropies described by the Hamiltonian

$$H = 2J\sum_i \left[\vec{S}_i \cdot \vec{S}_{i+1} + \delta (S_i^z)^2 + b(S_i^y)^2\right] . \qquad (1)$$

The classical ground state configuration of (1) is Ising like, $\vec{S} = (\pm S, 0, 0)$. For small anisotropies we expect that two neighbouring spins are almost antiparallel to each other at low temperatures. Here we make use of the angle variables introduced by Mikeska /1/ to treat one-dimensional classical antiferromagnets:

$$\vec{S}_i = (-1)^i S\Big\{\sin(\Theta_i + (-1)^i v_i)\cos(\Phi_i + (-1)^i \alpha_i),$$
$$\sin(\Theta_i + (-1)^i v_i)\sin(\Phi_i + (-1)^i \alpha_i) , \quad \cos(\Theta_i + (-1)^i v_i)\Big\} , \qquad (2)$$

Θ and Φ are the angles giving the sublattice magnetization, whereas v and α describe the deviations from perfect anti-alignment, and can be assumed to be small at low temperatures. Substituting (2) into the Hamiltonian (1) we can keep only the terms up to the second order in the small quantities v, α and the spatial variation of Θ and Φ. The variables v and α are then eliminated with $\partial\Theta/\partial t$ and $\partial\Phi/\partial t$, which can be derived from the equation of motion of the spins. This reduction from (1) finally results, using the continuum approximation, in

$$H = J S^2 \int dz \left\{ \frac{4}{S^2} P_\Theta^2 + \frac{4}{S^2} \frac{P_\Phi^2}{\sin^2\Theta} + \left(\frac{\partial\Theta}{\partial z}\right)^2 + \sin^2\Theta \left(\frac{\partial\Phi}{\partial z}\right)^2 + 2\delta \cos^2\Theta + \right.$$
$$\left. + 2b \sin^2\Theta \sin^2\Phi \right\} , \qquad (3)$$

---

1) Caixa Postal 702, 30.000, Belo Horizonte, MG, Brazil.

2) Work partially supported by CNPq and FINEP.

where the canonical momenta P_Θ and P_Φ are given by

$$P_\Theta = \frac{1}{8J}\frac{\partial\Theta}{\partial t}, \qquad P_\Phi = \frac{1}{8J}\sin^2\Theta\,\frac{\partial\Phi}{\partial t}, \tag{4}$$

and z is the coordinate along the chain. Equation (3) is the Hamiltonian, in the continuum limit, for a chain of coupled spherical pendula in an external potential given by

$$V(\Theta, \Phi) = \delta\cos^2\Theta + b\sin^2\Theta\sin^2\Phi . \tag{5}$$

The equations of motion can be obtained directly from the Hamiltonian (3). We obtain

$$\frac{\partial^2\Theta}{\partial z^2} - \frac{1}{c^2}\frac{\partial^2\Theta}{\partial t^2} = \sin\Theta\cos\Theta\left[\left(\frac{\partial\Phi}{\partial z}\right)^2 - \frac{1}{c^2}\left(\frac{\partial\Phi}{\partial t}\right)\right]^2 + 2\sin\Theta\cos\Theta(b\sin^2\Phi - \delta), \tag{6}$$

$$\frac{\partial^2\Phi}{\partial z^2} - \frac{1}{c^2}\frac{\partial^2\Phi}{\partial t^2} = -2\cot\Theta\left[\left(\frac{\partial\Theta}{\partial z}\right)\left(\frac{\partial\Phi}{\partial z}\right) - \frac{1}{c^2}\left(\frac{\partial\Theta}{\partial t}\right)\left(\frac{\partial\Phi}{\partial t}\right)\right] + 2b\sin\Phi\cos\Phi . \tag{7}$$

where c = 4JS.

' The static limit of (6) and (7) agrees with the corresponding limit for the equations of motion for a ferromagnet with two anisotropies and an antiferromagnet with one anisotropy in an external field /2/. The dynamics, however, is different from both models. As we can see $\Phi = 0$ and $\Theta = \pi/2$ satisfies the full equations (6) and (7) and complete dynamical solutions, which satisfy the ground state condition (i.e. are aligned along the x axis for $z = \pm\infty$), are obtained from

$$\Phi = 0, \quad \frac{\partial^2\Theta}{\partial z^2} - \frac{1}{c^2}\frac{\partial^2\Theta}{\partial t^2} = -2\delta\sin\Theta\cos\Theta , \tag{8}$$

$$\Theta = \frac{\pi}{2}, \quad \frac{\partial^2\Phi}{\partial z^2} - \frac{1}{c^2}\frac{\partial^2\Phi}{\partial t^2} = 2b\sin\Phi\cos\Phi . \tag{9}$$

These are sine-Gordon equations in the variables 2Θ and 2Φ, the fully dynamical xz and xy solitons are therefore given by

$$\text{xz: } \Theta = \frac{\pi}{2} + 2\arctan\left[\exp(\sqrt{2\delta}(z - vt)\gamma)\right], \tag{10}$$

$$E_{xz} = E^o_{xz}\gamma, \quad E^o_{xz} = 4JS^2/\sqrt{2\delta} ; \tag{11}$$

$$\text{xy: } \Phi = 2\arctan\left[\exp(\sqrt{2b}(z - vt)\gamma)\right], \tag{12}$$

$$E_{xy} = E^o_{xy}\gamma, \quad E^o_{xy} = 4JS^2\sqrt{2b} , \tag{13}$$

where $\gamma = (1 - v^2/c^2)^{-1/2}$. Depending on the ratio $\lambda = \delta/b$ the xz ($\lambda < 1$, easy xz plane) or the xy soliton ($\lambda > 1$, easy xy plane) has lower energy.

Let us consider the case $b < \delta$. The xy soliton is then the lowest energy soliton. The behavior of small oscillations in the presence of a single static soliton $\Phi_o(z)$ is determined by solutions to (6) and (7) of the form

$$\Theta(z, t) = \pi/2 + \tilde{\Theta}(z, t), \quad \Phi(z, t) = \Phi_o(z) + \xi(z, t) . \tag{14}$$

Substitution of (14) in (6) and (7), linearization in $\tilde{\Theta}$ and ξ and writting $\tilde{\Theta}$ and ξ as

$$\tilde{\Theta}(z, t) = s(z) \exp(i\omega_1 t), \qquad \xi(z, t) = r(z) \exp(i\omega_2 t) \tag{15}$$

leads to the following eigenvalue equations:

$$\frac{d^2 r}{dz^2} + \frac{\omega_2^2}{c^2} r = 2b(1 - 2 \operatorname{sech}^2 \sqrt{2b}\, z) r , \tag{16}$$

$$\frac{d^2 s}{dz^2} + \frac{\tilde{\omega}_1^2}{c^2} s = 2b(1 - 2 \operatorname{sech}^2 \sqrt{2b}\, z) s , \tag{17}$$

where

$$\tilde{\omega}_1^2 = \omega_1^2 - 2(\delta - b)c^2 .$$

The dispersion relation is determined by the behavior far from the soliton centre. We find

$$\omega_1^2 = (2\delta + q^2)c^2 , \quad \omega_2^2 = (2b + q^2)c^2 . \tag{18}$$

Equation (16) possesses a bound-state solution with $\omega_2 = 0$ as in the SG problem /3/. Equation (17) has the same form only that now $\tilde{\omega}_1 = 0$ gives a bound state with frequency

$$\omega_b^2 = 2(\delta - b)c^2 . \tag{19}$$

Both (16) and (17) have the same phase shift $\Delta(q)$ for the continuum states. We have /3/

$$\Delta(q) = \pi q/|q| - 2 \tan^{-1}(q/\sqrt{2b}) . \tag{20}$$

Now we are going to calculate the soliton density for the xy solitons using the phenomenological approach. We will take $b \ll \delta$ such that at low temperatures we can neglect the effects of the xz solitons to the partition function.

According to Currie et al. /3/ (see also /4/), the average number of both solitons and anti-solitons is given by

$$n = -\beta F = \frac{2}{L}\ln\int\frac{dz\,dp}{2\pi}\exp\left[-\beta(E(p)+\Sigma_q)\right], \tag{21}$$

where F is the free energy per unit length, L is the system size, E(p) the energy-dispersion of solitons and Σ_q the 'self-energy' of solitons given by /3/

$$\beta\Sigma_q = \int\frac{dq}{2\pi}\frac{d\Delta(q)}{dq}(q)\ln(\beta^2\omega_1\omega_2) + \ln(\beta\omega_b) \quad , \tag{22}$$

where ω_1, ω_2, $\Delta(q)$ (the frequencies and asymptotic phase shift of the linear scattering solutions of the equations of motion in the presence of a soliton) and ω_b (the frequency of the bound state) have been calculated above.

Inserting (18), (19), and (20) in (22) we obtain

$$\beta\Sigma_q = -\ln(\beta\sqrt{2b}\,c)^2 N_b - \frac{1}{\pi}\int_0^\infty\frac{\ln(1+x^2)}{1+x^2} - \frac{1}{\pi}\int_0^\infty\frac{\ln(\delta/b+x^2)}{1+x^2}dx + \ln(\beta\omega_b) \ , \tag{23}$$

where $x = q/\sqrt{2b}$ and N_b is the number of bound states. Evaluating the integrals we find

$$\exp(-\beta\Sigma_q) = \frac{2\beta c\sqrt{2b}\,(1+\sqrt{b/\delta})}{(1-b/\delta)^{1/2}} \ . \tag{24}$$

Inserting (24) in (21) we obtain

$$n \approx n_{SG}\frac{1+\sqrt{b/\delta}}{(1-b/\delta)^{1/2}} \quad , \tag{25}$$

where n_{SG} is the soliton density for the pure SG model given by

$$n_{SG} = 4\sqrt{\frac{2}{\pi}}\,J^{1/2}S\sqrt{2b}\left(\frac{\sqrt{2b}}{T}\right)^{1/2}\exp(-4JS^2\sqrt{2b}/T) \ . \tag{26}$$

The thermodynamical properties of Hamiltonian (1) can be obtained using an equivalent ferromagnet with two anisotropies /5/. For low temperature this latter system has been studied by Sasaki /6/ using the transfer integral technique. For example the inverse correlation length k_x (which is related to the soliton density by $k_x = 2n$) is given by (for $T \ll E^o_{xy}$),

$$k_x = 8J\delta S^2\tau_o/T \ , \tag{27}$$

where

$$\tau_o = 4\sqrt{2}\,\Delta^{3/4}(1+\sqrt{\Delta})^{1/2}\exp(-E^o_{xy}/T)\,e^{-\xi}\,I_o(\xi) \tag{28}$$

and

$$\xi \equiv \left[(1-\sqrt{\Delta})/2\sqrt{\Delta}\right](E^{o}_{xy}/T) \ , \quad \Delta = b/\delta \ , \qquad (29)$$

and I_o is the modified Bessel function.

To order $\sqrt{b/\delta}$ (25) agrees with an expansion of (27) for small Δ. We see that for small temperatures and small Δ the leading correction to the SG result comes from fluctuations out of the easy xy plane. The case $b \gg \delta$ is identical to the one studied above by just interchanging b and δ.

For $b = \delta$ (6) and (7) present a rotational degeneracy. That is, we have a dynamic SG soliton in any plane passing through the x axis. So if $b = \delta$ solitons with all possible phase value Φ_o may be excited, Φ_o is given by

$$\sin\Theta = (\cos^2\Phi_o + \sin^2\Phi_o \tanh^2\sqrt{2\delta}\ z)^{1/2} \ .$$

For low temperatures most of the solitons are then at rest and effects of the velocity can be neglected to lowest order. Now $\omega_1 = \omega_2$ and $\omega_b = 0$. We find for n

$$n = (32\,JS^2\,b/T)\exp(-4JS^2\sqrt{2b}/T) \ , \qquad (30)$$

in agreement with (27) for $\Delta = 1$. The difference in the temperature dependence of (26) and (30) is due to the fact that for $b = \delta$ the system has an "Ising-like" behaviour with the phase Φ_o delocalized. Decreasing b, F crosses over to the "planar-like" free energy where $\Theta \approx \pi/2$.

Thus the statistical mechanics of Hamiltonian (1) as obtained from the transfer integral method is reproduced by the phenomenological approach using solitons and phonons.

References

/1/ H.J. MIKESKA, J. Phys. C 13, 2913 (1980).
/2/ N. FLÜGGEN and H.J. MIKESKA, Solid State Commun. 48, 293 (1983).
/3/ J.F. CURRIE, J.A. KRUMHANSL, A.R. BISHOP, and S.E. TRULLINGER, Phys. Rev. B 22, 477 (1980).
/4/ K. OSANO, J. Phys. C 17, L511 (1984).
/5/ A.S.T. PIRES and M.E. GOUVEA, J. Phys. C 17, 4009 (1984).
/6/ K. SASAKI, Progr. theor. Phys. 68, 1518 (1982).

(Received August 19, 1985)

phys. stat. sol. (b) 133, K11 (1986)
Subject classification: 75.10

Department of Theoretical Physics,
University of Newcastle upon Tyne

The Combined Effect of Crystal-Field Anisotropy and Biquadratic Exchange in a Heisenberg Ferromagnet

By
ZHENG SIMING[1)]

Introduction A system described by the Hamiltonian

$$H = -\sum_{i,j} J_{ij} \vec{S}_i \cdot \vec{S}_j - \alpha \sum_{i,j} J_{ij} (\vec{S}_i \cdot \vec{S}_j)^2 - D \sum_i (S_i^z)^2 \tag{1}$$

will be considered. The first term in (1) is the usual Heisenberg-Dirac exchange interaction which takes into account the isotropic coupling between the highly localized electrons in a magnetic insulator. The second term represents biquadratic exchange and α is a corresponding parameter. The summations in these terms extend over all the lattice points. The last one is the crystal-field anisotropic term.

The effect of crystal-field anisotropy on a magnetically ordered system was studied by Devlin /1/. Some special Green functions were introduced for the Hamiltonian (1) when $\alpha = 0$. It is found that the anisotropic Green functions are unnecessary to be decoupled. Devlin's work overcomes the unphysical result of previous investigations which predicted that the Curie temperature T_C tends to infinity as D tends to infinity.

Later this method was used by Biegala /2/ to examine the effect of biquadratic exchange on the Curie temperature (i.e. with Hamiltonian (1) for D = 0).

Using standard basis operators, Micnas /3/ studied the Hamiltonian (1). The numerical result of the Curie temperature, however, was confined to D = 0. Therefore the combined effect of crystal-field anisotropy and biquadratic exchange in a Heisenberg ferromagnet on the Curie temperature was not examined.

Recently Tiwari and Srivastava /4/ studied the combined effect using the technique based on Devlin /1/. Their conclusion is that these two effects annul each other at $\alpha = 0.8$ and $D/J = 0.1$ and from then the Curie temperatures increase with the increase of α (up to $\alpha = 1$) instead of decreasing when

1) Permanent address: Physics Department, Luoyang Institute of Technology, Henan, People's Republic of China.

$\alpha < 0.8$. Unfortunately because of a series of approximations their final formula is obviously incorrect (e.g. Curie temperatures tend to infinity when $\alpha = 6/7$ and $D > 0$ and are irrelevant to D if $\alpha = 3/4$) and the graph of their conclusion is inconsistent with the final formula.

In this note we are interested in the combined effect on the Curie temperature in the frame of molecular field, RPA, and a Callen-type approximation. The discussion is restricted to spin-one systems and the exchange is between nearest neighbours only.

In the next part the molecular-field approximation is presented. Finally we use Green functions based on Devlin's paper /1/ and present a Callen-type decoupling procedure to deal with higher-order Green functions (it reduces to RPA under a certain condition). The equations for calculating the Curie temperature and the quadrupolar parameter are derived and solved.

Molecular field approximation (MFA) By some algebraics (1) reads

$$H = -\sum_{i,j} J_{ij}(S_i^+ S_j^- + S_i^z S_j^z) - \alpha \sum_{i,j} J_{ij}\Big[\frac{3}{2}(S_i^z)^2 (S_j^z)^2 - 2(S_i^z)^2 - \frac{1}{2} S_i^z S_j^z + \frac{1}{2}(S_i^+)^2 (S_j^-)^2 + S_i^z S_i^+ S_j^z S_j^+ + S_i^+ S_i^z S_j^- S_j^z\Big] - D\sum_i (S_i^z)^2 \quad . \qquad (2)$$

In MFA S_i^z is replaced by its ensemble average $\langle S_i^z \rangle$,

$$\sum_{i,j} J_{ij} S_i^z S_j^z \approx 2\sigma J_0 \sum_i S_i^z \quad ,$$

where $\sigma = \langle S^z \rangle$ is independent of the lattice-site index i and $J_0 = \sum_j J_{ij}$.

Now (2) reduces to

$$H^{MFA} = -(D + \alpha Q J_0)\sum_i (S_i^z)^2 - 2J_0\left(1 - \frac{\alpha}{2}\right)\sigma \sum_i S_i^z \quad , \qquad (3)$$

$$Q = 3\langle (S^z)^2 \rangle - 2 \qquad (4)$$

is the so-called quadrupolar parameter. The eigenvalues of H^{MFA} are

$$E(m) = -(D + \alpha Q J_0)\, m^2 - 2\sigma J_0\left(1 - \frac{\alpha}{2}\right) m \quad , \qquad (5)$$

where m is the azimuthal quantum number which has three values -1, 0, 1. By the definition of the ensemble average we have

$$\langle (S^z)^n \rangle = \frac{\sum_{m=-1}^{1} m^n \exp(-\beta E(m))}{\sum_{m=1}^{1} \exp(-\beta E(m))} \quad . \qquad (6)$$

To get the Curie temperature $T_C = 1/k\beta_C$, let $\sigma \to 0$. We obtain

$$\frac{k\,T_C^{MFA}}{J_o} = \frac{2(1-\alpha/2)}{1+\frac{1}{2}\exp\left[-(D-2\alpha J_o)/(kT_C^{MFA}) - \frac{3\alpha}{2-\alpha}\right]} \tag{7}$$

For very large anisotropy D we have

$$\frac{kT_C^{MFA}}{J_o} \xrightarrow{D\to\infty} 2\left(1-\frac{\alpha}{2}\right) \; ; \tag{8}$$

when $\alpha = 0$ this result is the same as that given in /1/.

RPA and Callen-type approximation Following Devlin's notations we define the following two Green functions:

$$G^i_{gh} = \langle\langle A^i_g ; S^-_h \rangle\rangle \quad (i = 1,\, 2)\,, \tag{9}$$

where

$$A^1_g = S^+_g\,, \qquad A^2_g = -\left[A^1_g,\, (S^z_g)^2\right] = S^+_g S^z_g + S^z_g S^+_g$$

which also satisfy the relation

$$\left[A^i_g\,,\, S^z_g\right] = -A^i_g \quad (i = 1,\, 2)\ .$$

The equations of motion of G^i_{gh} generated from (2) can be written as

$$E\,G^1_{gh} = \frac{2\sigma}{2\pi}\delta_{gh} + DG^2_{gh} + 2\sum_f J_{gf}\left(1-\frac{\alpha}{2}\right)\langle\langle S^z_f A^1_g - S^z_g A^1_f ; S^-_h\rangle\rangle + \alpha\sum_f J_{gf}\langle\langle Q_f A^2_g - Q_g A^2_f ; S^-_h\rangle\rangle +$$
$$+\alpha\sum_f J_{gf}\langle\langle (S^+_g)^2(S^-_f S^z_f + S^z_f S^-_f) - (S^+_f)^2(S^-_g S^z_g + S^z_g S^-_g);\, S^-_h\rangle\rangle\,, \tag{10}$$

$$E\,G^2_{gh} = \frac{2Q}{2\pi}\delta_{gh} + DG^1_{gh} + 2\sum_f J_{gf}\left[\left(1-\frac{\alpha}{2}\right)\langle\langle S^z_g A^2_f ; S^-_h\rangle\rangle - \frac{\alpha}{2}\langle\langle S^z_f A^2_g ; S^-_h\rangle\rangle\right] - \sum_f J_{gf}\langle\langle S^-_f (S^+_g)^2 +$$
$$+2\,Q_g A^1_f ; S^-_h\rangle\rangle + \alpha\sum_f J_{gf}\langle\langle Q_f A^1_g + Q_g A^1_f ; S^-_h\rangle\rangle - \alpha\sum_f J_{gf}\langle\langle (S^+_f)^2 (Q_g S^-_g + S^-_g Q_g) -$$
$$-(S^+_g)^2 S^-_f ;\, S^-_h\rangle\rangle \tag{11}$$

with $Q_g = 3(S^z_g)^2 - 2$ and $Q = \langle Q_g\rangle$.

The RPA decouples higher order Green functions as follows:

$$\langle\langle S^z_f A^i_g ;\, S^-_h\rangle\rangle \approx \sigma G^i_{gh}\,, \quad \langle\langle Q_f A^i_g ;\, S^-_h\rangle\rangle \approx Q\,G^i_{gh}\,, \quad \langle\langle (S^+_f)^2 \ldots,\, S^-_h\rangle\rangle \approx 0. \tag{12}$$

Instead of (12) the Callen decoupling is

$$\langle\langle S_f^z A_g^1 ; S_h^- \rangle\rangle = \langle\langle S_f^z S_g^+ ; S_h^- \rangle\rangle \approx \sigma G_{gh}^1 - \frac{\sigma}{2} \langle S_f^- S_g^+ \rangle G_{fh}^1 \ . \tag{13}$$

It was first used by Callen /5/ in an isotropic Heisenberg ferromagnet where other functions such as $\langle\langle A_g^2 S_f^z ; S_h^- \rangle\rangle$, $\langle\langle A_g^i Q_f ; S_h^- \rangle\rangle$ were not involved. There have been some attempts to decouple them in Callen type when $\alpha = 0$ /6 to 8/ but they all have a common shortcoming, that is, their results fail to reduce to Callen's when D = 0. Here we present a new Callen-type decoupling scheme and the result does reduce to Callen's when D = 0 and $\alpha = 0$.

Similar to (13) we write

$$\langle\langle S_f^z A_g^2 ; S_h^- \rangle\rangle \approx \sigma G_{gh}^2 - \frac{Q}{2} \langle S_f^- S_g^+ \rangle G_{fh}^1 \ , \tag{14}$$

$$\langle\langle Q_f A_g^i ; S_h^- \rangle\rangle \approx Q G_{gh}^i - \frac{Q}{2} \langle S_f^- A_g^i \rangle G_{fh}^1 \ . \tag{15}$$

Under the approximations (13) to (15) and applying Fourier transforms one can rewrite (10) and (11) as

$$\begin{pmatrix} E - 2\sigma J_o (1 - \frac{\alpha}{2})(1 + \frac{f}{2})(1 - \gamma_{\vec{k}}) - \frac{\alpha}{2} Q J_o F (1 - \gamma_{\vec{k}}) & -D - \alpha Q J_o (1 - \gamma_{\vec{k}}) \\ -D - Q J_o \left[\alpha - (2 - \alpha)\gamma_{\vec{k}} + f(1 - \gamma_{\vec{k}})\right] & E - 2\sigma J_o \left(1 - \frac{\alpha}{2} - \frac{\alpha}{2}\gamma_{\vec{k}}\right) \end{pmatrix} \begin{pmatrix} G_{\vec{k}}^1 \\ G_{\vec{k}}^2 \end{pmatrix} = \frac{1}{2\pi} \begin{pmatrix} 2\sigma \\ 2Q \end{pmatrix} \tag{16}$$

where

$$\gamma_{\vec{k}} = J_{\vec{k}}/J_o \ , \quad f = \frac{1}{N} \sum_{\vec{k}} \gamma_{\vec{k}} \Psi_{\vec{k}}^1 \ , \quad F = \frac{1}{N} \sum_{\vec{k}} \gamma_{\vec{k}} \Psi_{\vec{k}}^2$$

and $\Psi_{\vec{k}}^i$ are the Fourier transforms of $\langle S_h^- A_g^i \rangle$ (i = 1, 2).

If we put D = 0 and $\alpha = 0$ into (16) its solution can be easily written as

$$G_{\vec{k}}^1 = \frac{2\sigma}{E - E_o(\vec{k})} \ , \qquad G_{\vec{k}}^2 = \frac{2Q}{E - E_o(\vec{k})} \ ,$$

where

$$E_o(\vec{k}) = 2\sigma J_o (1 + \frac{1}{2} f)(1 - \gamma_{\vec{k}}) \ .$$

Using the spectral theorem one gets

$$\langle S_g^- S_g^+ \rangle = 2\sigma \Phi \ , \quad f = \frac{2\sigma}{N} \sum_{\vec{k}} \gamma_{\vec{k}} \Phi_{\vec{k}} \ ,$$

$$\Phi = \frac{1}{N} \sum_{\vec{k}} \Phi_{\vec{k}} = \frac{1}{N} \sum_{\vec{k}} \left[\exp(\beta E_o(\vec{k})) - 1\right]^{-1} \ .$$

These equations are the same as Callen's. Hence our decoupling procedure does give Callen's result when D = 0 and $\alpha = 0$.

Solving (16) and using the relation between Green function and its corresponding correlation function we obtain a set of equations satisfied by σ, Q, f, and F. In order to determine Curie temperature T_C and Q, letting $\sigma \to 0$, one gets

$$F = -\frac{\alpha}{2} Q^2 J_o F \frac{1}{N} \sum_{\vec{k}} \frac{\gamma_{\vec{k}}(1-\gamma_{\vec{k}})}{E_{\vec{k}}} \coth \frac{E_{\vec{k}}}{2kT_C} .$$

Its only solution is F = 0 because both sides should be equal for any values of α.

Now the equations to determine T_C, Q, and f are

$$\frac{4-Q}{3Q} = \frac{1}{N} \sum_{\vec{k}} \frac{D + \alpha Q J_o (1-\gamma_{\vec{k}})}{E_{\vec{k}}} \coth \frac{E_{\vec{k}}}{2kT_C} , \tag{17}$$

$$\frac{4(1-Q)}{3Q^2 J_o} = \frac{1}{N} \sum_{\vec{k}} \left[\frac{(1-\alpha)\gamma_{\vec{k}} - \frac{1}{2} f(1+\frac{\alpha}{2})(1-\gamma_{\vec{k}})}{E_{\vec{k}}} \coth \frac{E_{\vec{k}}}{2kT_C} + \right.$$
$$\left. + \frac{2-\alpha-\gamma_{\vec{k}} + \frac{1}{2} f(1-\frac{\alpha}{2})(1-\gamma_{\vec{k}})}{2kT_C} \operatorname{cosech}^2 \frac{E_{\vec{k}}}{2kT_C} \right] , \tag{18}$$

$$f = \frac{Q}{N} \sum_{\vec{k}} \frac{\gamma_{\vec{k}}}{E_{\vec{k}}} \left[D + \alpha Q J_o (1-\gamma_{\vec{k}}) \right] \coth \frac{E_{\vec{k}}}{2kT_C} , \tag{19}$$

$$E^2_{\vec{k}} = \left[D + \alpha Q J_o (1-\gamma_{\vec{k}}) \right] \left\{ D + Q J_o \left[\alpha + f - (2-\alpha+f)\gamma_{\vec{k}} \right] \right\} . \tag{20}$$

When f = 0 (17) and (18) are the same as those in /3/ and will further reduce to those in /1/ and /2/ when $\alpha = 0$ and D = 0, respectively.

Solving (17) to (19) we can get Curie temperatures T_C for various D and α. Numerical calculations are restricted in the range $0 \leq \alpha \leq 1$ for the b.c.c. lattice. Fig.1 shows T_C as a function of α for different D/J. It can be seen that for all values of D/J T_C are always decreasing functions of α. Therefore the effect of crystal-field anisotropy and biquadratic exchange never annul each other as far as $0 \leq \alpha \leq 1$ is concerned. It seems that the effect of biquadratic exchange is more important.

Fig. 2 shows $T_C/T_C(0)$ as function of D/J for different α. $T_C(0)$, the Curie temperatures for D = 0, can be evaluated from

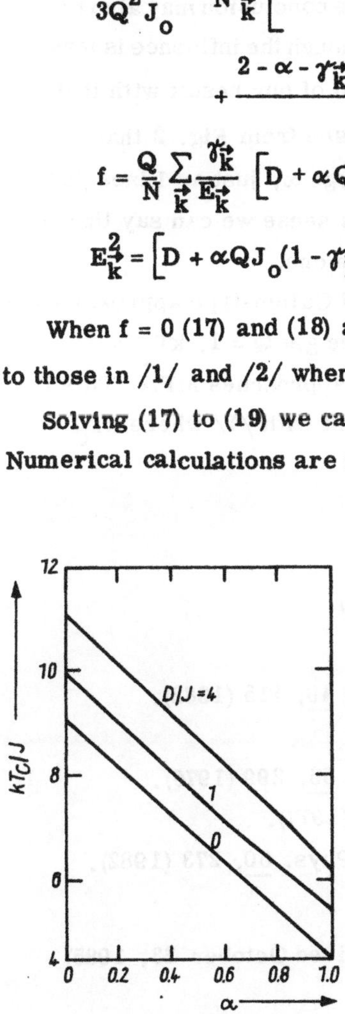

Fig. 1. The variation of Curie temperatures T_C as a function of α for D/J = 0, 1, and 4 is shown (Callen-type approximation)

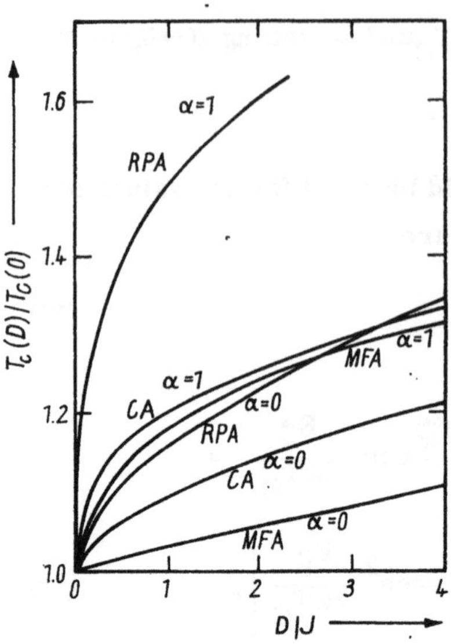

Fig. 2. It shows the reduced Curie temperatures $T_C(D)/T_C(0)$ as a function of D/J for $\alpha = 0$ and 1 for MFA, RPA and CA (Callen-type approximation)

$$\frac{kT_C(0)}{J_o} = \frac{4(5W - 2)}{qW^2}\left(1 - \frac{1}{2}\alpha\right)$$

which may be derived similar to /2, 5/, W is the Watson sum. For larger values of α the influence of D on T_C is much stronger, especially for small D. The conclusion may also be drawn in MFA although the influence is less.

By comparison of our result with that of RPA one may see from Fig. 2 that the influence is very strong in RPA especially for large α, just as Lines /9/ pointed out RPA overestimated this influence. In this sense we can say that our Callen-type decoupling approximation is a reasonable one.

For very large anisotropy D, $f \approx 0$, both RPA and Callen-type approximations reach the same result. From (17), (18) when $D \to \infty$ we get $Q = 1$, $kT_C = 2J_o(1 - \alpha/2)$. It means that for large D our result approaches MFA. This is similar to Devlin's conclusion but is in sharp contrast with /4/ where $kT_C = 2J_o(1 + \alpha/2)$ (when $D \to \infty$) can be derived.

References

/1/ J.F. DEVLIN, Phys. Rev. B 4, 136 (1971).
/2/ L. BIEGALA, Acta phys. Polon. A47, 129 (1975).
/3/ M. MICNAS, J. Phys. C 9, 3301 (1976).
/4/ M. TIWARI and R.N. SRIVASTAVA, Z. Phys. B 49, 115 (1982).
/5/ H.B. CALLEN, Phys. Rev. 130, 890 (1963).
/6/ V. KUMAR and K.C. SHARMA, Phys. Letters A 56, 299 (1976).
/7/ G. CHADDHA, J. Phys. Chem. Solids 38, 1218 (1977).
/8/ M. TIWARI and R.N. SRIVASTAVA, Canad. J. Phys. 60, 273 (1982).
/9/ M.E. LINES, Phys. Rev. B 12, 3766 (1975).

(Received October 23, 1985)

phys. stat. sol. (b) 133, K17 (1986)

Subject classification: 61.50 and 78.80; S8.11; S8.12; S8.13

Department of Physics, S.V. College, Aligarh[1)]

Modified PVV Theory and X-Ray K Absorption Edge Studies in Zinc Chalcogenides (Ionicity Determination and d Electron Effect on It)

By

O.P. SINGH and V.P. GUPTA

Introduction Various theories, i.e. thermo-chemical /1/, valence bond /2/, and spectroscopic /3/, have been put forward to study ionic and covalent effects in different compounds. The spectroscopic theory known as Phillips and Van Vechten (PVV) theory has been used for calculating ionic gap, homopolar gap, average gap, and ionicity of zinc chalcogenides. But the results are not in good agreement with X-ray absorption data /4/.

The PVV theory was modified in order to extend it to complex structures, transition metal compounds /5/, and layer-type materials /6, 7/. In the presented note the modified PVV theory has been used to evaluate ionic gap, hompolar gap, and average gap and hence ionicity of zinc chalcogenides.

Theory, results, and discussion According to the PVV theory the ionicity has been defined in terms of ionic gap (E_c) and average gap (E_g) as follows:

$$f_i = E_c^2/E_g^2 , \tag{1}$$

where $E_g^2 = E_c^2 + E_h^2$. E_h and E_c are contributions of homopolarity and heteropolarity of the bond A-B to the average gap. E_h and E_c are given in terms of crystallographic parameters as follows:

$$E_h = 40.468/r_{AB}^{2.5} , \tag{2a}$$

$$E_c = b\, e^2(Z_A/r_A - Z_B/r_B)\exp(-K_s r_{AB}/2) , \tag{2b}$$

where Z_A, Z_B and r_A, r_B are valence and covalent radii of atoms A and B, respectively, K_s is the Thomas-Fermi screening parameter, r_{AB} is the interatomic distance between atoms A and B, b is an adjustable parameter.

In order to extend the PVV theory to complex structures and transition metal compounds Lavin /5/ proposed the following modifications.

1) Aligarh 202001, India.

Table 1

material	N_F	E_c (eV)	E_h (eV)	E_g (eV)	f_i (present)
ZnS	8.5	5.325	4.73	7.112	0.56
ZnSe	8.75	4.54	4.31	6.26	0.53
ZnTe	9.0	3.458	3.61	4.999	0.48

Lavin introduced r_o ($r_o = r_{AB}/2$) instead of r_A and r_B in the equation for the ionic gap. The adjustable parameter b was defined in terms of the average coordination number $\bar{N}_c$ ($b = 0.089\,\bar{N}_c^2$).

In case of transition and noble metal compounds the contribution of d electrons is also considered along with the contribution of valence electrons. On the basis of the above modifications the relation for the ionic gap E_c has been taken in the following form:

$$E_c = \left[b\,e^2(Z_A - Z_B)/r_o\right]\exp(-K_s\,r_o) \quad . \tag{3}$$

Zn is neither a transition metal nor noble metal, however, the effect of d electrons has been observed by Van Vechten /8/, the value of the effective number of electrons (n_{eff}) has been given 4.3, 4.7, and 4.9 for ZnS, ZnSe, and ZnTe, respectively. The value of n_{eff} for II-VI compounds should be 4. Hence 0.3, 0.7, and 0.9 are the contributions from d electrons. Moreover it is a well known fact that in ionic compounds almost complete transfer of valence electrons from cation to anion takes place and the involvement of d electrons in bond formation is negligibly small while in case of covalent compounds more d electrons take part in bond formation. It means that increase in covalency increases the number of d electrons sharing in bond formation.

Now based on the above concept and following the modified PVV theory we determine E_h, E_c, E_g, and f_i. The evaluated data have been presented in Table 1. Table 2 gives a comparison between reported values of ionicity from spectroscopic theory (Phillips), X-ray absorption studies /4/, and evaluated values. It is evident that our results are in better agreement with experimental values (X-ray absorption studies) compared to that of Phillips, which support our potent of view of d electron sharing in bond formation.

Table 2

material	f_i (Phillips)	f_i (X-ray)	f_i (present)
ZnS	0.63	0.55	0.56
ZnSe	0.63	0.54	0.53
ZnTe	0.61	0.49	0.48

Conclusion The modified PVV theory has been used to evaluated the ionicity of zinc chalcogenides. The evaluated values were found in good agreement with experimental values which confirms the d electron contribution in bond formation.

References

/1/ L. PAULING, The Nature of Chemical Bond, Cornell Univ. Press, Ithaca (New York) 1960.

/2/ C.A. COULSON, L.B. REDEI, and D. STOCKER, Proc. Roy. Soc. (London) 270, 352 (1962).

/3/ J.C. PHILLIPS, Rev. mod. Phys. 42, 317 (1970).
J.C. PHILLIPS and J.A. VAN VECHTEN, Phys. Rev. 132, 2147 (1970).
J.A. VAN VECHTEN and J.C. PHILLIPS, Phys. Rev. 132, 2160 (1970).

/4/ S.V. ADHAPAK and A.S. NIGAVEKAR, J. Phys. Chem. Solids 37, 1037 (1976).

/5/ B.F. LAVIN, Phys. Rev. B 7, 2591 (1973).

/6/ V.P. GUPTA, A. GUPTA, and V.K. SHRIVASTAVA, phys. stat. sol. (b) 108, 323 (1981).

/7/ V.P. GUPTA and O.P. SINGH, phys. stat. sol. (b) 125, K1 (1984).

/8/ J.A. VAN VECHTEN, Phys. Rev. 187, 1007 (1969).

(Received October 14, 1985)

phys. stat. sol. (b) 133, K21 (1986)
Subject classification: 71.25; 72.15; S4

Department of Physics, Sardar Patel University, Vallabh Vidyanagar[1)]

Quantum Number Dependent Model Pseudopotential for Metallic Properties

By
A.R. JANI and H.K. PATEL

The pseudopotential theory is one of the methods to calculate various metallic properties. It has been observed, in general, that a pseudopotential yielding satisfactory results for one type (group) of metals fails many times to reproduce the results of other metals. Most recently, Gohel et al. /1/ have pointed out strongly that the wave vector $\vec{q}_0$, where the form factor becomes zero, plays a vital role and its correct determination is highly desirable.

Recently, Mǎsović and Zeković /2/ have suggested a form factor of pseudopotential (which is the extended form of earlier work /3 to 5/) to describe the effective electron-ion interaction of the following form:

$$\langle \vec{k} + \vec{q}|W|\vec{k}\rangle = \beta_1 \beta_2 j_0(x) + C_1 j_1(x) + C_2 j_2(x) , \tag{1}$$

where $x = 2\pi\beta_2(q/2k_F)$, j_0, j_1, and j_2 are spherical Bessel functions. β_1, β_2, C_1, and C_2 are the parameters of this potential.

In this note, we propose a following simple form factor to calculate metallic properties:

$$\langle \vec{k} + \vec{q}|W|\vec{k}\rangle = -\frac{2}{3} E_{F_0} \frac{\sin(\beta q)}{\beta q} \exp\left[-\frac{1}{2}\left(\frac{q}{2k_F}\right)^4\right] , \tag{2}$$

where β is the only parameter of the pseudopotential and E_{F_0} is the free electron Fermi level. The exponential function is chosen to achieve proper damping at high values of $\hat{q}$ /6/. This form factor satisfies the screening condition in the limit $q \to 0$.

For getting a proper trend of the potential and a reasonable reproducibility of q_0 by it, we propose the following dependence of the parameter β for its determination:

$$\beta = R_c(1 + s + z')^{1/3} , \tag{3}$$

where R_c is the usual core radius of the ion in metal, l and s are the orbital

1) Vallabh Vidyanagar 388120, Gujarat, India.

Table 1

$\frac{q_o}{2k_F}$ and parameter β for various metals

metal	β	$\frac{q_o}{2k_F}$ (present value)	$\frac{q_o}{2k_F}$ (reported values /6/)
Li	2.497	1.26	1.26, 1.13
Na	3.282	0.96	0.98, 0.97, 0.96, 0.89, 0.86
K	3.600	0.87	0.93, 0.89
Rb	3.770	0.83	0.94, 0.82, 0.78
Cs	3.869	0.81	1.07, 0.79
Mg	4.073	0.77	0.83, 0.78, 0.75
Zn	3.553	0.88	0.95, 0.87, 0.86, 0.83, 0.78
Cd	3.923	0.80	0.86, 0.84, 0.81
Hg	4.145	0.76	1.19, 0.88, 0.67
Al	4.165	0.75	0.76, 0.75, 0.73
Ca	3.390	0.93	0.94, 0.92, 0.91, 0.85
Sn	4.284	0.73	0.83, 0.82, 0.70
Pb	4.680	0.67	0.84, 0.64

and spin quantum numbers of the valence electrons, respectively. Here z' is the effective nuclear charge (Z_e) /7/ seen by an electron near the ion which does not possess d electrons in the last shell. In the case of ions which have d electrons in the last shell, z' is to be considered simply the valence of the element.

We have applied the proposed form factor (2) to various metals for estimating the q_o values. The values of $q_o/2k_F$ and the parameter β are given in Table 1 which are found to be highly satisfactory.

Aluminium, in spite of its complicated Fermi surface, is regarded as one of the simplest metals wherein nonlocality plays a very small role. Hence some other calculations are carried out for this metal as a test case. The computed values of the form factor for Al are given in Table 2 along with such values due to Animalu and Heine /8/. A comparison of these values justifies the present proposition. In addition to this, we note that the pseudopotential (2) has affirmed

Table 2

Computed values of form factor for aluminium (in Ryd)

$\frac{q}{2k_F}$	present value	reported value /8/
0.0	-0.5736	-0.5736
0.1	-0.5571	-0.5529
0.2	-0.5010	-0.4929
0.3	-0.4337	-0.4018
0.4	-0.3381	-0.2943
0.5	-0.2324	-0.1873
0.6	-0.1286	-0.0938
0.7	-0.0388	-0.0204
0.8	0.0268	0.0312
0.9	0.0630	0.0625
1.0	0.0714	0.0766

Table 3

Liquid phase electrical resistivity ($\mu\,\Omega$m) and thermoelectric power (μV/K) of aluminium at 976 K

property	present value	experimental value	previous theoretical values
electrical resistivity	0.2369	0.2484	0.2845 /1/, 0.1701 /1/, 0.1900 /2/, 0.6320 /5/, and 0.2920 /11/
thermoelectric power	-2.384	-2.10	-2.60 /1/, -3.92 /1/, -2.80 /12/, and -1.01 /13/

the experimental values /9/ of W(200) and W(111) within the error of 0.002 Ryd and reasonable damping at high q.

Furthermore, the proposed potential is also used to calculate the electrical resistivity and thermoelectric power of liquid Al according to Ziman's theory /1/

at 976 K. The results were obtained by means of the theoretical structure factor from the Percus-Yevick equation /10/ with the packing density $\eta = 0.46$. The theoretical values of resistivity and thermoelectric power of Al due to present work are given in Table 3. It can be seen therefrom that they are the best values among all the reported ones. This implies that the proposed pseudopotential (2) can be tried for other metals also because it yields satisfactory values of q_o.

The remaining study of Al and other metals is in progress. The results will be reported in due course.

References

/1/ V.B. GOHEL, C.K. ACHARYA, and A.R. JANI, phys. stat. sol. (b) 124, 137 (1984).

/2/ D.R. MAŠOVIĆ and S. ZEKOVIĆ, phys. stat. sol. (b) 89, K57 (1978).

/3/ V. VELJKOVIĆ and I. SLAVIĆ, Phys. Rev. Letters 29, 105 (1972).

/4/ V. VELJKOVIĆ, Phys. Letters A 45, 41 (1973).

/5/ F.R. VUKAJLOVIĆ, S. ZEKOVIĆ, and V. VELJKOVIĆ, Physica 92B, 66 (1977).

/6/ M.L. COHEN and V. HEINE, Solid State Phys. 24, 37 (1970).

/7/ V. HEINE and D. WEAIRE, Solid State Phys. 24, 418 (1970).

/8/ A.O.E. ANIMALU and V. HEINE, Phil. Mag. 12, 1249 (1965).

/9/ N.W. ASHCROFT, Phil. Mag. 8, 2055 (1963).

/10/ N.W. ASHCROFT and J. LEKNER, Phys. Rev. 145, 83 (1966).

/11/ C.B. SO and S. WANG, J. Phys. F 7, 35 (1977).

/12/ L.J. SUNDSTRÖM, Phil. Mag. 11, 657 (1965).

/13/ N.N. SINHA and P.L. SRIVASTAVA, phys. stat. sol. (b) 90, 369 (1978).

(Received November 26, 1985)

phys. stat. sol. (b) 133, K25 (1986)

Subject classification: 71.35 and 78.40; S8.16

Institute of Physics, Academy of Sciences of the Azerbaidzhan SSR, Baku[1)]

Special Features of Exciton Absorption Spectra of $A^{III}B^{III}X_2^{VI}$-Type Layer-Semiconductor Crystals

By

S.G. GUSEINOV, G.D. GUSEINOV, N.Z. GASANOV, and S.B. KYAZIMOV

The optical properties of $A^{III}B^{III}X_2^{VI}$-type layer-and-chain crystals are not clearly understood up to now. Published data /1 to 6/ are episodical, contradictory and have been discussed irrespective of the low-temperature structural phase transitions typical of this class of compounds.

In this report we confine ourselves to setting forth some new results obtained while studying the effects of intercalation, uniaxial pressure, temperature (5 to 180 K), composition variations, and structural phase rearrangements on exciton absorption spectra of single crystals of $A^{III}B^{III}X_2^{VI}$-type ternary layer semiconductors ($TlGaS_2$, $TlGaSe_2$, $TlInS_2$) and solid solutions based on them. As a result of thorough measurements it has been revealed that the nature of the thermal shift of the exciton absorption band is very complex: specifically, when the light is directed transversely to the cleavage plane of a $TlGaS_2$ crystal (d = 23 μm) the temperature coefficient for the ground state (n = 1) of a direct exciton has an unusual positive sign and varies in absolute value from 1×10^{-4} to 2.7×10^{-4} eV/K. As this takes place, the exciton absorption peak is displaced from 2.606 to 2.637 eV as the temperature increases from 4.2 to 180 K, but the trend of the $E_{ex}(T)$ dependence is quite different from that referred to in /4/. Actually, uniaxial compression perpendicular to the layers is essential, but in the opposite direction pointed out in /4/ it affects the energy positions of the exciton ground state and the values of their temperature coefficient, whereas in certain pressure and temperature regions it even leads to a reversal of the coefficient sign (see Tables 1 and 2).

As this takes place, the value and the sign of the differential baric displacement coefficient of the exciton absorption band depend on the temperature and pressure regions (see Table 3).

---

1) prospekt Narimanova 33, 370143 Baku, USSR.

Table 1

Dependence of the energetic position of the exciton ground state $E_{ex}(n = 1)$ (meV) in a $TlGaS_2$ single crystal (d = 23 μm) on temperature and uniaxial pressure perpendicular to the layers

P (mPa) \ T(K)	10	30	50	77	100
0	2610.0	2612.0	2617.0	2619.0	2626.0
28.6	2611.0	2612.7	2617.0	2618.4	2626.0
42.9	2611.5	2613.0	2618.0	2618.0	2626.0
57.2	2612.0	2613.0	2619.0	2618.3	2626.0
85.6	2612.5	2613.0	2619.0	2619.0	2626.0
114.4	2613.0	2613.7	2619.0	2621.0	2626.0
128.7	2613.5	2614.0	2622.0	2622.0	2626.0
143.0	2614.0	2613.7	2619.9	2621.0	2626.0
171.6	2615.8	2613.0	2619.9	2619.0	2626.0

Table 2

Variations of the temperature coefficient of the exciton absorption band $(\partial E_{ex}/\partial T)_p$ (10^{-4} eV/K) in a $TlGaS_2$ single crystal (d = 23 μm) over some temperature ranges, depending on uniaxial compression applied transversely to the layers

P (mPa) \ T(K)	10 to 30	30 to 50	50 to 77	77 to 100
0	+1	+1	+0.74	+3.04
28.6	+0.85	+2.15	+0.70	+3.30
42.9	+0.75	+2.50	0.00	+3.26
57.2	+0.50	+3.00	-0.259	+2.91
85.6	+0.25	+3.00	0.00	+2.07
114.4	+0.35	+2.65	+0.74	+2.17
128.7	+0.25	+2.70	+0.96	+2.17
143.0	-0.15	+3.10	+2.592	+3.04
171.6	-0.14	+3.45	-3.33	+3.04

Table 3

Baric coefficients of the positions of the exciton ground state in a $TlGaS_2$ single crystal (d = 23 μm) at constant temperature

T(K)	p (mPa)	$\left(\frac{\delta E_{ex}}{\delta P}\right)$ (10^{-6} eV/bar)
10	0 to 57.2	+2.273
	57.2 to 114.4	+1.75
	114.4 to 143	+3.496
	143 to 186	+4.545
30	0 to 42.9	+2.273
	42.9 to 85.8	0
	85.8 to 128.7	+2.273
	128.7 to 171.6	-2.273
	171.6 to 214.5	0
50	0 to 28.6	0
	28.6 to 57.2	+6.993
	57.2 to 114.4	0
	114.4 to 143	+3.496
	143 to 171.6	0
77	0 to 45.76	-2.273
	45.76 to 86.2	+2.273
	86.2 to 128.7	+6.993
	128.7 to 171.6	-6.993
	171.6 to 214.5	+2.273
100	0 to 28.6	0
	28.6 to 57.2	-3.496
	57.2 to 85.8	+3.496
	85.8 to 114.4	0
	114.4 to 143	+6.993
	143 to 171.6	+6.993

A comparative record was made of exciton absorption spectra of a non-intercalated (d = 34 μm) and a lithium-intercalated (d = 30 μm) $TlGaS_2$ single crystal in the temperature range from 5 to 180 K. In this temperature range both crystals exhibit the same temperature coefficient of the exciton absorption band (from +1.8 to 2.0×10^{-4} eV/K). The displacement of their energetic positions as a result of intercalation with lithium proved to be unessential. Consequently, the energetic position of the exciton absorption band of $TlGaS_2$ crystals with a more pronounced two-dimensionality proves to be more sensitive to uniaxial compression perpendicular to the layers, but less sensitive to intercalation. The reverse is observed when passing to an isostructural $TlGaSe_2$ compound with a comparatively lesser degree of two-dimensionality, in the case of which the usual thermal displacement trend of the exciton absorption band takes place.

Thus, with isostructural $TlGaSe_2$ as well as with $TlInS_2$ the differential coefficients of thermal displacement of the exciton absorption band have a normal negative sign, as distinct from $TlGaS_2$ crystals. As with $TlGaS_2$, their thermal and baric displacement coefficients of the exciton absorption band experience a jump in the course of structural phase rearrangements typical of crystals of the above semiconductor types and due to uniaxial compression (stretching) or temperature variations. With $TlGaSe_2$ crystals of certain thickness the existence of three types of exciton absorption bands has been revealed. One of them with the highest energy begins to show up at a temperature of 90 K ($5 \leqq T \leqq 90$ K) near which a temperature-dependent structural phase transition takes place, as shown by a set of other physical measurements. The energetic position of the ground state of the above-mentioned direct exciton which is likely to belong to a newly formed phase of $TlGaSe_2$ ($T \leqq 90$ K) varies from 2.3820 to 2.3937 eV as the temperature decreases from 90 to 10 K. Its temperature coefficient corresponds to $\partial E'_{ex}/\partial T = 1.8\times10^{-4}$ eV/K in the range of $20 \leqq T \leqq 90$ K, experiences a jump near 20 K, and amounts to -0.25×10^{-4} eV/K in the $T < 20$ K region.

Intercalation of a $TlGaSe_2$ single crystal with lithium ions (Li^+), unlike $TlGaS_2$ crystals, brings about a substantial displacement of the basic positions of the above exciton (E'_{ex}) towards low energies over the whole temperature range under study (from 55 to 110 K). Intercalation also results in a decrease of the temperature coefficient amounting to $(\partial E'_{ex}/\partial T)^{\text{interc.}} = -1.2\times10^{-4}$ eV/K in the $90 \geqq T \geqq 20$ K region. Upon intercalation, this exciton shows up in the higher-temperature region ($T \leqq 110$ K), too. In the regions below 20 K and above 90 K, however, its temperature coefficient drops sharply in absolute value, which is typical of second-order structural phase transitions.

Table 4

temperature range (K)	$\partial E''_{ex}/\partial T$ (10^{-4} eV/K)
120 to 90	3.65
90 to 50	2.85
50 to 20	1.3
20 to 5	0.6

With $TlGaSe_2$ single crystals the second exciton absorption band (E''_{ex}) is observed in the $120 \geqq T \geqq 5$ K region. In this case the positions of exciton absorption peaks are also shifted towards higher energies (from 2.153 to 2.178 eV) as the temperature decreases from 120 to 5 K. The temperature coefficient of this exciton state experiences a jump at temperatures of about 90, 50, and 20 K. In Table 4 these coefficients are presented with their respective temperature ranges.

The energetic positions of this exciton (E''_{ex}), just as those of the high-energy exciton (E'_{ex}) mentioned above, are shifted towards low temperatures as a result of intercalation of $TlGaSe_2$ crystals and correspond to E''_{ex} = = 2.1578 eV at 5 K and E''_{ex} = 2.1532 eV at 70 K.

The third exciton absorption band (E'''_{ex}) shows up simultaneously with the just mentioned E''_{ex} when there is an (induced) unilateral surface tension in the 90 to 5 K region.

The positions of the E'''_{ex} exciton absorption peaks in the 5 to 90 K range vary from 2.1592 to 2.1522 eV and the $E'''_{ex}(T)$ dependence exhibits a number of special features.

With $TlInS_2$ single crystals the temperature coefficients of the exciton ground state amount to -0.5×10^{-4}, -1.5×10^{-4}, -3×10^{-4}, and -6×10^{-4} eV/K in the temperature regions of 5 to 20 K, 20 to 30 K, 30 to 60 K, and (60 to 70) to 180 K, respectively.

A partial replacement of trivalent ions by iron in the $TlGaS_2$, $TlGaSe_2$, and $TlInS_2$ compounds (in $TlGa_{1-x}Fe_xS_2$, $TlGa_{1-x}Fe_xSe_2$, and $TlIn_{1-x}Fe_xS_2$ solid solutions) causes a displacement of the positions of exciton states towards low energies, changes the nature of their temperature dependence, and rearranges substantially their band spectrum as a whole.

References

/1/ G.D. GUSEINOV, E. MOOSER, E.M. KERIMOVA, R.S. GAMIDOV, I.V. ALEKSEEV, and M.Z. ISMAILOV, phys. stat. sol. (b) 34, 33 (1969).

/2/ G.D. GUSEINOV, S.G. ABDULLAEVA, E.M. GODZHAEV, A.SH. MEKHTIEV, L.A. ALIEVA, and M.Z. ISMAILOV, phys. stat. sol. (b) 81, K47 (1977).

/3/ G.I. ABUTALYBOV, S.G. ABDULLAEVA, and N.M. ZEINALOV, Fiz. Tekh. Poluprov. 16, 2086 (1982).

/4/ M.YA. BAKIROV, N.M. ZEINALOV, S.G. ABDULLAEVA, V.A. GAJIYEV, and E.M. GODZHAEV, Solid State Commun. 44, 205 (1982).

/5/ S.G. ABDULLAEVA, N.A. ABDULLAEV, G.L. BELENKII, N.T. MAMEDOV, andR.A. SULEIMANOV, Fiz. Tekh. Poluprov. 17, 2068 (1983).

/6/ S.G. ABDULLAEVA, G.L. BELENKII, and N.T. MAMEDOV, Fiz. Tekh. Poluprov. 15, 943 (1981).

(Received November 8, 1985)

phys. stat. sol. (b) 133, K31 (1986)
Subject classification: 76.40; S5.12

Institute of Physics of Solids and Semiconductors, Academy of Sciences of the Belorussian SSR, Minsk[1)]

Cyclotron Resonance Anisotropy of Charge Carriers in Irradiated Germanium

By
V.I. GATALSKAYA

The utilization of the cyclotron resonance (CR) method enables to obtain reliable quantitative data associated with the band structure of pure germanium /1, 2/. The effects of impurities and that of plastic deformation on the band parameters of electrons and holes in germanium were reported in /3, 4/. In /5/ we first measured the charge carrier (c.c.) effective masses in irradiated Ge. The objective of this note was to study further the radiation effect on the anisotropy of the effective mass of electrons and holes in germanium.

The studies were carried out on dislocation free n-type Ge single crystals with resistivity of 46 Ωcm cut into circular disks of 2 mm diameter with thickness of 0.3 mm, the flat surface of the disk always being in the (110) plane. The samples were irradiated by electrons at room temperature. The microwave frequency used in this note was 36 GHz under magnetic fields up to 0.7 T. The c.c. effective mass values were determined from the CR spectra at 4.2 K by the maximum positions of the resonance lines with an accuracy of 2 to 3% /6/.

Table 1 gives the results of the CR study on initial and irradiated Ge (m_t, m_l mean transverse and longitudinal electron masses, m_L, m_H the effective masses of light and heavy holes) as well as the c.c. effective mass values from /1/.

As follows from the CR data even at large irradiation doses the electron effective mass values do not change practically, i.e. the ellipsoidic character of the electronic valleys near the minimum of the conduction band remains with the electron effective mass ratio being $m_l/m_t = 19.5$. At the same time an anomaly in the behaviour of the hole effective mass is observed during irradiation. And though the resonance lines of the holes are greatly broadened when compared to the initial ones, the requirements of CR observation $\omega_c\tau \gg 1$

1) Podlesnaya 17, 220726 Minsk, USSR.

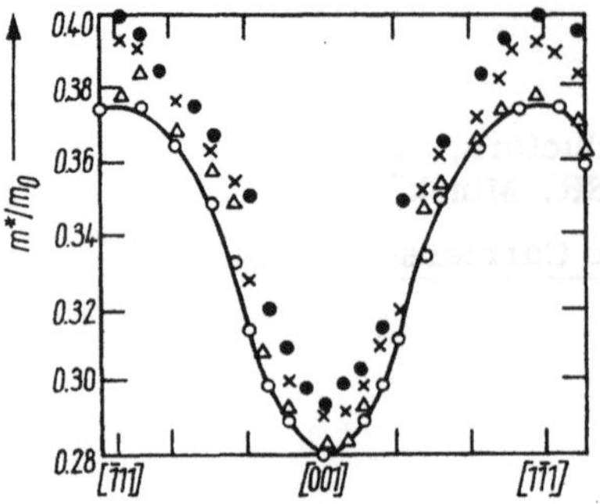

Fig. 1. Heavy hole effective mass anisotropy for the initial and irradiated Ge. $\Phi = 0$ (○), 5×10^{15} cm^{-2} (△), 2×10^{16} cm^{-2} (×), 7×10^{16} cm^{-2} (●)

are met and no marked overlapping is preserved in determining hole effective mass values. The light hole effective mass increases under irradiation by more than 10% for the largest irradiation doses, in this case it remains isotropic in character, which is indicative of the spherical energy contour of light holes. The increase in the heavy hole effective mass values with irradiation is accompanied by a change in character of their anisotropy.

Fig. 1 shows the angular dependence of heavy holes effective masses for different irradiation doses. The full line demonstrates the initial crystal. With enhancement of the irradiation dose, as follows from the figure, the increase in effective masses of heavy holes is greatest for the [111] and [001] axes for which the anisotropy of the band of heavy holes decreases.

It is known that the c.c. in the valence band of Ge in the K-space move along concentric isoenergetic surfaces, one of which is close to a spherical one (light holes), the other is a warping sphere (of heavy holes) /1/. Table 1

Table 1

$\Phi(cm^{-2})$	m_t/m_o	m_l/m_o	m_L/m_o	m_H/m_o [111]	m_H/m_o [001]	m_H/m_L [111]	m_H/m_L [001]
0	0.081	1.58	0.041	0.375	0.28	9.1	6.8
5×10^{15}	0.081	1.58	0.043	0.38	0.283	8.8	6.5
2×10^{16}	0.081	1.58	0.047	0.395	0.291	8.4	6.1
7×10^{16}	0.082	1.58	0.050	0.40	0.298	8.0	5.9
0 /1/	0.082	1.58	0.044	0.36	0.28	-	-

lists the hole mass ratios along [111] and [001]. After irradiation the values drop, i.e. the warping of the heavy holes is decreased along [111] and [001] axes.

For comparison we quote the CR data on plastically deformed germanium /4/, when an isotropic increase in the light hole effective mass and a preferable decrease in the anisotropy of the band of heavy holes in the [001] direction have been observed. In the case of Ge doped by isovalent impurities (for instance silicon /3, 7/) the CR data and those of optical measurements also give evidence for the decrease of the hole effective mass ratios with doping being rather heavy (37 mol% Si), this ratio tends to unity, i.e. in this case the curvatures of the energy bands of light and heavy holes are approaching each other.

References

/1/ V. DRESSELHAUS, A.F. KIP, and C. KITTEL, Phys. Rev. 98, 677 (1955).

/2/ R.N. DEXTER, H.Y. LEIGER, and B. LAX, Phys. Rev. 104, 637 (1956).

/3/ E. OTSUHA, S. NAGATA, and K. MURASE, J. Phys. Soc. Japan 20, 727 (1965).

/4/ L. RUSINIAK, phys. stat. sol. (b) 69, 619 (1975).

/5/ V.I. LUKASHEVA and L.A. KUROCHKIN, Dokl. Akad. Nauk Beloruss. SSR 28, 801 (1975).

/6/ V.I. LUKASHEVA, Avtoreferat kand. dissert., Minsk 1978.

/7/ R. BRAUNSTEIN, Phys. Rev. 130, 869 (1963).

(Received September 4, 1985)

phys. stat. sol. (b) 133, K35 (1986)
Subject classification: 61.80; S1.3

Istituto di Metodologie Avanzate Inorganiche del C.N.R.,
Istituto ci Chimica Generale ed Inorganica dell'Università, Roma[1)]

Electron-Energy-Loss Spectra of Ag Metal

By
MARISA SCROCCO

Following our XPS measurements on Ag /1/, we return to this metal stimulated by some observations of Eckardt and Fritsche /2/ on the interpretation of the satellites appearing in our XPS spectra. In the previous work we had assigned the satellites observed at the binding energies between 3.0 to 9.0 eV from the main peaks, as interband transitions, correlating them to the band structure as calculated by Chen and Segall /3/.

The alternative theoretical interpretation proposed by Eckardt and Fritsche is that some of these satellites are plasmon loss in nature. To clarify this point we now re-examine this problem making measurements in electron-energy-loss spectroscopy (EELS).

Plasmon satellites in X-ray photoemission spectroscopy differ from those excited in EELS experiments in having an intrinsic contribution during the photoemission process (the passage of the electron through the bulk solid).

The experiment was carried out in a UHV apparatus (VG ESCA-3 MK2) with the chamber equipped with an electron gun useful for Auger and energy-loss measurements. The sample was prepared by evaporating in situ a thick Ag layer on a Cu plate in $1x10^{-8}$ Torr vacuum. During the measurements the vacuum in the apparatus was about $1x10^{-9}$ Torr. Two different incident energies were adopted (E_p = 200 eV and 2000 eV) to differentiate between surface and bulk phenomena. Indeed, from the relationship of the intensity of the loss peaks with the incident-electron penetration depth, it is possible to distinguish bulk transitions from transitions between surface states. The spectra of Fig. 1 and 2 were recorded with a modulation of 1.0 V (peak-to-peak) and are reported in first and second derivative. The derivatives are useful because the former permits better resolution and the latter greater accuracy in the energy position

---

1) Città Universitaria, 00185 Rome, Italy.

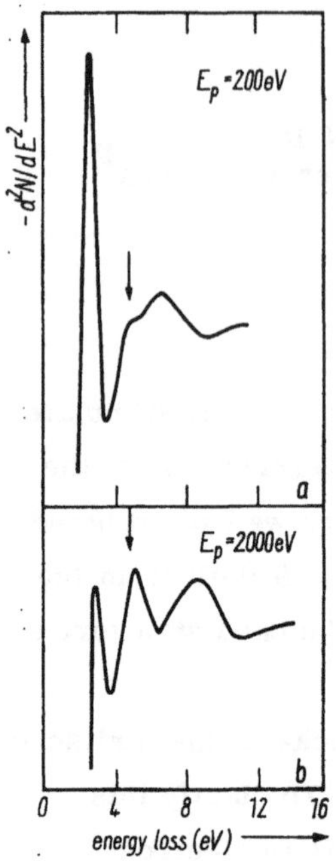

Fig. 1. Second-derivative electron energy-loss spectra of Ag at a) E_p = 200 eV and b) 2000 eV

of the energy-loss structure. The positions of the satellites with respect to the elastic peak are reported in Table 1. The more relevant assignments regard the transitions appearing at a binding energy < 10 eV.

In Fig. 1 we report the spectrum of this energy range, recorded in second derivative, at the incident energies of E_p = 200 eV and 2000 eV. Within the limits of the resolution selected, we observe four transitions at $\Delta\varepsilon$ = 3.5, 5.8, 7.4, and 9.0 eV which exhibit different relative intensities as a function of the primary beam energy E_p.

Comparing the two spectra it is possible to see that at E_p = 200 eV a very strong peak at 3.5 eV appears together with a doublet formed by a shoulder at about 5.8 eV and a peak at 7.4 eV. Conversely, at a primary energy of E_p = 2000 eV, the intensity of the peak at 3.5 eV is strongly decreased, the shoulder at 5.8 eV has grown into a distinct peak, the transition at 7.4 eV has disappeared and a new, very clear, peak appears at a binding energy of 9.0 eV.

From these results it is possible to assign the transitions that 3.5 and 7.4 eV as essentially surface transitions and those at 5.8 and 9.0 eV as essentially bulk transitions.

At 3.5 eV, the possibility exists of an overlap between this surface transition and a bulk transition observed at about the same energy /4 to 6/. The resolution of our apparatus does not allow a clear separation of these two transitions, although a very little shoulder is perhaps present in Fig. 1b.

For a free-electron metal such as Ag with one 5s electron per atom, the plasmon energy is predicted to lie at about 9.0 eV /7/. Thus, our peak at 9.0 eV can be assigned to this bulk plasmon with a surface plasmon consequently predicted at a binding energy of about 6.4 eV ($P_s = P_b/\sqrt{2}$). Our sur-

Table 1

Peak energy (eV) in the second derivative of the electron-loss spectra of Ag metal. The error limit can be evalued as ± 0.5 eV

E_p = 200 eV	E_p = 2000 eV
3.5	3.5
5.8 (shoulder)	5.8
7.4	-
	9.0
	15.0
	17.5
	20.0
	26.0
	30.5
	34.0
	42.5
	48.0
	55.0 Ag $4p_{3/2}$

face peak is at 7.4 eV and we presume that the shift of this peak is caused by the strong overlap with the bulk peak at 5.8 eV which can be ascribed to a shake-up transition. Having assigned the surface and bulk plasmons at 7.4 and 9.0 eV, we can ascribe the transitions observed (Fig. 2) at 15.0 and 17.5 eV to the double surface and bulk plasmon losses, respectively.

At higher binding energy, we find (Fig. 2) two pronounced peaks at 20.0 and 26.0 eV. Considering eleven "free" electrons per atom of Ag, Pines /7/ calculates the plasmon loss in this latter position (≈ 26 eV). Otto and Petri /8/ explain all the structure starting with the peak at about 25 eV as transitions from the higher d-shell electrons to the f-like

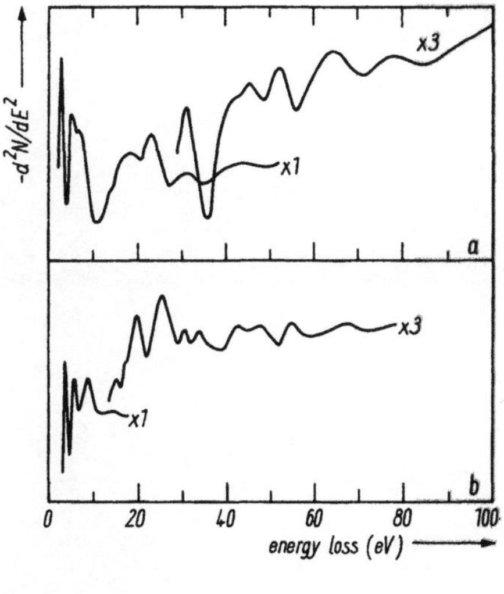

Fig. 2. a) First and b) second derivative electron energy-loss spectra of Ag metal at E_p = 2000 eV

free electron state with an energy modulation of these transitions as in the case of the extended X-ray absorption fine structure (EXAFS).

The structure at 55.0 eV is the Ag $4p_{3/2}$ core excitation.

References

/1/ MARISA SCROCCO, phys. stat. sol. (b) 125, 713 (1984).

/2/ H. ECKARDT and L. FRITSCHE, Solid State Commun. 54, 405 (1985).

/3/ AN-BAN CHEN and B. SEGALL, Phys. Rev. B 12, 600 (1975).

/4/ R. COURTH, V. BACHELIER, and S. HÜFNER, Solid State Commun. 38, 887 (1981).

/5/ J. GEIGER and K. WITTMAAK, Z. Phys. 195, 44 (1966).

/6/ J. DANIELS, Z. Phys. 203, 235 (1967).

/7/ D. PINES, Rev. mod. Phys. 28, 184 (1956).

/8/ A. OTTO and E. PETRI, Solid State Commun. 20, 823 (1976).

(Received October 22, 1985)

phys. stat. sol. (b) 133, K39 (1986)

Subject classification: 71.35 and 78.30; S5.11

Zentralinstitut für Elektronenphysik der Akademie der Wissenschaften der DDR, Berlin[1)]

Metastable Species of Oxygen-Related Donors in Silicon

By

D. WRUCK and F. SPIEGELBERG

In the last years, our knowledge about the electronic structure of oxygen-related donors formed after heat treatment at temperatures between 300 and 500 °C in oxygen-containing silicon, the so called thermal donors, has been greatly enlarged by a successful use of infrared spectroscopic techniques /1 to 4/.

Up to now it has been well established that these thermal donors give rise to a series of double donor species with closely spaced ionization energies. For explaining these properties, new microscopic models have been elaborated /5, 6/, assuming the formation of electrically active clusters with increasing number of oxygen atoms to be responsible for the occurrence of the different donor species. Especially in /5/ it was pointed out that in the "starting" phase metastable oxygen complexes would be stabilized by successively captured oxygen atoms.

In this note we report the observation of some unusual behaviour in the infrared absorption spectra of the high-energy series (D^+) of oxygen-related donors in silicon which gives some evidence for the metastable character of donor species in the initial stage of its formation.

The samples used were boron-doped Czochralski silicon crystals into which, by appropriate heat treatment at 450 °C, thermal donors have been introduced with concentrations of 1×10^{14} cm^{-3} (sample 1), 1×10^{15} cm^{-3} (sample 2), and 1.3×10^{15} cm^{-3} (sample 3). Infrared transmission measurements have been carried out with a commercial infrared grating spectrophotometer allowing the sample to be illuminated with polychromatic light with or without band gap excitation.

The full curve of Fig. 1 shows a part of the spectrum of the high-energy (D^+) series of oxygen-related donors of sample 3 measured at liquid neon temper-

1) Hausvogteiplatz 5-7, DDR-1086 Berlin, GDR.

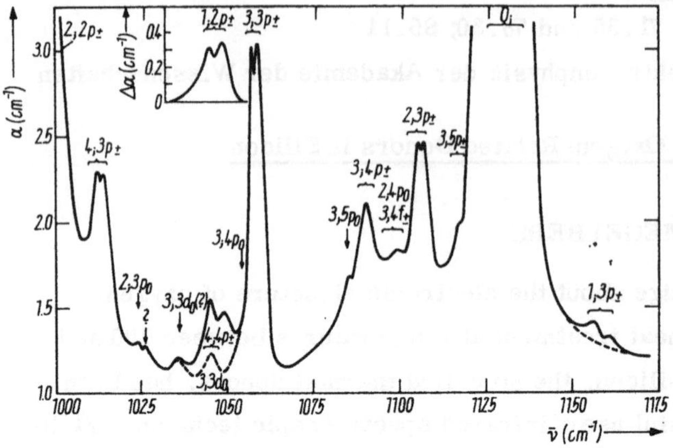

Fig. 1. Absorption spectrum oi sample 3 at LNeT measured with (full curve) and without (dashed curve) band gap illumination during cooling, the insert showing the difference spectrum in the vicinity of the 1(2p±) band

ature (LNeT) with the sample being illuminated with band gap light when cooling down from 300 K to LNeT. If the same sample is cooled in darkness (dashed curve), the 2p± and 3p± absorption bands of the donor species 1:D^+(1) (the notation "1" corresponds to "a" of /2/) are missing. The absorption bands belonging to the higher-order donor species remain unchanged.

If the temperature at which the band gap illumination is switched on in the cooling procedure is decreased, the intensities of the 1(2p±) and 1(3p±) bands measured at liquid nitrogen temperature (LNT) remain constant to a temperature of about 130 K but then rapidly drop to zero. This indicates a thermally activated process with relatively large activation energy.

On the other hand, after having formed by cooling from 300 K to LNT with illumination, the D^+(1) absorption bands can be decreased by raising the temperature above LNT (see Fig. 2). At about 140 K the D^+(1) absorption bands have disappeared while the corresponding bands of the higher-order donor species still remain.

Further, as seen from a comparison of the behaviour of the 1(2p±) band for three samples with different positions of the Fermi level at room temperature shown in Fig. 3, the observed occurrence or missing of the D^+(1) ab-

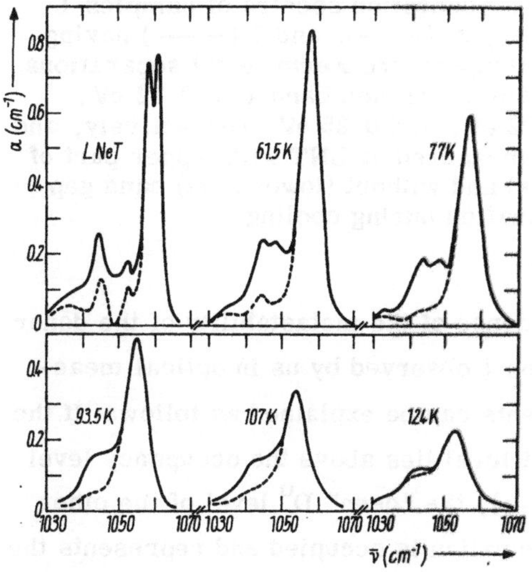

Fig. 2. Absorption spectra of sample 3 at different temperatures

sorption bands when cooling with or without band gap light is controlled by the position of the Fermi level at room temperature. The effect of removing the 1(2p±) band by cooling in darkness is seen for the samples 2 and 3 with $E_c - E_f < 0.32$ eV but not for sample 1 with $E_c - E_f \geq 0.32$ eV.

The above results suggest that the $D^+(1)$ level is caused by a metastable defect whose formation as well as its destruction is obvisouly controlled by thermally activated carrier capture processes via potential barriers of different heights and whose metastability depends on the position of the Fermi level.

If the formation of this metastable state would be caused by a temperature dependent electron capture via a potential barrier, one would need a doubly negatively charged centre, which is not available in our case of a D^{2+} donor level. Another interpretation, however, becomes possible if the $D^+(1)$ state originates from the transformation of another defect configuration being occupied by two electrons, for instance, the D^0 state of a deep double donor.

This at first glance somewhat speculative assumption becomes more founded if recent results of electrical measurements on the behaviour of thermal donors in silicon by Tkachev et al. /7/ and analogous data on germanium by Litvinov et al. /8/ are taken into consideration. Analyzing the temperature dependence of equilibrium carrier concentration together with the long-term relaxation of the non-equilibrium conductivity in silicon containing oxygen-related donors, the authors of /7/ concluded that the double donor levels at $E_c - 0.05$ eV and $E_g - 0.15$ eV are metastable states of an Anderson negative-U system /9/, whose $E(0|++)$ occupancy level should lie at $E_c - 0.32$ eV.

Obeying the usual precautions when relating the thermal ionization energies of the shallow double donors given in /7/ to our optical energies, the Fermi level

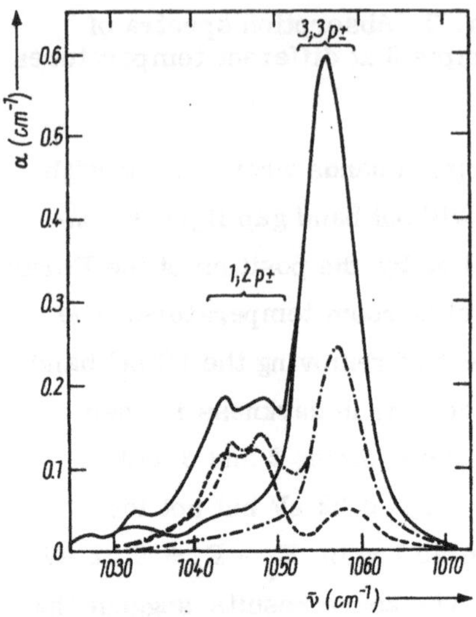

Fig. 3. Absorption spectra of samples 1 (1 ---), 2 (—·—), and 3 (——) having room temperature Fermi level separations from the conduction band at ≧ 0 32 eV, ≧ 0.22 eV, and 0.25 eV, respectively, and being measured at LNT with (upper part of curves) and without (lower part) band gap illumination during cooling

dependence of the metastability of the donor species 1 observed by us in optical measurements can be explained as follows: If the Fermi level lies above the occupancy level E(0 | ++), the "deep" D^0 level of the other configuration is occupied and represents the ground state of the system but may be converted into the "usual" D^+ configuration by a potential barrier controlled capture of a hole from the valence band created by band gap excitation. Its realization would necessitate a sufficiently high temperature (e.g. 300 K) when applying the band gap excitation and a freezing-in of the so formed D^+ state by cooling to a sufficiently low temperature. If the Fermi level lies below E(0 | ++), the "deep" D^0 level is not occupied and the only stable states are those of the "usual" double donor states which then would not be affected by the above cooling procedures. The position of the deep level connected with the E(0| ++) occupancy level can be estimated to E(0| +) = 2E(0| ++)- - E(+| ++) = 0.64 - 0 15 eV = 0.49 eV. Due to its negative-U character, for this level a large Franck-Condon shift and a correspondingly large optical ionization energy would be expected. Our attempts for detecting this level in optical absorption performed in a spectral range up to 1.24 eV failed, possibly due to the facts that the required optical energy exceeds the band gap and/or the absorption cross section will be too low.

In conclusion, we have observed optically a Fermi-level dependent metastable behaviour of the first species of oxygen-related double donors in silicon, whose temperature dependence indicates that potential-barrier controlled carrier capture mechanisms are involved. The above effect can be understood, if a mutual transformation of the D^+ state into another configuration occupied by two electrons is assumed. Linking our results with those of recent electrical measurements of Tkachev et al. /7/, this two-electron configuration can be associated with the E(0| ++) occupancy level of an negative-U system, whose metastable states are the "usual" shallow double donor levels of the species 1:D^+(1) and D^o(1).

References

/1/ H.J. HROSTOWSKI and R.H. KAISER, Phys. Rev Letters 1, 199 (1958).

/2/ D. WRUCK and P. GAWORZEWSKI, phys. stat. sol. (a) 56, 577 (1979).

/3/ B. PAJOT, H. COMPAIN, J. LEROUEILLE, and B. CLERJAUD, Physica 117/118 B, 110 (1982).

/4/ R. ÖDER and P. WAGNER, in: Defects in Semiconductors, Vol. II, Ed. S. MAHAJAN and J.W. CORBETT, North Holland, New York 1983 (p. 171).

/5/ G.S. OEHRLEIN and J.W. CORBETT, ibid. (p. 107).

/6/ A. OURMAZD, W. SCHRÖTER, and A. BOURRET, J. appl. Phys. 56, 1670 (1984).

/7/ V.D. TKACHEV, L.F. MAKARENKO, V.P. MARKEVICH, and L.I. MURIN, Fiz. Tekh. Poluprov. 18, 526 (1984).

/8/ V.V. LITVINOV, G.V. PALCHIK, and V.I. URENEV, Fiz. Tekh. Poluprov. 19, 1366 (1985).

/9/ G.A. BARRAFF, E.O. KANE, and M. SCHLÜTER, Phys. Rev. B 21, 5662 (1980).

(Received December 9, 1985)

phys. stat. sol. (b) 133, K45 (1986)

Subject classification: 78.55; S8.15

Institute of Semiconductor Physics,
Academy of Sciences of the Lithuanian SSR, Vilnius[1)]

Temperature Dependence of Photoluminescence Excitation Spectra in $Cd_{1-x}Mn_xTe$

By

G. AMBRAZEVIČIUS and G. BABONAS

Recent studies /1, 2/ have shown that photoluminescence (PL) bands B (2.0 eV), C (1.2 eV), and D (1.4 to 1.6 eV) in $Cd_{1-x}Mn_xTe$ are related to the Mn^{2+} ions. The existence of some temperature dependent PL excitation (PLE) mechanisms has been suggested /2/ from the PLE studies. In this note the temperature dependence of the PLE spectra of the B, C, and D bands in $Cd_{1-x}Mn_xTe$ ($0.2 \leq x \leq 0.7$) have been investigated in the range 10 to 150 K.

The shape of the PLE spectrum for the C band changes considerably with temperature (Fig. 1). At 10 K two peaks are revealed in the broad band extending in the range 2.00 to 2.25 eV. The higher-energy peak (2.18 eV at 80 K) agrees well with the peak in the piezoreflectance /3/ and photoresponse /4/ spectra. The characteristics of the lower energy peak (2.07 eV at 10 K for x = 0.70) are close to those of the E_1 structure in the wavelength-modulated reflectivity spectra /5/.

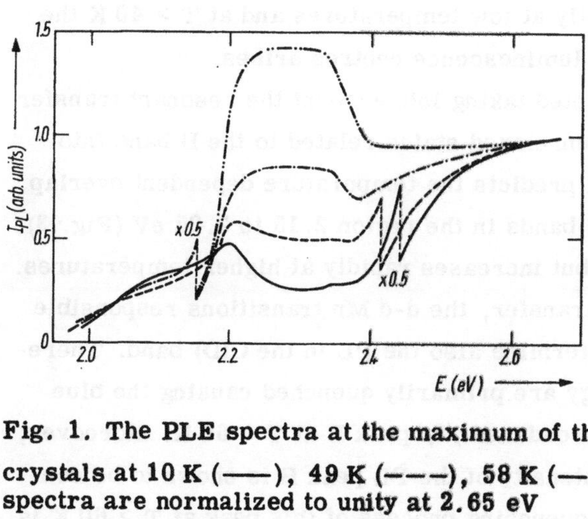

Fig. 1. The PLE spectra at the maximum of the PL band C of $Cd_{0.49}Mn_{0.51}Te$ crystals at 10 K (——), 49 K (– – –), 58 K (— · —), and 62 K (– - - –). The spectra are normalized to unity at 2.65 eV

---

1) 52 K. Pozhelos, 232600 Vilnius, USSR.

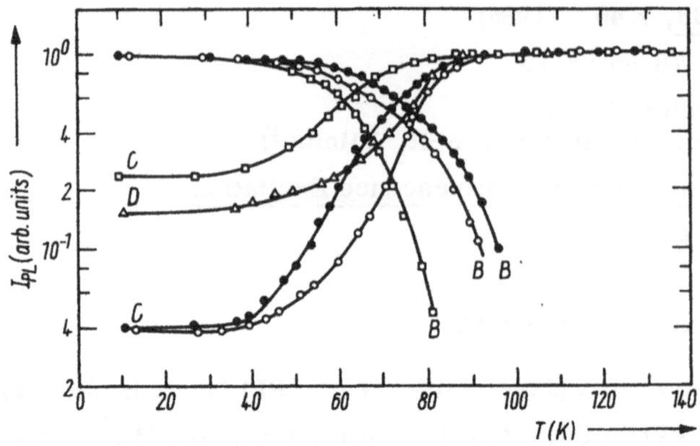

Fig. 2. Temperature dependence of peak intensity of B, C, and D PL bands at excitation into the plateau region (2.3 eV) for x = 0.51 (●), 0.62 (○ , △), and 0.70 (□). Solid lines are drawn to guide the eye. Curves are normalized to unity at the maximum

When the temperature increases, the shape of the PLE spectra for the C band approaches continuously to that for the B band /2/. However, in contrast to the B band, the intensities of the C and D bands show a sharp rise at T > 40K under excitation in the plateau (Fig. 2). The PLE spectra for the D band are similar to those of the C band /2/, though the band at 2.2 eV (Fig. 3) is narrower and cut at the low-energy side. The experimental results show that the PL bands B and C(D) are excited independently at low temperatures and at T > 40 K the connection between the respective luminescence centres arises.

This connection can be interpreted taking into account the resonant transfer of excitation energy /6/ from the unrelaxed states related to the B band into those of the C(D) band. The model predicts the temperature dependent overlap of the PLE spectra for B and C(D) bands in the region 2.15 to 2.25 eV (Fig. 3). The overlap is small at T < 40 K but increases rapidly at higher temperatures. Then, due to the efficient energy transfer, the d-d Mn transitions responsible for excitation of the PL band B determine also the PL in the C(D) band. Therefore, the B centres of lower energy are primarily quenched causing the blue shift /7/ of the inhomogeneously broadened /8/ peak B at T > 60 K. Moreover, the decay time of the integrated intensity of the PL peak B is shown to be constant up to T ≈ 80 K /8/, and the quenching process of this peak at T > 60 K is

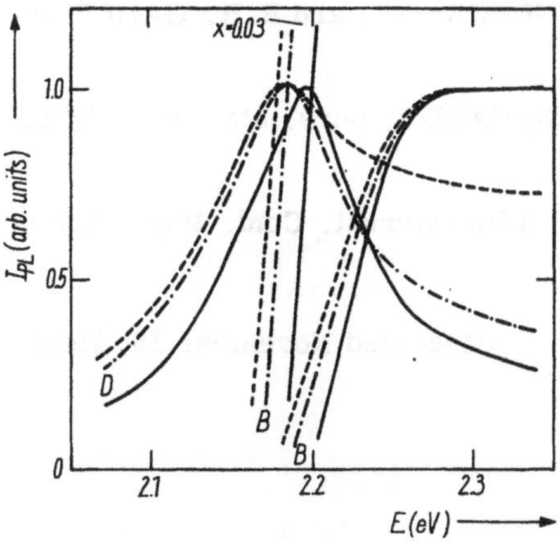

Fig. 3. The PLE spectra at the maximum of PL bands B and D of $Cd_{0.38}Mn_{0.62}Te$ crystals at 10 K (——), 52 K (—·—), and 71 K (- - -). The spectra are normalized to unity at the maximum

predicted /8/ to occur immediately after the PL excitation. This is in good agreement with the suggested model.

The short time of the excitation energy transfer indicates the centres responsible for B and C(D) PL bands to be sufficiently close together and, therefore, to be related to Mn ions. The correlation of the shape of PLE spectra for the C band /2/ with the photocurrent and photoresponse /4/ spectra shows that PL excitation involves the creation of free carries, while the PL band can be excited only by d-d Mn transitions.

The different properties of the states related to 3d impurities are expected from the recently developed spin-polarized tight binding model /9/. Two types of states are predicted for transitional metal impurities in tetrahedral semiconductors: the non-bonding states of e-symmetry and t_2-states, strongly hybridized with the host. Therefore, the different behaviour of Mn related bands in $Cd_{1-x}Mn_xTe$ is in agreement with the expected different manifestation of the transition metal ion states in crystals.

References

/1/ M.M. MORIWAKI, R.Y. TAO, R.R. GAŁĄZKA, W.M. BECKER, and J.W. RICHARDSON, Physica 117/118B, 467 (1983).

/2/ G. AMBRAZEVIČIUS, G. BABONAS, and YU.V. RUD, phys. stat. sol. (b) 125, 759 (1984).

/3/ Y.R. LEE and A.K. RAMDAS, Solid State Commun. 51, 861 (1984).

/4/ P. LEMASSON, BING-LIANG WU, R. TRIBOULET , and J. GAUTRON, Solid State Commun. 47, 669 (1983).

/5/ B. MONTEGU, A. LAUGIER, and R. TRIBOULET, , J. appl. Phys. 56, 3061 (1984).

/6/ D.L. DEXTER, J. chem. Phys. 21, 836 (1953).

/7/ M.M. MORIWAKI, W.M. BECKER, W. GEBHARDT, and R.R. GAŁĄZKA, Phys. Rev. B 26, 3165 (1982).

/8/ E. MÜLLER, W. GEBHARDT, and V. GERHARDT, phys. stat. sol. (b) 113, 209 (1982).

/9/ P. VOGL and J.M. BARANOWSKI, Proc. 17th Internat. Conf. Phys. Semicond., San Francisco 1984.

(Received November 18, 1985)

phys. stat. sol. (b) 133, K49 (1986)

Subject classification: 78.30; 71.55; S8.16

Institute of Physics of Solids and Semiconductors, Academy of Sciences of the Belorussian SSR, Minsk[1)]

Absorption Edge of Stoichiometric and Nonstoichiometric $CuInSe_2$ Crystals

By

N.N. BELEVICH, A.I. LUKOMSKII, L.A. MAKOVETSKAYA, and A.P. CHERNYAKOVA

$CuInSe_2$, one of the $A^IB^{III}C_2^{VI}$ semiconducting compounds, with a band gap at about 1 eV, can be obtained with both n- and p-type conductivity, and therefore, can be used in fabricating effective solar energy converters /1, 2/. The desired type of conductivity of the crystal and p-n junctions /3, 4/ could be obtained by a controlled deviation from stoichiometry of the quantities of the elements of the compound, since the electrical properties of the material, even if doped with elements which do not form it, are determined by native defects, i.e. by the vacancies of Cu, In, Se and by the interstitial atoms of these elements /5, 6/. The electrical properties of $CuInSe_2$, depending on the deviation from the stoichiometric quantities of the components and other technological factors, are studied in /7, 8/. The optical properties of crystals annealed in copper, indium, and selenium vapours are investigated in /9, 10/. In these papers the band gap and the ionization energy of some impurity states are determined.

In this note the effect of indium or selenium excess on the absorption spectra in the fundamental absorption band edge is investigated.

Single crystal ingots were grown by the vertical Bridgman technique. The specimens with an excess of 0.5% In had n-type conductivity, while those with an excess of 0.5% Se had p-type conductivity. The samples grown from the stoichiometric melt had, in general, n-type conductivity. The Hall concentration of charge carriers at room temperature was 10^{17} to $5x10^{18}$ cm^{-3} for p-type samples and 10^{16} to 10^{17} cm^{-3} for n-type samples, with the mobility ranging from 10 to 70 $cm^2 V^{-1} s^{-1}$ and from 200 to 700 $cm^2 V^{-1} s^{-1}$, respectively. From the electrical measurements the following activation energies have been obtained: for acceptors 0.020, 0.082, and 0.300 eV and for donors 0.013 eV.

---

1) Podlesnaya 17, 220726 Minsk, USSR.

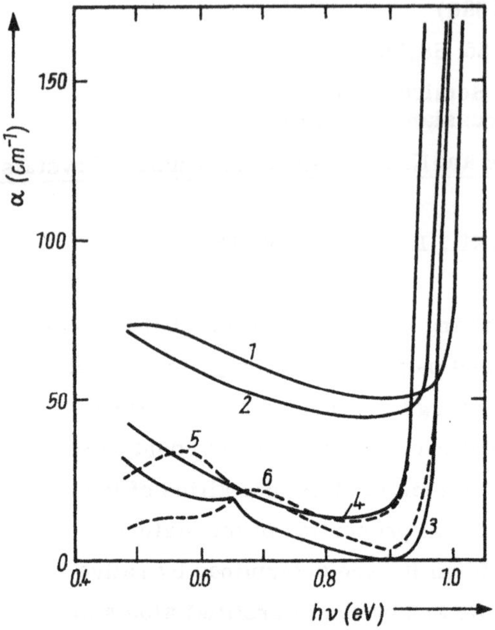

Fig. 1. Absorption spectra of $CuInSe_2$ crystals grown (1), (2) with an excess of In; (3), (4) with an excess of Se; and (5), (6) from the stoichiometric melt. (1), (3), (5) T = 77 K; (2), (4), (6) T = 293 K

Fig. 1 shows the dependences of absorption coefficients on the photon energy at 77 and 293 K for $CuInSe_2$ crystals grown in an excess of Se and In as well as from the stoichiometric melt. The absorption coefficient α was calculated from the measured transmission coefficient T and reflection coefficient R by using the well-known formula

$$\alpha = \frac{1}{d} \ln\left\{\frac{(1-R)^2}{2T} + \left[\frac{(1-R)^4}{4T^2} + R^2\right]^{1/2}\right\},$$

where d is the thickness of the plane-parallel sample. In the spectral range from 0.6 to 1.1 eV the reflection coefficient appeared to be equal to 0.24 for all the crystals at both temperatures.

The investigation of the obtained spectra allows to conclude that there is a dependence of the energetic position of the fundamental absorption edge and of the absorption peak in the range immediately adjacent to it at the long wavelength side on the elemental composition of the samples.

We determined the band gap E_g from the dependence of $(\alpha h\nu)^2$ on $h\nu$ employing the intersection point of its linear part with the abscissae. At 293 K the following E_g values were obtained: 0.959, 0.932, 0.934 eV for crystals grown with an excess of In and Se and from the stoichiometric melt, respectively. At 77 K these values appeared to be equal to 0.992, 0.958, and 0.954 eV. The measurement error was ±0.005 eV. The corresponding E_g values known in the literature /9 to 20/ differ widely and range from 1.055 eV /10/ to 0.961 eV /12/ at 77 K. The most reliable data seem to be given in /10/ and /11/, where the E_g value has been determined from luminescence /10/ and electroreflectance spectra /11/. At 77 K they are 1.055 and 1.038 eV, respectively. These values are the largest of those given in /9 to 20/.

For the $CuInSe_2$ crystals investigated by us the E_g values were in all cases smaller than those given in /10, 11/ and depended on the components ratio of the compound. Since all the investigated crystals of both conductivity types had high degrees of compensation (> 0.9) and, consequently, high concentrations of the ionized acceptor and donor levels, it is natural to assume that the position of the absorption edge should be determined by the acceptor-conduction band and(or) valence band-donor transitions. In this case the quantities and the ratios of the acceptor and donor levels, induced by the native defects, are given by the degree and direction of deviation from stoichiometry.

References

/1/ J.L. SHAY, S. WAGNER, and N.M. KASPER, Appl. Phys. Letters 27, 89 (1975).

/2/ L.L. KAZMERSKY, F.R. WHITE, and G.K. MORGAN, Appl. Phys. Letters 28, 269 (1976).

/3/ J. PARKES, R.D. TOMLINSON, and M.J. HAMPSHIRE, Solid State Electronics 16, 773 (1973).

/4/ I. GONZÁLEZ, C. RINCÓN, A. REDONDO, and P. NEGRETE, Japan. J. appl. Phys., 19, Suppl. 3, 29 (1980).

/5/ P.W. YU, J. appl. Phys. 47, 677 (1976).

/6/ P.W. YU, J.S. PARK, and J.T. GRANT, Appl. Phys. Letters 28, 214 (1976).

/7/ K.J. BACHMANN, M. FEARHEILEY, Y.H. SHING, and N. TRAN, Appl. Phys. Letters 44, 407 (1984).

/8/ R. NOUFI, R. AXTON, C. HERRINGTON, and S.K. DEB, Appl. Phys. Letters 45, 668 (1984).

/9/ C. RINCÓN and G. SANCHEZ PÉREZ, Solid State Commun. 50, 899 (1984).

/10/ G. MASSE and E. REDJAI, J. appl. Phys. 56, 1154 (1984).

/11/ J.L. SHAY, B. TELL, H.M. KASPER, and L.M. SCHIAVONE, Phys. Rev. B 7, 4485 (1973).

/12/ W. HÖRIG, H. NEUMANN, H.-J. HÖBLER, and G. KÜHN, phys. stat. sol. (b) 80, K21 (1977).

/13/ W. HÖRIG, H. NEUMANN, H. SOBOTTA, B. SCHUMANN, and G. KÜHN,

Thin Solid Films 48, 67 (1978).

/14/ L.Y. SUN, L.L. KAZMERSKY, A.H. CLARK, P.J. IRELAND, and D.W. MORTON, J. Vacuum Sci. Technol. 15, 265 (1978).

/15/ H. SOBOTTA, H. NEUMANN, V. RIEDE, G. KÜHN, J. SELTMANN, and D. OPPERMANN, phys. stat. sol. (a) 60, 531 (1980).

/16/ C. RINCÓN, J. GONZÁLEZ, and G. SÁNCHEZ PÉREZ, phys. stat. sol. (b) 108, K19 (1981).

/17/ G.K. AVERKIEVA, G.A. MEDVEDKIN, and A.A. YAKOVENKO, Fiz. Tekh. Poluprov. 11, 2081 (1983).

/18/ C. RINCÓN, J. GONZÁLEZ, and G. SÁNCHEZ PÉREZ, phys. stat. sol. (b) 117, K123 (1983).

/19/ C. RINCÓN, J. GONZÁLEZ, and G. SÁNCHEZ PÉREZ, J. appl. Phys. 54, 6634 (1983).

/20/ I.V. BODNAR, B.V. KORZUN, and A.I. LUKOMSKII, phys. stat. sol. (b) 105, K143 (1981).

(Received October 23, 1985)

phys. stat. sol. (b) 133, K53 (1986)
Subject classification: 71.30 and 72.80; S8.16

Institute of Physics of Solids and Semiconductors, Academy of Sciences of the Belorussian SSR, Minsk[1)]

Semiconductor-Metal Transition and Variable Range Hopping Conductivity in $CuInSe_2$

By
N.N. BELEVICH and L.A. MAKOVETSKAYA

The $CuInSe_2$ ternary semiconducting compound is of considerable interest because of its application in nonlinear optics, electroluminescence diodes, and photoelectric devices /1/and at present it attracts even growing attention. The electrical properties of this compound have been studied by many authors /2, 3/ and it has been established that $CuInSe_2$ could be obtained with both n- and p-type conductivity /4/. Since the study of the electrical properties of $CuInSe_2$ has been carried out mainly above 80 K, there is no detailed information on the mechanism of conductivity for the above mentioned compound.

In this note the measurements of electrical conductivity of $CuInSe_2$ single crystal samples of n- and p-type in the range from 4.2 to 300 K are reported. The single crystal ingots were grown by the vertical Bridgman technique. The n-type $CuInSe_2$ single crystal samples contain 0.5 at% excess of In and those of p-type - 0.5 at% excess of Se. Different impurity concentration of donors (acceptors) was obtained by annealing of the ingots after the growth at $\approx$ 1000 K.

The Hall concentrations of charge carriers at 290 K were 10^{17} to $5x10^{18}$ cm^{-3} for p-type samples and 10^{16} to 10^{17} cm^{-3} for n-type ones, the Hall mobilities were 10 to 70 cm^2/Vs and 200 to 700 cm^2/Vs, respectively.

The results obtained show that in the temperature range from 100 to 300 K the impurity activation mechanism of conductivity is dominant (Fig. 1, curves 1, 2). Activation energies of 0.01 to 0.05 eV, 0.23 to 0.26 eV for p-type and 0.008 to 0.013 eV for n-type samples, respectively, were obtained depending on the impurity concentration. Taking into account the expansion of impurity levels and the formation of impurity bands the calculated activation energies of acceptors for the case of "pure" samples were 0.02, 0.082, 0.3 eV. The activation energy of donors taking into account its increase with increasing

1) Podlesnaya 17, 220726 Minsk, USSR.

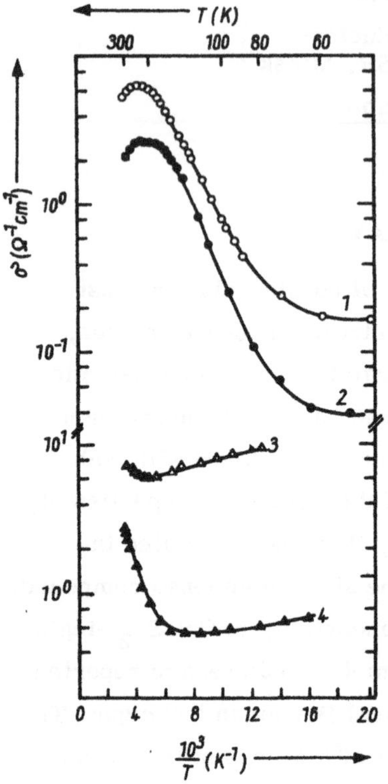

Fig.1

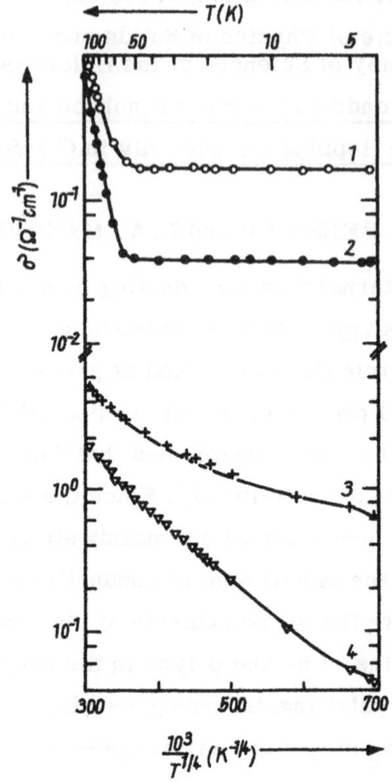

Fig.2

Fig. 1. Electrical conductivity as a function of temperature for p-type $CuInSe_2$: (1) $p_{290K} = 1.6x10^{17}$ cm^{-3}, $N_A = 7.85x10^{18}$ cm^{-3}; (2) $p_{290K} = 2.5x10^{17}$ cm^{-3}, $N_A = 4.05x10^{18}$ cm^{-3}; (3) $p_{290K} = 4.7x10^{18}$ cm^{-3}, $N_A = 5.7x10^{19}$ cm^{-3}; (4) $p_{290K} = 1.6x10^{18}$ cm^{-3}, $N_A = 3.8x10^{19}$ cm^{-3}

Fig. 2. Variable range hopping conductivity for p-type and n-type $CuInSe_2$: (1), (2) corresponds to the samples (1), (2) of Fig. 1; (3) $n_{290K} = 7.1x$ $x10^{16}$ cm^{-3}, $N_D = 1.5x10^{18}$ cm^{-3}; (4)$n_{290K} = 1.1x10^{17}$ cm^{-3}, $N_D = 5.7x$ $x10^{18}$ cm^{-3}

degree of compensation of donors is taken to be equal to 0.008 eV. The impurity concentration for p-type samples was calculated using the Brooks-Herring formula and for n-type ones using the results from /3/. Fig. 1

(curves 1 2 shows that the activation energy $\varepsilon = 0.082$ eV converts to ε · .02 eV with temperature rise and then saturation takes place when all the acceptors with $\varepsilon = 0.02$ eV are ionized. When the impurity concentration increases up to $N_A \approx 2.4 \times 10^{19}$ cm^{-3} and the degree of compensation $\varkappa \lesssim 0.96$, there occurs a semiconductor-metal transition /5/, with σ increasing weakly with the decrease of T (curves 3, 4). The transition takes place below $T \approx 200$ K, corresponding to the transition of the Fermi level across the percolation level /5/.

As follows from curves 1, 2 (Fig. 1) the dependence of σ on 1/T at $T < 100$ K fails to satisfy the activation mechanism of conductivity. The activation energy in this case starts to decrease which corresponds to the variable range hopping conductivity. In this temperature range, as is seen from Fig. 2, σ obeys the Mott equation /6/:

$$\sigma = \sigma_o \exp\left[-(T_o/T)^{1/4}\right], \tag{1}$$

$$kT_o = \beta\left[N(\varepsilon_F)\, a_o^3\right]^{-1}, \tag{2}$$

where k is the Boltzmann constant, $\beta \approx 16$ /7/, $N(\varepsilon_F)$ is the density of states at the Fermi level, a_o is the radius of the state localization. Curves 1, 2 in Fig. 2 are obtained for the same samples as curves 1, 2 in Fig. 1. At $T > 65$ K the parameter T_o calculated from (1) for n- and p-type samples varies from 7×10^2 to 3×10^7 K, that, according to (2), corresponds to state densities of 10^{18} to 10^{22} eV^{-1} cm^{-3}.

On the curves of hopping conduction as a function of temperature both for n-type and p-type $CuInSe_2$ at $T < 65$ K a break is revealed, which corresponds to a decrease of the characteristic temperature T_o. The largest break (up to nine orders of magnitude) is observed for p-type samples with dominant acceptor states with $\varepsilon = 0.082$ eV (curves 1, 2 in Fig. 2). From (2) one can conclude that the reason for the sharp drop of the characteristic temperature T_o at $T < 65$ K is the significant increase of the localization radius a_o for impurity states and, probably, some increase in the density of states.

This fact can probably be explained by the transformation of the valence band structure in this temperature range. As a consequence, the carriers with much smaller effective mass start to dominate in the conductivity. These carriers can be light holes of valence subbands split from the basic band by the crystal field and by spin-orbit interaction.

References

/1/ S. WAGNER, 3rd Internat. Conf. Ternary Compounds, Edinburgh 1977, Inst. Phys. Conf. Ser. No. 35, 205 (1977).

/2/ T. IRIE, S. ENDO, and S. KIMURA, Japan. J. appl. Phys. 18, 1303 (1979).

/3/ H. NEUMANN, R.D. TOMLINSON, N. AVGERINOS, and E. NOVAK, phys. stat. sol. (a) 75, K199 (1983).

/4/ P. MIGLIORATO, J.L. SHAY, H.M. KASPER, and S. WAGNER, J. appl. Phys. 46, 1777 (1975).

/5/ V.I. SHKLOVSKII and A.L. EFROS, Zh. eksper. teor. Fiz. 61, 816 (1971).

/6/ N.F. MOTT, Phil. Mag. 19, 835 (1969).

/7/ V. AMBEGAOKAR, B.I. HALPERIN, and J.S. LANGER, Phys. Rev. B 4, 2612 (1971).

(Received September 9, 1985)

phys. stat. sol. (b) 133, K57 (1986)

Subject classification: 73.40; S7.13

Sektion Physik der Humboldt-Universität zu Berlin[1)] (a) and International Laboratory of High Magnetic Fields and Low Temperatures, Wrocław[2)] (b)

Thermopower Measurements on the Two-Dimensional Electron Gas of a Grain Boundary in an InSb Bicrystal

By

R. HERRMANN (a), U. PREPPERNAU (a), and M. GLIŃSKI (b)

Quantum oscillations of the thermopower were measured on the two-dimensional electron gas of a grain boundary in an InSb bicrystal for the first time. Hitherto such thermopower measurements on the two-dimensional electron gas exist for GaAs-$Al_xGa_{1-x}As$ heterostructures only /1/.

The experimental arrangement is shown in Fig. 1. We have investigated the thermopower in an n-inversion layer adjacent to the grain boundary in a p-InSb bicrystal. The InSb bicrystal has been grown by means of a double-seed technique as described in /2/. Two [11$\bar{1}$] p-type InSb crystals ($p \approx 5\times10^{15}$ cm^{-3}) are tilted by an angle $\Theta = 13^o$ along the rotation axis [112]. A symmetrical β-grain boundary was obtained in [112] (1$\bar{1}$0) orientation with a ($\bar{1}$10) symmetric plane. Concentration and Hall mobility of electrons in the grain boundary layer are $n = 1.6\times10^{12}$ cm^{-2} and $\mu_H = 1.1\times10^4$ cm^2/Vs, respectively. Soldered contacts were fixed to the grain boundary layer as shown in Fig. 1. A temperature gradient ∇T is realized in the direction of the long axis of the bicrystal (x-direction). Therefore the temperature of contact 1 is T_1 whereas the temperature of contacts 2 and 3 is T_2. The temperature gradient was obtained by connecting one side of the sample to a small electric heater and the other to a heat sink (bath temperature $T_B = 4.2$ K). The temperature difference ΔT between contacts 1 and 2 was measured with AuFe-chromel thermocouples.

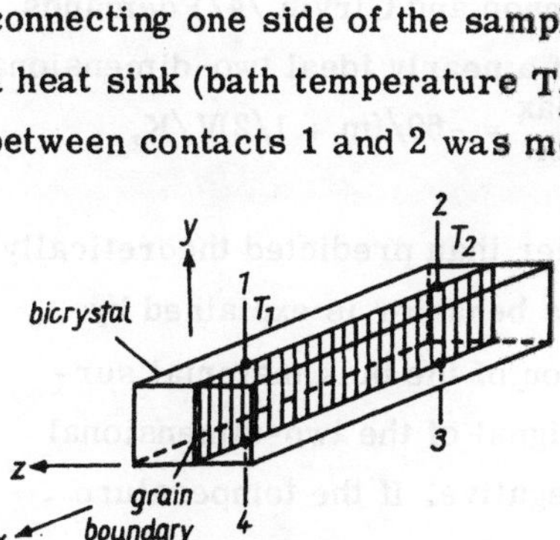

Fig. 1. Experimental arrangement. $\nabla T = ((\nabla T)_x, 0, 0)$; $\vec{B} = (0, 0, B)$; $\Delta T = T_2 - T_1$; $T_B = 4.2$ K

Measurements carried out at temperature gradients from $|\nabla T| = 4$ K/cm up to 8 K/cm (corresponding temperature differences

1) Invalidenstr. 110, DDR-1040 Berlin, GDR.

2) ulica A. Próchnika 95, 53-529 Wrocław, Poland.

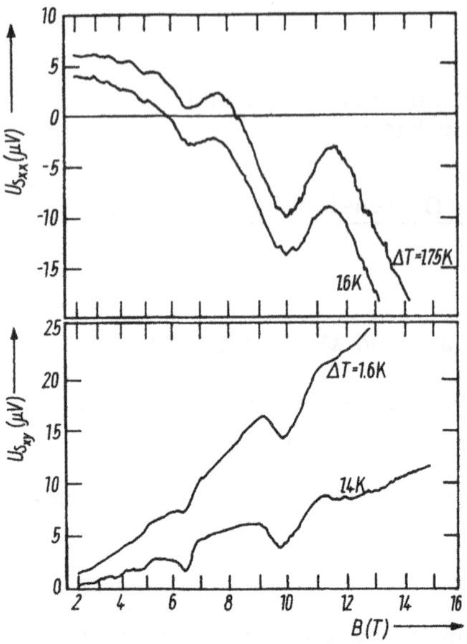

Fig. 2. $U_{S_{xx}}$ and $U_{S_{xy}}$ as a function of magnetic field B at different temperature gradients

from $\Delta T = 1$ K up to 2 K). These temperature gradients are considerably higher than those used by Obloh et al. /1/ for heterostructures.

Adding a magnetic field perpendicular to the plane of grain boundary the voltages $U_{S_{xx}}$ between contacts 1 and 2 and $U_{S_{xy}}$ between contacts 2 and 3 can be measured in dependence on magnetic field $\vec{B}$. Typical measured curves for field values up to B = 15 T are shown in Fig. 2. The thermopower components S_{xx} and S_{xy} followed (see Fig. 3), respectively, from voltages $U_{S_{xx}}$ and $U_{S_{xy}}$ measured and from the temperature difference $\Delta T = T_2 - T_1$ as well as distances between contacts (in the case of S_{xy}). The tensor of thermopower $\hat{S}$ is determined completely by S_{xx} and S_{xy} with regard to $S_{yy} = S_{xx}$ and $S_{yx} = -S_{xy}$.

The tensor component S_{xx} shows oscillatory behavior as a function of 1/B (cf. Fig. 3). The oscillation period is $\Delta(1/B) = 0.05\ T^{-1}$. From this followed a carrier concentration of $n_{so} = 1.02 \times 10^{12}\ cm^{-2}$ for the lowest occupied subband as expected in correspondence with the value from Shubnikov-de Haas measurements. The theory (e.g. Středa /3/, Jonson and Girvin /4/) demands that S_{xx} is always negative or zero in the case of a nearly ideal two-dimensional electron gas. The extremal values should be $S_{xx}^{max} = -60/(m + 1/2)V/K$, (m = 1, 2, 3, ...).

Our experimental values $|S_{xx}^{max}|$ are smaller than predicted theoretically. Additional experiments have to show whether this behavior is explained by properties of the grain boundary. The contribution of the bulk material surrounding the grain boundary superposes on the signal of the two-dimensional electron gas and effects that S_{xx} is not always negative. If the temperature gradient ∇T existing at the sample increases the influence of the bulk material

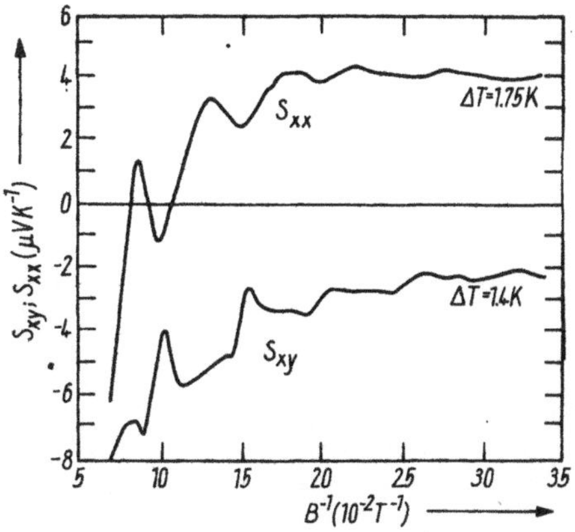

Fig. 3. The thermopower tensor components S_{xx} and S_{xy} as a function of reciprocal magnetic field 1/B at different temperature gradients

becomes relatively stronger. There are different reasons that S_{xx} is not zero in the regions between the minima, e.g. the occupation of several subbands in the InSb inversion layers. But Kraak et al. /5/ found recently that the component of magnetoresistance ρ_{xx} is not zero in the regions of Hall plateaus if one subband is occupied only.

The experimental results of the magnetic field dependence of the tensor component S_{xy} agree qualitatively with theoretical predictions. It becomes obvious from Fig. 3 that there is a superposition of the signal due to two-dimensional electron gas and the contribution of the bulk material surrounding the grain boundary.

Reducing the bath temperature below 4.2 K results in a decrease of the absolute values of S_{xx} and S_{xy}. The reasons of this phenomenon could be similar to those (mentioned above) producing too small signal values.

The experiments have been performed in the International Laboratory of High Magnetic Fields and Low Temperatures in Wrocław (Poland).

The authors would like to thank the Academy of Science of the GDR and the coworkers of the International Laboratory for support in these experiments.

References

/1/ H. OBLOH, K. VON KLITZING, and K. PLOOG, Surface Sci. 142, 236 (1984).

/2/ G. WORM, Dissertation A, Humboldt-Universität zu Berlin, 1984.

/3/ P. STŘEDA, phys. stat. sol. (b) 125, 849 (1984).

/4/ M. JONSON and S.M. GIRVIN, Phys. Rev. B 29, 1939 (1984).

/5/ W. KRAAK, G. NACHTWEI, and R. HERRMANN, phys. stat. sol. (b) 133, (1986).

(Received December 2, 1985)

phys. stat. sol. (b) 133, K61 (1986)
Subject classification: 72.80; S8

Institute of Applied Physics, Kishinev[1)]

Kondo-Type Behaviour of Conductivity in Tl-Doped PbTe Single Crystals

By

K.I. ANDRONIK, V.F. BANAR, V.G. KANTSER, and A.S. SIDORENKO

Group III elements, particularly In and Tl, form resonant type impurity levels in lead chalcogenides and can considerably change physical properties of these semiconductors. As it has been mentioned more than once in literature (see references in /1/), the impurities in PbTe can be single-electron as well as double-electron ones. The available experimental data for indium testify to the existence of double-electron states /2/. As for thallium the situation is more vague.

Since the maximum interstitial lattice sites in PbTe have sizes about 0.07 nm /3/ and ionic radii given by different authors to be Tl^{3+}: 0.095 to 0.105 nm and Tl^{+}: 0.136 to 0.149 nm /4/, it is accepted that the thallium impurity should substitute lead in the cation sublattice. If thallium atoms are in the charge state Tl^{2+} (as Pb^{2+}) they have one s-electron in the outer shell. The charge states Tl^{+} and Tl^{3+} are also possible; in these cases the outer shells of ions are filled and have more convenient configuration from the viewpoint of energy.

We have performed galvanomagnetic measurements with Tl-doped PbTe single crystals to investigate the thallium charge states in lead telluride.

The samples were prepared in the following way. The TlTe polycrystal had been synthesized in hydrogen atmosphere. Since pure Tl oxidizes in the air quickly and greatly, it is the use of thallium telluride that makes preparation of initial mixtures considerably easier. The polycrystal $(PbTe)_{1-x}(TlTe)_x$ was synthesized also in H_2 atmosphere in a quartz tube, and then a single crystal was grown in the same tube by the method of zone sublimation. The main advantage of this method is that the single crystal composition (except the small initial section of growth) is strictly conditioned by the charge composition. The rectangular parallelepipeds $1.5 \times 3 \times 6$ mm^3 oriented along {100} were cut from the obtained ingot. The faces were mechanically polished by diamond pastes.

1) ul. Grosul 5, 277028 Kishinev, USSR.

Sometimes we succeeded in cleaving a face of a crystal and in these cases mechanical polishing was not performed. Besides a number of samples were chemically etched (95 vol% HBr + 5 vol% Br_2) before making the contacts, but we have not observed any difference in results with respect to the unetched samples. No additional annealing was performed. We tested a few methods of preparation of ohmic contacts. The best results were obtained by deposition of thin gold or platinum layers on the corresponding parts of the sample faces using $HAuCl_4 \cdot 4H_2O$ or $H_2PtCl_6 \cdot 6H_2O$ with subsequent soldering of copper wires by a soft solder.

Investigations of the temperature dependence of resistivity down to liquid-helium temperature showed its anomalous character which is not typical for the classical scattering mechanism. The dependences $\varrho(T)$ (Fig. 1) are characterized by minima in the low temperature region, and their positions correlate with the concentration of the introduced thallium impurity (Table 1). The resistivity is typical for strongly degenerate semiconductors in the temperature region $T_{min} < T$, but when $T < T_{min}$ $\varrho(T)$ begins to rise. We could not identify this behaviour in analytical form because the temperature range $4.2\ K < T < T_{min}$ is rather narrow, although the low-temperature rising seems to be logarithmic with a small slope. The samples with x = 0.75 at% did not show the minimum, but the electrical resistivity practically did not change when $T < 14$ K. Since the hole density in these samples is comparatively high (Table 1), the absence of minima is possibly connected with the inferior quality of the ingot.

Samples of the composition with x = 0.005 showed negative magnetoresistance at helium temperatures. Despite that, investigated samples of other compositions did not show a negative magnetoresistance, the dependences $\Delta\varrho$ versus magnetic field differ in type from the classical ones. Fig. 2 shows the dependence $\Delta\varrho(B)/\varrho_0$ versus magnetic field induction for one of the samples with x = 0.005.

We have also measured field dependences of the Hall coefficient. But one could not detect any peculiarity in the low magnetic field region because of the Hall signal weakness.

Earlier such anomalies in the behavior of kinetic coefficients were observed in many semiconductors, doped by impurities /5, 6/. In this note they are presented for the first time for lead telluride doped by thallium.

Table 1

Tl doping	x = 0.005	0.0075	0.01	0.0125
T_{min}(K)	8	not observed	12	15
$p_{4.2}$(10^{18} cm^{-3})	8	67	43	70

Nowadays several mechanisms are known, that can lead to anomalous temperature and field dependences of the resistivity: quantum corrections due to interference of electron-electron and electron-phonon interactions /7/, including spin-dependent scattering mechanisms of this type /8/; resonant scattering mechanisms of this type /8/; resonant scattering /9, 10/; Kondo-type mechanism /11/ and other ones, connected with spin-flip in carrier scattering /12/. Quantum corrections seem to be unlikely because $\rho(T)$ minima have been observed on samples with mobilities of charge carriers $\mu \approx 10^2$ cm^2/Vs (x = 0.01, x = 0.0125) as well as on samples with mobilities $\mu \approx 10^4$ cm^2/Vs (x = 0.005). According to results of Kaidanov et al. and Chernik and Lykov (see /1/), the thallium impurity forms a resonant level in PbTe, which lies at 4.2 K $\approx$ 0.17 eV lower than the light valence band edge. According to

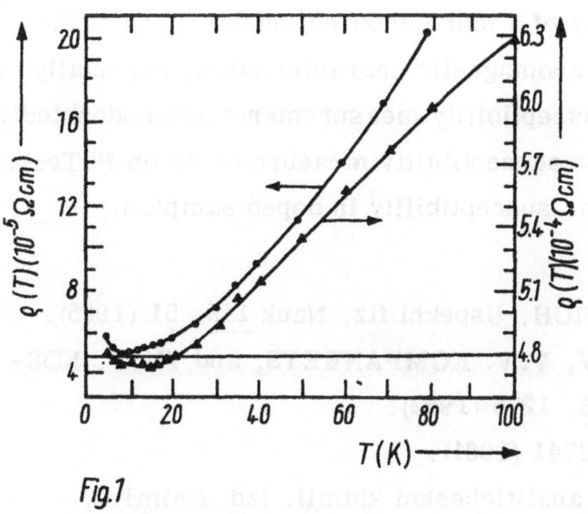

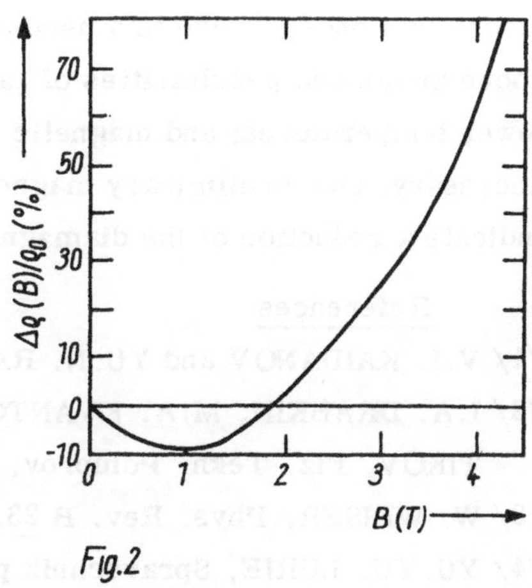

Fig. 1. Temperature dependence of the resistivity $\rho(T)$ of PbTe doped with Tl. x = ● 0.005, ▲ 0.0125

Fig. 2. The magnetoresistance $\Delta\rho/\rho_0$ versus magnetic field for PbTe:Tl with x = 0.005

/13/ in epitaxial films the impurity level of Tl lies somewhat nearer to the valence band top, $E_A \approx 0.09$ eV. With increasing temperature the level shifts away from the valence band edge /13, 14/. Computation of the Fermi energy in frames of the Volkov-Pankratov model /15/ gave $E_F = 0.09$ eV for samples with x = 0.005. Hence resonant scattering must be suppressed because of the relatively great difference between Fermi energy and impurity level.

Thus, the detected peculiarities in the temperature behaviour of resistivity and magnetoresistance of PbTe:Tl at low temperatures seem to be connected with Kondo scattering on localized magnetic moments with spin-flip. Thallium impurities not ionized with respect to the lead submatrix can be carriers of localized magnetic moments in PbTe:Tl. The results of papers /16, 17/ indirectly confirmed the existence of Tl^{2+} ions, as the efficiency coefficient of thallium was found to be less than unity (1/5 to 1/3).

The existence of localized magnetic moments can be connected with the presence of a thallium impurity resonant level, the existence of which is of no doubt /1/. There appear considerable difficulties in the theoretical treatment of definite mechanisms of localized magnetic moments formation in doped semiconductors. That is why there is up to now more or less no strict theory of these phenomena.

As for PbTe:Tl there are many experimental problems besides the theoretical ones. There is a necessity of a more detailed investigation of the above mentioned peculiarities of galvanomagnetic characteristics, especially at lower temperatures; and magnetic susceptibility measurements are undoubtedly necessary. Our preliminary magnetic susceptibility measurements on PbTe:Tl indicate a reduction of the diamagnetic susceptibility in doped samples.

References

/1/ V.I. KAIDANOV and YU.N. RAVICH, Uspekhi fiz. Nauk 145, 51 (1985).

/2/ I.A. DRABKIN, M.A. KVANTOV, V.V. KOMPANEETS, and YU.P. KOSTIKOV, Fiz. Tekh. Poluprov. 16, 1276 (1982).

/3/ W. WEISER, Phys. Rev. B 23, 2741 (1981).

/4/ YU.YU. LURIE, Spravochnik po analiticheskoi khimii, Izd. Khimiya, Moscow 1979.

/5/ YU.V. SHMARTSEV, E.F. SHENDER, and T.A. POLYANSKAYA, Fiz. Tekh. Poluprov. 4, 2311 (1970).

/6/ J. OOTUKA, S. KOBAYASHI, S. IKEHATA, W. SASAKI, and J. KONDO, Solid State Commun. 30, 169 (1979).

/7/ B.L. ALTSHULER and A.G. ARONOV, Zh. eksper. teor. Fiz. 77, 2028 (1979).

/8/ Y. ONO and J. KOSSUT, J. Phys. Soc. Japan 53, 1128 (1984).

/9/ V.I. KAIDANOV, S.A. NEMOV, YU.I. RAVICH, and A.M. ZAITSEV, Fiz. Tekh. Poluprov. 17, 1613 (1983).

/10/ I.V. KUCHEREBKO, O.A. PANKRATOV, A.E. SVISTOV, and A.P. SHOTOV, Fiz. Tekh. Poluprov. 18, 1583 (1984).

/11/ J. KONDO, Progr. theor. Phys. 32, 37 (1964).

/12/ V.F. GANTMAKHER and I.B. LEVINSON, Rasseyanie nositelei toka v metallakh i poluprovodnikakh, Izd. Nauka, Moscow 1984.

/13/ Z. FEIT, D. EGER, and A. ZEMEL, Phys. Rev. B 31, 3903 (1985).

/14/ A.N. VEIS and S.A. NEMOV, Fiz. Tekh. Poluprov. 13, 2384 (1979).

/15/ B.A. VOLKOV and O.A. PANKRATOV, Zh. eksper. teor. Fiz. 85, 1395 (1983).

/16/ K.G. GARTSMAN, T.B. ZHUKOVA, and S.A. NEMOV, Izv. Akad. Nauk SSSR, Ser. neorg. Mater. 21, 498 (1985).

/17/ P. GILLE and M. SCHENK, phys. stat. sol. (a) 84, K121 (1984).

(Received October 1, 1985)

phys. stat. sol. (b) 133, K67(1986)
Subject classification: 73.60; S8.16

Department of Physics,
University College of Science and Technology, Calcutta[1)]

On the Generalized Einstein Relation in Ultrathin Films of Ternary Chalcopyrite Semiconductors in the Presence of Crossed Electric and Magnetic Fields

By
M. MONDAL and K. P. GHATAK[2)]

The Einstein relation for the diffusivity-mobility ratio of the carriers in semiconductors (hereafter referred to as DMR) is a very useful relation since by being a thermodynamic relation, this is independent of any scattering mechanism and is, therefore, more accurate than any of the individual relations for diffusivity or mobility which are considered to be the two most widely used parameters of carrier transport in semiconductor physics. Furthermore, the connection of DMR with the velocity auto-correlation function /1/, its relation with the screening length /2/, and the different modifications of the same ratio under various physical conditions have been widely investigated in the literature /3 to 6/. Nevertheless, the DMR in ultrathin films of degenerate ternary chalcopyrite semiconductors has yet to be theoretically worked out for the more difficult case which occurs from the consideration of all types of anisotropies in the energy spectrum since the above class of semiconductors are being studied in /7, 8/ for their peculiar physical characteristics. It may be noted in this context that, in such ultrathin films where the film dimension is comparable to the de-Broglie wavelength of the carriers along the direction normal to the film (say, the z-direction) may be viewed as carrier confinement in a one-dimensional potential well leading to the quantization (known as quantum size effect) of the wave vector of the carriers along the direction of z-axis which produces a discrete energy spectrum. Besides, with the advent of liquid phase epitaxy and other techniques, such ultrathin films have been experimentally realized /9/. In the present communication, we shall derive a generalized expression of the DMR in ultrathin films of degenerate ternary chalcopyrite

---

1) 92, Acharya P.C. Road, Calcutta 700009, India.

2) Centre of Advanced Study in Radio Physics and Electronics, University of Calcutty, 1, Girish Vidyaratna Lane, Calcutta 700009, India.

semiconductors under crossed fields and study theoretically the doping dependence of it, taking degenerate n-$CdGeAs_2$ as an example.

The DMR can be expressed /3/ as

$$D/\mu = \frac{1}{e} n_o \Big/ \frac{d n_o}{d E_F} , \tag{1}$$

where e is the electron charge, n_o is the electron concentration, and E_F is the Fermi energy as measured from the edge of the conduction band. It appears, then, that the derivation of DMR using (1), requires the appropriate electron statistics which in turn is determined by the corresponding electron energy spectrum. Incidentally, using the same set of basis functions as proposed by Kildal /7/, the form of the $\vec{k}\cdot\vec{p}$ matrix can be written as

$$H = \begin{bmatrix} H_1 & H_2 \\ H_2^+ & H_1 \end{bmatrix} , \tag{2}$$

where

$$H_1 \equiv \begin{vmatrix} E_g & P_\parallel k_z & 0 & 0 \\ P_\parallel k_z & -\left(\delta + \frac{\Delta_\parallel}{3}\right) & \frac{\Delta_\perp \sqrt{2}}{3} & 0 \\ 0 & \frac{\Delta_\perp}{\sqrt{2}} & -\frac{2}{3}\Delta_\parallel & 0 \\ 0 & 0 & 0 & 0 \end{vmatrix} \qquad H_2 \equiv \begin{vmatrix} 0 & 0 & \frac{P_\perp z_{,-}}{\sqrt{2}} & \frac{P_\perp z_{,+}}{\sqrt{2}} \\ 0 & 0 & 0 & 0 \\ \frac{-P_\perp z_{,-}}{\sqrt{2}} & 0 & 0 & 0 \\ \frac{P_\perp z_{,+}}{\sqrt{2}} & 0 & 0 & 0 \end{vmatrix}$$

E_g is the bandgap, P is the momentum-matrix element along the direction of the c-axis, δ is the crystal field splitting parameter, $\Delta_\parallel$ is the spin-orbit splitting parameter along the c-axis, $\Delta_\perp$ is the same parameter in a plane perpendicular to the c-axis, $P_\perp$ is the momentum-matrix element perpendicular to the direction of the c-axis, $z_{,\pm} \equiv k_x \pm i k_y$ and $i \equiv \sqrt{-1}$. Thus, neglecting the contributions of the higher bands and the free electron energy, the diagonalization of the above matrix leads to the dispersion relation of the conduction electrons in ternary chalcopyrite semiconductors

$$U(E) = p_s^2/2M_\perp + \left[p_z^2 V(E)\right]\left[2 M_\parallel\right]^{-1} , \tag{3}$$

where
$$U(E) \equiv \left[E(1+\alpha E)\left\{(E+E_g)(E+E_g+\Delta_\parallel)+\delta(E+E_g+\frac{2}{3}\Delta_\parallel)+\frac{2}{9}\times\right.\right.$$
$$\left.\left.\times(\Delta_\parallel^2-\Delta_\perp^2)\right\}\right]\left[\zeta(E)\right]^{-1},$$

E is the energy as measured from the edge of the conduction band, $\alpha \equiv 1/E_g$,
$$\zeta(E) \equiv \left[(E+E_g)(E+E_g+\frac{2}{3}\Delta_\parallel)+\delta(E+E_g+\frac{\Delta_\parallel}{3})+\frac{1}{9}(\Delta_\parallel^2-\Delta_\perp^2)\right],\ p_s \equiv \hbar k_s ,$$

$\hbar \equiv h/2\pi$, h is the Planck constant, $M_\perp \equiv m_\perp^*(E_g+\frac{2}{3}\Delta_\perp)(E_g+\Delta_\perp)^{-1}$, $m_\perp^*$ is the transverse effective mass at the edge of the conduction band in the xy plane, $M_\parallel \equiv m_\parallel^*(E_g+\frac{2}{3}\Delta_\parallel)/(E_g+\Delta_\parallel)$, $m_\parallel^*$ is the longitudinal effective mass at the edge of the conduction band, $p_z \equiv \hbar k_z$, and $V(E) \equiv [\zeta(E)]^{-1}\left[(E+E_g)(E+E_g+\frac{2}{3}\Delta_\parallel)\right]$. Incidentally, under the substitutions $\Delta_\parallel = \Delta_\perp = \Delta$ (the isotropic spin-orbit splitting parameter), $m_\parallel^* = m_\perp^* = m^*$ (the isotropic effective electron mass at the edge of the conduction band), and $\delta = 0$, (3) assumes the form /10/

$$\frac{\hbar^2 k^2}{2m^*} = \gamma(E)\ , \qquad \gamma(E) \equiv \frac{E(E+E_g)(E+E_g+\Delta)(E_g+\frac{2}{3}\Delta)}{E_g(E_g+\Delta)(E+E_g+\frac{2}{3}\Delta)} \tag{4}$$

which is the well-known three-band Kane model. Thus, following Zawadzki and Lax /11/, the modified electron energy spectrum under crossed electric and magnetic fields can be expressed as

$$U(E) = (n+\frac{1}{2})\hbar\omega_0 + \left[\hbar^2 k_z^2/2a(E)\right] - \frac{E_0}{B}\,p_y\,\rho(E) - \frac{M_\perp E_0^2 \rho^2(E)}{2B^2}\ , \tag{5}$$

where n is the Landau quantum number, $\omega_0 \equiv eB/M_\perp$, B is the quantizing magnetic field along the z-direction, $a(E) \equiv M_\parallel/V(E)$, E_0 is the electric field along the x-direction,

$$\rho(E) \equiv U(E)\left[\{1+2\alpha E\}\{E(1+\alpha E)\}^{-1} - J(E)\{I(E)\}^{-1} + (E+E_g+\Delta_\parallel)^{-1} -\right.$$
$$\left.-(E+E_g+\frac{2}{3}\Delta_\parallel)^{-1}\right],$$
$$J(E) \equiv I(E)\left[c(E)\{1+A(E)\}^{-1} - H(E)\{1+G(E)\}^{-1}\right],$$
$$I(E) \equiv [1+A(E)]\,[1+G(E)]^{-1}\ ,$$
$$A(E) \equiv \left[(E+E_g+\Delta_\parallel)(E+E_g)\right]^{-1}\left[\frac{1}{9}(\Delta_\parallel^2-\Delta_\perp^2)+\delta(E+E_g+\frac{\Delta_\parallel}{3})\right],$$
$$G(E) \equiv \left[(E+E_g)(E+E_g+\frac{2}{3}\Delta_\parallel)\right]^{-1}\left[\delta(E+E_g+\frac{\Delta_\parallel}{3})+\frac{1}{9}(\Delta_\parallel^2-\Delta_\perp^2)\right],$$

$$c(E) \equiv A(E)\left[\left\{\delta(E+E_g+\frac{2}{3}\Delta_\parallel)+\frac{2}{9}(\Delta_\parallel^2-\Delta_\perp^2)\right\}^{-1}\delta - - (2E+2E_g+\Delta_\parallel)\left\{(E+E_g)(E+E_g+\Delta_\parallel)\right\}\right]^{-1} ,$$

and

$$H(E) \equiv G(E)\left[\left\{\delta(E+E_g+\frac{\Delta_\parallel}{3})+\frac{1}{9}(\Delta_\parallel^2-\Delta_\perp^2)\right\}^{-1}\delta - - (2E+2E_g+\frac{2}{3}\Delta_\parallel)\left\{(E+E_g)(E+E_g+\frac{2}{3}\Delta_\parallel)\right\}^{-1}\right] .$$

Using the same conditions, as used for obtaining (4) from (3), (5) may be simplified for the three-band Kane model as

$$\gamma(E) = (n+\frac{1}{2})\hbar\omega_0 + \frac{\hbar^2 k_z^2}{2m^*} - \frac{E_0}{B}\hbar k_y L(E) - \frac{m^* E_0^2 L^2(E)}{2B^2} , \quad (6)$$

where

$$\omega_0 \equiv \frac{eB}{m^*}$$

and

$$L(E) \equiv (E_g+\frac{2\Delta}{3})(E_g+\Delta)^{-1}(E+E_g+\frac{2\Delta}{3})^{-1} \times \times\left[1+2\alpha E+E(1+\alpha E)\left\{(E+E_g+\Delta)^{-1}-\left(E+E_g+\frac{2\Delta}{3}\right)^{-1}\right\}\right].$$

Besides, for $\Delta = 0$ as for the two-band Kane model (6) gets simplified to

$$E(1+\alpha E) = (n+\frac{1}{2})\hbar\omega_0 + \frac{\hbar^2 k_z^2}{2m^*} - \frac{eE_0\hbar k_y}{m^*\omega_0}(1+2\alpha E) - -\frac{1}{2}\frac{e^2E_0^2}{m^*\omega_0^2}(1+2\alpha E)^2 , \quad (7)$$

as derived for the first time by Zawadzki and Lax /11/.

In ultrathin films of semiconductors, when size quantization becomes operative, the wave vector in the direction normal to the film (i.e. along the z-direction here) gets quantized as it can be assumed that the carriers are in a 1-D potential well of width d_0. Thus the modified electron energy spectrum can be expressed, using (5), as

$$U(E) = (n+\frac{1}{2})\hbar\omega_0 + \frac{\hbar^2}{2a(E)}\left(\frac{l\pi}{d_0}\right)^2 - \frac{E_0}{B}\hbar k_y \varrho(E) - \frac{M_\perp E_0^2 \varrho^2(E)}{2B^2} , \quad (8)$$

where l (≡ 1, 2, 3, ...) is the size quantum number. Therefore, the use of (8) leads to the expression of the electron concentration in ultrathin films of degenerate ternary chalcopyrite semiconductors in the presence of crossed electric and magnetic fields as

$$n_o = \frac{eB}{\pi\hbar} \sum_{n=0}^{n_{max}} \sum_{l=1}^{l_{max}} F_{-1}(\eta_{nl}) \quad , \tag{9}$$

where $F_j(\eta_{nl})$ is the one parameter Fermi-Dirac integral /12/, $\eta_{nl} \equiv (k_B T)^{-1}(E_F - E_{nl})$, k_B is the Boltzmann constant, T is the temperature, and E_{nl} can be obtained by putting $k_y = 0$ and $E = E_{nl}$ in (8). Incidentally, under the substitutions $\Delta_\perp = \Delta_\parallel = 0$, $\delta = 0$, $\alpha = 0$, $E_o = 0$, and $m_\perp^* = m_\parallel^* = m^*$, (9) takes the form /13/

$$n_o = \frac{eB}{\pi\hbar} \sum_{n=0}^{n_{max}} \sum_{l=1}^{l_{max}} F_{-1}\left[(k_B T)^{-1}\left\{E_F - (n+\frac{1}{2})\hbar\omega_o - \frac{\hbar^2}{2m^*}\left(\frac{l\pi}{d_o}\right)^2\right\}\right]. \tag{10}$$

Thus combining (1) and (9), the generalized expression for the DMR in ultrathin films of ternary chalcopyrite semiconductors under crossed fields can be written as

$$\frac{D}{\mu} = \frac{k_B T}{e}\left[\sum_{n=0}^{n_{max}} \sum_{l=1}^{l_{max}} F_{-1}(\eta_{nl})\right]\left[\sum_{n=0}^{n_{max}} \sum_{l=1}^{l_{max}} F_{-2}(\eta_{nl})\right]^{-1} . \tag{11}$$

Besides for $\eta_{nl} \ll 0$, as for non-degenerate carrier concentrations, (11) takes the well-known form $D/\mu = k_B T/e$, as it should.

Using (9) and (11), we have computed the DMR versus n_o in ultrathin films of degenerate n-$CdGeAs_2$ at 4.2 K in the presence of crossed electric and magnetic fields as shown in Fig. 1 where the same dependence for $\delta = 0$ is also shown for the purpose of comparison. The parameter values used in these computations are as follows /7, 8/: $m_\parallel^* = 0.039\ m_o$, $m_\perp^* = 0.030\ m_o$, $\Delta_\parallel = 0.034$ eV, $\Delta_\perp = 0.36$ eV, $\delta = -0.21$ eV, $Eg = 0.57$ eV, $E_o = 10^3$ V/M, $d_o = 40$ nm, T = 4.2 K, and B = 2.3 T. It appears from the figure that the DMR increases with increasing electron concentration as expected for degenerate semiconductors and oscillates with it. Besides, the crystal field splitting parameter effectively enhances the values of the ratio in the whole range of the concentrations considered. It is also apparent that the value of the DMR becomes several orders of magnitude larger in the ultrathin films of ternary chalcopyrite semiconductors under crossed fields than in the bulk specimen of the same semiconductor as derived elsewhere /6/. Incidentally, the effects of the electron spin, the collision broadening, the formation of band tails, and the many-body effects have not been considered in the present derivation. However, the basic qualitative features of this simplified analysis will not be altered to a significant extent

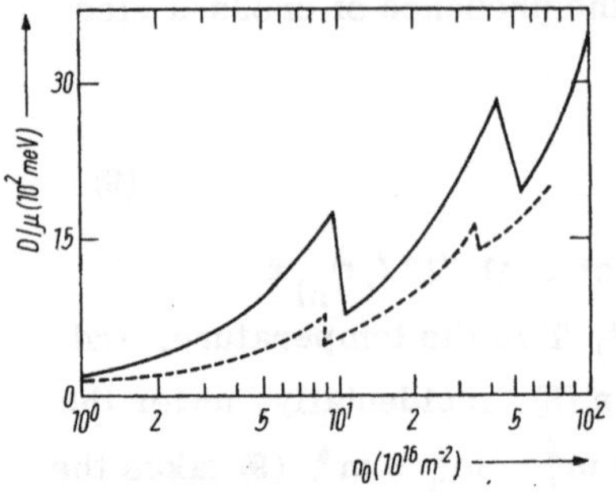

Fig. 1. Doping dependence of the DMR in ultrathin films of degenerate n-$CdGeAs_2$ at 4.2 K in the presence of crossed electric and magnetic fields. The dashed plot corresponds to $\delta = 0$

even after the necessary modifications. It may finally be noted that the conclusions made here would be of particular significance in view of the fact that the switching speed, the thermal noise, and the performance at the device terminals of the devices made of highly degenerate semiconductors can be related to the DMR /14 to 16/.

References

/1/ G.G. EMCH, J. math. Phys. 14, 1775 (1973).

/2/ R. KUBO, J. Phys. Soc. Japan 12, 537 (1957).

/3/ B.R. NAG, Electron Transport in Compound Semiconductors, Springer Series Solid State Sci., Vol. 11, Springer-Verlag, Heidelberg 1980 (p. 219).

/4/ P.T. LANDSBERG, Europ. J. Phys. 2, 213 (1981).

/5/ S.T.H. ABIDI and S.N. MOHAMMAD, Solid State Electronics 27, 1153 (1984).

/6/ M. MONDAL and K.P. GHATAK, Physica Scripta 30, 217 (1984).

/7/ H. KILDAL, Phys. Rev. B 10, 5082 (1974).

/8/ J.L. SHAY and J.H. WERNICK, Ternary Chalcopyrite Semiconductors: Growth Properties and Applications, Pergamon Press, London 1975.

/9/ L.E. VOROLEV, Electronic Properties of Ultrathin Films of Degenerate Semiconductors, Nauka, Leningrad/Moscow 1982 (p. 80) (in Russian).

/10/ V.A. VILLKOTSKII, D.S. DOMANEVSKII, R.D. KAKANOKOV, and V.V. KRASOVSKII, Soviet Phys. - Semicond. 13, 553 (1979).

/11/ W. ZAWADZKI and B. LAX, Phys. Rev. Letters 16, 1001 (1966).

/12/ J.S. BLAKEMORE, Semiconductor Statistics, Pergamon Press, London 1962 (p. 79).

/13/ M. MONDAL and K.P. GHATAK, phys. stat. sol. (b) 126, K41 (1984).

/14/ T.J. KAMINS and R.S. MULLER, Solid State Electronics 10, 423 (1967).

/15/ B.R. NAG and A.N. CHAKRAVARTI, phys. stat. sol. (a) 67, K113 (1981).

/16/ G.D. HATCHEL and P.E. RUCHLI, IEEE Trans. Electron Devices 15, 437 (1968).

(Received July 1, 1985)

phys. stat. sol. (b) 133, K73 (1986)
Subject classification: 72.15; S1.2

Department of Physics, Indian Institute of Technology, Delhi[1)]

The Width of the Scattering Resonance and the Electrical Transport Properties in Partially Filled d-Band Transition Metals

By
SADHANA GARG, H.C. GUPTA, and B.B. TRIPATHI

Recently, transport properties of liquid transition metals and alloys are actively investigated, both from experimental and theoretical view point /1 to 4/. Among these properties the electrical resistivity of the partially filled d-band transition metals and their alloys is of special interest for the theoreticians because till date none of the theories were really able to explain it. One of the reasons is the fact, that these transition metals have a very high resistivity at their melting point.

The electrical resistivity of simple liquid metals can be described by Ziman's theory /5/. This formulation was extended to liquid transition metals and their alloys /6, 7/, where d-electrons are treated as resonance states scattered by muffin-tin potentials in the conduction band. Evans et al. /6/ assumed that the d-phase shift dominates at the Fermi energy in transition metals. Therefore they have expressed the resistivity, because of the resonant nature and large magnitude of the l = 2 phase shift as

$$\varrho = \frac{30\pi^3\hbar^3}{m e^2 \Omega_o k_F^2 E_F} \sin^2\eta_2(E_F) a(2k_F) . \tag{1}$$

All the terms used in the above expression have the same meaning as given by Evans et al. /6/. The resistivity, ϱ, depends upon E_d and width Γ of the scattering resonance, because

$$\tan\eta_2 = \frac{\Gamma}{E_d - E} , \tag{2}$$

$$\varrho = \frac{30\pi^3\hbar^3}{m e^2 \Omega_o k_F^2 E_F} a(2k_F) \frac{\Gamma^2}{\Gamma^2 + (E_d - E_F)^2} . \tag{3}$$

The term involving Γ in the above expression corresponds to $N(E_F)$ in the Lloyd single site approximation to the density of d-states /8/. Further it has

1) Hauz Khas, New Delhi 110016, India.

been observed that the density of states per atom may be a reasonably universal function of the electron to atom ratio within a given structure /9/. Therefore, in the present treatment, we have considered the $N(E_F)$ dependence upon the valency.

In order to investigate this dependence, we find that the choice of the parameters Z, E_F, E_d, and Γ used is very important in calculating the electrical transport properties. We have used an approach in which we have initially estimated Γ, using the resistivity as input parameter for different valency of partially filled d-band transition metals such as Ni (Z = 2), Ti (Z = 4), V(Z = 5), and W(Z = 6), the numerical calculations having been made using expression (3) where $k_F^3 = 3\pi^2 Z/\Omega_0$ and $E_F = k_F^2 \hbar^2/2m$.

The liquid structure factor $a(2k_F)$ has been taken from the neutron studies of liquid metals by Waseda /10/ and E_d is taken from the work of Harrison /9/. The calculated values of Γ as shown in Table 1, suggest that Γ is the multiple of a constant which has a value equal to $\pi/8$. Hence a relation for Γ is proposed as

$$\Gamma = 2^Z \pi/8 \text{ eV} , \qquad (4)$$

where Z is the valency. The width of the scattering resonance (Γ) obtained from (4) suggests in transition metals Γ to be a rather universal function of valency. With this expression of Γ, the electrical resistivity is calculated for

Table 1

Data used for calculating the width of the scattering resonance Γ

system	Ni	Ti	V	W
Z	2	4	5	6
k_F (nm^{-1})	17.6	18.7	22.0	22.4
E_F (eV)	11.76	13.34	18.44	19.16
$a(2k_F)$	1.18	0.85	0.96	0.81*)
E_d (eV)	6.91	7.76	8.13	9.45
ρ_{exp} ($\mu\Omega$ cm)	85	167	135	135
Γ_{cal} (eV)	1.59	6.40	12.46	25.45

*) Using the $a(2k_F)$ exp data of Cr scaled to the W value of $2k_F$, because the structure factor for W has not yet been measured.

Table 2

Results of resistivity calculation for liquid transition metals

system	Fe	Ni	Pd	Co	Ti	V	W	Mo	Cr
Z	2	2	2	2	4	5	6	6	6
k_F (nm^{-1})	17.1	17.6	15.9	17.5	18.7	22.0	22.4	22.5	24.6
E_F (eV)	11.16	11.76	9.64	11.64	13.34	18.44	19.16	19.28	23.06
E_d (eV)[a]	7.64	6.91	4.52	7.36	7.76	8.13	9.45	7.12	8.01
ρ_{cal} ($\mu\Omega$cm)	137	82	84	91	164	135	135	124[*]	137
ρ_{exp} ($\mu\Omega$cm)	140[b]	85[b]	83[b]	100[b]	167[c]	135[c]	135[c]	97[c]	145[d]
ρ_{Ziman} ($\mu\Omega$cm)[e]	582	60	-	328	689	887	-	-	891
ρ_{Molt} ($\mu\Omega$cm)[e]	983	704	-	900	834	794	-	-	818
ρ_{Hirata} ($\mu\Omega$cm)[f]	182	55	58	83	356	281	-	-	120

a) Data from /9/, b) Results from /10/, c) Results from /3/, d) Results from /1/, e) Results from /2/, f) Results from /4/.

*) The disagreement between theory and experiment can be attributed to the value of $a(2k_F)$ taken in the present calculation from the table of Cr as given by Waseda /10/ as the value of $a(2k_F)$ for Mo was not experimentally available.

different partially filled d-band transition metals using (3), the results are shown in Table 2. Gallagher and Greig /2/ have discussed the Ziman formalism and found that it is difficult to see how any reasonable choice of E_F, k_F, and Z could bring the values of Ziman's formalism in line with experimental results, where one is tempted to use Z>1 for the transition metals. The present investigations deal with the actual valency of Z for most of the transition metals, thereby modifying the value of E_F and k_F as compared to previous calculations /4/, with Z = 1, but we observe that Z = 1 does not give a good agreement between theory and experiment. The values of E_d are the same as used by Gallagher and Greig /2/ and now if we see Table 2 for the excellent agreement obtained between theory and experiment, the reason only is that the expression for the value of the width of the scattering resonance Γ, which we have proposed as (4), is a universal function of valency.

In conclusion, we can say that the present formalism for Γ gives excellent agreement in resistivity for most of the partially filled d-band transition metals.

One of us (S.G.) is grateful to CSIR, India, for financial assistance.

References

/1/ J.B. VAN ZYTVELD, J. non-crystall. Solids 61/62, 1085 (1984).

/2/ B.L. GALLAGHER and D. GREIG, J. Phys. F 12, 1721 (1982).

/3/ V. SEYDEL and W. FUCKE, J. Phys. F 10, L 203 (1980).

/4/ K. HIRATA, Y. WASEDA, A. JAIN, and R. SHRIVASTAVA, J. Phys. 7, 419 (1977).

/5/ J.M. ZIMAN, Electrons and Phonons, Clarendon Press, 1960 (Chap. 5).

/6/ R. EVANS, D.A. GREENWOOD, and P. LLOYD, Phys. Letters A 35, 57 (1971).

/7/ O. DREIRACH, P. EVANS, H.J. GUNTHERODT, and H.U. KUNZI, J. Phys. F 2, 709 (1972).

/8/ P. LLOYD, Proc. Phys. Soc. 90, 207 (1967).

/9/ W.A. HARRISON, Electronic Structure and the Properties of Solids, Freeman, San Francisco 1980 (p. 491).

/10/ Y. WASEDA, The Structure of Non-Crystalline Materials, McGraw-Hill, 1980.

/11/ N.F. MOTT, Phil. Mag. 26, 249 (1972).

(Received October 10, 1985)

Pre-Printed Titles

Papers to be published
in "physica status solidi (b) - basic research"
Vol. 133, No. 2

Original Papers

TRAN THOAI DUY BAO, Ho Chi Minh City
Linear Density Response at the Metal Surface in the Hydrodynamic Approximation

I.I. SASOVSKAYA and V.P. KORABEL, Sverdlovsk
Optical Properties of α- and β-CuZn Brasses in the Region of Quantum Absorption

V.I. SHULGA, Moscow
Surface Semichanneling of Molecular Ions

J.Q. SHEN, C.W. LUNG, and K.L. WANG, Hefei and Trieste
Dislocation Core Models and Their Positron Annihilation Effects

S.O. CHOLAKH, V.A. PUSTOVAROV, and B.V. SHULGIN, Sverdlovsk
The Effects of Electron Excitation Energy Transfer to the Impurity Centre in LiH Crystals

BACH THANH CONG, Hanoi
The Functional Integral Method in the Theory of Ferromagnetism for a Magnetic System Containing Itinerant Electrons Interacting with Localized Spins

R. LENK, Karl-Marx-Stadt
The Density Matrix of an Infinite Muffin-Tin System and Mutually Avoiding Scattering Paths

H.C. GUPTA, G.S.N. REDDY, and B.B. ṬRIPATHI, New Delhi
Phonon Density of States in Amorphous $Si_{0.5}Ge_{0.5}$

A.A. MANENKOV, G.N. MIKHAILOVA, S.YU. SOKOLOV, and S.G. TIKHODEEV, Moscow
Phonon Wind Generation in Ge by CO_2 Laser Induced Heating of Electron-Hole Drops

R. LACROIX, Genève
The Ligand Field on d^n Ions in a Strained Corundum Crystal

ELEONORA TRIF and A. NICULA, Cluj-Napoca
EPR of Nd^{3+} Ions in Y-Type Zeolites

I.I. GEGUSIN, V.N. DATSYUK, A.A. NOVAKOVICH, L.A. BUGAEV, and R.V. VEDRINSKII, Rostov-on-Don
Multiple Scattering Approach to the XANES Theory of Alkali Halides Crystals (III)

P.V. SMITH and J.E. SZYMANSKI, Newcastle, Shortland and Heslington, York
Model Calculations of the LUC-CNDO Modulating Function

U. PIETSCH, Leipzig
Bond Charges and Electronic Charge Transfer in Ternary Semiconductors

H. BURGHARDT, T. FRAUENHEIM, and C. HAMANN, Karl-Marx-Stadt
The Influence of an Electric Field on the Mobility in Semiconductors

D.S. DOMANEVSKII, V.V. KRASOVSKII, M.V. PROKOPENYA, V.A. VILKOTSKII, and S.V. ZHOKHOVETS, Minsk
The Origin of the Minority Impurity States in Heavily Doped Semiconductors

ARUMA DHOBLE and M.P. VERMA, Ujjain
Thermodynamic Analysis of the Anderson-Grüneisen Parameters

L.I. LEVIN, Sverdlovsk
Semiphenomenological Theory of the Gd^{3+} S-State Splitting in Low-Symmetry Crystals

H. CONZELMANN, A. HANGLEITER, and J. WEBER, Stuttgart
Thallium-Related Isoelectronic Bound Excitons in Silicon

J. MYCIELSKI and A.M. WITOWSKI, Warsaw
Effect of Chemical and Spin Disorder on Shallow Impurity States in Semiconductor Alloys

G.V. IONOVA, Y.N. KOSTRUBOV, and A.V. NIKOLAEV, Moscow
Charge Ordering in Intermediate-Band Crystals

A. ROJEK, C. SUŁKOWSKI, A. ZYGMUNT, and G. KOZŁOWSKI, Wrocław
Low-Temperature Properties of Antiferromagnetic $GdRh_xSn_y$ Compound

T. STOBIECKI and M. PRZYBYLSKI, Cracow
Spontaneous Hall Effect and Magnetization of Fe-(B, Zr) and Co-Mo Amorphous Films

K. ŚWIDERCZAK, A. ZAGÓRSKI, and J. KRÓL, Warsaw
Amorphous Ferromagnets with Two Magnetic Components

P.K. KATANA, Kishinev
Multiphoton Interband Absorption of Light in Semiconductors in External Parallel Electric and Magnetic Fields

M. KOTERA, K. MURATA, and K. NAGAMI, Osaka
Monte-Carlo Studies of Energy Straggling of Electrons in Solids (II)

V.E. MASHCHENKO, V.F. KHARSIK, and S.V. BREZHNEVA, Moscow
Secondary Emission of Excitons in CuCl Polycrystals

J. SINKKONEN and S. ERÄNEN, Espoo
AC Transport in Short n^+nn^+ Layers

A.J. PAL and T.N. MISRA, Calcutta
Delayed Luminescence of Organic Mixed Crystals

T.M.A. KHAJIL and M. TOMAK, Ankara
The Resistivity of Liquid Metals with Finite Mean-Free Path

S.I. GUBAREV, Chernogolovka
Valence Band Splittings in Wurtzite-Type Semimagnetic Semiconductors

S.M. PILLAI and C.P.G. VALLABHAN, Karaikudi and Cochin
A Study on the Electroluminescence Spectrum of CaS:Sm and CaS:Dy

F. ENGELKE, D. MICHEL, W. WINDSCH, and J. PETERSSON, Leipzig and Saarbrücken
^{15}N-NMR Investigations of Dynamical Processes near the Ferroelectric Phase Transition of TSCC

S.M. RYABCHENKO and YU.G. SEMENOV, Kiev
EPR and Spin-Flip Raman Scattering in the n-Type Magnetically Mixed Semiconductors

G.V. SMIRNOV, YU.V. SHVYDKO, U. VAN BÜRCK, and R.L. MÖSSBAUER, Moscow and München
Experimental Study of the Influence of Lattice Vibrations on the Suppression Effect at Room Temperature

V. JANIŠ, Prague
A Self-Consistent Approximation for the Random-Field Ising Model

B.A. OLI and J.S. ONONIWU, Enugu
Effect of Non-Locality in the Calculation on Liquid Metal Resistivity of Transition Metals

A. KLAMT and H. TEICHLER, Stuttgart
Quantum-Mechanical Calculations for Hydrogen in Niobium and Tantalum (I)

A.N. BOTVICH, V.F. SHABANOV, A.N. VTYURIN, and T.A. POZDNAKOVA, Krasnoyarsk
Photon-Phonon Interaction Influence on the External Vibration Spectra of Molecular Crystals

JIAN-WEI HE and P.J. MØLLER, Copenhagen
On the Energy Dependence of the Very Low Energy Reflectance from Clean and Manganese-Doped MgO(100) Surfaces

R. KEIPER and O. ZIEP, Berlin
Theory of Quantum Hall Effect

Papers to be published
in "physica status solidi (a) - applied research"
Vol. 93, No. 1

Review Article

K.V.S. RAMA RAO and S. RAMAPRABHU, Madras
Nuclear Quadrupole Resonance Studies on Inorganic Solids

Original Papers

K.H. HEINIG and H.-D. GEILER, Dresden and Jena
Phenomenological Theory of Explosive Solid-Phase Crystallization of Amorphous Silicon (II)

A.B. AHMED and R.K. GARTIA, Imphal
Trapping Levels in $KAlSi_3O_8$

V.N. ABRAMOV, A.P. TYUTNEV, E.D. POZHIDAEV, V.S. SAENKO, and A.V. VANNIKOV, Moscow
Charge Carrier Mobility in Polyolefins

H. VOGT, R. ULM, P. WÜRFEL, and W. RUPPEL, Karlsruhe
Anomalous Charge Storage in $NaNO_2$ Layers

K. IWAUCHI and Y. IKEDA, Kyoto
Dielectric Properties of Hexagonal Ferrites

K. BIEDRZYCKI, Wrocław
New Aspects of Electron Emission from Virgin TGS Single Crystals

N. SAITO and N. TANAKA, Hamamatsu
Preparation and Properties of Magnetron Sputtered Fluorinated and Hydrogenated Amorphous Si-C Films

F. RETTENMEIER and H. KRONMÜLLER, Stuttgart
Reversible and Irreversible Structural Relaxations in Soft Ferromagnetic Amorphous Alloys (I)

H.-J. FITTING and A. VON CZARNOWSKI, Rostock
Electron Breakdown in Polar Insulating and Semiconducting Layers

J. DUTKIEWICZ, J. SALAWA, H. LÖFFLER, and E. KORNGIEBEL, Cracow and Halle
The Effect of Small Additions of Si on the Structure of an Al-2 at% Zn-1 at% Mg Mother Alloy

O. SIMMICH, Halle
Diffusion-Controlled Phase Transformations in Finite Regions

G. HÜBNER and H. LÖFFLER, Halle
The Growth of Guinier-Preston Zones in an Al-12 at% Zn Alloy during the Ageing at RT in Dependence on the Homogeneization Temperature

H. REUTHER, Rossendorf
Conversion Electron Mössbauer Spectroscopy of Nitrogen and Phosphorus Implanted Iron

R.V. GOPALA RAO and RATNA DAS, Calcutta
Structure Factor Studies of Like Charges in the One-Component Plasma

H.S. TAN and S.C. NG, Singapore
Thickness Dependence of Threshold Energy Density for Pulsed-Laser Melting of Si

V.K. YANOVSKII and V.I. VORONKOVA, Moscow
Ferroelectric Phase Transitions and Properties of Crystals of the $KTiOPO_4$ Family

NGUYEN VAN DEN, A. HEINRICH, K. KLOSTERMANN, and G. SOBE, Dresden
IR Absorption in CrSiO Thin Films

K.Z. BOTROS and S.S. SHEININ, Kuweit and Edmonton, Alberta
On Identifying Crystal Surfaces in Electron Microscope Images of Stacking Faults

M.E. ABU-ZEID, A.E. RAKHSHANI, A.A. AL-JASSAR, and Y.A. YOUSSEF, Kuwait
Determination of the Thickness and Refractive Index of Cu_2O Thin Film Using Thermal and Optical Interferometry

YU-JIN CHANG and XIANG-RONG QUIAN, Shanghai
A High Resolution Microscopic Study on Nd-Fe-B Permanent Magnetic Alloys

E.V. ZAROCHENTSEV, S.M. OREL, and A.YU. YAKOVETS, Donetsk
Elastic Contants of a Stressed Crystal (IV)

S.A. ABO-NAMOUS and R.W. FANE, Birmingham

Post-Annealing Effects on Doping Efficiency in Ne-Sputtered a-Si

I.I. FISHCHUK, Kiev

The AC Conductivity and Hall Effect in Inhomogeneous Semiconductors

G. SORGE, H. MAACK, and L.A. SHUVALOV, Halle and Moscow

Low-Frequency Dielectric Properties of Rb_2ZnCl_4

H. LASS, J. PEBLER, M. PEKALA, and A. POŁACZEK, Marburg and Warsaw

Magnetic and Mössbauer Effect Study of the Cluster Distribution Function in the Spin-Glass Phase

P. MËTSCH, F.H.M. SPIT, and H. BAKKER, Amsterdam

Computer Simulation of Simultaneous Bulk, Grain-Boundary, and Surface Diffusion

P. DORDOR, S. JOIRET, J.P. DOUMERC, J.C. LAUNAY, J. CLAVERIE, and P. HAGENMÜLLER, Talence

Electrical Properties of Lead-Doped $LaCoO_3$ Single Crystals

A. ZENTKO, P. DUHAJ, M. TIMKO, and V. KAVEČANSKÝ, Košice and Bratislava

The Effect of Neutron Irradiation on the Electrical Resistivity of Amorphous $Co_{47}Ni_{25}B_{18}Si_{10}$ Alloys

J. SZCZYRBOWSKI, K. SCHMALZBAUER, and H. HOFFMANN, Cracow and Regensburg

Optical Properties and Dynamic Current Density for Polycrystalline Metallic Films

O.A. ALEKSANDROVA, M.I. KAMCHATKA, M.S. MIROPOLSKII, and V.V. PASSYNKOV, Leningrad

Diffusion of Native Defects in $(Pb_{1-x}Sn_x)_{1-y}Te_y$ during Liquid Phase Epitaxy

N.V. MOSEEV, V.G. CHUDINOV, B.N. GOSHCHITSKII, and V.I. PROTASOV, Sverdlovsk and Ustinov

Molecular Dynamics Study of Atomic Configuration of Point Defects in Intermetallic Compound Mo_3Si

V.E. BORISENKO and V.P. PARKHUTIK, Minsk

RBS Study of Transient Thermal and Anodic Oxidation of Tantalum Films

I. PETR, Prague

Activation Energy of Trapping Centres in SiO_2 Gate Insulator

R. JAKUBAS, Z. CZAPLA, Z. GALEWSKI, L. SOBCZYK, O.J. ZOGAŁ, and T. LIS, Wrocław
Structure and Phase Transition in $(CH_3NH_3)_3Sb_2Cl_9$

E. SCHEID and P. CHENEVIER, Grenoble
1D Simulation of Oxidation-Enhanced and Oxidation-Retarded Diffusion in Silicon and Validity of the Physical Model

D. WOLF and S. KEMMLER-SACK, Tübingen
The Luminesence of Nd^{3+} in the Polymorphic Perovskites $Ba_6B_2W_3O_{18}$

J.P. ZHANG, H.Q. YE, K.H. KUO, and S. AMELINCKX, Antwerp
A High Resolution Electron Microscopy Study of the Domain Structure in Ni_3Nb (II)

O. GOEDE, W. HEIMBOLDT, and F. SITTEL, Berlin
Photoacoustic Spectroscopy of Isovalent Impurities in II-VI-Semiconductors

G.D. GUSEINOV, V.A. ALIEV, and E.F. BAGIRZADE, Baku
Recombination Levels in Non-Stoichiometric $TlGaSe_2$

T.N. MANONTOVA, KH.K. MUKANOV, G.Z. VINOGRADOVA, N.G. MAISASHVILI, and I.Z. BABIEVSKAYA, Leningrad and Moscow
A Study of the Optical Properties and Photoluminescence of Glasses in the Ge-P-S System

C. ZALDO and F. AGULLO-LOPEZ, Madrid
Optical Monitoring of the Precipitation-Induced Hardening in NaCl:Sr and NaCl:Ca

E. WENDLER, W. WESCH, and G. GÖTZ, Jena
Near-Edge Optical Absorption Behaviour in Weakly Damaged Ion-Implanted GaAs

U. PITTERMANN and SIGRID RIPPER, Darmstadt
Crystallization Behaviour of Thin NiP Alloy Films Prepared by Electroless or Galvanic Deposition

W. FERTIG and H. PAGNIA, Darmstadt
Photoemission of Electrons from Interface States in Silicon MOS Structures

V.I. MALYUTIN, V.E. OSUKHOVSKII, A.A. IVANOV, L.A. CHEBOTKEVICH, I.V. LOBOV, and YU.D. VOROBIEV, Moscow
The Effect of Volume and Surface Inhomogeneities on the Coercivity of Thin Magnetic Films

Substance Classification (according to physica status solidi)

S1. Transition metals and their alloys and compounds (for oxides see S.10)
- S1.1. Fe
- S1.2. Other 3d metals
- S1.3. Noble metals
- S1.4. Lanthanides
- S1.5. Actinides
- S1.61. Carbides, silicides, germanides
- S1.62. Nitrides, phosphides, arsenides
- S1.63. Sulfides, selenides, tellurides

S2. Alkali metals and their alloys and compounds (for alkali halides see S9.11)

S3. Alkaline earth metals and their alloys and compounds (for alkaline earth halides, see S9.12)

S4. Other metals and their alloys

S5. Nonmetallic elements
- S5.11. Silicon
- S5.12. Germanium
- S5.2. Solidified Gases

S6. Group IV compounds (for transition metal compounds, see S1.61)

S7. Group V compounds (for transition metal compounds, see S1.62)
- S7.1. III–V compounds
- S7.11. Phosphides
- S7.12. Arsenides
- S7.13. Antimonides
- S7.15. Mixed crystals
- S7.16. Higher compounds

S8. Chalcogenides (for transition metal compounds, see S1.63)
- S8.1. II–VI compounds
- S8.11. Sulfides
- S8.12. Selenides
- S8.13. Tellurides
- S8.15. Mixed crystals
- S8.16. Higher compounds

S9. Halides
- S9.11. Alkali halides
- S9.12. Alkaline earth halides
- S9.15. Mixed crystals
- S9.16. Higher compounds

S10. Oxides
- S10.1. Simple oxides
- S10.15. Mixed crystals

S11. Other higher compounds
- S11.1. Salts of oxy-acids
- S11.2. Spinels, garnets, ferrites

S12. Organic compounds
- S12.1. Quasi-one-dimensional systems

Attention

The *Classification Scheme* on cover two is reproduced and used with permission of International Council for Scientific and Technical Information (ICSTI).

Because of the lack of space only the main points of the Classification Scheme are printed here. The *detailed subdivided scheme* is to be found in the January issues of each annual set and in the Index Volumes from 1986 on.

The *Subject Classification* given at the head of each paper begins with the main subject matter of the paper, and this is followed, if necessary, by figures referring to subsidiary subject matter and the type of solids investigated.

The papers published in this issue are listed in the order of their subject classification in the *Systematic List* at the end of the table of contents.

32913

Manuscripts and letters for physica status solidi (b) — basic research and physica status solidi (a) — applied research

should be sent to

DDR-1086 Berlin, Leipziger Straße 3—4, P. B. 1233

or

Kunst und Wissen, Erich Bieber OHG, Postfach 330104, D-1000 Berlin 33.

The object of "physica status solidi (b)" and "physica status solidi (a)" is to publish rapidly important and topical results concerning fundamental research in the field of solid state physics and aspects of application and experimental techniques of fundamental solid state research, respectively. The issues of "physica status solidi (b)" and "physica status solidi (a)" will be published on the 1st and the 16th day of each month, respectively. Either series will contain the following categories:

1. Review articles written by arrangement with the editors. They will be remunerated.
2. Original papers selected from those which have not been submitted for publication elsewhere, after being reviewed. (The latest date for acceptance of contributions is the 8th and the 23rd day of each month for "physica status solidi (b)" and "physica status solidi (a)", respectively, 50 days before the delivery of the journal.)
3. Short notes published in the off-set part of the journal. They are, if necessary after being reviewed, selected by editorial staff. (The latest date for acceptance of manuscripts for this part is the 12th and the 27th day of each month for "physica status solidi (b)" and "physica status solidi (a)", respectively, 16 days before the delivery of the journal.)
4. Pre-printed titles of papers to be published in the next issue of both series.

Editorial Information

Manuscripts should be submitted preferentially in English. Besides, papers in German, French, and Russian are admitted. The final decision on the incorporation of the papers in the series (b) or (a) is taken by the editorial staff after consultation of the referee. Papers on liquids are only published in physica status solidi if the problems treated are also relevant for the solid state. Authors are asked to observe the following rules:

1. Manuscripts have to be written as concisely as possible. At most, 20 and 4 pages (28 lines per page; abstracts, tables, references, figures, and captions of figures included, where 3 figures are counted as about one page) are admitted for original papers and short notes, respectively. All manuscripts should be type-written on one side only. Two copies are required. Badly arranged manuscripts are returned.
2. The author's name, initials, and institution should be given.
3. One abstract of about ten lines in English and one in German, French, or Russian are required for original papers.
4. Because of lack of space conclusions (if any) must not have the character of a summary (i. e. substantially a repetition of the contents of the abstract).
5. Formulae should not be written too small and not with pencil. Separate lines for formulae are desirable. Indistinct writing will lead to a return of the manuscript.
6. Drawings should be enclosed on separate sheets of paper, not exceeding the maximum and minimum sizes of 20×30 and $10 \times 15\ cm^2$, respectively. Photographs should be submitted, on white paper, only for review articles and original papers. Duplicates are required.
7. Captions for the figures should be submitted on separate sheets of paper, and special instructions with regard to the size of illustrations should be given on these.
8. References should be numbered in order of appearance and should be given, on a separate sheet, in the form used in physica status solidi.
9. SI-units should be used.
10. Each paper must contain a proposal for the subject classification according to the classification scheme appearing in the January issues of each annual set.

The authors will be sent galley proofs of review articles and original papers. Proof reading of short notes and pre-printed titles will be done by the editorial staff.

For each original paper and short note a total of 75 reprints is available free of charge. The number of reprints free of charge per review article is 25. Requests for additional reprints (against payment) should be given together with the submitted manuscripts. **The deadline for orders of additional reprints is the 15th and the 30th of each month** for "physica status solidi (b)" and "physica status solidi (a)", respectively, two weeks before the delivery of the journal.

Bei Fragen zur Produktsicherheit wenden Sie sich bitte an:
If you have any questions regarding product safety,
please contact:

Walter de Gruyter GmbH
Genthiner Straße 13
10785 Berlin
productsafety@degruyterbrill.com